FOUNDATION ANALYSIS AND DESIGN

FOUNDATION ANALYSIS AND DESIGN

Second Edition

Joseph E. Bowles

Professor of Civil Engineering
Bradley University

McGRAW-HILL BOOK COMPANY

New York   St. Louis   San Francisco   Auckland   Bogotá   Düsseldorf
Johannesburg   London   Madrid   Mexico   Montreal   New Delhi
Panama   Paris   São Paulo   Singapore   Sydney   Tokyo   Toronto

FOUNDATION ANALYSIS AND DESIGN

1 2 3 4 5 6 7 8 9 0   K P K P   7 8 3 2 1 0 9 8 7

This book was set in Times New Roman. The editors were B. J. Clark and Richard S. Laufer; the cover was designed by Pencils Portfolio, Inc.; the production supervisor was Leroy A. Young. The drawings were done by A. E. Dini.
Kingsport Press, Inc., was printer and binder.

Library of Congress Cataloging in Publication Data

Bowles, Joseph E.
   Foundation analysis and design.

   Includes index.
   1. Foundations.   2.   Soil mechanics.   I.   Title.
TA775.B63   1977   624'.15   76-43303
ISBN 0-07-006750-3

Contents

PREFACE   xiii

CHAPTER 1 / INTRODUCTION   1
    1-1   Foundation—Definition and Purpose   1
    1-2   Foundation Classifications   2
    1-3   Foundation Site and System Economics   2
    1-4   General Requirements of Foundations   4
    1-5   Foundation Selection   4
    1-6   SI and Metric Units   4
    1-7   Computational Accuracy   6

CHAPTER 2 / SOIL MECHANICS IN FOUNDATION
               ENGINEERING   8
    2-1   Introduction   8
    2-2   Foundation Materials   9
    2-3   Soil Volume and Density Relationships   10
    2-4   Atterberg Limits   13
    2-5   Specific Gravity   14
    2-6   Unit Weight   14
    2-7   Water Content   15

2-8	Relative Density	15
2-9	Grain Size	16
2-10	Soil-Classification Terms	17
2-11	Soil Classification for Foundation Engineering	22
2-12	Shear Strength	23
2-13	Elastic Properties of Soil	33
2-14	Isotropic and Anisotropic Soil Masses	38
2-15	In Situ Stresses and K_0 Conditions	41
2-16	Consolidation Characteristics	43
2-17	Sensitivity and Thixotropy	51
2-18	Soil Water–Soil Hydraulics	52
2-19	Pore Pressure and Pore-Pressure Parameters	57
2-20	The Stress-Path Method	59

CHAPTER 3 / EXPLORATION, SAMPLING, AND IN SITU SOIL MEASUREMENTS — 69

3-1	Data Required	69
3-2	Methods of Exploration	70
3-3	Planning the Program	72
3-4	Soil Boring	73
3-5	Soil Sampling	78
3-6	The Standard Penetration Test	82
3-7	Other Penetration Methods	87
3-8	Core Sampling	87
3-9	Water-Table Location	90
3-10	Depth and Number of Borings	92
3-11	Presentation of Data	93
3-12	Field Load Tests	95
3-13	Field-Vane Testing of Soils	99
3-14	Measurement of In Situ Stresses and K_0 Conditions	101
3-15	Static Penetration Testing—Dutch-Cone Penetration Test (CPT)	105
3-16	The Borehole Shear Test	107
3-17	Seismic Exploration	108

CHAPTER 4 / BEARING CAPACITY OF SHALLOW FOUNDATIONS — 113

4-1	Introduction	113
4-2	The Terzaghi Bearing-Capacity Equation	113
4-3	The General Bearing-Capacity Equation	117
4-4	General Comments on Bearing-Capacity Computations	122
4-5	Footings with Eccentric or Inclined Loads	124
4-6	Effect of Water Table on Bearing Capacity	126
4-7	Bearing Capacity for Footings on Layered Soils	127
4-8	Bearing Capacity of Footings on Slopes	131

4-9   Bearing Capacity from Penetration Testing   133
4-10 Bearing Capacity of Foundations with Uplift or Tension
        Forces   137
4-11 Bearing Capacity Based on Building Codes
        (Presumptive Pressure)   139
4-12 Safety Factors in Foundation Design   141
4-13 Bearing Capacity of Rocks   143

CHAPTER 5 / FOUNDATION SETTLEMENTS   146
5-1   The Settlement Problem   146
5-2   Stresses in a Soil Mass due to Footing Pressure   147
5-3   Boussinesq Method for Evaluating Soil Pressure   148
5-4   Westergaard's Method for Evaluating Soil Pressures   153
5-5   Immediate (Elastic) Settlement Computations—Theory   157
5-6   Immediate Settlements—Application   159
5-7   Alternative Methods of Computing Elastic Settlements   165
5-8   Stresses and Displacements in Layered and Anisotropic
        Soils   168
5-9   Consolidation Settlements   169
5-10 Reliability of Settlement Computations   172
5-11 Proportioning Footings for a Given Settlement or Equal
        Settlements   172
5-12 Structures on Fills   175
5-13 Structural Tolerance to Settlement and Differential
        Settlements   176

CHAPTER 6 / IMPROVING SITE SOILS FOR FOUNDATION
              USE   181
6-1   Introduction   181
6-2   Compaction   181
6-3   Precompression to Improve Site Soils   183
6-4   Drainage Using Sand Blankets and Drains   185
6-5   Vibrating Methods to Increase Soil Density   186
6-6   Foundation Grouting and Chemical Stabilization   188
6-7   Altering Groundwater Conditions   189

CHAPTER 7 / FACTORS TO CONSIDER IN FOUNDATION
              DESIGN   191
7-1   Foundation Depth   191
7-2   Displaced Soil Effects   194
7-3   Net vs. Gross Soil Pressure—Design Soil Pressures   195
7-4   Erosion Problems for Structures Adjacent to Flowing
        Water   196
7-5   Corrosion Protection   197

7-6   Water-Table Fluctuation   197
7-7   Foundations in Sand Deposits   198
7-8   Foundations on Loess   198
7-9   Foundations on Expansive Soils   200
7-10 Foundations on Clay   202
7-11 Foundations on Sanitary Landfill Sites   203
7-12 Frost Depth and Foundations on Permafrost   204
7-13 Environmental Considerations   205

CHAPTER 8 / SPREAD-FOOTING DESIGN   207
8-1   Footing Classifications   207
8-2   Assumptions Used in the Design of Spread Footings   207
8-3   Reinforced-Concrete Design—USD   209
8-4   Structural Design of Spread Footings   215
8-5   Bearing Plates for Metal Columns   222
8-6   Pedestals   225
8-7   Rectangular Footings   228
8-8   Wall Footings and Footings for Residential
        Construction   232
8-9   Spread Footings with Overturning Moment   234

CHAPTER 9 / COMBINED FOOTINGS AND BEAMS ON
              ELASTIC FOUNDATIONS   240
9-1   Introduction   240
9-2   Rectangular Combined Footings   240
9-3   Design of Trapezoidal-Shaped Footings   248
9-4   Design of Strap or Cantilever Footings   253
9-5   Eccentrically Loaded Rigid Footings   255
9-6   Unsymmetrical Footings   263
9-7   Modulus of Subgrade Reaction   267
9-8   Classical Solution of Beam on Elastic Foundation   271
9-9   Finite-Element Solution of Beam on Elastic Foundation   276
9-10 Bridge Piers   285
9-11 Ring Foundations   287
9-12 General Comments on the Finite-Element Procedure   290

CHAPTER 10 / MAT FOUNDATIONS   294
10-1 Introduction   294
10-2 Types of Mat Foundations   294
10-3 Bearing Capacity of Mat Foundations   296
10-4 Mat Settlements   297
10-5 Design of Mat Foundations (Approximate Method)   298
10-6 Finite-Difference Method for Mats   304
10-7 Finite-Element Method for Mat Foundations   306
10-8 Mat-Superstructure Interaction   317
10-9 Circular Mats   318

CHAPTER 11 / LATERAL EARTH PRESSURE   321
11-1   The Lateral-Earth-Pressure Problem   321
11-2   Coulomb Earth-Pressure Theory   325
11-3   Rankine Earth Pressures   330
11-4   Active and Passive Earth Pressure Using Theory of
         Plasticity   334
11-5   Earth Pressure on Walls, Soil-Tension Effects, Rupture
         Zone   337
11-6   Reliability of Lateral Earth Pressures   340
11-7   Soil Properties and Lateral Earth Pressure   340
11-8   Earth-Pressure Theories in Retaining-Wall Problems   342
11-9   Graphical and Computer Solutions for Lateral Earth
         Pressure   345
11-10 Lateral Pressures by Theory of Elasticity for Surcharges   355
11-11 Other Causes of Lateral Pressure   361
11-12 Pressures in Silos, Grain Elevators, and Coal Bunkers   362

CHAPTER 12 / RETAINING WALLS   372
12-1   Introduction   374
12-2   Common Proportions of Retaining Walls   374
12-3   Soil Properties for Retaining Walls   377
12-4   Stability of Walls   379
12-5   Retaining-Wall Forces   381
12-6   Allowable Bearing Capacity   389
12-7   Settlements   389
12-8   Tilting   390
12-9   Design of Gravity and Semigravity Walls   393
12-10 Wall Joints   396
12-11 Drainage   397
12-12 Abutment Wing and Retaining Walls of Varying Height   398
12-13 Design of a Cantilever Retaining Wall   399
12-14 Design of a Counterfort Retaining Wall   405
12-15 Basement or Foundation Walls; Walls for Residential
         Construction   407
12-16 Reinforced-Earth Retaining Structures   408

CHAPTER 13 / SHEET-PILE WALLS—CANTILEVERED
                AND ANCHORED   412
13-1   Introduction   412
13-2   Soil Properties for Sheet-Pile Walls   413
13-3   Types of Sheetpiling   415
13-4   Safety Factors for Sheet-Pile Walls   417
13-5   Cantilever Sheetpiling   418
13-6   Anchored Sheetpiling: Free-Earth Support   426

13-7   Rowe's Moment Reduction Applied to the Free-Earth-
        Support Method   433
13-8   Finite-Element Analysis of Sheet-Pile Walls   436
13-9   Wales and Anchorages for Anchored Sheetpiling   444

CHAPTER 14 / BRACED COFFERDAMS FOR EXCAVATIONS   452
14-1   Construction Excavations   452
14-2   Soil Pressures on Braced Sheeting or Cofferdams   455
14-3   Conventional Design of Single-Wall (Braced)
        Cofferdams   458
14-4   Estimation of Ground Loss around Excavations   463
14-5   Finite-Element Analysis for Braced Excavations   465
14-6   Instability Due to Heave of Bottom of Excavation   472
14-7   Other Causes of Cofferdam Instability   474
14-8   Construction Dewatering   475
14-9   Slurry-Wall (or -Trench) Construction   479

CHAPTER 15 / CELLULAR COFFERDAMS   483
15-1   Cellular Cofferdams: Types and Uses   483
15-2   Cell Fill   487
15-3   Stability of Cellular Cofferdams   488
15-4   Practical Considerations in Cellular Cofferdam Design   497
15-5   Design of Diaphragm Cofferdam Cell   498
15-6   Circular-Cofferdam Design   501
15-7   Cloverleaf-Cofferdam Design   506

CHAPTER 16 / SINGLE PILES—STATIC CAPACITY AND
                INCLUDING LATERAL LOADS;
                PILE/POLE BUCKLING   508
16-1   Introduction   508
16-2   Timber Piles   509
16-3   Concrete Piles   511
16-4   Steel Piles   519
16-5   Corrosion of Steel Piles   522
16-6   Soil Properties for Pile Foundations   522
16-7   Pile Capacity in Cohesive Soils   523
16-8   Piles in Cohesionless Soils   530
16-9   Point-Bearing Piles   539
16-10   Bored or Cast-In-Place Piles   539
16-11   Static Pile Capacity Using Load-Transfer Load-Test Data   540
16-12   Tension Piles—Piles for Resisting Uplift Forces   542
16-13   Laterally Loaded Single Piles   543
16-14   Buckling of Fully and Partially Embedded Piles and
        Poles   550

CHAPTER 17 / SINGLE PILES—DYNAMIC ANALYSIS   556
    17-1   Dynamic Analysis   556
    17-2   Pile Driving   556
    17-3   The Rational Pile Formula   561
    17-4   Other Dynamic Formulas and General Considerations 565
    17-5   Reliability of Dynamic Pile-Driving Formulas   571
    17-6   The Wave Equation   573
    17-7   Pile Load Tests   579
    17-8   Pile-Driving Stresses   580
    17-9   General Comments on Pile Driving   583

CHAPTER 18 / PILE FOUNDATIONS—GROUPS   586
    18-1   Single Piles vs. Pile Groups   586
    18-2   Pile-Group Considerations   586
    18-3   Efficiency of Pile Groups   588
    18-4   Stresses on Underlying Strata   591
    18-5   Settlements of Pile Groups   598
    18-6   Pile Caps   602
    18-7   Batter Piles   605
    18-8   Negative Skin Friction   605
    18-9   Matrix Analysis for Pile Groups   608

CHAPTER 19 / CAISSONS INCLUDING DRILLED PIERS   618
    19-1   Types of Caissons   618
    19-2   Open-End Caissons   619
    19-3   Closed-End, or Box, Caissons   623
    19-4   Pneumatic Caissons   628
    19-5   Drilled Caissons   630
    19-6   Bearing Capacity and Settlements of Drilled Caissons 633
    19-7   Design of Drilled Caissons   637
    19-8   Laterally Loaded Caissons   641
    19-9   Inspection of Drilled Caissons   642

CHAPTER 20 / DESIGN OF FOUNDATIONS FOR
               VIBRATION CONTROL   644
    20-1   Introduction   644
    20-2   Elementary Vibrations   645
    20-3   Forced Vibrations for a Lumped Mass   650
    20-4   Approximate Solution of Vibrating Foundation—
           Theory of Elastic Half-Space   655
    20-5   Lumped-Mass Solution of the Vibrating Foundation 661
    20-6   Soil Properties—Elastic Constants   668
    20-7   Coupled Vibrations   671
    20-8   Effect of Piles to Reduce Foundation Vibrations   672
    20-9   Other Considerations for Machinery Foundations   674

APPENDIX A / GENERAL PILE-HAMMER AND PILE
　　　　　　　　　DATA TABLES　　　　　　　　　　　　676
　　　　A-1　H Piles　　　　　　　　　　　　　　　　676
　　　　A-2　Pile Hammers　　　　　　　　　　　　　677
　　　　A-3　Sheet Piles　　　　　　　　　　　　　　680
　　　　A-4　Pipe Piles　　　　　　　　　　　　　　　682
　　　　A-5　Prestressed-Concrete Piles　　　　　　　684

APPENDIX B / SELECTED COMPUTER PROGRAMS　　　685
　　　　B-1　Boussinesq Equation for Vertical Stresses　　686
　　　　B-2　Boussinesq Equation for Lateral Stresses　　687
　　　　B-3　Influence Factors F_1 and F_2 for Settlement of Footings
　　　　　　　on Stratum of Finite Thickness　　　　　688
　　　　B-4　Fox Equation for Footing-Depth Influence Factors　689
　　　　B-5　Beam, Ring, Sheet-Pile and Lateral-Pile Finite-Element
　　　　　　　Program　　　　　　　　　　　　　　690
　　　　B-6　Passive Pressure Using Plasticity Theory　　698
　　　　B-7　Mat Program　　　　　　　　　　　　　702
　　　　B-8　Culmann Program for Both Active and Passive Force　709
　　　　B-9　Retaining-Wall Program　　　　　　　　712
　　　B-10　Pile-Buckling Program　　　　　　　　　716
　　　B-11　Three-dimensional Pile-Group Program　　721

REFERENCES　　　　　　　　　　　　　　　　　723

INDEXES　　　　　　　　　　　　　　　　　　　741
　Name Index
　Subject Index

Preface

The intervening eight years since the first edition have produced considerable change in the approach to foundation-engineering problems. Changes have been most notable in computer applications, but many new ideas have been advanced in the application of soil-mechanics theories and principles to solving foundation problems. A veritable literature explosion has simultaneously occurred, with the various geotechnical publications and conferences producing well over 38,000 pages of material during this time.

In this period the Système International (SI) units have been introduced into science and engineering, and as of this writing the United States is the single country not officially using SI. This text uses both fps and metric/SI units, and most chapters present examples and problems in both systems of units. "Preferred usage" units are used with the SI problems. I have deliberately alternated the general discussion between fps and SI to assist the reader in the units transition.

This edition is the author's attempt to provide a textbook reflecting the latest developments in computer applications and soil-mechanics principles. I have not attempted a literature survey, although the bibliography is considerably larger than in the first edition. I have attempted to provide a recent reference for any necessary additional in-depth study, as a user convenience for every topic presented, but at the same time have attempted to make the text as self-contained as possible.

I am indebted to the many users of the first edition, including practitioners, educators, and students, who provided many helpful criticisms and suggestions.

I have kept the first-edition format of presenting both the theory and large numbers of illustrative examples. Users have indicated this as a particularly desirable feature. I have increased the number of chapters but decreased their length so that the text can be used more easily for a two-semester-course sequence at the senior-graduate level. This format also allows for greater ease of subject-matter identification.

More emphasis has been placed on soil properties than in the first edition. Generally a section of each design chapter is allocated to a discussion of the necessary soil properties/parameters.

I have included new computer-program listings to aid the user in developing and using the computational methods. The text programs are photocopies of listings made from operating computer programs in FORTRAN IV as used on an IBM 370 computer.

I have attempted where possible and practical to include actual field data in the worked examples and home problems. Selected answers for home problems are provided. In many cases where answers are not provided, this is because the answer depends on the problem assumptions, thus becoming too lengthy to provide in a textbook.

I wish to express especial appreciation to Dr. William D. Kovacs, Purdue University, for carefully reviewing the entire manuscript and making many helpful suggestions and criticisms.

I also wish to thank Marion J. Frobish of the computer center for assistance in computer programming, and Dr. Rathi K. Bhatacharya who gave some useful advice in the theoretical mechanics area of the text.

My wife, Faye J. Bowles, typed the entire manuscript and assisted in table preparations and in cross-checking references and figures; to simply express appreciation for this enormous contribution is hardly adequate.

Joseph E. Bowles

FOUNDATION ANALYSIS AND DESIGN

ONE

INTRODUCTION

1-1. FOUNDATION—DEFINITION AND PURPOSE

All engineered structures resting on the earth, including earth fills, dams (both earth and concrete), buildings, and bridges, consist of two parts, the upper or *superstructure*, and the lower or *foundation*. The foundation is the interfacing element between the superstructure and the underlying soil or rock. In the case of earth fills or earth dams there is no clear line of demarcation between the "superstructure" and the foundation.

Foundation engineering is the art and science of applying engineering judgment and the principles of soil mechanics to solve the interfacing problem. It is also concerned with solutions to problems of retaining earth masses by several types of structural elements such as retaining walls and sheet piles.

Foundation engineering is also the art and science of using engineering judgment and the principles of soil mechanics to predict the response of earth masses to changed conditions of geometry and/or loads.

It should be noted that foundation engineering has been defined as the "art and science" of applying engineering judgment and the principles of soil mechanics. The science of soil mechanics has progressed rapidly and considerably over the past fifty years. However, because of the natural variability of soil and the problems of testing (which will be elaborated upon in later chapters) the design of a foundation still depends to a large degree upon "art," or the application of engineering judgment.

This text will be concerned principally with foundation engineering in the context of those soil-mechanics principles particularly applicable to the design of interfacing elements and retaining structures and the estimating of probable soil response to load/geometry changes.

Soil mechanics as used here implies a body of knowledge embodying soil behavior as a by-product of engineering-geology considerations and other pertinent engineering disciplines including mechanics of materials, hydraulics, structures and structural vibrations, and theory of elasticity.

1-2. FOUNDATION CLASSIFICATIONS

Foundations for structures such as bridges and buildings, from the smallest residential to the tallest high-rise structures, are for the purpose of transmitting the building loads, which are usually of very high stress intensity [in the case of steel columns perhaps 20,000 pounds per square inch (1,400 kg/cm^2)], to the far lower supporting capacity of the soil. This transmission of stresses may be via use of:

1. Shallow foundations (footings or spread footings, $D < B$; see Chap. 4)
2. Deep foundations (pile or caisson foundations, $D > B$; see Chaps. 16 to 19)

Any structure whose purpose is to retain a soil or other similar mass in a geometric shape other than that occurring naturally under the influence of gravity will be termed a *retaining* structure. Any foundations not included in the above categories will be termed *special* foundations.

Normally, both shallow and deep foundations are buried or are beneath the superstructure in such a configuration that access will be difficult in case of any problems developing after part or all of the superstructure is in place. For this reason it is normal practice to be conservative in the design of these elements.

Typical foundation types are:

1. Foundations for buildings (shallow or deep)
2. Foundations for smokestacks, radio and television towers, bridge piers, etc. (shallow or deep)
3. Foundations for port or marine structures (may be shallow or deep, with extensive use made of retaining structures)
4. Foundations for machinery, turbines, generators, etc. (shallow or deep)
5. Foundation elements to support open cuts or retain earth masses or bridge abutments (retaining structures)

Foundations for buildings are built in vast quantities, foundations for the other categories in generally lesser numbers.

1-3. FOUNDATION SITE AND SYSTEM ECONOMICS

It is obvious that the foundation of a building must be adequate if the building is to perform satisfactorily and be safe for occupancy. Other foundations must be adequate to perform their intended functions in a satisfactory and safe manner; however, buildings usually have more stringent criteria for safety and performance than other structures—notable exceptions being nuclear-plant facilities, turbines for power generation, and certain types of radio-antenna equipment.

Almost any reasonable structure can be built and safely supported if there is enough money to spend. Unfortunately in the real situation this is seldom, if ever, the case, and the foundation engineer has the dilemma of making a decision under much less than the ideal condition. Also, even though his mistake may be buried, the evidence cannot be concealed but can show up relatively soon or, more likely, years later.

The designer is always faced with the question of what constitutes an economical design. Considering the possible and probable different soils at any site, and nowadays these may be partially man-made problems such as old fills or garbage dumps as land, especially near urban centers, becomes scarcer, the design becomes highly subjective. Not to be overlooked is the possibility that the act of construction may alter the soil properties considerably from those used in the initial analysis/design of the foundation. This type of design is so subjective and difficult to quantify that two design firms might come up with completely different designs that would perform equally satisfactorily. The costs may differ enormously, however. This problem and the widely differing solutions would depend, for example, on the following:

1. What constitutes satisfactory and tolerable settlement; how much extra should or could be spent to reduce the settlements by small amounts, for example, from say 2 to 1.5 cm?
2. How variable is the soil profile, and has the client been willing to authorize an adequate exploration program?
3. Can the building be supported by the underlying soil on spread footings (least cost), mats (intermediate in cost), piles (cost of several times spread footing cost)?
4. What is the likelihood of a lawsuit if the foundation does not perform adequately?
5. Is money available for the foundation portion of the construction? It is not unheard of that the foundation alone would cost so much the site is abandoned in favor of another site where the foundation costs are what the owner can afford.
6. What is the ability of the local construction force? It is hardly sensible to design an elaborate foundation if no one can build it or if it is so different in design that the contracting personnel, fearful of problems, make it uneconomical to build.
7. What is the engineering ability of the foundation engineer? While listed last, this is not necessarily of least importance in economical design. It should, of course, be realized that different persons have different capabilities, and just as there are excellent, intermediate, and poor lawyers, doctors, teachers, carpenters, mechanics, etc., there are also these categories of engineers.

If the foundation fails because of any cost shaving, no one will remember that the client had a temporary benefit for which he was grateful. He is at this point (facing heavy damages and/or a lawsuit) probably the least grateful of all the parties involved. Thus, one should always bear in mind that absolute dollar economics may not produce good foundation engineering.

The foundation engineer must look at the entire system: the building purpose, probable service-life loading, type of framing, soil profile, construction methods, and construction costs to arrive at a design consistent with the client/owner's needs and financial capability.

1-4. GENERAL REQUIREMENTS OF FOUNDATIONS

A foundation must be capable of satisfying several stability and deformation requirements such as:

1. Depth must be adequate to avoid lateral expulsion of material from beneath the foundation, particularly footings and mats.
2. Depth must be below seasonal volume changes such as freezing and thawing or the zone of active organic materials.
3. System must be safe against overturning, rotation, sliding, or soil rupture (shear-strength failure).
4. System must be safe against corrosion or deterioration due to harmful materials present in the soil.
5. System should be adequate to sustain some changes in later site or construction geometry or be easily modified should later changes be major in scope.
6. The foundation should be economical in terms of the method of installation.
7. Total earth movements (generally settlements) and differential movements should be tolerable for the foundation element and/or any superstructure elements.

   In general, foundation engineering is a soil-structure interaction problem and should be analyzed as such.

1-5. FOUNDATION SELECTION

The different types of foundations named in this section will be taken up in some detail in later chapters. It will be useful, however, at this point to enumerate the several types and their potential application. Where groundwater is present, it is understood that if it is below the depth of the footing (or other construction) it will not be a problem. If groundwater is within the construction zone it will be removed by lowering the water table by pumping, or by using grout curtain walls, steel shells (or compressed air) in the case of caissons, or other means as appropriate.

1-6. SI AND METRIC UNITS

This textbook will use both United States (foot-pound-second, or fps) and metric units. Metric units will generally comply with the SI format except where "preferred usage" appears to predominate. Problems will be either all fps or all SI-metric. The practice of placing the other system of units in parentheses as in many of the technical journals will rarely be done.

   Preferred usage will entail most weight and pressure units. The reason is that

Foundation type	Use	Applicable soil conditions
Spread footing, wall footings	Individual columns, walls, bridge piers	Any conditions where bearing capacity is adequate for applied load. May use on single stratum; firm layer over soft layer or soft layer over firm layer. Check immediate, differential, and consolidation settlements
Mat foundation	Same as spread and wall footings. Very heavy column loads. Usually reduces differential settlements and total settlements	Generally soil bearing value is less than for spread footings; over one-half area of building covered by individual footings. Check settlements
Pile foundations Floating	In groups (at least 2) to carry heavy column, wall loads; requires pile cap	Poor surface and near surface soils. Soils of high bearing capacity 20–50 m below basement or ground surface, but by distributing load along pile shaft soil strength is adequate. Corrosive soils may require use of timber or concrete pile material
Bearing	In groups (at least 2) to carry heavy column, wall loads; requires pile cap	Poor surface and near-surface soils; soil of high bearing capacity (point bearing on) is 8–50 m below ground surface
Caisson (shafts 75 cm or more in diameter) generally bearing or combination of bearing and skin resistance	Larger column loads than for piles but eliminates pile cap by using caissons as column extension	Poor surface and near-surface soils; soil of high bearing capacity (point bearing on) is 8–50 m below ground surface
Retaining walls, bridge abutments	Permanent retaining structure	Any type of soil, but a specified zone (Chaps. 11, 12) in back of wall usually of controlled backfill
Sheet-pile structures	Temporary retaining structures as excavations, waterfront structures, cofferdams	Any soil; waterfront structure may require special alloy or corrosion protection. Cofferdams require control of fill material

most soil laboratory equipment lasts for years (scales, pressure gages, calipers, ruling devices, etc.), and few laboratories (even in SI countries) have this equipment in SI units.

In this text, then, we will compute unit weights in g/cm^3. Pressures in laboratory testing, and allowable steel and concrete stresses will be in units of kg/cm^2 or lb/sq in (psi).

We will report all unit weights (soil, concrete, etc.) in kiloNewtons/cubic meter

Table 1-1. Useful SI and metric conversion factors

To convert from	To	Multiply by
inch	centimeter (cm)	2.54
square inch	square centimeter	6.45160
cubic inch	cubic centimeter	16.38706
kilogram force kg_f	Newton	9.807
pound force, lb_f	kg_f	0.45359
kip	kiloNewton	4.44747
kg/cm^2	kN/m^2 (kPa)	98.07
kip/sq in	kN/m^2 (kPa)	6,894.28
lb/sq in	kg/cm^2	0.07031
ton/sq ft	kg/cm^2	0.97650
foot-kip	kN-meter	1.35560
kcf	kN/m^3	157.09304
in^4	cm^4	41.62314

(kN/m^3) or lb/cu ft, also kips/cu ft (pcf, kcf). KiloNewtons will be used throughout, since a Newton is too small and a megaNewton, while permissible, is an extra unit.

Pressures such as bearing capacity will be in units of kiloNewtons/m^2 (kilo-Pascals or kPa) or kips/sq ft (ksf).

A conversion factor set very useful to remember is that to convert g/cm^3 to kN/m^3 or lb/cu ft (pcf)

$$1 \text{ g/cm}^3 = 9.807 \text{ kN/m}^3$$

$$1 \text{ g/cm}^3 = 62.5 \text{ lb/cu ft}$$

The unit weight of water is 62.5 lb/cu ft or 9.807 kN/m^3. Soil would normally vary from 90 to about 130 lb/cu ft in round numbers (also in SI) or from 14 to 21 kN/m^3.

Another unit (fps) which in the past has been widely used is tons/sq ft. Unconfined compression and consolidation testing has often used these units; pile load tests often used tons (the author prefers kips/sq ft or kips, primarily for consistency in units). Tons/sq ft and kg/cm^2 are very close to equal (see Table 1-1). Tons (2,000 lbs) in the fps system should not be confused with tons in the metric system (1,000 kg or 2,204 lbs and often spelled tonnes).

1-7. COMPUTATIONAL ACCURACY

The pocket and desktop electronic calculators with 10 to 14 digit computational accuracy tend to give a fictitiously high precision to computed quantities. The use of these devices is rapidly becoming universal and the author has used one in preparing example problems in this textbook. Numbers have been reported with two decimal places, generally, depending on the input, and are for use by the reader in problem verification.

The reader should realize that foundation loads and soil properties are not likely to be known to a precision closer than $\pm$ 10 to 20 percent. Nevertheless, for problem checking, or in design offices, where work is usually checked by others, the answers should be reported to sufficient digits that round-offs do not accumulate. This is certainly not likely using two decimal digits and is not excessively burdensome using the electronic calculator.

TWO

SOIL MECHANICS IN FOUNDATION ENGINEERING

2-1. INTRODUCTION

No construction material is more variable than the ground. It varies enormously both laterally and vertically, and quality control over any extended period of time is impossible, because of the large quantities involved. Its properties are difficult to determine, partly because of the variability but also because of the state of the art on sampling and testing methods. These problems will be considered and discussed as appropriate to the subject matter throughout this text.

Soil mechanics may also be defined as the engineering study of soil to obtain properties such as:

1. Strength parameters
2. Compressibility indexes
3. Permeability
4. Gravimetric-volumetric data (unit weight, specific gravity, void ratio)

This makes possible engineering predictions and estimates of:

1. Bearing capacity
2. Settlements
   Amount
   Rate
3. Earth pressures
4. Pore pressures and dewatering quantities

Not all branches of soil mechanics are of equal importance in the design and construction of foundations. This chapter will place emphasis on those soil-mechanics principles most applicable for foundation analysis and design.

Until 1925, when K. Terzaghi, often called the father of soil mechanics, published his book "Erdbaumechanik auf bodenphysikalischer Grundlage," soil mechanics, and foundation engineering, were largely empirical. From this point onward, and with the years from 1950 especially fruitful, soil engineering has developed into a rational approach, rather than putting total reliance on empirical rules.

Much soil-mechanics laboratory work has been done and reported in the literature in recent years. Most of this work has been done on idealized laboratory-prepared samples which tend to considerable homogeneity and uniformity. This type of testing is thought by its proponents to have considerable use in predicting what may happen in the field; however (owing to sampling technique, gravel, hydrostatic pressure, unloading of in situ stresses on sample upon recovery, etc.), the field-sample test results may bear little or no resemblance to the laboratory samples. Under these circumstances the elaborate laboratory testing programs may have less value than intended.

With the problems associated with testing, the testing of small quantities of soil or a few samples, and the sample-recovery problems and extrapolating the results to the soil mass in situ, a considerable amount of "art" or "engineering judgment" is required. The term "engineering judgment" is not to be equated with guessing at the solution, regardless of how it may appear to the casual observer.

In the application of engineering judgment to a foundation problem it is essential that the foundation engineer have available as much site, soil-profile, soil-property, and geological information as necessary to arrive at a decision. In some cases such as one-story load-bearing wall construction (department stores or service stations) where the soil is relatively homogeneous the necessary information may be simply the boring logs from four or five relatively shallow exploratory borings. For a 10-story building the necessary information would normally have to be more. For a 100-story building the amount of information necessary would be considerable and might cost several hundreds of thousands of dollars to obtain.

2-2. FOUNDATION MATERIALS

The foundation engineer is concerned with the construction of some type of engineering structure on the *earth*. The earth is composed of rock and soil which is weathered or degraded rock. Water in varying amounts is found in the pores or voids of the soil and in more limited quantities in the cracks and pores of the rock. Air is also present in many of the void spaces but is usually of no particular significance to foundation design. *Rock* will be defined as that naturally occurring material composed of mineral particles so firmly bonded together that relatively great effort is required to separate the particles (i.e., blasting, heavy crushing, or ripping forces). *Soil* will be defined as naturally occurring mineral particles which are fairly readily separated into relatively small pieces and in which the mass may contain *air*, *water*, or *organic* materials in varying amounts. The mineral particles of the soil mass are formed from decomposition of the rock by mechanical weathering (air, ice, wind, and water) and chemical processes.

The soil may be further described as a residual or transported soil. A *residual* soil is one which has been formed at a location by the decomposition of the parent material (the rock) and is found at the site of formation. A *transported* soil is one which was formed at one location and has been transported by wind, water, glacier, or gravity forces to the present location. The terms residual or transported must be taken in context, since many of the current residual soils are formed (or are being formed) from transported soil deposits of earlier geological periods which have later formed into rocks and/or undergone uplift as, for example, the Appalachian mountain and Piedmont regions of the Eastern United States.

2-3. SOIL VOLUME AND DENSITY RELATIONSHIPS

The more common soil definitions and mathematical relationships are presented in the following material. Figure 2-1 illustrates certain of the terms used in the following relationships.

Void ratio e. The ratio of the volume of voids V_v to the volume of solids V_s in a given volume of material, usually expressed as a decimal.

$$e = \frac{V_v}{V_s} \qquad (2\text{-}1)$$

Porosity n. The ratio of the volume of voids to the total volume V_t of a soil mass; may be expressed as a percentage or a decimal.

$$n = \frac{V_v}{V_t} \qquad (2\text{-}2)$$

Water content w. The ratio of the weight of water W_w in a given mass of soil to the

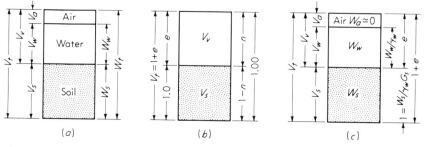

Figure 2-1. (*a*) Weight/volume relationships for a soil mass; (*b*) volume/void relationships; (*c*) volumes expressed in terms of weights and specific gravity.

weight of soil W_s in the same mass, expressed as a percentage.

$$w = \frac{W_w}{W_s} \times 100 \tag{2-3}$$

Unit weight γ (also termed density of the soil). The ratio of the weight of soil to the corresponding volume, with units of weight per unit volume. The general expression is

$$\gamma = \frac{W}{V_t} \tag{2-4}$$

Dry unit weight is based on using W_s, and *wet unit weight* uses W_t in Eq. (2-4).

Degree of saturation S. The ratio of the volume of water to the total volume of voids in a soil mass, expressed as a percentage.

$$S = \frac{V_w}{V_v} \times 100 \tag{2-5}$$

Specific gravity. The ratio of the unit weight of a material in air to the unit weight of water. The water should be distilled, and both materials should be at 4°C; however, no serious error is introduced if the weight and volume are used at the usual temperature ranges of 10 to 25°C. The average specific gravity of a mass of soil grains G_s is

$$G_s = \frac{W_s/V_s}{W_w/V_w} = \frac{W_s/V_s}{\gamma_w} = \frac{\gamma_s}{\gamma_w} \tag{2-6}$$

With the basic definitions given, other mathematical relations may be readily derived. For example, let the volume of solids be 1.00 (Fig. 2-1b). Since

$$e = \frac{V_v}{V_s} = \frac{V_v}{1} = V_v$$

From Fig. 2-1b, $V = V_s + V_v = 1 + e$. From Eq. (2-2) and substituting for V_v and V_s, one obtains

$$n = \frac{V_v}{V_t} = \frac{e}{1 + e} \tag{2-7}$$

Solving Eq. (2-7) for the void ratio, one obtains

$$e = \frac{n}{1 - n} \tag{2-8}$$

Also, from Fig. 2-1a the total weight of the soil W_t is

$$W_t = W_s + W_w$$

But by definition the weight of water W_w is

$$W_w = wW_s; \qquad \text{therefore} \qquad W_t = W_s + wW_s$$

Solving for the weight of solids, we obtain

$$W_s = \frac{W_t}{1 + w} \qquad (2\text{-}9)$$

and dividing through by V_t, we obtain

$$\gamma_{dry} = \frac{\gamma_{wet}}{1 + w} \qquad (2\text{-}9a)$$

Another useful relationship for a saturated soil ($S = 100$ percent) can be obtained as follows:

$$W_w = V_w \gamma_w G_w$$

But when $V_t = 1 + e$, $V_s = 1.0$ and $V_w = e$; therefore, substituting, we obtain

$$W_w = e \gamma_w G_w$$

Substituting this value for W_w and W_s from Eq. (2-6) into Eq. (2-3) and taking $G_w = 1.0$, we obtain

$$w = \frac{W_w}{W_s} = \frac{e \gamma_w}{G_s \gamma_w V_s}$$

and solving for the void ratio with $\gamma_w = V_s = 1.0$, we have

$$e = w G_s \qquad \text{(valid for } S = 100 \text{ percent)} \qquad (2\text{-}10)$$

From Fig. 2-1c and the definition of dry (no water present) unit weight $= W_s/V_t$, it is obvious that if $V_s = 1.0$, then $W_s = \gamma_w G_s$, and the dry unit weight is

$$\gamma_{dry} = \frac{\gamma_w G_s}{1 + e} \qquad (2\text{-}11)$$

Also, by definition, $S = V_w/V_v$, but $V_v = e$; thus $V_w = eS$, and from Eq. (2-6) for water

$$V_w = \frac{W_w}{\gamma_w G_w}$$

Therefore, $W_w = eS\gamma_w$ with $G_w = 1.0$, and the total weight is

$$W_t = W_s + W_w$$

and substituting from the above for W_w, taking $V_t = 1 + e$ and Eq. (2-6) for W_s,

$$\gamma_{wet} = \frac{W_t}{V_t} = \frac{\gamma_w G_s + eS\gamma_w}{1 + e} = \frac{\gamma_w(G_s + eS)}{1 + e} \qquad (2\text{-}11a)$$

Since $W_w = wW_s = w\gamma_w G_s$, the wet unit weight may also be written

$$\gamma_{wet} = \frac{\gamma_w G_s + w\gamma_w G_s}{1 + e} = \frac{G_s \gamma_w(1 + w)}{1 + e} \qquad (2\text{-}11b)$$

There are, of course, many other combinations of relationships, but these are left for the reader's ingenuity.

Example 2-1. A cohesive soil specimen (from a split spoon; see Chap. 3 for method) was subjected to laboratory tests, yielding the following data: The moisture content was 22.5 percent; G_s was found to be 2.60; and to determine the approximate unit weight, a sample weighing 224.0 g was placed in a 500-cm³ container with 382 cm³ of water required to fill the container.

REQUIRED. 1. The wet unit weight
2. The dry unit weight
3. The void ratio and porosity
4. The degree of saturation
5. The dry-bulk specific gravity

SOLUTION. 1. $\gamma_{\text{wet}} = \dfrac{W_{\text{wet}}}{V_t} = \dfrac{224}{500-382} = \dfrac{224}{118} = 1.9$ g/cm³

$$= 1.9 \times 62.5 = 118.8 \text{ pcf}$$

2. $\gamma_{\text{dry}} = \dfrac{\gamma_{\text{wet}}}{1+w} = \dfrac{1.9}{1+0.225} = \dfrac{1.9}{1.225} = 1.55$ g/cm³

3. $e = \dfrac{V_v}{V_s}$

$V_s = \dfrac{W_s}{G_s\gamma_w} = \dfrac{1.55}{2.60(1)} = 0.596$ cm³

$V_t = 1.0$ cm³ since we are working with weight of 1 cm³

$V_v = 1.0 - 0.596 = 0.403$ cu ft

$e = \dfrac{0.403}{0.596} = 0.677$

$n = \dfrac{V_v}{V_t} = \dfrac{0.403}{1} = 0.403$

4. $S = \dfrac{V_w}{V_v} \times 100$

$W_w = 1.9 - 1.55 = 0.35$ g

$V_w = \dfrac{W_w}{\gamma_w} = \dfrac{0.35}{1} = 0.35$ cm³

$S = 0.35 \times \dfrac{100}{0.403} = 86.8$ percent

5. Dry-bulk specific gravity

$G_b = \dfrac{\gamma_{\text{dry}}}{\gamma_w} = \dfrac{1.55}{1} = 1.55$

2-4. ATTERBERG LIMITS

The *Atterberg limits* are laboratory tests for arbitrary moisture contents to determine when the soil is on the verge of being a viscous liquid (liquid limit w_L) or nonplastic (plastic limit w_P). The plasticity index I_P:

$$I_P = w_L - w_P$$

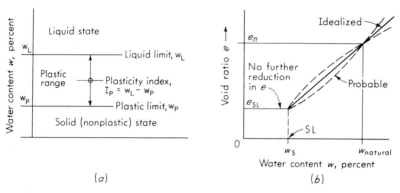

Figure 2-2. (*a*) Graphical significance of liquid and plastic limits; (*b*) no further reduction in void ratio *e* occurs at water contents below shrinkage limit (SL).

is the arbitrary range of water contents for which the soil is plastic. The *shrinkage limit* is the water content beyond which no further reduction of mass volume takes place with additional moisture loss. Two other "limits" proposed by Atterberg are the sticky limit (moisture content at which soil "sticks" to a metal surface) and the cohesion limit (moisture content below which soil crumbs no longer stick together). The sticky limit is occasionally used, but the cohesion limit is almost unheard of in the United States. The three common limits are displayed in Fig. 2-2.

The liquid and plastic limits are useful for classifying the soil in most classification systems (Fig. 2-5) and for empirically estimating settlements on cohesive soils.

2-5. SPECIFIC GRAVITY

Specific gravity G may be determined in a laboratory test with moderate difficulty. Table 2-1 presents values of specific gravity for selected soil minerals. A value of 2.65 to 2.70 is often assumed for sands (and other soils) without performing a test. A value is necessary for certain volumetric relationships.

2-6. UNIT WEIGHT

Unit weight γ is fairly easy to evaluate for a cohesive soil but is extremely difficult to impossible to determine in cohesionless soils unless they are located very near the surface. See Example 2-1 for an approximate method of determining unit weight for a cohesive soil. Tables 3-3 and 3-4 give empirical values for unit weight which may be used in the absence of laboratory tests.

Unit weight is necessary in computing in situ vertical and horizontal stresses and lateral pressures against retaining structures.

Table 2-1. Typical values of specific gravity for soil minerals

Mineral	Specific gravity	Mineral	Specific gravity
Bentonite	2.13–2.18	Muscovite (mica)	2.80–2.90
Gypsum	2.30	Dolomite	2.87
Gibbsite	2.30–2.40	Aragonite	2.94
Montmorillonite	2.40	Anhydrite	3.00
Orthoclase feldspar	2.56	Biotite (mica)	3.0 –3.1
Illite	2.60	Hornblende	3.00–3.47
Quartz	2.60	Augite	3.20–3.40
Kaolinite	2.60–2.63	Olivine	3.27–3.37
Chlorite	2.6 –3.0	Limonite	3.8
Plagioclase feldspar	2.62–2.76	Siderite	3.83–3.88
Talc	2.70–2.80	Hematite	4.90–5.30
Calcite	2.80–2.90	Magnetite	5.17–5.18

2-7. WATER CONTENT

This is a routine laboratory test, as part of the Atterberg limits, compaction tests, compression tests, shear tests, etc. If it is determined on soils obtained from exploration borings, it is termed the natural-moisture content.

The *natural moisture* w_N may be used as an aid in predicting settlements. A plot of w_N with depth for a field boring may detect height of capillary rise, perched water, or water-table location if all pertinent factors are taken into account. In the same type of soil and below the groundwater table, changes in water content would be indicative of density changes.

2-8. RELATIVE DENSITY

In cohesionless soils, *relative density* D_r is defined as

$$D_r = \frac{e_{max} - e_n}{e_{max} - e_{min}} \tag{2-12}$$

where e_{max} = void ratio of soil in loosest state
e_{min} = void ratio of soil in densest state
e_n = existing void ratio of soil

Equation (2-12) has been used to define the state of the soil.
Relative density can be defined in terms of unit weights as

$$D_r = \frac{\gamma_n - \gamma_{min}}{\gamma_{max} - \gamma_{min}} \frac{\gamma_{max}}{\gamma_n} \tag{2-12a}$$

where γ = unit weight, in situ (n), maximum (max), and minimum (min)

This equation is more widely used than Eq. (2-12) because of the normal laboratory procedures of using a compaction mold and making weight-volume measurements.

Holtz (1973) reports that relative-density determination is applicable for soils in the Unified Soil Classification system (see Fig. 2-5) as follows:

all GW, GP, SW, SP soils.

GW-GM, GP-GM, GW-GC, GW-GM if fines (− No. 200 sieve) are less than 8 percent.

SW-SM, SW-SC, SP-SM, SP-SC if fines are less than 12 percent, although some SW-SM and SP-SM soils with up to 16 percent fines may be used.

The relative density of soils has been used by some agencies to identify potential soil liquefaction under shock loads as in earthquakes [Durham and Townsend (1973), Finn et al. (1971)].

It is the author's opinion that the relative-density test is almost worthless and that simply using a soil density is as good a control criterion as practical. The reason is that it is very difficult to determine the three unit weights required for Eq. (2-12a). For example, a coarse, well-graded sand (SW) was tested a large number of times, yielding

$$\gamma_{max} = 120 \text{ pcf} - 3 \text{ pcf}$$

$$\gamma_{min} = 97 \text{ pcf} + 3 \text{ pcf}$$

and taking
$$\gamma_n = 110 \pm 2 \text{ pcf}$$

what is the *range* of D_r?

Note the error distribution, γ_{max} with the error tending low; that is, most values were *under* 120 pcf, and likewise most minimum-density values were *over* 97 pcf. Natural field density determinations would be expected to be ($\pm$)

$$D_r = \frac{108 - 100}{117 - 100}\left(\frac{117}{108}\right) = 0.51$$

A second possibility is
$$D_r = \frac{112 - 100}{117 - 100}\left(\frac{117}{112}\right) = 0.74$$

or in this case a range of $74 - 51 = 23$ percent. In spite of this shortcoming many persons still refer to and use relative density. Chapter 3 indicates indirect field methods of obtaining it. Bowles (1970) in Experiment 18 illustrates a simple laboratory procedure which requires a minimum of equipment and has been found to give values as good as those obtained by the ASTM D2049-69 test procedure.

2-9. GRAIN SIZE

The grain-size-distribution test is used for soil classification, although for much foundation analysis the soil classification is based on a visual inspection, supplemented by methods indicated later. The engineer is usually concerned only with whether the soil is gravel, sand, silt, or clay, or sandy clay, silty sand, etc., where the second term is the predominant material and the first indicates the filler; for example, sandy clay is a clay with some sand present. This type of visual classification can, with some experience, be determined quite reliably by one of the following procedures or tests:

1. To differentiate between gravel and sand, samples of each type of material can be prepared to include jars of fine, medium, and coarse sizes if the material is to be further subdivided, and can be kept in the laboratory. The engineer simply makes a visual comparison.
2. To differentiate between fine sand and silt, both materials may appear as dust when dry. By placing a spoonful of the soil in a test tube of water and shaking, sand or silt can be detected, since the sand settles out in $1\frac{1}{2}$ min or less, whereas the silt takes 10 or more minutes to settle (i.e., water clears). One can observe relative thickness of the sediments for subclassification, as silty sand, etc.
3. To differentiate between silt and clay:
   a. One moistens a spot on the soil and rubs a finger on it. If the rubbed spot appears smooth, the material is clay, but if it appears scratched, it is silt or silty.
   b. The dispersion test is made by mixing the material in water in a test tube and observing the time for the water to clear. Silt usually takes 1 hr or less, whereas clay may take several hours or more.
   c. Mix a small quantity of soil with water to form a plastic ball and place it in the palm of the hand and shake horizontally with jerking movements. If the material is silt or predominantly silt, the surface will become wet and shiny since water travels through silt particles relatively easily and the inertia forces cause the water to move to the surface. Clay, on the other hand, shows no change. This phenomenon is referred to as *dilatancy*.
   d. Crushing of dry clay lumps is relatively difficult, whereas silt lumps break quite easily.
   e. Clay can be rolled out into small threads, whereas silt is much more difficult to roll into small threads and generally requires more water.

Grain size is of considerable importance in seepage and soil-drainage problems.

2-10. SOIL-CLASSIFICATION TERMS

Identifying names, some of which are local in nature, are assigned to certain sizes or types of rock or soil formations, of which some of the principal ones are as follows:

1. Bedrock. Rock in its native location, usually extending greatly both horizontally and vertically. This material is generally overlain by soil of varying depths. If exposed, the outer portions may become weathered. Bedrock varies from igneous rocks, generally the hardest and formed from molten magma, to metamorphic rocks formed from metamorphizing sedimentary rocks under great heat and pressure, and sedimentary rocks formed from chemical action and pressure from overlying soil deposits. Rocks may be solid, but the interface with the overlying soil may be much fractured. Depending on the geologic history of the area, the rocks may be much fractured, folded, and faulted. Various textbooks on geology should be consulted for further information—especially concerning particular areas [e.g., Legget (1962), Thornbury (1965), both with extensive references].

2. Boulders. Smaller pieces of material which have broken away from the bedrock, usually 10 to 12 in or more in dimension. Pieces smaller than boulders may be called *cobbles* (2 to 3 in minimum size) or *pebbles* ($\frac{1}{8}$ to $\frac{1}{4}$ in minimum size).

3. Gravel. Common term used to describe pieces of rock from about 6 in maximum to less than $\frac{1}{4}$ in minimum size. May be *crushed stone* when man-made, *bank-run gravel* when excavated from a naturally occurring deposit and containing finer material, or *pea gravel* if it has been screened to sizes $\frac{1}{4}$ to $\frac{1}{8}$ in (pea size). Gravel is a cohesionless material; that is, it does not possess particle adhesion or attraction.

4. Sand. Mineral particles smaller than gravel, but larger than about 0.05 to 0.074 mm. May be *fine, medium,* or *coarse,* depending on the size of the majority of the particles. Sand is a *cohesionless* material; however, if it is damp or moist, the surface tension of the water may give an *apparent cohesion* which disappears when the material dries or becomes saturated. Sand is a favorable construction material. It has excellent bearing capacity *if confined.* Unconfined sand will flow from beneath foundations, pavements, etc., and this process can be accelerated by water flowing through and by wave or stream action eroding it. Since water flows easily through sand, any sand holding water must contain nonsand materials such as silts, clays, or mixtures of both. Excavations in sand will stand on slopes of about 1 to 1.5 or less. Excavations steeper than this are potentially unstable.

5. Silt. Mineral particles ranging in size from 0.05 to 0.074 mm maximum to 0.002 to 0.006 mm. It is called *organic silt* if it contains appreciable quantities of organic materials, and *inorganic silt* if no organic materials are present. Silt usually exhibits some *cohesion,* or particle attraction and adhesion, and may also have *apparent cohesion,* or loss of cohesion upon drying. Generally the cohesion in silt soils is due to the presence of clay particles dispersed through the mass. Often as little as 5 to 8 percent clay particles will give significant clay characteristics to a silt. Silt is generally not a very good foundation material unless dry or highly compressed into a sedimentary rock (siltstone). It is normally loose and quite compressible. As a construction material, it is difficult to compact unless the water content is carefully controlled. Material too wet is likely to weave (sometimes termed bull's liver) ahead of compaction equipment.

6. Clay. Mineral particles smaller than silt size (most authorities currently take clay particles as sizes 0.002 mm or smaller). If the particles are smaller than about 0.001 mm (1 μm), they may be called *colloids.* The *clay* particles are complex hydroaluminum silicates ($Al_2O_3 \cdot nSiO_2 \cdot kH_2O$, where n and k are numerical values of molecules attached). *Montmorillonite* is a term for the most common mineral of the montmorillonite group, and is the most active of the identified clay minerals. The particle consists of a gibbsite (approximately, $Al_2O_3 \cdot 3H_2O$) mass sandwiched between two silica sheets for a total thickness of approximately 10 Å (angstroms). This material has a strong affinity for water, and may take on as much as 200 Å of water for a total of 400 Å between clay particles. It is this affinity for water which accounts in a large part for the high shrinkage and swelling characteristics (activity) of the montmorillonite mineral. The usual thickness of water is probably 10 to 100 Å. Another clay mineral, the most commonly occurring of the illite group, *illite*, is less active than montmorillonite, since the adjacent silica layers are bonded with potassium ions, which provide a stronger bond than the water bond of the montmorillonite mineral. A third clay mineral, the most commonly occurring of the kaolin group, is *kaolinite* ($Al_2O_3 \cdot 2SiO \cdot 2H_2O$), the least active of the clay minerals. Some other clay minerals are *bentonite*, which contains large quantities of montmorillonite and is highly active, *halloysite, pyrophyllite, chlorite*, and *vermiculite*. Clay is also defined as a *cohesive* material, that is, a material in which the particles tend to stick together because of either particle attraction or adhesion. Kaolinite, illite, and montmorillonite are the most commonly occurring clay minerals. Clays tend to be named in certain locales. For example, London clay is a clay found in London, England, with certain characteristics. Boston Blue clay from Boston, Mass., has been reported by Casagrande and Fadum (1944) and others. Chicago Blue clay from Chicago, Ill., has been extensively studied by Peck [see Peck and Reed (1954)] and others. Leda clay found in large areas of Ottawa Province in Canada has been extensively studied and reported [Crawford (1961), Soderman and Quigley (1965)]. Typical profiles of the Boston, Chicago, and London clay are illustrated in Fig. 2-3.

7. Terms which tend to be localized:

a. *Adobe.* A clayey material found notably in the Southwest.

b. *Caliche.* A conglomeration of sand, gravel, silt, and clay bonded by carbonates. Usually found in arid areas.

c. *Glacial till* or *glacial drift.* A mixture of material which may include sand, gravel, silt, and clay, deposited by glacial action. Large areas of central North America, much of Canada, northern Europe, the Scandinavian countries, and the British Isles are overlain with glacial till or drift. Drift is usually used to describe any materials laid down by the glacier. Till is usually used to describe materials precipitated out of the ice, but the user must check the context of usage, as the terms are used interchangeably. Moraines are glacial deposits scraped or pushed ahead (terminal), or alongside the glacier (lateral). These deposits may also be called ground moraines if formed by seasonal advances and retreats of a glacier. The Chicago, Ill., area, for example, is underlain by three identifiable ground moraines.

d. *Gumbo.* A clayey or loamy material which is very sticky when wet.

e. *Hardpan.* This term may be used to describe caliche or any other dense, firm deposits which are excavated with difficulty.

f. *Loam.* A mixture of sand, clay, silt; an organic material; also called *topsoil*.

g. *Loess.* A uniform deposit of silt-sized material formed by wind action. Often

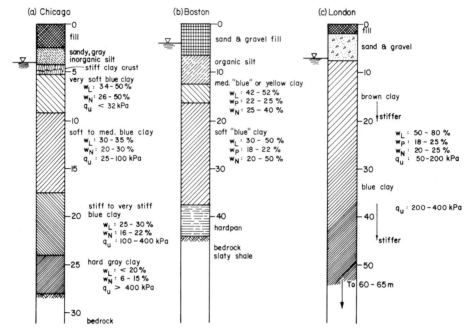

Figure 2-3. Typical soil profiles at locations indicated. Values for soil properties indicate order of magnitude—not to be used for design.

found along the Mississippi River, where damp air rising affects the density of the air transporting the material, causing it to deposit out. Such deposits are not, however, confined to the Mississippi Valley. Large areas of Nebraska, Iowa, Illinois, and Indiana are covered by loess deposits. Large areas of China and Russia (Siberia) and some areas of Europe are covered with loess deposits. Loess is considered to be a *transported* soil.

h. Muck. A thin watery mixture of soil and organic material.

i. Peat. Partly decayed organic matter.

8. Other terms used in soil classification:

a. Alluvial deposits. Soil deposits formed by sedimentation of soil particles from flowing water; may be lake deposits if found in lake beds; deltas at the mouths of rivers; marine deposits if deposited through salt water along and on the continental shelf. Alluvial deposits are found worldwide. For example, New Orleans is located on a delta deposit. The low countries of Holland and Belgium are founded on alluvial deposits from the Rhine River exiting into the North Sea. Lake deposits are found around and beneath the Great Lakes area of the United States. Large areas of the Atlantic coastal plain, including the eastern parts of Maryland, Virginia, the Carolinas, and the eastern part and most of South Georgia, Florida, South Alabama, Mississippi, Louisiana, and Texas consist of alluvial deposits. These deposits formed when much of this land was covered with the seas. Later upheavals such as that forming the Appalachian mountains have exposed this material. Alluvial deposits are fine-grained materials, generally silt-clay mixtures, silts, or clays and fine to medium sands. If the sand and clay layers alternate, the deposit is a *varved* clay. Alluvial deposits are usually soft and highly compressible.

b. *Black cotton soils.* Semitropical soils found in areas where the annual rainfall is 50 to 75 cm. They range from black to dark gray. They tend to become hard with very large cracks (large-volume-change soils) when dry and very soft and spongy when wet. These soils are found in large areas of Australia, India, and Southeast Asia.

c. *Laterites.* Another name for residual soils found in tropical areas with heavy rainfalls. These soils are typically bright red to reddish brown in color. These soils are formed initially by weathering of igneous rocks with the subsequent leaching and chemical erosion due to the high temperature and rainfall. The colloidal silica is leached downward, leaving behind aluminum and iron which become highly oxidized and are relatively insoluble in the high-pH environment (greater than 7). Well-developed laterite soils are generally porous and relatively incompressible. Lateritic soils are found in Alabama, Georgia, South Carolina, many of the Caribbean islands, large areas of Central and South America, and parts of India, Southeast Asia, and Africa.

d. *Residual soil.* Soil formed in place by mechanical and chemical weathering of rocks. This soil is found over much of the Eastern part of the United States east of the Appalachian mountains. It is also found over much of the Southeastern United States and large areas located in the Ozark and Rocky mountains. It is also found on most islands of the world, in large areas of South America, Australia, and parts of Europe.

e. *Saprolite.* Still another name for residual soils formed from weathered rock. Soil is often characterized by soil particles to large angular stones in the soil deposit. Check the context of use to see if the term is being used to describe laterite soils or residual soils.

f. *Shale.* A fine-grained, sedimentary rock composed essentially of compressed and/or cemented clay particles. It is usually laminated from the general parallel orientation of the clay particles as distinct from claystone or siltstone, which are indurated deposits of random particle orientation. According to Underwood (1967), shale is the predominant sedimentary rock in the earth's crust. It is often misclassified; layered sedimentary rocks of quartz or argillaceous materials such as argillite are not shale. Shale may be grouped as (1) compaction shale, and (2) cemented (rock) shale. The compaction shale is a transition material from soil to rock and can be excavated with modern earth-excavation equipment. Cemented shale can sometimes be excavated with excavation equipment but more generally requires blasting. Compaction shales have been formed by consolidation pressure and very little cementing action. Cemented shales are formed by a combination of cementing and consolidation pressure. They tend to ring when struck by a hammer, do not slake in water, and have the general characteristics of good rock. Compaction shales, being of an intermediate quality, will generally soften and expand upon exposure to weathering when excavations are opened. Shales may be clayey, silty, or sandy if the composition is predominately clay, contains consider-

Sieve size or number

System	> 3 in.	3 in.	3/4 in.	No. 4	No. 10	No. 40	No. 200	No. 270	0.005 in.	0.002 in.	0.001 in.
ASTM			Gravel			Sand		Silt		Clay	Colloid
FAA			Gravel			Sand			Silt	Clay	
Unified	Cobbles			Gravel		Sand			Silt or Clay		
AASHO	Boulders		Gravel			Sand		Silt		Clay	Colloids
USDA	Cobbles		Gravel			Sand			Silt		Clay
MIT		Gravel				Sand			Silt		Clay

| Size in millimeters | | 76.2 | 19.05 | 4.76 | 2.00 | 0.42 | 0.074 | 0.05 | 0.005 | 0.002 | 0.001 |

(a)

U.S. Sieve no.	mm	British (B.S.) Sieve no.	mm	German DIN Sieve no.	mm	French Sieve no.	mm
4	4.76	—		—		—	
10*	2.00	8*	2.057	—		34*	2.000
20	0.841	16	1.003	—		31	1.000
30	0.595	30	0.500	500	0.500	28	0.500
		36†	0.422	400†	0.400	27†	0.400
40†	0.420	—		—		—	
50	0.297	52	0.295	—		—	
60	0.250	60	0.251	250	0.250	25	0.250
80	0.177	85	0.178	160	0.160	23	0.160
100	0.149	100	0.152	125	0.125	22	0.125
200	0.074	200	0.076	80	0.080	20	0.080
270	0.053	300	0.053	50	0.050	18	0.050

* Breakpoint between sand and gravel.
† Use for Atterberg Limits.

(b)

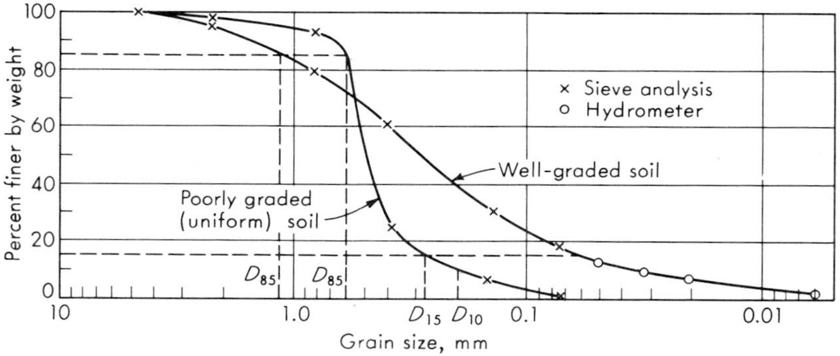

Figure 2-4. (a) Range of grain sizes for various classifications; (b) various standard sieve numbers and screen openings; (c) grain-size-distribution curves.

able silt or sand, respectively. Dry unit weight of shale may range from about 1.28 g/cm^3 for poor-quality compaction shale to 2.56 g/cm^3 for high-quality cemented shale.

Figure 2-4a illustrates the grain-size distribution for several soil-classification systems proposed or in current use. The figure also displays the most commonly used numbered sieves and their openings in millimeters. Figure 2-4c displays typical grain-size distribution curves for a well-graded and a poorly graded soil.

2-11. SOIL CLASSIFICATION FOR FOUNDATION ENGINEERING

It is necessary for the foundation engineer to classify the soils at any site investigated for use as a foundation. This is for several reasons:

1. To be able to use data of others in predicting foundation performance.
2. To build the engineer's background in the "art" application of the design.
3. To maintain a permanent record which can be understood by others should problems later develop and outside parties be required to investigate the original design.

Normally in foundation-design work the Unified Soil Classification system with slight modifications is used (Fig. 2-5). For example, in much foundation work it is academic whether a sand is well graded or poorly graded, but its density and the presence of gravel would be of much more interest. Whether a fine-grained cohesive soil is actually a clayey silt rather than a silty clay is not as important as identifying its strength and settlement characteristics. Because of these reasons, grain-size analyses are often not done, specific-gravity tests are seldom done, and few liquid- and plastic-limit tests may be run. Other tests such as shrinkage, hydrometer compaction, and permeability are run only for special problems.

The foundation engineer relies heavily on a written description of the soil to assist him. Where a soil may simply classify as a SM or CH in the Unified soil system, the foundation engineer would say

"Reddish-brown, dense, silty sand, low plasticity, SM"
"Gray-blue stiff clay with trace of sand and gravel; high plasticity, CH"

The coloring and other distinguishing features such as "dense," "stiff," and "trace of sand" are self-explanatory, but note the ease of cross-referencing to the next job the engineer might have in this area where the exploration program encounters these materials. The terms SM and CH would not convey much information alone, and it is altogether possible that another SM or CH could exist in the same boring at a different depth with very different foundation-design properties (i.e., loose, gravelly, very dense, less silty, higher or lower w_L, soft or hard, etc.).

2-12. SHEAR STRENGTH

The shear strength is generally the property of primary interest when working with soil. Occasionally the tensile strength may be of interest; however, it is usually low, being on the order of 0.15 to 0.6 kg/cm^2 or less [Ramanathan and Raman (1974), Fang and Chen (1971), Bishop and Garga (1969)] as intuitively suspected and sometimes observed.

The shear strength involves two strength parameters and has been presented in a linear equation termed the Mohr-Coulomb equation as

$$s = c + \sigma' \tan \phi_e \qquad (2\text{-}13)$$

where s = shear strength

c = cohesion

σ' = *effective* normal pressure $(\sigma - u)$. The total normal pressure σ is often used instead of the effective pressure

ϕ = angle of internal friction; use effective angle ϕ_e with σ'

Equation (2-13) is the rupture envelope for a soil. It has been found that this envelope and the parameters ϕ and c depend heavily on the following factors [Bowles (1974a, p. 48), Banks and MacIver (1969), Hvorslev (1960), also current literature]:

ϕ angle

1. Method of performing test (UU, CU, CD)
2. Previous stress history (increasing OCR reduces ϕ)
3. Cell or normal pressure (increasing cell or normal pressure tends to reduce ϕ)
4. Grain size and shape (larger grains and increasing angularity tend to increase ϕ)
5. Density (increasing density on the order of only 0.05 g/cm^3 or 3 to 4 pcf may increase ϕ by two or three degrees)
6. Water content (depending on test and water content, ϕ can range from 0° to the correct value)
7. Mineral composition (uncertain effects)

Cohesion

1. Water content
2. Test method (from $\phi = 0$ and maximum cohesion to $c = 0$ in a CD test)
3. Previous stress history (in overconsolidated clays, $c \neq 0$ in all CD tests)
4. Mineral composition

With two unknown quantities (c, ϕ) in Eq. (2-13), at least two tests must be performed in which the shear strength is measured in order to obtain the parameters. It is usual to test three or perhaps four samples and solve Eq. (2-13) graphically as shown in Fig. 2-6 for the cohesion intercept and the slope of the straight line defined by the equation. Note, as shown in Fig. 2-6d, that the "rupture" line may be broken if the soil is preconsolidated. A portion of the rupture line up to the point where the

Figure 2-5. Unified Soil Classification. [Casagrande (1948)].

Major divisions			Group symbols	Typical names	Laboratory classification criteria
Coarse-grained soils (More than half of material is larger than No. 200 sieve size)	Gravels (More than half of coarse fraction is larger than No. 4 sieve size)	Clean gravels (Little or no fines)	GW	Well-graded gravels, gravel-sand mixtures, little or no fines	$C_u = \dfrac{D_{60}}{D_{10}}$ greater than 4; $C_c = \dfrac{(D_{30})^2}{D_{10} \times D_{60}}$ between 1 and 3
			GP	Poorly graded gravels, gravel-sand mixtures, little or no fines	Not meeting all gradation requirements for GW
		Gravels with fines (Appreciable amount of fines)	GM* (d / u)	Silty gravels, gravel-sand-silt mixtures	Atterberg limits below "A" line of I_p less than 4 / Above "A" line with I_p between 4 and 7 are borderline cases requiring use of dual symbols.
			GC	Clayey gravels, gravel-sand-clay mixtures	Atterberg limits above "A" line with I_p greater than 7
	Sands (More than half of coarse fraction is smaller than No. 4 sieve size)	Clean sands (Little or no fines)	SW	Well-graded sands, gravelly sands, little or no fines	$C_u = \dfrac{D_{60}}{D_{10}}$ greater than 6; $C_c = \dfrac{(D_{30})^2}{D_{10} \times D_{60}}$ between 1 and 3
			SP	Poorly graded sands, gravelly sands, little or no fines.	Not meeting all gradation requirements for SW
		Sands with fines (Appreciable amount of fines)	SM* (d / u)	Silty sands, sand-silt mixtures	Atterberg limits below "A" line or I_p less than 4 / Limits plotting in hatched zone with I_p between 4 and 7 are borderline cases requiring use of dual symbols
			SC	Clayey sands, sand-clay mixtures	Atterberg limits above "A" line with I_p greater than 7

Determine percentages of sand and gravel from grain-size curve.
Depending on percentage of fines (fraction smaller than No. 200 sieve size), coarse-grained soils are classified as follows:

Less than 5 percent	GW, GP, SW, SP
More than 12 percent	GM, GC, SM, SC
5 to 12 percent	*Borderline* cases requiring dual symbols†

26

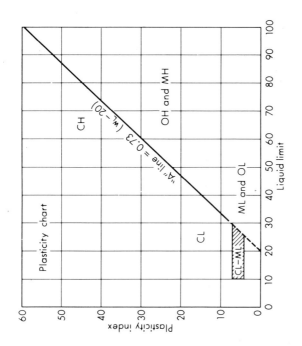

Plasticity chart

A line $= 0.73 (w_L - 20)$

CH

OH and MH

CL

CL-ML

ML and OL

Plasticity index

Liquid limit

(axis values: 0 10 20 30 40 50 60 70 80 90 100 ; 10 20 30 40 50 60)

ML	Inorganic silts and very fine sands, rock flour, silty or clayey fine sands, or clayey silts with slight plasticity	Silts and clays (Liquid limit less than 50)	Fine-grained soils (More than half of material is smaller than No. 200 sieve)
CL	Inorganic clays of low to medium plasticity, gravelly clays, sandy clays, silty clays, lean clays.		
OL	Organic silts and organic silty clays of low plasticity		
MH	Inorganic silts, micaceous or diatomaceous fine sandy or silty soils, elastic silts	Silts and clays (Liquid limit greater than 50)	
CH	Inorganic clays of high plasticity, fat clays		
OH	Organic clays of medium to high plasticity, organic silts		
Pt	Peat and other highly organic soils	Highly organic soils	

* Division of GM and SM groups into subdivisions of d and u are for roads and airfields only. Subdivision is based on Atterburg limits; suffix d used when w_L is 28 or less and the I_P is 6 or less; suffix u used when w_L is greater than 28.

† Borderline classifications, used for soils possessing characteristics of two groups, are designated by combinations of group symbols. For example: GW-GC, well-graded gravel-sand mixture with clay binder.

27

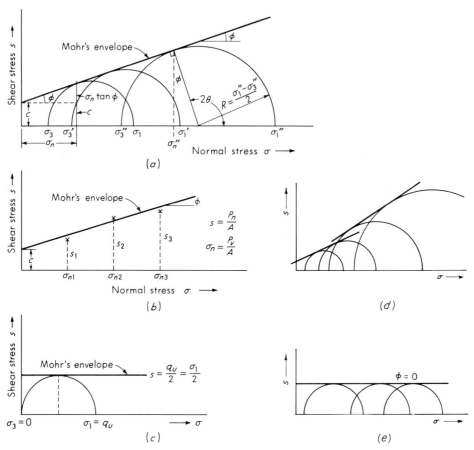

Figure 2-6. Mohr's circle. (*a*) Typical plot from triaxial test data; (*b*) plot from direct-shear data; (*c*) plot from unconfined-compression-test data; (*d*) plot of triaxial tests on overconsolidated clay; (*e*) undrained triaxial tests on any saturated soil.

laboratory test consolidation (or confining) pressure σ_3 equals the in situ normal or preconsolidation pressure will have a smaller ϕ angle and a cohesion intercept (probably even for consolidated-drained tests if the soil is preconsolidated). The location of the "break" in the rupture line may also depend on how the sample is consolidated, i.e., anisotropically ($\sigma_1 \neq \sigma_3$) or isotropically ($\sigma_1 = \sigma_3 = \sigma_c$) consolidated [Ladd (1966)]. When the testing pressures exceed the in situ pressures, the slope of the ϕ–c curve increases as shown, with an intercept at the origin of zero for normally consolidated soils in consolidated-undrained tests and consolidated-drained tests if pore pressure is accounted for.

The strength parameters c and ϕ are usually obtained from:

1. Direct shear tests [laboratory tests, see Bowles (1970)].
2. Triaxial tests [laboratory tests, see Bishop and Henkel (1962)] and may be
    a. Compression with confining pressure constant and increasing the deviator stress to failure.

b. Compression with confining pressure increasing.

c. Extension with confining pressure constant and decreasing the axial (or deviator) stress.

d. Extension with confining pressure increasing and axial (or deviator) stress constant [for a short assessment of extension test errors, see Barden and Khayatt (1966, p. 343)].

e. Other tests, depending on local equipment and technician ability [Sowers (1963) lists several variations with references].

3. Borehole shear tests (in situ test)—this is a recent innovation which involves placing a shear device into a borehole and performing essentially a consolidated-undrained or consolidated-drained direct shear test [Wineland (1975)].

4. Vane shear tests [laboratory or, more generally, in situ, see Arman *et al.* (1975), with references].

The direct shear test is often called a *plane-strain* test in that the shear box configuration tends to limit deformations to the plane of failure. It appears that the angle of internal friction from plane-strain shear tests varies from 0 to about 8° higher than values obtained from triaxial tests [Lee (1970) with extensive sources of tests cited]. General details of the direct shear and triaxial test are illustrated in Fig. 2-7.

Both direct shear and triaxial tests may be done as follows:

1. Unconsolidated-undrained (UU or quick) tests—these tests are done immediately upon applying the normal load or cell pressure and so rapidly that no pore water can escape from the failure plane during the test. This test always gives $\phi = 0°$ for cohesive soils unless the soil is partially saturated and the pore pressure is measured. Strain rates vary from about 0.5 to 1.5 (0.02 to 0.05 in/min) mm/min.

2. Consolidated-undrained (CU or consolidated-quick) tests—these tests are done after the volume change due to applying the normal load or confining cell pressure has halted. The test is then performed similarly to the UU test above. This test usually gives $\phi = 0°$ to small angle values for cohesive soils. If the soil is partially saturated and pore pressure is measured, the test gives the true soil parameters c and ϕ.

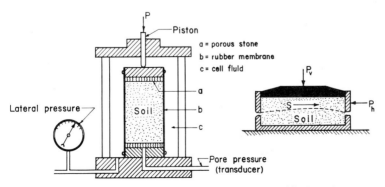

Figure 2-7. Typical laboratory test setup for (*a*) triaxial test; (*b*) direct shear test.

3. Consolidated-drained (CD or slow) tests—these tests are done after applying the normal load or confining cell pressure, waiting for complete consolidation, then testing the sample to failure at a slow enough strain rate (perhaps 1 to 3 percent/24 hr) that no excess pore pressure develops on the failure plane within the sample. This test gives the "effective" or true soil parameters c and ϕ_e. In normally consolidated soils this test would be expected to give $c \simeq 0$.

For dry, damp, or saturated cohesionless soils, the above three test methods in the direct shear test give essentially the same results. In the triaxial test, the water content (or degree of saturation) will influence the results considerably, as will the test method. This is because it is difficult to control drainage in the direct shear device, but with the sample confined inside a rubber triaxial membrane it is easy to control external drainage. The lubrication effect of water on the angle of internal friction is on the order of 1 to 2° (see Table 2-2).

The *unconfined-compression test* is a special triaxial test in which the confining pressure is atmospheric (or zero). This test is often termed an "unconsolidated-undrained" or simply an "undrained" test, and this text will always use the symbol q_u for the unconfined-compression strength. The UU test is also often called a $\phi = 0°$ test, since the test conditions of the unconfined-compression test produce the conditions of Fig. 2-6c, in which $\phi = 0°$. The UU test in general produces the conditions shown in Fig. 2-6e. The unconfined-compression test can be used only for cohesive soil.

Table 2-2. Representative values for angle of internal friction ϕ

Soil	Type of test*		
	Unconsolidated-undrained UU	Consolidated-undrained CU	Consolidated-drained CD
Gravel			
Medium size	40–55°		40–55°
Sandy	35–50°		35–50°
Sand			
Loose dry	28.5–34°		
Loose saturated	28.5–34°		
Dense dry	35–46°		˙43–50°
Dense saturated	1–2° less than dense dry		43–50°
Silt or silty sand			
Loose	20–22°		27–30°
Dense	25–30°		30–35°
Clay	0° if saturated	14–20°	20–42°

* See a laboratory manual on soil testing for a complete description of these tests, e.g., Bowles (1970).

There is some opinion [Skempton and Sowa (1963)] that the unconfined-compression test gives as good soil parameters as a triaxial test. This has been attributed to:

1. The fact that as the sample tends to expand during unloading, recovery causes a negative pore pressure (or soil suction) to build up which may be equal to the in situ stresses
2. The fact that surface-tension effects in the lower-humidity laboratory will tend to "confine" the specimen during testing by values on the order of the in situ stresses
3. The disturbance associated with inserting a triaxial specimen into the rubber membrane

It would appear, however, that 1 and 2 above are severely limited by the maximum surface tension of the pore water. Ladd and Lambe (1963), with extensive references and data on nine soils, refute at least items 1 and 2 above [see also discussion by Ladd and Bailey (1964) of Skempton and Sowa] and show that the UU test considerably underestimates the shear strength of a soil. They show that the measured shear strength may be 40 to 80 percent of the true shear strength. The opinion of most investigators is that sample disturbance results in laboratory strength tests being generally about 50 percent smaller than in situ determinations. Actual values tend to range from 30 to 80 percent and are principally dependent on overconsolidation ratio (OCR), water content, sensitivity, and sampling methods.

The fact that different persons obtain wide-ranging values of ϕ and c from shear tests, which is attributable to material and test differences, data interpretation, and sampling disturbance, has had the result of generating a tremendous amount of both substantiating and conflicting (and confusing) literature in the recent past and continuing. Two relatively recent conferences have been devoted entirely to shear strength of soils: Research Conference on Shear Strength of Cohesive Soils at Boulder, Colo. (1960 with 1,164 pages) published by ASCE and Laboratory Shear Testing of Soils, STP 361 (1964 with 505 pages) published by ASTM. All the geotechnical engineering journals regularly carry papers on shear strength, more tests on different soils, and new interpretations of data from tests and field observations.

Practical problems arise in obtaining strength parameters because of the soil type, previous (geological) stress history, and sample disturbance during recovery. Samples of sensitive soils are difficult to recover, and the vane shear is an attempt to overcome this problem. Soils (especially residuals and glacial tills) with varying amounts of gravel are difficult to recover and to produce test samples with the ends adequately squared and of sufficient length for testing. Recovery of undisturbed cohesionless samples for a triaxial test is nearly impossible. Preconsolidated soils will yield different results depending on test orientation with the bedding plane. Sample rebound due to unloading of in situ stresses is always a problem. Some of the latter problems are claimed to be solved using the borehole shear device; however, bedding-plane orientation will of necessity always be perpendicular to the vertical boreholes most commonly used.

The best samples for laboratory testing are obtained from block samples carefully hand-cut from the side or bottom of a test pit. This may not be extremely expensive if the sample is not at a great depth and the pit can be excavated by a backhoe or similar mechanical device. The next samples in order of quality are thin-walled tube samples, with quality deteriorating as the tube diameter decreases. Last in quality are the samples obtained from the penetration test described in Chap. 3.

Engineering judgment must be applied to interpret correctly and use the shear-strength and/or the appropriate test to obtain the shear-strength parameters. Generally the unconsolidated-undrained test will provide satisfactory data where the foundation loadings are so rapid that UU conditions are obtained in the field, such as rapidly constructed clay embankments over soft clay deposits, the clay cores of rapidly constructed earth dams where little water-content change can occur, spread footings placed on a clay or clayey deposit and rapidly loaded, or spread footings placed on a soil overlying a clay deposit at a shallow depth and rapidly loaded.

A consolidated-undrained test may be more appropriate when an earth embankment is slowly constructed in the cases cited above for UU analyses or the footings are more slowly loaded. Consolidated-undrained tests would be appropriate for sites where the UU strength is so low the soil is incapable of supporting the proposed loads but by preloading the site to increase the shear strength the soil might be usable. CU tests could be used to indicate the expected soil strength after preloading.

Consolidated-drained tests may be used for the true soil parameters as in research studies. The true parameters may be of some use in certain long-time-strength evaluations. Owing to the cost of these tests, it is doubtful if there are many actual site conditions which would warrant this expense—especially when the soil loadings are likely to go through either UU or CU (or both) conditions prior to attaining CD stress conditions.

Shear strength may be estimated from penetration tests (N) as indicated in Tables 3-3 and 3-4.

Graphically, shear strength is plotted on a Mohr's circle, as illustrated in Fig. 2-6, to arrive at a "Mohr's envelope," or the line for which the shear-strength equation applies. Definitions of the symbols used in Fig. 2-6 are as follows:

σ_3 = confining-chamber pressure for triaxial test
σ_1 = maximum stress, or stress at 20 percent strain from stress-strain curve of a compression test (20 percent being an estimate of ultimate stress when the stress-strain curve exhibits large amounts of residual soil strength)
s = maximum shearing stress; from a direct shear test this is a measured value
σ_n = normal stress from normal load applied to a direct shear specimen or normal stress on the failure plane in the compression test

By inspection of Fig. 2-6a,

$$s = c + \sigma_n \tan \phi \qquad (a)$$

$$= \frac{\sigma_1 - \sigma_3}{2} \sin 2\theta \qquad (b)$$

$$\sigma_n = \frac{\sigma_1 + \sigma_3}{2} + \frac{\sigma_1 - \sigma_3}{2} \cos 2\theta \qquad (c)$$

From a simultaneous solution of the above equations, one obtains the following widely used equations:

$$\sigma_1 = \sigma_3 \tan^2 \left(45° + \frac{\phi}{2}\right) + 2c \tan \left(45° + \frac{\phi}{2}\right) \tag{2-14}$$

$$\sigma_3 = \sigma_1 \tan^2 \left(45° - \frac{\phi}{2}\right) - 2c \tan \left(45° - \frac{\phi}{2}\right) \tag{2-15}$$

The lower limit of the angle of internal friction of a dry, cohesionless soil can be estimated by carefully pouring the dry material into a pile and measuring the *angle of repose*, or slope the pile forms with the horizontal.

For fine-grained soils which undergo marked changes in shearing stresses when disturbed, a *vane shear* may be used. Laboratory versions of this device exist as well as devices used in field testing (Chap. 3). The vane can be inserted in a laboratory (or field) sample, and the applied torque related to the shear strength as

$$T = s\pi \left(\frac{d^2 h}{2} + \frac{d^3}{6}\right) \tag{2-16}$$

where T = applied torque
   d, h = respectively, diameter and height of vane
   s = shear strength of soil

This equation assumes a uniform stress distribution on the two ends of the vane. For a triangular or parabolic distribution of end shear stresses the $d^3/6$ term must be adjusted (see Sec. 3-13).

The laboratory device can be calibrated for a fixed-size vane, so that the shear strength is related to the angle through which the known torque is applied. With the angle measured, a chart may be entered and the shear strength read directly. The laboratory test can be performed on undisturbed tube samples (Chap. 3), as well as the standard compaction-test samples, as long as the soil is cohesive and does not contain large grains (pebbles) of soil. A pocket vane device is available which is pushed into a sample about 6 mm. It is self-reading, and the user simply rotates the device until a shear failure occurs. The shear strength of the soil is read directly from the device.

2-13. ELASTIC PROPERTIES OF SOIL

The stress-strain modulus (also modulus of elasticity) and Poisson's ratio are the principal elastic properties of interest. Both the stress-strain modulus E_s and Poisson's ratio μ are of use in evaluating settlements and may be used to compute the modulus of subgrade reaction. The modulus of subgrade reaction is a useful concept in soil-structure interaction problems in that the soil medium can be replaced by a system of springs with a spring constant which is dependent on the value of subgrade reaction.

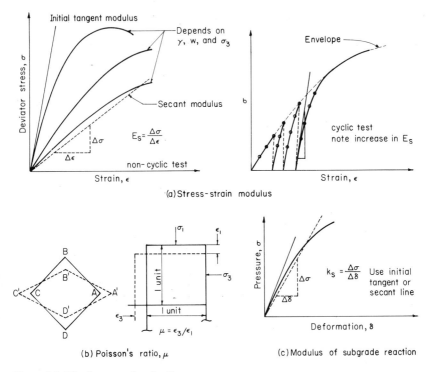

Figure 2-8. Elastic properties of soil.

The stress-strain modulus is computed from mechanics of materials (refer to Fig. 2-8a; typical values in Table 2-3) as

$$E_s = \frac{\text{stress}}{\text{strain}} = \frac{\sigma}{\varepsilon} \qquad (d)$$

Poisson's ratio is defined as the ratio of lateral strain ε_3 to longitudinal strain ε_1 when the applied stress is uniaxial (refer to Fig. 2-8b; typical values in Table 2-4).

$$\mu = \frac{\varepsilon_3}{\varepsilon_1} \qquad (e)$$

The modulus of subgrade reaction is defined as the ratio of stress to deformation (refer to Fig. 2-8c)

$$k_s = \frac{\sigma}{\delta} \qquad (2\text{-}17)$$

The shearing modulus G, defined as the ratio of shear stress to shear strain, is related to E_s and μ as follows:

$$G = \frac{\text{shear stress}}{\text{shear strain}} = \frac{s}{\varepsilon_s} = \frac{E_s}{2(1 + \mu)} \qquad (f)$$

Table 2-3. Typical range of values for the static stress-strain modulus E_s for selected soils. Field values depend on stress history, water content, density, etc.

Soil	E_s ksi	E_s kg/cm^2
Clay		
Very soft	0.05–0.4	3–30
Soft	0.2–0.6	20–40
Medium	0.6–1.2	45–90
Hard	1–3	70–200
Sandy	4–6	300–425
Glacial fill	1.5–22	100–1,600
Loess	2–8	150–600
Sand		
Silty	1–3	50–200
Loose	1.5–3.5	100–250
Dense	7–12	500–1,000
Sand and gravel		
Dense	14–28	800–2,000
Loose	7–20	500–1,400
Shales	20–2,000	1,400–14,000
Silt	0.3–3	20–200

Table 2-4. Typical range of values for Poisson's ratio μ

Type of soil	μ
Clay, saturated	0.4–0.5
Clay, unsaturated	0.1–0.3
Sandy clay	0.2–0.3
Silt	0.3–0.35
Sand (dense)	0.2–0.4
Coarse (void ratio = 0.4–0.7)	0.15
Fine-grained (void ratio = 0.4–0.7)	0.25
Rock	0.1–0.4 (depends somewhat on type of rock)
Loess	0.1–0.3
Ice	0.36
Concrete	0.15

From Fig. 2-8 the shearing strain is the change in the right angle at the point such that

$$\varepsilon_s = \text{angle } BAD - \text{angle } B'A'D' \qquad (g)$$

Another concept occasionally used is volumetric strain, defined as

$$\varepsilon_v = \frac{\Delta V}{V} \qquad (h)$$

The bulk modulus E_b is defined as the ratio of hydrostatic stress to the volumetric strain ε_v.

$$E_b = \frac{2}{3} G \frac{1 + \mu}{1 - 2\mu} = \frac{E_s}{3(1 - 2\mu)} \qquad (i)$$

Since G and E_b cannot be negative, Eqs. (f) and (i) set the following limits of μ: $-1 \leq \mu \leq 0.5$. In soils it appears the range of Poisson's ratio is 0.00 to 0.50.

Hooke's generalized stress-strain law in terms of principal strains and stresses is

$$\varepsilon_1 = \frac{1}{E}(\sigma_1 - \mu\sigma_2 - \mu\sigma_3)$$

$$\varepsilon_2 = \frac{1}{E}(\sigma_2 - \mu\sigma_1 - \mu\sigma_3) \qquad (2\text{-}18)$$

$$\varepsilon_3 = \frac{1}{E}(\sigma_3 - \mu\sigma_1 - \mu\sigma_2)$$

which simplifies to

$$\varepsilon_1 = \frac{1}{E}(\sigma_1 - 2\mu\sigma_3) \qquad (2\text{-}19)$$

for the case of triaxial tests where only the confining pressure σ_3 and the deviator stress are acting (and assuming consolidation under σ_3). If we "smooth" out the results of a triaxial test by plotting the stress-strain curve and run a smooth curve through the points, we should be able to solve Eq. (2-19) for the unknown value of stress-strain modulus and Poisson's ratio by taking two adjacent curve points (close enough that we may assume a linear variation between the two points) to obtain values of strain ε_i and stress. Doing this on the computer so that a large number of points that are very close together may be used, and starting with ε very close to 0.0, it has been found consistently that Poisson's ratio exceeds 0.5 at very small strain values. This indicates that there is really only a very small region in which soil is "elastic," since Poisson's ratio greater than 0.5 would indicate "plastic" behavior. It is also found that Poisson's ratio depends on both stress level and strain.

Since soil is "elastic" for only small strain values, the initial tangent modulus is generally the best estimate of the stress-strain modulus. It also appears that some caution must be exercised in applying the finite-element method to soil masses in that beyond certain strains one may only be manipulating numbers.

In using the triaxial test to obtain the stress-strain modulus, some persons suggest that the sample be cyclic loaded-unloaded three to five times or more and the stress-strain modulus is then computed as the stress-strain modulus of the last reload [Barden et al. (1969), Sangrey et al. (1969), Leonards (1968), Soderman et al. (1968), Makhlouf and Stewart (1965), Larew and Leonards (1962)]. This method of testing is to allow sample seating, grain readjustment to the original in situ condition, strain hardening, and pore-pressure stabilization. The initial tangent modulus by this method tends to be more than on the first cycle but depends on soil, confining pressure, and ratio of deviator stress/confining pressure (usually with deviator stresses of about one-third to one-half estimated failure value) on the load-unload cycles (see Fig. 2-8a).

The most common method of computing the stress-strain modulus is to use the initial tangent value, or the slope of the stress-strain curve at the origin. Values may also be obtained as the initial secant modulus using the origin and the secant-line intercept at a stress level of, say, one-third to one-half of the deviator stress at ultimate or failure (σ_f). Alternatively, the secant modulus using intercepts of $\frac{1}{4}$ and $\frac{3}{4}$ σ_f could be used depending on the user, problem, and estimated working-stress level. One must be careful, however, to use stresses in the strain region where the soil is at least pseudo-elastic; otherwise the stress-strain modulus is simply a number.

The stress-strain modulus of soils (and Poisson's ratio) is heavily dependent on

1. Method of performing the compression test (unconfined, triaxial, and whether UU, CU, CD, also whether "compression" or "extension")
2. Confining pressure in cell (increases with increasing cell pressure at least to somewhat above any preconsolidation pressure)
3. Overconsolidation ratio (OCR), defined as the preconsolidation pressure/presen overburden pressure
4. Water content of soil—low water contents tend to higher moduli values (and brittle fracture at low strain values)
5. Density
6. Strain rate—at low rates the modulus value is less in general than at higher rates

The stress-strain curve for all soils is curved for the full region except possibly a small portion near the origin for very small strain values (which are not normally obtained in a test). Kondner (1963) [see also Duncan and Chang (1970)] proposed that the stress-strain curve (Fig. 2-9a) could be represented by a hyperbolic equation of the form

$$\sigma_1 - \sigma_3 = \frac{\varepsilon}{a + b\varepsilon} \tag{2-20}$$

which could be rewritten as

$$\frac{\varepsilon}{\sigma_1 - \sigma_3} = a + b\varepsilon \tag{2-20a}$$

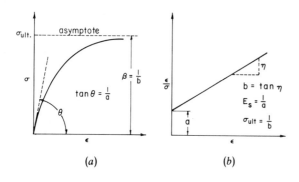

Figure 2-9. (*a*) Usual stress-strain plot —hyperbolic-curve approximation; (*b*) transformed stress-strain representation of stress strain—gives approximate linear curve as shown. [*After Kondner (1963)*.]

The left side of Eq. (2-20a) is known for various values of deviator stress and the corresponding strain; thus a linear plot can be made as shown in Fig. 2-9b which gives the values of the constants *a* and *b*. While Kondner proposed this procedure specifically for clay soils, it should be equally applicable for any soil (or material) with a nonlinear stress-strain response. This method of presenting the stress-strain data is useful in computer applications where the computer program develops the deviator stress and by using Eq. (2-20) can obtain the corresponding strain; with the strain and deviator stress, Eq. (*a*) can be used to compute a stress-strain modulus number for that stress level in the soil mass.

2-14. ISOTROPIC AND ANISOTROPIC SOIL MASSES

An *isotropic* material is one in which the elastic properties (stress-strain modulus E_s and Poisson's ratio μ) are the same in all directions. In *anisotropic* materials the elastic properties are different in different directions. A material is *homogeneous* when the properties are the same throughout the volume of interest. A soil mass is often assumed to be isotropic and homogeneous as a computational convenience.

Two elastic constants E_s and μ are necessary to define the stress-strain relation of an isotropic material. The matrix representation of the generalized Hooke's law for an isotropic material is shown in Eq. (2-21):

$$\varepsilon = D\sigma \qquad (2\text{-}21)$$

where the matrix D has the following form:

$$D = \frac{1}{E_s}$$

$\varepsilon \,\backslash\, \sigma$	1	2	3
1	1	$-\mu$	$-\mu$
2	$-\mu$	1	$-\mu$
3	$-\mu$	$-\mu$	1

For the most general case of anisotropic material, 21 elastic constants appear in the stress-strain relation. But the number of elastic constants is reduced if the medium is elastically symmetric in certain directions. Transversely isotropic (cross-anisotropic) materials have five elastic constants. In general, soil is deposited in horizontal layers, and it is reasonable to assume that the elastic properties of soil are symmetrical about the vertical axis and different from the elastic properties in the vertical direction. Thus the theory of cross-anisotropic materials may very well be applied to soil deposits. Seven elastic constants for cross-anisotropic materials are defined as follows (actually, of course, only five of these seven constants are independent): The xz plane is the plane of isotropy, which is horizontal, and the y axis is vertical.

E_V = stress-strain modulus in the vertical direction
E_H = stress-strain modulus in the horizontal plane, i.e., in the plane of isotropy
$\mu_1 = \varepsilon_z/\varepsilon_x$ when the applied stress is σ_x
$\mu_2 = \varepsilon_x/\varepsilon_y$ when the applied stress is σ_y
$\mu_3 = \varepsilon_y/\varepsilon_x$ when the applied stress is σ_x
G_H = shear modulus in the horizontal plane
G_V = shear modulus in the vertical plane

But
$$G_H = \frac{E_H}{2(1 + \mu_1)}$$

and
$$\frac{\mu_2}{E_V} = \frac{\mu_3}{E_H} \tag{a}$$

So the five elastic constants for a cross-anisotropic material are G_V, E_V, E_H, μ_1, and μ_2. A more detailed discussion on cross-anisotropic behavior of soil deposits can be found in Bhatacharya (1968).

The generalized Hooke's law for cross-anisotropic material takes the following form:

$$\varepsilon_x = \frac{\sigma_x}{E_H} - \mu_2 \frac{\sigma_y}{E_V} - \mu_1 \frac{\sigma_z}{E_H}$$

$$\varepsilon_y = \frac{\sigma_y}{E_V} - \mu_3 \frac{\sigma_x}{E_H} - \mu_3 \frac{\sigma_z}{E_H}$$

$$\varepsilon_z = \frac{\sigma_z}{E_H} - \mu_1 \frac{\sigma_x}{E_H} - \mu_2 \frac{\sigma_y}{E_V}$$

$$\gamma_{xy} = \frac{\tau_{xy}}{G_V} \quad \gamma_{xz} = \frac{\tau_{xz}}{G_H} \quad \gamma_{yz} = \frac{\tau_{yz}}{G_V}$$

$$\tag{b}$$

For problems of plane strain (when $\varepsilon_z = \gamma_{xz} = \gamma_{yz} = 0$)

$$\sigma_z = \mu_1 \sigma_x + \mu_2 \frac{E_H}{E_V} \sigma_y \tag{c}$$

Substituting Eq. (c) in Eqs. (b), using Eq. (a) to obtain μ_3, and noting that $\gamma_{xz} = \gamma_{yz} = 0$, the following form of the generalized Hooke's law for cross-anisotropic material in plane strain is obtained:

$$\varepsilon_x = A\sigma_x + B\sigma_y$$

$$\varepsilon_y = B\sigma_x + C\sigma_y$$

$$\gamma_{xy} = \frac{\tau_{xy}}{G_V}$$

where
$$A = \frac{1 - \mu_1^2}{E_H} \qquad\qquad B = \frac{-\mu_2 - \mu_1 \mu_2}{E_V}$$

$$C = \frac{1 - n\mu_2^2}{E_V} \qquad \text{and} \qquad n = \frac{E_H}{E_V}$$

(d)

Hence, the D matrix for plane-strain problems of cross-anisotropic materials is

ε \ σ	1	2	3
1	A	B	0
2	B	C	0
3	0	0	$\dfrac{1}{G_V}$

$D =$ (e)

Thus for plane-strain problems of cross-anisotropic materials it is only necessary to know the four parameters A, B, C, and G_V, which can be determined [Chowdhury (1972)] as follows:

1. Perform a set of plane-strain triaxial tests with a constant cell pressure on a sample with the plane of isotropy horizontal.
2. Plot the deviator stress vs. axial strain.
3. Plot the deviator stress vs. lateral strain. The lateral strain can be computed from the axial strain and volume-change measurements.
4. Compute: $1/B$ = slope of curve of step 3.
         $1/C$ = slope of curve of step 2.
5. Perform a plane-strain triaxial test with a constant cell pressure on a sample with the plane of isotropy vertical such that the direction of plane strain is parallel to the plane of isotropy.
6. Plot steps 2 and 3 above to obtain a second set of curves.
7. Compute: $1/B$ = slope of curve of step 3 (should check reasonably with step 4).
         $1/A$ = slope of curve of step 2.
8. Test a sample with the plane of isotropy inclined at 45° to the horizontal (samples may be difficult to obtain except from a test pit).

9. Plot the deviator stress vs. axial strain. The slope $d\sigma/d\varepsilon$ of the curve is related to G_V by the following equation:

$$G_V = \frac{1}{4/\text{slope} - (A + 2B + C)} \qquad (f)$$

Thus the four constants required to solve the plane-strain problems of cross-anisotropic soil can be obtained from three sets of plane-strain triaxial tests; one set of tests is on soil samples with the plane of isotropy horizontal; the second set is on samples with the plane of isotropy vertical; and the third set is on samples with the plane of isotropy 45° inclined to the horizontal. Effective or total stresses may be used as appropriate, but all values should be consistent. Since the value of G_V is particularly critical [Raymond (1970)], all four constants A, B, C, and G_V must be correctly determined if one wants to consider the cross anisotropy of the soil. If the correct evaluation of each of the four constants is not possible, the soil should be treated as an isotropic material.

2-15. IN SITU STRESSES AND K_0 CONDITIONS

When soil is formed at a site, either residual from in situ rock weathering or transported via deposition of sediments, any element within the soil mass is confined by the surrounding soil. After the deposit is completely formed, the soil structure reaches equilibrium conditions with respect to both stresses and strains. Each element within the soil mass is subjected to both total and effective vertical and lateral stresses of such magnitude as to result in a condition of zero strain. This condition is referred to in soil-mechanics literature as the K_0 *condition*. This is an important condition of state, and any new stresses in the soil mass caused either by applying building loads or from removal of loads, as along the bottom or sides of excavations, will use the K_0 state as the starting reference point.

Triaxial tests in laboratory testing to determine the stress-strain modulus and the shear-strength parameters ϕ and c should involve using initial confining, σ_3, and σ_1 stresses which reproduce the K_0 stress state in the test specimen. Since ϕ and c are determined from the test, one must usually estimate K_0 or the triaxial cell pressure/sample stress conditions in advance. One of the early proposals of the at-rest earth-pressure coefficient K_0 was made by Jaky (1948):

$$K_0 = 1 - \sin \phi \qquad (2\text{-}22)$$

Later investigators [Wroth (1972), Myslivec (1972), with several equations, Brooker and Ireland (1965)] have shown that this equation is reasonably satisfactory for cohesionless soils (and for various materials such as corn and wheat in silo storage) but the use of

$$K_0 = 0.95 - \sin \phi \qquad (2\text{-}22a)$$

would be more appropriate for normally consolidated clay. In both the above equa-

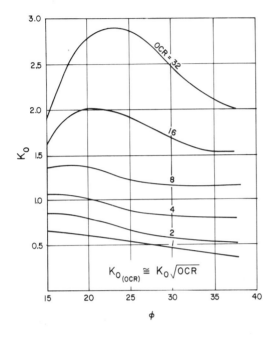

Figure 2-10. Relationships between overconsolidation ratio (OCR), K_0, and *drained* angle of internal friction. [*After Brooker and Ireland (1965)*.]

tions the effective or CD value of angle of internal friction ϕ should be used. Brooker and Ireland attempted to relate K_0 to the plasticity index I_p of normally consolidated clays. Over the region from $I_p = 0$ to 40 percent the relationship is approximately linear with the following approximate equation:

$$K_0 = 0.40 + 0.007I_p \tag{2-23}$$

In the region $I_p = 40$ to 80, K_0 is approximately

$$K_0 = 0.68 + 0.001(I_p - 40) \tag{2-23a}$$

Figure 2-10 illustrates the effect of OCR and angle of internal friction on K_0. As one would expect, anisotropy, where the soil is heavily preconsolidated, results in larger values of K_0. One would also expect that the upper limit on K_0 would be the passive earth-pressure coefficient of Chap. 11. It should be emphasized that the K_0 state is a condition of zero lateral strain and is not to be confused with the concept of active and passive pressures considered later in Chaps. 11 and 12. The latter values may be limiting cases of the K_0 state depending on the past stress history of the soil mass. Since the K_0 state is a condition of no lateral strain, Poisson's ratio is uniquely related. From the general Hooke's law relationships in Eqs. (2-18) and setting the strain $= 0$ in the second of the equations, we have

$$\frac{\sigma_2 - \mu\sigma_1 - \mu\sigma_3}{E} = 0$$

Also let $\sigma_2 = \sigma_3 = K_0 \sigma_1$ and substituting, we have

$$K_0 = \frac{\mu}{1 - \mu}$$

also, rearranging, we have

$$\mu = \frac{K_0}{1 + K_0}$$

Methods of obtaining the at-rest earth-pressure coefficient K_0 in situ are presented in Chap. 3.

2-16. CONSOLIDATION CHARACTERISTICS

The settlement characteristics of saturated cohesive soils may be estimated using laboratory consolidation (also oedometer) tests. The consolidation test is performed by placing a carefully trimmed "undisturbed" sample into a ring ranging from about 4.5 to 11.28 cm (1.8 to $4\frac{1}{2}$ in) in diameter, and 2 to 3 cm (0.75 to 1.5 in) in height with a porous stone on the top and/or bottom to facilitate sample drainage. A compressive load is then applied to the sample. If the ring is not attached to the base of the machine, compression takes place on both faces of the sample, and the device is termed a *floating-ring consolidometer*. The floating ring theoretically reduces side-ring friction on the sample, but the fixed-ring device (settlement from one face only, with the ring attached to the machine base) is required if rates of pore-pressure dissipation are to be determined. Rings may be lined with Teflon to reduce friction effects. The use of larger rings should provide relatively smaller errors, since friction and sample disturbance should be about the same for both large and small samples.

Consolidation of a soil will be defined as a void-ratio reduction which takes place as a function of time. It is considered to be taking place during the time excess pore pressure exists in the consolidating stratum owing to an increase in pressure within the stratum from exterior loading. This consolidation is also termed *primary consolidation*. *Secondary consolidation* is that additional void-ratio change (settlement) taking place after the excess pore pressure has essentially dissipated. This consolidation is sometimes termed *creep* and may continue for some considerable time after primary consolidation is complete. This is especially true in organic or peat soils. For most soils secondary compression is much less than "consolidation" settlements, except for the organic or peat soils mentioned.

The samples are loaded with increasing load increments (usually a load-increment ratio $\Delta p/p = 1$), and settlements as a function of time are recorded. Studies by Leonards and Girault (1961) and Leonards and Ramiah (1959) indicate that load-increment ratios less than 1 do not generally give reliable results. Smaller load-increment ratios may be necessary for soft soils—at least in the initial-loading stage to avoid soil-structure failure. Duration of load increment does not seem important as long as the increment is of sufficient duration to allow most of the primary consolidation to take place in the case of clays and most of the secondary compression to take place in the case of organic or peat soils.

These data are plotted to settlement (dial reading) vs. log time (minutes), or $\sqrt{t}$, as illustrated in Fig. 2-11a and b. This operation is necessary to estimate the time when 100 percent primary (excess pore pressure is zero) consolidation takes place. On the dial reading (DR) vs. log t plot, the time and dial reading at 100 percent consolidation is found by drawing two tangents, as illustrated in Fig. 2-11a. For the dial reading vs. $\sqrt{t}$ plot, Taylor (1948) recommends drawing a straight line through the first few points, extended as required. A second line with an abscissa 15 percent larger is then laid off from the same origin. Additional points are established on the deformation curve until the curve crosses the 1.15 offset line. This settlement is taken as 90 percent of the primary consolidation. By extrapolating the settlement curve and increasing the 90 percent settlement by 10 percent and plotting to the extrapolated settlement curve, the time of 100 percent primary consolidation is found. This method is illustrated in Fig. 2-11b. The initial dial reading D_0 must also be established. On the DR vs. log t curve this is done by inspecting the curve; and if it is approximately parabolic proceeding as follows: A difference in ordinates between any two points with time ratios of 4 : 1 in the parabolic region is marked off. This distance is then laid off above the upper point (lowest time). Several points should be used, and the ordinate defined by a horizontal line drawn through the average of these points may be taken as D_0. If the early part of the curve is not parabolic, use the initial dial reading as D_0. For the $\sqrt{t}$ method, where the straight line intersects, $\sqrt{t} = 0$ is taken as D_0. In both cases

$$D_{50} = \frac{D_0 + D_{100}}{2}$$

from which t_{50} may be found (D_{50} and t_{50} refer to the dial reading and corresponding time for 50 percent of primary consolidation to occur).

An expression to attempt to relate primary consolidation to the total consolidation (primary + secondary consolidation) can be mathematically expressed as

$$r = \frac{D_a - D_{100}}{D_0 - D_f}$$

where D_a is the actual dial reading at the beginning of the test (a load increment) and D_f will be the final dial reading of the current load increment. Of course, the closer r is to unity, the less the theoretical secondary consolidation is to be expected. However, the validity of this expression may be suspect if the test is of short duration [Barber (1961)], as, for example, using the Taylor square-root-of-time fitting technique.

From laboratory time data for a percent of consolidation of the sample, the time for primary consolidation can be estimated in terms of a *consolidation coefficient* as

$$c_v = \frac{TH^2}{t} \tag{2-24}$$

where T = the time factor; Table 2-5 gives values of T for various percentages of consolidation for the usually assumed distributions of water pressures through depth of sample

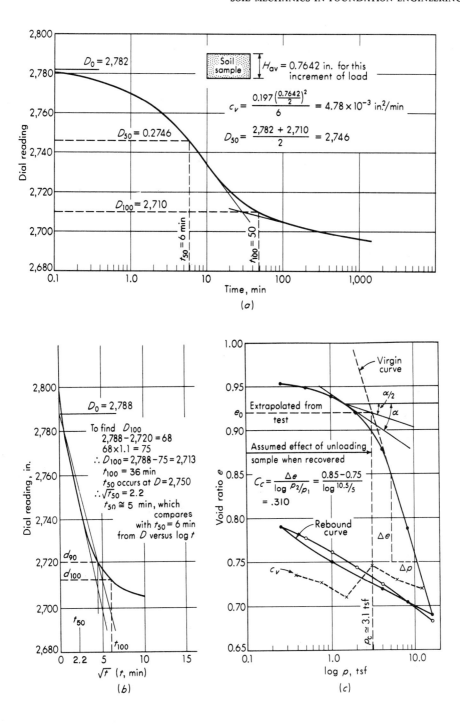

Figure 2-11. (a) Dial reading vs. log time curve to find t_{50}, D_{100}, and c_v; (b) dial reading vs. $\sqrt{t}$ to find D_{90} and D_{100}; (c) e vs. log p curve. Note finding p'_0 by the Casagrande method.

Table 2-5. Time factors for indicated pressure distribution

$U, \%$	Case I	Case II
0	0.000	0.000
10	0.008	0.048
20	0.031	0.090
30	0.071	0.115
40	0.126	0.207
50	0.197	0.281
60	0.287	0.371
70	0.403	0.488
80	0.567	0.652
90	0.848	0.933
100	∞	∞

Case I $\qquad$ Case Ia $\qquad$ Case II

$$u_i = u_0 \qquad u_i = u_0 - u_1 \frac{z}{H} \qquad u_i = u_0 \sin \frac{\pi z}{H}$$

Pore-pressure distribution for case I
usually assumed for case Ia
[Taylor (1948)]

Pore-pressure distribution
for case II

The following approximations hold for case I of Table 2-5: [Means and Parcher (1964)]:

$$T \simeq \begin{cases} KU^2 & \text{where } K \text{ is taken as } 1/12{,}700 & \text{for } U \leq 60\% \\ 1.781 - 0.93 \log (100 - U\%) & & \text{for } U > 60\% \end{cases}$$

H = length of *longest drainage* path a particle of water must take to escape normal pressure. In the laboratory it is half the thickness of the sample when floating-ring consolidometers are used

t = time for a percent of consolidation to take place, such as t_{10}, t_{50}, t_{90}, etc., where 10, 50, 90 are percents of consolidation at the corresponding time t

Since t_{50} is frequently used, Eq. (2-24) becomes

$$c_v = \frac{0.197 H^2}{t_{50}} \qquad (2\text{-}24a)$$

For each loading, c_v is computed and plotted, commonly on the same graph as indicated in Fig. 2-11c. This curve is usually found to be erratic because of limitations on the consolidation theory such as:

1. Saturation less than 100 percent.
2. Coefficient of permeability not constant—probably one of the most critical fac-

tors, since any change in height of a sample 2 to 3 cm in height for a load increment will represent a considerable percentage change in void ratio, whereas in the field a 2- to 3-cm change in stratum thickness is a negligible reduction in void ratio. The coefficient of permeability (and consequently drainage) is heavily dependent on the void ratio.

3. Temperature of test not constant—this is probably the second most critical factor in that the viscosity of water is heavily dependent on temperature. In the field the groundwater temperature will be nearly constant at around 10 to 12°C with a viscosity of around 0.0134 poise. For a laboratory temperature of 25°C the viscosity is around 0.009 poise. The resulting percentage change is on the order of 35 percent *faster in the laboratory.*
4. Ring friction on sample.
5. Other factors which may be peculiar to the soil under consideration such as presence of stones, or nonhomogeneity.

The amount of settlement to be expected is related to a plot of void ratio vs. load, usually plotted to a logarithmic scale as in Fig. 2-11c. The void ratio is computed from weight/volume relationships of the soil specimen and the measured settlement under load.

All consolidation-test specimens are under a pressure in situ at least equal to the present overburden pressure. Since the specimen is unloaded ($p_0 = 0$) when the sample is removed from the ground, an initial portion of the curve is obtained, with characteristics somewhat as indicated on the curve shown in Fig. 2-11c. This early portion of the curve can be approximately duplicated (at a lower void ratio) by loading the specimen in load increments to some point, unloading in increments (allowing rebound), and then reapplying the load increments.

Casagrande (1936) proposed that an approximate method for determining the effective preconsolidation pressure p_c is as follows (illustrated in Fig. 2-11c):

1. On the sharpest part of the curve, as estimated by eye, draw a tangent to the curve.
2. Through the point of tangency, draw a horizontal line forming an angle α with the tangent line.
3. Bisect α with a line through the point of tangency.
4. Extend the straight-line portion of the e vs. log p curve back until it intersects the α bisector in step 3.
5. From the intersection established in step 4, drop a vertical to read on the log p scale the magnitude of p_0.

This method should be recognized as an approximate method, and the shape of the e vs. log p curve should be used as a guide to determine whether to use this technique or some other approximations for establishing the preconsolidation pressure p_0. This is especially true if the e vs. log p plot is not straight in the region beyond the point of known preconsolidation pressure (so-called virgin curve).

Sample disturbance is a critical factor in consolidation testing. Work by Rutledge (1944), Schmertmann (1955), and Hamilton and Crawford (1959), among others, indicates that it may be difficult to establish the preconsolidation pressure

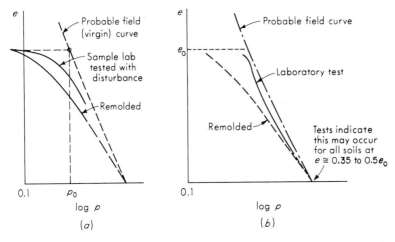

Figure 2-12. Qualitative e log p curves indicating effect of sample disturbance. (a) Nonsensitive clay; (b) sensitive clay, with void ratio $e > 1.5$.

from the e log p curve unless the degree of disturbance is small. Tests show that disturbed (or remolded) samples (see Fig. 2-12a) yield a lower C_c value as well as lower apparent values of preconsolidation pressure p_0. If the in situ void ratio is greater than about 1.5, the "straight"-line part of the e log p curve tends to be concave up, rather than straight, indicating a soil-structure collapse above some critical pressure increment. This load increment would be of some considerable significance in the field loading of a soil with this characteristic. Schmertmann's (1955) tests also tend to indicate that virgin-curve, remolded, and "undisturbed" field samples merge at void ratios of around 35 to 50 percent of the in situ void ratio e_0 (Fig. 2-12b).

If p_0, as established from the e vs. log p curve, is equal to the present overburden pressure, the clay is said to be *normally* consolidated. If p_0 is greater than the present overburden pressure, the clay is said to be *overconsolidated*, or *preconsolidated*, and the symbol p_c is used. If p_c is less, or only somewhat more, than the present overburden pressure, the material is probably only normally consolidated and the effect is caused by disturbance in recovery, trimming, and placing in the testing machines, as well as by the previously indicated limitations on the theory. If the samples have been carefully removed, disturbance due to unloading does not seem to be critical, according to Noorany and Poormand (1973), who compared soft bay mud samples with "perfect" samples made in the laboratory by anisotropically consolidating remolded samples in a triaxial cell with p_0' vertical and $K_0 p_0'$ lateral stresses.

The assumptions generally made in consolidation theory are as follows:

1. Soil stratum is homogeneous.
2. Soil is saturated ($S = 100$ percent).
3. Deformation is due to change in volume (voids) and can take place only as free water is forced out of the soil voids.
4. Deformation is in the direction of applied load (vertical).

5. Flow of water is in the direction of applied load (vertical).
6. The coefficient of consolidation c_v is constant during consolidation.

If the e vs. load curve is plotted to a natural scale, the slope of the curve is termed the *coefficient of compressibility*

$$a_v = \frac{\Delta e}{\Delta p} \qquad \text{area/unit weight} \qquad (2\text{-}25)$$

Strictly speaking, the sign of a_v is negative, but it is of academic importance for settlement computations.

The in situ porosity is

$$n = \frac{e}{1 + e_0}$$

But $1 + e_0$ is a constant, and e_0 is the initial void ratio of the soil in situ; therefore, the change in porosity is

$$\Delta n = \frac{\Delta e}{1 + e_0}$$

Substituting Δe, from Eq. (2-25),

$$\Delta n = \frac{a_v \, \Delta p}{1 + e_0} = m_v \, \Delta p$$

where m_v is the *coefficient of volume compressibility*, with units of 1/stress-strain modulus,

$$m_v = \frac{a_v}{1 + e_0} \qquad \text{units of area/force} \qquad (2\text{-}26)$$

From Fig. 2-13, by proportion,

$$\frac{S}{H} = \frac{\Delta e}{1 + e_0}$$

Solving for S and substituting Δe from Eq. (2-25) results in the following expression for the settlement:

$$S = m_v \, \Delta p \, H \qquad (2\text{-}27)$$

where H is the total thickness of the strata.

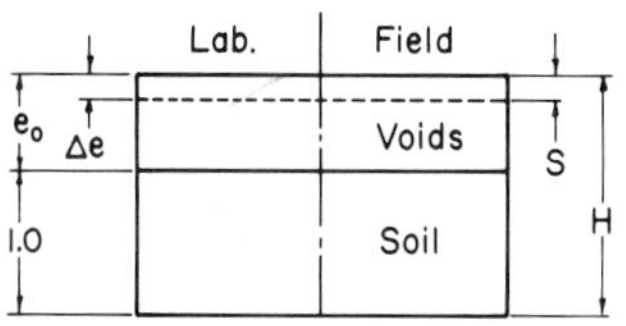

Figure 2-13. Soil relationships for settlement equations. Left side is laboratory, right side field relationships.

One other equation, occasionally needed, but offered without proof [see any textbook on soil mechanics, e.g., Taylor (1948)] to show that the rate of consolidation is dependent on the coefficient of permeability and void ratio, is

$$c_v = \frac{k(1 + e)}{\gamma_w a_v} \tag{2-28}$$

From the straight-line portion of the plot of e vs. log p curve the *compression index* C_c is

$$C_c = \frac{\Delta e}{\log (p_2/p_1)} \tag{2-29}$$

For *preconsolidated* soils we may obtain a *recompression index* from the average slope of the recompression (or from averaging both the unload and recompression branch) of the e log p

$$C_r = \frac{\Delta e}{\log (p_c/p_0)} \tag{2-29a}$$

where Δe should be obtained as the difference in void ratio between the in situ overburden pressure p_0 and the preconsolidation pressure as estimated from the e log p curve.

Rearranging terms and defining Δe as $e_0 - e$,

$$e = e_0 - C_c \log \left(\frac{p_2}{p_1}\right) \tag{2-29b}$$

which is a useful expression for computing the void ratio at various depths in the soil stratum, and shows that as one goes deeper in any soil stratum, the void ratio decreases.

Since

$$p_2 = p_1 + \Delta p$$

and

$$S = \frac{\Delta e H}{1 + e_0}$$

we may substitute the value of Δe from Eq. (2-29) and obtain the following widely used equation for settlement:

$$S = \frac{C_c H}{1 + e_0} \log \frac{p_0 + \Delta p}{p_0} \tag{2-30}$$

where p_0 is the present overburden pressure.

Since the consolidation test is expensive and time-consuming, considerable effort has been devoted to relating the compression index to some soil property which is more easily determined. Using the liquid limit as the property of interest, an equation of the form

$$C_c = a(w_L - b)$$

has been proposed. Values of the a parameter range from 0.005 to 0.009, with the b parameter ranging from 10 to 14 [see Bowles (1974a), p. 31]. Terzaghi and Peck (1967) suggested a revision of Skempton's (1944) work $(a = 0.007, b = 10)$ to the following:

$$C_c = 0.009(w_L - 10) \tag{2-31}$$

which may be used for clays of medium to low sensitivity (many glacial clays found in Middle Western United States). This equation has a reliability on the order of ± 30 percent, which it might be noted in passing is not too good but is actually intended as an estimate in preliminary work and not for final design predictions.

Another method to estimate C_c has been proposed by Hough (1969, p. 134) in the form

$$C_c = c(e_0 - d)$$

Hough provides several values for the parameters c and d for different soils. Sowers and Sowers (1970, p. 102) suggest $c = 0.75$ and $d = 0.2$ for porous rock to $d = 0.8$ for highly micaceous soils. There are wide divergences between Hough's (northeastern soils) and Sower's (principally southeastern laterites) parameters, which illustrates that local soils may behave markedly differently; and also of the preceding Eq. (2-31) with its ± 30 percent reliability the conclusion is that one should use these "short-cuts" with considerable caution.

The conventional method of consolidation testing uses a constant ratio of σ_3 to σ_1 because of the confining ring. This is also termed a K_0 test, since there is no lateral displacement. The pressure-increment ratio $\Delta p/p = 1$. A nonstandard test may, for a given problem, provide the designer with more realistic design data; for example, in soft clay a ratio less than 1 may avoid a structure collapse. Various procedures including constant-rate-of-stress, constant-rate-of-strain, and load-increment ratios other than 1 are reported by Hamilton and Crawford (1959). Lowe (1974), and Lowe et al. (1964, 1969) have proposed consolidation testing with back pressure and/or controlled pore-pressure gradients. Aboshi et al. (1970) have tested at constant loading rate. Wissa et al. (1971) and Smith and Wahls (1969) have reported consolidation tests at constant strain rate.

2-17. SENSITIVITY AND THIXOTROPY

Cohesive soil may lose a portion to all or nearly all the shear strength on remolding or disturbance. Therefore, where piles, caissons, excavations, fill, or footings may disturb the soil, the sensitivity S_t of the soil may need to be determined. Sensitivity is defined as the ratio

$$S_t = \frac{\text{undisturbed strength}}{\text{remolded strength}} = \frac{q_u}{q_u \text{ remolded}} \tag{2-32}$$

where q_u is the unconfined-compression-test strength and should be at constant water content.

The sensitivity of most clays is from 2 to 4, and termed *insensitive*. A *sensitive* soil has values from 4 to 8, and an *extrasensitive* soil has values greater than 8. Some soils may be so sensitive, in conjunction with a high water content, that they have no measurable remolded strength.

Remolded clays of sensitivity less than 16 usually regain a portion to all of the original shear strength with elapsed time. This regain of shear strength with time is termed *trixotropy*. Piles driven in clay, for example, may immediately after driving possess little to no load-carrying capacity, but after a few hours to several days the load capacity may build up to an adequate amount. Remolded quick ($S_t > 16$) clays will usually recover very little of their original strength [Skempton and Northey (1952)], at least within reasonable (say less than 2 to 4 months) time lapses.

2-18. SOIL WATER–SOIL HYDRAULICS

The foundation engineer is concerned with water in the soil voids primarily in terms of *effective pressure*, pore or *neutral pressure*, *permeability*, and *drainage*.

Effective Pressure

All natural soil deposits contain free water in their voids. Near the ground surface, after prolonged dry periods, the amount of water may be quite small, whereas at or near the surface immediately after a rain, or below the capillary zone, the voids may be completely filled.

Effective pressure σ' is defined as the pressure the individual soil grains exert on each other. It is this pressure multiplied by the corresponding effective coefficient of friction which provides the shear strength of granular materials,

$$s = \sigma' \tan \phi_e$$

and a component of shear strength in cohesive soils,

$$s = c + \sigma' \tan \phi_e$$

Considering Fig. 2-14a, neglecting the shearing resistance along the sides, and considering an element with an area of 1 sq ft, as shown, the nominal grain-to-grain contact pressure at the surface of the water table is

$$\sigma' = \frac{\gamma_1 h_1 + \gamma_2 h_2}{1 \times 1} = \gamma_1 h_1 + \gamma_2 h_2 \qquad \text{(psf or kN/m}^2\text{)}$$

Note that although most of the pores in the capillary zone are filled with water ($S = 100$ percent), this is contributing to the weight, since this water is not free to move.

The *effective* pressure at point A (Fig. 2-14a) can be established as

$$\sigma' = \gamma_1 h_1 + \gamma_2 h_2 + \gamma_3 h_3 - \gamma_w h_3$$

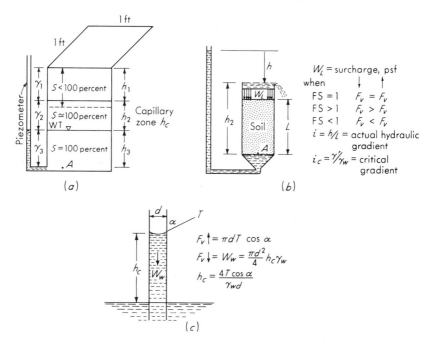

Figure 2-14. (*a*) Soil/water relationships for effective and pore-pressure concepts; (*b*) critical hydraulic gradient and concept of safety factor against a "quick" condition; (*c*) computation of height of capillary rise in a capillary tube.

If we let that soil below the water table $\gamma_3 - \gamma_w$ be defined as the *buoyant* weight of a soil γ', then

$$\sigma' = \gamma_1 h_1 + \gamma_2 h_2 + \gamma' h_3$$

But $\gamma_1 h_1 + \gamma_2 h_2$ may be interpreted as a surcharge q applied to the surface of the soil stratum under consideration; thus the effective pressure may be rewritten

$$\sigma' = q + \gamma' h_3$$

Reconsidering the general expression for effective pressure,

$$\sigma' = q + \gamma_3 h_3 - \gamma_w h_3$$

and setting the surcharge q equal to zero,

$$\gamma_3 h_3 = \sigma' + \gamma_w h_3$$

or

$$\sigma = \sigma' + u$$

and rearranging and solving for the effective stress σ', we obtain

$$\sigma' = \sigma - u \tag{2-33}$$

where σ = total vertical pressure on a plane

$u = \gamma_w h$ = pore pressure

and for changes in pore pressure Δu, a study of Eq. (2-33) indicates that the following is valid (with due regard to the sign of Δu):

$$\sigma_e = \sigma - (u + \Delta u) \tag{2-33a}$$

In general, Eq. (2-33a) is expressed [Bishop and Blight (1963), Skempton (1960)] as follows:

$$\sigma' = \sigma - [u_a - \chi(u_a - u_w)] \tag{2-34}$$

where u_a = air pressure in soil pores
$\quad\;\; u_w$ = water pressure in soil pores
$\quad\;\; \chi$ = coefficient which depends on the degree of saturation S
$\quad\quad\;$ = 1 for saturated soils
$\quad\quad\;$ = between 0 and 1 for partially saturated soils and depends on both soil type, percent clay, etc., and the degree of saturation

In normal foundation-engineering work, the soil is either saturated where $\chi = 1$ and Eq. (2-33) is used or the pore pressure is not evaluated and the effective pressure is taken as the total pressure.

Equation (2-33a) indicates that in increasing the neutral pressure $(u + \Delta u)$ the effective pressure is decreased by the same amount. If the neutral pressure is sufficiently increased, the effective pressure reduces to zero; i.e., the soil will, if granular, possess no shear strength. This condition is referred to as a "quick" condition, and conditions for its occurrence may be approximately evaluated as follows.

From Fig. 2-14b, equating the upward and downward pressures at point A,

$$(h_2 - L)\gamma_w + L\gamma + W_L \text{ (downward)} = (h + h_2)\gamma_w \text{ (upward)}$$

At a safety factor F of 1, this system of pressures is in equilibrium; canceling terms and simplifying, we obtain

$$\frac{h}{L} = \frac{\gamma'}{\gamma_w} + \frac{W_L}{\gamma_w L} \tag{2-35}$$

but from basic soil mechanics, h/L is defined as the hydraulic gradient i across an element of soil of length L. The surcharge on top of the soil element is W_L in appropriate pressure units.

If the surcharge in Eq. (2-35) is zero, and since most sands have a G_s of approximately 2.65, with void ratios from 0.25 to 1.0, and if the γ'/γ_w term is interpreted as the critical hydraulic gradient i_c, then

$$i_c = \frac{\gamma'}{\gamma_w} = \frac{G_s - 1}{1 + e} \tag{2-36}$$

and the critical hydraulic gradient i_c is found to be near 1.0, with values ranging from about 0.8 to 1.3.

Table 2-6. Order of magnitude values for permeability k, based on description of soil and by Unified Classification, cm/sec

10^2		10^0		10^{-3}		10^{-7}		10^{-9}
	Clean gravel GW, GP		Clean gravel and sand mixtures GW, GP SW, SP GM		Sand-silt mixtures SM, SL, SC		Clays	

Permeability

Flow of soil water, for nonturbulent conditions, has been expressed by Darcy as

$$v = ki \qquad (2\text{-}37)$$

where i = hydraulic gradient h/L, as previously defined
k = coefficient of permeability as proposed by Darcy, length/time. Table 2-6 lists typical order-of-magnitude values for various soils.

The quantity of flow q is

$$q = kiA \qquad \text{volume/time}$$

Two tests commonly used in the laboratory to determine k are the *constant-head* and *falling-head* methods. See Fig. 2-15 for a schematic diagram of each and the significance of the terms used.

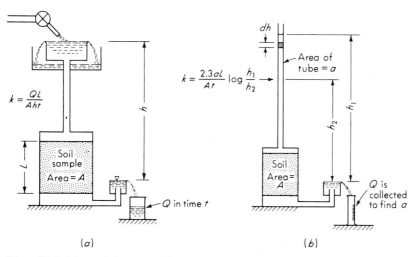

$$k = \frac{QL}{Aht}$$

$$k = \frac{2.3aL}{At} \log \frac{h_1}{h_2}$$

(a) (b)

Figure 2-15. Schematic for permeability determination. (a) Constant-head permeameter; (b) falling-head permeameter.

For the constant-head permeability apparatus

$$k = \frac{QL}{Aht} \tag{2-38}$$

For the falling-head permeability apparatus

$$k = \frac{2.3aL}{At} \log \frac{h_1}{h_2} \tag{2-39}$$

Capillary Water

Capillary rise in a soil may be estimated (referring to Fig. 2-14c) by summing forces in the vertical direction ($\sum F_v = 0$) to obtain

$$\frac{\pi d^2}{4} h_c \gamma_w = \pi \, dT \cos \alpha$$

Solving for the height of capillary rise,

$$h_c = \frac{4T \cos \alpha}{\gamma_w d} \tag{2-40}$$

where h_c = height of capillary rise
$\quad d$ = pore diameter (may use $\frac{1}{5}D_{10}$ grain size as an estimation for soils)
$\quad T$ = surface tension of water (approximately 0.075 g/cm for usual soil-water temperatures of 50 to 60°F)
$\quad \alpha$ = angle of intersection of surface film and conduit, usually taken as 0°

Equation (2-40) generally overestimates the height of capillary rise considerably because the soil pore system is not regular. Few laboratory observations of capillary rise have been found to exceed 1 or 2 m.

Flow Nets

The flow of water through soil under an energy potential can be mathematically expressed by a Laplace equation as

$$k_x \frac{\partial^2 h}{\partial x^2} + k_y \frac{\partial^2 h}{\partial y^2} = 0$$

where k_x, k_y = coefficients of permeability parallel to the x, y axes, respectively
$\quad h$ = energy potential

The above equation is for two-dimensional flow, which with appropriate axis rotation, will apply to most seepage problems. A graphical solution of this equation results in families of intersecting orthogonal curves which are called a flow net. One set of the curves represents equipotential lines (lines of constant piezometric head) and the other set intersecting at right angles represents flow paths. The flow net consists of squares of varying dimension if $k_x = k_y$ and rectangles otherwise. In general, for reasonably homogeneous soil a graphical solution of the Laplace equation provides seepage quantities which are at least as correct as one is likely to obtain the coefficients of permeability.

Seepage quantity from a flow net can be computed as

$$Q = kH\frac{n_f}{n_d}Wt \qquad \text{(cu ft or m}^3 \text{ in time } t) \qquad (2\text{-}41)$$

where k = transformed coefficient of permeability when $k_x \neq k_y$ and so the resulting flow net consists of squares, $k = \sqrt{k_x k_y}$

H = differential head of fluid across system

n_f, n_d = numbers of flow paths and equipotential drops, respectively, in system

W = width of seepage flow

t = time base (1 hour, 1 day, 1 week, etc.)

Figure 2-16a illustrates a flow net for one side of a cofferdam-type structure which will be of most interest in this text. We may use the flow net to estimate how much drawdown may be allowed on the construction side of the wall or how much excavation can be performed before the construction side becomes "quick."

For other seepage problems the user is referred to any text on soil mechanics [e.g., Wu (1966), Leonards (1962, Chap. 3)].

Example 2-2. From Fig. 2-16a assume the following data:

$$H = 6.0 \text{ m} \qquad k_x = k_y = 4 \times 10^3 \text{ cm/sec}$$

$$\gamma = 19.8 \text{ kN/m}^3 \text{ (saturated sand)}$$

Distances: $AB = 2$ m, $BC = 2$ m, $CD = 1.5$ m, $DE = 1$ m

REQUIRED. What is the effective pressure at point C? Is point C "quick"?

SOLUTION. With the drawdown even with the soil on the left side, $H = 6 + 2 = 8$ m. The excess pore pressure at C: $\Delta u = 8 - \frac{7}{8}(8) = 1$ m. Total pore pressure at C:

$$u = (1 + 2)9.807 = 29.42 \text{ kPa}$$

Total pressure at C: $\sigma = 2(19.8) = 39.6$ kPa
Effective pressure at C:

$$\sigma' = \sigma - u = 39.6 - 29.4 = 10.2 \text{ kPa}$$

With $\sigma' > 0.00$ some effective pressure exists and point "C" is not "quick".

Note in this problem that excess pore pressure at point C is the height of a column of water in a piezometer at C above the static water level at the surface of the ground.

2-19. PORE PRESSURE AND PORE-PRESSURE PARAMETERS

A method of evaluating effective pressures was presented in Eq. (2-33) and Example 2-2 which is widely used for saturated ($S = 100$ percent) soils. One may also measure the pore pressure (usually at the base of the sample) in laboratory triaxial

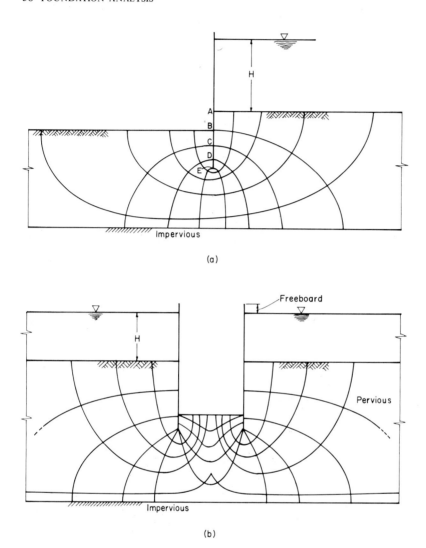

Figure 2-16. Typical flow nets as used for sheet-pile or cofferdam structures. (*a*) Single sheet-pile wall or other wall too far to influence net; (*b*) double-wall cofferdam as used for bridge piers, etc.

tests using a pressure transducer as illustrated in Fig. 2-7. For both saturated and partially saturated soils, it appears that the change in pore pressure due to applied-stress changes is related to the material, the degree of saturation, and whether the soil mass is normally consolidated or preconsolidated. In saturated soils the various pore-pressure-measuring devices (pressure transducers, null devices, etc., in the lab-oratory and various types of piezometers in the field) enable reasonably good meas-urements of changes in pore pressure to be made. Field measurements of laboratory predictions are also reasonably good. When partially saturated soils are encountered, however, the laboratory predictions become just that—predictions. In many cases, a prediction is better than guessing, and in an attempt to provide some aid, Skempton

Table 2-7. Typical values for the A parameter* for Eq. (2-42) (at failure)

Type of material	Approximate range of values
Highly sensitive clay	1.2–2.5
Normally consolidated clay	0.7–1.3
Compacted sandy clay	0.75–0.25
Lightly preconsolidated clays	0.3–0.7
Heavily preconsolidated clays	−0.5–0.0
Very loose fine sand	2–3

* Skempton (1954), Lambe and Whitman (1969).

(1954) proposed an equation of the following form:

$$\Delta u = B[\Delta\sigma_3 + A(\Delta\sigma_1 - \Delta\sigma_3)] \qquad (2\text{-}42)$$

where A, B = pore-pressure parameters

Δu = change in pore pressure due to increased stresses

$\Delta\sigma_1, \Delta\sigma_3$ = change in principal stresses caused by soil loading or stresses induced in a soil sample in a triaxial test

$\Delta\sigma_1 - \Delta\sigma_3$ = deviator stress

The B parameter ranges from 0 to 1.0, being zero for a dry soil ($S = 0$ percent) and 1.0 (practically, in the laboratory, the soil is probably saturated at any measured B above about 0.95) for a fully saturated soil ($S = 100$ percent). There does not seem to be, however, a linear relationship between percent saturation and the B parameter. Inspection of Eq. (2-42) indicates that one may evaluate the B parameter with a triaxial cell and pore-pressure apparatus by applying a chamber pressure $\Delta\sigma_3$ and measuring the resulting pore pressure, since the deviator stress ($\Delta\sigma_1 - \Delta\sigma_3$) is zero until a vertical load is applied to the specimen.

The A parameter tends to be heavily dependent on the type of soil and the previous stress history, with a typical range of values as indicated in Table 2-7.

2-20. THE STRESS-PATH METHOD

The stress-path method can be used to provide insight into the stresses developed in a soil mass under field loadings, or the most appropriate laboratory-test method to develop the needed soil parameters for the field loading conditions. It can be used also as an alternative analytical approach to the Mohr's-circle method to obtain the soil parameters. The basic concept of stress paths has been in use for some time by various investigators; however, various methods for tracing the "stress path" have been proposed including using $\sigma'_v(\sigma'_{axial})$ vs. $\sigma'_h\sqrt{2}$ (or $\sigma'_{radial}\sqrt{2}$) by Henkel (1959) and Rendulic (1936). Roscoe et al. (1958) and Roscoe and Poorooshasb (1959) used the deviator stress ($\sigma'_1 - \sigma'_3$) vs. mean normal stress [$(\sigma'_1 + 2\sigma'_3)/3$]. Henkel (1970) used σ'_1 vs. σ'_3 (or the equivalent σ'_v, σ'_h), with the recommendation that this would be satisfactory for plane-strain problems. The prime superscript indicates effective stresses.

The p, q coordinate system proposed by Lambe (1964, 1967), and Lambe and Whitman (1969) is simple, straightforward, and has a unique relationship between the p, q system and Mohr's failure envelope. The p and q coordinates are obtained as follows:

$$p = \frac{\sigma_1 + \sigma_3}{2} \qquad q = \frac{\sigma_1 - \sigma_3}{2} \qquad (2\text{-}43)$$

The stress coordinates may be either p, q for total stresses or p', q for effective stresses, i.e., using either σ_1, σ_3 or σ'_1, σ'_3 in Eq. (2-43).

Since

$$\sigma'_1 = \sigma_1 - u$$

$$\sigma'_3 = \sigma_3 - u$$

subtracting

$$\sigma'_1 - \sigma'_3 = \sigma_1 - \sigma_3$$

and dividing through by 2, it is evident that

$$q' = q$$

$$p' = p - u$$

or the x coordinate for effective stresses differs from the total stress coordinate by the pore water pressure. The coordinates p, q are the origin and radius of Mohr's circle, respectively.

The stress path for a triaxial test considering effective stresses is idealized in Fig. 2-17. Figure 2-17a is the conventional Mohr's-circle representation of one stage of sample loading under conditions of $\sigma_c = \sigma_3$ and a deviator stress component to give $\sigma_1 = \sigma_3 +$ deviator stress. If, at the instant of the value of σ_1 shown, the ratio of

$$\frac{q}{p} = \frac{1 - K_0}{1 + K_0}$$

line OB through the origin and the coordinates p, q is the K_0 line. The K_0 line will be approximately through the origin for all normally consolidated soils when considering effective stress conditions. The K_0 line for preconsolidated soils may not pass through the origin. If the above q/p ratio is not satisfied at failure, the locus of q/p points will be termed a K_f line (for "failure" as Fig. 2-18).

Figure 2-17b is a set of Mohr's circles and p, q coordinates from a triaxial test as follows:

1. Begins with a confining cell pressure σ_c with p, q coordinates $p = q = \sigma_3 = \sigma_c$.
2. Point 2 is the p, q coordinates on the corresponding Mohr's circle for an intermediate deviator stress increment as is point 3.
3. Point 4 is obtained by holding the major principal, σ_1, stress constant at some point during the test and decreasing σ_3.
4. That part of the stress path 1, 2, 3 is always inclined at an angle of 45° to the horizontal axis.

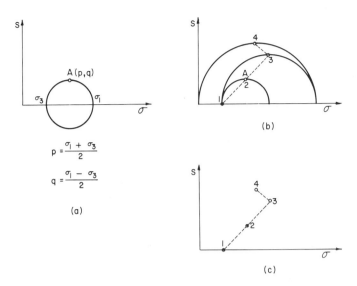

Figure 2-17. (*a*) Coordinates of a point on a stress path; (*b*) stress path on a set of Mohr's circles; (*c*) stress path drawn using coordinates *p, q*.

Figure 2-17*c* is the stress-path plot of the *p, q* coordinates without the accompanying Mohr's circles to provide greater clarity.

Additional information from the stress-path method and a display of the four basic triaxial tests is shown in Fig. 2-18. The K_f lines (locus of *p, q* at failure) shown in the figure are obtained from the following data:

For Fig. 2-18*a*: (all pressures in kg/cm²)

Test	σ_1	σ_3	p	q
1	4.8	1.2	3.0	1.8
2	12.0	4.0	8.0	4.0

For Fig. 2-18*c*:

Test No. 1: Initial horizontal pressure = 3.00 = initial vertical pressure. Decrease horizontal pressure to 0.55 (compression test); $p = 1.78$; $q = 1.22$.

Test No. 2: Initial horizontal pressure = initial vertical pressure = 3.00. Increase vertical pressure to 9.4 (usual compression test); $p = 6.2$; $q = 3.2$.

Test No. 3: Initial horizontal pressure = initial vertical pressure = 3.00. Decrease vertical pressure to 0.55 (extension test); $p = 1.78$; $q = -1.22$.

Test No. 4: Initial horizontal pressure = initial vertical pressure = 3.00. Increase the horizontal pressure to 9.4 (extension test); $p = 6.2$; $q = -3.2$.

Several observations can be made in the plots in Fig. 2-18*a* and *c*:

1. Lines 1 and 2 in Fig. 2-18*a* and lines 1, 2, 3, and 4 of Fig. 2-18*c* are all at slopes of 45°.

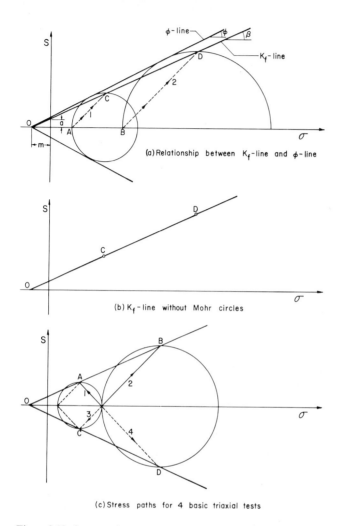

(a) Relationship between K_f-line and ϕ-line

(b) K_f-line without Mohr circles

(c) Stress paths for 4 basic triaxial tests

Figure 2-18. Stress paths.

2. The point 0 is common to both the K_f line and the ϕ line.
3. A relationship exists between the ϕ line and the K_f line as

$$m = a/\tan \beta = c/\tan \phi$$

and
$$q/(m + p) = \tan \beta = q/(m + p) = \sin \phi$$

$$c = a/\cos \phi$$

4. The slope of the K_f line is common to all four tests. In this example $m = 1.00$ and the ratio of the positive q values$/(m + p)$ is $1.22/2.78 = 3.2/7.2 = 0.44$ (within normal round-off). From $\sin \phi = 0.44$, $\phi = 26°$.

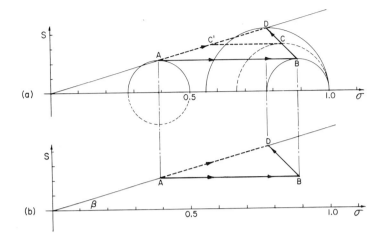

Figure 2-19. Stress path for consolidation test from load increment 0.5 to 1 kg/cm² plotted on a Mohr's circle; (b) same plot using p, q coordinates.

The stress path for a consolidation test can be established (refer to Fig. 2-19) as follows:

1. Take point A as the state of effective stress at the end of primary consolidation for any load increment.
2. Suddenly apply the next load increment. Until drainage can occur after some time elapses, there is no change in volume and correspondingly no increase in any shearing stress including the maximum value q. Thus, $q = (\sigma_1 - \sigma_3)/2$ is unchanged, causing the diameter of the circle to be unchanged but displaced to the right an amount equal to the change in vertical pressure. In the case of saturated soil $\Delta\sigma_1 = \Delta u$ according to consolidation theory.
3. As drainage occurs, the pore pressure decreases and the Mohr's circle of effective stress increases as indicated at some elapsed times by circles C and D. Circle D is the effective stress state when primary consolidation under this load increment has occurred.

Example 2-3. A normally consolidated clay soil with $K_0 = 0.95 - \sin \phi = 0.56$ is consolidated in an oedometer (consolidation) test. At the end of primary consolidation under a 0.5 kg/cm² total load, the principal stresses (neglecting any ring-side friction effects) are

$$\sigma_1' = 0.5 \text{ kg/cm}^2 \qquad \sigma_3 = K_0\sigma_1 = 0.28 \text{ kg/cm}^2$$

and the corresponding p, q coordinates are

$$p' = 0.39 \qquad q = q' = (0.50 - 0.28) = 0.11$$

These coordinates plot point A on the Mohr's circle as shown in Fig. 2-19a.

The next load increment of $\Delta p/p = 1$ gives a total stress on the sample of 1.0 kg/cm². This load increment, at $t = 0$, moves circle A horizontally a distance of $1.0 - 0.5 = 0.5$ kg/cm² (this is also the instantaneous pore pressure) to circle B with p, q coordinates of

$$p = \frac{(0.5 + 0.5) + (0.28 + 0.5)}{2} = 0.89 \qquad \text{(also } 0.39 + 0.5)$$

$q = 0.11$ as before

After primary consolidation is complete, the effective pressure σ_1' is equal to the initial total pressure:

$$\sigma_1' = 1.0 \text{ kg/cm}^2$$

$$\sigma_3' = K_0 \sigma_1 = 0.56 \text{ kg/cm}^2$$

and $\qquad p' = 0.78 \text{ and } q = q' = 0.22 \text{ kg/cm}^2$

The latter p, q coordinates represent point D of Fig. 2-19. Curve C is at some arbitrary intermediate consolidation stress state. Line $ABCD$ represents the total stress path of the soil in consolidating under the 0.5 kg/cm² load increment. Line AD represents the effective stress path. The horizontal distance between AD and BCD represents the excess pore pressure at some instant in time; CC' is the excess pore pressure at a time when the total stress path is at p, q = point C (0.835, 0.165) with a value

$$\Delta u = 0.835 - 0.585 = 0.25 \text{ kg/cm}^2$$

or 50 percent of the excess pore pressure has dissipated.

Line AD is a segment of the K_0 line for this soil. If this line is extended, it will intercept the origin at $p = q = 0$, since the clay is normally consolidated and at the beginning of the test $p = q = \sigma_1 = \sigma_3 = 0$. The slope of this K_0 line is

$$\tan \beta = \frac{1 - K_0}{1 + K_0}$$

PROBLEMS

2-1. One cubic foot of soil weighs 124.5 lb. If $G_s = 2.72$ and $w = 12.0$ percent, find: γ_{dry}, e, n, S.
   Answer: 111.2, 0.53, 0.35, 61.5%.

2-2. One cubic meter of soil weighs 19.54 kN. If $G_s = 2.72$ and $w = 12.0$ percent, find: γ_{dry}, e, n, S.
   Answer: 17.44, 0.53, 0.35, 61.9%.

2-3. A soil has a natural void ratio e_0 of 1.87, $w_N = 60.0$ percent, and $G_s = 2.75$. What is the wet unit weight and S?
   Answer: 15.04 kN/m³ (95.8 pcf), $S = 88.3$ percent.

2-4. A sample of *saturated* clay weighs 1,853.5 g wet and 1,267.4 g dry and has a $\gamma_{dry} = 1.50$ g/cm³. What is (a) wet unit weight; (b) void ratio; (c) specific gravity; (d) wet unit weight if $S = 50$ percent? Comment on what type of clay this might be.

2-5. Plot a curve of $e = f(n)$ with e on the abcissa range of values. How would this plot differ from $n = f(e)$?

2-6. In a consolidation test a specimen was 2.00 cm thick (ring diameter = 6.27 cm) at the beginning of the test. At the end of the test the sample was 1.32 cm thick. The dry weight of the 1.32-cm soil cake was 98.72 g. The G_s of the soil solids is 2.60. Compute (a) initial void ratio; (b) final void ratio.

2-7. The void ratio at the beginning of a consolidation test was 1.18. The initial specimen height was 0.800 in, and the final height was 0.514 in. What is the final void ratio?
   Answer: $e = 0.40$.

2-8. A triaxial (CU) test on a $\phi - c$ soil yielded the following data:

Test No.	σ_3, kg/cm²	Deviator stress (failure)
1	0.8	1.2
2	1.6	1.7
3	2.4	2.0

Find the angle of internal friction and c.

    Answer: $c \simeq 0.3$ kg/cm².

2-9. A triaxial test was performed on a saturated clay as a UU test. Confining pressure was 10 psi, and the total vertical pressure ($\sigma_1 + \sigma_3$) was 30 psi. Compute the soil parameters c and ϕ.

    Answer: $\phi = 0$; $c = 1,440$ psf.

2-10. The stress conditions on an element of sand in a soil stratum are

$$\sigma_v = 2 \text{ ksf (computed)} \qquad \sigma_h = 0.8 \text{ ksf (computed)} \qquad s = 0.9 \text{ ksf}$$

Find the principal stresses and orient the principal plane.

    Answer: $\sigma_3 \simeq 0.35$ ksf; $\theta = 28.15°$.

2-11. A direct shear test (CU) was performed on a $2\frac{1}{2}$-in-diameter specimen with the following data:

Test No.	P_v (normal force), lb	P_h (shear force), lb
1	2.2	1.3
2	8.8	2.6
3	11.2	3.1

Determine the cohesion and angle of internal friction for the soil.

2-12. A triaxial test was performed on a silty clay with the following data:

Test No.	σ_3, psi	$\sigma_1 - \sigma_3$, psi
1	5	18
2	10	24
3	15	28

Plot the data and determine the soil parameters. Comment on the test.

2-13. Plot the data of Prob. 2-12 using a p, q diagram (stress paths). Using the relationship between β and ϕ (Fig. 2-18), determine the soil parameters ϕ and c.

2-14. Classification tests were performed on a light-brown sandy soil with the following results:

Sieve No.	Percent passing
10	100
40	38
200	15
$LL = 26.2\%$,	$PL = 14.1\%$

Classify the soil in the Unified system.

2-15. Given the data of Table P2-15 from a consolidation test:

Table P2-15. Dial readings ($\times$ 0.0001)

Time, min	1/4 tsf	1/2 tsf	1 tsf
0	2,240	2,188	2,127
0.25	2,234	2,180	2,119
0.50	2,230	2,172	2,113
1.0	2,227	2,162	2,105
2.0	2,222	2,153	2,094
4.0	2,218	2,144	2,083
8.0	2,213	2,139	2,073
16.0	2,208	2,135	2,062
30.0	2,204	2,132	2,055
60.0	2,200	2,131	2,050
120.0	2,197	2,130	2,047
240.0	2,193	2,129	2,046
480.0	2,190	2,128	2,045
1,440.0	2,188	2,127	2,045

Required for each load increment:
1. Plot dial reading vs. log time and find t_{50}.
2. Plot dial reading vs. $\sqrt{t}$ and find t_{50}.
3. If the initial specimen height was 0.800 in, and using two-way drainage, find c_v.
   Answer: $\simeq$ 3.0, 1.1, 3.0 min, respectively.

2-16. Plot Example 2-3 in Sec. 2-20, and verify that the next standard load increment (2.0 kg/cm^2) will fall on the K_0 line. For this load increment make a plot of σ' vs. Δu for several arbitrarily selected σ' points.

2-17. Three consolidation tests were performed on separate layers of *soft* clay underlying a site (refer to Fig. P2-17).

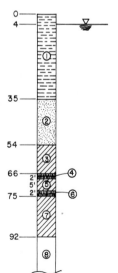

Figure P2-17

Test load, tsf	Void ratio e		
	Soil No. 3	Soil No. 5	Soil No. 7
0.08	1.395	1.190	1.140
0.16	1.393	1.187	1.135
0.32	1.390	1.180	1.130
0.64	1.385	1.175	1.105
1.4	1.380	1.125	1.080
2.8	1.360	1.050	1.060
5.6	1.180	0.925	0.965
6.8	1.110		
10.2	0.925	0.760	0.805
20.4	0.725	0.625	0.680

$w_L = 77.0\%$	53.0	69.0
$w_P = 24.0\%$	26.0	24.0
$G_s = 2.74$	2.70	2.66
$e_0 = 1.20$	1.18	1.06
$p_0 = 5.1$ tsf	4.2	3.9

Soil:	1	2	3	4	5	6	7
γ, pcf:	100.0	116.5	*	118.5	*	119.2	*

* Compute from G_s, e_0.

Soil No: 1—organic silt and clay
2—medium dense sand
3, 5, 7—clay
4, 6—thin silt seams

Required: Plot e vs. log p curves as assigned. Find C_c, p_0, p_c; comment on tests and test results.

2-18. For Fig. P2-18: (a) Estimate h at which the sand would be expected to become quick; (b) if $h' = 25$ cm, what is effective pressure at point A?
Answer: (a) 44.8 cm; (b) 1.95 kPa (40.7 psf).

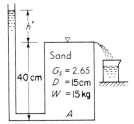

Figure P2-18

2-19. A triaxial test with pore-pressure-measuring equipment was used on a clay sample with the following data:

$\Delta\sigma_1$, psi	$\Delta\sigma_3$, psi	Δu, in Hg
10	10	1.22
15	15	1.84
25	10	2.33
35	15	3.56

Compute the pore-pressure parameters A and B. Is this a saturated soil?

2-20. Plot the triaxial test data of Prob. 5-12; draw a smooth curve and for each pair of strain values *from the curve* starting with 0.005, compute E_s and μ using Eq. (2-19). Note the strain and stress level when $\mu > 0.5$. Make appropriate comments.

THREE

EXPLORATION, SAMPLING, AND IN SITU SOIL MEASUREMENTS

3-1. DATA REQUIRED

The review material presented in Chap. 2 indicated the basic data required to enable one to classify and establish the parameters of a soil. The basic essentials of consolidation theory, settlement, and soil hydraulics were also presented. In Chaps. 4 and 5 we will consider the problems of soil reaction and stresses developed in the soil from imposed loads and the relationship of the soil parameters and soil type in the solution of these problems. This leaves for discussion the problem of the collection of soil samples for laboratory testing and the determination of the arrangement of the materials underlying a site. Although this topic appears later in the discussion sequence, it is actually the first operation performed in the analysis and design of a foundation. The field and laboratory investigations to obtain the required information are usually termed *soil,* or *site, exploration.*

Knowledge of the underground conditions at a site is prerequisite to the economical design of the substructure elements. To attempt to save a few dollars by bypassing a site exploration only to find, after the design is completed and construction has started, that the foundation conditions encountered necessitate a new design is a false economy. It is doubtful if any major structures are constructed at present without site exploration being undertaken, but for smaller structures there is a wide practice of little or no exploration. For a new structure the soil-site investigation should provide data on the following items:

1. Location of groundwater level (at least to the extent of whether it is in the problem zone)

2. Bearing capacity of the soil
3. Selection of alternative types and/or depth of foundation
4. Data on soil parameters and properties so that earth pressures and construction methods may be evaluated
5. Settlement predictions
6. Potential problems concerning adjacent property

It may be necessary to institute an exploratory program on existing structures to investigate the ability of the foundation to carry contemplated additions, or the current status of the safety of the structure, if the foundation behavior is not what the original designer may have expected. Settlement predictions and necessary remedial measures for the control of settlements may also be determined from investigations on the existing structures.

Soil exploration is also used for road, airfield, and other extended sites, as contrasted with the compact-site[1] exploration of the preceding paragraphs. The extended-site program is useful to locate line and grade, locate groundwater level and rock, delineate zones of poor-quality soil, and establish sources of construction materials (borrow and sand and/or gravel pits).

3-2. METHODS OF EXPLORATION

The most widely used method of subsurface investigation for compact sites and most extended sites is boring holes into the ground from which samples may be collected for either visual inspection or laboratory testing. Several procedures are commonly used to drill the holes and to obtain the soil samples. These will be taken up in more detail later.

Aerial photographs, in conjunction with drill holes or test pits, may be especially useful for a soil-exploration program encompassing large areas. Color photographs are an added cost, but they tend to be more sensitive to soil color changes and are generally more useful. Minard and Owens (1962, with a bibliography) presents several articles on using aerial photographs for large-scale exploration programs. For large areas, statistical techniques in conjunction with aerial photographs may prove helpful in reducing the amount of sample collection and testing. This technique was recently applied to a large-scale mapping project in South Africa by Kantey and Morse (1965).

Geophysical methods are also used with primary application on extended-site exploration. These methods fall in two general categories, namely, *seismic* and *resistivity* methods. The use of geophysical methods is usually limited to establishing the location of bedrock underlying softer materials or locating gravel (or sand) deposits.

Visual inspection and sample collection of the exposed subsoil in erosion ditches, construction excavations, or test pits are sometimes satisfactory. Load tests in test

---

[1] A compact site may be considered as one in which the construction cost per unit of area is relatively high compared with an extended-site construction cost.

Table 3-1. Summary of soil-exploration methods

	No samples taken	
Method	Depths	Applicability
Geophysical seismic resistivity	Usually less than 35 m	Locating firmer material underlying softer material. Certain equipment is adapted to determination of density and soil moisture
Vane shear	Limited by torque stresses on rod	In situ shear strength of sensitive cohesive nongravelly soils
Sounding	Limited by encountering rock and driving equipment	Locating soft material and rock by probing, using solid rods as opposed to a split spoon
Dutch cone	Same as "sounding"	
Pressuremeter	Usually less than 35 m	In situ E_s, μ, K_0
Glötzl cell	Same as pressure meter	In situ K_0
Fracture apparatus	Same as pressure meter	In situ K_0
	Disturbed samples taken	
Auger boring	Depends on equipment and time available, practical depths being up to about 35 m	All soil where hole will maintain wall without casing
Rotary drilling Wash boring Percussion drilling	Depends on equipment, most equipment can drill to depths of 70 m or more	All soils. Some difficulty may be encountered in gravelly soils. Rock requires special bits, and wash boring is not applicable. *Penetration testing* is used in conjunction with these methods, and disturbed samples are recovered in the split spoon. Penetration counts are usually taken at 1- to 2-m increments of depth
Test pits and open cuts	As required, usually less than 6 m; use power equipment	All soils
	Undisturbed samples taken	
Rotary drilling, percussion drilling, wash boring	Depends on equipment, as for disturbed-sample recovery	Thin-walled tube samplers and various piston samplers are used to recover samples from holes advanced by these methods. Commonly, 5- to 10-cm-diam. samples can be recovered
Test pits	Same as for disturbed samples	Hand-trimmed samples. Careful trimming of sample should yield the least sample disturbance of any method

pits (or in the site excavation) can be used to establish soil bearing capacity, but are relatively costly. Table 3-1 summarizes various soil-exploration methods available and discussed in the following sections.

3-3. PLANNING THE PROGRAM

It has been indicated that a soil-exploration program can be used to obtain a considerable amount of information on the subsoil. Some methods obviously will yield more information than other methods; e.g., a plate-load test in a test pit 1 m below the ground surface may indicate to the designer that the soil can carry 200 kPa. However, this does not preclude the possibility that at a greater depth (perhaps only 1 or 2 more m) there is a thick layer of compressible clay underlying the building site on which the proposed 200 kPa would induce disastrous settlements. A boring would be likely to disclose this situation.

The soil engineer should constantly keep in mind, when planning the exploration program, the purpose of the program and the relative costs involved. It may be more economical to provide a conservative foundation design than to expend large quantities of money on an elaborate boring and testing program. Often an indication of the extent of an exploration program can be estimated from the history of foundation successes and failures in an area. In this phase of the program, experience in an area is very helpful. Also, for planning the program, the engineer should be well acquainted with the current methods of soil boring, sampling, and testing and have some idea of the limitations on both the field and laboratory equipment and methods.

The actual planning of a subsurface exploration program includes some or all of the following steps:

1. *Assembly of all available information* on dimensions, column spacing, type and use of the structure, basement requirements, and any special architectural considerations of the proposed building. Foundation regulations in the local building code should be consulted for requirements which may be a local peculiarity. For bridges the soil engineer should have access to type and span lengths as well as pier loadings. This information will indicate probable soil loadings and the building tolerance to settlement.
2. *Reconnaissance of the area.* This may be in the form of a field trip to the site which can reveal information on the type and behavior of adjacent structures such as cracks, noticeable sags, and possibly sticking doors and windows. The type of local existing structure may influence, to a considerable extent, the exploration program and the best foundation type for the proposed adjacent structure. Since the existing structures must be maintained in an "as is" condition, nearby excavations or construction vibrations will have to be carefully controlled. Erosion in existing cuts (or ditches) may also be observed, but this information may be of limited use in the foundation analysis of buildings. For highways, however, runoff patterns, as well as soil stratification to the depth of the erosion or cut, may be observed. Rock outcrops may give an indication of the presence or the depth of bedrock.

The reconnaissance may also be in the form of a study of the various sources of information available, some of which are:

Geological maps. Either U.S. government or state geological survey maps.
Agronomy maps. Published by the Department of Agriculture (state or U.S.).
Aerial photographs. May require special training to interpret soil data, but terrain features are easily recognized by the nonspecialist.
Water-well logs (and oil wells).
Hydrological data. Data collected by the U.S. Corps of Engineers on stream-flow data, tide elevations, and flood levels.
State highway department soil manuals.
State university publications. These are usually engineering experiment station publications. Information can be obtained from the state university if it is not known whether a state study has been undertaken and published.

3. *A preliminary site investigation.* This is usually in the form of a few borings or a test pit to establish the types of materials, stratification of the soil, and possibly the location of the groundwater level. For small projects this step may be sufficient to establish foundation criteria, in which case the exploration program is finished.
4. *A detailed site investigation.* For complex projects or where the soil is of poor quality and/or erratic, a more detailed investigation may be undertaken in which samples are collected for shear-strength determination and settlement analysis. Table 3-1 summarizes the various methods of soil exploration.

3-4. SOIL BORING

Exploratory holes into the soil may be made by hand tools (Fig. 3-1), but more commonly mounted power tools (Figs. 3-2 and 3-3) are used.

Hand Tools

The earliest method of obtaining a test hole was to excavate a test pit using a pick and shovel. Because of economics, the current procedure would be to use power-excavation equipment such as a backhoe to excavate the pit, then use hand tools to remove a block sample or shape the site for in situ testing. This is the best method at present for obtaining quality *undisturbed* samples or samples for testing at other than vertical orientation. For small jobs, where the sample disturbance is not critical, hand or powered augers held by one or two persons can be used. Hand-augered holes can be drilled to depths of about 35 m, although great depths (say, greater than 6 to 10 m) are usually not practical. Commonly, the depths run 2 to 5 m, as on roadway, airport, or small-structure investigations.

Mounted Power Drills

For numerous borings to greater depths and to collect samples that are *undisturbed,* the only practical method is to use power-driven equipment. *Wash boring* is a term used to describe one of the most common methods of advancing a hole into the

(c)

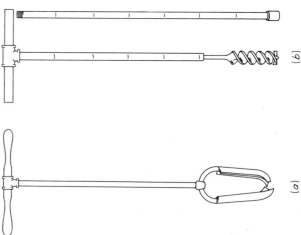

(b)

(a)

74

ground. A hole is started by driving *casing* (Fig. 3-2) to a depth of 2 to 3.5 m. Casing is simply a pipe which supports the hole, preventing it from caving in. The casing is cleaned out by means of a chopping bit fastened to the lower end of the drill rod. Water is pumped through the drill rod, and exits at high velocity through holes in the bit. The water rises between the casing and drill rod, carrying suspended soil particles, and overflows at the top of the casing through a T connection into a container, from which the effluent is recirculated back through the drill rod. The hole is advanced by raising, rotating, and dropping the bit onto the soil at the bottom of the hole. Drill rods, and if necessary casing, are added as the depth of the boring increases. Usually 6 m or less of casing is required at a hole site. This method is quite rapid for advancing holes in all but the very hard soil strata.

Rotary drilling is another method of advancing test holes. This method uses rotation of the drill bit, with the simultaneous application of pressure to advance the hole. Rotary drilling is the most rapid method of advancing holes in rock unless it is badly fissured; however, it can also be used for any other type of soil. If this method is applied in soils where the sides of the hole tend to cave in, a *drilling mud* may be used. The drilling mud is usually a water solution of a thixotropic clay (such as bentonite), with or without other admixtures, which is forced into the sides of the hole by the rotating drill. This provides sufficient strength in conjunction with the hydrostatic pressure of the mud suspension ($\gamma \simeq 1.1$ to $1.2 \ g/cm^3$) against the soil so that it maintains the hole. The mud pressure also tends to seal off the water flow into the hole from the permeable water-bearing strata. Various drill heads are available, such as auger heads for shallow highway and borrow-pit exploration, grinding heads for soil and rock, and coring bits for taking cores from rock, as well as from concrete and asphalt pavements.

Continuous-flight augers are probably the most popular method of soil exploration at present (Fig. 3-3). The flights act as screw conveyors to bring the soil to the surface. The method is applicable in all soils, although in saturated sand under several feet of hydrostatic pressure, the sand tends to flow into the lead sections of the auger, requiring a washdown prior to sampling. Borings up to nearly 100 m can be made with these devices, depending on the driving equipment, soil, and auger size. The augers may be *hollow-stem* or *solid*, with the hollow-stem type generally preferred as penetration testing or tube sampling may be done through the stem. Borings do not have to be cased using continuous-flight augers for obvious reasons, and this is a decided economic advantage over other boring methods. Continuous-flight augers are available in 3- to 5-ft sections and in several diameters including the following (in inches):

Solid stem OD	$2\frac{5}{8}$	$3\frac{1}{4}$	4	$4\frac{1}{2}$	$5\frac{1}{2}$	6	7
Hollow stem ID × OD	$2\frac{1}{2} \times 6\frac{1}{4}$	$2\frac{3}{4} \times 7$	3×8	$3\frac{1}{2} \times 9$	4×10	5×10	6×12

Figure 3-1. Hand tools for soil exploration. (*a*) Posthole, or iwan, auger; (*b*) small helical auger; (*c*) gasoline-engine-powered hand auger equipped with a continuous-flight auger. Note additional auger flights in the foreground.

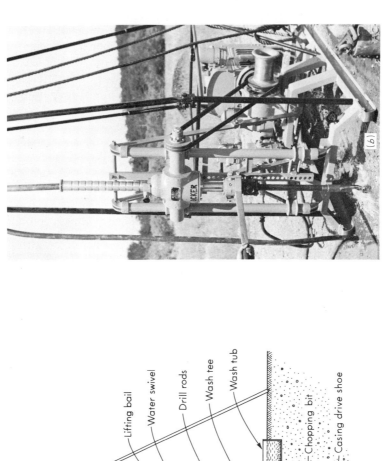

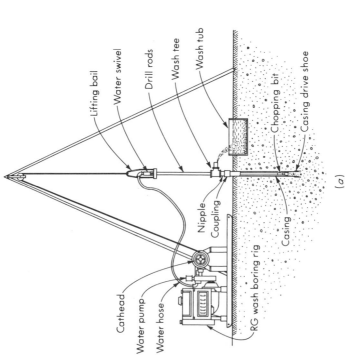

Figure 3-2. (*a*) Schematic of wash boring operations; (*b*) photograph of wash boring operation. Note weight in lower right foreground to advance the casing and to take penetration numbers when the chopping bit is replaced with the split spoon. (*The Acker Drill Company.*)

Figure 3-3. Rotary drilling using a continuous-flight auger. (*The Acker Drill Company.*)

From these sizes it is evident that almost any practical size of tube sample may be recovered through the stem for laboratory testing. Holes will be about $\frac{1}{2}$ in larger than the auger size. In practice a cutting head is attached with or without a plug (depending on soil) and the hole is advanced. At the desired depth the plug is removed and a sample or SPT test is performed. Flights are added as required. If a plug is not used, it may be necessary to wash the several feet of soil from the bottom of the auger stem which has entered during the advancing of the hole prior to sampling.

Percussion drilling is still another method of forming a hole. In this method the drill is lifted and dropped onto the bottom of the hole. Water is circulated to bring the soil particles to the ground surface; casing is required as well as a pump to circulate the water.

3-5. SOIL SAMPLING

Materials recovered from soil borings are classified as *disturbed* or *undisturbed* samples. Strictly speaking, there is no such thing as an undisturbed sample; however, with the use of proper recovery equipment and precautionary field methods, the degree of disturbance can be kept to a minimum.

Where classification of the soil is the prime object, disturbed samples are more satisfactory, since they are cheaper to obtain. The principal requirement of the sample is that it be truly representative of the stratum. Samples from wash borings termed *wash samples* (soil brought up suspended in water) are generally not satisfactory because the finer material tends to separate from the coarse fraction. Auger borings are reasonably representative, and are usually considered satisfactory. A drawback of both auger and wash borings is inability to detect accurately changes in soil stratification.

For stress-strain relationships, undisturbed samples are required. Much effort has been and is being expended in attempts to obtain undisturbed samples suitable for these tests. The major problems associated with this endeavor are as follows:

1. The sample is unloaded of its in situ overburden, thus tending to expand in the sampler.
2. The sample is collected in a container through the drill hole, and as the collection device displaces the surrounding soil, disturbance of the sample occurs.
3. Friction on the sides of the recovery container creates disturbance.
4. Samples collected below the water table tend to drain during the recovery process.
5. Changes in pore or air pressures will disturb the sample an unknown amount.

In sands or gravelly materials the problem is even more pronounced, since it is difficult to predict the internal effect on the sample due to a surface dislocation of a sand grain or pebble.

Samples both disturbed and "undisturbed" are usually obtained by driving or pushing into the soil an open-end device, termed a *sampling spoon*. The *split spoon* (Fig. 3-4) is made up of a cutting shoe, a barrel consisting of a length of tube split lengthwise, and a coupling to attach the spoon to the drill rod. For cohesionless soils and thin muds, inserts (Fig. 3-4d) are available to retain the sample in the tube. When a sample has been recovered, the shoe and coupling are unscrewed, and the two halves of the barrel are opened to expose the soil. The field technician makes a visual soil classification, and a portion of the sample is then placed in a small glass jar; on the lid is inscribed the job, depth of the boring, the hole number, and the penetration number (see Sec. 3-6). Larger quantities of material for compaction tests, sieve analysis, Atterberg limits, etc., may be placed in a sack or a larger jar. In deposits of

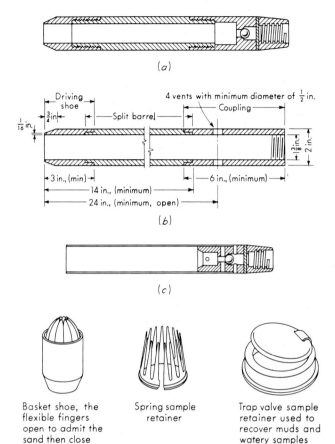

Figure 3-4. Soil-sampling tools. (a) Standard split-spoon sampler; (b) dimensions of the standard split-sampler assembly; (c) thin-wall (Shelby tube) sampler; (d) split-spoon sampler inserts.

Basket shoe, the flexible fingers open to admit the sand then close when the tube is withdrawn

Spring sample retainer

Trap valve sample retainer used to recover muds and watery samples

(d)

fine sand a slot sampler (Fig. 3-5) may be required. Samples are recovered by pushing this sampler into the deposit, and then rotating it, so that the slot trimmings fall inside the device.

Recovery of "undisturbed" samples in cohesive soils is accomplished by replacing the split spoon on the drill rod with specially constructed thin-wall (16- to 20-gage) seamless brass or steel tubing which is driven, but preferably pushed, into the soil. The term "Shelby tube" is widely used to describe any thin-wall tube used in this manner. Actually, Shelby tube was the trade name for hard-drawn seamless-steel tube manufactured by the National Tube Division of the U.S. Steel Corporation until several years ago. The tube is slightly rotated, or a special cutting device is attached, to cut the sample off. Friction holds the sample in the tube as the sample is withdrawn; however, there are also special valve or piston (Fig. 3-6) arrangements which use a pressure differential (suction) to retain the sample in the tube.

A special sampler termed a *foil sampler* (Fig. 3-6c) was developed in Sweden [see Hvorslev (1949, p. 269), Kjellman (1948)] to overcome two principal deficiencies of the usual sampling tubes and piston samplers. These deficiencies are short sample

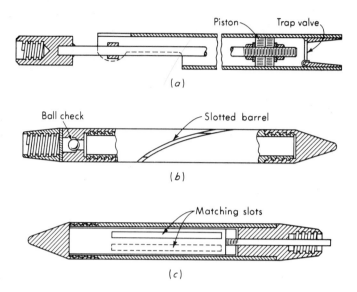

Figure 3-5. Special sampling tools. (*a*) Sand-pump sampler which utilizes pumping action to recover a sample; (*b*) spiral-slot sampler that is filled by rotation of device; (*c*) slot sampler that is filled by rotation of device.

length and side friction between wall and soil as it is forced into the sampler. Reducing friction required reducing the sampler length. If one is in a soil suspected of being particularly troublesome, it may be necessary to take continuous samples. This is not practical with samplers of, say, 1 m maximum length because of continually pulling the drill rods to attach a new tube. The foil sampler is a means of allowing sample recovery of from 10 to 20 m with a minimal friction effect. The interested reader should consult the cited references for exact details, but essentially the sampler operates by first placing it on the bottom of the borehole. Next it is pushed into the soil; as the sample enters the tube it is surrounded by 16 foil (thin metal strips about 13 mm wide by 0.5 to 1.0 mm thick) strips which carry the sample up the tube. Friction between soil and foils results in reducing the compressive stress in the sample as the length of recovered sample increases in the tube.

Liners are available with certain split-spoon samplers which are simply thin-walled tubes placed inside of the barrel portion. The sample is collected, the exterior barrel opened, and the liner containing the sample is removed. Liner-recovered samples are of doubtful quality if an undisturbed sample is the intent, the major reason being that the sample is obtained using the driving shoe, which has a rather large volume displacement, to trim the sample.

Although sample disturbance depends on factors such as rate of penetration of spoon, whether the cutting force is obtained by pushing or driving, and presence of gravel, it also depends on the volume of soil displacement to the volume of the collected sample, *area ratio* A_r.

$$A_r = \frac{D_o^2 - D_i^2}{D_i^2} \times 100 \qquad (3\text{-}1)$$

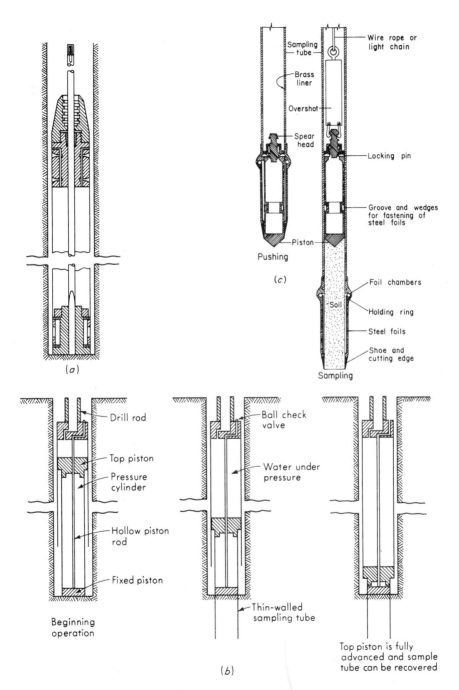

Figure 3-6. Typical piston samplers. (*a*) Stationary-piston sampler for recovery of "undisturbed" samples of cohesive soils. Piston remains stationary on soil and tube is pushed into the soil; piston is then clamped and sample is recovered; (*b*) Osterberg piston sampler; (*c*) Swedish foil sampler [*Hvorslev* (1949)].

where D_o = outside diameter of tube
$\quad\quad D_i$ = inside diameter of cutting edge

Well-designed sample tubes should have an area ratio less than about 10 percent [the widely used 2-in thin-wall (Shelby) tube has an A_r of about 13 percent as computed in Example 3-1].

Another term used in estimating the degree of disturbance of a cohesive or rock-core sample is the *recovery ratio* L_r:

$$L_r = \frac{\text{actual length of recovered sample}}{\text{theoretical length of recovered sample}} \quad\quad (3\text{-}2)$$

A recovery ratio of 1 (recovered length of the sample/the length sampler was forced into the stratum) indicates that, theoretically, the sample did not become compressed from friction on the tube. A recovery ratio greater than 1.0 would indicate a loosening of the sample from rearrangement of stones, roots, removal of preload, or other factors within the recovered specimen.

In the final analysis, however, engineering judgment must be relied upon to extrapolate the results of tests on "undisturbed" samples to the prediction of field behavior.

Undisturbed samples in cohesionless deposits are extremely difficult to obtain. Samples have been recovered by either freezing the soil totally [Corps of Engineers (1938)] or just freezing the material at the lower end [Fahlquist (1941)] of the sampler, so that the tube is plugged. Asphaltic material [Van Bruggen (1936)] may be injected into the sand; the sample is then recovered, and the asphalt later removed with a suitable solvent. Fortunately, it is seldom necessary to recover undisturbed cohesionless samples, since the standard penetration test of Sec. 3-6 provides fairly reliable data at modest expense. Also, it appears that if a laboratory sample of sand is prepared to the in situ density, the soil structure (void ratio) is reasonably well duplicated for most engineering purposes.

Example 3-1. What is the area ratio of the 2-in thin-wall (Shelby) tube?

SOLUTION. Using nominal dimensions from the supplier's catalog,
OD = 2 in.

ID = $1\frac{7}{8}$ in (the actual ID of the tube is slightly larger than the ID of the cutting edge to reduce side friction on the sample as the tube is pushed into the soil).

$$A_r = \frac{D_o^2 - D_i^2}{D_i^2} \times 100 = \frac{2^2 - 1.875^2}{1.875^2} \times 100 = 13.6 \text{ percent}$$

3-6. THE STANDARD PENETRATION TEST

It is impossible to obtain undisturbed cohesionless soil samples and highly desirable to have a continuous record of soil conditions in a borehole for all soils. Various methods to accomplish the above two objectives are in use. One of the earliest, and currently the most popular and economical in terms of cost/amount of information supplied, is the *standard penetration test* (SPT). Some persons feel this procedure should not be dignified by calling it a "test." Others feel that the data obtained should be called an index (SPI). Nonetheless the test has been "standardized" by ASTM D1586-67 (1976) and consists in:

1. driving the standard split-tube sampler of dimensions shown in Fig. 3-4
2. a distance of 18-in (45.5 cm) into the soil at the bottom of a borehole, and
3. counting the number of blows to drive the device the last 12 in (30.5 cm) (this is the N number of the SPT test),
4. using a 140-lb (63.5-kg) drive weight
5. falling 30 in (76.2 cm).

The blow count for the first 6 in (15 cm) is not used, as this is to seat the sampler below the disturbed soil in the bottom of the borehole. Some early authorities recommended that a blow count be taken of each 6-in increment and the values summed from the two least counts for N. This is not currently used or recommended.

The number of blows to drive the sampler the final 12 in is the penetration resistance N. If it is not practical to drive the sampler the full 12 in, because of very dense, cemented hardpan, rock, etc., the boring log will indicate the blow count and penetration as a ratio, i.e., 70:4 for 70 blows and 4 in of penetration. A premium price is generally charged for blow counts greater than 50 or 60/ft owing to the greatly increased borehole time and excessive equipment wear—especially the driving shoe on the sampler. Normal boring operations are usually discontinued when the blow counts become on the order of $N = 100$. "Refusal" is usually shown on the boring log if the blow count is on the order of 50:1.

There are numerous factors in performing the SPT test which make it very difficult to reproduce the N values—even in adjacent borings—and nearly impossible in similar soils at other locations. Schmertmann (1975), from considerable work with various types of penetration test devices, also cites considerable difficulty with reproducing N values. The state-of-the-art presentation by De Mello (1971) presents a considerable summary of efforts to "standardize" the SPT. Fletcher (1965) discussed the difficulties associated with the SPT. Sanglerat (1972) made an extensive study of the penetration test in sands to improve the Terzaghi and Peck (in 1948 and 1967 editions) method of using the SPT value. In spite of this tremendous amount of literature, most highly critical of the SPT, it is not expected that the test will be discontinued for several very important reasons:

1. Economic, as cited earlier, of cost/information obtained.
2. The test results in recovery of soil samples which, even though rather highly disturbed, can be tested or visually examined.
3. The test does give some indication of soil state in the borehole.
4. Long service life of the enormous amount of this type of equipment in use.
5. The accumulation of files in the engineering offices of successful (and empirical) usage of SPT data (which will continue to increase because of item 4).
6. The fact that the SPT can be used *with* more refined methods, in both the field and the laboratory when it is evident, as the exploration program progresses, that alternatives to the SPT are needed.

Some of the factors which tend to make the SPT nonreproducible are as follows:

1. Variations in the 30-in drop height of the drive weight, since this is often done by eye.
2. Interference with the free fall of the drive weight by guides and/or the rope used to hoist the drive weight for successive blows.
3. Use of a drive shoe on the split-tube sampler which has been distorted from driving onto rock.
4. Failure to seat the sampler on undisturbed material for the blow count.
5. Inadequate cleaning of loosened material from the bottom of the hole.
5. Failure to maintain sufficient hydrostatic pressure in the boring so that the test zone becomes "quick." Too large a hydrostatic pressure as with use of drilling mud and/or head greater than static ground level may also influence N.
7. Use of too light or too long a string of drill rods.
8. Driving a stone ahead of the sampler.
9. Careless work on the part of the drill crew.

Studies have been made which indicate that the use of three instead of two turns of rope around the cathead to hoist the drive weight could increase N by as much as 40 percent [see Kovacs et al. (1975)]. Various boring-equipment manufacturers have now put equipment on the market which uses mechanical hoisting trip devices to avoid the energy loss due to rope friction and looping interference and maintain a more uniform height of free fall. Studies by Gibbs and Holtz (1957) indicate that drill-rod lengths of the E and A sizes (Table 3-2) did not affect N for lengths up to about 20 m. Later analytical studies by McLean et al. (1975) using the wave equation indicate that the longer drill rods may increase the blow count by as much as 14 blows/ft.

In sand (coarse, cohesionless soil) extensive studies by Gibbs and Holtz (1957) and Bazaraa [(1967), see also Peck and Bazaraa (1969)] indicate that the penetration number N should be corrected for overburden pressure. Bazaraa makes the following proposal based on analysis of large numbers of borings by others:

$$N = \frac{4N'}{1 + 2p_0} \qquad p_0 < 1.5 \text{ ksf} \tag{3-3}$$

$$N = \frac{4N'}{3.25 + 0.5p_0} \qquad p_0 > 1.5 \text{ ksf} \tag{3-3a}$$

The following comments can be made regarding these equations:

1. N is increased from the actual blow count N' when $p_0 < 1.5$ ksf (effective-pressure value).
2. N is decreased from the actual blow count when $p_0 > 1.5$ ksf.
3. When the blow count indicates a relative density $D_r < 0.5$ (see Table 3-3), do not use these equations.
4. Do not adjust N to more than $2N'$ (author's recommendation).
5. Use these equations cautiously.

Table 3-2. **Standard designation and sizes for drill rods and casing**

Drill rod	OD, in	Casing and core barrel	Core-barrel-bit OD, in	Approx. diam of borehole,* in	Diam of core sample, in
E	$1\frac{5}{16}$	EX	$1\frac{7}{16}$	$1\frac{1}{2}$	$\frac{7}{8}$
A	$1\frac{5}{8}$	AX	$1\frac{7}{8}$	2	$1\frac{1}{8}$
B	$1\frac{7}{8}$	BX†	$2\frac{3}{8}$	$2\frac{1}{2}$	$1\frac{5}{8}$
N	$2\frac{3}{8}$	NX	$1\frac{15}{16}$	3	$2\frac{1}{8}$

  * Diameter of borehole is very nearly the ID of the casing.
  † In soft or fractured rock, BX or larger cores are preferred.

   The SPT was originally developed for cohesionless soils so that samples would not have to be taken. The test has evolved to the current practice of routinely determining N for all soils. In the zones of particular interest from about 2.5 ft or 1 m below ground surface to considerable depth below the estimated base of the foundation the test is performed every 2.5 ft or 1 m depth increment. At considerable depths where the boring becomes more informational the depth increment for testing is often increased to 5 ft or 2 m.
   Empirical correlations between N and various soil properties have been attempted for cohesionless soils (Table 3-3). Table 3-3 should be used cautiously; for example, a "loose" soil with a range of D_r between 15 and 35 percent places rather arbitrary numbers on a rather tenuous description of a soil.

Table 3-3. **Empirical values for ϕ, D_r, and unit weight of granular soils based on the standard penetration number with corrections for depth and for fine saturated sands**

Description	Very loose		Loose		Medium		Dense		Very dense
Relative density D_r*	0		0.15	0.35		0.65		0.85	1.00
Standard penetration no. N			4	10		30		50	
Approx. angle of internal friction ϕ°†	25°–30°		27–32°	30–35°		35–40°		38–43°	
Approx. range of moist unit weight, (γ) pcf (kN/m³)	70–100‡ (11–16)		90–115 (14–18)	110–130 (17–20)		110–140 (17–22)		130–150 (20–23)	

   * USBR [Gibbs and Holtz (1957)].
   † After Meyerhof (1956). $\phi = 25 + 25D_r$ with more than 5 percent fines and $\phi = 30 + 25D_r$ with less than 5 percent fines. Use larger values for granular material with 5 percent or less fine sand and silt.
   ‡ It should be noted that excavated material or material dumped from a truck will weigh 70 to 90 pcf. Material must be quite dense and hard to weigh much over 130 pcf. Values of 105 to 115 pcf for nonsaturated soils are common.

Table 3-4. Empirical values for q_u* and consistency of cohesive soils based on the standard penetration number

Consistency	Very soft		Soft	Medium	Stiff		Very stiff	Hard
q_u, ksf	0	0.5		1.0	2.0	4.0		8.0
N, standard penetration resistance	0	2		4	8	16		32
$\gamma_{(sat)}$, pcf (kN/m³)		100–120 (16–19)			110–130 (17–20)		120–140 (19–22)	

* These values should be used as a guide only. Local cohesive samples should be tested, and the relationship between N and the unconfined compressive strength q_u established as $q_u = KN$.

Correlation between unconfined compressive strength and penetration number of cohesive coils has also been published, as in Table 3-4. The indicated values of unconfined compressive strength correlated to penetration number in Table 3-4 should be used cautiously, however. Experiences in different areas indicate that it is not unusual to get penetration numbers of 6 to 10 on soils where the unconfined compressive strength may be from 6 to 12 ksf, which is considerably different from data shown in the table, where q_u is approximately one-fourth of the penetration number. A valid correlation between the penetration number and the shear strength of cohesive (ϕ-c) soils can be made only on a local basis, and then the validity may be suspect, unless large enough quantities of tests are made to allow a statistical analysis.

In general, to correlate penetration number to the shear strength of a cohesive (ϕ-c) soil involves evaluating the following expression for the constant of proportionality:

$$q_u = KN \tag{3-4}$$

where K = proportionality constant
N = penetration number

The penetration test applied to gravel or gravelly soils and silty sands yields results which require careful interpretation. In loose gravel the voids formed when the gravel is displaced by the driving shoe of the split spoon may yield low penetration numbers. On the other hand, if the spoon pushes a large piece of rock, the number may be too high.

In very fine, or silty, saturated sand Terzaghi and Peck (1st ed., p. 426) recommended that the penetration number be adjusted if N is greater than 15 as

$$N = 15 + \tfrac{1}{2}(N' - 15) \tag{3-5}$$

where N' is to be used in further computations or is for use in Table 3-3. Bazaraa (1967) also studied Eq. (3-5) for numerous sites. From a plot of N' vs. N a better fit is obtained using

$$N = 0.6N' \tag{3-5a}$$

than the original Terzaghi and Peck proposal—especially in the region where the actual field blow counts N' are less than 15.

Example 3-2. The penetration test was performed on a sand deposit for which the unit weight of the soil is 17.26 kN/m³ (average for first 3.5 m). At elevation -3.0 m the blow count was 8. According to Eq. (3-3), what is the value of N to use in the bearing-capacity equations of Chap. 4?

SOLUTION. $p_0 = 3.0(17.26) = 51.78$ kPa < 71.82; use Eq. (3-3). From Table 3-3 it is estimated that $D_r = 0.35$. In fps, 51.78 kPa $= 1.08$ ksf

$$\frac{4(8)}{1 + 2(1.08)} = 10$$

The adjusted N value is larger than the original N value, therefore, the user would have to make a decision of whether to use this adjusted value or the original value. Most persons would be likely to use the original value because the Bazaraa adjustment has not been proposed for a long enough time that many organizations have any substantial empirical files of the advantage from using this method.

3-7. OTHER PENETRATION METHODS

Cone-type penetrometers are quite popular in Europe. The Dutch-cone penetrometer (Fig. 3-7) is one form that is rather widely used, especially in the Low Countries, and is becoming widely used worldwide. This device will be discussed in some detail in Sec. 3-15. Palmer and Stuart (1957) report correlations on a British device (Fig. 3-7g) for which the penetration number is about the same as for the standard method. This particular device is primarily for testing in gravels, where the problem of pushing a large stone is found.

Other penetrometers and *sounding* devices are also used in the United States. The split-spoon samplers are available from 2 to $4\frac{1}{2}$ in in diameter by $\frac{1}{2}$-in increments. A device termed a peat sampler, which is primarily for hand insertion into very soft deposits (under 35 ft), is available for collecting $\frac{3}{4}$-in-diameter samples. Sounding rods are also used to probe for rock. In the simplest form these are drill rods equipped with a special point.

For the larger split spoons and sounding rods, drop weights weighing 250 to 375 lb are available. These weights may also be used to drive the casing into the borehole. Sowers (1954) published a correlation factor for converting the penetration obtained by nonstandard procedures to the standard test. This factor was 1.0 for all samplers and drive weight combinations except a 1.3-in-OD sampler with a 140-lb weight where the factor was $N = 1.5N'$. In light of recent studies concerning hammer input energy, the factors may no longer be 1, and new evaluations may be required.

3-8. CORE SAMPLING

In rock, except for very soft or partially decomposed sandstone or limestone, very high blow counts are necessary to get any penetration, and the term *refusal* is commonly used to describe this condition. For these cases core samples may be

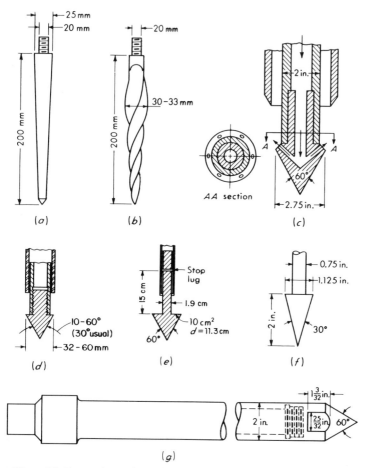

Figure 3-7. Penetration and sounding soil-exploration devices. With the exception of (*g*) the devices utilize static methods. (*a*) Danish penetration device. Penetration is recorded for 25, 50, 75, and 100 kg; then device is rotated, and the penetration is noted for each 25 half-turns. [*After Godskesen (1936)*.] (*b*) Swedish penetrometer. Same essential method of operation as the Danish penetrometer. [*Dahlberg (1974)*.] (*c*) Terzaghi wash-point penetrometer. The cone is jacked 10 in into the soil, and the penetration resistance is measured. Water is then pumped through the system and circulated to the surface, where the soil is collected. The sleeve is then pushed down to the top of the cone, and the process is repeated. [*After Hvorslev (1949)*.] (*d*) Swiss penetrometer. [*Crettaz and Zeindler (1974)*.] (*e*) Early Dutch-cone penetrometer. [*Barentsen (1936)*.] (*f*) Waterways Experiment Station penetrometer for shallow soundings. The device is pushed into the soil by one man, and the force necessary to penetrate 2 in is observed. The device has a load ring which is used to obtain the penetration resistance. (*g*) Modified split spoon. The conical point is for use in granular soil so that large pieces of stone are not trapped, thus causing an increase in the blow count. [*After Palmer and Stuart (1957)*.]

necessary, for several reasons, one principal reason being that the boring operation is usually charged on the basis of per foot of hole drilled, and high blow counts reduce the daily footage considerably. High blow counts also tend to damage the drilling equipment, and in rock or very hard soil the impact tends to reduce the sample to chips, so that the quality of the material may not be determined.

If rock is encountered close to the ground surface, it may be necessary to determine if it is bedrock or a suspended boulder, unless prior geological knowledge in the area can be relied upon. In this case cores may be taken, although probing in the vicinity of the drill hole may indicate the size of the rock.

Rock cores are necessary if the soundness of the rock is to be established; however, cores smaller than BX size (see Table 3-2) tend to break up in the boring process. Cores larger than BX size also tend to break up (rotate inside the core barrel and degrade)—especially if the rock is soft or fissured. Drilling small holes and injecting colored grout (water-cement mixture) into the seams can sometimes be used to recover relatively intact samples. Colored grout present in the recovered core indicates fissures, as well as fissure size, and with cores from several adjacent borings indicates the fissure orientation.

Unconfined and high-pressure triaxial tests can be performed on recovered cores to determine the elastic properties of the rock. These tests are performed on pieces of sound rock from the core sample and may give much higher compressive strengths in laboratory testing than the "effective" strength available from the rock mass. This is because the rock mass may be fractured or fissured.

Figure 3-8 illustrates several commonly used drill bits, which are attached to a piece of hardened steel tube (core barrel) 0.6 to 3 m long. In the drilling operation the bit and core barrel rotate while pressure is applied, thus grinding a groove around

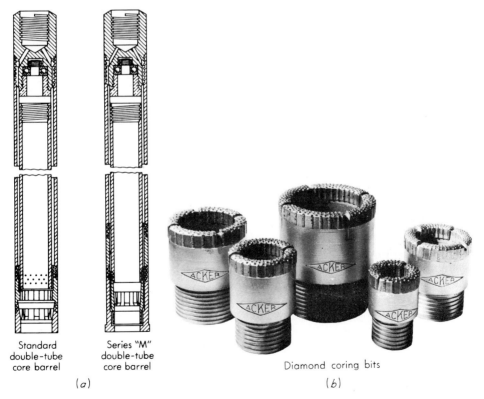

Standard
double-tube
core barrel

Series "M"
double-tube
core barrel

(a)

Diamond coring bits

(b)

Figure 3-8. (a) Core barrels to which (b) coring bits are attached to obtain rock cores. (*The Acker Drill Company.*)

the core. Water under pressure is forced down the barrel and into the bit to carry the rock dust out of the hole as the water is circulated.

The recovery-ratio term used earlier also has significance for core samples. A recovery ratio near 1.0 usually indicates good-quality rock. In badly fissured or soft rocks the recovery ratio may be 0.5 or less.

Rock quality designation (RQD) is an index or measure of the quality of a rock mass [Stagg and Zienkiewicz (1968)] used by many engineers. RQD is computed from recovered core samples as

$$RQD = \frac{\Sigma \text{ lengths of intact pieces of core} > 10 \text{ cm}}{\text{length of core advance}}$$

For example, a core advance of 150 cm produced a sample length of 131 cm consisting of dust, gravel, and intact pieces of rock. The sum of lengths of pieces 10 cm or larger (pieces vary from gravel to 28 cm) in length is 89 cm. The recovery ratio $L_r = 131/150 = 0.87$ and $RQD = 89/150 = 0.59$.

The rating of rock quality may be used to approximately establish field reduction of modulus of elasticity and/or compressive strength and the following may be used as a guide.

RQD	Rock Description	E_f/E_{lab}^*
<0.25	very poor	0.15
0.25–0.50	poor	0.20
0.50–0.75	fair	0.25
0.75–0.90	good	0.3–0.7
>0.90	excellent	0.7–1.0

* Approximately for field/laboratory compression strengths also.

3-9. WATER-TABLE LOCATION

Since groundwater affects many elements of foundation design and construction, its location should be established as accurately as possible if it is within the probable construction zone; otherwise, it is necessary only to determine where it is not. This can be done with considerably less accuracy. It is generally determined by measuring to the water level in the borehole after a suitable time lapse. A period of 24 hr is a value which is widely used. In soils with high permeability, such as sands and gravels, the lapse of several hours is usually sufficient unless the hole has been sealed with drilling mud. In soils with low permeability, such as silts, fine sands, and clays, it may take several days to weeks to determine precisely the water level. For these cases it may be necessary to resort to indirect means to establish the approximate location of the water table, as follows:

1. One obvious method is to plot natural water contents with depth. The major drawback is in collecting reliable natural-water-content data.

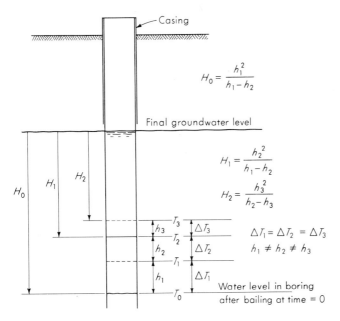

Figure 3-9. Method of computation of location of the stabilized groundwater level by measuring rise of water in the borehole for equal time intervals of ΔT. [*After Hvorslev (1949)*.]

2. Another method is to fill the hole and bail it out. After bailing a quantity, observe if the water level in the hole is rising or falling. The true level is between the bailed point where the water was falling and the bailed point where it was rising.
3. Apply a computational method proposed by Hvorslev (1949, p. 77). In this method measure the rise (or fall) for two or more equal time intervals ΔT.

$$T_1 - T_0 = \Delta T \qquad \text{for height of rise } h_1$$

$$T_2 - T_1 = \Delta T \qquad \text{for height of rise } h_2$$

$$T_3 - T_2 = \Delta T \qquad \text{for height of rise } h_3$$

The distance to the rising water surface from the stabilized groundwater level is H_0 to T_0; H_1 to $h_1 + h_2$; H_2 to $h_1 + h_2 + h_3$; etc. (refer to Fig. 3-9 and Example 3-3). The distances H are computed from the measured changes in water level:

$$H_0 = \frac{h_1^2}{h_1 - h_2} \qquad H_1 = \frac{h_2^2}{h_1 - h_2} \qquad H_2 = \frac{h_3^2}{h_2 - h_3}$$

If reliable data are necessary, a piezometer should be installed in a borehole and periodically inspected over a longer period of time until the groundwater level stabilizes.

Artesian pressures and perched water levels can create an interpretation problem for the unwary. If the groundwater is under pressure (artesian water), deeper borings tend to raise the water level. Perched water may be indicated if the water level tends

to disappear when the boring extends deeper, especially through a relatively imper-
meable material into an underlying material, e.g., a clay stratum overlying a sand
deposit.

     Example 3-3. It is desired to establish the location of the groundwater table in a clayey material. The
borehole was bailed to a depth of 35 ft below the ground surface, and the water rise was recorded on three
successive days as follows:

$$h_1 = 2.1 \text{ ft} \qquad \text{at 24 hr}$$

$$h_2 = 1.9 \text{ ft} \qquad \text{at 24 hr}$$

$$h_3 = 1.7 \text{ ft} \qquad \text{at 24 hr}$$

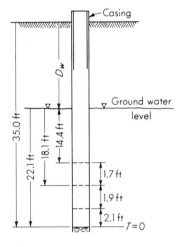

Figure E3-3

SOLUTION. $H_0 = \dfrac{(2.1)^2}{2.1 - 1.9} = 22.1$ ft

$$H_1 = \dfrac{(1.9)^2}{2.1 - 1.9} = 18.1 \text{ ft}$$

$$H_2 = \dfrac{(1.7)^2}{1.9 - 1.7} = 14.4 \text{ ft}$$

Referring to the figure, on which the H values just computed
have been placed, the depth to the water table is as follows:

First day: $\qquad\qquad\qquad\qquad\qquad D_w + 22.1 = 35$ ft

$$D_w = 12.9 \text{ ft}$$

Second day: $\qquad\quad D_w + 18.1 + 2.1 + 1.9 = 35$ ft

$$D_w = 12.9 \text{ ft}$$

Third day: $\qquad\quad D_w + 14.4 + 1.7 + 1.9 + 2.1 = 35$ ft

$$D_w = 14.9 \text{ ft}$$

Averaging results, we obtain a depth to the water table of

$$D_w = 13.6 \text{ ft}$$

3-10. DEPTH AND NUMBER OF BORINGS

Generally, a project should be approached from the view of what data are needed,
the information already available, the funds available for an exploration program,
and the possible costs associated with inadequate subsurface information. This pre-
cludes some rule-of-thumb criteria for the number of borings. One boring is in
general not adequate, although it is better than no boring at all. Probably a minimum of three borings should be taken in a building site if the soil is reasonably
uniform; five borings are preferable—one at each corner and one in the center. On
the other hand, if the soil is erratic, large numbers of borings will provide little
additional information. If the soil goes from one type to another, extra borings to
delineate the two zones are required. Additional borings in the vicinity of heavily
loaded columns and beneath delicate machinery are usually desirable.

Depths of borings are also difficult to estimate. Here a rule of thumb that the borings should extend to a depth such that, from a Boussinesq analysis (Chap. 5), the increase in pressure is 10 percent of the contact pressure is often used. Since this is near 1.5 to 2 times the width of the structure, this criterion is also sometimes used. Again, the project should be analyzed as an individual problem. For example, a one-story warehouse, say, 20 × 40 m in plan, would not generally require borings 30 m or more deep. On the other hand, for a 20-story building 20 m wide, it would be desirable to have at least one boring to bedrock, which may be quite a bit below 30 to 40 m into the ground. Here, however, the cost of a deep boring is more than offset by the possible damage due to a design based on inadequate subsurface information.

3-11. PRESENTATION OF DATA

The boring data obtained for a site are usually plotted graphically on a sheet of drawing paper. The plot data are obtained from the drilling record kept by the boring foreman. A typical drilling record for one hole of a project is shown in Fig. 3-10. The graphical presentation, termed a *boring log*, should include a plot plan, showing the hole location and identification number, and a separate plot, showing the soil profile as established from the drilling record. Obviously, some extrapolation is necessary to define strata boundaries in the zones between borings; however, unless the soil contains lenses, or is erratic, the extrapolation errors are usually minor. The adjacent borings should be plotted next to each other to establish the soil profile regardless of the identification number of the hole. The penetration number, unconfined-compression strengths, Atterberg limits, natural water content, and other appropriate laboratory data may be shown for each boring on the soil profile. A plot of a boring project is shown in Fig. 3-11. Exact details of a plot differ somewhat in various engineering offices. On small projects the drilling records (as Fig. 3-10) may be all that are made other than recommendations from their analysis.

For projects where the number of borings may be larger than three or four, or if the soil profile seems erratic, a plot which progresses with the site boring may provide the best means to determine the need for, and location of, additional borings.

From an analysis of the boring log, which includes consideration of the elevation of footings, column location and loads, location of compressible strata, water-level location, laboratory data, the extent of extrapolation of soil profile, and any additional factors peculiar to a particular project, the foundation engineer makes a decision as to the probable safe soil bearing capacity, considering both settlements and bearing failure. This value may be submitted to the client as a single value (as used in Chap. 4) for all-size footings, or in the form of curves, which indicate the bearing capacity as a function of the footing size [$q = f(B)$], as illustrated in Example 4-7, and should also state whether the value is *gross* or *net* pressure. This information is sent to the client in the form of a report, together with several copies of the boring log and the boxes of soil samples collected from the split spoon as the borings were taken.

FIELD LOG

Project Name _____
Site _____
Project Number 03-55
Boring Number B-5

Date of Boring 9-12-75
Water Level 11'-6" at 0 Hr.
Drillers A.B. & C.D.

A&H ENGINEERING CORPORATION

ELEV	DESCRIPTION	DEPTH	SAMPLE	6"	12"	N		q_u	q_p	W_c	γ_d
0 – 3'	Dark brown silty clay with organic matter(fill)	2½	1 AU								
			2 SS	7	8	15	56/20	2.5			
3' – 5'	Brown clayey silt (fill)	5	3 SS	4	6	10	36/20	1.0			
		7½	4 SS	5	4	9	19/15	0.7			
5' – 7'	Light brown clayey silt to silty clay	10	5 SS	3	4	7	20/15	0.7			
		12½	6 SS	4	3	7	24/15	1.0			
7' – 15'-6"	Light brown silt	15	7 SS	4	3	7	36/20	1.2			
		20	8 SS	4	4	8		1.2			
15'-6" – 20'	Brown silty clay, trace sand	25	9 SS	3	5	8	21/20	1.0			
20' – 25'-6"	Light brown sandy silty clay, some gravel	30	10 SS	7	9	16	48/15	1.7			
25'-6" – 31'	Gray clayey silt, some sand and gravel										

(a)

RECORD OF SUBSURFACE EXPLORATION

Boring B-5

Project Name: _____ High School
Site: _____, Illinois
Date of Boring: September 12, 1975
Project No.: 03-55_

DEPTH	DESCRIPTION	SAMPLE	N	q_u	q_p	M_c	REMARKS
	SURFACE Elev. 103.7						
	Dark brown Silty CLAY with organic matter (fill)	1 AU	–	–	–	23	
	Brown Clayey SILT (fill)	2 SS	15	2.2	2.5	24	
5	Light brown Clayey SILT to Silty CLAY	3 SS	10	1.4	1.0	28	LL = 30.5 PI = 3.7
		4 SS	9	0.8	0.7	24	0 Hr.
10	Light brown SILT	5 SS	7	0.8	0.7	24	
		6 SS	7	1.0	1.0	23	
15		7 SS	7	1.4	1.2	23	
	Brown Silty CLAY with trace sand						
20		8 SS	8	–	1.2	25	
	Light brown Sandy Silty CLAY with some gravel						
25		9 SS	8	0.8	1.0	15	
	Gray Clayey SILT with some sand and gravel						
30		10 SS	16	2.0	1.7	13	
	END OF BORING						

(b)

2-1

94

3-12. FIELD LOAD TESTS

A semidirect method to estimate the bearing capacity of a soil in the field is to apply a load to a model footing and measure the amount of load necessary to induce a given amount of settlement. Round plates from 6 to 30 in in diameter by 6-in increments are available, as well as square plates of 1 sq ft area. These are carefully machined 1-in-thick steel plates, and where 18, 24, and 30 in diameters are used, the plates are usually stacked so that bending effects are reduced.

   Because of the small size of the plate compared with the size of the footing, and since the stress zone of influence is dependent on the size of the loaded area (Fig. 5-10), careful extrapolation of load-test results is required. There is, additionally, the problem of providing an adequate reaction load on the plate, this being the principal reason for using the small plate dimensions.

   The plate-load test, as with the standard penetration test, is subject to individual variations, although ASTM has attempted to standardize it. The method of performing the plate-load test as prescribed by ASTM D1194-72 is essentially as follows:

1. Decide on the type of load application, and if it is to be a reaction against piles, they should be driven first to avoid excessive vibration and loosening of the soil in the excavation where the load test will be performed.
2. Excavate a pit to the depth the test is to be performed. The test pit should be at least four times as wide as the plate and to the depth the foundation is to be placed. If it is specified that three sizes of plates are to be used for the test, the pit should be large enough so that there is an available spacing between tests of 3D of the largest plate.
3. A load is placed on the plate, and settlements are recorded from a dial gage accurate to 0.25 mm. Observations on a load increment should be taken until the rate of settlement is beyond the capacity of the dial gage. Load increments should be approximately one-fifth of the estimated bearing capacity of the soil. Time intervals of loading should not be less than 1 hr and should be approximately of the same duration for all the load increments.
4. The test should continue until a total settlement of 25 mm is obtained, or until the capacity of the testing apparatus is reached. After the load is released, the elastic rebound of the soil should be recorded for a period of time at least equal to the time duration of a load increment.

   Figure 3-12 presents the essential features of the load test, and Fig. 3-13 displays typical load-settlement data from a test. In Fig. 3-13, the ultimate soil pressure is taken as that load which is on the vertical part of the settlement curve (or 1 in if this is as far as the test progresses).

   For extrapolating the load-test results to full-size footings, it will be shown in

---

Figure 3-10. (a) Field boring record; (b) boring log to present the data from the field boring record of (a), together with any laboratory tests. (*A and H Engineering and Testing Corporation.*)

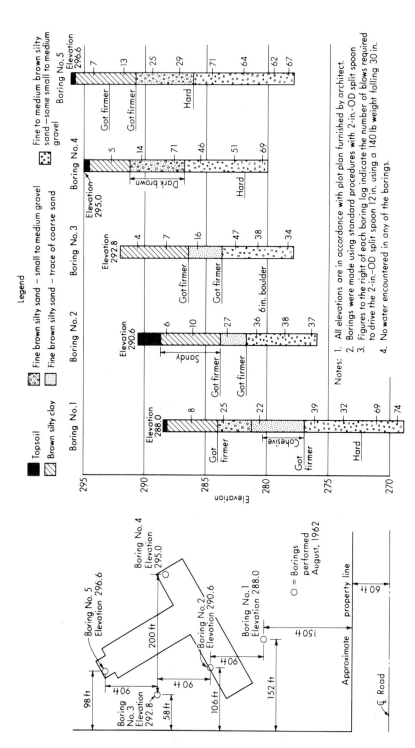

Figure 3-11. An alternative method of presenting the boring information on a project.

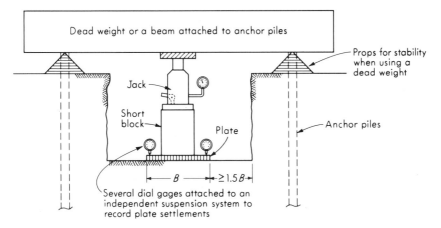

Figure 3-12. Plate-load testing. The method of performing this test is outlined in some detail as ASTM Standard Procedure D1194-57 (Part 19 of ASTM Book of Standards).

Chap. 4 that the bearing capacity of clay is essentially independent of the footing size, or

$$q_{\text{footing}} = q_{\text{load test}} \tag{3-6}$$

In sands and gravels the bearing capacity increases linearly with the size of the footing:

$$q_{\text{foundation}} = M + N \frac{B_{\text{foundation}}}{B_{\text{plate}}} \tag{3-7}$$

where the reader may easily recognize[1] that the M term includes the N_c and N_q terms, and the N term includes the N_y portion of Eqs. (4-1) to (4-3). By using plates of more than one size, Eq. (3-7) may be solved graphically. Practically, for extrapolating plate-load tests we may use

$$q_f = q_p \frac{B_f}{B_p} \tag{3-7a}$$

From Examples 4-1 and 4-2 it would appear that the NB_f/B_{plate} term is not a major part of the soil foundation pressure; however, in those examples the footing ratio B_2/B_1 was near 1, whereas a footing/plate ratio B_f/B_p may be 7 to 15 or more.

Housel (1929) has suggested that the bearing capacity of a footing on cohesive (ϕ-c) soils be written

$$V = Aq + Ps \tag{3-8}$$

where V = total load on a bearing area A
    A = contact area of footing (or plate)
    P = perimeter of footing
    q = bearing pressure beneath A
    s = perimeter shear

---

[1] The settlement of the footing will also increase using this equation, as outlined in Sec. 5-5.

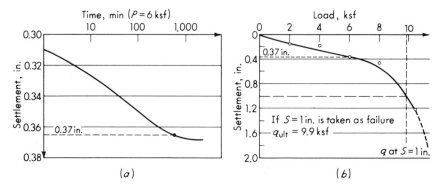

Figure 3-13. Plate-load-test data. (a) Plot of settlement vs. log time to determine the maximum settlement for a load increment (6 ksf in this case); (b) load vs. settlement plot to establish the maximum design pressure.

This procedure requires data from at least two plate-load tests so that Eq. (3-8) may be solved for q and s (for a given settlement). Once the values of q and s are known, the size of a footing to carry a given load may be established as in Example 3-4.

Settlement predictions may also be estimated from plate-load tests. On clay, if it is homogeneous and saturated to a depth of 2 to $2.5B$ of the full-size footing, one may [making reference to Fig. 5-10 and by proportion from Eq. (5-10)] write

$$S_{\text{foundation}} = S_{\text{plate}} \frac{B_{\text{foundation}}}{B_{\text{plate}}} \qquad (3\text{-}9)$$

Settlement predictions on granular soils may be extrapolated using Eq. (5-15).

Since a load test is of short duration, consolidation settlements usually cannot be predicted. If the test is performed on a material overlying a saturated compressible stratum, the test may give highly misleading information. As a precautionary measure against this event, it is good practice to have borings performed at the load-test site.

Example 3-4. Two plate-load tests were performed using plates 1×1 ft and 1.5×1.5 ft. For a $\frac{1}{2}$-in settlement, the loads were 8,350 and 17,100 lb, respectively. What size square footing is required to carry a 72-kip column load?

SOLUTION. $V = Aq + Ps$

$q + 4s = 8{,}350$

$2.25q + 6s = 17{,}100$

$q = 6{,}090 \text{ psf}$

$s = 565 \text{ psf}$

For full-size footing,

$B^2q + 4Bs = 72{,}000$

$B^2(6.09) + 4B(0.565) = 72$

$B^2 + 0.371B = 11.82$

$B \simeq \sqrt{11.82} = 3.4$ ft $0.371B$ being relatively insignificant

Use $B = 4 \times 4$ ft   and $S \simeq \frac{1}{2}$ in

3-13. FIELD-VANE TESTING OF SOILS

There is considerable difficulty in recovering an undisturbed sample from very soft, sensitive clay deposits, as with the recovery of cohesionless samples. As previously stated, one major problem of all undisturbed-sample recovery is that the effect of overburden pressure is lost. Additionally, pore-pressure changes may take place because of loss of hydrostatic pressure or a large area ratio of the sampler. Vibrations from the sampling operations may affect the soil structure. In laboratory testing, there is the problem of application of a realistic lateral pressure in the triaxial cell for performing the confined-compression tests necessary to define the soil parameters.

To overcome some of these difficulties, in situ shear testing using a *vane-shear* apparatus was developed. Figure 3-14 illustrates the U.S. Bureau of Reclamation [Gibbs et al. (1960)] version of this instrument in use since around 1954. The vane-testing apparatus seems to have been developed in Europe around 1948 [Skempton (1948), Carlson (1948)]. The Bureau of Reclamation device uses standard A-size drill rods in 5-ft lengths, to which the vane is attached, thus allowing progressive depth tests in multiples (or fractions) of 5-ft increments. Standard BX casing pipe is used to enclose the vane rods. A watertight bearing through which the vane rod fits is made a part of the BX casing point, or first section of casing just above the vane. The vane dimensions used by the Bureau of Reclamation are 2 in in diameter by 4 in long; 3 in in diameter by 6 in long; and 4 in in diameter by 8 in long. Commercial vanes $(L/d = 2)$ 2.2 × 4.32 in long and 2.56 × 5.13 in long are also available. Bushings may be added on the drill rod to maintain vertical alignment and reduce whip of the rod under torque.

In use a hole is drilled to the desired depth (above the point the actual vane test is made), the vane is carefully pushed into the soil, and the torque necessary to shear the cylinder of soil defined by the vane blades is measured. An alternative method is to push the vane assembly, using a special point, into the soil without the benefit of a borehole. At the desired depth the vanes are extended and the torque measured, as in the method using a borehole. Equation (2-16) may be used to compute the shear strength of the soil, or charts may be prepared showing the relationship of torque (for a particular vane dimension) to the shear strength of the soil. Equation (2-16) is repeated here for convenience, and it may be noted that if the d/h ratio is known, the equation can be somewhat simplified.

$$T = s\pi\left(\frac{d^2h}{2} + \frac{ad^3}{4}\right) \tag{2-16}$$

where a = 2/3 for uniform end shear [usual and for Eq. (2-16) in Chap. 2]
      3/5 for parabolic end-shear distribution
      1/2 for triangular end-shear distribution

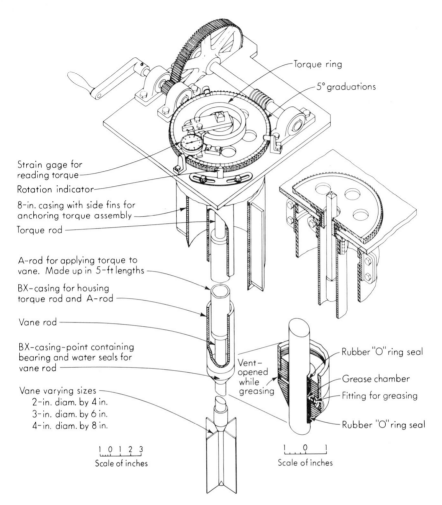

Torque ring

5° graduations

Strain gage for
reading torque

Rotation indicator

8-in. casing with side fins for
anchoring torque assembly

Torque rod

A-rod for applying torque to
vane. Made up in 5-ft lengths

BX-casing for housing
torque rod and A-rod

Vane rod

BX-casing-point containing
bearing and water seals for
vane rod

Vane varying sizes
  2-in. diam. by 4 in.
  3-in. diam. by 6 in.
  4-in. diam. by 8 in.

Vent—
opened
while
greasing

Rubber "O" ring seal

Grease chamber

Fitting for greasing

Rubber "O" ring seal

1 0 1 2 3
Scale of inches

1 0 1
Scale of inches

Figure 3-14. The Bureau of Reclamation vane-shear test apparatus. [*Gibbs et al. (1960), courtesy of Gibbs and Holtz of the USBR.*]

The in situ vane-shear test gives values of the consolidated-undrained shear strength, since the measurement is made with the average effective stress acting on the sample. It was initially the opinion of users that the test would give reliable values of the undisturbed shear strength. The test probably does give a reliable value of the strength if all factors are taken into account, although some persons feel the values are unconservative [Bjerrum (1972)] and require correction (Fig. 3-15); others take the opposite extreme. La Rochelle and Lefebvre (1970) in tests on a highly sensitive, preconsolidated, Leda clay found that the field vane gave values of s_u on the order of 900 psf where laboratory undrained shear tests on carefully obtained hand-cut block samples gave values of about 1,400 psf. Their laboratory vane-test values were on the order of 1,050 psf. Arman et al. (1975) in a very extensive test program on normally consolidated Mississippi Delta type soils found the vane s_u to be about two times the

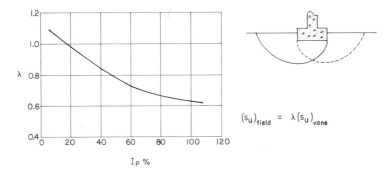

$$(s_u)_{field} = \lambda (s_u)_{vane}$$

Figure 3-15. Correction factor for vane-shear test. [*Bjerrum (1972)*.]

laboratory undrained test values. Eden (1965) in tests on Leda clay found vane strengths on the order of two times unconfined compression values in agreement with Arman et al.

There is some indication that the field vane may, because of its size, cause more disturbance on insertion than originally thought. Field-vane tests are commonly conducted by

1. Working from the bottom of a borehole.
2. Pushing the vane from the ground surface to some depth, then making the test. The operator must separate rod-friction torque from vane torque.
3. Pushing the vane, protected by a cover, to the proper depth, then extending the vane below the cover and performing the test, taking precaution to isolate rod friction from vane torque.

No doubt anisotropy is a significant factor in the vane-shear test as well as fissuring. There is some indication that the diameter of the shear zone is about 5 percent larger than the vane diameter, but this would reduce the computed shear strength not more than 10 to 15 percent. There is also some evidence that progressive failure may start from the blade intersection with the rupture cylinder, but use of thinner blades could reduce this effect.

3-14. MEASUREMENT OF IN SITU STRESSES AND K_0 CONDITIONS

It is difficult to nearly impossible to measure K_0 stresses in a soil mass, since they disappear on the sides of boreholes and on the sides of excavations as the soil is removed. One method which gives a K_0 measurement of some accuracy as well as a measure of the stress-strain modulus is the *borehole pressuremeter*. This device was originally developed by Menard (1956, 1965) and has been modified by Wroth (1975) and others at Cambridge and by Jezequel and others in France [Baguelin et al. (1974)] to be self-boring (Fig. 3-16*b* and *c*). The Menard-type borehole pressuremeter operated on the principle of expanding a cylinder in the borehole. By observing the

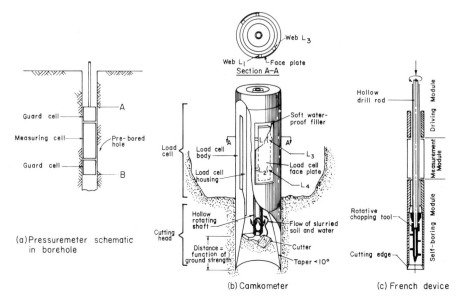

Figure 3-16. Pressuremeter testing. (*b*) and (*c*) above are "self-boring" or capable of advancing the distance *AB* of (*a*) so that in situ lateral stress is not lost.

amount of expansion and the pressure to obtain this deformation, one may use the theory of an infinitely thick cylinder (the soil around the pressuremeter) subjected to an internal pressure to obtain the desired elastic constants. The fundamental equation is

$$\Delta r = \frac{pr_1(1 + \mu)}{E_s} \tag{3-10}$$

In practice it is easier to measure the initial volume V_0 of the device and the change in volume by monitoring the amount of fluid (and the pressure) added to the pressuremeter. A plot can be made from test data similar to Fig. 3-17. From this plot [and rearrangement of Eq. (3-10)] the stress-strain modulus is

$$E_s = 2(1 + \mu)V_0 \frac{\Delta p}{\Delta V} \tag{3-11}$$

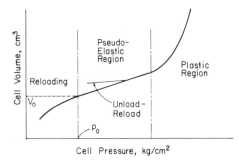

Figure 3-17. Pressure-volume curve for a pressuremeter test.

where Δp = pressure increment producing the corresponding change
in volume ΔV

V_0 = initial volume of the measuring cell of pressuremeter

The in situ lateral pressure is taken as the cell pressure necessary to obtain the initial volume of the pressuremeter as shown in Fig. 3-17. The lateral earth-pressure coefficient can then be computed as

$$K_0 = \frac{\sigma_3}{\sigma_1} = \frac{\sigma_3}{p_0'}$$

The pressuremeter operation depends upon the assumption that the soil is isotropic, linearly elastic, and homogeneous. It should be self-evident that the stress-strain modulus obtained is parallel to the bedding plane and if the material is anisotropic the lateral stress-strain modulus may be considerably larger than the vertical modulus.

It is necessary to assume a value of Poisson's ratio to obtain E_s from Eq. (3-11) unless points outside the linear range are also used so that two equations with two unknowns can be used. It is a complicated process to use the pressuremeter test to obtain the soil parameters ϕ and c [Gibson and Anderson (1961), Ladanyi (1963)] and involves making some additional assumptions.

Other methods to measure in situ conditions include a total-pressure cell (Fig. 3-18) (Glötzl cell) which is inserted into the ground with or without aid of a borehole. If a borehole is not used, the insertion of the pressure cell results in initial conditions of larger than K_0 pressures which will generally dissipate within about 7 to 10 days to nearly K_0 values [Massarsch (1975), Massarsch et al. (1975)]. If the total-pressure cell is inserted into a borehole, the initial lateral stresses are too low; however, with careful backfilling and time the lateral pressure will build up to approximately the K_0 level. Careful interpretation of the pressure vs. time plot is necessary to approximate the K_0 conditions.

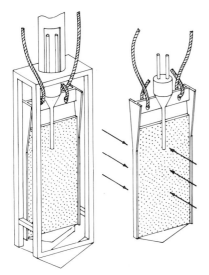

Figure 3-18. Glötzl earth-pressure cell with protection frame to measure lateral earth pressure after the protection frame is withdrawn.

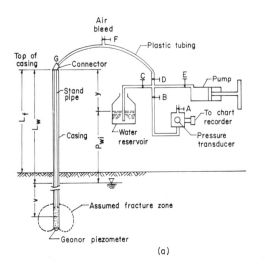
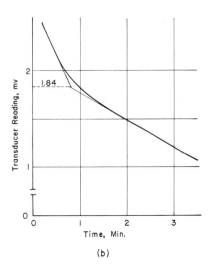

Figure 3-19. (a) Schematic of hydraulic-fracture test setup; (b) qualitative-data plot from fracture test. Value of 1.84 is used in Example 3-5. [*After Bozozuk (1974).*]

Hydraulic fracture is a method which may be used in both rock and clay soils. It cannot be used in cohesionless soils with large coefficients of permeability. Water under pressure is pumped through a piezometer which has been carefully installed in a borehole to sustain considerable hydraulic pressure, before a breakout occurs, out through the piezometer point. At a sufficiently high hydraulic pressure a crack will develop in the soil at the level of water injection. At this time the water pressure will drop and level off at an approximate constant value and flow rate. Closing the system will cause a rapid drop in pressure as the water flows out through the crack in the wall of the boring. The crack will then close as the pressure drops to some value with a resulting decrease in flow from the piezometer system. By close monitoring of the system and making a plot as in Fig. 3-19b one can approximate the pressure at whic the crack opens.

Steps in performing a fracture test are as follows (refer to Fig. 3-19a):

1. Prepare a saturated piezometer with a $\frac{1}{4}$-in standpipe tube filled with deaired water with the top plugged and pushed to the desired depth L using a series of drill rods of about size E (Table 3-2). The plug will keep water from flowing into the system under excess pore pressure developed by pushing the piezometer into the ground.
2. Measure L_w, L_f and compute depth of embedment of piezometer and V.
3. After about $\frac{1}{2}$ hr unplug the $\frac{1}{4}$-in tube standpipe and connect the fracture apparatus to the standpipe using an appropriate tube connector at G.
4. Fill the fracture system with deaired water from the 1- or 2-liter reservoir bottle by opening and closing the appropriate valves. Use the hand-operated pump/metering device to accomplish this.
5. Take a zero reading on the pressure transducer.
6. Apply system pressure using the hand pump at a slow rate until fracture occurs as observed by a sudden drop-off in pressure.

7. Quickly close valve E.
8. From the plot of pressure vs. time, the break in the curve is interpreted as relating to σ_3.

The following example [from Bozozuk (1974)] illustrates the method of obtaining K_0 from a fracture test.

Example 3-5. Data from a hydraulic-fracture test are as follows (refer to Fig. 3-19a):
Length of casing used $= L_w + V = 20.50$ ft
Distance top of casing to ground $L_f = 5.10$ ft
Distance L_w measured inside drill rods with a probe $= 6.62$ ft
Saturated unit weight of soil assuming groundwater nearly to ground surface and $S = 100$ percent for full depth $= 106.9$ pcf
Y measured when fracture apparatus connected $= 0.87$ ft
The pressure-transducer output is calibrated to 0.254 ksf/mV

REQUIRED. Find the at-rest earth-pressure coefficient K_0

SOLUTION. $\qquad V = 20.50 - 6.62 = 13.88$ ft

$$P_{wi} = L_w - Y = 6.62 - 0.87 = 5.75 \text{ ft}$$

$$P_{wi} = 5.75(0.0625) = 0.359 \text{ ksf}$$

Next computing the in situ effective overburden pressure p'_0:

$$p_0 = 15.40(0.1069) = 1.646 \text{ ksf}$$

$$u = 13.88(0.0625) = 0.868 \text{ ksf}$$

$$p'_0 = 0.778 \text{ ksf}$$

Using the transducer reading (Fig. 3-19b) × constant to convert to pressure, the fracture pressure is

$$FP = 1.84(0.254) = 0.468 \text{ ksf}$$

$$P_{wi} = 0.359$$

$$\text{Total pressure} = 0.827 \text{ ksf}$$

$$K_0 = 0.827/0.778 = 1.06$$

3-15. STATIC PENETRATION TESTING— DUTCH-CONE PENETRATION TEST (CPT)

The CPT test is a simple test which is becoming widely used whereby a cone is pushed into the soil stratum of interest and the corresponding resistance is measured. The resistance may be of the cone alone or, more commonly, a cone resistance and the side or skin resistance of a short segment of pipe pushed along with the cone. The cone resistance is related to the undrained shear strength, since the test is so rapid undrained conditions are developed.

The two most common cone configurations are shown in Fig. 3-20; in a is one of the earliest cone versions to measure point and point and side resistance. Figure 3-20b is an electronic version which measures the point resistance using a strain-gage bridge. For this cone the operator observes the total resistance necessary to push the

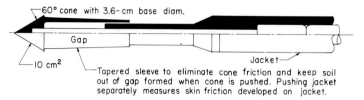

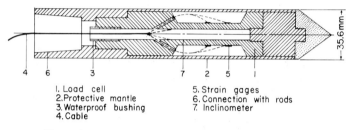

(a) Dutch cone modified to measure both point resistance C_R and skin friction

1. Load cell
2. Protective mantle
3. Waterproof bushing
4. Cable

5. Strain gages
6. Connection with rods
7. Inclinometer

(b) Electric strain gage penetrometer (De Ruiter, 1971)

Figure 3-20. Dutch cones.

system. The point resistance is obtained from the strain-gage bridge and the side resistance is easily computed. A cone version similar to this can be used for ocean-floor tests [Hirst et al. (1972)]. In the earlier version of Fig. 3-20a it is necessary to observe the resistance of the point and when the cone advance is sufficient to engage the friction sleeve a second resistance is observed. The friction resistance q_f is easily computed as the difference between the two resistances.

The data are usually presented as cone resistance q_c and a friction ratio F_R, where

$$F_R = \frac{\text{friction resistance}}{\text{cone resistance}} = \frac{q_t - q_c}{q_c} = \frac{q_f}{q_c} \qquad (3\text{-}12)$$

This test was developed for fine sands, silty fine sands, and clay deposits, as found especially in Holland, Belgium, and in the marine sediments in the fjords of Scandinavia (and also as would be found along large areas of the Gulf Coast of the United States, around and in large lakes and other offshore marine deposits characterized by being soft and homogeneous).

The equipment for a CPT can be truck-mounted with a hole made in the truck bed for pushing the cone and drill rods via use of a hydraulic ram system and using the truck, with or without additional anchorage, for a reaction. Alternatively a system of anchor piles or screw anchors can provide the reaction. Reactions generally do not exceed 10,000 kg, with most on the order of 3 to 5,000 kg. Proponents of the CPT cite the fact it can be used to obtain a continuous or nearly so record. It can be pushed through stones, but the results, except in fine-grained deposits, tends to be misleading. Gages on the hydraulic jack used to push the cone indicate the pushing resistance. These gages can also be connected to pressure transducers and chart recorders to record the resistance vs. depth electronically.

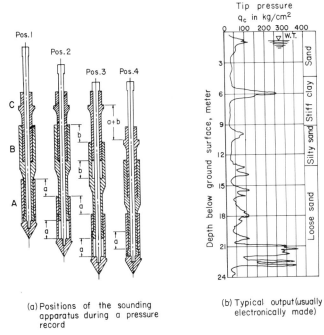

(a) Positions of the sounding
apparatus during a pressure
record

(b) Typical output (usually
electronically made)

Figure 3-21. Sequence of operations in Dutch-cone test and typical plot as obtained using electrical recording equipment.

The cone is forced into the ground at a rate varying from 1 to 2 cm/sec. Figure 3-21a illustrates the pushing sequence for the nonelectronic cone. Figure 3-21b illustrates a typical data output from a CPT. These data would be used as in Chaps. 4 and 5 to estimate bearing capacity and settlements or in Chap. 16 to estimate static pile capacity. The state-of-the-art paper by Schmertmann (1975, with extensive bibliography) summarizes the current status of CPT.

3-16. THE BOREHOLE SHEAR TEST

This test consists in lowering a serrated cylinder which has been split lengthwise into a borehole. The cylinder is constructed so that it can be expanded into the opposite sides of the borehole by applying pressure through from the surface through a pipe system. The test proceeds by applying a known pressure and then pulling the cylinder vertically until slip occurs. The cylinder pressure is incremented and the test repeated several times until sufficient data are obtained to plot a Mohr's rupture envelope as Fig. 2-6b. The cylinder pressure is the "normal" force and the pull necessary to cause slip and is related to the shear stress since we know the dimensions of the shear faces. This test is likely to produce CD shear strengths, since the serrations on the device are of a shallow depth such that excess pore pressure can dissipate rather rapidly.

This device is described in some additional detail by Wineland (1975). Figure 3-22 illustrates the essential features of the borehole shear device. Note that it is

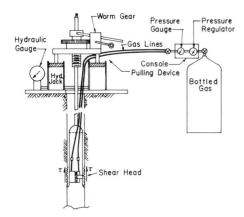

Figure 3-22. Borehole shear device. [*After Wineland* *(1975)*.]

relatively compact and light. The worm-gear pulling force is on the hydraulic gages to obtain the shear stress. The test consists in augering a hole slightly under 76 mm, then reaming it with a special hand auger to "exactly" 76 mm, then inserting the device for the test series.

3-17. SEISMIC EXPLORATION

Seismic methods are fast, reliable means of establishing rock profiles or the location of dense strata underlying softer materials. The method proceeds by inducing impact or shock waves into the soil by exploding small charges in the soil or striking a plate placed on the soil with a hammer. The shock waves are picked up through listening devices termed *geophones*. The velocity of the wave in the surface soil can then be determined by recording the time lapse of the wave traveling to the geophone. As the shock point is moved away from the geophone, some of the waves pass from the surface strata into the underlying layer and then back into the surface strata and into the geophone, enabling the computation of wave velocity in the underlying material.

Depth to rock underlying a single overlying stratum by the seismic method is found essentially as follows:

1. Move the shock-producing device along the desired line, and at given intervals of distance induce a shock and record the time of shock-wave travel and distance readings from the impulse point to the geophone.
2. Plot the results as distance vs. time (ordinate). The plot will "break" when the strike plate is far enough from the geophone so that some of the shock waves enter the denser underlying layer, and thus arrive at the geophone earlier than waves traveling through the upper material. The slope of the initial portion of the graph is

$$\frac{t}{d} = \frac{1}{\text{velocity}}$$

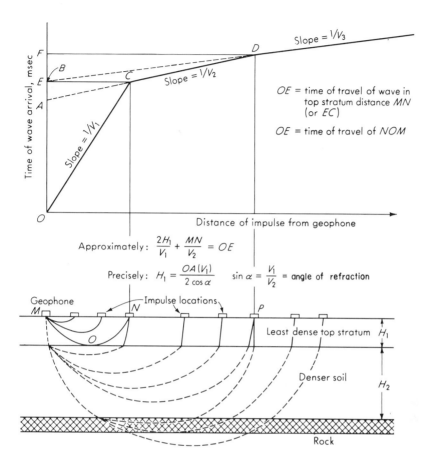

Figure 3-23. Time-distance graph plotted with respect to the physical location of strata, impulse, and sensing device.

This is the velocity of the shock wave in the overlying soil. The slope of the curve beyond the break is the velocity, in the denser soil. Actually, one recording would give a value of soil velocity, but by plotting d versus t, an averaged curve may be obtained which is somewhat more precise than a single value.

3. By measuring the distance $\overline{EC}$ to the break (Fig. 3-23) and using the computed velocities of the shock waves, the depth to the dense layer can be computed as

$$H_1 = \frac{\overline{EC}}{2} \sqrt{\frac{V_2 - V_1}{V_2 + V_1}} \qquad (3\text{-}13)$$

The distance $\overline{EC}$ is the distance from the geophone to the impulse point such that the time for the sound to travel through the two soil strata along $\overline{MON}$ is the same time that it takes the sound to travel through the overlying soil $\overline{MN}$. Table 3-5 gives typical velocities of waves for shallow depths.

Table 3-5. Range of velocities of seismic waves in soil near the surface or at shallow depths*

Material	Velocity, m/sec
Sand	200–2,000
Loess	300– 600
Alluvium	500–2,000
Loam	800–1,800
Clay	1,000–2,800
Marl	1,800–3,800
Sandstone	1,400–4,300
Limestone	1,700–6,400
Slate and shale	2,300–4,600
Granite	4,000–5,700
Quartzite	6,100

* After Leet (1950)

Figure 3-23 indicates three strata, for which simplified forms [Shepard and Haines (1944)] of the seismic equations are

$$H_1 = \frac{\overline{OA}(V_1)}{2\cos\alpha} \qquad (3\text{-}13a)$$

$$H_2 = \frac{\overline{AB}(V_2)}{2\cos\beta} \qquad (3\text{-}13b)$$

where $\overline{OA}$ and $\overline{AB}$ are times from the plot of d versus t; H_1 and H_2 are depths of the strata; and the angles of refraction of stratum interface, α and β, are defined as

$$\sin\alpha = \frac{V_1}{V_2} \qquad \sin\beta = \frac{V_2}{V_3}$$

For more advanced problems, such as three or more different strata, or sloping strata, see Shepard and Haines (1944) or Leet (1950).

When $V_2 < V_1$, the seismic method does not yield useful results. If there is gradual transition from soil to bedrock, that is, the soil becomes denser so that there is not a clear contrast in wave velocities, or if the interface of the two materials is irregular, the method does not yield satisfactory results. It should also be recognized that the method does not tell the operator the type of materials through which the wave is traveling. Thus borings or soundings are usually required to supplement the seismic data. However, with adequate velocity contrast and careful data collection this method can yield bedrock or strata changes within 5 percent of depths established by boring.

Seismic techniques may be used between two or more boreholes using a shock device in one hole and a geophone in the adjacent hole(s). Data obtained are the time for the compression (P wave) and shear (S wave) to arrive at the geophone and the distance traveled. Ballard and McLean (1975) consider the cross-hole seismic method and interpretation of data in some detail beyond the scope of this text.

Example 3-6. Referring to Fig. 3-23, the graphical plot yielded the following data. $V_1 = 2,000$ fps, and $V_2 = 14,000$ fps (the soil was a two-layer system and no V_3 was detected). The distance MN was found $(\overline{EC})$ to be 300 ft. The following data are required: OE, OA, and H_1.

SOLUTION. OE is readily found from the velocity equation.

$$d = vt \qquad \text{but} \qquad t = OE$$

$$300 = 2,000t \qquad \text{therefore} \qquad t = \tfrac{3}{20} \text{ sec}$$

The time OA is found to be

$$OA = \frac{300}{14,000} = \frac{3}{140} \text{ sec}$$

Approximately, H_1 can be found by considering the following paths for the same time interval:

$$\frac{2H_1}{V_1} + \frac{\overline{MN}}{V_2} = OE$$

$$\frac{2H_1}{2,000} + \frac{300}{14,000} = \frac{3}{20} \qquad H_1 = 128.5 \text{ ft} \qquad \text{approx.}$$

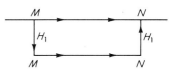

Precisely,

Figure E3-6

$$H_1 = \frac{\overline{MN}}{2}\sqrt{\frac{V_2 - V_1}{V_2 + V_1}} = \frac{300}{2}\sqrt{\frac{14,000 - 2,000}{14,000 + 2,000}} = 129.5 \text{ ft}$$

PROBLEMS

3-1. Referring to Fig. 3-4b, compute the area ratio for the standard split spoon. Based on your computed value of A_r, would you expect a sample recovered in this manner to be undisturbed?

3-2. From a supplier's catalog, the following two Shelby tube dimensions were taken. Compute A_r for both sizes. From these values and Example 3-1 can you draw any conclusions concerning sample disturbances?

OD, in	ID, in	Length, in
3	2.875	24
$4\tfrac{1}{2}$	4.375	24

Partial Answer: $A_r = 8.8$ percent.

3-3. A Shelby tube was pushed into the firm soil at the bottom of a borehole a distance of 28 in. The length of recovered sample was 23.6 in. What is the recovery ratio? Is this a good quality sample?
Answer: 0.84.

3-4. For Fig. P3-4, compute adjusted N values. Use Table 3-3 and make a reasonable estimate of D_r. Assume the sand increases linearly in density from 95 to 115 pcf. Estimate the allowable bearing capacity at elevation -5 for a building with column loads of 100 to 150 kips.

3-5. From the plate-load-test data shown in Fig. P3-5, which was on a damp sand, compute the allowable bearing pressure (SF = 3) for a footing 8×8 ft. Take the ultimate bearing of the plate test at $S = 1$ in, and the plate is 2.5×2.5 ft.

3-6. Assume the plate-load-test data (1×1 ft) of Fig. P3-5 are on a cohesive material. What is the allowable bearing capacity of an 8×8 ft footing using SF = 3? Take the ultimate pressure of the plate as that value where the curve is becoming vertical.
Answer: $q = 1.33$ ksf.

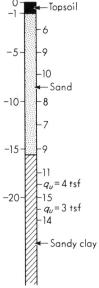

Figure P3-4

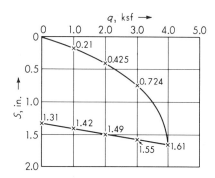

Figure P3-5

3-7. Given the following pressuremeter test data:

ΔV, cm^3	88	130	175	195	230	300	400	600
p, kg/cm^2	0.30	1.92	2.90	3.25	3.90	4.30	4.60	4.90

Assume $V_0 = 88$ cm^3 and $\mu = 0.2, 0.4$.
The soil weighs 1.80 g/cm^3 and the pressuremeter is 1.60 m below the ground surface. Plot these data; find σ_3, E_s for both values of Poisson's ratio and K_0. Make appropriate comments, especially concerning Poisson's-ratio effects.

    Answer: E_s for $\mu = 0.2$ is 800 kPa.

3-8. Referring to Fig. 3-23, V_1 has been found to be 1,200 fps, $V_2 = 3,800$ fps, and $V_3 = 8,100$ fps. If the distance $\overline{MN}$ was 125 ft, compute OE. Using the approximate expression given in Fig. 3-23, compute H_1.

    Answer: $H_1 = 42.8$ ft.

3-9. Use the data of Prob. 3-8 to compute H_1 by exact methods. (*Hint:* cos α is defined from sin α.)

    Answer: $H_1 = 47.5$ ft.

3-10. If NP of Fig. 3-23 is 65 m, $\overline{MN}$ is 30 m, OE is 0.08 sec, OF is 0.10 sec, and V_3 is 3,500 m/sec, find H_2.

    Answer: AB = 0.0184 sec, cos $\beta = 0.866$, $H_2 = 18.6$ m.

3-11. Two plate-load tests yielded the following data:

Plate size, ft	V, lb	s, in
1 × 1	8,500	0.5
2 × 2	21,500	0.5

What should the size of a square footing be to resist a load of 120 kips?

    Answer: 6.0 ft.

FOUR

BEARING CAPACITY OF SHALLOW FOUNDATIONS

4-1. INTRODUCTION

Foundation design is based on providing a means of transmitting the loads from a structure to the underlying soil without a soil shear failure (i.e., a plastic flow and/or a lateral expulsion of soil from beneath the foundation) or causing excessive settlements of the soil under the imposed loads. If both these requirements for a structure are not satisfied, the structure will, in general, perform unsatisfactorily. That is, it will settle excessively, tilt, and form unsightly cracks, and it may even collapse if the differential settlements induce sufficient overstress in critical members.

This chapter presents the currently accepted theories of computing bearing capacity. Chapter 5 will present several methods of computing stresses at a depth in a soil mass due to vertical loading (Chap. 11 considers lateral loading), and methods of computing settlements for soils both by consolidation (settlements as a function of time) and immediate-settlement theories.

Foundations are often classified as *shallow* and *deep* foundations, depending on the depth of the load-transfer member below the superstructure. Thus a deep, as compared with a shallow, foundation becomes a somewhat relative term. Terzaghi (1943) defined a shallow foundation as one in which the depth to the bottom of the footing is less than or equal to the least dimension of the footing (Fig. 4-1a). Others have considered a shallow foundation to be $D \leq 4$ or $5B$.

4-2. THE TERZAGHI BEARING-CAPACITY EQUATION

Terzaghi, using the shallow foundation of $D \leq B$, derived a general bearing-capacity equation from a modification of equations based on soil being perfectly plastic and using plasticity theory proposed by Prandtl (ca. 1920). This equation was derived

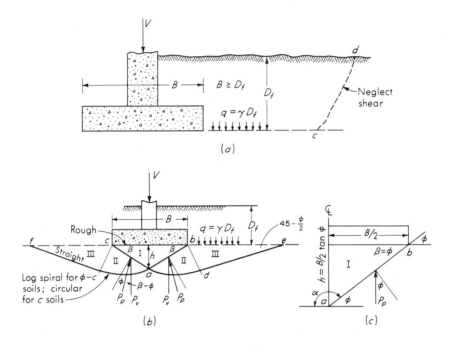

Figure 4-1. Development of the Terzaghi bearing-capacity equations. (*a*) Conditions for a "shallow foundation" as shear along plane *cd* is neglected; i.e., replace soil by a surcharge $q = \gamma D_f$; (*b*) assumptions for the Terzaghi bearing-capacity solution; (*c*) geometry if $\beta = \phi$. Note $\beta = 45 - \phi/2$ for perfectly smooth to $45 + \phi/2$ for perfectly rough base.

based on neglecting the shear resistance of the soil above the horizontal plane through the base of the footing (Fig. 4-1*a*) and replacing the soil above this plane with a surcharge ($q = \gamma D_f$). The application of the load V in Fig. 4-1*b* tends to push the wedge of soil *abc* into the ground with a tendency to lateral displacement of zones II (a sector of a logarithmic spiral if ϕ and γ are greater than zero and a circular sector if ϕ or $\gamma = 0$, as arc *adb*) and III, a triangular passive Rankine zone *bde*. This lateral displacement is resisted by shear stresses developed along the slip plane *ade* and *acf* and by the weight of soil in these zones and the equivalent effective surcharge pressure q along lines *be* and *cf*. From static equilibrium conditions relating vertical forces P_v and V and simplifying assumptions, including $\beta = \phi$, the following equations were obtained:

$$\text{Continuous footing:} \quad q_{\text{ult}} = cN_c + qN_q + \tfrac{1}{2}\gamma BN_\gamma \qquad (4\text{-}1)$$

$$\text{Square footings:} \quad q_{\text{ult}} = 1.3cN_c + qN_q + 0.4\gamma BN_\gamma \qquad (4\text{-}2)$$

$$\text{Round footings:} \quad q_{\text{ult}} = 1.3cN_c + qN_q + 0.3\gamma BN_\gamma \qquad (4\text{-}3)$$

where q_{ult} = ultimate soil bearing pressure (use consistent units for
c, q, and B)

c = cohesion of soil

ϕ = angle of internal friction of soil

$q = \gamma D_f$
B = least lateral dimension of footing (use diameter = B for round footings)

γ = unit weight of soil (use submerged weight for soil below water table or a correction as given in Sec. 4-6 with the saturated unit weight)

$$N_c = \text{bearing-capacity factor} = \cot \phi \left[\frac{a^2}{2 \cos^2 (\pi/4 + \phi/2)} - 1.0 \right]$$

$$N_q = \frac{a^2}{2 \cos^2 (45 + \phi/2)}$$

$$N_\gamma = \frac{\tan \phi}{2} \left(\frac{K_{p\gamma}}{\cos^2 \phi} - 1.0 \right)$$

$a = e^{(\pi 3/4 - \phi/2) \tan \phi}$
$K_{p\gamma}$ = a term relating the passive pressure of the soil in zones II and III on zone I

The Prandtl solution can also consider the condition of the base of the footing as to whether it is *smooth* or *rough* and whether a *local-shear* or a *general-shear* failure is possible. A smooth-footing base gives a smaller value of q_{ult} (which can also be computed from a rough base and a larger safety factor). The equations given are for a rough base, which is the usual condition developed regardless of the compacted condition of the soil at the interface when pouring a concrete footing onto the soil. Figure 4-2 gives the conditions for both a local-shear and a general-shear failure. A local-shear failure is associated with considerable vertical soil movement before soil bulging takes place (soil is loose relative to a general-shear failure, and this failure is sometimes called "punching" shear). Local-shear-failure soil parameters were proposed by Terzaghi as

$$c' = 2/3c \qquad (4\text{-}4)$$

$$\tan \phi' = 2/3 \tan \phi \qquad (4\text{-}5)$$

It would be expected that foundations would never be designed for a local-shear-failure condition because (1) there would be no point in determining ϕ and c except in the in situ soil state and hence applying Eqs. (4-4) and (4-5) is pointless, and (2) local shear results in such a large reduction in bearing value that the soil would be compacted to a general-shear condition.

Table 4-1 presents values of N_c, N_q, and N_γ for selected angles of internal friction for use in Eqs. (4-1) to (4-3). Use interpolation for intermediate values not in the table. A study of Eqs. (4-1) to (4-3) leads to the following conclusions:

1. The ultimate bearing capacity increases with depth of footing.
2. The ultimate bearing capacity of a cohesive soil ($\phi = 0$) is independent of footing size; i.e., at the ground surface ($D_f = 0$) $q_u = 5.7c$.

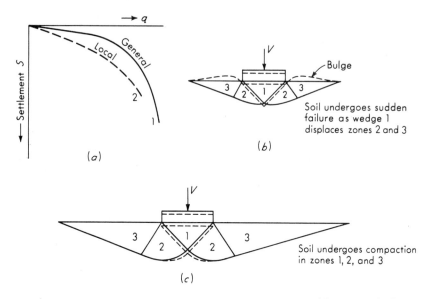

Figure 4-2. Illustration of general-shear and local-shear failures. (*a*) Curve 1 is for a general-shear failure where the soil resists load with small settlements until suddenly soil fails; (*b*) displacements for general-shear failure; (*c*) conditions for a local shear failure. [*After Terzaghi* (1943).]

3. The ultimate bearing capacity of a cohesionless soil ($c = 0$) is directly dependent on footing size, but the depth of footing is more important than size.

The Terzaghi bearing-capacity equations are not much used at present, even though the bearing-capacity factors are not greatly different from factors proposed by others, e.g., by comparing Tables 4-1 and 4-2. The principal reason is that these equations are based on an incorrect failure pattern including the angle β, which has

Table 4-1. Bearing-capacity factors for use in Eqs. (4-1) to (4-3) for general-shear conditions N_c, N_q, N_γ and local-shear conditions N'_c, N'_q, N'_γ

ϕ	N_c	N_q	N_γ	N'_c	N'_q	N'_γ
0	5.7	1.0	0.0	5.7	1.0	0.0
5	7.3	1.6	0.5	6.7	1.4	0.2
10	9.6	2.7	1.2	8.0	1.9	0.5
15	12.9	4.4	2.5	9.7	2.7	0.9
20	17.7	7.4	5.0	11.8	3.9	1.7
25	25.1	12.7	9.7	14.8	5.6	3.2
30	37.2	22.5	19.7	19.0	8.3	5.7
34	52.6	36.5	35.0	23.7	11.7	9.0
35	57.8	41.4	42.4	25.2	12.6	10.1
40	95.7	81.3	100.4	34.9	20.5	18.8
45	172.3	173.3	297.5	51.2	35.1	37.7
48	258.3	287.9	780.1	66.8	50.5	60.4
50	347.5	415.1	1,153.2	81.3	65.6	87.1

been found to be closer to $45 + \phi/2$ than to ϕ. Also these equations do not have provision for including depth effects and other boundary conditions as, for example, the equations of the next section.

Example 4-1. Compute the allowable soil bearing pressure for the footing and soil parameters shown in the accompanying figure. Use Eq. (4-2) and a safety factor $F = 3.0$. Data are obtained from a series of triaxial UU tests. Is the soil saturated?

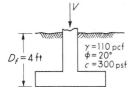

$D_f = 4$ ft

$\gamma = 110$ pcf
$\phi = 20°$
$c = 300$ psf

Figure E4-1

SOLUTION. 1. The soil is not saturated since tests are UU tests and $\phi \neq 0°$.
2. From Table 4-2 and $\phi = 20°$,

$$N_c = 17.7, \quad N_q = 7.4, \quad \text{and} \quad N = 5.0.$$

3. From Eq. (4-2), $q_{ult} = 1.3cN_c + qN_q + 0.4BN$, and substituting values and $q = \gamma D_f$, we have

$$q_{ult} = 1.3(0.3)(17.7) + 0.110(4.0)(7.4) + 0.4(0.110)(5.0)B$$

$$= 10.1 + 0.22B$$

4. The allowable soil pressure $q_a = q_{ult}/F$

$$q_a = 10.1/3.0 + 0.22B/3.0 = 3.4 + 0.07B$$

For all practical purposes and since B must have some kind of value, we may take $q_a = 4.0$ ksf (rounding up slightly). An alternative value of q_a would be 3.5 ksf.

Normally in bearing-capacity computations if the bearing pressure is over 1.0 ksf we would be justified in reporting the value to the nearest 0.5 ksf. For values under 1.0 ksf we would generally report the value to the nearest 0.1 ksf. The rationale for this is that for the small bearing values (under 1.0 ksf) the soil properties would be determined more carefully and rounding can result in a rather large percentage error. For higher bearing-capacity values the soil capacity becomes less critical and, probably, the soil parameters are not as well defined. In any case, it should be evident from soil-mechanics principles, test limitations, and material from Chaps. 2 and 3 that bearing-capacity values are not highly precise numbers.

4-3. THE GENERAL BEARING-CAPACITY EQUATION

In recent years there have been numerous proposals for computation of the ultimate bearing capacity of foundations including those of Meyerhof (1953, 1951), Hansen (1970), Hu (1964), Chen and Davidson (1973), and Balla (1962). These proposals make various assumptions concerning the formation of the β angle of the triangular wedge (Fig. 4-1c) beneath the footing and of the shape of the failure surfaces of

Table 4-2. Bearing-capacity factors N_c, N_q, and N_γ and ϕ-dependent terms for use in the shape factor s_c and d_q of Eq. (4-6)

ϕ	N_c	N_q	N_γ	N_q/N_γ	$2 \tan \phi \, (1 - \sin \phi)^2$
0	5.14	1.0	0.0	0.19	0.000
5	6.49	1.6	0.1	0.24	0.146
10	8.34	2.5	0.4	0.30	0.241
15	10.98	3.9	1.2	0.36	0.294
20	14.83	6.4	2.9	0.43	0.315
25	20.72	10.7	6.8	0.51	0.311
26	22.25	11.9	7.9	0.53	0.308
28	25.80	14.7	10.9	0.57	0.299
30	30.14	18.4	15.1	0.61	0.289
32	35.49	23.2	20.8	0.65	0.276
34	42.16	29.4	28.8	0.70	0.262
36	50.59	37.8	40.1	0.75	0.247
38	61.35	48.9	56.2	0.80	0.231
40	75.31	64.2	79.5	0.85	0.214
45	133.87	134.9	200.8	1.01	0.172
50	266.88	319.1	563.6	1.20	0.130

Table 4-3. Shape, depth, inclination, and other factors for Eq. (4-6). Table combined from Hansen (1970). De Beer (1970), and Vesic (1973). Primed factors are used for $\phi = 0°$ or undrained (UU) conditions

Shape factors	Depth factors	
$s_c' = 0.2B/L$	$d_c' = 0.4D/B$	$D \le B$
$s_c = 1 + N_q B/N_c L$	$d_c' = 0.4 \tan^{-1} \dfrac{D}{B}$	$D > B$
$s_q = 1 + (B/L) \tan \phi$	$d_c = 1 + 0.4 \dfrac{D}{B}$	$D \le B$
$s_\gamma = 1 - 0.4B/L$	$d_c = 1 + 0.4 \tan^{-1} \dfrac{D}{B}$	$D > B$
	$d_q = 1 + 2 \tan \phi(1 - \sin \phi)^2 \dfrac{D}{B}$	$D \le B$
	$d_q = 1 + 2 \tan \phi(1 - \sin \phi)^2 \tan^{-1} \dfrac{D}{B}$	$D > B$
	$d_\gamma = 1.00$ for all ϕ	

Table 4-3 (*Cont.*)

Inclination factors

$i'_c = 0.5 - 0.5\sqrt{1 - H/A_f c}$

$i_c = i_q - (1 - i_q)/(N_q - 1)$

$i_q = \left(1 - \dfrac{0.5H}{V + A_f c \cot \phi}\right)^5$

$i_\gamma = \left(1 - \dfrac{0.7H}{V + A_f c \cot \phi}\right)^5$ horizontal ground

$i_\gamma = \left(1 - \dfrac{(0.7 - \eta°/450°)H}{V + A_f c \cot \phi}\right)^5$ sloping ground

---

Ground factors (for horizontal ground use $g'_c = 0.0$ and $g_q = g_\gamma = 1.00$)

$g'_c = \psi°/147°$
$g_c = 1 - \psi°/147°$
$g_q = g_\gamma = (1 - 0.5 \tan \psi°)^5$

Base factors (for horizontal ground use $b'_c = 0.0$ and $b_q = b_\gamma = 1.00$)

$b'_c = \eta°/147°$
$b_c = 1 - \eta°/147°$
$b_q = \exp(-2\eta \tan \phi)$
$b_\gamma = \exp(-2.7\eta \tan \phi)$

---

where A_f = effective footing contact area $B'L'$
      L' = effective footing length = $L - 2e_L$
      B' = effective footing width = $B - 2e_B$
      D = depth of footing in ground
   e_B, e_L = eccentricity of load with respect to center of footing area
       c = cohesion of base soil
       ϕ = angle of internal friction of soil
   H, V = load components parallel and perpendicular to footing, respectively
  $\tan \delta$ = coefficient of friction between footing and base soil {use $\delta = \phi$ for
         concrete poured on ground [Schultze and Horn (1967)]}
   η, ψ = as shown in accompanying figure with positive directions shown

---

Notes: L and B are interchangeable depending on failure direction.
        Normally the width B is used.
        Limitations: $H \leq V \tan \delta + cA_f$
                  $i_q, i_\gamma > 0.0$
                  $\psi \leq \phi$
                  $\eta + \psi \leq 90°$

Fig. 4-1b. Either the depth of the footing (the development of shear stress along cd of Fig. 4-1a) is considered directly in the computations, or factors to account for the depth of the footing, the shape of the footing, and the inclination or eccentricity of the loads are introduced as applicable.

Considerable quantities of test results of both models [Ko and Davidson (1973), Chummar (1972), both with references] as well as full-size footings are becoming available [Milovic (1965)], from which the various theories may be compared. In passing, it should be mentioned that Terzaghi (1943) had only limited experimental data on which to base his theory. Two of the proposed theories seem to provide values that compare well with the experimental results, but their use tends to be restricted to a soil type.

One theory (which is somewhat similar to the Terzaghi method) has been proposed and later extended by Hansen (1970) and gives better test vs. computed bearing capacities than the Terzaghi equations (see Table 4-4). According to Hansen, the ultimate soil pressure can be computed as

$$q_{ult} = cN_c s_c d_c i_c g_c b_c + \bar{q}N_q s_q d_q i_q g_q b_q + \tfrac{1}{2}\gamma BN_\gamma s_\gamma d_\gamma i_\gamma g_\gamma b_\gamma \qquad (4\text{-}6)$$

For undrained conditions ($\phi = 0°$) the equation is to be modified to

$$q_{ult} = 5.14c(1 + s_c' + d_c' - i_c' - b_c' - g_c') + \bar{q} \qquad (4\text{-}6a)$$

where s = shape factors to account for the shape of the foundation in developing a failure surface

d = depth factors to account for embedment depth and the additional shearing resistance along line cd of Fig. 4-1a

i = inclination factors to allow for both horizontal and vertical foundation loads

$\bar{q}$ = effective surcharge pressure = γD as shown in Fig. 4-1b

g, b = ground and base factors, respectively, in Table 4-3

The bearing-capacity factors are computed as

$$N_q = \tan^2 (45 + \phi/2) \exp (\pi \tan \phi) = K_p \exp (\pi \tan \phi)$$
$$N_c = (N_q - 1) \cot \phi \qquad (4\text{-}7)$$
$$N_\gamma = 1.50(N_q - 1) \tan \phi$$

Bearing-capacity factors for selected angles of internal friction ϕ are shown in Table 4-2. Table 4-3 provides a list of shape, depth, inclination, and geometry factors for use with Eq. (4-6). The user may wish to approximate these values where the soil parameters are not accurately known. Hansen considered the bearing capacity as a *plane-strain* problem which is correct for the infinitely long strip footing considered by Prandtl. As a plane-strain problem the ϕ angle from triaxial tests tends to be too low based on Hansen (1970) and Bishop's (1961) suggestions and the extensive literature survey made by Lee (1970). Hansen suggests that

$$\phi_{\text{plane strain}} = 1.1 \, \phi_{\text{triaxial}}$$

since the average increase in plane strain ϕ angle tends close to 15 percent. The author suggests that this increase be used only on ϕ angles greater than 25°.

Example 4-2. A full-scale load test made by H. Muhs, Berlin [Hansen (1970)] with problem data as follows:

$$D = 0.5 \text{ m} \qquad \gamma' = 9.31 \text{ kN/m}^3 \text{ water-table level with ground surface}$$

$$B = 0.5 \text{ m}$$

$$L = 2.0 \text{ m} \qquad \phi = 42.7° \text{ (triaxial)} \qquad c = 0$$

The measured failure load,

$$P_{\text{ult}} = 1{,}863 \text{ kN}$$

REQUIRED. Compute the ultimate bearing capacity and compare to the measured value.

SOLUTION. $\phi_{ps} = 1.1\phi_t = 1.1(42.7) = 47°$

From which

$$N_q = 187$$

$$N_\gamma = 1.5(186) \tan \phi = 299$$

$$d_q = 1 + 2 \tan \phi(1 - \sin \phi)^2 D/B = 1.15$$

$$s_q = 1 + \tan \phi(B/L) = 1.27$$

$$d_\gamma = 1.00$$

$$s_\gamma = 1 - 0.4B/L = 0.9$$

with $c = 0$ and all factors $= 1.00$ except those shown, Eq. (4-6) becomes

$$q_{\text{ult}} = qN_q d_q s_q + \tfrac{1}{2}\gamma' BN_\gamma d_\gamma s_\gamma$$

and inserting values

$$q_{\text{ult}} = 0.5(9.31)(187)(1.15)(1.27) + \tfrac{1}{2}(9.31)(0.5)(299)$$

$$(1)(0.9) = 1{,}271.3 + 626.3 = 1{,}897.7 \text{ kPa}$$

The measured $q_{\text{ult}} = P/A = 1{,}863/(0.5 \times 2) = 1{,}863 \text{ kPa} \simeq 1{,}898$. In this case the computed and measured values agree extremely well. What would have been the computed vs. measured agreement if $\phi = 42.7°$ had been used?

$$N_q = 94.7 \qquad N_\gamma = 1.5(93.7) \tan 42.7 = 129.6$$

$$q_{\text{ult}} = 1{,}271.3(94.7/187) + 626.3(129.6/299) = 915.3$$

with 915.3 kPa vs. 1,863 kPa measured, the computed agreement without adjusting ϕ for plane-strain conditions is very poor.

Example 4-3. A series of full-scale footing tests were performed on soft Bangkok clay [Brand et al. (1972)]. One of the tests consisted in a 1.05-m-square footing with $D = 1.5$ m. At a 1-in immediate settlement the load was approximately 14.1 tons from interpreting the given load-settlement curve. Unconfined compression, triaxial, and vane-shear tests gave UU values as

$$q_u = 3.0 \text{ t/m}^2 \qquad s_u = 2.0 \text{ t/m}^2 \qquad \text{and} \qquad \text{vane } s_u = 2.4 \text{ t/m}^2$$

These data were obtained by inspection of plots in the zone within 1 to 2 m from bottom of footing and the considerable scatter with q_u and triaxial tests. Plasticity data: $w_P = 35$ percent; $w_L = 80$ percent. These load-settlement data are in Prob. 4-14. The units of this problem are to match those used in the test ($\text{t/m}^2 \times 9.807 = \text{kPa}$).

REQUIRED. Compute the ultimate bearing capacity and compare with the load-test values.

SOLUTION. With a UU test, $\phi = 0°$; $N_c = 5.14$ and $N_q = 1.00$; we will neglect N_q effects. From Fig. 3-15 and I_p of 45, the reduction factor to apply to the vane-shear strength is 0.8 and $s_u = 0.8(2.4) = 1.92$ t/m². From Eq. (4-6a) and a vertical load, we have

$$q_{ult} = 5.14s_u(1 + s_c' + d_c')$$

and $\qquad\qquad s_c' = 0.2B/L = 0.2(1) = 0.2 \qquad$ (Table 4-3)

$$d_c' = 0.4 \tan^{-1}(D/B) = 0.4 \tan^{-1}(1.5/1.05) = 0.38$$

Substituting these values we obtain

$$q_{ult} = 5.14(1.92)(1 + 0.2 + 0.38) = 15.6 \text{ t/m}^2$$

From the load test we have $q_{ult} = 14.1/(1.05)^2 = 12.8$ t/m²

Parry (1971) in a study of embankment failures on soft clay (also a bearing-capacity problem) concluded that isotropy, progressive failure, and strain rate tend to reduce s_u values considerably. Considering this reduction on the order of 15 percent let us recompute the ultimate bearing capacity from the triaxial test value of $2.0(0.85) = 1.70$ t/m² and substituting we have

$$q_{ult} = 5.14(1.7)(1.58) = 13.8 \text{ t/m}^2$$

which is quite close to the measured value above; if we use $c = q_u/2$, $q_{ult} = 12.2$ t/m². If we put the usual safety factor of $F = 3.0$ for clay, we have 4.6 t/m², corresponding to a settlement of less than $\frac{1}{8}$ in.

4-4. GENERAL COMMENTS ON BEARING-CAPACITY COMPUTATIONS

Balla (1962) considered bearing-capacity analysis and presented figures and curves necessary for making the computations. Table 4-4 indicates the Balla method provides " good " solutions for soil with little cohesion. Since the Hansen equations using plane-strain values are equally good, the Balla method is not presented in this text edition.

Meyerhof (1951, 1955, 1963) has also considered the bearing capacity of spread footings and has presented equations and charts for the designer's use in obtaining bearing-capacity factors. An examination of the charts indicates that Meyerhof's values lie somewhat between the general- and local-shear values of Terzaghi when the depth of footing D_f is zero, but Meyerhof also considered depth effects, and as indicated in Table 4-4, where several theories are compared, this increases the bearing capacity considerably.

With several methods available to compute the ultimate bearing capacity which can provide rather widely divergent values, the question naturally arises of which method to use? In fact one can use any of the methods with reasonable assurance that the results will be conservative for design use (by applying a suitable safety factor). The Terzaghi method has been widely used but is not much used at present for reasons discussed earlier. Meyerhof's values are used principally in England. His values as shown in Table 4-4 tend to swing widely between being conservative and

Table 4-4. Comparison of theoretical bearing capacities and experimental values*. ϕ **values are triaxial values**

Method of determining bearing capacity	Type of soil			
		Soil with little to no cohesion q, kg/cm²		Cohesive soil q, kg/cm²
	No. 1	$D_f = 0$ m $\phi = 37°$ $c = 6.37$ kPa	No. 5	$D_f = 0.40$ m $\phi = 22°$ $c = 12.75$ kPa
Terzaghi		7.62		4.47
Meyerhof	$D = 0.0$‡	6.68	$D = 0.4$	6.58
B. Hansen†	$B = 0.5$	9.70	$B = 0.71$	3.98
Balla	$L = 2.0$	10.34	$L = 0.71$	6.74
Muhs	$\gamma = 15.69$	10.80§	$\gamma = 17.65$	
Milovic				4.1§
	No. 2	$D_f = 0.50$ $\phi = 35\frac{1}{2}°$ $c = 3.92$ kPa	No. 6	$D_f = 0.50$ $\phi = 25°$ $c = 14.71$ kPa
Terzaghi		7.80		5.77
Meyerhof	$D = 0.5$	16.84	$D = 0.5$	8.84
B. Hansen†	$B = 0.5$	14.10	$B = 0.71$	5.74
Balla	$L = 2.0$	14.11	$D = 0.71$	10.18
Muhs	$\gamma = 16.38$	12.0§	$\gamma = 17.65$	
Milovic				5.50§
	No. 3	$D_f = 0.50$ $\phi = 38\frac{1}{2}°$ $c = 7.84$ kPa	No. 7	$D_f = 0$ $\phi = 20°$ $c = 9.81$ kPa
Terzaghi		15.23		2.51
Meyerhof	$D = 0.5$	34.86	$D = 0.0$	2.51
B. Hansen†	$B = 0.5$	29.76	$B = 0.71$	1.98
Balla	$L = 2.0$	25.18	$D = 0.71$	2.93
Muhs	$\gamma = 17.06$	24.20§	$\gamma = 17.06$	
Milovic				2.20§
	No. 4	$D_f = 0.50$ $\phi = 38\frac{1}{2}°$ $c = 7.84$ kPa	No. 8	$D_f = 0.30$ $\phi = 20°$ $c = 9.81$ kPa
Terzaghi		18.55		2.90
Meyerhof	$D = 0.5$	46.96	$D = 0.3$	4.10
B. Hansen†	$B = 1.0$	40.87	$B = 0.71$	2.57
Balla	$L = 1.0$	32.50	$D = 0.71$	4.40
Muhs	$\gamma = 17.06$	33.0§	$\gamma = 17.06$	
Milovic				2.57§

* After Milovic (1965).
† Values computed by author increasing ϕ triaxial to 1.1ϕ.
‡ Dimensions in m and weight as kN/m³.
§ Experimental results.

123

unconservative. Meyerhof developed his theory using very small model footing tests (many were only 2.5 cm wide). For this reason and the fact the theory does not provide reliable agreement with the full-scale footing tests the author does not re-commend their use. Hansen's equations are shown to have good agreement with measured values for both cohesionless and cohesive soils as shown in Table 4-4. Example 4-2 with an adjustment for $\phi_{ps} = 1.1\phi_t$ also shows excellent agreement for cohesionless soils. Hansen's method does not provide both static and kinematic equilibrium—nor do the Terzaghi or Meyerhof procedures. Balla's method is an attempt to provide a failure zone which is both statically and kinematically admis-sible. As shown in Table 4-4, this method provides reasonable agreement for cohesionless soils.

The author suggests using Hansen's equations for all soils as a reasonable and simple means of estimating bearing capacity.

Seldom does bearing capacity control the design of foundations. Settlement is usually the controlling factor, and bearing pressures to limit settlement are usually less than the allowable value obtained from any of the bearing-capacity equations. This is the principal reason for the many bearing-capacity theories in existence and being used.

The author would suggest some caution in using bearing-capacity values much over 8 to 10 ksf (380 to 470 kPa) unless the soil parameters are known with a high degree of reliability or if a failure is not catastrophic.

4-5. FOOTINGS WITH ECCENTRIC OR INCLINED LOADS

A footing may be eccentrically loaded, as with axial load and overturning moments about either or both the x and y axes as shown in Fig. 4-3. Alternatively, the vertical load does not coincide with the center of the footing area as when part of the footing is later cut off during remodeling or installation of new equipment. Research and observations [Meyerhof (1953), Hansen (1970, p. 6)] indicate that the effective foot-ing area should be computed using footing dimensions of

$$L' = L - 2e_y \quad \text{and} \quad B' = B - 2e_x \quad \text{and the area } A'_f = B'L'$$

where the terms are identified in Fig. 4-3. This places the location of the resultant vertical forces in the centroid of the reduced foundation area A'_f. These reduced footing dimensions should be used in Eq. (4-6) for the B term and in the appropriate foundation-factor equations given in Table 4-2 except for the D/B ratio in the depth factors.

The use of this concept reduces the total allowable load on the footing through a reduced ultimate bearing pressure from use of B' in Eq. (4-6) and in the foundation-factor equations in Table 4-2. The total allowable load is also reduced since

$$V_{ult} = q_{ult}(B'L') \text{ instead of } q_{ult}(BL)$$

Inclined loads will have horizontal and vertical components parallel and perpen-dicular to the base, respectively. The inclination factors allow for reducing the ulti-mate bearing pressure as given in Table 4-2. It is evident from inspection of the

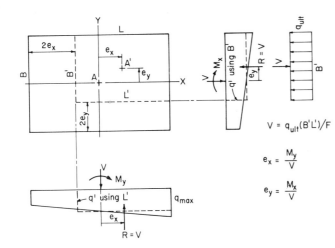

$$V = q_{ult}(B'L')/F$$

$$e_x = \frac{M_y}{V}$$

$$e_y = \frac{M_x}{V}$$

Figure 4-3. Method of computing effective footing dimensions when footing is eccentrically loaded.

inclination factors that placing the base perpendicular to the inclined load will result in the largest bearing capacity and for modest η angles (Table 4-2), the reduction in bearing capacity due to the resulting sloping-ground condition will probably be less than 10 percent.

Example 4-4. A footing 6 × 6 ft square is loaded with an axial load of 400 kips and $M_x = 200$ ft-k, $M_y = 120$ ft-k. Undrained triaxial tests (soil not saturated) give $\phi = 32.7°$ and $c = 200$ psf. The footing depth $D = 6.0$ ft; the soil unit is 115 pcf; the water table is -20.0 ft.

REQUIRED. What is the allowable soil pressure if $F = 3.0$? Will the footing have to be resized?

SOLUTION. Compute eccentricity and adjusted dimensions

$$e_x = M_y/V = 120/400 = 0.3 \qquad \text{and} \qquad 2e_x = 0.6 \text{ ft}$$

$$e_y = 200/400 = 0.5 \qquad \text{and} \qquad 2e_y = 1.0 \text{ ft}$$

$$B' = 6 - 1 = 5.0 \text{ ft} \qquad L' = 6 - 0.6 = 5.4 \text{ ft}$$

Compute bearing-capacity factors

$$\phi_{ps} = 1.1\phi_t = 1.1(32.7) = 36°$$

and

$$N_c = 50.6 \qquad N_q = 37.8 \qquad N_\gamma = 40.1 \qquad \text{(from Table 4-2)}$$

$$D/B = 6/6 = 1$$

$$s_c = 1 + N_q/N_c(B'/L') = 1.69 \qquad d_c = 1 + 0.4D/B = 1.4$$

$$s_q = 1 + B'/L' \tan \phi = 1.67 \qquad d_q = 1 + 2 \tan \phi(1 - \sin \phi)^2 \frac{D}{B} = 1.25$$

$$s_\gamma = 1 - 0.4B'/L' = 0.63 \qquad d_\gamma = 1.00$$

all $i = g = b = 1.0$

$$q_{ult} = cN_c s_c d_c + \bar{q}N_q s_q d_q + \tfrac{1}{2}\gamma B' N_\gamma s_\gamma d_\gamma$$

$$= 0.2(50.6)(1.69)(1.4) + 6(0.115)(37.8)(1.67)(1.25)$$

$$+ 0.5(0.115)(5.0)(40.1)(0.63)(1.0) = 23.9 + 54.4 + 7.2$$

$$= 85.5$$

with

$$F = 3, \quad q_a = 85.5/3 = 28.5 \text{ ksf}$$

$$P_a = 28.5(5)(5.4) = 769 \text{ kips} \gg 400 \text{ kips}$$

The footing would probably be left at 6×6 ft even though it is apparently too large. It might have to be increased in size to reduce settlements. A load of 400 kips is rather large to put on any footing under 6×6, since the nominal soil pressure is now on the order of $q = 400/36 = 11.1$ ksf or, using the reduced area, $q' = 400/(5 \times 5.4) = 14.8$ ksf, both extremely high soil bearing values.

4-6. EFFECT OF WATER TABLE ON BEARING CAPACITY

The *effective* unit weight of the soil is used in developing the bearing-capacity equations. It is this value which should be used to compute

$$\bar{q} = \gamma D \text{ in the } N_q \text{ term}$$

$$\tfrac{1}{2}\gamma B \text{ in the } N_\gamma \text{ term}$$

Where the water table is level with the base of the footing, it is easy to compute the effective unit weight for the zone abc of Fig. 4-1 [and using a recommended β angle of $45 + \phi/2$ (and not ϕ as proposed by Terzaghi)]. When the water level is at some distance below the footing, the effective unit weight in that zone can be computed by averaging the unit weights of the wet and submerged values as in Example 4-5. A satisfactory reduction factor can be computed as $F_w = 0.5 + d_w/B \tan (45 + \phi/2)$ to apply to N_γ. The term d_w is the depth from the footing base to the water table. $F_w = 1$ when $d_w = 0.5B \tan (45 + \phi/2)$.

It is unlikely that the water table will be above the base of the footing except by accident (a rising water table). If it is, however, the $\bar{q}$ term will have to be adjusted similarly to the case of water below the base. If it is expected that the water table will rise, both the effect of submergence on reducing the bearing capacity and the effect on reducing the cohesion and angle of internal friction (and any buoyancy on the base) should be analyzed.

It would be expected that the soil parameters are based on a stable-water-table condition. That is, the cohesion and angle of internal friction are determined consistent with field conditions.

Example 4-5. A square footing vertically and concentrically loaded is to be placed on a cohesionless soil with the following properties and dimension:

$$\phi = 35° \qquad \gamma = 115 \text{ pcf} \qquad D = 4 \text{ ft} \qquad D_w = 6 \text{ ft} \qquad B = 8 \text{ ft}$$

REQUIRED. What is the allowable bearing capacity using Eq. (4-6) and $F = 2.0$?

SOLUTION. Note that B would in the general problem not be known but would depend on the column load and the allowable soil pressure. We could, however, compute q_{ult} for several values of B and plot a curve of q_a versus B. Here we will obtain a single value of q.

The effective soil pressure in $\bar{q}$ is based on $\gamma = 115$ pcf. We must compute the effective pressure in the soil beneath the footing in a depth $D_1 = B/2 \tan (45 + \phi/2) = 8/2 \tan 62.5 = 7.68'$.

The saturated unit weight will require an assumption of $G_s = 2.68$, and we will estimate the water content of the wet soil at about 10 percent.

$$\gamma_{dry} = \gamma_{wet}/(1 + w) = 115/(1.10) = 104.5 \text{ pcf}$$

$$V_s = 104.5/(62.5)(2.68) = 0.624 \text{ cu ft}$$

$$V_v = 1 - 0.624 = 0.376 \text{ cu ft}$$

$$\gamma_{sat} = 104.5 + 0.376(62.5) = 128.0 \text{ pcf}$$

The soil in zone D_1 is made up of

To water table $(6 - 4)(0.115) =$ 0.23
Below water table $(7.68 - 2)(0.128 - 0.0625) = \underline{0.372}$
                                   0.602 ksf

The effective weight across the zone 7.68 ft deep is

$$\gamma' = 0.602/7.68 = 0.078 \text{ kcf}$$

Alternatively: $$\gamma = F_w \gamma_{sat}$$

$$\gamma = [0.5 + 2/15.37](0.128) = 0.081 \text{ kcf}$$

From Table 4-2

$$N_q = 33.29 \qquad N_\gamma = 33.91,$$

and from Eq. (4-6) and omitting factors which are 1.0,

$$q_{ult} = qN_q s_q d_q + \tfrac{1}{2}\gamma BN_\gamma s_\gamma d_\gamma .$$

From Table 4-3 the appropriate foundation factors are

$$d_q = 1 + 2 \tan 35(1 - \sin 35)^2 4/8 = 1 + 0.13 = 1.13 \qquad \text{and} \qquad d_\gamma = 1.00$$

$$s_q = 1 + \tan \phi = 1.70 \qquad s_\gamma = 1 - 0.4 = 0.6$$

Substituting

$$q_{ult} = 4(0.115)(33.3)(1.13)(1.70) + \tfrac{1}{2}(0.081)(8)(33.9)(0.6)(1)$$

$$= 29.4 + 6.6 = 36.0$$

$$q_a = q_{ult}/F = 36.0/2 = 18.0 \text{ ksf}$$

It is unlikely that a soil with this foundation geometry would be loaded to this high a pressure; settlements would probably be the controlling factor.

4-7. BEARING CAPACITY FOR FOOTINGS ON LAYERED SOILS

It may be necessary to place footings on stratified deposits where the thickness of the top stratum is insufficient to fully enclose the rupture zone (surface *ade* of Fig. 4-1*a*). If the rupture zone extends into the lower layer (or layers if very thin) the ultimate bearing capacity will be modified to some extent. One would expect it to increase if the lower soil is of better quality and decrease for poorer soil. Numerous solutions have been proposed to enable one to make an estimate of the bearing capacity. One of the first proposals was by Button (1953) for a $\phi = 0°$ analysis. With $\phi = 0°$ the curved sector ad of Fig. 4-1b becomes circular, and Button actually used a circular segment $cade$ with the center above and to the right of point b. The solution was limited to a two-layer system with $\phi = 0°$ in both soils. The solution was obtained by finding the minimum value of the pressure ratio

$$N_c = \frac{q}{c}$$

where q = applied footing contact pressure across the width B and c is the cohesion of the soil stratum immediately underlying the footing.

Reddy and Srinivasan (1967) extended the Button solution to anisotropic soils defined by a coefficient of anisotropy of the soil immediately underlying the footing as

$$K = \frac{q_1}{q_3}$$

where q_1 = vertical shear strength
q_3 = horizontal shear strength

A value of $K < 1$ indicates overconsolidation; $K = 1$ = isotropic, and $K > 1$ = normally consolidated. Charts from Reddy and Srinivasan are shown for three K values of 0.8, 1.0, and 1.2 in Fig. 4-4c. The values for $K = 1.0$ are identical to those of Button (1953). The charts are made up for the angle between the failure plane and the minor principal stress of 35° which was used from an earlier summary of tests by Lo (1965). Button, and as shown in Fig. 4-4 for $K = 1.0$, used an average value of $N_c = 5.5$ for the case of a single layer which is an approximate average of the values 5.14 and 5.7 shown in Tables 4-1 and 4-2. Meyerhof and Brown (1967) show that the additional charts for various K values are not needed; the chart for $K = 1$ is sufficient if one uses an average $q_u = (q_1 + q_3)/2$ and the corresponding cohesion. Errors associated with this would be not more than 10 percent and are less than the assumption of a circular failure surface. A circular failure surface is obviously an erroneous assumption with anisotropy, since there is a ϕ angle.

The bearing capacity for the layered soil conditions of Fig. 4-4a is computed from Eq. (4-6) as

$$q_{ult} = c_1 N_c (1 + s'_c + d'_c) + q N_q \tag{4-8}$$

where N_c is now obtained from Fig. 4-4c, and c_1 is the cohesion of the soil immediately underlying the footing. Figure 4-4b illustrates the limiting d/b and c_2/c_1 values to have a failure zone all in the top layer with the rupture circle just tangent to the second layer. This figure can be obtained from the curves of $K = 1$ from (c) at the breaks in the d/b lines beyond which N_c = constant. Any c_2/c_1 ratios larger than the "limiting" values for that d/b ratio will result in the bearing-capacity failure occurring entirely within the top soil layer.

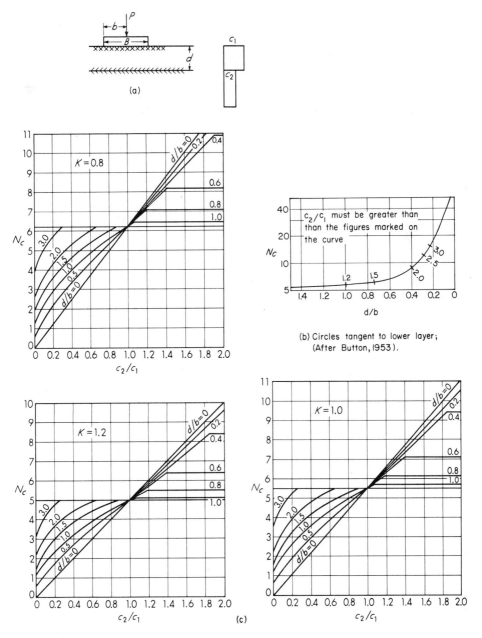

Figure 4-4. Bearing-capacity factors for a footing located on a two-layer cohesive soil system, with cohesion in each layer as shown. [*After Reddy and Srinivasan (1967).*]

Example 4-6. A footing 6.1×3.05 m is to be placed at $D = 1.83$ m into a two-layer clay deposit. Soil data: $c_1 = 76.60$ kPa; $c_2 = 114.91$ kPa; $\gamma = 17.26$ kN/m³; thickness of top layer $= 3.05$ m.

REQUIRED. Estimate the ultimate bearing capacity.

SOLUTION. We will assume UU tests and that anisotropy was not investigated (a common practice). Therefore,

$$c_2/c_1 = 114.91/76.60 = 1.50 \qquad d/b = (3.05 - 1.83)/1.525 = 0.80$$

From the curves for Fig. 4-4c for $K = 1$ we have $N_c = 6.1$.
From Table 4-3 the depth and shape factors are computed as follows:

$$s'_c = 0.2B/L = 0.2(3.05/6.1) = 0.1$$

$$d'_c = 0.4\frac{D}{B} = 0.4\left(\frac{1.83}{3.05}\right) = 0.24$$

$$N_q = 1.00 \text{ for } \phi = 0°$$

Substituting into Eq. (4-8)

$$q_{ult} = 76.60(6.1)(1 + 0.1 + 0.24) + 1.83(17.26)(1)$$
$$= 626.1 + 31.6 = 657.7 \text{ kPa}$$

It should be pointed out that one could use a slope-stability program such as in Bowles (1974a) to solve this problem. The program as given in that reference would require slight modification to treat the footing as a surcharge. Increase in shear strength with depth can be approximated by using several additional soil layers of varying shear strength. One also could approximate parts of the failure surface in the area of acb (Fig. 4-1a) with appropriate boundary lines and solve $\phi \neq 0°$ problems.

One may also obtain solutions for the bearing capacity of layered soils, or where soil properties increase with depth using the finite element of the elastic continuum [see Zienkiewicz (1971)]. To use this method and computer program it would be necessary to express the soil properties in terms of stress-strain modulus E_s and Poisson's ratio μ instead of ϕ and c. It will be necessary to define "failure" as either a specified stress level or a specified deflection. One would proceed in the solution by obtaining values of E_s and μ as necessary to define the strata. The next step would be to load sufficient nodes to define the footing (usually one-half the footing on the corresponding amount of the elastic continuum) with either pressure increments and plot resulting deflections or with deflection increments and plot the resulting stress. It will be necessary to use a nonlinear stress-strain modulus if it is desired to obtain a curved pressure vs. deflection plot. This method may be used for layered soils of dissimilar materials, i.e., sand over clay, clay over sand, sand over sand, clay over clay, etc.

Meyerhof (1974a) and Yamagughi (1963) have presented studies of bearing capacity on two-layered soils, and both references present some bearing-capacity charts for sand overlying clay. Purushothamaraj et al. (1974) presents an extension of the two-layer system of clay to clay with both c and ϕ. With the rapid development of the finite-element method which can handle larger numbers of boundary conditions, Fig. 4-4 should be sufficient to enable the designer to approximate his solution and if the approximation is not sufficient then have recourse to the cited references, the finite-element method, or to the following described approximation.

We may approximate the bearing capacity of two-layered soils of any type by investigating the worse of two cases:

1. The footing on the top layer.
2. A pseudo-footing loading the second layer with a reduced pressure intensity obtained from the Boussinesq or Westergaard pressure-bulb concept (or a 2 on 1) as displayed in Fig. 5-4 of Sec. 5-2.
3. The bearing capacity will be based on the contact pressure and the upper layer if the reduced pressure of step 2 can be carried by the lower layer. The bearing capacity will have to be back-figured from the contact pressure which the lower layer can carry if that pressure controls.

4-8. BEARING CAPACITY OF FOOTINGS ON SLOPES

Another special problem that may be encountered occasionally is that of a footing located on or adjacent to a slope (Fig. 4-5). From the figures it can be seen that the lack of soil on the slope side of the footing will tend to reduce the stability of the footing.

The author developed Table 4-5 to solve the footing on or adjacent to a slope as follows:

1. Develop the exit point E for a footing as shown in Fig. 4-5. The angle of the exit is taken as $45 - \phi/2$ since the slope line is a principal plane.
2. Compute a reduced N_c based on the failure surface $cade = L_0$ of Fig. 4-1 and the failure surface $cadE = L_1$ of Fig. 4-5a to obtain

$$N'_c = N_c \frac{L_1}{L_0}$$

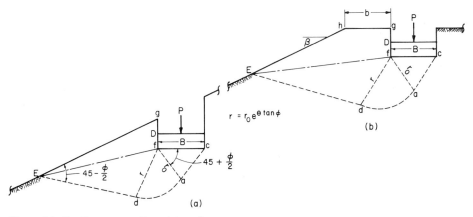

Figure 4-5. Footings on or adjacent to a slope.

Table 4-5. Bearing-capacity factors N'_c, N'_q for footings on or adjacent to a slope. Refer to Fig. 4-5 for variable identification. Base values of N_c, N_q are used when length or area ratios > 1. Base values of N_c, N_q apply for $b/B > 2$ (approximately)

D/B = 0.0, b/B = 0.0

B		φ 0	10	20	30	40
0	$N_c =$	5.14	8.34	14.83	30.14	75.31
	$N_q =$	1.03	2.47	6.40	18.40	64.20
5		5.01	8.07	14.19	28.43	69.67
		1.03	2.47	6.40	18.40	64.20
10		4.89	7.80	13.57	26.80	64.42
		1.03	2.47	6.40	18.40	64.20
20		4.63	7.28	12.39	23.78	55.01
		1.03	2.47	6.40	18.40	64.20
30		4.38	6.77	11.28	21.05	46.88
		1.03	2.47	6.40	18.40	64.20
60		3.62	5.33	8.33	14.34	28.56
		1.03	2.47	6.40	18.40	64.20

D/B = 0.50, b/B = 0.50

	0	10	20	30	40
5	5.14	8.34	14.83	30.14	75.31
	0.83	1.85	4.41	11.67	37.38
10	5.14	8.34	14.93	30.14	74.12
	0.94	2.03	4.73	12.17	37.80
20	5.14	8.34	14.83	29.79	65.74
	1.03	2.34	5.19	12.66	36.91
30	5.14	8.34	14.83	27.46	58.32
	1.03	2.47	5.36	12.40	34.02
60	5.14	8.07	12.37	20.74	40.00
	1.03	2.13	3.92	7.83	18.04

D/B = 0.0, b/B = 0.50

	0	10	20	30	40
5	5.14	8.25	14.45	28.84	70.40
	1.03	2.47	6.40	18.40	64.20
10	5.14	8.15	14.08	27.52	65.87
	1.03	2.47	6.40	18.40	64.20
20	5.12	7.96	13.40	25.39	57.87
	1.03	2.47	6.40	18.40	64.20
30	5.10	7.77	12.75	23.40	51.07
	1.03	2.47	6.40	18.40	64.20
60	4.86	7.07	10.89	18.40	35.81
	1.03	2.47	6.40	18.40	64.20

D/B = 0.50, b/B = 1.00

	0	10	20	30	40
5	5.14	8.34	14.83	30.14	75.31
	0.86	1.95	4.70	12.53	40.33
10	5.14	8.34	14.83	30.14	75.31
	1.01	2.26	5.35	13.94	43.57
20	5.14	8.34	14.83	30.14	68.61
	1.03	2.47	6.40	16.29	47.76
30	5.14	8.34	14.83	29.80	62.51
	1.03	2.47	6.40	17.81	48.90
60	5.14	8.34	14.83	24.80	47.25
	1.03	2.47	6.40	15.66	35.82

D/B = 0.0, b/B = 1.00

	0	10	20	30	40
5	5.14	8.34	14.71	29.25	71.13
	1.03	2.47	6.40	18.40	64.20
10	5.14	8.34	14.60	28.43	67.33
	1.03	2.47	6.40	13.40	64.20
20	5.14	8.34	14.41	26.99	60.74
	1.03	2.47	6.40	18.40	64.20
30	5.14	8.34	14.23	25.74	55.26
	1.03	2.47	6.40	18.40	64.20
60	5.14	8.34	13.45	22.46	43.06
	1.03	2.47	6.40	18.40	64.20

D/B = 1.00, b/B = 0.0

B		φ 0	10	20	30	40
0	$N_c =$	5.14	8.34	14.83	30.14	75.31
	$N_q =$	1.03	2.47	6.40	18.40	64.20
5		5.14	8.34	14.83	30.14	75.31
		0.99	2.13	4.92	12.55	38.59
10		5.14	8.34	14.83	30.14	75.31
		1.03	2.19	4.91	12.16	36.24
20		5.14	8.34	14.83	30.14	70.75
		1.03	2.17	4.63	10.88	30.54
30		5.14	8.34	14.83	29.17	61.39
		1.03	1.98	4.04	9.05	24.02
60		5.06	7.33	11.29	19.03	36.93
		0.45	0.77	1.39	2.74	6.27

D/B = 0.50, b/B = 0.0

B		φ 0	10	20	30	40
0	$N_c =$	5.14	8.34	14.83	30.14	75.31
	$N_q =$	1.03	2.47	6.40	18.40	64.20
5		5.14	8.34	14.83	30.14	75.31
		0.76	1.69	4.03	10.65	34.14
10		5.14	8.34	14.83	30.14	72.67
		0.79	1.70	3.94	10.15	31.61
20		5.14	8.34	14.83	28.19	62.88
		0.79	1.62	3.58	8.79	25.88
30		5.14	8.34	13.83	25.11	54.13
		0.72	1.43	3.03	7.09	19.79
60		4.34	6.33	9.81	16.68	32.75
		0.28	0.50	0.95	1.98	4.78

D/B = 1.00, b/B = 0.50

	0	10	20	30	40
5	5.14	8.34	14.83	30.14	75.31
	1.03	2.25	5.19	13.22	40.59
10	5.14	8.34	14.83	30.14	75.31
	1.03	2.44	5.47	13.52	40.14
20	5.14	8.34	14.83	30.14	73.61
	1.03	2.47	5.80	13.56	37.73
30	5.14	8.34	14.83	30.14	65.57
	1.03	2.47	5.77	12.81	33.55
60	5.14	8.34	13.84	23.09	44.19
	1.03	2.04	3.63	7.01	15.55

Table 4-5. (*Cont.*)

	D/B = 1.00		b/B = 1.00				D/B = 1.50		b/B = 0.50		
	0	10	20	30	40		0	10	20	30	40
5	5.14 / 1.03	8.34 / 2.34	14.83 / 5.41	30.14 / 13.81	75.31 / 42.45	5	5.14 / 1.03	8.34 / 2.47	14.83 / 6.02	30.14 / 14.95	75.31 / 44.56
1C	5.14 / 1.03	8.34 / 2.47	14.83 / 5.95	30.14 / 14.75	75.31 / 43.83	10	5.14 / 1.03	8.34 / 2.47	14.83 / 6.33	30.14 / 15.26	75.31 / 43.94
20	5.14 / 1.03	8.34 / 2.47	14.83 / 6.40	30.14 / 16.11	75.31 / 44.80	20	5.14 / 1.03	8.34 / 2.47	14.83 / 6.40	30.14 / 15.20	75.31 / 41.04
3C	5.14 / 1.03	8.34 / 2.47	14.83 / 6.40	3J.14 / 16.61	69.76 / 43.38	30	5.14 / 1.03	8.34 / 2.47	14.83 / 6.40	30.14 / 14.23	72.83 / 36.18
60	5.14 / 1.03	8.34 / 2.47	14.83 / 6.36	27.14 / 12.28	51.44 / 27.10	60	5.14 / 1.03	8.34 / 2.19	14.83 / 3.85	25.43 / 7.28	48.37 / 15.77

	D/B = 1.50		b/B = 0.0				D/B = 1.50		b/B = 1.0		
B φ	0	10	2C	30	40		0	10	2J	3J	40
0	N_c = 5.14 / N_q = 1.03	8.34 / 2.47	14.83 / 6.40	30.14 / 18.40	75.31 / 64.20	5	5.14 / 1.03	8.34 / 2.47	14.83 / 6.21	30.14 / 15.45	75.31 / 46.05
5	5.14 / 1.03	8.34 / 2.47	14.83 / 5.79	30.14 / 14.40	75.31 / 42.97	10	5.14 / 1.03	8.34 / 2.47	14.83 / 6.40	3J.14 / 16.30	75.31 / 46.93
10	5.14 / 1.03	8.34 / 2.47	14.83 / 5.85	30.14 / 14.13	75.31 / 40.81	20	5.14 / 1.03	8.34 / 2.47	14.83 / 6.40	30.14 / 17.39	75.31 / 46.85
20	5.14 / 1.03	8.34 / 2.47	14.83 / 5.65	30.14 / 12.93	75.31 / 35.14	30	5.14 / 1.03	8.34 / 2.47	14.83 / 6.40	30.14 / 17.48	75.31 / 44.32
3C	5.14 / 1.03	8.34 / 2.47	14.83 / 5.04	3J.14 / 10.99	68.64 / 28.23	60	5.14 / 1.03	8.34 / 2.47	14.83 / 6.17	29.49 / 11.69	55.63 / 25.26
6C	5.14 / 0.62	8.34 / 1.04	12.76 / 1.83	21.37 / 3.52	41.12 / 7.80						

3. Compute a reduced N_q based on the ratio of area $D_f(be) = A_0$ of Fig. 4-1 and the area Efg of Fig. 4-5a or the alternative (Fig. 4-5b) $Efgh = A_1$ to obtain

$$N_q' = N_q \frac{A_1}{A_0}$$

where the slope is such that $A_1 > A_0$, $N_q' = N_q$

4. The overall slope stability should be checked for the effect of the footing load. This can be done using the slope-stability program in Bowles (1974a) or one the user may already have.

Bearing capacity is computed using Eq. (4-6) and the reduced bearing-capacity factors as

$$q_{ult} = cN_c' s_c d_c i_c + \bar{q} N_q' s_q d_q i_q + \tfrac{1}{2}\gamma BN_\gamma s_\gamma d_\gamma i_\gamma$$

Since N_γ depends on the soil wedge cfa of Fig. 4-5a, this requires no modification for slope effects.

4-9. BEARING CAPACITY FROM PENETRATION TESTING

The standard penetration test (SPT) of Chap. 3 can be used to establish the bearing capacity of a cohesionless soil such as sand, sand-gravels, and sands with some silt. The method is largely empirical but is widely used. The earliest published relationship seems to be the curves of N vs. allowable soil pressure q_a for various widths of footings by Terzaghi and Peck (1967, p. 491; in the 1948 edition p. 423). These curves approximately fit an equation presented by Teng (1962)

$$q_a = 0.720(N - 3)\left(\frac{B + 1}{2B}\right)^2 \qquad (a)$$

According to Bazaraa (1967, p. 133) the Terzaghi and Peck chart was largely developed on an intuitive basis. Meyerhof (1956, 1974b) published equations similar to Eq. (a) of the form

$$q_a = \frac{N}{4} \qquad\qquad B \le 4 \qquad (b)$$

$$q_a = \frac{N}{6}\left(\frac{B + 1}{B}\right)^2 \qquad B > 4 \qquad (c)$$

where q_a = allowable *net* increase in soil pressure in kips/sq ft for an estimated maximum settlement of 1 in

N = penetration number corrected for overburden pressure if necessary

The original Terzaghi and Peck N value was to be the lowest N found in any of the site borings in the footing stress-influence zone (within about $2B$ below footing depth). This is too rigid a criterion for the reasons cited in Sec. 3-6 (and not likely to be done at present anyway), and a more representative value of N would be obtained from a *rough statistical analysis* of the most probable value. Note again that N is the representative penetration number in the zone of stress influence and should not be obtained from using all the N values obtained from the borings.

Both these equations may be corrected for depth as

$$K_d = 1 + \frac{0.2D}{B} \le 1.20 \text{ for Eq. } (a)$$

$$K_d = 1 + \frac{0.33D}{B} \le 1.33 \text{ for Eqs. } (b) \text{ and } (c)$$

The effect of water is already in the equations as an effect on N, and additional adjustments are not recommended.

Considerable observation of footings using the above equations indicates that they are too conservative for design and should be increased at least 50 percent based on the currently used methods of obtaining N. If we do this, the Meyerhof equations become

$$q_a = \frac{N}{2.5} K_d \qquad\qquad B \le 4 \qquad (4\text{-}9)$$

$$q_a = \frac{N}{4}\left(\frac{B + 1}{B}\right)^2 K_d \qquad B > 4 \qquad (4\text{-}9a)$$

Figure 4-6 is a plot of Eqs. (4-9) for rapid determination of the allowable bearing capacity.

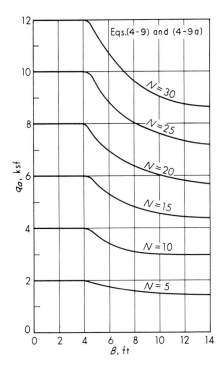

Figure 4-6. Allowable bearing capacity from Eq. (4-9) for surface loaded footings, 1-in (2.5-cm) settlement, and B and N values shown.

For mat foundations the allowable pressure can be computed according to Meyerhof (1956, 1965), and with the author's 50 percent increase from the originally proposed equation, as

$$q_a = \frac{N}{4}(K_d) \tag{4-10}$$

For all these equations, q_a is the net allowable bearing pressure for an assumed 1-in settlement. In general, the net allowable bearing pressure for any settlement S_j is

$$q_{a(S_j)} = q_a S_j \tag{4-11}$$

Example 4-7. Given the data and corrected N value of Example 3-2. The footing will be at -1.6 m (blow count is at -3 m, which is in stress area).

REQUIRED. What is the allowable bearing capacity for a 4-cm settlement?

SOLUTION. With a 4-cm settlement we will report the allowable pressure in kPa and use metric values of B for plotting fps units for q_a to avoid revising Eq. (4-9)

$$N = 10 \text{ corrected value} \qquad D = 1.6 \text{ m} \qquad D_r \cong 0.35$$

Since the allowable soil pressure depends on the size of the footing, we will prepare a table of values and present the results in a plot of q_a vs. B (Fig. E4-7).

B, m (ft)		D/B	K_d	q_a, kPa	q_a (4 cm), kPa
1.5	(4.92)	1.06	1.33	153.69	242.03
2.0	(6.66)	0.80	1.26	133.55	210.36
3.0	(9.84)	0.53	1.17	113.47	178.69
4.0	(13.13)	0.40	1.13	104.85	165.12
5.0	(16.4)	0.33	1.11	100.06	157.57
6.0	(19.7)	0.27	1.09	95.76	150.80

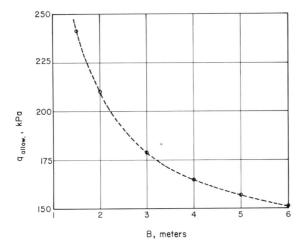

Figure E4-7

A method of presenting the allowable bearing pressure, since it depends on B, is with a plot as shown. The above method illustrates the method of using the standard-penetration-test data. It should be pointed out, however, that it would not be good practice to put a foundation of any importance on a soil with a $D_r = 0.35$. The soil would probably be made more dense, in which case the bearing capacity would be substantially different from that above.

The bearing capacity of sand using the Dutch-cone penetration test can be computed [Schmertmann (1975)] by back computations for

$$N_\gamma \cong \frac{q_c}{0.80} \tag{4-12}$$

where q_c = cone resistance, kg/cm²
$\quad N_\gamma$ = bearing-capacity factor in Table 4-1

With N_γ one can obtain ϕ and N_q to use in Eqs. (4-1) through (4-3).

Meyerhof (1956) suggested an easier and more direct procedure for the allowable bearing capacity for a 1-in settlement as

$$q_a = \frac{q_c}{15} \quad \text{(ksf)} \qquad\qquad B \leq 4 \qquad\qquad \text{(4-13)}$$

$$q_a = \frac{q_c}{25}\left(\frac{B+1}{B}\right)^2 \quad \text{(ksf)} \quad B > 4 \qquad\qquad \text{(4-13a)}$$

where q_c = cone-point resistance, ksf
$\quad B$ = footing width, ft

For rafts Meyerhof suggests doubling q_a in Eq. (4-13a).

In cohesive soils one may obtain [see Schmertmann (1975)] the undrained shear strength s_u as

$$s_u = \frac{q_c - \gamma D}{N_f} \qquad\qquad \text{(4-14)}$$

where q_c = cone resistance
$\quad \gamma D$ = total overburden pressure at cone tip
$\quad N_f$ = bearing-capacity factor ranging from about 5 to 70 but probably in the range of 10 for electrical cones and 16 for mechanical cones and clays with OCR < 2 and I_p > 10 percent

4-10. BEARING CAPACITY OF FOUNDATIONS WITH UPLIFT OR TENSION FORCES

Footings beneath elevated water tanks, and for the anchor cables of radio and television towers, or supporting the legs of power-transmission towers and other tall structures on isolated columns subject to overturning forces will require an analysis for stability against the tension forces in the foundation.

Footings to develop tension resistance are idealized in Fig. 4-7. Balla (1961) considered this problem. He assumed a failure surface (the dashed line ab in Fig. 4-7) as circular and developed some highly complicated mathematical expressions which were verified on model tests in a small glass jar and by some larger tests of others. The only footings considered were circular. Meyerhof and Adams (1968) also considered the problem and proposed the conditions of Fig. 4-7, namely, that footings should be considered as either shallow or deep since deep footings could develop only to some limiting pull-out force. Circular and rectangular footings were considered in both cohesive and cohesionless soils. They compared the theory (following equations) with models as well as full-scale tests on circular footings and found considerable scatter; however, with a factor of safety of around 2 to 2.5 these equations should be satisfactory.

These equations are developed by neglecting the larger pull-out zone observed in tests and using an approximation defined by line ab' in Fig. 4-7. If we take the cohesion developed on this cylinder, it is πBcD. The passive earth-pressure friction

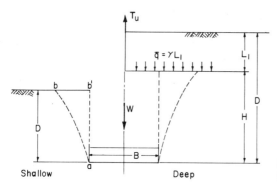

Figure 4-7. Footings subject to uplift or tension forces.

resistance developed on this cylinder is $(\pi/2)B\gamma D^2 K_u \tan \phi$ and should include a shape factor s_f. The remaining resistance is the weight of the soil and footing material within the cylinder W, and the ultimate uplift resistance T_u is

$$T_u = \pi BcD + s_f(\pi/2)B\gamma D^2 K_u \tan \phi + W \qquad (4\text{-}15)$$

For deep footings where $D > H$, Eq. (4-15) becomes

$$T_u = \pi cBH + s_f(1.57)\gamma B(2D - H)HK_u \tan \phi + W \qquad (4\text{-}16)$$

These equations can be used for both circular and square footings.

The shape factor is approximately

$$s_f = 1 + \frac{mD}{B}$$

with a maximum value for deep footings of

$$s_f = 1 + \frac{mH}{B}$$

Values of m, s_f, and H/B for various ϕ angles are as follows:

	ϕ	20°	25	30	35	40	45	48
Limiting	H/B	2.5	3	4	5	7	9	11
	m	0.05	0.10	0.15	0.25	0.35	0.50	0.60
Maximum	s_f	1.12	1.30	1.60	2.25	3.45	5.50	7.60

The passive earth-pressure coefficient K_u is computed from K_p where $K_p = \tan^2(45 + \phi/2)$ or values from Table 11-2, and $K_u = K_p \tan(2\phi/3)$. This value of earth-pressure coefficient is used to account for the actual shape of the rupture surface in terms of the angle of inclination of the passive earth resistance and the resulting effect on the friction resistance.

For rectangular footings at shallow depths the uplift resistance can be computed as

$$T_u = 2cD(B + L) + \gamma D^2(2s_f B + L - B)K_u \tan \phi + W \qquad (4\text{-}17)$$

and for deep rectangular footings

$$T_u = 2cH(B + L) + \gamma(2D - H)(2s_f B + L - B)HK_u \tan \phi + W \qquad (4\text{-}18)$$

For square footings use $L = B$ in these equations.

These equations are based on Meyerhof's assumption that the shape factor is acting on the two end portions of $B/2$ and in the interior part $L - B$ the passive pressure is the same as for strip footings ($s_f = 1.0$).

For footings founded in very poor soils, Robinson and Taylor (1969) found a satisfactory design resistance could be obtained by using only the weight term W in Eqs. (4-15) through (4-18), and a safety factor slightly greater than unity for transmission-tower anchorages.

Example 4-8. A footing $4 \times 4 \times 2$ ft is placed at a depth of 6 ft in a soil of $\gamma = 110$ pcf; $\phi = 20°$; $c = 400$ psf undrained.

REQUIRED. Estimate the allowable uplift force for a $F = 2.5$.

SOLUTION. $D/B = 6/4 = 1.5 < 2.5$ for $\phi = 20°$; therefore, the footing is classed as shallow and we will use Eq. (4-17).

$$T_u = 2cD(B + L) + \gamma D^2(2s_f B + L - B)K_u \tan \phi + W$$

$$s_f = 1 + mD/B = 1 + 0.05(1.5) = 1.075$$

$$K_u = \tan^2 (45 + 20/2) \tan [2(20)/3] = 0.483$$

$$W = \text{weight of concrete} + \text{weight of soil}$$

$$W = 4(4)(6 - 4)(0.150) + 4(4)(6 - 2)(0.110) = 4.8 + 7.04$$

$$= 11.84 \text{ kips}$$

Substituting values

$$T_u = 2(0.4)(6)(4 + 4) + 0.110(6)^2[2(1.075)(4) + 4 - 4](0.483) \tan 20° + 11.84$$

$$= 38.4 + 5.99 + 11.84 = 56.2 \text{ kips}$$

$$T_a = 56.2/2.5 = 22.5 \text{ kips}$$

The anchor footing would have to be able to carry structurally at least 22.5 kips with an appropriate safety factor.

4-11. BEARING CAPACITY BASED ON BUILDING CODES (PRESUMPTIVE PRESSURE)

In many cities the local building code stipulates values of allowable soil pressure to use when designing foundations. These values are usually based on years of experience, although in some cases they are simply used from the building code of another city. Values such as these are also found in engineering and building-construction handbooks. These arbitrary values of soil pressure are often termed *presumptive* pressures. Most building codes now stipulate that other soil pressures may be applied if laboratory testing and engineering considerations can justify the use of alternative values. Presumptive pressures are based on a visual soil classification.

Table 4-6 indicates representative values of building-code pressures. These values are primarily for illustrative purposes, since it is generally conceded that in all but minor construction projects some soil exploration should be undertaken. Major drawbacks to the use of presumptive soil pressures are that they do not reflect the depth of footing, size of footing, location of water table, or potential settlements.

Table 4-6. Presumptive bearing capacities from indicated building codes, psf

Soil description	Chicago, 1975	Natl. Board of Fire Underwriters, 1967	Atlanta,* 1973	New York City, 1968	Uniform Bldg. Code, 1964†
Clay, very soft	500				
Clay, soft	1,500	3,000	1,000	2,000	1,500
Clay, ordinary	2,500				
Clay, medium stiff	3,500	5,000	2,000		2,500
Clay, stiff	4,500		3,000	4,000	
Clay, hard	6,000			10,000	8,000
Sand, compact and clean	5,000		6,000	6,000–16,000	
Sand, compact and silty	3,000				
Inorganic silt, compact	2,500				
Sand, loose and fine		4,000		<4,000	1,500
Sand, loose and coarse, or sand-gravel mixture, or compact and fine		6,000		8,000	2,500
Gravel, loose, and compact coarse sand		8,000		12,000	8,000
Sand-gravel, compact		12,000	8,000	16,000–20,000	8,000
Hardpan, cemented sand, cemented gravel	12,000	20,000		24,000	
Soft rock				16,000	
Sedimentary layered rock (hard shale, sandstone, siltstone)		30,000		16,000	
Bedrock	200,000	200,000		40,000–120,000	

* Use of presumptive pressures limited to structures not over four stories.

† Maximum values; however, code allows consideration of depth of footing, as, for example, soil classed as hard clay, $q_a = 400D_f \leq 8,000$ psf.

4-12. SAFETY FACTORS IN FOUNDATION DESIGN

The buildings are designed on the basis of introducing into the analysis a safety factor which may be defined as the ratio of the resistance R of the structure to the applied loads L or $F = R/L$. The magnitude of the safety factor depends mainly on the reliability of the design data and the assessment of the structural resistance, and applied loads. Accuracy of the structural analyses, the quality of construction and maintenance, and the probability and seriousness of a failure during the service life of the structure also influence the value of F.

In the design of foundations, there are more uncertainties and approximations than in the design of other structures because of the complexity of soil behavior and incomplete knowledge of the subsoil conditions. These uncertainties and approximations have to be evaluated for each case and a safety factor assigned which should be reasonable and sufficient but taking into account some or all of the following:

1. Magnitude of damages (loss of life, lawsuits, property damage)
2. Relative cost of increasing or decreasing F
3. Relative change in probability of failure by changing F
4. Reliability of soil data
5. Construction tolerances
6. Changes in soil properties due to construction operations
7. Accuracy (or approximations used) in developing design/analysis methods.

The customary procedure for applying a safety factor in foundation design is similar to the superstructure design, namely, an overall value. Typically, values of F are as in Table 4-7. The upper values are normally used for normal-design-load

Table 4-7. Values of customary safety factors

Failure mode	Foundation type	F
Shear	Earthworks	
	Dams, fills, etc.	1.2–1.6
Shear	Retaining structure	
	Walls	1.5–2.0
Shear	Sheetpiling, cofferdams	1.2–1.6
	Braced excavations,	
	(temporary)	1.2–1.5
Shear	Footings	
	Spread	2–3
	Mat	1.7–2.5
	Uplift	1.7–2.5
Seepage	Uplift, heaving	1.5–2.5
	Piping	3–5

service conditions and the lower values for maximum or transient loading conditions. For example, it is customary to use a bearing capacity with a $F = 3.0$ for the design load as follows:

$$\text{Design load} = DL + R_L(L) + R_s(S) \pm \text{hydrostatic pressure}$$

Wind and earthquake loads are considered temporary, and one might use $F = 2.0$ for the following combination of loads:

$$\text{Temporary load} = DL + R_L(L) + R_s(S) + R_w(\text{wind}) \pm \text{hydrostatic pressure}$$

Note that it would not be likely that both wind and earthquake loading would be acting simultaneously. The factors R_i are code reduction factors for both load condition and load combination. The load terms are obtained from Table 4-8.

Partial safety factors have been proposed by some investigators [see Hansen (1967)] using different values on cohesion and ϕ. Others propose that statistics and probability theory be applied to the problem [for example, Meyerhof (1970), Wu and Kraft (1967), Lumb (1966, 1974)]. The concept of partial safety factors is somewhat academic, since we are concerned with failure vs. no failure and whether we assign a single number or several numbers to make a composite we are still concerned with a possible failure.

For statistics and probability theory, a few observations will be made:

1. Few persons use this method at present (1976).
2. It may be useful to quantify certain parameters and to obtain a probability number of say, 0.95 or 0.99 as the end result of step 5 following.
3. It is difficult to apply statistics in soil mechanics—some parameters such as ϕ may obey normal frequency-distribution laws approximately when the sand is *dry* whereas ϕ (and c) for cohesive soils depends on water content, density, and test procedure.
4. It is necessary for the statistician to estimate the variation of the soil parameters and other geological, climate, etc., effects.

Table 4-8. Foundation loads

Load	Includes
Dead load (DL)	Weight of structure and all permanently attached material
Live load (LL)	Any load not permanently attached to the structure, but to which the structure may be subjected
Snow load (S)	Acts on roofs; value to be used generally stipulated by codes
Wind load (W)	Acts on exposed parts of structure
Earthquake (E)	A lateral force (usually) which acts on the structure

Table 4-9. Range of properties for selected rock groups

Type of rock	Typical unit wt. pcf	Modulus of elasticity E, ksi	Poisson's ratio μ	Compressive strength, ksi
Basalt	178	7,000–13,000	0.27–0.32	25–60
Granite	168	4,000–7,000	0.26–0.30	10–40
Schist	165	2,000–5,000	0.18–0.22	5–15
Limestone	165	2,000–6,000	0.24–0.25	5–25
Porous limestone				1–5
Sandstone	145–150	1,000–3,000	0.20–0.30	4–20
Shale	100–140	500–2,000	0.25–0.28	1–6
Concrete	100–150	Variable		2–6

5. When one combines a series of variation estimates (step 4) into any equation to arrive at the net overall effect (in this case F), the result is now a quantifiable estimate.

4-13. BEARING CAPACITY OF ROCKS

Although the topic of rock mechanics is beyond the scope of this book, a few comments on bearing capacity of rocks will be made.

With the exception of a few porous rocks, such as some shales, limestones, sandstones, and volcanic rocks, the strength of bedrock in situ is probably as great as the concrete of a footing. This, of course, may not be true if the rock is not sound. The problem of the engineer, then, is to determine the quality (soundness) of the rock and whether it is a small stone suspended in the soil mass or is in fact bedrock (or a stone large enough to have the same structural effect).

To allow for the possibility of unsound rock, it is common practice to use high safety factors (say, 6 to 10 on the unconfined compression strength q_u) or, in the case of placing concrete on the rock, to use no more than the compressive strength of the concrete as the working compression strength of the rock. In general

$$q_a = Cq_u$$

where C may be on the order of 0.2 to 0.3 of q_u. For sandstone in Table 4-9, $q_a \cong 115$ to 173 ksf for $q_u = 4$ ksi.

Table 4-9 provides data ranges for some commonly occurring rocks.

PROBLEMS

4-1. A square footing is as shown in Fig. P4-1. Find the required B for the soil properties and load given. (Note both ϕ and q_u.) Use $F = 3.0$ and Eq. (4-2).

    Answer: $q_a = 5$; $B = 6.5$ ft.

4-2. If the water table is at the base of the footing in Prob. 4-1, what should the size of a square footing be?

    Answer: $B = 6.5$ ft.

4-3. Redo Prob. 4-1 using Eq. (4-6)

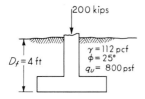

Figure P4-1

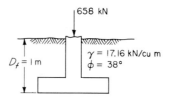

Figure P4-4

4-4. A footing is located as shown in Fig. P4-4. For a square footing, what should be the size, using bearing-capacity factors in Table 4-2? Use $F = 3$.

4-5. Verify the Terzaghi and Hansen values of q_{ult} for the assigned test number of Table 4-4.

4-6. Redo Prob. 4-4; (a) water table at base of footing; (b) increase D to 1.5 m. What is the percent increase in bearing capacity between (a) and (b)?

4-7. A footing is as shown in Fig. P4-7. Use $F = 3.0$.
   () Find the size of square footing to carry the load, using Eq. (4-6).
   (b) Find the most economical size footing to carry the load.
   (c) Compute percentage of area savings between (a) and (b).
   (d) What is the allowable bearing pressure in (a) and (b)?
   Answer: (a) 6.0×6.0; (b) 4×7 ft; (c) $\simeq 13\%$; (d) part a: $q_a = 7$ ksf; part b: $q_a = 7.5$ ksf.

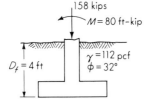

Figure P4-7

4-8. A set of plate load-test data on a thick sand deposit is as follows [Bazaraa (1967)]:

			S, in	
q, tsf	1 × 1 ft	1 × 1	4 × 4	4 × 4
2	0.04	0.06	0.10	0.06
4	0.08	0.12	0.18	0.19
6	0.14	0.18	0.34	0.47
8	0.25	0.30	0.53	0.94
10	0.37	0.50	0.78	1.53
SPT =	N = 34	N = 32	N = 25	N = 26

The depth of tests was between 7 and 9.5 ft. The water table is not a significant factor for the problem. Plot the specified load test, and using the method of Secs. 3-12 and 4-9, estimate the allowable soil pressure.

4-9. A portion of a cone penetration test is in Fig. P4-9. Estimate the allowable bearing pressure at (a) −2 m; (b) −4 m.

4-10. For the portion of the cone penetration test shown in Fig. 3-21b, estimate the allowable bearing pressure at −3 m.

4-11. For the penetration-test data shown in Fig. 3-10, estimate the allowable bearing pressure at −6.0 ft. Will water be a problem?

4-12. For the boring logs shown in Fig. 3-11: What is the recommended soil pressure for footings located in the vicinity of elevation 284 for the part of the building area covered with boring No. 2? The building is a four-story (five on the low side) office building with load-bearing walls and column loads around 160 kips. State your general recommendations and reasons.

4-13. Redo Prob. 4-12 for boring No. 4, but the footings will be around elevation 290.

4-14. The following load-test data are obtained from Brand et al. (1972). The footings are all square of dimensions given and located approximately 1.5 m below the ground surface. Plot the assigned load test and estimate the "failure" load. Compare the failure load with Eq. (4-2) and Eq. (4-6). Comment on your assumptions and results. For the 1.05-m footing see Example 4-3.

	Square plate size, m			
Load, tons	1.05	0.9	0.75	0.60
0	0.000	0.000	0.000	0.000
2	0.030	0.043	0.062	0.112
3				0.212
4	0.075	0.112	0.175	0.406
5			0.243	0.631
6	0.134	0.187	0.325	0.912
7			0.450	1.456
8	0.212	0.306	0.606	
9		0.394	0.862	
10	0.331	0.500	1.293	
11		0.625		
12	0.537	0.838		
13		1.112		
14	0.706	1.500		
15	1.143			
16	1.425			

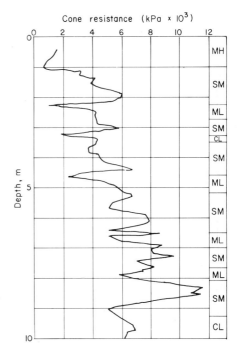

Figure P4-9

FOUNDATION SETTLEMENTS

5-1. THE SETTLEMENT PROBLEM

Foundation settlements must be assessed as accurately as possible for most structures including buildings, bridges, towers, power plants, etc. It is often necessary to
obtain an indication of the probable magnitude of settlement for other structures
such as fills, dams, braced sheeting, and retaining walls. Settlement computations are
particularly difficult because they tend to be a problem in elasticity and

1. Soils are elastic only for very small strains.
2. It is very difficult to obtain the elastic properties of a soil for reasons presented in
   Chap. 2.

   In spite of the above shortcomings, it is convenient to treat the soil as an elastic,
isotropic, homogeneous mass for estimating settlements. More recent developments
can allow for layered soils and to a limited extent anisotropy. Some of these will be
taken up later in this chapter.
   Settlements are of two general types

1. *Immediate* or those which take place as soon (0 to less than about 7 days) as the
   load is applied.
2. *Consolidation*, or those which are time-dependent.

   Immediate-settlement analyses are appropriate for all fine-grained soils (silts and
clays) with a degree of saturation less than about 90 percent. All soils with a large

coefficient of permeability (rapid draining under a hydraulic gradient) including all cohesionless soils would undergo immediate settlements.

Consolidation-settlement analyses are appropriate, for all saturated, or nearly saturated, fine-grained soil deposits (k on the order of 10^{-4} cm/sec or less). Consolidation-settlement analyses are actually a modified form of elasticity computations. They are done in the alternative form, however, because in the development of the "elastic constant" C_c, one also obtains a time-rate constant (the coefficient of consolidation c_v) so that one obtains estimates of both amount of settlement and time duration.

5-2. STRESSES IN A SOIL MASS DUE TO FOOTING PRESSURE

Several methods are currently used to estimate the increased pressure on an element of soil at some depth in the strata below the foundation member. The simplest method is to use a $2:1$ slope. Some persons prefer to use a stress zone defined by some angle (say, 30 to 45°) with the vertical, as indicated in Fig. 5-1. If the stress zone is defined by the $2:1$ slope, the change in pressure Δq at a depth z beneath the footing is, simply,

$$\Delta q = \frac{V}{(B + z)(L + z)} \tag{5-1}$$

which simplifies for a square footing to

$$\Delta q = \frac{V}{(B + z)^2} \tag{5-1a}$$

where V = total load applied to foundation member
 B, L = footing dimensions, ft
 z = depth from footing base to elevation in soil, where increase in stress is desired

This method is an approximation and becomes even more approximate for layered soils.

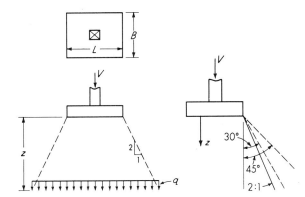

Figure 5-1. Approximate methods of evaluating stress increase in the soil at a depth z beneath the footing.

5-3. THE BOUSSINESQ METHOD FOR EVALUATING SOIL PRESSURE

The Boussinesq and Westergaard theories are more mathematically oriented methods for estimating soil pressures at various points in a soil stratum. Both these theories are based on elastic methods. Boussinesq's equation considered a point load on the surface of an infinitely large homogeneous, isotropic, elastic half-space to obtain

$$q_v = \frac{3Q}{2\pi z^2} \cos^5 \theta \qquad (5\text{-}2)$$

with symbols as identified in Fig. 5-2a. From this figure we can also write $\tan \theta = r/z$ and define a new term $R^2 = r^2 + z^2$ and take $\cos^5 \theta = (z/R)^5$. With these terms inserted in Eq. (5-2), we obtain

$$q_v = \frac{3Qz^3}{2\pi R^5} \qquad (5\text{-}3)$$

which can also be written

$$q_v = \frac{3Q}{2\pi z^2} \frac{1}{[1 + (r/z)^2]^{5/2}} = \frac{Q}{z^2} A_b \qquad (5\text{-}4)$$

Since the A_b term is a function of only the ratio of r/z, values may be tabulated as in Table 5-1, once and for all.

A foundation is not a point load, making the use of the Boussinesq equation somewhat impractical. If, however, a contact pressure q_0 is considered to be applied to a circular area as shown in Fig. 5-2b, the total load Q can be written $Q = \int_0^A q_0 \, dA$.

The stress on the soil element from the contact pressure q_0 on the surface area dA of Fig. 5-2b is

$$dq = \frac{3q_0}{2\pi z^2} \frac{1}{[1 + (r/z)^2]^{5/2}} \, dA \qquad (a)$$

Table 5-1. Values of A_b and A_w for selected r/z ratios for use in the Boussinesq (Eq. 5-4) or Westergaard (Eq. 5-8)

| r/z | A_b | A_w | | r/z | A_b | A_w | |
		$\mu = 0.0$	$\mu = 0.4$			$\mu = 0.0$	$\mu = 0.4$
0.0	0.4775	0.3183	0.9549	1.5	0.0251	0.0247	0.0173
0.1	0.4657	0.3090	0.8750	2.0	0.0085	0.0118	0.0076
0.2	0.4329	0.2836	0.6916	2.5	0.0034	0.0064	0.0040
0.5	0.2733	0.1733	0.2416	3.0	0.0015	0.0038	0.0023
0.8	0.1386	0.0925	0.0897	3.5	0.0007	0.0025	0.0015
1.0	0.0844	0.0613	0.0516	4.0	0.0004	0.0017	0.0010

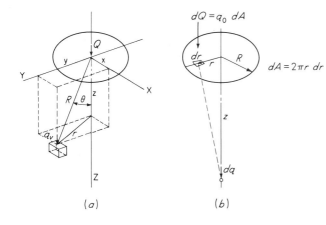

Figure 5-2. (a) Intensity of pressure q based on Boussinesq approach; (b) pressure at a point of depth z below the center of a circular area acted on by intensity of pressure q_0.

but $dA = 2\pi r\, dr$, and Eq. (a) becomes

$$q = \int_0^r \frac{3q_0}{2\pi z^2} \frac{1}{[1 + (r/z)^2]^{5/2}} 2\pi r\, dr \tag{b}$$

Performing the integration and inserting limits,

$$q = q_0\left\{1.0 - \frac{1}{[1 + (r/z)^2]^{3/2}}\right\} \tag{c}$$

If we rearrange this equation, solve for r/z, and take the positive root,

$$\frac{r}{z} = \sqrt{\left(1 - \frac{q}{q_0}\right)^{-2/3} - 1} \tag{5-5}$$

The interpretation of this equation is that the r/z ratio is the relative size of a circular bearing area such that, when loaded, it gives a unique pressure ratio q/q_0 on the soil element at a depth z in the stratum. If values of the q/q_0 ratio are put into this equation, corresponding values of r/z may be obtained, as in Table 5-2. The results of Table 5-2 may be used to draw a series of concentric circles termed an influence chart, as shown in Fig. 5-3. This concept and the influence charts were presented by Newmark (1942). The use of the chart is based on a factor termed the *influence value*, determined from the number of units into which the chart is subdivided. For example, if the series of rings are subdivided so that there are 400 units, often made approximate squares, the influence value is $1/400 = 0.0025$. In making a chart it is necessary that the sum of the units between two concentric circles multiplied by the influence value be equal to the change in the q/q_0 of the two rings (i.e., if the change in two rings is $0.1q/q_0$, then the influence value I multiplied by the number of units M should equal 0.1). This concept enables one to construct a chart of any influence value. Figure 5-3 is subdivided into 200 units; therefore, the influence value is $1/200 = 0.005$. Smaller influence values increase the number of squares and the amount of work involved, since the sum of the squares used in a problem is merely a mechanical integration of Eq. (a). It is doubtful if much accuracy is gained using very small influence values, although the amount of work is increased considerably.

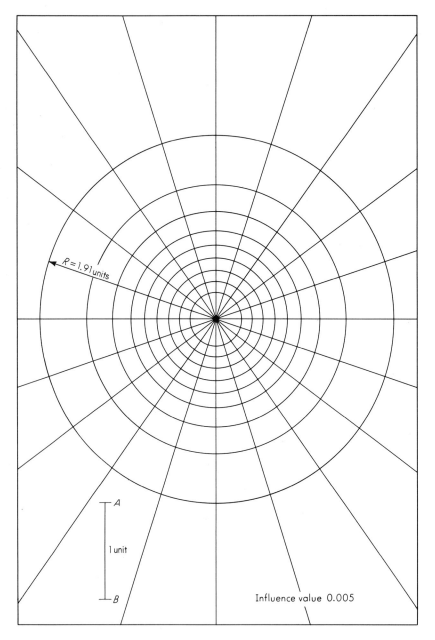

$R = 1.91$ units

A

1 unit

B

Influence value 0.005

Figure 5-3. Influence chart for vertical pressure. [*After Newmark (1942).*]

Table 5-2. Values of r/z for selected values of q/q_0 for construction of the influence (Newmark) chart of Fig. 5-3

q/q_0	r/z	q/q_0	r/z
0.0000	0.0000	0.5000	0.7664
0.1000	0.2698	0.6000	0.9176
0.2000	0.4005	0.7000	1.1097
0.3000	0.5181	0.8000	1.3871
0.4000	0.6370	0.9000	1.9084
		1.0000	∞

The influence chart may be used to compute the pressure on an element of soil beneath a footing, or pattern of footings, and for any depth z below the footing. It is only necessary to draw the footing pattern to a scale of z = length AB of the chart. Thus, if z = 15 ft, the length AB becomes 15 ft; if z = 20 ft, the length AB becomes 20 ft; etc. Now if AB is 1 in, scales of 1 in = 15 ft and 1 in = 20 ft, respectively, will be used to draw the footing plans. These footing plans will be placed on the influence chart with the point for which the stress Δq is desired at the center of the circles. The units (segments or partial segments) enclosed by the footing or footings are counted, and the increase in stress at the depth z is computed as

$$\Delta q = q_0 M I \qquad (5\text{-}6)$$

where Δq = increased intensity of soil pressure due to foundation loading at depth z, psf
$\qquad q_0$ = foundation contact pressure, psf
$\qquad M$ = number of units counted (partial units are estimated)
$\qquad I$ = influence factor of a particular chart used

The influence chart is especially useful for several footings, a mat or raft foundation, footings with different contact pressures, or other type of problem where the stress in the soil at a point may be caused from loadings at several other points. For a single circular, square, or long footing, the concept of the *pressure bulb* is useful. Figure 5-4 gives pressure bulbs (isobars) for the two common types of footings. These isobars were constructed by plotting a footing of dimension B to scale at various z values or ratios of footing dimension and using the influence chart to find pressure intensities at various points both beneath and outside the footing prism. The pressure isobars for a circular foundation are not given since these are simple to obtain by proper interpretation of the influence chart.

Lateral pressure on an element of soil in a stratum, q_h, may also be computed by a Boussinesq equation as

$$q_h = \frac{Q}{2\pi z^2} \left[3 \sin^2 \theta \, \cos^3 \theta - \frac{(1 - 2\mu)\cos^2 \theta}{1 + \cos \theta} \right] \qquad (5\text{-}7)$$

where the terms are the same as previously defined, in Fig. 5-2 and μ = Poisson's ratio.

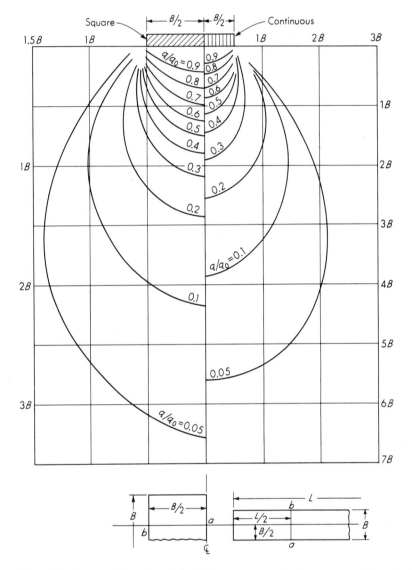

Figure 5-4. Pressure isobars based on the Boussinesq equation for square and long footings. Applicable only along line *ab* as shown.

The assumptions for using the Boussinesq equations are as follows:

1. The soil is weightless.
2. The soil is elastic, homogeneous, semi-infinite, and isotropic and obeys Hooke's law.
3. The change in soil volume is neglected.
4. The soil is unstressed before the application of the load Q.
5. There is stress continuity.
6. Stress distribution is symmetrical with respect to the vertical (z) axis.

It may be observed that vertical stesses are independent of both stress-strain modulus and Poisson's ratio. The lateral pressure depends on Poisson's ratio but is independent of stress-strain modulus. Values of Poisson's ratio of 0.5 for saturated cohesive soils and 0.2 to 0.3 for other soils will generally be satisfactory for use in Eq. (5-7).

The Boussinesq method of obtining the increase in vertical stresses is widely used for all types of soil masses, normally consolidated, overconsolidated, anisotropic, layered, etc. The computed stresses have been found to be in quite good agreement with measured (or measured settlement and back-computed stresses) stresses in most cases.

The easiest method to obtain the solution of Eqs. (5-3) and (5-7) is by numerical integration on the digital computer. The procedure is as follows:

1. Draw the soil and footing plan to approximate scale.
2. Decide on the various depths (z coordinates) to which it is desired to evaluate the vertical stress [Eq. (5-3)] or horizontal stress [Eq. (5-7)].
3. Divide the footing into unit areas, depending on size of footing and establish x, y coordinates for the centroid of each unit area. The computer should be programmed to do this.
4. Compute the stress at each z coordinate as the sum of the stresses contributed by the appropriate equation from each unit area.

The author has found that subdividing a footing to smaller unit areas than 1 sq ft or 0.10 m^2 does not improve the stress value, but it does increase the amount of computation considerably. Round foundations can be analyzed after converting to an equivalent square foundation of $B = \sqrt{A_f}$. Computer programs are in Appendix B to solve Eqs. (5-2) and (5-7).

5-4. WESTERGAARD'S METHOD FOR EVALUATING SOIL PRESSURES

When the soil mass consists of layered strata of finer and coarser materials (typified by stratified sedimentary deposits) or nonisotropic soils, the Boussinesq equations given above may not provide reliable solutions. To formulate a mathematical model for this situation, Westergaard (1938) presented the following equation:

$$q_v = \frac{Q}{2\pi z^2} \frac{\sqrt{(1 - 2\mu)/(2 - 2\mu)}}{[(1 - 2\mu)/(2 - 2\mu) + (r/z)^2]^{3/2}} \tag{5-8}$$

where q_v = intensity of stress at a point in the soil due to surface loading Q
$\quad\quad \mu$ = Poisson's ratio, for which the minimum value of zero simplifies the problem considerably
$\quad z, r$ = same as in Eq. (5-3)

When Poisson's ratio μ is taken as zero, Eq. (5-8) simplifies to

$$q_v = \frac{Q}{\pi z^2} \frac{1}{[1 + 2(r/z)^2]^{3/2}} \tag{d}$$

As with the Boussinesq equation, we may take the latter part of this equation as a term involving only a ratio and rewrite it (see Table 5-1)

$$q_v = \frac{Q}{z^2} A_w \qquad (5\text{-}8a)$$

The load Q is generally not a point load; so, using Fig. 5-2b for the significance of terms, since the figure is also applicable to the Westergaard problem, and from Eq. (5-8), let

$$a = \frac{1 - 2\mu}{2 - 2\mu}$$

The load Q will be written as the sum of the contact pressures on differentials of area.

$$Q = \int_0^A q_0 \, dA$$

where q_0 is the contact pressure

$$dA = 2\pi r \, dr$$

With these substitutions Eq. (5-8) is rewritten

$$q = \frac{q_0 \sqrt{a}}{2z^2} \int_0^A \left[a + \left(\frac{r}{z}\right)^2 \right]^{-3/2} 2r \, dr \qquad (e)$$

Integrating this expression and solving for the positive root of r/z,

$$\frac{r}{z} = +\sqrt{\frac{a}{(1 - q/q_0)^2} - a} \qquad (5\text{-}9)$$

If this equation is solved for selected values of Poisson's ratio and incremental quantities of q/q_0, as was done with the Boussinesq equation, values to plot a Westergaard influence chart may be computed. A chart from the values in Table 5-3 (with $\mu = 0.0$) is shown in Fig. 5-5. This chart is used in the same manner as the Boussinesq (Newmark) influence chart. The influence value of 0.005 is derived from using 200 units as $I = 1/200$. Note that in this case concentric rings were drawn to include only $q/q_0 = 0.8$, using the radius of $3.464AB$ from Table 5-3, according to the

Table 5-3. Values of r/z for selected values of q/q_0 for construction of a Westergaard influence chart

$\mu = 0.00$				$\mu = 0.4$			
q/q_0	r/z	q/q_0	r/z	q/q_0	r/z	q/q_0	r/z
0.000	0.000	0.500	1.227	0.000	0.000	0.500	0.707
0.100	0.343	0.600	1.620	0.100	0.198	0.600	0.935
0.200	0.530	0.700	2.249	0.200	0.306	0.700	1.298
0.300	0.721	0.800	3.464	0.300	0.417	0.800	2.000
0.400	0.943	0.900	7.036	0.400	0.544	0.900	4.062

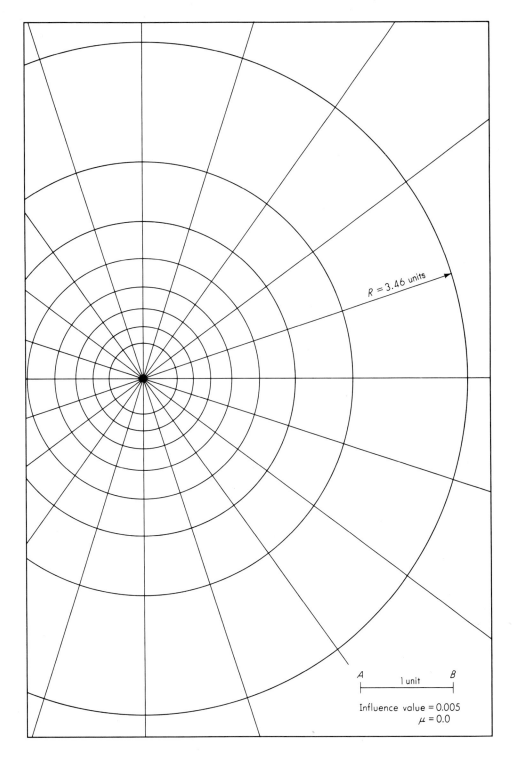

R = 3.46 units

A   1 unit   B

Influence value = 0.005
$\mu = 0.0$

Figure 5-5. Influence chart for vertical pressure at any depth $z = AB$ in the soil, based on the Westergaard theory. Chart is constructed from values given in Table 5-3.

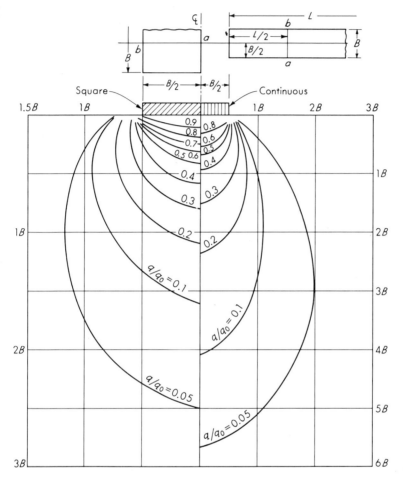

Figure 5-6. Pressure isobars based on the Westergaard equation for square and long continuous footings of dimension $B \times B$ or $B \times L$. Values for the continuous footings are at the point $L/2$ from the end. Chart contructed from Fig. 5-5 ($\mu = 0$). Applicable along line ab as shown.

AB scale shown on the chart, and thus the chart includes only 160 of the 200 units used to establish the influence value.

Figure 5-6 is a set of isobars prepared from Fig. 5-5 to obtain stresses at a point beneath continuous and square foundations. Figure 5-5 may be used for round contact areas.

A numerical procedure may be used to solve Eq. (5-8) on the computer in a similar manner to the Boussinesq equation.

In comparing the Boussinesq and Westergaard methods for vertical stresses, one finds that

1. At small r/z (close to footing) the Boussinesq equation gives larger stress intensity.
2. At r/z about 1.8 both values are approximately equal.
3. At r/z greater than 1.8 the Westergaard equation gives larger stresses.

5-5. IMMEDIATE (ELASTIC) SETTLEMENT COMPUTATIONS—THEORY

The settlement of soils such as nonsaturated clays and silts, sands and gravels both saturated and unsaturated, and clayey sands and gravels can be computed from the following equation:

$$S = qB \frac{1 - \mu^2}{E_s} I_w \tag{5-10}$$

where S = settlement
q = intensity of contact pressure
B = least lateral dimension of footing
I_w = influence factor which depends on shape of footing and its rigidity (typical values are given in Table 5-4)
E_s, μ = elastic properties of soil (typical values in Tables 2-3 and 2-4)

For flexible footings, the influence factor can be computed from an equation proposed by Steinbrenner (1934) for the corner of a rectangle area of dimensions $B \times L$ as

$$I_w = \frac{1}{\pi} \left\{ \frac{L}{B} \ln \left[\frac{1 + \sqrt{(L/B)^2 + 1}}{L/B} \right] + \ln \left[\frac{L}{B} + \sqrt{\left(\frac{L}{B}\right)^2 + 1} \right] \right\} \tag{5-11}$$

The influence value for the "center" of a footing is obtained by summing two corner values from the above equation with due attentionn paid to the B and L dimensions of the contributing area. Two areas are used instead of four because B = footing

Table 5-4. Influence factors for various-shaped members, I_w, I_m, for flexible and rigid footings

Shape	Flexible			Rigid	
	Center	Corner	Average	I_w	I_m *
Circle	1.00	0.64 (edge)	0.85	0.88†	6.0
Square	1.12	0.56	0.95	0.82	3.7
Rectangle:					
$L/B = 0.2$					2.29
0.5					3.33
1.5	1.36	0.68	1.15	1.06	4.12
2	1.53	0.77	1.30	1.20	4.38
5	2.10	1.05	1.83	1.70	4.82
10	2.54	1.27	2.25	2.10	4.93
100	4.01	2.00	3.69	3.40	5.06

* Lee (1962).
† Others have used the value $0.79 = \pi/4$ for the rigid-footing influence factor for circular footings.

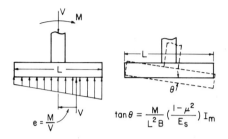

$$\tan \theta = \frac{M}{L^2 B} \left(\frac{1 - \mu^2}{E_s}\right) I_m$$

$$e = \frac{M}{V}$$

Figure 5-7. Rotation of a rigid footing.

width in Eq. (5-10) and not $b = B/2$. Influence factors for several footing ratios obtained from Eq. (5-11) are given in Table 5-4.

According to Schleicher (1926) and others, the influence factor for rigid footings is about 7 percent smaller than for flexible foottings; thus we have a simple means to obtain the values of Table 5-4 for the rigid footing.

Influence values for rotation have been proposed by Lee (1962), some of which are included in Table 5-4, and are used in the following equation:

$$\tan \theta = \frac{Ve}{BL^2} \frac{1 - \mu^2}{E_s} I_m \qquad (5\text{-}12)$$

where terms are identified in Fig. 5-7.

The theory-of-elasticity equation for immediate settlement [Eq. (5-10)] was derived for a footing on the surface of a semi-infinite, isotropic, homogeneous, elastic half-space. Most footings are located at some depth D in the soil mass, and it is often layered and of finite depth.

To approximate the surface deflection of a footing on an elastic layer of finite depth H, Steinbrenner computed the surface settlement S' and the settlement S'' at depth H at the corner of a rectangle $B \times L$. He reasoned that the net surface settlement at the corner would be

$$S = S' - S''$$

which resulted in the following equation:

$$S = qB \frac{1 - \mu^2}{E_s} \left(F_1 + \frac{1 - 2\mu}{1 - \mu} F_2\right) \qquad (5\text{-}13)$$

where $F_1 = \frac{1}{\pi} \left[M \ln \frac{(1 + \sqrt{M^2 + 1})\sqrt{M^2 + N^2}}{M(1 + \sqrt{M^2 + N^2 + 1})} + \ln \frac{(M + \sqrt{M^2 + 1})\sqrt{1 + N^2}}{M + \sqrt{M^2 + N^2 + 1}} \right]$

$F_2 = \frac{N}{2\pi} \tan^{-1} \left(\frac{M}{N\sqrt{M^2 + N^2 + 1}}\right)$ ($\tan^{-1}$ in rads)

$M = \frac{L}{B}$ $N = \frac{H}{B}$

The settlement at the center of a footing would be *two* times the value obtained for a corner as before with Eq. (5-11). Equation (5-11) is a special case of F_1 with $H/B = \infty$, and the same values can be obtained when H/B is around 1,000.

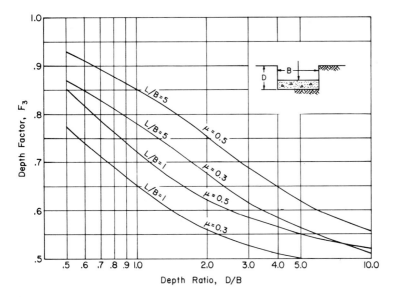

Figure 5-8. Influence factor for footing at a depth D. Use actual footing width and depth dimension for D/B ratio.

Footings are almost never placed on the ground surface (see also Sec. 7-1). They are usually at some depth D in the ground. Fox (1948) proposed a correction to apply to the surface deflection which depends on both the depth of embedment and Poisson's ratio. The corrected settlement S_f becomes

$$S_f = S(F_3)$$

where F_3 is obtained from Fig. 5-8 and S is obtained from Eqs. (5-10), (5-13), or (5-14). Figure 5-8 has been obtained by programming the Fox equations (which are too lengthy to be presented here) on the computer to obtain the settlement ratio vs. D/B and for selected L/B and Poisson's ratios. The computer program is in the Appendix so that the user can obtain values for any ratios not shown. It might be added that the program in the Appendix can be used to compute elastic settlements for a footing on a semi-infinite, elastic, homogeneous half-space if the computer variable WC is multiplied as follows:

$$S = WC \frac{(q)(1 + \mu)}{4\pi E_s(1 - \mu)} \tag{5-14}$$

5-6. IMMEDIATE SETTLEMENTS—APPLICATION

Equation (5-10) can be interpreted in terms of the deformation equation from mechanics of materials ($e = PL/AE = \sigma L/E$) where the relationships are

$$q = \sigma \qquad B(1 - \mu^2)I_w = L \qquad \text{and} \qquad E_s = E$$

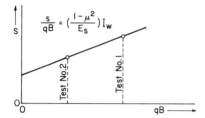

Figure 5-9. Graphical solution to find $[(1 - \mu^2)/E_s]I_w$ using plate-load tests with two or more plates of the same shape. Generally, Test No. 1 uses smaller plate.

It is evident, therefore, that we now have two problems in applying Eq. (5-10):

1. Correctly evaluating L
2. Correctly evaluating E_s

One way to evaluate E_s is as shown in Fig. 5-9, using several different size plates (of same shape) in a series of plate-load tests. For some load the several plates will undergo settlements for that qB which can be plotted to some scale. The slope of the resulting plot is $(1 - \mu^2)I_w/E_s$, as shown in the figure. For example, using the data from the 0.6- and 1.05-m-square plates of Prob. 4-14 at 4 tons, we have

$$S = 0.003 \text{ m and } qB = 3.81 \text{ for the 1.05-m-square plate}$$

$$S = 0.010 \text{ m and } qB = 6.67 \text{ for the 0.6-m plate}$$

from which (without plotting)

$$\frac{1 - \mu^2}{E_s} I_w = \frac{1}{408.6}$$

If we assume $I_w = 0.82$ and Poisson's ratio $= 0.5$ (saturated clay)

$$E_s = (1 - 0.5^2)(0.82)(408.7) = 251.4 \text{ tons/m}^2$$

which are awkward units but are consistent with the problem data. One should use a test load in the region where the S vs. load plot is reasonably linear.

Stress-strain modulus values can be obtained from stress-strain plots from triaxial tests, either from the initial tangent modulus on a single cycle or from the initial tangent modulus after several cycles. This was discussed in Sec. 2-13. One may estimate E_s from the undrained shear strength s_u from UU tests as

$$E_s = (250 \text{ to } 500)s_u$$

for normally consolidated, sensitive clay; and for insensitive clays of OCR less than 2, one may estimate

$$E_s = 1,000s_u$$

When using penetration tests, Mitchell and Gardner (1975) presented extensive tables of equations used by various persons. For the SPT it appeared that a reasonable estimate is

$$E_s = 10(N + 15) \qquad \text{(ksf) for sands}$$

$$E_s = 6(N + 5) \qquad \text{(ksf) for clayey sand}$$

These values are intermediate in Mitchell's table between equations giving much higher and some giving considerably lower values. For the cone penetration test (CPT) an estimate is

$$E_s = 3q_c \text{ (units of } q_c) \text{ for sands (several sources)}$$

$$E_s = (2 \text{ to } 8)q_c \text{ (units of } q_c) \text{ for clays and depends on both}$$
$$q_c \text{ and the soil}$$

The stress-strain modulus should be representative in the stressed zone beneath the footing (Fig. 5-10). It generally increases with depth in all soils owing to a decrease in e and water content, and this coupled with the decrease in Δq with depth causes the computed settlements to be conservative more often than not. This would not be serious if the amount conservative was small, but it is often conservative by a factor of 2 or 3. At one time it was thought $E_s = $ constant for normally consolidated clays; however, if the samples are carefully recovered, it is found E_s generally increase with depth.

There is considerable evidence that the modulus of elasticity determined by the unconfined-compression test on soils with cohesion may be in error by a considerable amount. Liepens (1957) compared results of field settlements on several buildings in the Boston area with unconfined-compression tests and found that the field E_s was four to five times the laboratory E_s. Crawford and Burn (1962) compared data in Canada and found for unconfined-compression data that the field E_s was four to thirteen times the laboratory value, and for triaxial tests (unconsolidated-undrained) the field E_s was 1 to 1.5 times the laboratory value. In cohesionless soils E_s is underestimated owing to impossible recovery of undisturbed samples. Cementation of the grains, bedding planes, in situ stress duplication, etc., tend to reduce E_s from laboratory tests. In situ testing will overcome many of these problems.

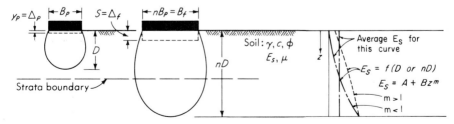

Figure 5-10. Influence of footing size on the depth of the stress zone and E_s. Note that, with an underlying stratum of different soil, the plate settlement does not reflect stresses in this material; thus, the settlement of the full-size footing can be seriously underestimated.

Figure 5-10 illustrates why plate-load tests may be misleading as to the supporting capacity of soil. First the depth of significant stress penetration may not enclose underlying strata, and secondly, the increase in stress-strain modulus is not identified—especially where 1- or 2-ft-square plates are used to estimate capacity of 6- to 10-ft footings.

Size effects have some influence in settlement studies as for bearing capacity. Terzaghi and Peck (1967, p. 489) in the 1948 edition proposed, from the work of Koegler and others in Berlin [see curves in Bond (1961)] that for extrapolating a load test from a 1-ft-square plate to a footing of size B, the following equation [note the inverse of this in Eq. (a) of Sec. 4-9]:

$$S = S_1 \left(\frac{2B}{B + 1} \right)^2 \tag{5-15}$$

where S, S_1 = settlements for footing of width B and for the 1-ft plate, respectively. This curve provides a best fit for Koegler's tests on 3- to 50-cm plates on dense sand. The fit is not very good for extrapolating large B/B_p ratios (say, greater than 4 or 5) and may either over- or underestimate the settlements considerably. An extensive testing program by D'Appolonia et al. (1968) compared 12-ft footings with 1-ft plates and found the actual settlements were about three times the predicted values. This equation will greatly underestimate settlements on loose sands; however, few footings are placed on loose sand—it is densified prior to use.

Bond (1961) proposed that one could estimate settlement as

$$\frac{S_a/B_a}{S_p/B_p} = \left(\frac{B_a}{B_p} \right)^n \tag{5-16}$$

where S_a, S_p = settlement of footing (desired quantity) and plate
    (measured), respectively
    B_a, B_p = full-size footing and load plate dimension, respectively
    n = coefficient depending on soil, with the following
    values in the absence of test values:
    For clay: 0.03 to 0.05
    For sandy clay: 0.08 to 0.10
    For dense sand: 0.40 to 0.50
    For medium dense sand: 0.25 to 0.35
    For loose sand: 0.20 to 0.25

Bond tested the above equation against a large number of plate-load tests, and the equation provides a reasonably good fit. It should be evident that one can determine the exponent by performing two or more plate-load tests and solving Eq. (5-16) for n.

Example 5-1. For the 6 × 6-ft-square footing shown in the accompanying figure (Fig. E5-1), plot the intensity of pressure with depth, (a) at the center; and (b) at the corner. Find the average Δp for the clay stratum beneath the center of the footing. Assume that the 120-kip load is the net increase on the soil.

Table E5-1A

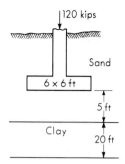

Center of footing			Corner of footing		
Elev.	D/B	Δq	Scale	M	$MI = \Delta q$
-5	0.83	$0.42q_0$	$1'' = 5'$	38.5	$0.193q_0$
-10	1.67	$0.16q_0$	$1'' = 10'$	21	$0.105q_0$
-15	2.50	$0.08q_0$	$1'' = 15'$	12	$0.06q_0$
-20	3.3	$0.05q_0$	$1'' = 20'$	8	$0.04q_0$
-25	4.2	$0.025q_0$	$1'' = 25'$	5.5	$0.02q_0$

Footing plan for Fig. 5-3 (x placed at center of influence chart).

Figure E5-1a

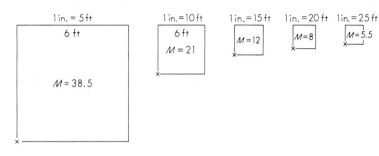

Figure E5-1b

SOLUTION. We will ignore footing depth effects and treat the stratum as isotropic and homogeneous. To treat as a layered mass would require E_s and μ of both the clay and sand.

For the center of the footing we will set up Table E5-1A and use the pressure-bulb chart of Fig. 5-4 to obtain the data shown.

For the corner of the footing it will be necessary to draw several footings to the scale $D = \overline{AB}$ of Fig. 5-3, place the scale drawing with the corner on point O, and count the squares enclosed in the scale footing M.

Table E5-1B is obtained using the 2 : 1 method which gives the same pressure intensity at all points beneath the footing.

Table E5-1B

For 2 : 1 method	Elev.	Δq
$q_0 = \dfrac{120}{36} = q_0$	0	q_0
$q_{-5} = \dfrac{36q_0}{(6+5)^2} = 0.297q_0$	-5	$0.297q_0$
	-10	$0.141q_0$
$q_{-10} = \dfrac{36q_0}{(6+10)^2} = 0.141q_0$	-15	$0.082q_0$
	-20	$0.053q_0$
	-25	$0.037q_0$

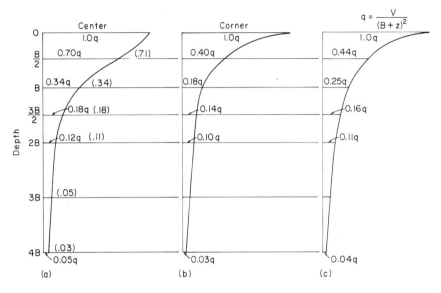

Figure E5-1c

Figure E5-1c is a plot of the data with depth. Also shown is a numerical integration (computer) of Eq. (5-2) for the center using 36 unit areas of 1 sq ft.

The average increase in pressure in the clay stratum is obtained from Eq. (5-18) of Sec. 5-8.

$$\Delta qH = 5\left(\frac{0.42 + 0.025}{2} + 0.16 + 0.08 + 0.05\right)q_0 = 2.56q_0$$

$$\Delta q = 2.56q_0/20 = 0.128q_0$$

For Fig. E5-1c:
  (a) Boussinesq theory for center of square footing (q) = computer numerical integration.
  (b) Boussinesq theory for corner of square footing.
  (c) Approximation using 2 : 1 method.
Note that stresses computed by this method do not depend on the location, i.e., center, edge, corner, etc.

Example 5-2. A footing is as shown in Fig. E5-2.

$$E_s = 340 \text{ ksf} \qquad \mu = 0.3$$

REQUIRED. Estimate the average settlement
(a) For a footing on the ground surface without depth effects
(b) For footing as shown
(c) Corner of footing

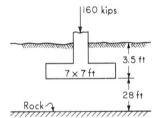

Figure E5-2

SOLUTION. Assume flexible footing ($I_w = 0.95$ from Table 5-4).

(a)
$$S = qB\frac{1 - \mu^2}{E_s}I_w = 3.27(7)\frac{1 - 0.3^2}{250}(0.95)$$

$$= 0.079 \text{ ft} \qquad \text{say, 1 in}$$

(b) Computing the settlement in two parts for center $B = L = \frac{7}{2}$ for F_1 and F_2

$$S = qB\frac{1 - \mu^2}{E_s}\left(F_1 + \frac{1 - 2\mu}{1 - \mu}F_2\right)$$

$$L/B = 1 \qquad H/B = \frac{28}{3.5} = 8$$

From a computer output $F_1 = 0.482$, $F_2 = 0.0196$ and substituting and multiplying by 2, since these are corner values, and using B, not $B/2$,

$$S = 3.27(7)\frac{1 - 0.3^2}{250}\ 2\left[0.482 + \frac{1 - 2(0.3)}{1 - 0.3}(0.0196)\right] = 0.082\ \text{ft}$$

The footing depth-effect factor F_3 for $D/B = 4.0$ and $\mu = 0.3$ is approximately 0.57 from Fig. 5-8, and

$$S = 0.082(0.57) = 0.047\ \text{ft}$$

(c) For corner settlement, $B = L = 7$ and

$$L/B = 1 \qquad H/B = \frac{28}{7} = 4$$

$$S = 3.27(7)\frac{1 - 0.3^2}{250}\left[0.408 - \frac{1 - 0.6}{1 - 0.3}(0.0375)\right] = 0.036\ \text{ft}$$

Note that S is not doubled since only one corner contribution is made whereas in (b) two corner contributions are made (actually four, but using B instead of $2B/2$ allows for remainder). For off-center or irregular shapes the foundation is gridded through the point desired and "corner" computations made for each abutting corner and summed to obtain total settlement, S.

Note that considering a finite instead of infinite stratum thickness and the depth of the footing reduces the settlement approximately 48 percent.

5-7. ALTERNATIVE METHODS OF COMPUTING ELASTIC SETTLEMENTS

From the preceding section it was stated that the settlement can be computed as

$$S = \int_0^L \varepsilon\ dL = \frac{\Delta qL}{E_s}$$

Thus, if one can evaluate the strain and apply an integration process over the depth of stress influence (approximately $2B$ to $2.5B$ as illustrated in Fig. 5-2) or the stratum depth, whichever applies, the settlement can be computed as

$$S = (L)\varepsilon_{\text{average}}$$

Alternatively, one may accomplish the same objective by integration of the stress (or stress difference) over the depth of stress influence to find the average value of stress on soil length L. A similar process must be applied to find the average value of stress-strain modulus E_s. With these two values known, the settlement can be computed as

$$S = (L)\left(\frac{\Delta q}{E_s}\right)_{\text{average}}$$

In both these computations it will be necessary to adjust the computed settlement for the effect of footing embedment in the soil mass. This may be done by making allowance for embedment depth when computing either ε or Δq or treating the footing as if it is on the surface of the elastic half-space and

$$S_{\text{depth}} = (F_3)S_{\text{surface}}$$

where F_3 is obtained from Fig. 5-8.

The average strain can be obtained using approximately the method proposed by Lambe (1964) using the stress-strain plot from a triaxial test with the cell pressure at the estimated in situ value (K_0 condition). Now one may assume that the field value of deviator stress $\Delta\sigma_1$ or the vertical stress Δq resulting from the imposed foundation loads causes similar strains as in the laboratory test. The use of either stress value to obtain the soil strain tends to give the same order of magnitude of settlement; however, the use of the deviator stress may be more realistic and is generally recommended.

These two methods of computing elastic settlements will be illustrated in the following example.

Example 5-3. Compute the immediate elastic settlement for the soil-footing system shown in E5-3a.

PRELIMINARY WORK. A series of triaxial (or direct-shear) tests must be run to establish ϕ. With ϕ the K_0 soil pressure can be computed so that the triaxial tests are performed at that value of cell pressure σ_3. Plot the initial part of the stress-strain curve to a large scale as shown in E5-3b. For cyclic tests plot the last cycle and shift the ordinate so the curve passes through the origin. For this example take:

$$\phi = 35° \qquad \gamma_1 = 17.3 \qquad \gamma_2 = 19.1 \text{ kN/m}^3$$

$K_0 = 1 - \sin 35° = 0.426$ (use single value of ϕ even though it has been previously shown that ϕ varies with soil density)

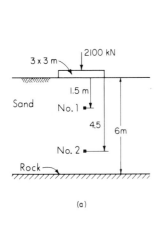

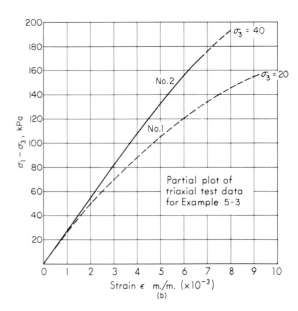

Figure E5-3

Test 1: $p_0 = 2(17.3) = 34.6$ kPa $\qquad \sigma_3 = 0.426(34.6) = 14.7$ kPa
Use cell pressure $= 20$ kPa (approx. 3 psi)

Test 2: $p_0 = 3(17.3) + 1.5(19.1) = 80.6$ kPa (estimating density)
Use cell pressure $= 40$ kPa

It is not a simple matter to test reliably at very low cell pressures. It is not generally easy to build sand samples to specific densities.

REQUIRED. Estimate footing settlement using:
(a) Strain integration.
(b) $S = \Delta\sigma_1 L/E_s$. Use a secant modulus of elasticity passing through the origin and stress point. The tangent modulus is more correct but is considerably more laborious to compute.

SOLUTION. Divide the 6-m stratum into four increments and make the following table. Obtain q/q_0 from Fig. 5-2; $q_h = q_v K_0$; ε from stress-strain plot at $\Delta\sigma_1$; and $\Delta\sigma_1 = q_v - q_h$.

$$q_0 = 2,100/9 = 233.3 \text{ kPa}$$

At D/B 0.0; $\Delta\sigma_1 = q_v(1 - K_0) = 233.3(1 - 0.426) = 133.9$ kPa
From the stress-strain plot (curve 1) $\varepsilon_1 = 7 \times 10^{-3}$
The corresponding secant modulus $E_s = 133.9/0.007 = 19,130$ kPa

Curve	D	D/B	q/q_0	q_v	$\Delta\sigma_1$	$\varepsilon \times 10^{-3}$	$E_s \times 10^3$ (kPa)
1	0	0	1	233.3	133.9	7.0	19.13
1	1.5	0.5	0.7	163.3	93.7	4.6	20.4
1	3.0	1.0	0.33	77.0	44.1	1.8	24.5
2	4.5	1.5	0.19	44.0	25.3	1.0	25.3
2	6.0	2.0	0.12	28.0	16.1	0.6	26.8

Using the strain-integration method the settlement is computed using numerical integration as

$$S = \int_0^6 \varepsilon \, dL = 1.5\left(\frac{7 + 0.6}{2} + 4.6 + 1.8 + 1.0\right)10^{-3} = 16.8 \times 10^{-3} \text{ m}$$

$$S = 1.68 \text{ cm}$$

Using the modulus-of-elasticity concept the settlement is obtained as follows:

$$E_s = \frac{1.5}{6}\left(\frac{19.13 + 26.8}{2} + 20.4 + 24.5 + 25.3\right)10^3 = 23.29 \times 10^3 \text{ kPa}$$

A similar computation for $\Delta\sigma_1$ gives 59.525

$$S = \frac{\Delta\sigma L}{E_s} = \frac{(59.525)(6)}{23,290} = 15.3 \times 10^{-3} \text{ m} = 1.53 \text{ cm}$$

This small discrepancy between the two methods is principally due to using the secant instead of the tangent modulus of elasticity.
Using Eq. (5-13) to compute the settlement with $\mu = 0.3$ and $F_3 = 1$ for a surface footing gives

$$S = 233.3(3)\left(\frac{1 - 0.3^2}{23,290}\right)0.285 = 0.009 \text{ m} = 0.9 \text{ cm}$$

5-8. STRESSES AND DISPLACEMENTS IN LAYERED AND ANISOTROPIC SOILS

There are numerous elastic solutions for special cases of stresses and displacements in layered or anisotropic soils. Special cases are sometimes useful to obtain an indication of probable (or possible) magnitude of error from using an idealized soil mass (isotropic, homogeneous, etc.). Generally, the special cases in the literature [Poulos and Davis (1974) summarizes a large amount of curves, charts, and tables] are not found in nature, or by the time one makes the necessary interpolations from curves and tables one could solve his own problem.

   The author proposes that one of the best uses of the finite-element method is to solve this type of problem. The computer program is available in Chap. 20 of Zienkiewicz (1971). One solves this type of problem as follows:

1. Model a reasonable size of half-space, once for all, and use a card generator to develop the cards to define the x, y coordinates of the nodes and the node numbers defining each element and the soil for that element. The model should have provision for about five different layers of soil (for fewer layers one simply uses the same soil properties for more than one layer).
2. Solve the problem for a point load at one node where the footing is placed. This is either in the ground or at the ground surface depending on if it is desired to obtain depth effects. For all layers of the same soil identify all soils in the model with the same properties.
3. Re-solve the problem with the point load at the same location but with the correct soil stratification.
4. From the Boussinesq pressure bulbs obtain the stress at the desired point beneath the footing (now we are incorporating the shape and three-dimensional effect of the load into the problem).
5. From steps 2 and 3 above find the point-load stress at the same point as obtained in step 4.
6. Compute the stress due to stratification as a proportion to obtain

$$q_{fL} = q_b \left(\frac{q_L}{q_h} \right) \tag{5-17}$$

   where q_b = Boussinesq value for a footing of same dimension and applicable corrections for depth, etc.
   q_{fL} = stresses due to footing in layered soil at the depth of interest
   q_L, q_h = stresses from the FEM solutions for the layered and homogeneous cases, respectively

This solution is at least as good as the soil parameters E_s and μ used in the FEM. This method allows using a two-dimensional plane-strain or plane-stress solution and avoids the tremendous computer core capacity for a three-dimensional solution. Deflections can be computed in an analogous manner.

5-9. CONSOLIDATION SETTLEMENTS

The settlements of fine-grained, saturated cohesive soils will be time-dependent, and consolidation theory is usually used, although elastic methods can be, and sometimes are, used. Equation (2-30) is usually used for consolidation settlements. If Eq. (2-27) is used,

$$S = m_v \, \Delta p \, H \qquad (2\text{-}27)$$

but m_v = units of $1/E_s$; therefore, when we use that equation we have the form of $S = \sigma L/E = \Delta p(H)/E_s$ which is the elastic method used in mechanics of materials. Some authorities use $1/m_v$ as a routine method of evaluating the stress-strain modulus using the consolidation test. With a sample on the order of 2 to 2.5 cm thick, however, the results may not be very representative.

Equation (2-27) is applicable for normally consolidated soils and for preconsolidated soils or for that part of the applied pressure which added to p_0 exceeds the preconsolidation pressure. When the soil is preconsolidated, and $p_0 + \Delta p$ does not equal the preconsolidation pressure p_c as determined either by the Casagrande method or inspection of the e log p curve, one should use either C_r from Eq. (2-29a) for C_c and/or p_c in Eq. (2-27) as in Example 5-4.

In addition to inspection of the e log p curve for the preconsolidation pressure, an indication of preconsolidation in a clay is that if the natural moisture content is closer to the liquid limit, the clay is probably normally consolidated. If the natural water content is closer to the plastic limit, the soil is preconsolidated. If the natural water content is higher than the liquid limit, the soil will be highly sensitive. This use of the natural water content is based on the fact that the void ratio will be less for a preconsolidated than for a normally consolidated material.

The average increase in pressure Δp in the stratum of interest is determined by one of the methods presented (2 : 1, Boussinesq, FEM, etc.). The average void ratio can be computed by rearranging Eq. (2-30) to obtain

$$e = e_0 - C_c \, \log \frac{p_0 + \Delta p'}{p_0}$$

where e_0 is the void ratio at the sample depth, p_0 the effective overburden pressure at sample location, and $\Delta p'$ the effective pressure due to increase in soil weight above the selected depth increment.

When the stratum is over 1.5 to 2 m thick, it should be broken up into several increments and the individual settlements obtained and summed to obtain the total settlement. This usually involves considerable computation, and the preferred method is to refine the average stratum stress increase, which can be done as follows:

1. For clay at a depth of B or more beneath the footing, use the 2 : 1 method and integration to obtain

$$H(\Delta p) = \int_{H_1}^{H_2} \frac{V}{(B+z)^2} \, dz$$

2. For other cases or when Boussinesq or FEM method is preferred, obtain the increase in pressure using Fig. 5-4 at several locations in the clay stratum including top, bottom, and one or more interior locations with ΔH = constant. Use the trapezoidal-rule formula to integrate numerically the area

$$A = H(\Delta p) = \Delta H \left(\frac{p_1 + p_n}{2} + p_2 + p_3 + \cdots + p_{n-1} \right) \qquad (5\text{-}18)$$

One may question the validity of using the Boussinesq method based on an isotropic, homogeneous mass when we are now considering soils overlying consolidating clay layers. While the method is certainly not exact, unless there is a significant difference say, by a factor of five times or more in the stress-strain modulus of the two materials more refined computations will improve the computed stress increase very little [see Morgan and Gerrard (1971)].

Example 5-4. Given the consolidation test, soil profile, and other data shown in the accompanying sketches.

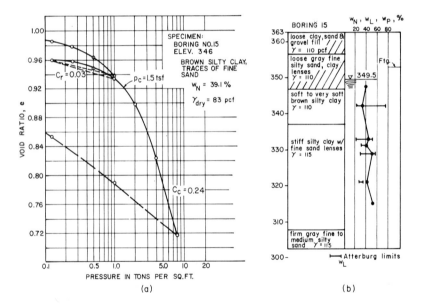

Figure E5-4

REQUIRED. Estimate the settlement of a 8 × 8 ft footing carrying 375 kips at elevation 353 on the "soft to very soft brown silty clay" (elevation 347 to 337).

SOLUTION. Note that p_c is estimated from the reload shape of the e log p curve of the second reload cycle, since it did not become linear at 1 tsf. The Casagrande method is not used, but if it were p_c would be approximately the same. Compute C_r as the slope of the rebound curve

$$C_r = \frac{\Delta e}{\log p_2/p_1} = \frac{0.960 - 0.937}{\log 1/0.14} = \frac{0.023}{0.854} = 0.027 \qquad \text{say, } 0.03$$

Compute C_c as the slope of the curve beyond p_c

$$C_c = \frac{0.821 - 0.719}{\log 8/4} = \frac{0.102}{0.426} = 0.24$$

from Eq. (2-31)

$$C_c = 0.009(78 - 10) = 0.612 \qquad \text{(differs by about 255 percent but clay is overconsolidated)}$$

Compute the average stress increase

(a) Using the 2 : 1 method with the footing at elevation 353; the depth to the top of the clay layer is $353 - 347 = 6$ ft; to the bottom $= 353 - 337 = 16$ ft.

$$\Delta pH = \int_6^{16} \frac{375}{(8 + z)^2} \, dz = -\frac{375}{8 + z} \Big]_6^{16}$$

$$\Delta p = \frac{1}{10}\left(-\frac{375}{24} + \frac{375}{14}\right) = 1.12 \text{ ksf}$$

(b) Using the Boussinesq pressure bulbs (Fig. 5-4)

Elev.	D/B	$\Delta q/q_0$	Δq
-6	0.75	0.5	2.93
8.5	1.06	0.33	1.93
11.0	1.375	0.23	1.35
13.5	1.68	0.16	0.94
-16	2.0	0.12	0.70

$$\Delta qH = 2.5\left(\frac{2.93 + 0.70}{2} + 1.93 + 1.35 + 0.94\right)$$

$$\Delta qH = 15.09$$

$$\Delta q = 15.09/10 = 1.51 \text{ ksf}$$

Computing the settlement

p_0 = in situ effective stress at center of stratum (elevation 342)

$$p_0 = 0.110(363 - 349.5) \qquad\qquad = 1.485 \text{ ksf} \qquad \text{(top down to W.T., not}$$
$$+ (0.110 - 0.625)(349.5 - 342) = 0.356 \qquad\qquad \text{from footing down)}$$
$$\overline{1.841 \;\text{ksf}}$$

$$p_c = 1.5 \text{ tsf} = 3 \text{ ksf} \qquad \text{(Note also } w_N \text{ closer to } w_p \text{ than to } w_L)$$

$$p_0 + \Delta p = 1.84 + 1.51 = 3.35$$

$$p_{\text{top}} = 1.84 - 5(0.110 - 0.625) = 1.60 \text{ at top of clay layer (approx. sample location)}$$

The average e is often difficult to estimate; we compute it approximately as

$$e = 0.955 - 0.47 \log \frac{1.84}{1.60} = 0.955 - 0.03 = 0.925$$

which we could have estimated in this case reasonably well because of p_c. Therefore, the settlement is made of two parts: Δp of $3.00 - 1.84 = 1.16$ ksf with C_r and $\Delta p = 1.51 - 1.16 = 0.35$ ksf with C_c. Up to p_c the settlement is

$$S_1 = \frac{0.030(10)}{1 + 0.925} \log \frac{1.84 + 1.16}{1.84} = 0.03 \text{ ft}$$

and from p_c on the settlement is (and no adjustment in e_0)

$$S_2 = \frac{0.24(10)}{1 + 0.925} \log \frac{3.0 + 0.35}{3.0} = 0.06 \text{ ft}$$

$$S = S_1 + S_2 = 0.03 + 0.06 = 0.09 \text{ ft or approximately } 1.1 \text{ in}$$

5-10. RELIABILITY OF SETTLEMENT COMPUTATIONS

Settlements in general are made up of immediate settlements S_i, consolidation settlement S_c, and secondary compression (also called consolidation or creep) S_s (see Sec. 6-3 for equation), or in general

$$S = S_i + S_c + S_s$$

In cohesionless soils, the immediate-settlement term would predominate, with possibly some creep effects. As a matter of fact it appears that settlement predictions on cohesionless soils are usually conservative with current methods. This is probably due to the high degree of conservatism in the empirical method whereby the prediction is made, and due to the fact that the undisturbed sand is considerably stiffer than the disturbed laboratory specimen. It is also highly likely that most of the settlements occur as the building is constructed owing to the combination of construction vibrations and loads and are largely built out of the structure during floor and column alignment.

In other soils the settlements are probably made up of all three settlement effects. In inorganic clays and silts the consolidation settlement probably predominates. In highly organic deposits the secondary-compression effects are likely to predominate, and in these soils it may be necessary to run the test longer to obtain a secondary-compression index [see Wahls (1962), Barden (1968)].

Consolidation-settlement predictions are reasonably reliable if care is taken in obtaining "undisturbed" samples and in testing and interpreting the data. Studies by MacDonald and Skempton (1955) and Skempton et al. (1955) indicate that on the average the settlement prediction is valid. Settlement rates are not nearly as well predicted. This is probably due to the fact that in the laboratory a small amount of sample compression results in a large change in the coefficient of permeability, whereas in the field the change is negligible. One of the basic assumptions in developing consolidation theory is that $k = \text{constant}$.

5-11. PROPORTIONING FOOTINGS FOR A GIVEN SETTLEMENT OR EQUAL SETTLEMENTS

The allowable pressure from penetration testing in Sec. 4-9 was for a 1-in settlement. Example 4-7 illustrated modification of the bearing capacity for a larger allowable settlement. Settlements from Eq. (5-10) are in direct proportion to the contact pressure. One can estimate the settlement ratio of two footings B_1 and B_2 using Eq. (5-10) to obtain

$$\frac{S_1}{S_2} = \frac{(qB)_1}{(qB)_2}$$

and (a) if the contact pressures are equal, $S_1/S_2 = B_1/B_2$; (b) if the settlements are equal, $q_1 B_1 = q_2 B_2$.

The problem with this is that the effect of the larger footing is to a greater depth in the ground with the resulting increase in E_s and change in μ. If these factors were to exactly compensate, then this method would be exact. In general, the factors do not compensate and the scaling effect from using this method becomes more of an estimate.

Proportioning footings for a given or equal settlement on consolidating clay deposits becomes a trial procedure, as illustrated in the following example.

Example 5-5. Proportion a footing such that the consolidation settlement is not over 1.5 in for the given conditions.

Table E5-5

	B = 8		B = 16		B = 24	
D	D/B	q/q_0	D/B	q/q_0	D/B	q/q_0
−10	1.25	0.25	0.625	0.6	0.417	0.77
−15	1.87	0.13	0.936	0.4	0.625	0.60
−20	2.5	0.08	1.25	0.25	0.833	0.40
−25	3.12	0.06	1.56	0.17	1.04	0.34

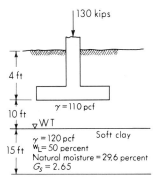

Figure E5-5(a)

SOLUTION. Assume that the net increase in soil pressure due to the concrete displacement of the soil is negligible. Since the settlement depends on the contact pressure and footing size and is nonlinear, several trials will be required, and it will be most convenient to use the average stress increase in the stratum Δp. The results will be plotted of S versus B to find the required footing size.

$$p_0 = 14(0.110) + 7.5(0.120 - 0.0625) = 1.97 \text{ ksf}$$

Take $C_c = 0.009(w_L - 10) = 0.009(50 - 10) = 0.36$

$$e_0 = wG_s = \frac{29.6}{100}(2.65) = 0.784 \qquad \text{assuming } S = 100\%$$

$$S = \frac{C_c H}{1 + e_0} \log \frac{p_0 + \Delta p}{p_0} = \frac{0.36(15)}{1.784} \log \frac{p_0 + \Delta p}{p_0} = 3.03 \log \frac{1.97 + \Delta p}{1.97}$$

Use the Boussinesq method (Fig. 5-4, and obtain data in the above table.
Computing the average stress Δp by the trapezoidal rule [Eq. (5-18)],

$$\Delta p = \frac{1}{15}\left(\frac{15}{3}\right)\left(\frac{0.25 + 0.06}{2} + 0.13 + 0.08\right) = \frac{0.36}{3} = 0.12q_0 \qquad \text{for } 8 \times 8 \text{ ft}$$

$$= \frac{1}{3}\left(\frac{0.6 + 0.17}{2} + 0.4 + 0.25\right) = 0.35q_0 \qquad \text{for } 16 \times 16 \text{ ft}$$

$$= \frac{1}{3}\left(\frac{0.77 + 0.34}{2} + 0.6 + 0.40\right) = 0.52q_0 \qquad B = 24 \times 24 \text{ ft}$$

$$q_8 = 0.12\left(\frac{130}{64}\right) = 0.243 \text{ ksf} \qquad q_{16} = 0.35\left(\frac{130}{256}\right) = 0.178 \text{ ksf}$$

$$q_{24} = 0.52\left(\frac{130}{576}\right) = 0.117 \text{ ksf}$$

$$S_{8ft} = 3.03 \log \frac{1.97 + 0.243}{1.97} = 3.03(0.05) = 0.1515 \text{ ft} = 1.82 \text{ in}$$

$$S_{16ft} = 3.03 \log \frac{1.97 + 0.178}{1.97} = 3.03(0.037) = 0.112 \text{ ft} = 1.35 \text{ in}$$

$$S_{24ft} = 3.03 \log \frac{1.97 + 0.117}{1.97} = 3.03(0.026) = 0.0788 \text{ ft} = 0.95 \text{ in}$$

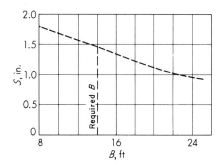

Figure 5-5(b)

From the plot the required width of footing is $B \cong 14^-$, say, 14 ft.
While it may appear that $B = 10, 16, 20$ would be better values to try, initially this is not known; thus the three values shown are satisfactory.

5-12. STRUCTURES ON FILLS

It is often advantageous, and sometimes necessary, to place the structure or parts of it on filled-in areas. These sites may be sanitary land fills, rubble dumps from old torn-down buildings, or fills constructed according to engineering criteria. In the situations where sanitary fills or rubble dumps are used, it is doubtful if a structure can be placed on this material and not undergo detrimental settlement unless the fill has had time to decompose and fully consolidate. For most cases of foundations on fills the loads will have to be carried through the fill material utilizing piles or caissons of a noncorrosive material (usually concrete or treated wood).

A well-constructed earth fill, using quality control with regard to both material and compaction, often produces a better foundation base than the original material underlying the fill. Many persons have been reluctant to place a footing on or in fills because of two main factors:

1. Unpleasant results from placing footings on poorly placed fills. With no quality control it is not unusual to get a fill with a hard crust over several feet of loose fill as a result of compacting only the last lift, or from placing a lift too thick to be compacted with the available equipment.
2. Placing a footing in the fill with unpleasant results obtained not from the fill settlement but from settlement of the underlying soil due to the weight of both the fill and the structure.

There are precautions one must take with a fill, in addition to exercising compaction control, such as eliminating soils of large volume change, providing adequate drainage, and making sure, if construction is to proceed relatively soon after the fill is placed, that consolidation settlements have been considered. Under consolidation processes the structure and fill will subside because of just the weight of the fill, and this will take place whether the footings are placed on the natural soil or in the fill. Excessive differential settlements may also result from consolidation in the underlying soft strata if the fill varies considerably in thickness or if part of the structure is on virgin soil and part is on fill. A poorly constructed fill will also undergo settlements with time, and there is no theory available which will indicate reliably the magnitude or the length of time for the settlement to be complete.

The determination of the bearing capacity (and settlements) proceeds as with the virgin soil. If the fill is placed before exploration takes place, the usual exploration methods of Chap. 3 (standard penetration tests and tests on recovered samples) are applicable. When the field exploration has already been performed, the bearing capacity of the fill may be determined by performing laboratory tests on specimens compacted to the proposed in situ density. Building-code values, coupled with successful experience on soils of similar properties and density, may also be used as a guide.

5-13. STRUCTURAL TOLERANCE TO SETTLEMENT AND DIFFERENTIAL SETTLEMENTS

Theoretical settlements can be computed for various points such as a corner center or beneath lightest and heaviest loaded footings to obtain the total settlement and the differential settlement between adjacent points. If the entire structure moves vertically some amount or rotates as a plane rigid body, this will not generally cause structural or architectural distress. For example, if a structure settles 2 cm on one side and 10 cm on the other and undergoes a linear settlement between the two points, except for aesthetic and public-confidence considerations structural damage is not likely to be developed. The building will have settled 2 cm and tilted an amount $\xi = (10 - 2)/L$. Local settlements below the tilt line between the two sides of the structure will cause any building distress. These local settlements below either the settlement or tilt line are the differential settlements which the foundation designer must control, since they will determine the acceptability of the structure. The total settlements (some of which occur during construction) can be landscaped into concealment when the building is completed or later. A cracked wall or warped roof is much more difficult to conceal.

Differential settlement can be computed as the difference in settlement between two adjacent points. It may be estimated as $\frac{3}{4}$ of the computed maximum total settlement; i.e., maximum total settlement = 4 cm; expected differential settlement, $\Delta s = 3/4(4) = 3$ cm.

MacDonald and Skempton (1955) made a study of 98 buildings, mostly older buildings of load-bearing wall, steel, and reinforced-concrete construction to provide the data of Table 5-5. This study was recently substantiated by Grant et al. (1974) from a study of 95 additional buildings of more recent construction (some were constructed after 1950). Feld (1965) cites a rather large number of specific structures with given amounts of settlement and structural response which might be of interest in considering a specific problem. Combining all sources, one can conclude that:

Table 5-5. Tolerable differential settlements of buildings, in inches*, recommended maximum values in parentheses

Criterion	Isolated foundations	Rafts
Angular distortion (cracking)	1/300	
Greatest differential settlement:		
Clays	$1\frac{3}{4}(1\frac{1}{2})$	
Sands	$1\frac{1}{4}(1)$	
Maximum settlement:		
Clays	$3(2\frac{1}{2})$	$3-5(2\frac{1}{2}-4)$
Sands	$2(1\frac{1}{2})$	$2-3(1\frac{1}{2}-2\frac{1}{2})$

* MacDonald and Skempton (1955).

1. The values in Table 5-5 should be adequate most of the time. The values in brackets being recommended for design, others are the range of settlements found for satisfactory structural performance.
2. One must carefully look at the differential movement between two adjacent points in assessing what constitutes an acceptable slope.
3. Residual stresses in the structure may be important, as some structures can tolerate much larger differential settlements than others.
4. Construction material—steel, being more ductile, can tolerate larger movements than either concrete or load-bearing walls.
5. Time interval during which settlement occurs—long time spans allow the structure to adjust and better resist differential movement.

If computed differential settlements are kept within the values in parentheses in Table 5-5, statistically the structure should adequately resist that deformation. Values of acceptable slopes between two adjacent points from the U.S.S.R. building code as presented by Polshin and Tokar (1957) and Mikhejev et al. (1961) are in Table 5-6.

Table 5-6. Permissive differential building slopes by the U.S.S.R. code on both unfrozen and frozen ground. All values to be multiplied by L = length between two adjacent points under consideration. H = height of foundation above foundation*

Structure	On sand or hard clay	On plastic clay	Average max. settlement, cm	
Crane runway	0.003	0.003		
Steel and concrete frames	0.0010	0.0013	10	
End rows of brick-clad frame	0.0007	0.001	15	
Where strain does occur	0.005	0.005		
Multistory brick wall			8	$L/H \geq 2.5$
L/H to 3	0.003	0.004	10	$L/H \leq 1.5$
Multistory brick wall				
L/H over 5	0.005	0.007		
One-story mill buildings	0.001	0.001		
Smokestacks, water towers,				
Ring foundations	0.004	0.004	30	
Structures on Permafrost				
Reinforced concrete	0.002–0.0015		15 at 4 cm/year†	
Masonry, precast concrete	0.003–0.002		20 at 6 cm/year	
Steel frames	0.004–0.0025		25 at 8 cm/year	
Timber	0.007–0.005		40 at 12 cm/year	

† not to exceed this rate per year.
* From Mikhejev et al. (1961) and Polshin and Tokar (1957).

PROBLEMS

Problems 5-1 to 5-7 are to be assigned by the instructor from the following table by key number, which provides the strata thickness of the soil profile given in Fig. P5-1, in feet or meters.
$B = 10 \times 10$ or 3×3 m

Key number	z_s	z_c	x	y
1	5	5	5	0
2*	2.5	2	2.5	0
3	15	5	10	5
4*	5	6	3	2
5	15	10	10	5
6*	4	3	3	1
7	10	15	6	4
8*	5	4.5	2	3
9	2	18	2	0
10*	1	6	1	1

* dimensions in meters.

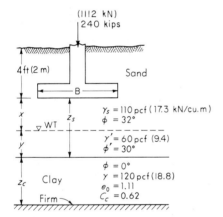

(1112 kN)
240 kips

4 ft (2 m)

Sand

B

$\gamma_s = 110$ pcf (17.3 kN/cu.m)
$\phi = 32°$

WT

$\gamma' = 60$ pcf (9.4)
$\phi' = 30°$

$\phi = 0°$
$\gamma = 120$ pcf (18.8)
$e_0 = 1.11$
$C_c = 0.62$

z_s

x

y

z_c Clay

Firm

Figure P5-1

5-1. Referring to Fig. P5-1, compute the average increase in stress for the clay strata (Δp) for the assigned key number from the table of strata thickness by (a) the Boussinesq method; (b) the Westergaard method; (c) the 2 : 1 slope method.

   Answer: 5-1(1): (a) 1.21; (b) 0.76; (c) 0.80 (ksf).
   5-1(10): (a) 38.9; (b) 31.1; (c) 27.8 (kPa).

5-2. Compute the consolidation settlement from the Δp obtained in Prob. 5-1, using the unit weights and other soil data as given in Fig. P5-1. Compare the settlements obtained by the three methods of stress computation.

   Answer: 5-1(1): (a) 5.56 in; (b) 3.93 in; (c) 4.09 in.
   5-1(10): (a) 30.7 cm; (b) 25.4 cm; (c) 23.1 cm.

5-3. A foundation plan is as shown in Fig. P5-3. The soil profile corresponds to the key numbers of Prob. 5-1. For the assigned key number for the profile and the following problem condition as assigned:

   (a) Consider that all footings have the same contact pressure, $q_0 = 3$ ksf.
   (b) Footing contact pressure is as follows: $q_1 = 5$ ksf; $q_2 = 3$ ksf; $q_3 = 6$ ksf.
   (c) The area is covered by a mat foundation with $q = 2.0$ ksf.

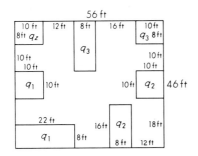

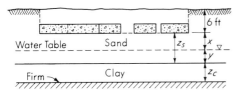

Figure P5-3

Find the footing with the maximum and minimum pressure increase in the clay stratum beneath the footing center and the magnitude of the pressure increases.

(*Note:* In all problems, assume the pressure given is the *net* increase in pressure. For metric problems use a direct conversion of dimensions and pressures to nearest 0.1 from Fig. P5-3 and above.)

5-4. Compute the approximate maximum differential consolidation settlement for the soil-profile key number used in Prob. 5-3*a*.

5-5. Compute the approximate maximum differential consolidation settlement for the soil-profile key number used in Prob. 5-3*b*.

5-6. Compute the approximate differential consolidation settlement for the soil-profile key number used for the mat foundation of Prob. 5-3*c*.

5-7. Use an average value of $E_s = 800$ ksf or 32,000 kPa as appropriate and compute the expected immediate settlement on the sane underlying the footing of Fig. P5-1 for the appropriate key number given in the soil-profile table. If the key number is the same as the analysis made in Prob. 5-2, what is the total expected settlement?

5-8. Use the load-test data of Prob. 4-14 and estimate the settlement of a 2-m-square footing with a contact pressure of 4 tons/m^2.

5-9. Estimate the settlement of a 6-ft-square footing loaded to the allowable pressure obtained from Prob. 4-11.

5-10. Estimate the settlement for 8-ft footings loaded to 4 ksf at $D = 4$ ft using the data of Prob. 2-17 (Chap. 2) for:
    (*a*) Soil no. 3
    (*b*) Soil no. 5
    (*c*) Soil no. 7
    (*d*) Total settlement of the three soils

5-11. Using the data of Prob. 2-17 and a footing 2×3 m loaded to 200 kPa at $D = 1$ m, estimate the settlement for:
    (*a*) Soil no. 3
    (*b*) Soil no. 5
    (*c*) Soil no. 7
    (*d*) Total settlement of the three soils

5-12. Two consolidated undrained triaxial tests (actual data shown) were performed on a light brown, silty clay. Footing loads are 200 to 300 kips placed 6 ft below the ground surface. The water table is 31.5 ft below the ground surface in a medium dense sand. The triaxial samples were at 16.5 ft below the ground surface. Find the allowable bearing capacity and an estimate of the settlements both total and differential. State all assumptions, and briefly discuss your results.

ε	Test No. 1 $\sigma_3 = 10$ psi $\Delta\sigma_1$, psi	Test No. 3 $\sigma_3 = 20$ psi $\Delta\sigma_1$, psi
0	0	0
0.01	3.8	2.5
0.014	5.6	5.6
0.02	13.5	13.5
0.03	19.5	19.0
0.04	20.6	21.8
0.05	24.4	28.6
0.07	26.8	32.1
0.09	29.8	33.8
0.12	34.1	33.0
0.14	34.6	35.6
0.16	34.9	37.6
0.19	38.4	35.4
0.21	38.6	33.1

5-13. Write a computer program to solve for values of F_1 and F_2 of Eq. (5-13) and plot the factors for D/B vs. factors for various L/B ratios.

IMPROVING SITE SOILS FOR FOUNDATION USE

6-1. INTRODUCTION

It is evident from soil-mechanics considerations as well as the bearing-capacity and settlement considerations that significant increase in bearing capacity may be obtained by altering the properties of the in situ soil mass by some means. Various means are currently available by which this may be achieved such as:

Compaction—usually the cheapest
Preloading—primarily to reduce future settlement
Drainage—primarily to speed up settlements under preloading, but also may increase shear strength
Vibroflotation-type densification using a vibroflot or vibrating piles
Grouting
Chemical stabilization

These several methods of soil improvement will be taken up in some additional detail in the following sections.

6-2. COMPACTION

This is usually the cheapest method of improving site soils. It may be accomplished by excavating some depth, then carefully backfilling in controlled lift thickness and compacting the soil with the appropriate compaction equipment. The standard compaction tests (ASTM, part 19) which may be used are:

ASTM D698-70 consists in: 5.5-lb hammer
12-in drop
1/30 cu ft mold
3 layers of soil at 25 blows/layer

ASTM D1557-70 consists in: 10-lb hammer
18-in drop
1/30 cu ft mold
5 layers at 25 blows/layer

Figure 6-1 presents typical compaction curves for several soils.

In building construction, the modified compaction test is not much used as there is little effective increase in soil properties for the additional compaction effort and necessary quality control. Around building foundations it is usually necessary to backfill in limited quarters, requiring the use of smaller equipment and close attention to compacting the soil so that later subsidence around column and wall footings does not form cavities beneath floor slabs and potential cracking when loads exceed the floor slab bending capacity.

Fills which will later support buildings should be placed using compaction control/criteria. With compaction control, the fill is often better quality than the underlying soil. The underlying soil will undergo settlements of varying magnitude depending on its characteristics and the depth of fill. Settlements will be nonuniform if the fill depth varies or if the site consists of both cut and fill. Settlements may be of long duration unless special steps are, or can be, taken to speed up the process such as overfill (to increase the pressure) or drainage methods to speed consolidation.

Specific details of compaction tests and methods and equipment necessary to compact various soils and methods of quality control are beyond the scope of this text. The interested reader should consult publications such as the following:

"Criteria for Compacted Fills," Building Research Advisory Board, National Academy of Sciences, Washington, D.C.
Symposium on Compaction of Earthwork and Granular Bases, *Highway Research Record* no. 177, National Academy of Sciences, Washington, D.C.
Compacted Clay; A Symposium, *Transactions ASCE*, vol. 125, 1960, pp. 681–756

These publications contain extensive references for additional study.

Compaction of cohesionless soils using vibratory rollers can be accomplished to depths of about $1\frac{1}{2}$ m, although Moorhouse and Baker (1968) report being able to obtain a depth of about 2.5 m. Where an ample supply of water is available compaction using vibratory rollers can be improved by flooding the site (saturation). It may be of little value to densify the top 1 or 2 m of a site if the underlying soil remains loose, as the potential for large settlements still exists. On the other hand this method is particularly attractive if the top 1 to 2 m of soil is all that is in a loose state.

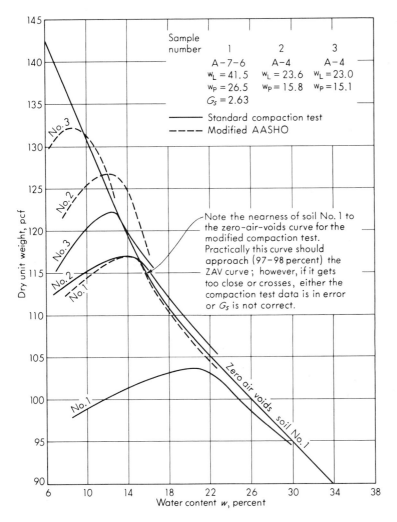

Figure 6-1. Typical compaction curves for three soils classified as indicated on the graph and by both standard and modified methods. The zero-air-voids curve is shown only for soil sample No. 1.

6-3. PRECOMPRESSION TO IMPROVE SITE SOILS

A relatively inexpensive, effective method to improve poor foundation soils in advance of construction of permanent facilities is to use preloading. The preload may consist of soil or sometimes rock, and in the case of oil or water tanks, gradual filling of the tanks may obtain preload goals. Sometimes the preload may be accomplished by lowering the groundwater table. Aldrich (1965) [see also Johnson (1970)] conducted a survey among several organizations and reported in a state-of-the-art paper several preload practices. How or what to use to accomplish preloading will be determined by relative economics.

Precompression (or preloading) is to accomplish two major goals:

1. Temporary surcharge loads are used to eliminate settlements what would otherwise occur after the structure is completed.
2. Improve the shear strength of the subsoil by increasing the density, reducing the void ratio, and decreasing the water content.

Preloading is normally used for normal to lightly overconsolidated silts, clays, and organic deposits. If the deposits are thick and do not have alternating sand seams, the preloading may necessitate using sand drains of the next section to reduce the time necessary to effect consolidation.

The amount of settlement eliminated by using preloading should be 100 percent of primary consolidation. As much secondary compression is removed as practical with the combination of eliminated settlement such that any remaining after completion of construction will be a tolerable amount. The primary consolidation can be computed by obtaining the stress increase using the Boussinesq method of Chap. 5 for several points beneath the loaded area and using Eq. (2-30). The secondary compression may be estimated from using the slope of the secondary-compression branch of the dial reading vs. log time curve (slope of curve to right of t_{100} in Fig. 2-11a), as $C_a = \Delta DR/\log t_2/t_1$. This value is used in the following equation:

$$ S_{\text{secondary}} = \frac{C_a H}{1 + e_0} \log \frac{t_{\text{total}}}{t_{\text{primary}}} $$

where H = thickness of stratum in field, t_{total} = service life, t_{primary} = time to the end of primary consolidation. The slope of the secondary-compression branch is approximately independent of load increment. The total settlement for the preload is the sum of the primary and secondary settlements.

Shear-strength tests before and after preloading will be necessary to evaluate the improvement in strength with consolidation.

Normally one would try to apply a surcharge greater than the estimated weight of the proposed structure so that postconstruction settlements are negligible. There will be some rebound and recompression as the preload is removed and the new load applied.

The principal difficulty with preloading a site is obtaining sufficient material to provide the necessary load. The preload material must be brought in, placed, and removed after settlements have occurred. This will require both a source and possibly a waste area.

An alternative form of preload was used in a housing-site development near San Francisco on bay mud [Garbe and Tsai (1972)]. The undrained shear strength was on the order of 300 to 500 psf and thus unsuitable for even the light loads of residential construction. The site was carefully backfilled so that the soft soil was not remolded by construction equipment bringing in the fill and spreading it. Fill was placed to varying depths of 2 to 6 ft. This preload accomplished an increase in soil strength, a decrease in settlement of the houses, and a spreading of the foundation loads to a reduced stress intensity on the underlying mud. In this case, the foundation loads were on the order of only 300 to 400 psf. For heavier foundation loads, the site would have been unsuitable and either pile or mat foundations would have been required.

Ponding may be used where the soil in contact with the water is of sufficiently low permeability. A perimeter dike can be constructed and the site ponded to the necessary depth to achieve the desired settlement.

6-4. DRAINAGE USING SAND BLANKETS AND DRAINS

The time for consolidation is computed from rearranging Eq. (2-24) to obtain the time as

$$t = \frac{TH^2}{c_v}$$

The dimensionless factor T depends on the percent consolidation (Table 2-5) and is about 0.848 and 0.197 for 90 and 50 percent consolidation. The coefficient of consolidation c_v is usually back-computed from a consolidation test by solving the above equation for c_v. The coefficient is also

$$c_v = \frac{k}{\gamma_w m_v}$$

where all terms have been defined in Chap. 2. For radial drainage as in sand drains, the coefficient of permeability k in the above equation would be the horizontal value, which is often four or five times as large as the vertical value.

The theory of radial drainage into sand drains, including allowance for "smear" effects on the sides of the holes reducing inflow, has been presented by Barron (1948) and more recently by Richart (1959). Since one is fortunate to determine the order of magnitude of k (the exponent of 10), for practical purposes the time for consolidation of a layer can be computed as:

1. Take $H = \frac{1}{2}$ distance between sand drains.
2. Compute c_v using the horizontal coefficient of permeability.
3. Use T from Table 2-5 for appropriate percent consolidation.

The time will be in some error owing to vertical drainage within the consolidating layer, depending on whether thin sand seams are present, drainage is from one or both faces, how the distance H compares with the clay thickness, etc.

Sand drains are installed by several procedures [see Landau (1966)] in diameters ranging from 6 to 30 in.

1. Mandrel-driven pipes—the pipe is driven with the mandrel closed. Sand is put in the pipe which falls out the bottom as the pile is withdrawn, forming the drain. Air pressure is often used to ensure continuity and densify the sand.
2. Driven pipes—the soil inside is then jetted. Rest of procedure is same as method 1.
3. Rotary drill—then filling boring with sand.
4. Continuous-flight hollow auger.

Figure 6-2 illustrates methods 1 and 3 above.

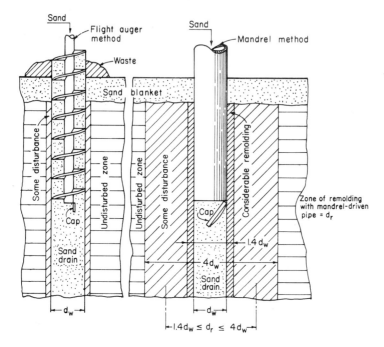

Figure 6-2. Two methods of constructing sand drains. [*Landau (1966).*]

When using earth for the surcharge, the length of the drainage path can be reduced by placing a sand blanket between the soil to be precompressed and the preload. Water can flow vertically through the sand drains to the sand blanket, then laterally to a ditch or other disposal means.

6-5. VIBRATORY METHODS TO INCREASE SOIL DENSITY

The allowable bearing capacity of sands depends heavily on the soil condition. This is reflected in the penetration number or cone resistance value as well as in the angle of internal friction. It is usually not practical to place a footing on loose sand because the allowable bearing capacity (based on settlements) will be too low to be economical. Additionally in earthquake analyses the local building code may not allow construction unless the relative density is above a certain value. Table 6-1 gives liquefaction-potential relationships between magnitude of earthquake and relative density for a water table 1.5 m below ground surface. This table can be used for a water table up to 3 m with slight error. The relative density may be related to penetration testing as in Table 3-2 after making suitable corrections to N for overburden effects.

The methods most commonly used to densify cohesionless deposits of sand and gravel with not over 20 percent silt or 10 percent clay are vibroflotation and insertion

Table 6-1. Approximate relationship between earthquake magnitude, relative density, and liquefaction potential for water table 1.5 m below ground surface*

Earthquake acceleration g	High liquefaction probability	Potential for liquefaction depends on soil type and earthquake acceleration	High liquefaction probability
0.10g	$D_r < 33$	$33 < D_r \leq 54$	$D_r > 54$
0.15g	< 48	$48 < D_r \leq 73$	> 73
0.20g	< 60	$60 < D_r \leq 85$	> 85
0.25g	< 70	$70 < D_r \leq 92$	> 92

* From Seed and Idriss (1971).

and withdrawing a vibrating pile [Terra-Probing, see Janes (1973)]. Vibroflotation (patented by the Vibroflotation Foundation Co.) utilizes a cylindrical penetrator about 17 in in diameter, 6 ft long, weighing about 4,000 lb. An eccentric weight inside the cylinder develops a horizontal centrifugal force of about 10 tons at about 1,800 rpm. The device has water jets top and bottom with a flow rate of between 60 and 80 gal/min at a pressure of 60 to 80 psi. Figure 6-3 illustrates the procedure for

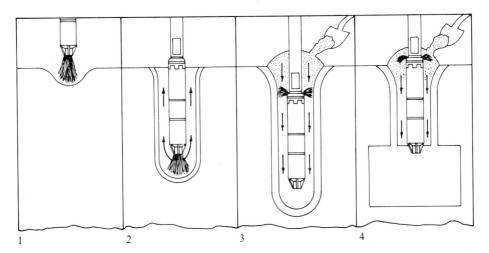

Figure 6-3. Vibroflotation.

1. Vibroflot is positioned over spot to be compacted and its lower jet is then opened full.
2. Water is pumped in faster than it can drain away into the subsoil. This creates a momentary "quick" condition beneath the jet which permits the Vibroflot to settle of its own weight and vibration. on typical sites, the Vibroflot can penetrate 15 to 25 ft in approximately 2 min.
3. Water is switched from the lower to the top jets and the pressure is reduced enough to allow water to be returned to the surface, eliminating any arching of backfill material and facilitating the continuous feed of backfill.
4. Compaction takes place during the one-foot-per-minute lifts which return the Vibroflot to the surface. First, the vibrator is allowed to operate at the bottom of the crater. As the sand particles densify, they assume their most compact form. By raising the vibrator step by step and simultaneously back-filling with sand, the entire depth of soil is compacted into a hard core.

vibroflotation. The device sinks at a rate of between 3 and 6 ft/min into the ground into the "quick" zone under the point caused by a combination of excess water and vibration. When the vibroflot reaches the desired depth, depending on footing size and stratum thickness, say, $2B$ to $3B$, after a few moments of operation the top jet is turned on and the vibroflot is withdrawn at the rate of about 1 ft/min. Sand is added to the crater formed at the top from densification as the device is withdrawn—typically about 10 percent of the compacted volume. Compaction rates of 10,000 to 20,000 cu yd in an 8-hr work shift are common. The probe is inserted on 4- to 10- or 15-ft centers depending on densification desired—maximum densification being in the immediate vicinity of the probe hole. Bearing capacities of 5,000 to 8,000 psf can be obtained using this method.

The Terra-Probe (patented by the L. B. Foster Co.) method involves mounting a vibratory pile driver on a probe (pile) and vibrating it into and out of the soil to be densified. This device is applicable to soils in which the vibroflotation method is applicable. This device is also applicable in underwater applications, e.g., shoreline construction. The probe is inserted on spacings of 4 to 15 ft depending on the amount of densification required.

The vibroflotation device can be used in soft, cohesive soils to insert granular columns to effect an increase in both strength and stiffness. The column is put into the ground by first sinking the device, then backfilling and simultaneously raising and lowering the device through the backfill about 1 ft at a time. The granular material is packed both vertically and rammed laterally into the soil by the weight of the vibroflot and the vibration effect.

The sand (or stone) columns are placed in a triangular grid on a 5- to 10-ft spacing and in an area 6 to 10 ft greater than the foundation plan. The process produces columns 2.5 to 3.5 ft in diameter and is particularly adapted to raft or storage-tank foundations. Additional details are described by Watt et al. (1967) and Hughes et al. (1975).

Sand piles (compaction piles) can be used in loose sand deposits to increase the density. They are formed by driving a pipe pile with a false bottom, sometimes called a hollow mandrel, which is closed during driving but drops open on withdrawal. The pipe is then filled with sand, closed on top, and air pressure is applied and the pipe withdrawn. The air pressure and gravity force the sand column out of the bottom of the pipe, which opens up on withdrawal. The same general grid layout is used as previously cited for sand columns in cohesive soils.

6-6. FOUNDATION GROUTING AND CHEMICAL STABILIZATION

Grouting is a technique of inserting some kind of stabilizing agent into the soil mass under pressure. The pressure forces the agent into the soil voids in a limited space around the injection tube. The agent reacts with the soil and/or itself to form a stable mass. The most common grout is a mixture of cement with or without sand and water. In general, grouting is one of the most expensive methods of treating a soil. It has a large number of applications such as:

1. Control of water problems by filling cracks and pores
2. Prevention of sand densification beneath adjacent structures due to pile driving
3. Underpinning using compaction (displacement) grouting
4. Reducing vibrations by stiffening the soil
5. Reducing settlements by filling voids and cementing the soil structure more firmly

Generally grout can be used if the permeability of the deposit is greater than 10^{-3} cm/sec. One of the principal precautions with grouting is that the injection pressure should not be sufficient to lift the ground surface. In using compaction grouting where a very stiff displacement volume is injected into the ground under high pressure, however, lifting of the ground surface as a grout lens forms is of minor consequence.

Various chemicals can be used as grouting agents. Chemical agents are the most expensive foundation treatment of all. They do have advantages in that they have low viscosity and absence of particulate material, and the setting time can often be controlled. A discussion of the advantages and disadvantages and of available types of chemical stabilizing agents is beyond the scope of this text. The reader is referred to ASCE (1966) for the most recent bibliography by the ASCE Committee on Grouting.

Chemical stabilization in the form of

Lime
Cement
Fly ash
Combinations of the above

is widely used in soil stabilization for road and street work. It can also be used for building construction to improve the soil. Lime, for example, will reduce the plasticity of most clays, which in return reduces volume-change potential (Secs. 7-1 and 7-9). Mixtures of cement, cement and fly ash, or lime, cement, and fly ash can improve the bearing capacity of the soil considerably. The optimum benefit of using these agents in stabilization must be determined by laboratory testing. Beyond a certain optimum percent of stabilizing material the effect may be detrimental.

6-7. ALTERING GROUNDWATER CONDITIONS

From the concept of submerged unit weight it is evident that the intergranular pressure can be increased by removing the buoyant effect of water. This can be accomplished by lowering the water table. In many cases this may not be feasible or perhaps only as a temporary expediency. Where it is possible, one obtains the immediate increase in intergranular pressure of $\gamma_w z$, where z is the change in water-table elevation.

With the increase in effective pressure, unwanted settlements may result, and as it is impossible to lower the water table exactly within the confines of ones' own property the effect of lowering the water table on adjacent property will need to be investigated.

PROBLEMS

6-1. The penetration number N of a loose sand varies from 8 at elevation -1.5 m to 15 at elevation -7.0 m. It is necessary to have a D_r of at least 0.75 for this soil. The area to be covered is 40 × 50 m. Vibroflotation or Terra-Probing will be used. What will be the expected N values after densification? How many cubic meters of sand will be required to maintain the existing ground elevation? (*Note:* Your answer depends on your assumptions.)

Answer: $N = 35$ to 40.

6-2. What is the additional settlement due to lowering the water table of Example 5-4 from 349.5 to 344.0? Comment on the effect of the water table rising to elevation 354.5 ft.

6-3. For soil No. 1 of Fig. 6-1 estimate the specific gravity G_s.

Answer: $G_s = 2.74$.

6-4. A soft clay deposit is 8.2 m thick overlying a dense sandy gravel. The following soil data are available: $k_h = 5 \times 10^{-4}$ cm/sec; $w_L = 62\%$; $w_P = 31\%$; $w_N = 58\%$; $G_s = 2.61$; $c_v = 3.1 \times 10^{-4}$ cm^2/min.

(a) Estimate the depth of preload for a site 60 × 50 m for a settlement of 6 cm using soil of $\gamma = 16.5$ kN/m^3.

(b) For sand drains at 2 m center to center, how long will it take the 6 cm of settlement to occur?

SEVEN

FACTORS TO CONSIDER IN FOUNDATION DESIGN

7-1. FOUNDATION DEPTH

Footings should be carried below:

1. The frost line
2. Zones of high volume change due to moisture fluctuations
3. Topsoil or organic material
4. Peat and muck
5. Unconsolidated material such as abandoned garbage dumps and similar filled-in areas

Footings should be placed below the frost line because of possible frost heave of the buildings and because alternate freezing and thawing of the soil tends to maintain it in an unconsolidated (loose) state. However, aside from the consideration that the soil may be loose, interior footings may be placed at convenient depths since the building warmth should control frost. Figure 7-1 presents approximate maximum frost depths for various parts of the United States; however, local building codes should be consulted for design values, which may be based on local experience and therefore be more realistic. Recent weather extremes may be obtained from weather records as a check that possible cold-weather cycles are not increasing the frost depth.

Clayey soils tend to shrink on drying and expand when wet. Generally, the lower the shrinkage limit and the wider the range of plasticity index, the most likely is

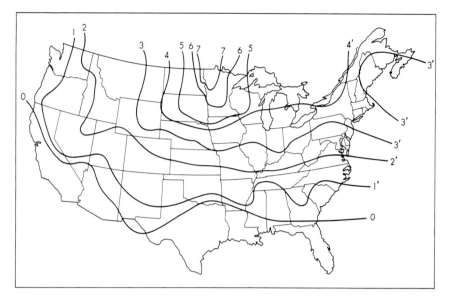

Figure 7-1. Approximate frost-depth contours for the United States, based on a survey of a selected group of cities.

volume change to occur (Table 7-1) and the greater the amount of such change. These changes in volume can be brought about by the drying of the soil after the structure is built, as, for example, in the protection of the soil from natural moisture. Loss in soil moisture by evaporation through heating the building or beneath or adjacent to heating units such as boilers may also create volume changes. Volume change may be induced by desiccation of the soil from vegetation, such as trees and shrubs used for landscaping the structure.

Volume change may also occur from artificially increasing the moisture in the soil beneath the structure. Cold-storage buildings with an uninsulated thermal gradient may condense the water vapor in the soil or create an upward flow of water vapor from the water table. Ice lenses may form if the temperature is sufficiently low.

Volume change is particularly troublesome in large areas of the Southwestern United States, India, and Australia, and in parts of Africa and the Middle East which

Table 7-1. Potential soil volume change as related to the plasticity index, I_P, and the liquid limit, w_L^*

Potential for volume change	Plasticity index I_P	Shrinkage limit w_S, %	Liquid limit w_L, %
Low	< 18	> 15	20–35
Medium	15–28	10–15	35–50
High	25–41	7–12	50–70
Very high	> 35	< 11	> 70

* *From* Holtz (1959) and Dakshanamurthy and Raman (1973).

are subject to long dry periods and periodic heavy rains of short duration. The dry periods tend to desiccate the soil; then the rains cause large amounts of swelling. There is not enough rainfall to leach and weather the troublesome clay minerals; thus they remain unaltered near the ground surface and are rapidly wetted during the rainy periods. Soils in these areas are particularly troublesome to build on as water vapor migrating from the water table which may be at a depth of many meters condenses on the bottom sides of the floor slabs and footings. The soil in the interior portions of the building eventually becomes approximately saturated from the condensate and swells unless the building provides sufficient weight to restrain the swelling pressure—buildings seldom provide the huge restraining pressures required to control swelling. A second difficulty arises in that in the arid climate the soil around the periphery remains in a much drier state than the interior soil and large differential movements result. Table 7-1 may be used as a guide in evaluating the potential for volume change of soils based on easy-to-determine index properties. This table is a summary of Holtz's (1959) data on several soils and the correlation of some 50 soils from other areas, including a large number of Indian Black Cotton soils by Dakshanamurthy and Raman (1973). In terms of relative values "low" volume change might be taken as not more than 5 percent where "very high" could be interpreted as over 25 percent. Structures founded on expansive soils require special construction techniques for the foundations. Some of the methods will be taken up in some detail in Sec. 7-9.

When placing footings adjacent to an existing structure, as indicated in Fig. 7-2a and b, the line from the base of the new footing to the edge of the old footing should be 45° or less with the horizontal plane. From this it follows that the distance m of Fig. 7-2a should be greater than the difference in elevation of the two footings, z_f. This is an approximation for reducing the pressure overlap of the two footings, which is considerably more liberal than the 2 : 1 ratio used in Sec. 5-2 for computing soil pressures at a depth in the soil.

Conversely, Fig. 7-2b indicates that if the new footing is lower than the existing

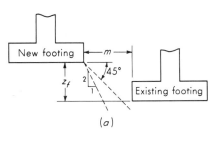

(a)

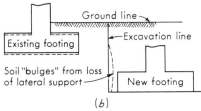

(b)

Figure 7-2. Location considerations for spread footings. (a) An approximation for spacing footings to avoid interference between old and new footings. If "new" footing is in relative position of "existing" footing, interchange words "existing" and "new." Make $m > z_f$. (b) Possible settlement of "existing" footing because of loss of lateral support of soil wedge beneath existing footing.

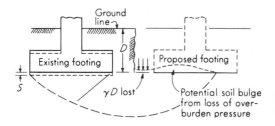

Figure 7-3. Potential settlement from loss of overburden pressure.

footing, there is a possibility that the soil may flow out from beneath the old footing, increasing the quantity of excavation and introducing settlement into the existing structure. This problem is difficult to analyze; however, an approximation can be obtained from Eqs. (2-14) and (2-15).

Figure 7-3 illustrates how one may get into trouble by excavating so close to an existing building that the qN_q term of the bearing-capacity equations of Chap. 4 is lost.

It is difficult to compute how close one may excavate as in either Fig. 7-2b or 7-3 before the adjacent structure is distressed. The problem may be avoided by constructing a wall (sheetpiling or other material, Chaps. 13 and 14) to retain the soil in place outside the excavation. This is one of the major concerns in open excavations for new buildings in built-up areas—building without causing damage, either imagined or real, to adjacent property owners.

Underground defects or utilities may affect the foundation depth, for example, limestone caverns, old mine tunnels, soft material, sewer tunnels, telephone-cable conduits, and possible flaws created by the pumping out of soil fluids (oil, water). Bridging action may be adequate for some cavities or across soft spots but should be relied upon only after a careful study of the conditions. In other cases, the solution may require a different type of foundation (such as piles or caissons) or even an abandonment of the site.

7-2. DISPLACED SOIL EFFECTS

Soil is always displaced by installing a foundation. In the case of spread footings the displacement is the volume of the footing pad and the negligible amount from the column resting on the footing. In cases where a basement is involved, the basement floor slab usually rests directly on top of the footing pad. In other cases, a hole is excavated for the footing, the footing and column is poured, and the remainder of the hole is backfilled to the ground surface as illustrated in Fig. 7-4a. When the footing is below ground, a concrete pedestal is used to connect to steel columns because of corrosion; for concrete columns, the column is simply attached to the footing with dowels at the footing level. Figure 7-4b illustrates the condition of footings beneath basements and walls. Figure 7-4c illustrates the situation of a mat foundation. The backfill soil should be carefully compacted over the footing of Fig. 7-4b if a floor slab is to rest on the ground surface. Select free-draining backfill is carefully placed around basement walls as shown in Fig. 7-4b and c, usually with a system of perimeter drainage to control any hydrostatic pressure.

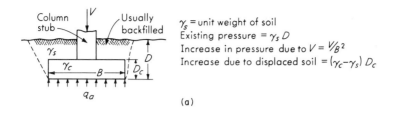

γ_s = unit weight of soil
Existing pressure = $\gamma_s D$
Increase in pressure due to $V = V/B^2$
Increase due to displaced soil = $(\gamma_c - \gamma_s) D_c$

(a)

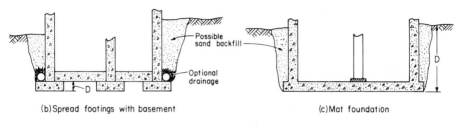

(b) Spread footings with basement   (c) Mat foundation

Figure 7-4. Footing placement and significance of q_a.

7-3. NET VS. GROSS SOIL PRESSURE—DESIGN SOIL PRESSURES

When the soil engineer gives an allowable bearing pressure (or a set of curves of q_a vs. B) to the structural designer, as is often the practice, what is the significance of these pressures? Is this a

1. *Net* pressure—pressure in excess of the existing overburden pressure which can be safely carried at the foundation depth D or a
2. *Gross* pressure—the total pressure which can be carried at the foundation depth, including the existing overburden pressure?

The bearing-capacity equations are based on gross soil pressure q_{ult}, which is everything above the foundation level. Settlements are caused only by net increases in pressure over the existing overburden pressure. Therefore, if the allowable pressure is based on the bearing-capacity equations of Chap. 4 (Terzaghi or Hansen), the pressure is a gross pressure. If the allowable pressure is based on settlement considerations, the pressure is net pressure, and computations proceed accordingly. Whether the pressure is a gross or net value should be stated in the foundation report submitted to the designer; however, it often is not.

There is a tendency of most designers to treat the bearing pressure provided by the soils engineer as a precise value which cannot be exceeded. In fact, from material presented in Chaps. 2, 3, and 4 relating to sampling, testing, and computation difficulties, the bearing pressure is hardly an exact value at all. Thus, it makes little difference if we exceed by 0.05 to about 0.2 ksf (2 to 10 kPa) the given values unless the given values are under 1.0 ksf (50 kPa) where we might not wish to exceed the values by over about 0.05 ksf. Recognition of the inherent limitations on precisely determining the bearing capacity can considerably reduce the computations in footing design of the next two chapters.

In this text footings will be proportioned for the soil pressure with no allowance for the footing weight, for the following reasons:

1. If one uses the given pressure as the *gross* soil pressure, the total increase q_i in pressure will be approximately (Fig. 7-4a)

$$q_t = \frac{V}{B^2} + \gamma_s D + \gamma_c D_c - \gamma_s D_c$$

but $\gamma_s D$ = existing pressure; thus $q_i = q_t - \gamma_s D$ or

$$q_i = \frac{V}{B^2} + D_c(\gamma_c - \gamma_s)$$

If we neglect $(\gamma_c - \gamma_s)D_c$, then $q_i = V/B^2$, which is recommended because of simplicity. For nominal values of concrete density of 150 pcf and soil density of 115 pcf, the neglected term amounts to about $35D_c$ psf, and soil bearing pressures are just not known to this precision.
2. If the given pressure is taken as the net soil pressure, which by definition is pressure in excess of the existing soil pressure, one may use the same reasoning as in point 1; that is, an error of $(\gamma_c - \gamma_s)D_c$ will be introduced if the term is neglected.
3. It is usual practice to use multiples of even inches, preferably to the nearest larger 3-in multiple for setting the footing forms. This rounding reduces the actual soil pressure and the mathematical errors of 1 and 2.

For footings, except rafts, and considering the factors listed above, it is recommended that proportioning be on the basis of the given q_a, with the limitation that where the depth of the resulting footing D_c is more than 3 ft (with a potential resulting error of $35 \times 3 = 105$ psf), the size will be increased slightly to compensate for the weight of the footing, if the size has not already been adjusted up for placing the forms. This procedure takes into consideration a backfill over the top of the footing, or it is flush with the ground surface, so that no backfill is required. If the soil is permanently removed, as for a basement, the effect is too large to ignore and must be considered in establishing the net (or gross) increase in soil pressure.

7-4. EROSION PROBLEMS FOR STRUCTURES ADJACENT TO FLOWING WATER

Bridge piers, abutments, bases for retaining walls, and footings for other structures adjacent to or located in flowing water must be located at a depth such that erosion or scour does not undercut the soil and cause a failure. The scour depth will depend on the geological history of the site (depth of prior erosion to bedrock and subsequent redeposition of sediments, stream velocity, and area runoff). Where the redeposition of sediments in the stream bed is on the order of 30 to 50 m, a careful analysis of borings into the sediments to predict the depth of maximum scour is necessary in order to provide a foundation that is economical. It may be possible to use spread footings if they can be placed at sufficient depth, but normally piles are required to support the foundation and an accurate prediction of scour depth is necessary to use as short pile lengths as possible. If careful records of driving resistances are kept, one may predict the scour depth as being where the penetration (SPT or cone) resistance increases substantially [see Kuhn and Williams (1961)].

An NCHRP (1970) report lists some 13 equations proposed by several authorities including those of Laursen and Toch (1956), and later proposals by Laursen (1962). This report indicates that engineering judgment is used more than any other method for estimating scour depth. The equations by different authorities for the same problem will compute scour depths differing by as much as 1,000 percent.

Scour occurs principally during floods. Some scour may occur at other times, and it usually leaves a scour hole. Scour holes forming during floods are usually refilled as high water falls. Scour is accelerated if the foundation creates channel obstruction; thus to reduce scour the foundation should create a minimum obstruction to normal stream-flow patterns.

Normally the approach to scour is as follows:

1. Design the foundation.
2. Estimate the scour effects, depth, etc.
3. Estimate the cost of foundations for normal and various scour conditions.
4. Determine the cost vs. risk, and revise the design accordingly.

7-5. CORROSION PROTECTION

In polluted ground areas such as old sanitary land fills, shorelines near outfall sewer lines from older industrial plants, or backwater areas where water stands over dead vegetation, there can be corrosion problems with metal foundation members as well as with concrete. Concrete is resistant to corrosion; however, if sulfates are present, it may be necessary to use sulfate-resistant concrete. It may occasionally be necessary to use air-entrained concrete for foundation members. It may be necessary to use treated timber piling instead of metal piling where the soil has a pH much more or less than 7 (7 being neutral).

7-6. WATER-TABLE FLUCTUATION

A lowered water table increases the effective pressure and may cause additional settlements. A raised water table may create instability from

1. Floating the structure
2. Reducing the effective pressure

It may create a wet basement if the basement walls are not watertight.

7-7. FOUNDATIONS IN SAND DEPOSITS

Foundations on sand will require consideration of the following:

1. Bearing capacity.
2. Settlement—loose deposits must be densified to control the settlement.
3. Placing the footing at a sufficient depth that the soil beneath the footing is confined. If the soil is not confined, it will roll out from the footing perimeter with a loss of density and bearing capacity.

Foundations in sand may consist in spread footings, mats, or piles, depending on the density, thickness, and cost of densifying the deposit, and the building loads. Solid-section, large-volume piles may be used both to carry load to a greater depth in the deposit and as a means of compacting the deposit. Small-volume piles are normally used to carry near-surface loads through loose sand deposits to firm underlying strata. Spread footings are used if the deposit is dense enough to support the loads without excessive settlements. Settlements on sand deposits are classified as immediate settlements, and many of them will be built out as the construction progresses because of loads and site vibration. Unfortunately, it is impossible to compute the built-out settlements. It is poor practice to place foundations on sand deposits where the relative density is not at least 60 percent or to a density of about 90 percent or more of the maximum density possible to obtain in the laboratory. This dense state reduces the possibility of both load settlements and possible settlement damage due to vibrations from passing equipment, earthquakes, etc.

The allowable bearing capacity for footings and mats is computed from methods given in Chaps. 4 and 5.

7-8. FOUNDATIONS ON LOESS

Loess is a fine-grained, uniform soil deposit formed by deposition of wind-borne (aeolian) particles. It is yellow to reddish-brown, with buff being the most common color. This soil covers approximately 17 percent of the United States (see Fig. 7-5). Loess covers about 17 percent (1.8×10^6 km^2) of Europe also, including the low countries, parts of France, Germany, and eastern Europe. Tremendous areas of Russia and Siberia (about 15 percent) are also covered by loess, as are large areas of China. Loess is found in New Zealand and the plains regions (pampas) of Argentina and Uruguay, principally between latitudes 30 and 40°. Very little loess is found in Canada and none in Australia or Africa, although both areas do have wind-borne soil deposits [Flint (1971)]. Loess appears to be formed by wind-borne deposits traveling over glacial outwash, with the higher humidity of the outwash causing precipitation of the soil particles. Since Australia, South America, Africa, and much of southern Europe were not glaciated, the deposits are not present. Depths of loess deposits range from less than 1 to more than 50 m in depth. Depths of 2 to 3 m are extremely common.

Figure 7-5. Location of major loess deposits in the United States. [*Gibbs and Holland (1960)*.]

Loess is characterized by a complete absence of gravel or pebbles and nearly 90 percent passing the No. 200 sieve (0.074 mm) with some 0 to about 15 percent smaller than 0.005 mm. The specific gravity ranges from about 2.60 to 2.80, with most values between 2.65 and 2.72 producing in situ dry density from about 66 to 104 pcf. The Atterberg limits vary considerably depending on the amount of clay present. Liquid limits commonly range from 25 to 55 and plastic limits from 15 to 30 percent. Standard compaction tests produce dry densities of 100 to 110 pcf at optimum moisture contents of 12 to 20 percent [Sheeler (1968)]. The in situ void ratio is in the range of 0.67 to 1.50 [Drannikov (1967)].

The density of loess is the most significant parameter of its usefulness as a foundation. Generally if the density is greater than 1.44 g/cm^3 (90 pcf), the settlements will be rather small. The principal difficulty with using loess is that the bearing capacity changes considerably upon saturation. Clevenger (1958) indicated that bearing capacities on the order of 10 ksf are possible on dry loess but may be only 0.5 ksf upon saturation if settlement is to be kept low, say around 0.5 in. Studies indicate that rainfall does not contribute to saturation of loess because of its low coefficient of permeability; i.e., 1 to 1.5 m below the surface it may be permanently dry. Ponding can cause saturation and is sometimes used to saturate and cause the structure to collapse so that settlement (preconsolidation) occurs prior to construction.

The typical loess structure consists in relatively uniform soil grains cemented together with clay particles present either during deposition or from later weathering of loess particles. Additional cementation is caused by carbonates, principally calcium and the presence of root holes or tubules which often carbonate up. This seems to account for the distinguishing characteristic for loess to stand on vertical cuts and to slough away on vertical banks. Wetting of the mass destroys or weakens the cement bonds sufficiently so that external loading or self-weight causes a large settlement or collapse depending on the density of the deposit. Loess below the permanent water table is in a stable condition. Loess which is eroded away and redeposited is commonly a "silt" deposit and loses the loess character.

Compacted loess is a satisfactory foundation material for spread footings or mats if the density is on the order of 1.6 g/cm³ or more. Additional stability can be obtained by using lime, lime fly ash, or cement as a stabilizing agent. Normally compaction would be necessary above the permanent water table or the full depth of the deposit of shallow (1 to 2 m) deposits to ensure satisfactory performance.

Pile foundations are necessary to control settlements when the in-place density is under about 1.44 g/cm³ unless the soil moisture can be controlled or the soil can be compacted. Piles should be driven through the loess layer into the underlying soil, as the capacity will be determined by this material. If the loess becomes wet, it will lose adhesion and/or point capacity if the pile is founded completely in the loess unless, of course, it terminates below the water table.

7-9. FOUNDATIONS ON EXPANSIVE SOILS

Soils which undergo volume changes upon wetting and drying are termed expansive soils. These soils are mostly found in arid areas and contain large amounts of clay minerals. The low rainfall has not enabled the montmorillonite clay minerals to weather to less active clay types nor has it allowed sufficient leaching to carry the clay particles far enough into the strata to reduce its effect. Expansive soils are found in large areas of the Southwest and Western United States including Oklahoma, Texas, Colorado, Nevada, California, Utah, and others. These soils are also found in large areas of India and Australia (sometimes called Black Cotton soils), South America, Africa, and the Middle East.

Table 7-1 can be used to give an indication of the potential for volume change caused by alternate wetting and drying.

When the problem is identified, one may

1. Alter the soil—for example, addition of lime, cement, or other admixture will reduce or eliminate the volume change on wetting or drying. Compaction to low densities at water contents on the wet side of optimum may also be used [Gromko (1974), with large number of references].
2. Control the direction of expansion—by allowing the soil to expand into cavities built in the foundation, the foundation movements may be reduced to tolerable amounts. A common practice is to build "waffle" slabs (see Fig. 7-6) so that the ribs hold the structure, the waffle voids allow soil expansion [BRAB (1968),

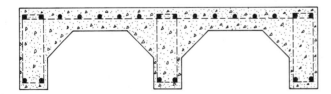

Figure 7-6. Waffle slab.

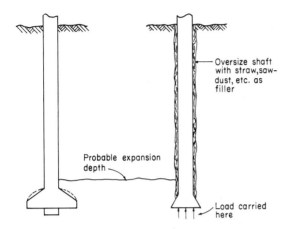

Figure 7-7. Belled piers in expansive soils.

Dawson (1959)]. It may be possible to build foundation walls to some depth into the ground using tiles placed such that the soil can expand laterally into the tile cavity.

3. Control the water—the soil may be excavated to a depth such that the weight of soil will control heave, lay a plastic membrane, and then backfill. The rising water vapor is collected at a depth such that volume change is controlled by the weight of overlying material and construction. The moisture above will also have to be controlled by paving, grading, etc.

4. Ignore the heave—by placing the footings at a sufficient depth and/or leaving an expansion zone between the ground surface and the building, swell can take place without causing detrimental movement. The procedure is to use belled piers with the bell at sufficient depth in the ground that tension on the shaft does not pull it out and that volume change does not heave the whole system. One could use small pipes and plates for smaller structures to accomplish this solution. The pier or pile shaft should be as small as possible to avoid high-tension stresses due to expansion pressure and adhesion. Sometimes the pier shaft is surrounded by straw or other porous material to reduce adhesion, as the primary purpose of the shaft is to transmit the building load below the zone of volume change. Figure 7-7 illustrates this solution.

5. Load the soil to sufficient pressure intensity to balance swell pressures—this method is used in many fills where the fill weight balances the swell pressure. This can also be used beneath buildings either by using spread footings of high-pressure intensity or excavating several feet of the clay and backfilling with granular backfill. This combined with foundation pressures may contain the swell. This method may not be practical for one-story commercial buildings and residences because of the small soil pressures developed.

The pressure to control soil expansion may be estimated from an equation proposed by Komornik and David (1969) based on a statistical analysis of some 200 soils. The equation is

$$\log P_s = \bar{2}.132 + 2.08(w_L) + 0.665(\gamma_d) - 2.69(w_N)$$

where log P_s is the logarithm to base 10 of the swelling pressure in kg/cm^2 and other terms are as previously defined. For example, if the liquid limit is 60 percent, the natural moisture content 25 percent, and the dry density 1.2 g/cm^3,

$$\log P_s = \bar{2}.132 + 2.08(0.6) + 0.665(1.2) - 2.69(0.25)$$

$$= \bar{2}.132 + 1.3735 = \bar{1}.5055$$

and taking the antilog of 0.5055 and moving the decimal 1 place to the right for the -1, we obtain the estimated swell pressure as

$$P_s = 0.32 \ kg/cm^2 \ (0.66 \ ksf)$$

A footing pressure (or other surcharge) in excess of 0.66 ksf should control the heave to tolerable limits.

7-10. FOUNDATIONS ON CLAY

Foundations on clay must be adequate to avoid a shear failure and excessive settlements. If the soil is not saturated, the settlement is computed as immediate settlements using elastic methods. The bearing capacity is normally obtained using the unconsolidated-undrained, or for partially saturated soils a quasi-consolidated-undrained shear strength. The 2-in tube samples are often used. If the engineer has experience in an area, he may use only q_u values from SPT samples.

Preconsolidated clays frequently contain fissures or slickensides, and it may be necessary to use triaxial tests to determine the undrained shear strength. The confining pressure tends to squeeze the fissure together so that the test failure more closely resembles field behavior.

The *net* ultimate bearing capacity for vertical loads is normally computed as

$$q_{ult} = cN_c s_c d_c + \bar{q} N_q s_q d_q - \bar{q}$$

which is often written as

$$q_{ult} = cN_c s_c d_c + \bar{q}(N_q - 1)$$

The combined effect of $N_c s_c d_c$ has a limiting value [Skempton (1951)] of 9.0 for square footings and about 7.6 for strip footings for $\phi = 0$ conditions. For example,

	D/B			
	0.0	1.0	2.0	5.0
Strip footing $N_c s_c d_c = 5.14$		6.4	7.2	7.6
Square footing $N_c s_c d_c = 6.2$		7.6	8.4	9.0

If the soil is saturated, a principal concern will be the consolidation settlement and it will be necessary to estimate if the soil is preconsolidated. This may be done by methods presented in Chap. 2. If the soil is preconsolidated, it will be necessary to apply considerable engineering judgment to determine the probable settlement, or excessive overdesign may result.

7-11. FOUNDATIONS ON SANITARY LANDFILL SITES

As land becomes scarce near urban areas, it may be necessary to use a former sanitary landfill. A sanitary landfill is an esoteric name for a garbage dump. Landfills which are likely to be used at present were often placed at some convenient location, generally where the ground was uneven so that the material could be placed in the depressions and later covered. Present landfills require location such that ground-water pollution is controlled and generally require daily covering of the refuse accumulation with a layer of earth. Good practice requires 0.6 to 1 m of disposal material covered with 0.15 to 0.2 m of compacted earth in alternating layers. This may not be achieved, owing to the practice of dumping old bedding, refrigerators, auto parts, demolition and construction refuse, broken-up pavements, metal cans, and tires as well as smaller materials. In the past the dumps were often uncovered for days at a time, creating both odor and pest nuisances. Currently, except in smaller communities, health authorities require covering the material daily. When the landfill is closed down, the surface is covered with up to 2/3 m of earth and sometimes compacted, or the equipment makes a few passes to consolidate the fill, and is then landscaped. As the refuse decays, the surface may become uneven or the underlying material may cavitate, depending on the rapidity of action, refuse materials, and thickness of fill cover.

In using a landfill for later construction, it may be extremely difficult to avoid settlements as the refuse decomposes and/or consolidates. It is certain that the settlements will be uneven owing to the varied character of the refuse material and the method(s) used to construct the fill. Yen and Scanlon (1974) reported several studies of landfill settlements, but no conclusive method of predicting settlement could be made.

Bearing capacity of the fill will consist in evaluation of the surface cover for a punching or rotational-shear failure as shown in Fig. 7-8a. It may be possible to add additional fill to reduce the pressure from the foundation on the refuse zone; this will speed up and increase the fill consolidation and may be desirable for relieving future settlements if sufficient time is available. The additional fill would be on the order of 1.5B thick to accomplish the stress reduction on the refuse zone. For light structures such as one- or two-story residential buildings, apartments, office buildings, and stores, where the bearing capacity may only be 0.5 to 1 ksf, the use of continuous foundations may provide adequate bearing and bridging capacity over local soft spots or cavities. If this is not sufficient, or the owner does not wish to take a chance on building damage, the only recourse is to use piles or piers (caissons) through the landfill into the underlying soil.

In using piles or piers, it will be necessary to use noncorroding materials, as any moisture in the fill will be corrosive to metal and may damage concrete. Generally only treated wood or concrete piles can be used. The concrete for the piles and caissons may be a considerable problem if the driving or boring encounters old paving rubble, auto bodies, or tires. The foundation construction may create odor problems as the ground cover is penetrated, and this should be investigated prior to construction as adjacent property owners may be able to obtain a court injunction

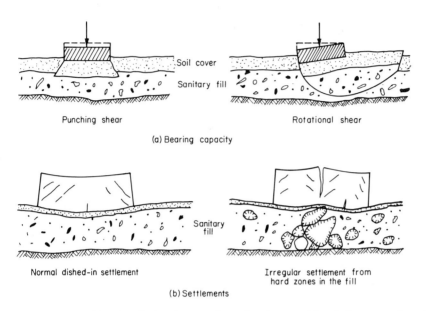

Figure 7-8. Considerations for bearing capacity and settlement of foundations on sanitary landfills with thin soil covers. [*Sowers (1968).*]

against the air pollution. Gas escaping from the fill, in addition to being malodorous, may be a health hazard requiring investigation as injured workmen may seek damage from everyone connected with the project including the foundation engineers.

7-12. FROST DEPTH AND FOUNDATIONS ON PERMAFROST

Figure 7-1 shows frost-depth contours based on a survey of selected cities for frost-depth practices at that location. Building codes may stipulate the depth the footing must be placed in the ground so that differential movement does not take place owing to water freezing in the ground beneath the foundation. Differential movement is difficult to evaluate, as it depends on the amount of water, whether an ice lens forms, and the soil density. Water expands approximately 10 percent on freezing; thus an ice lens in dense soil could cause a considerable amount of differential movement. It will not be practical to contain the ice expansion as is sometimes possible with soil expansion (Sec. 7-9), as the ice forces are quite large (on the order of 2,000 kg/cm^2). Ice adhesion and resulting uplift can be avoided by using granular backfill around the foundation walls or footing pedestals. With insulation [see Robinsky and Bespflug (1973)] it may be possible to reduce the depth of foundation.

Permafrost is a condition of permanently frozen ground at temperatures no higher than 0°C. This condition covers large areas of northern Canada, Alaska, and most of Siberia, northern parts of Scandinavia, and Antarctica. In many areas an "active" zone overlies the permafrost which thaws in season leaving a trapped mass of water-saturated bog, peat, and mud overlying the ice. Construction in these areas requires that the foundations be placed in the permafrost and insulated sufficiently that thawing does not occur—either from building heat or because of changed environmental conditions.

Where the soil is considered thaw-stable, the foundation design is the same as in temperate regions. Thaw-stable soils are granular materials as coarse sands and gravels. These soils will, of course, have to be of sufficient thickness that the active zone will not penetrate the permafrost. Spread and continuous footings, mats, and beam-and-post construction can be used. Sometimes these foundations can be used for thaw-unstable soils as well [Linnel and Johnston (1973)]. Often the use of these foundations may require the use of a thaw-stable fill or ducting to reduce heat transfer into the underlying permafrost.

Pile foundations are more reliable for permafrost areas, are much more expensive, but may be necessary where large differential settlements cannot be tolerated. Piles are commonly wood, steel pipe, or H piles. Concrete piles are less common for several reasons including transport cost, problem of curing if cast in place, and the high tensile stresses developed because of adfreezing in the active zone. Tensile stresses on the order of 10 ksf have been measured [Linnel and Johnston (1973)]. Piles may be driven or inserted into predrilled holes and using a soil- or sand-water slurry which freezes around the pile to fix it in place. The interested reader should consult Linnel and Johnston for additional details and for a source list of some 190 additional references cited.

7-13. ENVIRONMENTAL CONSIDERATIONS

The foundation engineer has the responsibility to ensure that his portion of the total design does not have a detrimental effect on the environment. The responsibility may be enforceable by the courts if a laxity on the part of any parties can be ascertained. While it may not be readily apparent, the foundation engineer does have some effect or potential effect on the environment, for example:

1. Soil borings through sanitary landfills (which may have been constructed on impervious soil to avoid groundwater pollution) can pollute the groundwater via seepage through the bore-holes.
2. Soil boring logs should be checked for indication of effect of site excavation on the environment in terms of runoff, pollution in runoff, odor problems, dust, and noise.
3. Investigate means to salvage topsoil for landscaping.
4. Pile-driver noise and vibration.
5. Alternatives to cutting trees either for site work or where trees cause seasonal volume changes from soil desiccation during the growth season and wetting during the dormant season.
6. Effect of soil borings on perched water tables.

7. That soil borings near streams do not cause piping problems during high-water periods. This may be avoided by careful plugging of the boreholes.
8. That hydraulic fill for cofferdams, roadway approaches, and retaining structures, usually obtained from river bottoms, does not cause pollution of the ground-water through loss of the relatively impervious silt layer in the stream bed.
9. That earth removal for fill from hillsides does not cause landslides which may destroy scenic areas.
10. Effect of river and marine structures on aquatic life.
11. Effect of river and marine structures on pollution of groundwater by either river or salt-water intrusions.

Particular sites and potential site development may create environmental considerations supplementing the above list. But this list may be considered as a reasonable starting point.

PROBLEMS

7-1. A soil has the following average properties: $w_L = 62.5\%$; $w_P = 28.0\%$; $\gamma_{dry} = 1.66$ g/cm³. The profile contains 28 ft of clay overlying a medium dense sand; q_u varies linearly from 2.5 ksf at -5 ft to 3.5 ksf at -25 ft. Assume this site is near Dallas, Tex.
   (a) Estimate the swell pressure.
   (b) A two-story load-bearing (concrete-block) wall apartment building consists in 8 units and a plan of 32 × 124 ft. State how this building can be constructed to have no differential settlement. Write your recommendations in a short report. Consider with and without a basement and include consideration of floor slabs on grade or basement slab.

7-2. What is the allowable point bearing capacity of a 18-in-diameter pier founded at elevation -25 ft in the above soil?

7-3. What is the allowable bearing capacity of a 5-ft-square spread footing at elevation -5, -10, and -15 using the equation given in Sec. 7-10 and using the soil of Prob. 7-1?

7-4. A soil investigation in an old sanitary landfill indicates 1 ft of cover; SPT blow counts to -15 ft range from 1 to 8 except one boring where $N = 50$ at -10. At elevation 15 approximately 2 ft of topsoil and organic material was encountered, and beginning at elevation -17 blow counts ranged up to from 12 to 18 to elevation -32. This soil was a silty, stiff gray clay with traces of sand and gravel. At elevation -32 the soil becomes a medium dense sand with blow counts ranging from 25 to 35. At elevation -50 this soil becomes very dense and the blow counts range from 40 to 45. Boring is terminated at -65. A one-story discount store consisting in 31,500 sq ft is proposed for this site. Assume the site is near Chicago, Ill. Draw the "typical" boring log and write a set of recommendations for foundations.

7-5. Use the soil boring data of Prob. 7-4. A 6-story office building using a steel-frame and curtain-wall construction is proposed for the site. Draw the typical boring log and write a set of foundation recommendations.

7-6. A series of boring logs for a site revealed that the top 32 ft was loose sand with blow counts ranging from 4 to 8 for the first 15 ft and from 6 to 12 for the remainder. Underlying this is a 31.5-ft stiff clay deposit with $w_L = 35.8$, $w_P = 21.3$, $w_N = 24\%$ and unit weight of 124.5 (all average). The water table is at elevation 20. Assume the site is near Memphis, Tenn.
   A two-story manufacturing plant is proposed with column loads averaging 85 to 100 kips and wall loads of 2.0 k/ft. Draw the profile and make foundation recommendations. Settlements should be limited to 1 in, and there will be machinery vibrations.

SPREAD-FOOTING DESIGN

8-1. FOOTING CLASSIFICATIONS

When a footing carries a single column, it is called a *single footing* (Fig. 8-1a); it may be *stepped*, or *sloped*, if the column loads are large or if no tensile reinforcement is used and if the savings in material will justify the extra forming costs. A single footing is also often termed a *spread footing*, since it "spreads" the load from the column (which is loaded to a high stress intensity) to the soil, which can take only a relatively low stress. A single footing using reinforcing steel may also be called a *two-way footing*. The "two-way" refers to using tensile steel in both directions. A spread footing may be used under a wall as a *wall footing* (Fig. 8-1d), or it may have a *pedestal* (Fig. 8-1e) which transmits the column load to the footing, if the footing is more than a nominal depth below the floor.

8-2. ASSUMPTIONS USED IN THE DESIGN OF SPREAD FOOTINGS

Theory of elasticity analysis and observations [Barden (1962), Schultze (1961), Borowicka, (1936)] indicates that the stress distribution beneath footings, symmetrically loaded, is not uniform. The actual stress distribution depends on the type of material beneath the footing and the rigidity of the footing. For footings on loose cohesionless material, the soil grains tend to displace laterally at the edges from under the load, whereas in the center the soil is relatively confined. This results in a pressure diagram somewhat as indicated in Fig. 8-2a. For the general case of rigid

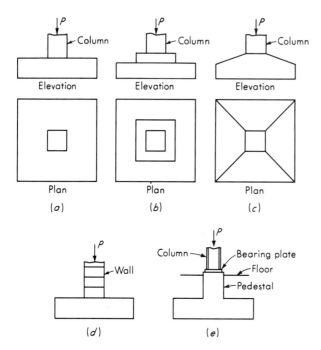

Figure 8-1. Typical footings. (*a*) Single or spread footing; (*b*) stepped footing; (*c*) sloped footing; (*d*) wall footing; (*e*) footing with pedestal.

footings on cohesive and cohesionless materials, Fig. 8-2*b* indicates the probable theoretical pressure distribution. The high edge pressure may be explained by considering that edge shear must take place before settlement can take place.

Because the pressure intensities beneath the footing depend on the rigidity of the footing, the soil type, and the condition of the soil, the problem is generally indeterminate. It is common practice to use a linear pressure distribution beneath the

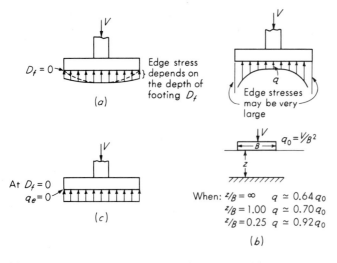

Figure 8-2. Probable pressure distribution beneath a rigid footing (*a*) on a loose cohesionless soil; (*b*) generally for cohesive soils; (*c*) usual linear pressure distribution.

footings, and this procedure will be followed in this text. Measurements by Schultze (1961) of pressure intensities beneath existing foundations indicate that unless the foundation is massive, as some bridge piers, the higher edge pressures of Fig. 8-2b are not developed. This is probably due to the low rupture strength of soil. In any case little difference in design results by using a linear pressure distribution [Bowles (1975, p. 490)].

8-3. REINFORCED-CONCRETE DESIGN—USD

The latest revision of the ACI Standard Building Code Requirements for Reinforced Concrete (ACI 318-71), hereinafter termed the Code, places almost total emphasis on strength-design (USD) methods, and that will be the only procedure used in this text. According to ACI this is the last major revision; the code will be periodically reprinted with errata and minor revisions made between printings. All notation pertaining to concrete design used in this text will conform to ACI Code notation. Where this conflicts with notation previously used, the user should take note. Strength design requires converting working design loads to *ultimate* loads through the use of load factors as

$$P_u = 1.4D + 1.7L \qquad (a)$$

$$P_u = 0.75(1.4D + 1.7L + 1.7W) \qquad (b)$$

$$P_u = 0.9D + 1.3W \text{ alternative with wind} \qquad (c)$$

For earthquake loading substitute E for W as applicable. Other load combinations may be used, but the user is referred to Art. 9.3 of the Code for their application.

Strength design considers workmanship and other uncertainties by use of ϕ factors as follows:

Design consideration	ϕ
Moment	0.90
Diagonal tension, bond, and anchorage	0.85
Compression members, spiral	0.75
Compression members, tied	0.70
Unreinforced footings	0.65
Bearing on concrete	0.70

Concrete strain at ultimate stress is taken as 0.003 in/in. Generally the yield strength of reinforcing steel is limited to 80 ksi per Art. 9.4. The most popular grade of reinforcing steel in current use is $f_y = 60$ ksi.

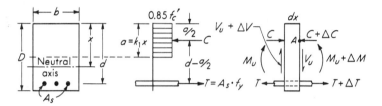

Figure 8-3. Assumptions used for the development of the ultimate-stress-design equations.

Elements of USD Design

For the partial development of the USD equations which follow, refer to Fig. 8-3.

From Fig. 8-3a the summing of horizontal forces, $\sum F_H = 0$, yields $C = T$, and taking the compressive-stress block as a rectangle of dimensions shown:

$$C = 0.85f'_c ba$$

The tensile force T is

$$T = A_s f_y$$

Equating the latter quantities yields an expression for the depth of the compression block as

$$a = \frac{A_s f_y}{0.85f'_c b} \tag{8-1}$$

From statics and summing moments at a convenient point (either T or C) we obtain

$$T\left(d - \frac{a}{2}\right) = M_u = C\left(d - \frac{a}{2}\right)$$

and solving for the ultimate resisting moment on a section and inserting the workmanship ϕ factor

$$M_u = \phi A_s f_y\left(d - \frac{a}{2}\right) \tag{8-2}$$

Alternatively, if terms p and q are defined as follows:

$$p = \frac{A_s}{bd} \qquad q = \frac{pf_y}{f'_c}$$

Equation (8-2) can be written

$$M_u = \phi bd^2 f'_c q(1 - 0.59q) \tag{8-2a}$$

The percentage of steel at a cross section has been defined as $p = A_s/bd$ and the percent at *balanced* design will be designated as p_b. To ensure a tensile failure rather than a sudden concrete compression failure p_d is taken as not over $0.75p_b$ (Art. 10.3.2)

Table 8-1. Maximum allowable percent of steel*

f_c', ksi (kg/cm²)		f_y, ksi (kg/cm²)		
f_c', ksi (kg/cm²)	k_1	45,000 (3,164)	50,000 (3,515)	60,000 (4,219)
3,000(211)	0.850	0.0278	0.0206	0.0160
3,500(246)	0.850	0.0325	0.0241	0.0187
4,000(281)	0.850	0.0371	0.0275	0.0214
5,000(352)	0.800	0.0437	0.0324	0.0252
6,000(422)	0.750	0.0491	0.0364	0.0283

* Table includes 25 percent reduction for bending using strength design ACI 318-71, Art. 8.1. Note that k_1 is reduced for $f_c' > 4$ ksi.

where the balanced percent reinforcement is computed based on the concrete strain at failure of 0.003 as

$$p_b = \frac{0.85k_1 f_c'}{f_y} \frac{87,000}{f_y + 87,000} \qquad (8\text{-}3)$$

The factor k_1 is taken as 0.85 for f_c' to include 4,000 psi and is reduced by 0.05 for each 1,000 psi in excess of 4,000; thus, for 4,500 psi, $k_1 = 0.825$ and for 5,000 psi concrete, $k_1 = 0.80$. Table 8-1 provides typical values of $p_a = 0.75p_b$ in both fps and metric units.

If tension-steel requirements exceed p_a, it will be necessary to use compression steel (Art. 10.3.5). Generally for foundations, p_a will be adequate; for compression-steel design consult a textbook on reinforced-concrete design.

Bond stresses can be obtained from Fig. 8-3, and summing moments at point A we have

$$(V_u + \Delta V_u)\, dx - \Delta T\left(d - \frac{a}{2}\right) = \Delta M \qquad \text{but} \qquad \Delta T = V_u(dx)S_p u$$

with S_p taken as the sum of the bar perimeters of the tension steel. Dropping second-order differentials, maximizing, and including the ϕ factor, one obtains

$$u = \frac{V_u}{\phi(d - a/2)S_p} \qquad (d)$$

The current Code considers an alternative to Eq. (d) by specifying the embedment length L_d (Art. 12.5) as follows:

For No. 11 or smaller bars	$0.04A_b f_y/\sqrt{f_c'}$
but not less than	$0.0004d_b f_y$
For No. 14 bars	$0.085f_y/\sqrt{f_c'}$
For No. 18 bars	$0.11f_y/\sqrt{f_c'}$

The lengths just given are to be multiplied by the applicable factor

Top reinforcement (more than 12 in of concrete below member)	1.4
Bars with f_y greater than 60,000 psi	$2-60,000/f_y$
For lightweight concrete	1.33
If spacing is at least 6 in on center and 3 in in from sides	0.8
Reinforcement in excess of that required	$A_{s(reqd)}/A_{s(furn)}$

In all cases $L_d \geq$ 12-in.

The development length for bars in compression is the largest of the following:

$$\frac{0.02f_y d_b}{\sqrt{f'_c}} \quad \text{or} \quad 0.0003 f_y d_b \quad \text{or} \quad 8 \text{ in}$$

If the compression bar is enclosed in a spiral of at least $\frac{1}{4}$ in bar diameter and 4-in pitch, the length may be reduced 25 percent, and if excess compression steel is furnished it may be reduced in the ratio as given for tension bars.

Standard hooks can be used to reduce the required value of L_d but are not usually required in foundation problems (Art. 12.8).

The shear equation can be developed from inspection of Fig. 8-3, which indicates that the force C must be carried across the neutral axis by shear; thus, summing horizontal forces

$$C + \text{shear} - C - \Delta C = 0$$

but shear $= b(dx)v_u$, neglecting second-order differentials as before, and observing that $\Delta T = \Delta C$, that $\Delta T(d - a/2) = V_u \, dx$, and making appropriate substitutions

$$v_u = \frac{V_u}{(d - a/2)b} \tag{e}$$

The ACI-ASCE Committee 326 Report on Shear and Diagonal Tension in 1962 indicated that the distance $(d - a/2)$ could be replaced by d, and we obtain

$$v_u = \frac{V_u}{bd} \tag{8-4}$$

to compute the actual concrete shear stress. This computed value is to be compared with the allowable *wide-beam* shear-stress value of Art. 11.4.1

$$v_c = 2\phi\sqrt{f'_c}$$

or the diagonal-tension shear value of Art. 11.10.3

$$v_c = 4\phi\sqrt{f'_c}$$

note that the ϕ factor has been put with the allowable value and not the computed value as is done in the Code.

Table 8-2. Allowable wide-beam and diagonal-tension shear by ACI 318-71 Code

	f_c, psi (kg/cm²)			
$\phi = 0.85$	3,000 (211)	3,500 (246)	4,000 (281)	5,000 (352)
Wide beam $2\phi\sqrt{f'_c}$:				
psi (kg/cm²)	93.1(6.5)	100.6(7.1)	107.5(7.6)	120.2(8.5)
ksf (kN/m²)	13.41(642)	14.49(694)	15.48(741)	17.31(829)
Diagonal tension $4\phi\sqrt{f'_c}$:				
psi (kg/cm²)	186.2(13.1)	201.1(14.1)	214.0(15.0)	240.4(17.0)
ksf (kN/m²)	26.81(1,283.7)	28.96(1,386.4)	30.96(1,475.4)	34.62(1,657.4)

The Code allows the use of shear reinforcement in foundation members (mats, spread footings, combined footings), but its use is not to be recommended, primarily because of the usual assumption of rigid members.

The bearing pressure of the column member on the footing may require dowels or increasing the depth of the footing if the contact stress is higher than the allowable compressive stress of the footing. The column contact stress is not to exceed $0.85\phi f'_c$ unless the footing surface is wider on all sides than the loaded area. If this is the case, then the permissible bearing stress becomes

$$f_c = 0.85\phi f'_c \sqrt{\frac{A_2}{A_1}} \qquad (f)$$

The ratio $\sqrt{A_2/A_1}$ cannot exceed 2, and the ϕ factor is taken as 0.7. The area A_2 is computed as the base of a frustum which can be placed in the footing with a height of d, an upper base of A_1, and sides of 1 vertical to 2 horizontal (Art. 10.14).

Table 8-2 gives allowable wide-beam and diagonal-tension shear values for several f'_c values. Table 8-3 summarizes the principal Code requirements applicable to concrete foundations (mats, footings, and retaining walls).

Table 8-3. Summary of foundation-member requirements ACI 318-71 (as of 1976)

Design factor	ACI Code article	General requirements
Spacing of reinforcement	7.4	Not less than D or 1 in or 1.33 × (max aggregate size); not more than 3 × depth of footing or 18 in
Lap splices	7.5.2	Not for bars > No. 11
In tension	7.6	See section in Code
In compression	7.7	See section in Code
Temperature	7.13	0.002 for $f_y = 40$ to 50 ksi
Shrinkage	10.5.2	$p = 0.0018*$ for $f_y = 60$ ksi or welded-wire fabric

(Continued)

Table 8-3 (*Cont.*)

Design factor	ACI Code article	General requirements
Minimum reinforcement cover	7.14	3 in against earth
Design-methods flexure	8.1	$M_u = \phi A_s f_y(d - a/2)$ $a = A_s f_y/0.85f'_c b$
Maximum reinforcement	10.3.2	$p_d = 0.75 \times$ Eq. (8-3) $p = A_s/bd \leq p_d$
Minimum reinforcement	10.5.1	$p \geq 200/f_y$ if footing of variable thickness; for slabs of uniform thickness use shrinkage and temperature percentage
k_1 factor	10.2.7	$k_1 = $ 0.85 for $f'_c \leq 4{,}000$ psi $0.85 - 0.05$ for each 1,000 psi over 4,000 psi
Limits of compression reinforcement	10.9	$0.01 \leq A_{st}/A_g \leq 0.08$
Modulus of elasticity	8.3	$E = w^{1.5}33\sqrt{f'_c}$ psi for w between 90 and 155 pcf $E_c = 57{,}000\sqrt{f'_c}$ psi for $w = 140$ to 150 pcf Take $n = E_s/E_c$ to nearest integer > 6
Load factors ϕ	9.2	Flexure $= 0.90$; shear $= 0.85$; bearing $= 0.70$; flexure plain concrete $= 0.65$
Load	9.3.1	$1.4 \times$ dead load; $1.7 \times$ live load
Bearing	10.14	$q_{brg} \leq \psi 0.85\phi f'_c \quad \psi \leq 2$
Shear, wide-beam	11.10.1a 11.2.1 11.4.1	$v_u = V_u/bd \quad v_c = 2\phi\sqrt{f'_c}$
Diagonal (punch tension)	11.10.1b 11.10.2 11.10.3	$v_u = V_u/bd \quad v_c = 4\phi\sqrt{f'_c}$
Shear reinforcement	11.11.1	For footings only 50% effective
Development of reinforcement	12.5 12.6	See values given in text
Grade beams	14.3 15.10	
Footings	15.1	General footing considerations
Location of bending moments	15.4.2	See Fig. 8-5
Distribution of reinforcing in rectangular footings	15.4.4	Percent in zone of width $B = 2/[L/(B + 1)]$
Shear	15.5 11.10.1a 11.10.1b	See Fig. 8-4

(*Continued*)

Table 8-3 *(Cont.)*

Design factor	ACI Code article	General requirements
Transfer of stress at base of column	15.6.5 10.14.1	At least 4 dowels with total $A_s \geq 0.005 A_g$
Unreinforced pedestals, footings	15.7	$f_c = 0.85\phi f'_c \quad \phi = 0.70$ $f_t = 5\phi\sqrt{f_c} \quad \phi = 0.65$
Round columns	15.8	Equivalent square column side, $w' = \sqrt{A_c}$
Minimum edge thickness	15.9	8 in for unreinforced footing; 6 in above reinforcement; 12 in on piles
Maximum tensile stress in unreinforced footings	15.7.2	$f_t \leq 5\phi\sqrt{f'_c} \quad \phi = 0.65$
Minimum wall thickness	14.2(k)	8 in
Moments	15.4.2	

* Recommend using 0.002 for all grades of steel

8-4. STRUCTURAL DESIGN OF SPREAD FOOTINGS

Shear stresses usually control the design of spread footings. Wide-beam shear may control the depth for rectangular footings when L/B is greater than about 1.2. Diagonal tension controls the depth for all square footings. The depth for diagonal tension results in a quadratic equation developed as follows:

Referring to Fig. 8-4b, the perimeter in shear for a square column is $4(w + d)$ and the depth to the center of the steel area is d. The soil pressure on the base of the footing is $q = P_u/BL$ in general. Summing vertical forces on the diagonal-tension zone (the footing weight cancels)

$$P_u = (w + d)^2 q - 4d(w + d)v_c = 0$$

rearranging

$$d^2\left(v_c + \frac{q}{4}\right) + d\left(v_c + \frac{q}{2}\right)w = \frac{BL - w^2}{4}q \qquad (8\text{-}5)$$

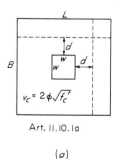

$v_c = 2\phi\sqrt{f_c'}$

Art. II.10.1a

(a)

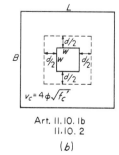

$v_c = 4\phi\sqrt{f_c'}$

Art. II.10.1b
II.10.2

(b)

Figure 8-4. *(a)* Section for wide-beam shear at a distance d from the column or wall face; *(b)* section for two-way diagonal (punching) shear at a distance of $d/2$ from the face of the column or wall. This shear analysis will generally control for square footings.

For round columns and using the diameter $= a$,

$$d^2\left(v_c + \frac{q}{4}\right) + d\left(v_c + \frac{q}{2}\right)w = \frac{BL - A_{col}}{\pi}q \tag{8-6}$$

One should not convert a round column to an equivalent square column and use Eq. (8-5) to obtain the depth, as this will be unsafe.

    Steps in footing design are:

1. Compute footing dimensions using the allowable soil pressure; for square footings

$$B = \sqrt{\frac{D + L}{q_a}}$$

2. Convert q_a to q_{ult} as follows (note L here is footing length):

$$q_{ult} = \frac{P_u}{BL}$$

3. With v_c from Table 8-2 and q_{ult} from the preceding step use Eqs. (8-5) or (8-6) as appropriate to find the effective concrete depth d.
4. For rectangular footings immediately check d for wide-beam shear, as it may control.
5. Compute the required steel for bending (and use the same amount each way) using q_{ult} and the critical section shown in Fig. 8-5. Treat a unit strip as a cantilever beam where the bending moment is computed as

$$M = \frac{ql^2}{2}$$

This value is used in Eq. (8-2) to find A_s. Check the resulting percentage of steel to satisfy minimum shrinkage or the maximum percent from Table 8-1.

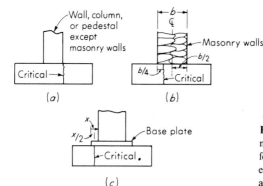

Figure 8-5. Sections for computing bending moment (ACI Art. 15.4.2). Bond is computed for section indicated in (a) for all cases; however, for convenience use bond at same section as moment.

This step is modified slightly from the Code, which requires taking the full footing width. It is self-evident that if the footing has a nonuniform but linear pressure distribution, the bending-moment computation will require modification to obtain the most critical bending-moment value.

The current Code procedure is based on tests by Richart (1948) which showed larger bending moments at the column face for column strips and lesser values on other strips. Bowles (1974a, Chap. 7), using finite-difference and finite-element analytical procedures, found that while the bending moment is higher in the column area, for finite-difference methods the average bending moment across the footing at the section taken in Fig. 8-5 is the same as the Code procedure (see also Example 8-2 reworked in Sec. 10-5). The maximum computed moment will exceed the average moment by about 30 percent for the finite-difference method and by more than 40 percent using the finite-element method and assuming column fixity—which is close to reality for concrete columns attached to the footing. It is implicit that readjustment will take place to reduce the cracking effect of the column-zone moment in the Code requirement. It may be questionable whether the 40 percent larger moment can be adequately readjusted without possible cracking and long-term corrosion effects. This problem was less severe when working-stress design methods were more popular than at present. The problem is such that it may be necessary in USD to use larger load factors than $1.4D$ and $1.7L$ for footing design in the near future. It is, of course, now permissible to use larger factors since the (any) Code provides only minimum values.

6. Compute column bearing and use dowels if bearing stresses are exceeded. It will be necessary to use a minimum of four dowel bars with an area of at least 0.5 percent of the area of the column or pedestal (Art. 15.6.5) regardless of the allowable bearing stress. If dowels are required to transfer column load, they must be of a length to satisfy compression requirements.
7. Detail the design.

Example 8-1. Design a plain concrete spread footing for the following conditions:

$$D = 222 \text{ kN} \qquad L = 311 \text{ kN} \qquad \text{Column} = 30 \text{ cm diam}$$

$$q_a = 145 \text{ kPa} \qquad f_y = 3,515 \text{ kg/cm}^2 \qquad w/6 \text{ bars at 20 mm}$$

$$f_c' = 211 \text{ kg/cm}^2 \text{ (column and footing)}$$

SOLUTION. *Step 1.* Find B

$$B = \sqrt{\frac{222 + 311}{145}} = 1.92 \text{ m} \qquad \text{Use 2 m}$$

Step 2. Find q_{ult}

$$q_{ult} = \frac{1.4(222) + 1.7(311)}{(2)(2)} = 209.9 \text{ kPa}$$

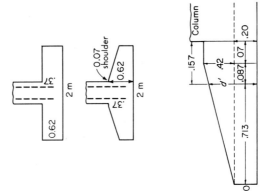

Figure E8-1c

Figure E8-1b

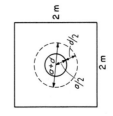

Figure E8-1a

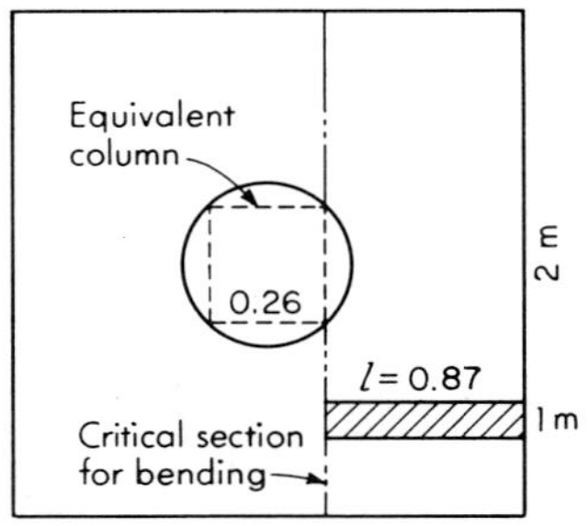

Step 3. Find d

$$d^2(v_c + q/4) + d(v_c + q/2)w = \frac{B - A_{col}}{\pi}q$$

$$v_c = 4\phi\sqrt{f'_c} = 1.06(0.85)\sqrt{211} \text{ in metric}$$

$$v_c = 13.088 \text{ kg/cm}^2 = 1,283.5 \text{ kPa}$$

$$v_c + q/4 = 1,335.98$$

$$(v_c + q/2)w = 416.53$$

$$\frac{B^2 - A_{col}}{\pi}q = \frac{4 - 0.7854(0.3)^2}{\pi}(209.9) = 262.5$$

$$1,335.98d^2 + 416.53d = 262.5$$

$$d^2 + 0.312d = 0.196$$

$$(d + 0.156)^2 = 0.2203$$

$$d = \pm 0.469 - 0.156 = 0.313 \text{ m}$$

Step 4. Check depth for bending
The equivalent side of a square column is

$$w' = \sqrt{0.7854(0.3)^2} = 0.26 \text{ m}$$

The bending moment is $M = qL^2/2$

where $L = (2 - 0.26)/2 = 0.87$ m
$M = 209.9(0.87)^2/2 = 79.44$ kN-m

The allowable tension stress according to Art. 15.7.2 is $5\phi\sqrt{f'_c}$, but modifying for metric

$$f_t = 1.33(0.65)\sqrt{211} = 12.56 \text{ kg/cm}^2$$

$$f_t = 12.56(98.07) = 1,231.5 \text{ kPa}$$

$$f_t = M/S$$

where S = section modulus = $bd^2/6$
Solving for d and taking $b = 1$-m strip

$$d = \sqrt{\frac{6(79.44)}{1(1,231.5)}} = 0.62 \text{ m} \qquad \text{(controls)}$$

Use $\qquad\qquad\qquad d = 0.62$ m

Step 5. Check bearing Arts. 10.14.1 and 15.6.5

$$A_1 = 0.7854(0.3)^2 = 0.071 \text{ m}^2$$

The equivalent diameter of the frustum entirely within the footing is

$$0.3 + 2(0.62)(2) = 2.78 \text{ m} \qquad \text{and} \qquad A_2 = 0.7854(2.78)^2 = 6.07 \text{ m}^2$$

$$\sqrt{A_2/A_1} = \sqrt{6.07/0.071} = 9.2 > 2 \qquad \text{Use 2}$$

The allowable bearing pressure is

$$f_c = 0.85\phi f'_c\sqrt{A_2/A_1} = 0.85(0.7)(211)(2) = 251 \text{ kg/cm}^2$$

$$= 24,616 \text{ kPa}$$

$$f_a = \frac{1.4(222) + 1.7(311)}{0.071} = 11,824 < 24,616$$

therefore, we need only minimum dowels per Art. 15.6.5. Use 6 dowels since there are 6 column bars

$$A_s = 0.005(0.071)(10,000) = 3.55 \text{ cm}^2$$

Use 6 bars at 20 mm to match column steel

$$L_d = 0.0755 f_y d_b / \sqrt{f_c'} \qquad \text{Appendix C of Code,}$$

$$\text{or } 0.00427 f_y d_b \text{ or } 20 \text{ cm}$$

Resulting in the following computations:

$$L_d = 0.0755(3,515)(2.0)/\sqrt{211} = 36.5 \text{ cm}$$

or

$$L_{d'} = 0.00427(3,515)(2.0) = 30 \text{ cm} > 20$$

the dowels must be 36.5 cm, say, 37 cm or 0.37 m < 0.62 furnished. We may use a constant-depth footing or either

(a) step it, or
(b) slope it

neither of which may be economical owing to extra labor costs, however, in checking a sloped footing. The edge must be 8 in or 20 cm minimum (Art. 15.9). d for shear at 0.157 m from column face is 0.313 m. The depth furnished is (leaving a 7-cm shoulder in order to place column forms)

$$d' = \frac{71.3}{80}(42) + (20) = 57.4 \text{ cm} \gg 31.3 \text{ required}$$

The bending requirement is

$$M = 209.9 \left(\frac{0.713}{2}\right)^2 = 53.35 \text{ kN-m}$$

$$d = \sqrt{\frac{6(53.35)}{1(1,231.5)}} = 0.51 \text{ m} < 57.4 \text{ furnished}$$

The sloped footing is satisfactory.

Example 8-2. Design a square footing by ultimate-strength design.

DATA. $D = 71$ kips $L = 100$ kips $q_a = 4$ ksf

$$f_c' = 3,000 \text{ psi (column and footing)} \qquad f_y = 50,000 \text{ psi}$$

Column $= 14 \times 14$ in with 4 No. 8 bars

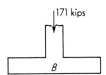

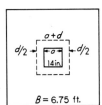

$B = 6.75$ ft.   **Figure E8-2a**

SOLUTION. *Step 1.* Find footing size

$$B = \sqrt{171/4} = 6.54 \qquad \text{Use 6.75 ft}$$

Step 2. Find depth for shear. We can use the same equation for depth based on punching shear d as for Example 8-1, using v_c ultimate and an ultimate soil pressure q_u.

$$d^2(v_c + q/4) + d(v_c + q/2)w = (B^2 - w^2)q/4$$

$$P_u = 1.4(71) + 1.7(100) = 269.4 \text{ kips}$$

$$q_{\text{ult}} = 269.4 = 5.91 \text{ ksf} \qquad v_c = 4\phi\sqrt{f'_c}$$

$$v_c = 4(0.85)\sqrt{3,000} = 186 \text{ psi or } 26.8 \text{ ksf}$$

$$v_c + q/4 = 26.8 + \frac{5.91}{4} = 28.28 \qquad (v_c + q/2)w = 34.81$$

$$(B^2 - w^2)q/4 = (6.75^2 - 1.17^2)\frac{5.91}{4} = 65.30$$

$$28.28d^2 + 34.81d = 65.30$$

$$d^2 + 1.23d = 2.31$$

Completing the square

$$(d + 0.615)^2 = 2.31 + 0.615^2$$

$$d = \pm 1.640 - 0.615 = 1.025 \text{ ft (12.3 in)} \qquad \text{Use } d = 13 \text{ in}$$

Not necessary to check wide-beam shear.

Step 3. Find required steel for bending; refer to sketch for L'

$$L' = \frac{6.75 - 1.17}{2} = 2.79 \text{ ft}$$

$$M = \frac{qL'^2}{2} = \frac{5.91(2.79)^2}{2}(12) = 276.02 \text{ in-kips}$$

$$M_u = \phi A_s f_y(d - a/2) \qquad \text{Eq. (8-2)} \qquad \phi = 0.9$$

$$a = \frac{A_s f_y}{0.85 f'_c b} = \frac{A_s(50)}{0.85(3)(12)} = 1.634 A_s$$

$$A_s\left(13 - \frac{1.634}{2} A_s\right) = \frac{276.02}{0.9(50)}$$

$$0.82 A_s^2 - 13 A_s = -6.13$$

$$A_s = 0.49 \text{ sq in/ft}$$

$$P = \frac{0.49}{(12)(13)} = 0.003 > 0.002 \qquad O.K.$$

$$< 0.0206 \qquad \text{Table 8-1 also } O.K.$$

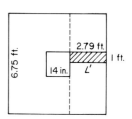

Figure E8-2b

Use 12 No. 5 bars at 6.75-in spacing· $A_s = 3.41 > 3.31$ sq in

$$L_d = 0.04 A_b f_y/\sqrt{f'_c} = 0.04(0.31)(50,000)/\sqrt{3,000} = 11.3 \text{ in} < 12 \qquad N.G.$$

$$L_d = 0.0004 d_b f_y = 0.0004(0.625)(50,000) = 12.5 > 12 \qquad O.K.$$

$$L_d \text{ required} = 12.5 \text{ in} \qquad L_d \text{ furnished} = 33.5 - 3 = 30.5 \text{ in}$$

Use $L_d = 30.5$ in \qquad allows 3-in cover (Art. 7.14)

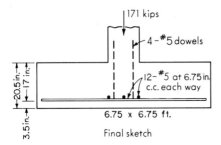

6.75 x 6.75 ft.

Final sketch   **Figure E8-2c**

Step 4. Check bearing Arts. 10.14 and 15.6.5

$$A_1 = 14^2 = 196 \qquad A_2 = (14 + 2(13)(2))^2 = 4,356$$
$$\psi = \sqrt{A_2/A_1} = 4.7 > 2 \qquad \text{Use 2}$$
$$f_c = 0.85(0.7)(3)(2) = 3.57 \text{ ksi}$$

The actual pressure is

$$f = 171/14^2 = 0.87 \text{ ksi} \ll 3.57$$

Use dowels per Art. 15.6.5

$$A_s = 0.005(196) = 0.96 \text{ sq in} \qquad \text{Use 4 No. 5 bars}$$
$$A_s = 4(0.31) = 1.24 \text{ sq in}$$
$$L_d = 0.02(50,000)(0.625)/\sqrt{3,000} = 11.4 > 8 \qquad O.K.$$

Check alternate

$$L_d = 0.0003(50,000)(0.625) = 9.4 < 11.4$$

Run dowels to flexure steel as in sketch.

Step 5. Design sketch
It will be necessary to provide at least 3 in of cover between bottom bars and soil.
It is necessary to place A_s both ways, and with such a shallow footing the top steel needs to be about 14 in down.
The final depth D_c is at least $13 + 0.625 + 3 = 16.625$ in taking d to center of bar crossing each way.
For convenience take $D_c = 17$ in and other dimensions as shown in sketch. This allows some bar misalignment.

8-5. BEARING PLATES FOR METAL COLUMNS

When the column is made of metal, the contact area will be small because of the higher allowable metal stresses. To transmit this stress to the footing, a bearing plate will be required.

$$P = A_p f_c \qquad\qquad (8\text{-}7)$$

Supplement 3, Art. 1.5.5 of AISC (1970, 1974) gives f_c as follows:

On full support area $0.35f'_c$

On less than full area $0.35f'_c\sqrt{\dfrac{A_2}{A_1}}$

But not more than $0.7f'_c$

one uses working loads with the above stresses. The $\sqrt{A_2/A_1}$ ratio is the same as used previously in Eq. (f), but the A_1 term is now taken as the area of the bearing plate.

The following design procedure for concentrically loaded column baseplates is adopted from the AISC Steel Construction Manual (1970, Supplement 3, 1974).

1. Find area of footing as $A_f = P/q_a$ if not already established.
2. Compute f_c as $0.35f'_c$ or $0.35f'_c\sqrt{A_2/A_1}$ but not more than $0.7f_y$.
3. Find area of plate, $A_p = P/f_c$.
4. Check A_p to see if correct f_c is used and adjust if necessary.
5. Establish B and C (Fig. 8-6). For maximum efficiency of the metal, in bending, make $m = n$.
6. Find actual bearing pressure q, psi.
7. Use largest value m or n to compute bending moment M.

$$M = \frac{q}{2}(m \text{ or } n)^2$$

8. Then $F_b = Mc/I \leq 0.75f_y$

and $$t = \sqrt{\frac{3q(m \text{ or } n)^2}{F_b}}$$ (8-8)

where t is the thickness of the bearing plate.

If there is axial load and a moment (eccentrically loaded plate), it will be necessary to adjust the steps to account for the moment. For example, in step 4,

$$q = \frac{P}{A} + \frac{Mc}{I} \leq f_c$$

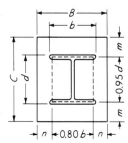

Figure 8-6. Dimensions for bearing plate. [After AISC (1973).]

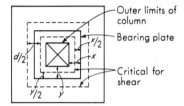

Figure 8-7. Critical section for shear in footing when column rests on a bearing plate.

With the bearing plate designed, the footing design proceeds as for a concrete column in that the depth for shear is next analyzed. To find the critical section for shear requires interpretation of the Code. The author's interpretation is shown in Fig. 8-7. This is based on reasoning that under compressive stress the edges of the bearing plate will tend to curl up and the diagonal tension crack will probably start at about the same place as the critical location for bond and moment.

Example 8-3. Design a reinforced-concrete footing with a metal column. Given the design data of Example 8-2 and that the 171-kip total load is carried by a 10 WF 45.

OTHER DATA. A-36 steel $f_y = 36$ ksi

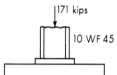

$q_a = 4$ ksi $f'_c = 3,000$ psi

$f_y = 50,000$ psi for reinforcing steel

$F_b = 0.75 f_y$ for baseplate

10 WF 45: $b = 8.022$ in

$d = 10.12$ in

SOLUTION. *Step 1.* Find footing area, Since loads and soil pressure are the same as in Example 8-2,

$B = 6.75$ ft   and $q = 3.75 +$ about 0.035 ksf

Step 2. Find dimensions of bearing plate. d may be larger than 13 in used because loads are the same but the effective plate area is less. Estimating that A_p will be about 10×12, then

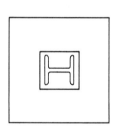

Figure E8-3a

$$A_1 = 120$$

$$A_2 = [10 + 4(13)][12 + 4(13)] = 3,968$$

$$\psi = \sqrt{3,968/120} = 5.75 \gg 2 \qquad \text{Use 2}$$

this is so large it will not have to be revised later

$$f_c = 0.85(0.70)f'_c(2) = 3.57 > 0.7f'_c$$

so we will use $f_c = 0.7f'_c = 2.1$ ksi

$$A_p = \frac{171}{2.1} = 81.42 < \text{projected area of column}$$

Use plate 9×11 with $2\frac{3}{4}$-in anchor bolts of $f_y = 50$ ksi steel. Anchor-bolt embedment is (Arts. 12.5 and 12.8)

$$L_d = 0.04(0.44)540 = 9.5 \text{ in} \qquad \text{Use 12 in with standard hook.}$$

Step 3. Check plate t for bending

$$m = \frac{11 - 0.95(10.12)}{2} = 0.69$$

$$n = \frac{9 - 0.8(8.02)}{2} = 1.29$$

$$q = \frac{171}{9(11)} = 1.73$$

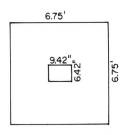

Figure E8-3b

The allowable bending stress will be taken as $0.75f_y = 27$ ksi. From Eq. (8-8)

$$t = \sqrt{\frac{3(1.73)1.29^2}{27}} = 0.565 \qquad \text{Use } \tfrac{5}{8}\text{-in plate}$$

Step 4. Find depth of footing for shear. From Fig. 8-7 the "column dimensions" are (refer also to Fig. E8-3b)

$$C = 11 - 2(0.69) = 9.62 \text{ in}$$

$$B = 9 - 2(1.29) = 6.42 \text{ in}$$

$$v_c = 0.186 \text{ ksi}$$

Because of the nonsquare column and usual rounding of depths, we will try values to find a satisfactory d. Try $d = 14$ in and the resulting perimeter shear resistance for diagonal tension is

$$2(13)v_c[(9.62 + 13) + (6.42 + 13)] = 203.3 > 171$$

Since this computation neglects soil pressure on this block, also resisting the 171-kip load, use $d = 14$ in.

Step 5. Compute required steel for bending. From sketch used in step 4 the largest moment arm is

$$L = \frac{6.75 - 0.54}{2} = 3.105$$

$$q_{\text{ult}} = 5.91 \qquad \text{(Example 8-2)}$$

$$M = \frac{5.91(3.105)^2(12)}{2} = 341.9$$

From Example 8-2, $a = 1.633A_s$

$$A_s\left(13 - \frac{1.633}{2}A_s\right) = \frac{341.9}{(0.9)(50)}$$

$$A_s - 15.92A_s = -9.31$$

$$A_s = \pm7.35 + 7.96 = 0.61 \text{ sq in/ft}$$

$$P = \frac{0.61}{(12)(13)} = 0.0039 > 0.002$$

$$< 0.0206 \qquad \text{Table 8-1}$$

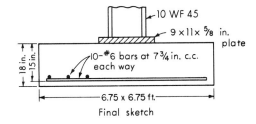

Figure E8-3c

Use 10 No. 6 bars at $7\tfrac{3}{4}$-in spacing each way. $A_s = 4.4 > 4.12$.

8-6. PEDESTALS

When it is necessary to place a footing at a depth in the ground and metal columns are used, a *pedestal* is used to carry the column load through the floor to the footing. This avoids possible corrosion of the metal from the soil. Careful backfill over the

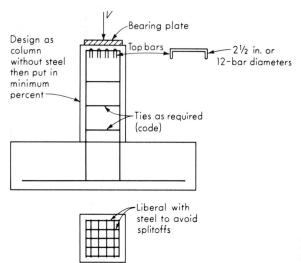

Figure 8-8. Pedestal details (approximate only).

footing and around the pedestal will be necessary if a floor slab is placed over the ground so that ground subsidence does not cause the floor to crack. Pedestals may be used for concrete columns as stage construction but are not usually economical.

The Code allows both reinforced and unreinforced pedestals (Art. 15.7.1). Generally the minimum percent steel for columns of $0.01A_g$ of Art. 10.9.1 should be used even if the pedestal is designed on the basis of no reinforcing steel. When steel bearing plates are used, the design should be on the basis of no steel to avoid point loads from the steel bars acting on the plate and increasing the bending stress.

Steel should be liberally added at the top to avoid spalls and to keep the edges from cracking as in Fig. 8-8. Room must be left, however, to place anchor bolts to hold the bearing plate (and column) in correct position.

Pedestals (and footings) are usually overdesigned because the overdesign is a small cost factor for the enormous potential benefit.

Pedestals can be designed as short columns ($L/r < 60$) because of the lateral support of the surrounding soil. They may be designed for moment, but this is beyond the scope of this text, and the user is referred to textbooks on reinforced design or design aids as published by ACI and others.

For the rather common condition of "simply supported" columns connecting to a pedestal via a baseplate in a nonmoment connection the pedestal may be designed on the basis of the following formula:

$$P_u = \phi(0.85f'_c A_c + A_s f_y) \tag{8-9}$$

where A_c = net area of pedestal concrete $(A_g - A_s)$; A_s = total steel area furnished; $\phi = 0.70$ for tied and 0.75 for spiral columns (pedestal). The use of this procedure will be illustrated by the following example.

Example 8-4. Design a pedestal and bearing plate for the following conditions:

W8 × 67 column of grade 50 steel ($f_y = 50$ ksi)

$D = 180$ kips $L = 140$ kips

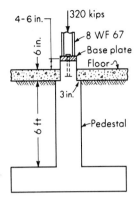

Figure E8-4a

The bearing plate is also grade 50 steel

$$f'_c = 3.75 \text{ ksi} \qquad f_y = 50 \text{ ksi (rebars)}$$

$$q_a = 4 \text{ ksf}$$

COLUMN DATA. $d = 9$ in, $b_f = 8.29$ in (AISC Handbook)

SOLUTION. *Step 1.* Assume the bearing plate is approximately the same dimensions as the top of the pedestal. We may revise it down later. The pedestal will be expanded 3 in as shown in the sketch to fit the floor slab. With the plate the same size as the pedestal top

$$f_c = 0.35 f'_c \text{ (AISC)}$$

$$f_c = 0.35(3.75) = 1.31 \text{ ksi}$$

and the baseplate must be

$$A_p = \frac{180 + 140}{1.31} = 244.3 \text{ sq in}$$

The area of the pedestal from Eq. (8-9), using no steel (later put in $0.01A_g$) and assuming "tied," is

$$A_c = \frac{1.4(180) + 1.7(140)}{0.85(3.75)(0.70)} = 219.6 \text{ sq in}$$

Use a pedestal 16 × 16 which will enlarge to 19 × 19 at the bottom of the floor slab.

Step 2. Design pedestal steel. Minimum steel $= 0.01(16)^2 = 2.56$ sq in. Use 4 No. 8 rebars with A_s furnished $= 4(0.79) = 3.16$ sq in. Use No. 3 bars for ties (Art. 7.12.3)

$$s = 16 \times 1 = 16 \text{ in}$$

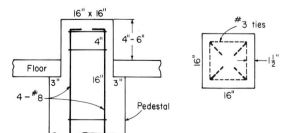

Figure E8-4b

or
$$s = 48 \times \tfrac{3}{8} = 18 \text{ in}$$

or
$$s = \text{column dimension}$$

Use
$$s = 16 \text{ in}$$

Place top tie with $1\tfrac{1}{2}$-in top cover and bend top bars toward center for at least 6 in.

Step 3. Design anchor bolts. Use a plate 16×16 with two $1\tfrac{1}{4}$-in anchor bolts $\times L_d$ of $f_y = 50$ ksi steel. These can be tightened to 29.07 kips each and should hold the column sufficiently.

The bearing plate will be shop-welded to the column and prepunched for the anchor bolts.

$$L_d = (0.75)0.085 f_y / \sqrt{f'_c} \qquad \text{(using value for No. 14 bar and considering that anchor is enclosed by tie)}$$

$$f_h = 330\sqrt{f'_c} \quad \text{(Art. 12.8)}$$
$$= 20{,}208 \text{ psi}$$

and
$$L_d = 0.75(0.085)(20208) / \sqrt{f'_c} = 21 \text{ in}$$

Use a standard hook on the anchor bolt of $6 \times 1\tfrac{1}{4} = 7.5$ in

Step 4. Design bearing plate. Plate $= 16 \times 16$

$$m = \frac{16 - 0.95d}{2} = \frac{16 - 8.55}{2} = 3.73 \text{ in}$$

$$n = \frac{16 - 0.8b}{2} = \frac{16 - 6.63}{2} = 4.68$$

$$q = \frac{320}{256} = 1.25 \text{ ksi} \qquad F_b = 0.75 f_y$$

$$t = \sqrt{\frac{3qn^2}{F_b}} = \sqrt{\frac{3(1.25)(4.68)^2}{37.5}} = 1.48, \quad \text{say, 1.5 in}$$

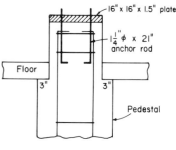

Figure E8-4c

Use a bearing plate $16 \times 16 \times 1.5$ in thick.

8-7. RECTANGULAR FOOTINGS

Rectangular footings are necessary where square footings cannot be used because of space limitations. The design is quite similar to that for a square footing. The depth will be controlled by shear, except that wide-beam action will probably control if the length exceeds the width by a ratio much greater than 1.

One other special consideration for rectangular footings is in the placement of the reinforcement. The reinforcement in the long direction is computed in the same manner as for a square footing. For the short direction a conservative approach would be to use the same steel area per foot of width as computed for the long direction. Since this is generally too conservative, the Code recommends that the required steel area be computed separately in the short direction by taking a section as illustrated in Example 8-5.

The area of footing adjacent to the column, being more effective in resisting bending, is the basis for an adjustment of the steel in the short direction. This adjustment places a percentage of the steel in a zone centered on the column or

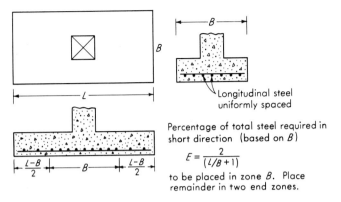

Longitudinal steel
uniformly spaced

Percentage of total steel required in
short direction (based on B)

$$E = \frac{2}{(L/B + 1)}$$

to be placed in zone B. Place
remainder in two end zones.

Figure 8-9. Placement of steel in short direction of a rectangular beam based on Code Art. 15.4.4.

pedestal with a width equal to the length of the short side B of the footing computed as follows:

$$E = \frac{2}{S + 1} \qquad (8\text{-}10)$$

where E = percentage of total steel required in short direction to be
placed in zone of width B
S = ratio of long side to short side, L/B

This adjustment of steel in the short direction, using Eq. (8-10), is illustrated in Fig. 8-9.

Example 8-5. Design a rectangular reinforced-concrete footing, given the following design data and metric units:

f'_c (column) = 352 kg/cm^2 f_y (column and footing) = 4,219 kg/cm^2

Square column 45 × 45 cm with eight 25-mm bars

$D = 1,110$ kN $L = 1,022$ kN

f'_c (footing) = 211 kg/cm^2 $q_a = 240$ kPa

SOLUTION. *Step 1.* Find footing dimensions

$$BLq_a = P$$

$$B = \frac{1,110 + 1,022}{2.2(240)} = 4.04 \qquad \text{Use 4.1 m}$$

and

$$q_{ult} = \frac{1.4(1,110) + 1.7(1,022)}{(4.1)(2.2)} = 364.9 \text{ kPa}$$

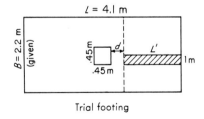

Trial footing

Figure E8-5a

Step 2. Find footing depth for shear. Since $L/B = 4.1/2.2 = 1.86$, check wide-beam first.
For a strip 1 m wide as shown and d from the column, summing forces: $d(1)v_c = (1.825 - d)(364.9)$

$$v_c = 642 \text{ kPa}$$

$$642d = 665.94 - 364.9d$$

$$1,006.9d = 665.94$$

$$d = 0.66 \text{ m}$$

Check diagonal tension (approximate)

$$\text{Perimeter} = (0.45 + 0.66)4 = 4.44$$

$$P_s = 4.44(1,283.7)(0.66) = 3,761.7 > 3,291.4 \text{ kN}$$

more refined analysis is not necessary. Final depth for shear, $d = 0.66$ m.

Step 3. Find steel required in long direction

$$L = 1.825 \text{ m} \qquad q_{ult} = 364.9 \text{ kPa}$$

$$M_u = \frac{364.9}{2} 1.825^2 = 607.7 \text{ kN-m}$$

$$a/2 = \frac{4,219A_s}{0.85(211)(1)(2)} = 11.76A_s$$

$$A_s(0.66 - 11.76A_s) = \frac{607.7}{(0.9)(4,219)(98.07)}$$

$$0.66A_s - 11.76A_s^2 = 0.00163$$

$$A_s^2 - 0.056A_s = -0.000139$$

$$(A_s - 0.028)^2 = -0.000139 + 0.000787$$

$$A_s = \pm0.0255 + 0.028$$

$$A_s = 0.00258 \text{ m}^2 = 25.8 \text{ cm}^2/\text{m}$$

This problem is quite sensitive to the number of decimal places used.

$$P = \frac{0.00258}{1(0.66)} = 0.0039 > 0.002 \qquad \text{instead of Code 0.0018}$$

$$< 0.016 \qquad \text{Table 8-2} \qquad O.K.$$

$$A_s \text{ (total)} = 2.2(25.8) = 56.76 \text{ cm}^2$$

From Table 8-4 using metric reinforcing bars use twelve 25-mm bars

$$A_s = 58.91 \text{ cm}^2 > 56.8$$

$$L_d = 84.7 < 175 \text{ cm} \qquad \text{available}$$

space at 2.2/12 = 18.3 cm with 7.5-cm side cover.

Step 4. Find steel in short direction

$$L' = \frac{2.2 - 0.45}{2} = 0.875$$

$$M = 364.9\left(\frac{0.875}{2}\right)^2 = 139.7$$

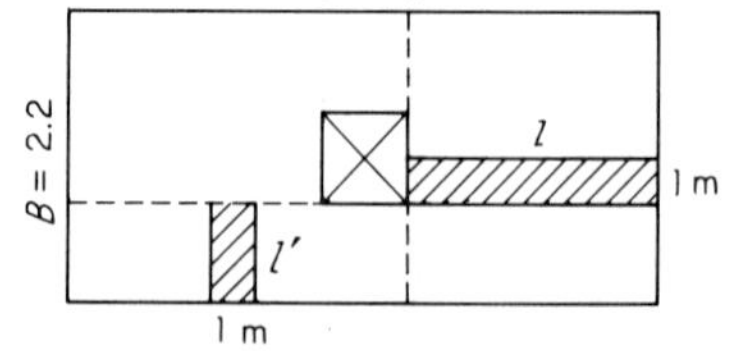

Figure E8-5b

use $d = 0.66 - 0.025 - 0.025/2 = 0.62$ to allow placing short side steel on top of longitudinal steel.

$$A_s(0.62 - 11.76A_s) = \frac{139.7}{(0.9)(4,219)(98.07)}$$

solving

$$A_s = \pm0.0257 + 0.0263 = 0.00057 \text{ m}^2/\text{m}$$

or $\qquad$ $A_s = 5.7 \text{ cm}^2/\text{m}$

$$p = \frac{0.00057}{(1)(0.66)} = 0.0008 < 0.002 \qquad \text{shrinkage controls}$$

Note that Art. 7.13 allows $p = 0.0018$; however, the author recommends 0.002 for all grades. Since the p to satisfy shrinkage controls

$$A_s = 0.002(0.66) = 0.0012 \text{ m}^2/\text{m} \qquad \text{or} \qquad 12 \text{ cm}^2/\text{m}$$

$$A_s(\text{total}) = 12(4.1) = 49.2 \text{ cm}^2$$

since shrinkage controls, do not use Eq. (8-10). Use at least $4.1/0.3 = 13^+$ bars for reasonable spacing.
    Use sixteen 20-mm bars $A_s = 50.27 > 49.2$; $L_d = 54.2 < 80$ cm furnished; space at 25.6 cm with end bars in 12.8 cm > 7.5.

    Step 5. Check bearing and design dowels.

$$A_1 = 0.45^2 = 0.2025 \text{ m}^2$$

$$A_2 = (0.45 + 4(0.66))^2 = 9.548$$

$$\psi = \sqrt{A_2/A_1} = \sqrt{9.548/0.2025} = 6.9 > 2 \qquad \text{Use 2}$$

$$f_c = 0.85(0.7)(211)(2)(98.07) = 24{,}624 \text{ kPa}$$

$$f_a = \frac{1.4(1{,}110) + 1.7(1{,}022)}{0.2025} = 16{,}253.8 \ll 24{,}624 \qquad O.K.$$

Use minimum dowels of $0.005A_c$

$$A = 0.005(45)^2 = 10.125 \text{ cm}^2$$

Use four 20-mm bars $A_s = 4(3.142) = 12.57 \text{ cm}^2$

$$L_d = 0.0755 \frac{f_y d_b}{\sqrt{f_c'}} = \frac{0.0755(4{,}219)(2)}{\sqrt{211}} = 43.8 \text{ cm} \qquad \text{(controls)}$$

or $\qquad L_d = 0.00427 f_y d_b = 0.00427(4{,}219)(2) = 36.0 \text{ cm}$

or $\qquad L_d = 20 \text{ cm}$

Use dowels with $L_d \cong 45$ cm

    Step 6. Design sketch

Make $\qquad\qquad\qquad\qquad D_c = 0.66 + \dfrac{0.025}{2} + 0.075 = 0.75 \text{ m}$

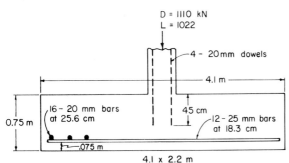

4.1 x 2.2 m

Final sketch

Figure E8-5c

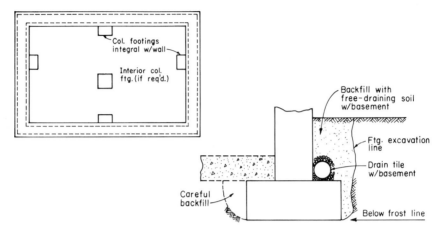

Figure 8-10. Footings for residential construction.

8-8. WALL FOOTINGS AND FOOTINGS FOR RESIDENTIAL CONSTRUCTION

Wall footings are continuous-strip footings for load-bearing walls. Sometimes they are corbeled out to accommodate columns integral with the wall. In these cases the columns support a major portion of the interior floor loads; the walls carry self-weight and perimeter floor loads. Figure 8-10 illustrates typical wall footings.

Design of a wall footing consists in providing a depth adequate for wide-beam shear (which will control as long as d is less than or equal to $2/3 \times$ footing projection). The remainder of the design consists in providing sufficient reinforcing steel for bending requirements of the footing projection. Longitudinal steel is required to satisfy shrinkage requirements. Longitudinal steel will, in general, be more effective in the top of the footing than in the bottom, as shown in Fig. 8-11.

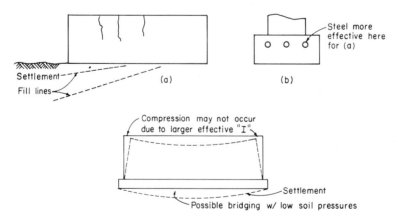

Figure 8-11. Settlements of residences.

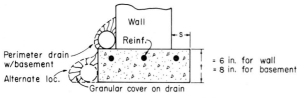

Wall

Reinf. ⊣ s ⊢

Perimeter drain w/basement

Alternate loc.

= 6 in. for wall
= 8 in. for basement

Granular cover on drain

	s, inches	
	Frame construction	Masonry or veneer
1-story, basement	3	4
no basement	2	3
2-story, basement	4	5
no basement	3	4

Figure 8-12. Federal Housing Administration (FHA) minimum wall-footing dimensions. Recommend use of at least 2 No. 5 reinforcing bars (author's, not FHA). Always use perimeter drains with a basement. [Further details in Bowles (1974*b*).]

Wall footings for residential construction are usually of dimensions to satisfy local building codes or Federal Housing Administration (FHA) requirements or of dimension to allow placing foundation walls. The contact pressure is usually on the order of 350 to 500 psf including the wall weight. The FHA requirement is shown in Fig. 8-12. Again longitudinal steel, if used, should be placed in the top rather than the bottom for maximum effectiveness in crack control when the foundation settles.

Interior footings for residential construction are usually nonreinforced and sized to carry not over 6 to 10 kips, resulting in dimensions on the order of 1.5 to 5 ft square or rectangle foundations. Often these footings can be placed in predrilled auger holes which are taken to a depth below seasonal volume change. Additional information on foundations for residential construction can be found in Bowles (1974*b*).

Example 8-6. Design a wall footing for the given data. Wall load consists in 70.1 kN/lin m ($D = 50$, $L = 20.1$) including wall, floor, and roof contribution.

$$f'_c = 211 \text{ kg/cm}^2 \qquad f_y = 4,219 \text{ kg/cm}^2$$

$$q_a = 200 \text{ kPa} \qquad \text{wall of concrete block } 20 \times 30 \times 40 \text{ cm}$$

SOLUTION. From Table 8-2 v_c wide-beam = 642 kPa
v_c diagonal tension = 1,283.7 kPa

Step 1. Find footing width

$$B = \frac{70.1}{200} = 0.35 \text{ m}$$

since this is only 5 cm wider than the concrete block (30 cm), we will arbitrarily make the footing project 30 cm on each side of wall or

$$B = 90 \text{ cm}$$

we will arbitrarily make the depth 40 cm deep overall ($d = 32$ cm)

$$q_{ult} = \frac{1.4(50) + 1.7(20.1)}{0.90} = 115.7 \text{ kPa}$$

Step 2. Check shear for $d = 32$ cm; since $d > 2/3$ projection, check diagonal tension $(0.30 - 0.32/2)115.7 = 16.198$ kN; shear resistance $= 0.32(1)(1,283.7) = 410.8 \geqslant 16.2$ *O.K.*

Step 3. Find steel for bending

$$L = 0.30 + 0.30/4 = 0.375 \text{ m} \qquad \text{(Fig. 8-5}b\text{)}$$

$$M = \frac{115.7(0.375)^2}{2} = 8.14 \text{ kN-m}$$

$$a = 11.76A_s \qquad \text{(from Example 8-5)}$$

$$A_s(0.32 - 11.76A_s) = \frac{8.14}{(0.9)(4,219)(98.07)}$$

$$A_s^2 - 0.0272A_s = -1.859 \times 10^{-6}$$

$$A_s = \pm 0.01353 + 0.0136 = 0.00007 \text{ m}^2 = 0.7 \text{ cm}^2$$

For shrinkage $A_s = 0.002(40)(100) = 8 \text{ cm}^2/\text{m}$ (note 0.002, not 0.0018)
 Use six 14-mm bars/meter $\qquad s = 16.7 \text{ cm}$

$$A_s = 6(1.539) = 9.2 \text{ cm}^2 > 8 \qquad O.K.$$

Step 4. Longitudinal steel; from step 3 shrinkage controls, use eight 14-mm bars with six bars at 4 cm from top, two bars at 8.0 cm from bottom

$$A_s \text{ furnished} = 12.3 \text{ cm}^2 > 8 \text{ required} \qquad O.K.$$

This is rather arbitrary; however, the quantity of steel is small and the bars are easy to place. The bottom bars will serve as support for the horizontal bars and result in some labor savings.

Step 5. Design sketch

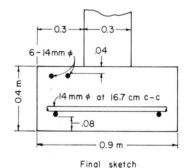

Final sketch   **Figure E8-6**

8-9. SPREAD FOOTINGS WITH OVERTURNING MOMENT

Gravity loads in combination with unsymmetrical frames and wind (or earthquake) analyses may produce column bending moments to be carried by footings. A carefully backfilled footing 1.2 or more meters in the ground can carry a substantial moment before rotation is sufficient to cause the moment capacity to dissipate. When columns are adequately doweled to carry the moment (using the same amount of steel into the footing as in the column) or if the baseplate of steel columns is designed to transmit moment, then it may be appropriate to analyze the footing for both axial load and bending moment.

Gravity-load moments due to unsymmetrical frames will be of known direction.

Wind and earthquake moments will be subject to reversal, which must be taken into account in the design.

Although much has been written in the past on footings with eccentricity and special methods for footings with eccentricity—especially for the resultant outside the middle one-third (kern limit), it would be poor practice to design footings on the basis of only part of it effective (see also Sec. 9-5).

The design will proceed as for a footing without moment. Design for shear, considering that it may now be possible for wide-beam shear to control owing to higher pressures on one side of the footing. Depending on the nature of the moment, reinforcing steel will be the same in both directions or heavier on the resisting side of the footing. Example 8-7 illustrates part of the design of a footing with moment.

Example 8-7. Given the following data including a clockwise moment of 25 ft-kips.

DATA. $D = 71$ $L = 125$ kips $q_a = 4$ ksf $f'_c = 3,000$ $f_y = 50,000$ psi
Moment due to loads, not wind.

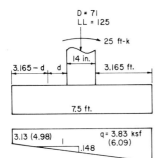

Figure E8-7

SOLUTION. *Step 1.* Find footing dimension B

Referring to the sketch, we will attempt to keep maximum soil pressures less than 4 ksf. We could write an expression and solve for B, but it will be easier to try for a value

$$B = \sqrt{\frac{71 + 125}{4}} = 7 \text{ ft, try } 7.5 \text{ ft}$$

Check soil pressure: Section 9-5 gives an equation of

$$q = \frac{P}{A}\left(1 \pm \frac{6e}{B}\right) \qquad e = \frac{M}{P} = \frac{25}{196} = 0.128$$

$$q = \frac{196}{7.5^2}\left[1 \pm \frac{6(0.128)}{7.5}\right] = 3.83 < 4 \qquad O.K.$$

$$= 3.48(1 - 0.1) \qquad = 3.13 \qquad O.K. \qquad (\text{use } 7.5 \text{ ft})$$

Compute q_{ult}:

$$P_u = 1.4(71) + 1.7(125) = 311.9$$

$$q = 3.83\left(\frac{311.9}{196}\right) = 6.09 \text{ ksf}$$

$$= 3.13\left(\frac{311.9}{196}\right) = 4.98 \text{ ksf}$$

Step 2. Find depth for shear. Since the pressure is linear, we may use an average value of $q_{ult} = 5.54$ in the column zone for diagonal tension, for which the depth is found to be

$$d = 1.15 \text{ ft} \qquad \text{say 14 in}$$

Check wide-beam shear (referring to sketch)

$$V_u \text{ (wide-beam)} = (7.5)(3.165 - d)\left[\frac{6.09 + 6.09 - (3.165 - d)0.148}{2}\right]$$

$$= (23.74 - 7.5d)(5.86 + 0.074d)$$

$$= 139.12 - 42.19d - 0.555d^2$$

This is equal to the resisting shear developed by the concrete of

$$V_c = 7.5(13.41)d$$

and equating

$$0.555d^2 + 142.76d = 139.12$$

solving, $d = 0.965 \text{ ft} < 1.15$ Use $d = 14$ in
    If these had been wind moments, the values with wind would be compared as

$$P_{u(g)} = 1.4D + 1.7L < P_{u(w)} = 0.75[1.4D + 1.7(L + W)]$$

then v_c of 26.81 and 13.41 would have been increased 25 percent to 33.51 and 16.76, respectively, and q_{ult} would have been based on $P_{u(w)}$.

Step 3. Find bending moment for design of A_s (for strip 1 ft wide).

$$M = \int_0^x V \, dx = \int_0^{3.165} \left(6.09x - \frac{0.148x^2}{2}\right) dx$$

$$= 6.09 \frac{x^2}{2} - 0.148 \frac{x^3}{6}\Big]_0^{3.165}$$

There is no integration constant because at $x = 0$, $M = 0$; inserting limits at the column face

$$M = 6.09 \frac{(3.165)^2}{2} - 0.148 \frac{(3.165)^3}{6} = 29.72 \text{ ft-kips/ft}$$

The remainder of the problem is similar to earlier examples.

Table 8-4. Standard U.S. reinforcing bars and metric bars as elsewhere

Bar No.	Diameter, in	Area, sq in	Perimeter, in	Size, mm	Area, cm^2	Perimeter, cm
2	0.250	0.05	0.786	6	0.283	1.88
3	0.375	0.11	1.178	8	0.503	2.51
4	0.500	0.20	1.571	10	0.785	3.14
5	0.625	0.31	1.963	12	1.131	3.77
6	0.750	0.44	2.356	14	1.539	4.40
7	0.875	0.60	2.749	16	2.011	5.03
8	1.000	0.79	3.142	18	2.545	5.65
9	1.128	1.00	3.544	20	3.142	6.28
10	1.270	1.27	3.990	25	4.909	7.85
11	1.410	1.56	4.430	32	8.042	10.05
14S	1.693	2.25	5.32	35	9.621	11.00
18S	2.257	4.00	7.09	40	12.566	12.57

PROBLEMS

8-1. Design the assigned problem of Table P8-1 (refer to Fig. P8-1). Both columns and footings are square.

Table P8-1

		Column data					Footing data		
	w, in	f_y	f'_c	No. of bars	D, kips	L, kips	f_y	f'_c	q_a
a	18	60	4.0	15 No. 11	420	320	50	3.0	4.5
b	21	60	3.5	8 No. 10	250	285	50	3.0	3.5
c	13	50	3.5	4 No. 11	130	100	60	3.5	2.5
d	13	50	3.5	4 No. 10	100	110	60	4.0	4.5
e	18	60	4.0	8 No. 11	280	250	50	4.0	3.0
f	15	60	4.0	8 No. 9	155	200	60	3.0	2.0

Partial Answer: ( ) = p

	B, ft	d, in	A_s/ft
a	13.00	29.2	1.04 (0.003)
c	9.75	14.8	0.537(0.003)
e	13.5	22.2	0.98 (0.004)

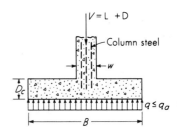

Figure P8-1

8-2. Design the assigned problem of Table P8-2 (refer to Fig. P8-1). Both columns and footings are square. Note column bars are in mm, f'_c and f_y in kg/cm²

Table P8-2

		Column data					Footing data		
	w	f_y	f'_c	No. of bars	D, kN	L, kN	f_y	f'_c	q_a, kPa
a	45	4,219	281	16 at 35	1,870	1,400	3,515	246	215
b	52	4,219	246	10 at 30	1,110	1,250	3,515	281	168
c	33	3,515	246	6 at 30	560	440	4,219	246	120
d	33	3,515	246	4 at 30	445	490	4,219	281	210
e	45	4,219	281	12 at 30	1,245	1,112	3,515	281	144
f	38	4,219	281	8 at 30	690	889	4,219	211	90

Partial Answer: ( ) = p

	B, m	d, cm	A_s/m
a	3.90	70.91	22.84(0.003)
c	2.90	37.11	11.23(0.003)
e	4.10	56.63	21.06(0.004)

8-3. Redo Prob. 8-1 if $L = 6.0$ ft.

Partial Answer: $d\text{-}t$ = diagonal tension; $w\text{-}b$ = wide-beam; $(\ ) = p$

	B, ft	d, in	A_s/ft (longit.)	A_s/ft (short)
b	25.50	41.21(w-b)	2.63(0.005)	0.99(0.002)
d	8.00	13.19(d-t)	0.72(0.005)	0.36(0.002)
f	29.75	32.28(w-b)	2.35(0.006)	0.77(0.002)

8-4. Redo Prob. 8-2 if $L = 2$ m.

Partial Answer: $d\text{-}t$ = diagonal tension; $w\text{-}b$ = wide beam; $(\ ) = p$

	B, m	d, cm	A_s/ft (longit.)	A_s/ft (short)
b	7.10	85.3(w-b)	55.68(0.007)	17.05(0.002)
d	2.30	33.4(d-t)	12.76(0.004)	9.08(0.003)
f	8.80	75.7(w-b)	47.78(0.006)	15.15(0.002)

8-5. Redo Prob. 8-1 if w = diam.

Partial Answer: $(\ ) = p$

	B, ft	d, in	A_s/ft
a	13.00	34.0	0.92(0.002)
c	9.75	17.4	0.47(0.002)
e	13.50	25.9	0.87(0.003)

8-6. Redo Prob. 8-2 if w = diam.

Partial Answer: $(\ ) = p$

	B, m	d, cm	A_s/m
a	3.90	82.5	20.09(0.002)
d	2.90	43.5	9.78(0.002)
e	4.10	66.2	18.40(0.003)

8-7. Design a wall footing for a concrete-block-wall building. The building has a 16-ft wall; the footing is 3.5 ft in the ground and has a plan of 40 × 120. The roof will weigh about 15 psf, and snow is 20 psf. The allowable soil pressure is 2 ksf, and about one-half of the building is on a fill of varying depth from 0 to 3 ft.

8-8. Design a wall footing for a two-story office building of concrete block and brick veneer. The building is 14 × 30 m in plan. The footing is 1.4 m below ground. The first-floor slab rests directly on the ground. Assume the floor dead load averages 1.5 kPa and live load 4.4 kPa. The roof is about 0.75 kPa, and snow is 0.9 kPa. Concrete blocks are 20 × 30 × 40 cm and weigh 4.2 kPa (wall surface). Brick (10 × 20 × 9 cm) will weigh 1.9 kPa (wall surface). The undrained shear strength q_u may be taken as 60 kPa.

8-9. Design the foundation for a residence with approximately 1,400 sq ft of floor area. A perimeter wall will be used and a single interior post-on-pad. Assume wood frame, aluminum siding, and brick trim. Take snow load at 15 psf. The plan is semirectangular with 32 × 45. Draw a building plan and place the post at a convenient location. Comment on the design as appropriate. (You must assume or specify any missing data needed for your design.)

8-10. Design a baseplate for a 10 WF 45 column carrying a 150-kip load. Use A36 steel, f'_c of footing of 3,500 psi, and an allowable soil pressure of 3 ksf.

8-11. Design a column baseplate for a 10 WF 72 column carrying a 210-kip load. Use grade 50 steel for column and baseplate, f'_c of the footing of 3,000 psi, and an allowable soil pressure of 4 ksf.

8-12. Design Prob. 8-10 with a pedestal 8 ft long. Design the footing, pedestal, and baseplate; use an average load factor of 1.55.

8-13. Design Prob. 8-11 for a pedestal 6 ft long. Design the footing, pedestal, and baseplate; use an average load factor of 1.60.

NINE

COMBINED FOOTINGS AND BEAMS ON ELASTIC FOUNDATIONS

9-1. INTRODUCTION

The preceding chapter presented elements of the design of spread and wall footings. This chapter considers some of the more complicated shallow-foundation problems. Among these are footings supporting more than one column in a line (combined footings), which may be rectangular or trapezoidal in shape, or two pads connected by a beam, as for a strap footing. Eccentrically loaded footings and unsymmetrically shaped footings will also be considered.

9-2. RECTANGULAR COMBINED FOOTINGS

When property lines, equipment locations, column spacing, or other considerations limit the footing clearance at the column locations, a possible solution is the use of a rectangular-shaped footing. This type of footing may support two columns, as illustrated in Fig. 9-1 and 9-2, or more than two columns with only slight modification of the design procedure. Bridge piers are also founded on combined footings. These footings are commonly designed by assuming a linear stress distribution on the bottom of the footing, and if the resultant of the soil pressure coincides with the resultant of the loads (and center of gravity of the footing), the soil pressure is assumed to be uniformly distributed. The linear pressure distribution implies a *rigid* footing on a homogeneous soil. The actual footing is generally not rigid, nor is the pressure uniform beneath it, but it has been found that solutions using this concept

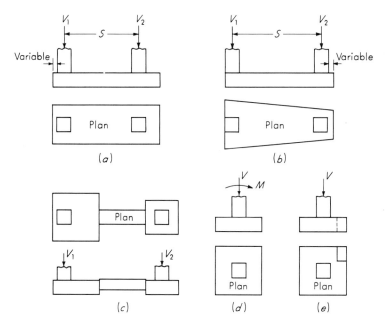

Figure 9-1. Typical special footings considered in this chapter. (*a*) Combined footing; (*b*) trapezoidal footing; (*c*) cantilever or strap footing; (*d*) footing with moment; (*e*) footing with a notch cut out.

are adequate. This concept also results in a rather conservative design. Consideration of the footing as a nonrigid member with nonuniform soil pressure is considered in Sec. 9-9.

The design of a rigid rectangular footing consists in determining the location of the center of gravity (cg) of the column loads and using length and width dimensions such that the centroid of the footing and the center of gravity of the column loads coincide. With the dimensions of the footing established, a shear and moment diagram can be prepared, the depth selected for shear (again it is conventional to make

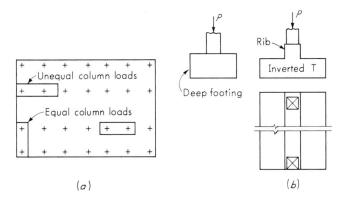

Figure 9-2. (*a*) Typical layout of combined footings for column loads as shown; (*b*) deep footings for heavy loads and the use of a rib or inverted T beam to reduce footing weight.

the depth adequate for shear without using shear reinforcement to satisfy rigidity requirements implicitly), and reinforcing steel selected for bending requirements. Critical sections for shear, both diagonal-tension and wide-beam, should be taken as indicated in Chap. 8. The maximum positive and negative moments are used to design the reinforcing steel and will result in steel in both bottom and top of the beam.

In the short direction, obviously, the entire length is not going to be effective in resisting bending. That zone closest to the column will be most effective for bending, and it is recommended that this approach be used. This is basically what the ACI Code specifies in Art. 15.4.4 for rectangular footings.

If it is accepted that the zone which includes the columns is most effective, what should this zone width be? Certainly, it should be something greater than the width of the column. Probably it should be no greater than the column width plus d to $1.5d$, depending on the column location based on the author's analytical work, lack of Code guidance, and recognizing that extra steel will "stiffen" the zone and increase the moments in this zone and reduce the moment out of the zone. An effective width using this method is illustrated in Fig. 9-3 (see also Example 9-1). For the remainder of the footing in the short direction, the ACI Code requirement for minimum-percentage steel (Art. 10.5 or 7.13) should be used.

In selecting dimensions for the combined footing, the length dimension is some-what critical if it is desired to have shear and moment diagrams mathematically close as an error check. This means that unless the length is exactly the computed value from the location of the cg of the columns, an eccentricity will be introduced into the footing, resulting in a nonlinear earth-pressure diagram. The actual as-built length, however, should be rounded to a practical length, say, to the nearest 0.25 or 0.5 ft (7.5 to 15 cm).

The column loads may be taken as concentrated loads for computing shear and moment diagrams. For design the shear and moment values at the edge (face) of the column should be used. The resulting error, using this approach, is negligible (Fig. 9-4).

If the footing is loaded by more than two columns, the problem is still statically determinate; the reactions (column loads) are known as well as the distributed loading, i.e., the soil pressure.

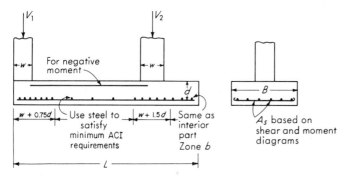

Figure 9-3. Steel placement for rectangular combined footing.

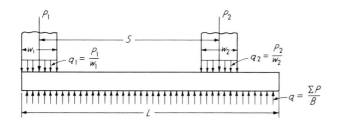

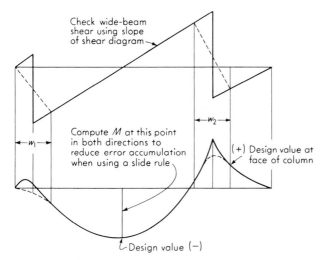

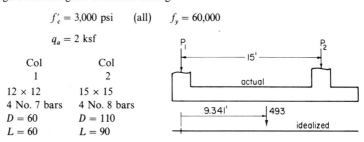

Figure 9-4. Shear and moment diagrams (qualitative) for a combined footing considering the column loads as point loads and as distributed loads (dashed line). It can be seen that in the design areas it makes no difference how the diagrams are drawn, and the point-load case is much simpler.

Example 9-1. Design of a rectangular combined footing.

Given:

$$f'_c = 3{,}000 \text{ psi} \quad \text{(all)} \quad f_y = 60{,}000$$

$$q_a = 2 \text{ ksf}$$

Col 1	Col 2
12 × 12	15 × 15
4 No. 7 bars	4 No. 8 bars
D = 60	D = 110
L = 60	L = 90

Figure E9-1a

SOLUTION. *Step 1.* Convert loads to ultimate

$$P_{u1} = 1.4(60) + 1.7(60) = 186 \text{ kips}$$

$$P_{u2} = 1.4(110) + 1.7(90) = 307 \text{ kips}$$

$$P_1 + P_2 = 60 + 60 + 110 + 90 = 320$$

$$\text{Ultimate ratio} = \frac{186 + 307}{320} = 1.54$$

$$q_{ult} = 2(1.54) = 3.08 \text{ ksf}$$

This is necessary so that eccentricity is not introduced in finding L using working loads and then switching to "ultimate" values.

Step 2. Find *L* and *B*.

$$493\bar{x} = 15(307)$$

$$\bar{x} = 9.341 \text{ ft}$$

To make 493 fall at *L*/2, we have

$$L = (9.341 + 0.5)2 = 19.682$$

We will use 19.75, but to make the moment diagram close, use 19.68 ft.

Step 3. Find *B*

$$BLq = 493$$

$$B = \frac{493}{(19.68)(3.08)} = 8.13 \text{ ft}$$

use *B* = 8.17 ft

L/B = 19.75/8.17 = 2.42

Step 4. Draw shear and moment diagrams [author used computer program from Bowles (1974*a*)] by hand.

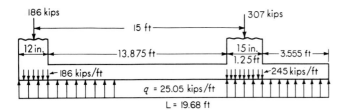

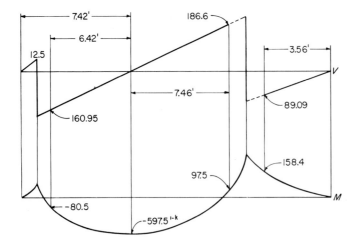

Figure E9-1*b*

Shear between columns 1 and 2 using integration

$$dV = \int q \, dx$$

$$V = 25.05x - 186(x - 0.5)$$

$$V = 0 \text{ at } x = 186/25.05 = 7.425 \text{ ft}$$

Moment between column 1 and 2

$$dM = \int_0^x dV \, dx$$

$$M = 25.05 \frac{x^2}{2} - 186(x - 0.5)$$

at $\qquad\qquad x = 1$ (right face of column 1)

$$M = \frac{25.05}{2} - 186(0.5) = -80.48 \text{ ft-kips}$$

Maximum M at $V = 0$ is

$$M = \frac{25.05(7.425)^2}{2} - 186(7.425 - 0.5) = 597.5 \text{ ft-kips}$$

other values would be similarly obtained.

Step 5. Select depth based on analysis for both wide-beam and diagonal tension.
a. Critical location for wide-beam is readily obtained.
b. Diagonal tension may have to be investigated for three conditions:
    1. 3 side zone, column 1
    2. 4 side zone, column 2
    3. 3 side zone, column 2
Check wide-beam first (slope of shear diagram = constant)

$$\text{Max. } V = 186.60 \text{ at column 2}$$

$$Bv_cd = 186.60 - 25.05d$$

$$v_c = 13.41 \text{ ksf (allowable)}$$

$$8.17(13.41)d = 186.60 - 25.05d$$

$$d = \frac{186.6}{134.58} = 1.39 \text{ ft (16.7 in)}$$

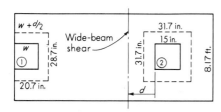

Figure E9-1c

Checking diagonal tension at column 1 using d just obtained

$$\text{Perim.} = (12 + 8.3)(2) + 12 + 16.7 = 69.3 \text{ in} = 5.78 \text{ ft}$$

$$A = \frac{(12 + 8.3)(28.7)}{144} = 4.05 \text{ sq ft}$$

$$V = P_{col} - P_{soil} = 186.0 - 4.05\left(\frac{25.05}{8.17}\right) = 173.6 \text{ kips}$$

$$V_c = 2(13.41) = 26.8 \text{ ksf} \qquad \text{(allowable)}$$

$$v = \frac{173.6}{(5.78)(1.39)} = 21.63 < 26.8 \qquad O.K.$$

At column 2

$$V = 307 - 6.98\left(\frac{25.05}{8.17}\right) = 285.6 \text{ kips}$$

$$v = \frac{285.6}{(10.57)(1.39)} = 19.4 < 26.8 \qquad O.K.$$

By inspection a 3-side diagonal tension is not critical.

Step 6. Design negative steel (between columns 1 and 2).

$$a = 1.96 A_s$$

$$A_s(d - a/2) = \frac{M_u}{\phi(60)}$$

$$A_s(16.7 - 0.98 A_s) = \frac{597.6(12)}{(0.9)(60)(8.17)}$$

$$A_s^2 - 17.03 A_s = -16.584$$

$$A_s = 1.037 \text{ sq in/ft}$$

$$p = \frac{1.037}{(16.7)(12)} = 0.005 > 0.003$$

$$< 0.016$$

Use 12 No. 8 bars at 8.2-in spacing

$$A_s = 12(0.79) = 9.48 > 8.47 \text{ sq in}$$

Run 1/3 of bars full length of footing (less 3 in each end).

$$L_d \text{ is O.K.}$$

Step 7. Find steel in short direction at column 1.

$$B' = 12 + 16.7(0.75) = 24 \text{ in}$$

$$q = \frac{186}{(8.17)(2)} = 11.38 \text{ ksf}$$

$$L' = \frac{8.18 - 1}{2} = 3.585$$

$$M = \frac{11.38(3.585)^2}{2}(12) = 877.5 \text{ in-kips}$$

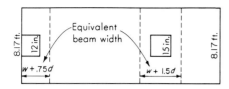

Figure E9-1d

Take $d = 1$ in less

$$A_s(15.7 - 0.98A_s) = \frac{877.5}{(0.9)(60)} = 16.25$$

$$A_s^2 - 16.02A_s = -16.58$$

$$A_s = \pm 6.90 + 8.01 = 1.11 \text{ sq in/ft}$$

p is O.K. from previous calculation

Use 4 No. 7 bars at 6 in

$$A_s = 2.40 > 2.22 \text{ sq in}$$

$$L_d = 0.04(0.60)60,000/\sqrt{3,000} = 26.3 \text{ in}$$

or $0.0004(0.875)(60,000) = 21$ in

$$L_d \text{ furnished} = 3.585(12) - 3 = 40 \text{ in}$$

At column 2

$$B' = 15 + 1.5(16) = 39 \text{ in} = 3.25 \text{ ft}$$

$$L' = \frac{8.17 - 1.25}{2} = 3.46$$

$$q = \frac{307}{(3.25)(8.17)} = 11.56$$

$$M = \frac{11.56(3.46)^2(12)}{2} = 830.35 \text{ in-kips}$$

$$A_s^2 - 16.02A_s = 15.69$$

$$A_s = \pm 6.96 + 8.01 = 1.05 \text{ sq in/ft}$$

Use 6 No. 7 bars at 6.5 in

$$A_s = 3.6 > 3.4 \text{ sq in}$$

Step 8. Check dowel requirements of column to footing. At column 1 the supporting area is not on *all* sides; therefore, the bearing stress is limited to

$$f_c = 0.85(0.7)(f'_c) = 1.785 \text{ ksi}$$

$$P = 12(12)1.785 = 257 > 186 \text{ kips}$$

Use 4 dowels to provide at least

$$A_s = 0.005(144) = 0.72 \text{ sq in}$$

Use 4 No. 6 for $4(0.44) = 1.76$ sq in

At column 2 with concrete all around

$$\psi = \sqrt{A_2/A_1} > 2 \qquad \text{Use } 2$$

$$f_c = 0.85(0.7)(f_c')(2) = 3.57 \text{ ksi}$$

$$P = 15(15)3.57 = 803 \gg 307$$

Use 4 dowels same size as column 1

Step 9. Steel in cantilever portion of footing is found to be 0.28 for moment and 0.67 sq in/ft (0.0033 per $200/f_y$) for minimum requirement.
Use 10 No. 7 bars $A_s = 6.00 > 5.47$ sq in
Run 5 bars full length to use as chairs for short-direction steel.

Step 10. Final sketch

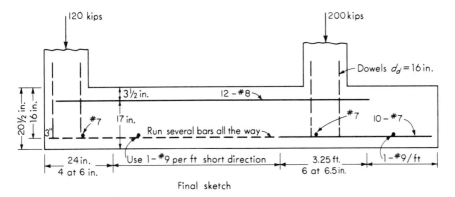

Final sketch

Figure E9-1e

9-3. DESIGN OF TRAPEZOIDAL-SHAPED FOOTINGS

This type of footing may be used to carry two column loads when space outside the structure is too limited for a spread footing and the exterior column carries the largest load. The location of the resultant force will then be closer to the large column, and doubling the centroid distance will not provide a length sufficient to reach the other column, as illustrated in Fig. 9-5, without introducing an eccentricity into the soil-pressure diagram.

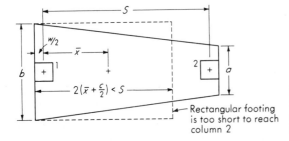

Figure 9-5. A trapezoid footing is required in this case unless the distance S is so great that a cantilever (or strap) footing would be more economical.

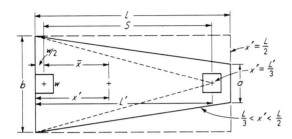

Figure 9-6. Conditions for valid trapezoidal-footing solution.

From the geometry of the trapezoid shown in Fig. 9-6, it can be shown that the area and the distance from the *large end* to the center of area are

$$A = \frac{a+b}{2}L \tag{9-1}$$

and

$$x' = \frac{L}{3}\frac{2a+b}{a+b} \tag{9-2}$$

From Eq. (9-2) it can be seen that if $a = 0$, we have a triangle; if $a = b$, we have a rectangle.

Therefore, it follows that a trapezoid solution exists between the limits:

$$\frac{L}{3} < x' < \frac{L}{2}$$

with the minimum value of L as out-to-out of the column faces.

For the most economical solution, simultaneous equations based on the minimum required area of the trapezoid and for the location of the centroid of area can be used to find the width dimensions a and b.

With the dimensions a and b established, the footing may be treated similarly to a combined footing in drawing the shear and moment diagrams, except that, in this case, the shear diagram will be a second-degree curve and the moment diagram a third-degree curve. For simplicity the column loading will be considered as point loads.

Example 9-2. Proportion and partially design a trapezoid footing.
Given data: columns = 0.46 m²

$D_1 = 1,200$ $L_1 = 816$ kN

$D_2 = 900$ $L_2 = 660$ kN

$f'_c = 211$ $f_y = 4,219$ kg/cm²

$q_a = 190$ kPa

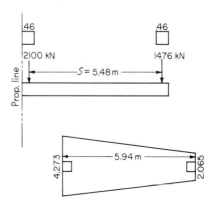

Figure E9-2a

SOLUTION. *Step 1.* Find q_{ult}

$$\text{Ultimate ratio} = \frac{1.4(1{,}200 + 900) + 1.7(816 + 660)}{1{,}200 + 900 + 816 + 660} = \frac{5{,}449}{3{,}576} = 1.52$$

$$q_{ult} = 190(1.52) = 289.5 \quad \text{kPa}$$

Step 2. Find end dimensions

$$5{,}449.2\bar{x} = 5.48[1.4(900) + 1.7(660)]$$

$$\bar{x} = \frac{13{,}053.4}{5{,}449.2} = 2.395 \text{ m}$$

and
$$x' = 2.395 + 0.46/2 = 2.625 \text{ m}$$

and from Eq. (9-1)

$$A = \frac{a+b}{2} L = \frac{a+b}{2} \quad (5.94)$$

but based on q_{ult} and the footing loads

$$A = \frac{5{,}449.2}{289.5} = 18.823 \text{ m}^2$$

equating
$$\frac{a+b}{2} (5.94) = 18.823$$

$$a + b = 6.338 \text{ m}$$

From Eq. (9-2)

$$x' = \frac{L}{3} \frac{2a+b}{a+b}$$

$$\frac{2a+b}{a+b} = \frac{3(2.625)}{5.94} = 1.326$$

but $a + b = 6.338$, from which $b = 6.338 - a$ and substituting for b

$$\frac{2a + 6.338 - a}{6.338} = 1.326$$

and solving $a = 2.065$ m and from back substitution

$$b = 6.338 - 2.065 = 4.273 \text{ m}$$

Step 3. Draw shear and moment diagrams

$$\text{Pressure big end} = 4.273(289.5) = 1{,}237.03 \text{ kPa/m}$$

$$\text{Pressure small end} = 2.065(289.5) = 597.82 \text{ kPa/m}$$

the slope of the pressure line $s = (1{,}237.0 - 598.1)5.94 = 107.5$

$$V = \int_0^x q \, dx$$

$$V = 1{,}237.0x - 107.5\frac{x^2}{2} + C$$

at $x = 0.23$ m, $C = 0$

$$V = 1{,}237.0(0.23) - 53.75(0.23)^2 = 282 \text{ kN}$$

at $x = 0.23 + dx$, $C = -3{,}067$

$$V = 282 - 3{,}067 = -2{,}785 \text{ kN}$$

at column 2 $x = 5.71$ $C = -3{,}067$

$$V = 2{,}242 \quad \text{and at } 5.71 + dx \quad V = -140 \text{ kN}$$

values of shear at faces of columns 1 and 2 are 2,509.7 and 2,095.5 kN, respectively

$$V = 0 \quad \text{at} \quad 1.237.0x - 53.81x^2 = 3{,}067$$

solving $x = 2.828$ m. Moments are computed similarly

$$M = \int_0^x V \, dx = 1{,}237.0\frac{x^2}{2} - 107.5\frac{x^3}{6} - Cx'$$

at $x = 0.23$, $x' = 0$

$$M = 32.0 \text{ kN-m}$$

at face of column 1, $M = -576.0$ kN-m. Maximum M is at $x = 2.828$

$$M = 4{,}538.0 - 3{,}067(2.828 - 0.23) = -3{,}430 \text{ kN-m}$$

at face of column 2, $M = 16.0$ kN-m. This is sufficient to draw the following shear and moment diagrams.

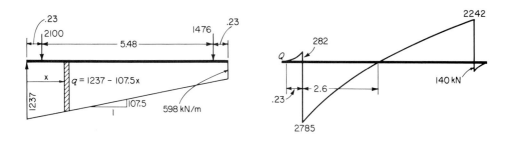

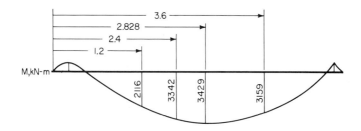

Figure E9-2b

Step 4. Find depth for wide-beam shear at small end and check diagonal tension at large end.

Reasoning:
$$V_b/V_s = \frac{2,509.7}{2,095.5} = 1.2$$

$$\frac{b}{a} = \frac{4.27}{2.06} = 2.07 \gg 1.2$$

$$V = 1,237.0x - 107.5\frac{x^2}{2} - 3,067 \qquad \text{where} \quad x = 5.48 - d$$

$$= 2,095.5 - 647.5d - 53.75d^2 \qquad \text{(net shear at section)}$$

$$\text{width} = 2.065 + \frac{4.27 - 2.06}{5.94}(d + 0.46)$$

$$= 2.065 + 0.372d + 0.17 = 2.24 + 0.372d$$

$$v_c = 642 \text{ kPa}$$

equating concrete shear to external shear $(2.24 + 0.372d)d(642) = 2,095.5 - 647.5d - 53.75d^2$

$$292.6d^2 + 2,085.6d = 2,095.5$$

$$d^2 + 7.1d = 7.2$$

$$d = 0.89 \text{ m}$$

Checking diagonal tension at large end (not possible at small end) shows need $d = 0.75$ m.

Step 5. Design reinforcing steel. Since width varies, one should check A_s for several locations, resulting in the following table:

x	V, kN	M, kN-m	w, m	A_s cm^2/m
0	0	0	4.27	0
0.6	− 2,344.6	− 916.1	4.05	8.2
1.2	− 1,660.6	− 2,115.8	3.83	20.4
1.8	− 1,015.4	− 2,916.6	3.60	30.2
2.4	− 408.9	− 3,342.0	3.38	37.2
2.828(max)	0.0	− 3,428.7	3.22	42.0
3.0	+ 159.0	− 3,415.0	3.16	41.2
3.6	688.1	− 3,159.0	2.94	40.6
4.8	1,630.3	− 1,752.4	2.49	26.1
5.94	0.0	0.0	2.07	0.0

The max steel ='156.3 cm^2/m
The min steel = 32.48 cm^2/m (based on 200/f_y or 14/f_y metric)

Step 6. Steel in short direction. Treat same as rectangular footing using appropriate zone of $w + 0.75d$, since columns are at end of footing. Use the average width of footing in this zone for bending, for example, at large end:

$$w + 0.75d = 0.46 + 0.75(0.89) = 1.13 \text{ m}$$

$$B_1 = 4.27 \qquad B_2 = 4.27 - 1.13\frac{4.27 - 2.07}{5.94} = 3.85$$

average
$$w = \frac{4.27 + 3.85}{2} = 4.06 \text{ m}$$

$$L' = \frac{4.06 - 0.46}{2} = 1.8 \text{ m}$$

$$M = \frac{289.5}{2} 1.8^2 = 469 \text{ kN-m}$$

The remainder of the problem is left for the reader.

9-4. DESIGN OF STRAP OR CANTILEVER FOOTINGS

A strap footing may be used where the distance between columns is so great that a combined or trapezoid footing becomes quite narrow, with resulting high bending moments, or where $\bar{x} < L/3$, as in Sec. 9-3.

A strap footing consists in two column footings connected by a member termed a strap, beam, or cantilever which transmits the moment from the exterior footing. Figure 9-7 illustrates a strap footing. Since the strap is designed for moment, either it should be formed out of contact with the soil or the soil should be loosened for several inches beneath the strap so that the strap has no soil pressure acting on it. For simplicity of analysis, if the strap is not very long, the weight of the strap may be neglected.

In designing a strap footing, it is first necessary to proportion the footings. This is done by assuming a uniform soil pressure beneath the footings; that is, R_1 and R_2 (Fig. 9-7) act at the centroid of the footings.

By taking moments about V_2 and neglecting the weight of the footings,

$$R_1 = V_1 \frac{S}{S'} \tag{9-3}$$

But from $\sum F_v = 0$ we have

$$R_2 = V_1 + V_2 - R_1 \tag{9-4}$$

It can be seen that this is somewhat of a trial-and-error solution, with several possible answers, depending on the selected value of e, since the length of the footing is

$$L = 2\left(e + \frac{w}{2}\right)$$

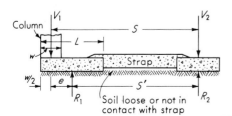

Figure 9-7. Assumed loading and reactions for a strap-footing design. With concentric locations of R_1 and R_2 it necessarily follows that the soil-pressure distribution is uniform.

The width of the footing can then be computed as

$$B = \frac{R_1}{Lq_a}$$

The strap must be a massive member for this solution to be valid. The development of Eq. (9-3) implies rigid body rotation; thus, if the strap is not able to transmit the eccentric moment from column 1 without rotation, the solution is not valid. Some work by the author indicates $I_{strap}/I_{footing} > 2$ to avoid exterior footing rotation. Maximum rigidity is obtained by running the strap from column to column as in the example rather than as in Fig. 9-7.

It is desirable to proportion both footings so that B and q are as nearly equal as possible to control differential settlements.

Example 9-3. Proportioning of a strap footing. Given the foundation conditions shown. Allowable soil pressure is 2 ksf. Both columns are 16 in square.

SOLUTION. Try $e = 4$ ft

$$16R_1 = 130(20)$$

$$R_1 = 162.5 \text{ kips}$$

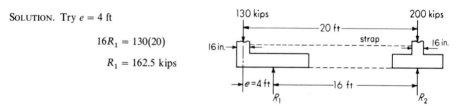

Figure E9-3a

moments about R_1 gives

$$R_2 = V_2 - V_1 \frac{e}{S'} = 200 - 130 \frac{4}{16} = 167.5$$

$$R_2 = V_1 + V_2 - R_1 = 330 - 162.5 = 167.5 \qquad Check$$

$$A_1 = \frac{R_1}{q} = \frac{162.5}{2} = 81.25 \text{ sq ft}$$

To make this value of eccentricity work,

$$L = (4 + 0.67)(2) = 9.33$$

$$B = \frac{81.25}{9.33} = 8.7 \text{ ft} \qquad \text{Use 8.75 ft}$$

Other combinations of L and B, for example, 9 × 9 ft, will also work.

$$q = \frac{162.5}{81.7} = 1.99 \text{ ksf}$$

sizing footing under V_2

$$A_2 = \frac{167.5}{2} = 83.75$$

$$B^2 = 83.75 \qquad B = \sqrt{83.75} = 9.15 \qquad \text{Use 9.25 ft}$$

$$q = \frac{167.5}{(9.25)^2} = 1.95 \simeq 1.99$$

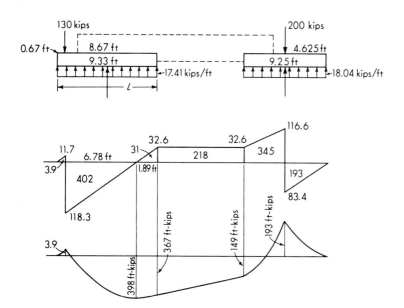

Figure E9-3b

Settlement should be nearly uniform, since both soil pressures are nearly equal.

    Design footing steel along L based on 398 ft-kips.

    Design strap based on $V = 33$ kips and $M = 367$ ft-kips.

    Design footing depths to satisfy wide-beam and punching shear.

9-5. ECCENTRICALLY LOADED RIGID FOOTINGS

Footings are frequently loaded with overturning moments as well as axial loads, in which case the soil-pressure resultant will not coincide with the centroid of the footing, as illustrated in Fig. 9-8. In these cases it is assumed that the soil-pressure resultant coincides with the axial force V, but not with the centroid of the footing, which results in a linear nonuniform stress distribution on the footing. Possible tilting of the footing due to the higher intensity of soil pressure at the toe may occur. This can be reduced by using a larger safety factor when computing the allowable soil pressure. The footing of Fig. 9-8a has a built-in eccentricity in that the load is not collinear with the center of area. One must investigate the column fixity of this situation as part of the analysis. The footing of Fig. 9-8b must be rigidly attached to the footing in order to transmit moment. For all cases where the column can or does transmit moment the apparent eccentricity $(e = M/V)$ resulting from the moment will be counteracted by the attempted footing rotation, which will always be in the direction of the load moment. If the footing rotation is θ, due to the applied moment, a resisting (opposite direction to the load moment) moment is developed of $M_r = 4EI\theta/L$ at the footing and $M_r' = 2EI\theta/L$ at the far end of column; these moments may require additional frame analysis. This concept is illustrated in an approximate manner in Example 9-5.

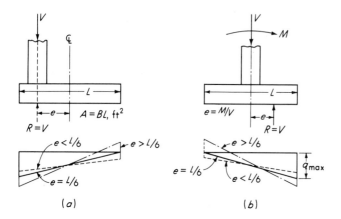

Figure 9-8. (a) Column eccentrically placed with respect to center of footing area; (b) column centrally located on footing, but with overturning moment.

Since the footing is considered to be rigid, the soil pressure can be computed from principles of mechanics of materials for combined bending and axial stresses.

$$q = \frac{V}{A} \pm \frac{Vec}{I} \tag{9-5}$$

Since

$$\frac{I}{c} = \frac{BL^2}{6} \quad \text{and} \quad A = BL$$

Eq. (9-5) may be rewritten

$$q = \frac{V}{BL}\left(1 \pm \frac{6e}{L}\right) \tag{9-6}$$

where V = vertical load, or resultant force
e = eccentricity of vertical load, or resultant force
BL = dimensions of footing (width B, length L)
q = intensity of soil pressure ($+$ = compression) and must be less than or equal to allowable soil pressure

From Eq. (9-6), if e is sufficiently large, the soil pressure becomes negative, signifying soil tensile stresses on the footing as it attempts to separate from the soil. Tensile stresses are not possible; thus they are neglected, resulting in a reduced effective footing area to carry the load. Solving Eq. (9-6) for $q = 0$, we obtain $e = L/6$, which is the maximum eccentricity of the soil pressure for the entire footing area to be effective and with no soil tensile stresses beneath it; i.e., as long as the resultant is in the "middle one-third," the entire footing is effective.

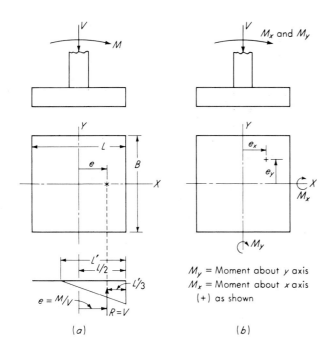

M_y = Moment about y axis
M_x = Moment about x axis
(+) as shown

Figure 9-9. (*a*) Pressure diagram when $e > L/6$; (*b*) location of resultant when there is a moment about both the x and y axes. If both $e > L/6$, only part of the footing is effective.

For the special case of eccentricity $e > L/6$, with respect to one axis only, one may easily derive equations for the maximum intensity of soil pressure as follows (Fig. 9-9):

$$(a) \quad \frac{L'}{3} = \frac{L}{2} - e \quad \text{and} \quad (b) \quad V = \frac{q}{2}(L'B)$$

Substituting L' from (*a*) into (*b*) and solving for q, we obtain

$$q = \frac{2V}{3B(L/2 - e)} \tag{9-7}$$

Best utilization of materials (smaller area) will occur for this case when the footing is a rectangle, since both L and the moment of inertia I are increased.

For footings with moments or eccentricity about both axes (Fig. 9-9*b*), we may write

$$q = \frac{V}{A} \pm \frac{M_y x}{I_y} \pm \frac{M_x y}{I_x} \tag{9-8}$$

where I_x = moment of inertia of footing about x axis
I_y = moment of inertia of footing about y axis
x = distance from y axis to point
y = distance from x axis to point

which is valid, for the assumptions used, as long as q is not in tension ($-$). When q is in tension, the method must be modified to take into account the reduced footing area. This modification is not easy, and in the past curves have been presented which

Figure 9-10. Model footing tests to check footing-soil separation.

attempted to solve for the line of zero pressure so that the soil pressure could be found. Bowles (1975*b*, p. 514) and Dunham (1962, p. 248) have provided curves for this problem. An easier solution and one which gives at the same time the soil pressure, deflections, bending moments, and the approximate pressure line is to use the mat computer program described in Chap. 10 appropriately coded.

It is the opinion of the author based on a series of model footing tests using 6-, 8-, and 10-in-square plates on both sand and clay beds that soil-footing separation does not occur except for rather large footings. Rather the footing punches into the ground and the soil yields in the zones of high pressure, resulting in a redistribution of soil pressure into the zones where normally one computes "tension." A photograph of one of the tests is shown in Fig. 9-10. These tests used a $\frac{1}{4}$-in aluminum plate rigidly attached to a 1.5 × 1.5-in square column to simulate 6-, 8-, and 10-ft footings connected to a 18-in column.

Factors to consider in analyzing footings subject to large eccentricities:

1. One must always investigate the column connection. A concrete column rigidly attached as common practice or a metal column in a moment-resisting connection will cause readjustment of the eccentricity as discussed earlier.
2. One must avoid large differential soil pressures which will fail the soil in the zone of highest pressure intensity and could lead to a foundation failure, as the remolded soil in the failure zone may behave considerably differently from the soil used to develop allowable bearing capacity. It may be necessary to strengthen the highly stressed soil by grouting or replacement. Investigate the possibility of changing the footing geometry so that the resultant coincides with the center of area.
3. If the maximum soil pressure is reduced, there will be far less likelihood of soil-footing separation.
4. By using the mat program as suggested, it is possible to include the footing weight and the effect of the column connection, which will help reduce the computed

tendency of the footing to separate and the footing eccentricity. One can also evaluate the effect of stiffening the soil in the highly stressed zone.

It is not recommended that the maximum soil pressure exceed the allowable soil pressure under any footing—including eccentrically loaded footings. If this recommendation is followed, it is doubtful if any eccentrically loaded footing will have the resultant outside the middle one-third (kern limit). It should be noted that with the resultant exactly at the kern limit the maximum soil pressure is two times the average value and the minimum pressure is zero, an undesirable condition, in general. The conventional method of analysis (but not generally recommended) is illustrated in the following two examples.

Example 9-4. Proportion a footing for a concentric column load and an overturning moment. Given data:

$$P = D + L = 1{,}600 \text{ kN} \qquad M = 800 \text{ kN-m}$$

$$q_a = 200 \text{ kPa}$$

SOLUTION. A rectangular footing will provide a minimum area of footing solution if L is taken in the direction to resist moment.

Using Eq. (9-6)

$$q = \frac{V}{BL}\left(1 \pm \frac{6e}{L}\right)$$

$$e = \frac{M}{V} = \frac{800}{1{,}600} = 0.5 \text{ m} \qquad \text{and is independent of footing dimensions}$$

$$\frac{1{,}600}{BL}\left[1 + \frac{(6)0.5}{L}\right] = 200 \quad \Rightarrow \quad \frac{8}{L}\left(1 + \frac{3}{L}\right) = B$$

Trial and error: (Assume L and solve B)

L, m	B, m	BL, m^2
2	10	20
3	5.33	16.0
4	3.5	14.0
5	2.56	12.8
6	2.0	12.0
100	0.08	8.0

From the data it is readily seen that a rectangle provides minimum area. Although it is not necessary to compute so many values of B, it illustrates the change in area requirements with L. Use footing 2.46×5 m to avoid the appearance of the column on a beam.

Example 9-5. Find soil pressure beneath a footing with eccentricity about both axes. Given data: footing $= 6 \times 6$ ft,

$$P = 60 \text{ kips, column } 15 \times 15, \qquad L = 14 \text{ ft}, \qquad M_x = M_y = 120 \text{ ft-kips},$$

$$k_s = 100 \text{ kcf}$$

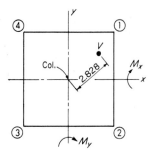

Figure E9-5a

SOLUTION. *Step 1.* Find e

$$e_x = e_y = \frac{120}{60} = 2 > \frac{L}{6} \text{ ft}$$

Therefore, Eq. (9-6) is not valid, and we must use a trial-type solution. Referring to the figure:

Step 2. Make initial zero-soil-pressure line estimate

$$q = \frac{V}{A} + \frac{M_x y}{I_x} + \frac{M_y x}{I_y}$$

and

$$I_x = I_y = \frac{bh^3}{12} = \frac{6(6)^3}{12} = 108 \text{ ft}^4$$

Compute soil pressure at corners:

Point	$\dfrac{V}{A}$	$\dfrac{M_x y}{I_x}$	$\dfrac{M_y x}{I_y}$	q, ksf $+$ = compression
1	1.67	3.33	3.33	8.33
2	1.67	−3.33	3.33	1.67
3	1.67	−3.33	−3.33	−5.01
4	1.67	3.33	−3.33	1.67

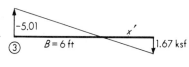

Figure E9-5b

Find the point of zero pressure along line 2-3 and line 3-4, referring to the accompanying figure.

$$x' = \frac{1.67}{6.67}(6) = 1.50 \text{ ft}$$

Refer to Fig. E9-5c for the position of the *zz* axis.

Step 3. Compute the new moment of inertia:

For area 2

$$I_{(2)} = \frac{bh^3}{3} = \frac{7.78(1.414)^3}{3} = 7.33 \text{ ft}^4$$

assuming average width of $b = 7.78$ ft.

For area 1 take $I_0 + Ad^2$

$$I_{(1)} = \frac{bh^3}{36} + Ad^2, \text{ triangle area}$$

$$= \frac{7.07(3.54)^3}{36} + 12.5(2.59)^2$$

$$= 8.7 + 83.9 = 92.6 \text{ ft}^4$$

$$I_{\text{total}} = 92.6 + 7.3 = 99.9 \text{ ft}^4$$

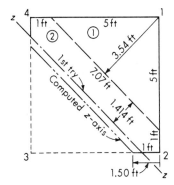

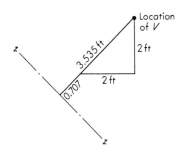

Figure E9-5c

Step 4. Compute q at points 1, 2, and 4 of Fig. E9-5c.

$$M_{zz} = V\bar{z} = 60(3.535) = 212.1 \text{ ft-kips}$$

$$q_1 = \frac{M_c}{I} = \frac{212.1(4.95)}{99.9} = 10.5 \text{ ksf}$$

$$q_{2,4} = \frac{212.1(0.707)}{99.9} = 1.51 \text{ ksf}$$

Step 5. Compute V'_q based on these soil pressures.

$$V_2 = 1.414(7.78)(1.51) = 16.6 \text{ kips}$$

$$V_1 = \int_{1.414}^{4.95} \frac{2M}{I}(4.95 - x)x \, dx = \frac{2(212.1)}{99.9}\left[4.95\frac{x^2}{2} - \frac{x^3}{3}\right]_{1.414}^{4.95}$$

$$= 4.25(16.2) = 69.0 \text{ kips}$$

and

$$V'_q = 16.6 + 69.0 = 85.6 \gg 60 \text{ kips} \qquad Redo$$

Step 6. Redo to find zz axis. From step 5 it is evident that we may solve directly for the z axis.
For a triangle the moment of inertia with respect to the base is $I = bh^3/12$. Here $b = 2h$; $I = h^4/6$, the location of V with respect to the zz axis is

$$\bar{x} = 2.828 - (6 \sin 45° - h) = h - 1.414$$

$$M_{zz} = 60(\bar{x}) = 60(h - 1.414) = 60h - 84.84$$

The soil resistance $V'_q = 60$ kips for equilibrium; thus,

$$q = \frac{Mx}{I} \qquad dA = \frac{a \times b}{2} \qquad \text{but} \qquad a = h - x \qquad b = 2(h - x)$$

$$V'_q = \int_0^h q \, dA = \int_0^h \frac{Mx}{I}(h - x)^2 \, dx$$

integrating using power rule

$$V'_q = \frac{2M}{I}\left(\frac{hx^2}{2} - \frac{x^3}{3}\right)\Big]_0^h = \frac{Mh^3}{3I}$$

Step 7. Find h and check.
Substitute M and I into the expression for V'_q

$$2(6)\frac{60h^2 - 84.84}{h^4}\left(\frac{h^3}{2} - \frac{h^3}{3}\right) = 60$$

$$12\frac{60h - 84.84}{h}\frac{1}{6} = 60$$

$$60h - 84.84 = 30h$$

$$h = 2.83 \text{ ft}$$

Check to see if h is correct

$$M'_{zz} = V'_q x = \int_0^h \frac{M}{I}(h - x)^2 x^2 \, dx = \frac{2M}{I}\left(\frac{hx^3}{3} - \frac{x^4}{4}\right)\Big]_0^h = \frac{Mh^4}{6I}$$

From step 6, $V'_q = \dfrac{Mh^3}{3I}$; thus

$$\frac{Mh^3}{3I}x' = \frac{Mh^4}{6I}$$

$$x' = \frac{h}{2} = \frac{2.83}{2} = 1.414 \text{ ft}$$

which is the location of the effective column load; thus $\sum F_v$ and $\sum M$ is satisfied.

Step 8. Find maximum soil pressure q.

$$q = \frac{Mc}{I} = \frac{60(h/2)(h)}{h^4/6} = \frac{60(3)}{2.83^2} = 22.5 \text{ ksf}$$

Step 9. Check reverse moment due to footing rotation. Approximate slope and taking $k_s = q/\Delta$ is

$$\Delta = q/k_s = \frac{22.5}{100} = 0.225 \text{ ft}$$

this varies from 0.225 ft at point 1 to zero at 2.83 ft from point 1

$$\theta = \frac{0.225}{2.83} = 0.079 \text{ rad}$$

$$\theta_x = \theta_y = 0.079 \cos 45 = 0.056 \text{ rad}$$

$$I_{col} = \frac{bh^3}{12} = \frac{15(15)^3}{12} = 4,218.8 \text{ in}^4$$

Take $E = 3{,}144$ ksi (3,000-psi concrete) and the developed moment resistance is

$$M_{r(x)} = M_{r(y)} \cong \frac{4(3{,}144)(4{,}219)(0.056)}{14(12)(12)} = 1{,}473.8 \text{ ft-kips}$$

since this value of resisting moment is far in excess of 120 ft-kips, the footing remains in contact with the soil and the stress pattern is probably very close to uniform.

9-6. UNSYMMETRICAL FOOTINGS

There are occasions when a footing must be built with a hole or notch for special purposes, or an existing footing may have a part removed to provide clearance for some reason. In problems such as this for *rigid* footings, and assuming linear stress distribution, one may write a general equation [Cross (1963)] for the stress at any point beneath a footing (Fig. 9-11):

$$q = ax + by + c \tag{a}$$

From statics the following equations can also be written:

$$V = \int_0^A q \, dA \tag{b}$$

$$M_x = \int_0^A y(q \, dA) \tag{c}$$

$$M_y = \int_0^A x(q \, dA) \tag{d}$$

These latter three equations can be substituted into Eq. (a), noting that

$$\int_0^A dA = A$$

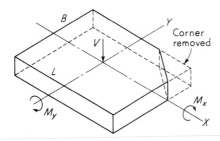

Figure 9-11. Unsymmetrical footing.

and with respect to the centroid of the footing,

$$\int_0^A x \, dA = 0$$

$$\int_0^A y \, dA = 0$$

Further, from statics it can be shown that

$$\int_0^A x^2 \, dA = I_y \qquad \int_0^A y^2 \, dA = I_x \qquad \int_0^A xy \, dA = I_{xy}$$

and by rewriting and substituting terms, the following expressions are obtained for V, M_x, and M_y:

$$V = \int_0^A q \, dA = (ax + by + c) \, dA = cA$$

$$M_x = \int_0^A q(y \, dA) = (ax + by + c)y \, dA = aI_{xy} + bI_x$$

$$M_y = \int_0^A q(x \, dA) = (ax + by + c)x \, dA = aI_y + bI_{xy}$$

Solving these expressions for coefficients a, b, and c, we get, by inspection,

$$c = \frac{V}{A} \tag{e}$$

and the following pair of equations containing the a and b coefficients

$$aI_{xy} + bI_x = M_x$$
$$aI_y + bI_{xy} = M_y$$

A simultaneous solution of this set of equations gives

$$a = \frac{M_y - M_x(I_{xy}/I_x)}{I_y(1 - I_{xy}^2/I_x I_y)} \tag{f}$$

and

$$b = \frac{M_x - M_y(I_{xy}/I_y)}{I_x(1 - I_{xy}^2/I_x I_y)} \tag{g}$$

Finally, the desired equation is obtained from Eq. (a):

$$q = \frac{M_y \pm M_x(I_{xy}/I_x)}{I_y(1 - I_{xy}^2/I_x I_y)}(x) + \frac{M_x - M_y(I_{xy}/I_y)}{I_x(1 - I_{xy}^2/I_x I_y)}(y) + \frac{V}{A} \tag{9-9}$$

where I_{xy} = product of inertia, may be + or −
   M_x = moment about x axis, may be + or −
   M_y = moment about y axis, may be + or −; positive directions for M_x, M_y,
      and V shown in Fig. 9-11
   x, y = respective distances to points under consideration, may be + or −
   q = intensity of soil pressure (+ = compression)

The sign convention outlined must be strictly followed to arrive at the correct answers.

Example 9-6. The square footing shown is subjected to a column load of 540 kips. The allowable soil pressure is 6 ksf. What is the actual maximum soil pressure when the corner is notched as shown? Compute new section properties.

$$f'_c = 3,000 \text{ psi}, \qquad f_y = 60,000, \qquad E_s = 452,700 \text{ ksf},$$

$$\text{column} = 2 \times 2 \text{ ft}, \qquad D = 1.67 \text{ ft}$$

Part	Area	x'	y'	Ax'	Ay'
Notch	−4.5	3.5 ft	4.25	−15.75	−19.13

SOLUTION. (note that solution is independent of material properties and D)

Step 1. Find new x, y axis,

$$\bar{x} = \frac{-15.75}{95.5} = -0.165 \text{ ft}$$

$$\bar{y} = \frac{-19.13}{95.5} = -0.20$$

which gives the location of new axes X' and Y' as in the figure.

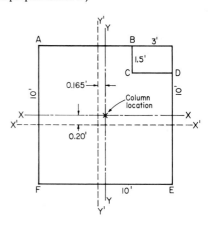

Figure E9-6

Step 2. Compute new properties

$$I_x, I_y, I_{xy}$$

Part	Area	x	y	Ax^2	Ay^2	I_{ox}	I_{oy}
Uncut	100	−0.165	−0.20	2.72	4.00	833.3	833.3
Notch	−4.5	+3.66	+4.45	−60.3	−89.1	−0.84	−3.38
Total	95.5						

$$I_{x'} = I_{ox} - I_{ox\,notch} + Ay^2$$

$$= 833.3 - 0.84 + 4.0 - 89.0 = 747.5 \text{ ft}^4$$

$$I_{y'} = I_{oy} - I_{oy\,notch} + Ax^2$$

$$= 833.3 - 3.38 + 2.73 - 60.5 = 772.15 \text{ ft}^4$$

$$I_{xy} = I_{oxy} + A\bar{x}\bar{y}$$

$$= 0 + (-4.5)(3.66)(4.45) + 100(-0.20)(-0.165) = -70.0 \text{ ft}^4$$

Step 3. Compute moments.

$$M_y = 540(0.165) = 89.1 \text{ ft-kips} \leftarrow \text{note that sign is } +$$

$$M_x = 540(0.20) = 108 \text{ ft-kips}$$

Step 4. Compute soil pressure at selected locations.

$$q = \frac{P}{A} + \frac{M_y - M_x(I_{xy}/I_x)}{I_y(1 - I_{xy}^2/I_xI_y)} x + \frac{M_x - M_y(I_{xy}/I_y)}{I_x(1 - I_{xy}^2/I_xI_y)} y \qquad \text{Eq. (9-9)}$$

$$M_y - \frac{M_xI_{xy}}{I_x} = 89.1 - 108\left(\frac{-70.0}{747.5}\right) = 99.2 \text{ ft-kips}$$

$$M_x - \frac{M_yI_{xy}}{I_y} = 108 - 89.1\left(\frac{-70.0}{772.15}\right) = 116.1 \text{ ft-kips}$$

$$I_y - \frac{I_{xy}^2}{I_x} = 772.2 - \frac{(-70.0)^2}{747.5} = 765.6 \text{ ft}^4$$

$$I_x - \frac{I_{xy}^2}{I_y} = 747.5 - \frac{(-70.0)^2}{772.15} = 741.1 \text{ ft}^4$$

taking $+$ = compression and noting at $x = y = 0$ from X', Y' axes $q = 5.65$ ksf, and in general

$$q = \frac{540}{95.5} + \frac{99.2}{765.6} x + \frac{116.1}{741.1} y = +5.65 + 0.13x + 0.157y$$

Table E9-6

Point (see figure)	x	y	P/A	$0.13x$	$+0.157y$	q
A	-4.84	$+5.20$	$+5.65$	-0.63	$+0.82$	$+5.84$
B	$+2.16$	$+5.20$	$+5.65$	$+0.28$	$+0.82$	$+6.75$
C	$+2.16$	$+3.70$	$+5.65$	$+0.28$	$+0.58$	$+6.51$
D	$+5.16$	$+3.70$	$+5.65$	$+0.67$	$+0.58$	$+6.90 > 6.00$
E	$+5.16$	-4.80	$+5.65$	$+0.67$	-0.75	$+5.57$
F	-4.84	-4.80	$+5.65$	-0.63	-0.75	$+4.27$

Percent increase of q with notch = 6.90/5.40 = 1.28, or 28 percent increase. Footing must be redesigned since 6.90 > 6.0. This problem is reworked in Example 10-3 to illustrate the effect of column fixity on soil pressure.

9-7. MODULUS OF SUBGRADE REACTION

The ratio between the unit soil pressure and the corresponding settlement is termed the *modulus of subgrade reaction* (also *coefficient of subgrade reaction*) and is mathematically expressed as (Eq. 2-17).

$$k_s = \frac{q}{\Delta} \tag{2-17}$$

where k_s = modulus of subgrade reaction (units of force $\times L^{-3}$)
q = intensity of soil pressure (FL^{-2})
Δ = average settlement for an increment of pressure (L)

This value can be determined by performing a plate-load test and plotting a curve of q versus Δ, as illustrated in Fig. 9-12.

Since it is difficult to load the plate uniformly when performing the test, the soil will tend to deflect more in the zone of higher stress intensity; thus the stresses and settlements beneath the plate are nonuniform. It is for this reason that, when using 18-, 24-, and 30-in load plates, they are stacked to increase the rigidity, to attempt to distribute the essentially point load more uniformly. The graph is obtained by plotting the average soil pressure computed as $q = V/A$ vs. the average measured settlement across the plate.

The value of subgrade reaction also depends on the size and shape of the plate used in the test. The plate settlement is nonlinear in that it increases not as a constant, as Eq. (2-17) implies, but will have a different value, depending on which pair of coordinates are taken (e.g., along the curve shown in Fig. 9-12). When utilizing the

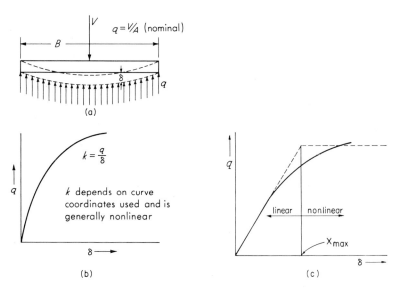

Figure 9-12. Determination of modulus of subgrade reaction k.

concept of modulus of subgrade reaction, therefore, values of q and Δ should be used (if test results are available) which approximate the actual loading to be expected on the soil.

From Sec. 5-6 it was noted that the deeper the stress influence penetrates into the soil, the larger the settlement will be. Therefore, if data from a plate-load test are to be used, the effect of k of using a different-size footing must be estimated.

Terzaghi (1955) proposed for footings on clay that one could extrapolate plate-load tests for a plate 1×1 ft to a footing as

$$k_{sf} = k_{s1} B \tag{9-10}$$

and for footings on sand and using the size-effect ratio used earlier for bearing capacity

$$k_{sf} = k_{s1} \left(\frac{B+1}{2B} \right)^2 \tag{9-11}$$

and for a rectangular footing of dimensions $b \times mb$

$$k_{sf} = k_{s1} \frac{m + 0.5}{1.5\, m} \tag{9-12}$$

with the limiting value of k_{sf} in Eq. (9-12) being $0.667 k_{s1}$. In these equations $k_{sf} = $ modulus of subgrade reaction for the footing and $k_{s1} = $ plate-load value of modulus of subgrade reaction.

Vesic (1961a, 1961b) proposed that the modulus of subgrade reaction could be computed using the stress-strain modulus from triaxial test data as

$$k'_s = 0.65 \sqrt[12]{\frac{E_s B^4}{E_f I_f}} \frac{E_s}{1 - \mu^2} \tag{9-13}$$

where E_s, $E_f = $ modulus of elasticity of soil and footing, respectively

B, $I_f = $ footing width and moment of inertia, respectively

To obtain k_s

$$k_s = \frac{k'_s}{B}$$

In this equation, $E_f I_f$ is the flexural rigidity of the footing, i.e., depends on footing thickness to resist bending.

It should be observed that the 12th root of any value will be close to 1, and this coupled with the inherent problems with determining the stress-strain modulus E_s means that for all practical purposes k'_s becomes relatively independent of B and is simply computed as

$$k'_s = \frac{E_s}{1 - \mu^2} \tag{9-13a}$$

Bowles (1974a) computes a value of k_s from the allowable bearing capacity as

$$k_s = 36 q_a \qquad \text{kcf} \tag{9-14}$$

Table 9-1. Range of values of modulus of subgrade reaction k_s. These values should be used for
a guide. Local values may be higher or lower. To
convert to kN/m^3, multiply by 157.09

Soil	Range of k_s, kcf
Loose sand	30–100
Medium sand	60–500
Dense sand	400–800
Clayey sand (medium)	200–500
Silty sand (medium)	150–300
Clayey soil:	
$q_u < 4$	75–150
$4 < q_u < 8$	150–300
$8 < q_u$	> 300

which has been found to give about as reliable values as any method. This equation is developed by reasoning that q_a (ksf) is valid for a settlement of about 1 in, and safety factor $F = 3$; hence, inserting both values and converting to feet, one obtains Eq. (9-14). In metric units and rounding slightly, Eq. (9-14) becomes

$$k_s = 120q_a \qquad kN/m^3 \qquad (9\text{-}15)$$

where $q_a = kPa$. Typical values of k_s are given in Table 9-1.

If it is necessary to know the correct value of deflection when using the concept of modulus of subgrade reaction, it is necessary to have the correct value of k_s. In many problems, the order of magnitude of deflection is sufficient and hence the use of Eq. (9-14) can be justified. In most problems, the interaction of the structural element with the soil mass is such that the EI (flexural rigidity) of the structure predominates and thus the value of k_s becomes of less importance. An increase of 100 to 200 percent in k_s may change the structural behavior only 15 to 25 percent. Deflections will increase in direct proportion to k_s, but often any deflection is tolerable.

It would appear that the most general equation to describe k_s is

$$k_s = A + BZ^n \qquad (9\text{-}16)$$

where A = constant as applicable for horizontally placed footings and mats
  B = coefficient for depth
  Z = depth below ground
  n = exponent and may be either greater or less than 1

For the use of the bearing-capacity equation of Chap. 2 and dropping the N_γ term and depth factors d_i

$$q_{ult} = cN_c s_c + qN_q s_q$$

and using Eq. (9-14) without $F = 3$, since we have q_{ult}, not q_a

$$k_s = 12q_{ult}$$

we have Eq. (9-16) developed with the terms defined as follows:

$$A = 12(cN_c s_c) \qquad B = 12(\gamma N_q s_q) \qquad \text{and} \qquad \text{depth} = Z^1$$

Bearing capacity does not increase without bound with depth; therefore, it is reasonable to expect that k_s will behave similarly.

As the pressure vs. deflection curve defining the modulus of subgrade reaction is nonlinear except for possibly a very small region near the origin, it is convenient to linearize it into one or more regions as illustrated in Fig. 9-12c. Probably a single value of k_s defined for all deflections less than X_{max} is satisfactory (at least the author has found this to be the case) when considering the uncertainty of the concept. From plasticity considerations one would replace the soil effect on the foundation where deflections exceed X_{max} by a force (or pressure)

$$q = k_s(X_{max})$$

The value of X_{max} would be determined by inspection of the load-settlement curve, by arbitrarily using, say 1 in or 2.5 cm, or by using the strain at failure in a compression test together with the foundation width to obtain

$$X_{max} \cong \varepsilon_f(1.5 \text{ to } 2B)$$

since strain times length of stress zone = deflection. The principally stressed zone will be on the order of 1.5 to 2B.

For footings, mat foundations, and similar structures one may use Eqs. (9-13a) or (9-14) to obtain rapid values for analysis. As will be seen in the computer programs where this elastic soil property is used, it is easy to change k_s and observe effects and attempt to bracket the possible solution range. This concept will be considered further in Chaps. 13, 18, and 19.

The use of a constant value of k_s beneath mats and footings (spread, combined, or trapezoidal) is usually adequate owing to the effect of superstructure rigidity contributing to the foundation element and the fact that the element flexural rigidity (EI) decidedly predominates. For many foundations such as oil and water tanks and any other foundation element supporting a superstructure of negligible rigidity and where the foundation-element rigidity is often not so predominating, it may be necessary to adjust the k_s value somewhat. This adjustment may be done as follows:

1. From inspection of Fig. 5-4 (Boussinesq pressure bulbs) it is seen that beneath the footing the soil-pressure influence is greater at the center than the edges. Therefore, program Eq. (5-3) to obtain the pressure profile at selected points beneath the footing. Halt the depth increment when the pressure contribution is less than $0.1q_0$. The pressure contribution does not have to be based on actual contact pressure of the foundation.
2. Numerically integrate the pressure profile from step 1, and find the average pressure increase. Also find the depth of stress influence (which is needed to find the average pressure increase).
3. Compute the resulting settlement as

$$S_i = \frac{\Delta q L}{E_s}$$

4. Compute the resulting modulus of subgrade reaction to use for design at the different points as

$$k_{s_i} = \frac{\Delta q_i}{S_i}$$

This will give "softer" soil springs near the center of large flexible mats, since S will be larger near the center than at the edge. While this may give the correct qualitative deflection profile, the correct quantitative values will depend on true values of k_s. Note that this method makes use of E_s, which may introduce some further difficulty. This procedure is illustrated in the following example.

Example 9-7. Evaluate k_s beneath a footing assuming the pressure profile for two points as given in Fig. E5-2 of Example 5-1. Additional data: $E_s = 650$ ksf, $q_{ult} = 4$ ksf (nominal as V/A)

SOLUTION. Take $q_0 = 1$ ksf and $L = 4B$ for both

Step 1. For the center of footing.

$$4B\Delta q = \frac{B}{2}\left(\frac{1.0 + 0.03}{2} + 0.71 + 0.34 + 0.18 + 0.11 + 0.09 + 0.05 + 0.04\right)$$

$$\Delta q = \frac{B}{2}\frac{2.035}{4B} = 0.254 \text{ ksf (center)}$$

For the corner a similar computation yields $\Delta q = 0.184$

Step 2. Find S

$$S_c = \frac{\Delta q L}{E} = \frac{0.254(4B)}{650} = 0.0016B$$

$$S_{cor} = \frac{0.184(4B)}{650} = 0.0011B$$

Step 3. Find k_s

$$k_{s(c)} = \frac{1(4)}{0.0016B} = \frac{2,500}{B}$$

$$k_{s(cor)} = \frac{1(4)}{0.0011B} = \frac{3,636}{B}$$

From this it is evident that the center will settle more than the edge for equal contact pressures (or forces).

9-8. CLASSICAL SOLUTION OF BEAM ON ELASTIC FOUNDATION

When flexural rigidity of the footing is taken into account, a solution is used that is based on some form of a beam on an elastic foundation. This may be of the classical Winkler solution of about 1867 in which the foundation is considered as a bed of springs ("Winkler foundation") or a finite-element procedure of the next section.

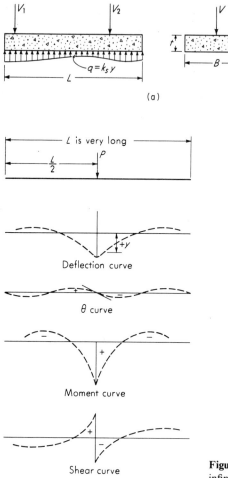

Figure 9-13. (a) Beam on elastic foundation; (b) beam of infinite length on an elastic foundation with concentrated load at the center.

The classical solutions, being of closed form, are not of general application as is the finite-element method. The basic differential equation is (see Fig. 9-13)

$$EI\frac{d^4y}{dx^4} = q = -k'_s y$$

where $k'_s = k_s B$. In solving the equations, a variable is introduced

$$\lambda = \sqrt[4]{\frac{k'_s}{4EI}} \qquad \text{or} \qquad \lambda L = \sqrt[4]{\frac{k'_s L^4}{4EI}} \qquad (9\text{-}17)$$

Table 9-2 gives the closed-form solution of the basic differential equations for several loadings as in Fig. 9-13c utilizing the Winkler concept. It is convenient to express the

Table 9-2. Closed-form solutions of infinite beam on elastic foundation (Fig. 9-13c)

Concentrated load at end	Moment at end
$y = \dfrac{2V_1\lambda}{k_s'}D_{\lambda x}$	$y = \dfrac{-2M_1\lambda^2}{k_s'}C_{\lambda x}$
$\theta = \dfrac{-2V_1\lambda^2}{k_s'}A_{\lambda x}$	$\theta = \dfrac{4M_1\lambda^3}{k_s'}D_{\lambda x}$
$M = \dfrac{-V_1}{\lambda}B_{\lambda x}$	$M = M_1 A_{\lambda x}$
$Q = -V_1 C_{\lambda x}$	$Q = -2M_1\lambda B_{\lambda x}$

Concentrated load at center	Moment at center	
$y = \dfrac{P\lambda}{2k_s'}A_{\lambda x}$	$y = \dfrac{M_0\lambda^2}{k_s'}B_{\lambda x}$	deflection
$\theta = \dfrac{-P\lambda^2}{k_s'}B_{\lambda x}$	$\theta = \dfrac{M_0\lambda^3}{k_s'}C_{\lambda x}$	slope
$M = \dfrac{P}{4\lambda}C_{\lambda x}$	$M = \dfrac{M_0}{2}D_{\lambda x}$	moment
$Q = \dfrac{-P}{2}D_{\lambda x}$	$Q = \dfrac{-M_0}{2}A_{\lambda x}$	shear

The A, B, C, and D coefficients are:

$$A_{\lambda x} = e^{-\lambda x}(\cos \lambda x + \sin \lambda x)$$
$$B_{\lambda x} = e^{-\lambda x}\sin \lambda x$$
$$C_{\lambda x} = e^{-\lambda x}(\cos \lambda x - \sin \lambda x)$$
$$D_{\lambda x} = e^{-\lambda x}\cos \lambda x$$

trigonometric portion of the solutions separately as shown in the bottom of Table 9-2 and where λ is defined by Eq. (9-17). Select values of these coefficients are tabulated in Table 9-3 for debugging a computer program so that an extensive user table can be prepared by the reader.

Hetenyi (1946) developed equations for a load at any point along a beam (see Fig. 9-14) measured from the left end as follows:

$$y = \frac{P\lambda}{k_s'(\sinh^2 \lambda L - \sin^2 \lambda L)}\{2\cosh \lambda x \cos \lambda x(\sinh \lambda L \cos \lambda a \cosh \lambda b$$
$$- \sin \lambda L \cosh \lambda a \cos \lambda b) + (\cosh \lambda x \sin \lambda x$$
$$+ \sinh \lambda x \cos \lambda x)[\sinh \lambda L(\sin \lambda a \cosh \lambda b - \cos \lambda a \sinh \lambda b)$$
$$+ \sin \lambda L(\sinh \lambda a \cos \lambda b - \cosh \lambda a \sin \lambda b)]\} \tag{9-18}$$

Table 9-3. Coefficients for the solution of an infinite beam on an elastic foundation

λx	$A_{\lambda x}$	$B_{\lambda x}$	$C_{\lambda x}$	$D_{\lambda x}$
0.0	1.0000	0.0000	1.0000	1.0000
0.1	0.9907	0.0903	0.8100	0.9003
0.5	0.8231	0.2908	0.2415	0.5323
1.0	0.5083	0.3096	−0.1108	0.1988
1.5	0.2384	0.2226	−0.2068	0.0158
2.0	0.0667	0.1231	−0.1794	−0.0563
3.0	−0.0423	0.0070	−0.0563	−0.0493
4.0	−0.0258	−0.0139	0.0019	−0.0120
5.0	−0.0045	−0.0065	0.0084	0.0019
6.0	0.0017	−0.0007	0.0031	0.0024
7.0	0.0013	0.0006	0.0001	0.0007
8.0	0.0003	0.0003	−0.0004	0.0000
9.0	0.0000	0.0000	−0.0001	−0.0001

$$M = \frac{P}{2\lambda(\sinh^2 \lambda L - \sin^2 \lambda L)}\{2 \sinh \lambda x \sin \lambda x(\sinh \lambda L \cos \lambda a \cosh \lambda b$$
$$- \sin \lambda L \cosh \lambda a \cos \lambda b) + (\cosh \lambda x \sin \lambda x - \sinh \lambda x \cos \lambda x)$$
$$\times [\sinh \lambda L(\sin \lambda a \cosh \lambda b - \cos \lambda a \sinh \lambda b)$$
$$+ \sin \lambda L(\sinh \lambda a \cos \lambda b - \cosh \lambda a \sin \lambda b)]\} \tag{9-19}$$

$$Q = \frac{P}{\sinh^2 \lambda L - \sin^2 \lambda L}\{(\cosh \lambda x \sin \lambda x + \sinh \lambda x \cos \lambda x)$$
$$\times (\sinh \lambda L \cos \lambda a \cosh \lambda b - \sin \lambda L \cosh \lambda a \cos \lambda b)$$
$$+ \sinh \lambda x \sin \lambda x[\sinh \lambda L(\sin \lambda a \cosh \lambda b - \cos \lambda a \sinh \lambda b)$$
$$+ \sin \lambda L(\sinh \lambda a \cos \lambda b - \cosh \lambda a \sin \lambda b)]\} \tag{9-20}$$

The equation for the slope θ of the beam at any point is not presented since it is of little value in the design of a footing. The value of x to use in the equations is from the end of the beam to the point for which the deflection, moment, or shear is desired. If x is less than the distance a, use the equations as given, and measure x from C. If x

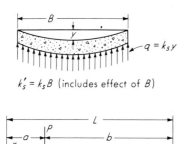

Figure 9-14. Beam of finite length on an elastic foundation.

Table 9-4. Coefficients for use in solving for deflections, moments, and shear for a beam of finite length on an elastic foundation

Distance	Deflection	Moment	Shear
$\lambda_L = 1.0$			
	Load at 0.0L		
0.0	4.0378	0.0000	0.0000
1.0	−1.9716	0.0000	0.0000
	Load at 0.5L		
0.0	0.9814	0.0000	0.0000
1.0	1.0124	0.2486	0.4999
$\lambda_L = 2.0$			
	Load at 0.0L		
0.2	1.4015	−0.4837	−0.2678
0.8	−0.3688	−0.1051	0.2340
	Load at 0.1L		
0.2	1.2296	−0.1395	−0.3887
0.8	−0.2051	−0.0735	0.1584
	Load at 0.4L		
0.2	0.6715	0.1074	0.2688
0.8	0.3015	0.0266	−0.0801

is larger than a, replace a with b in the equations, and measure x from D (Fig. 9-14). These equations may be rewritten

$$y = \frac{P\lambda}{k'_s} A' \qquad M = \frac{P}{2\lambda} B' \qquad \text{and} \qquad Q = PC'$$

where the coefficients A', B', and C' are the values for the hyperbolic and trigonometric remainder of Eqs. (9-18) to (9-20). Select values of A', B', and C' are given in Table 9-4 to aid in debugging a computer program.

It has been proposed that one could use λL from Eq. (9-17) to determine if a foundation should be analyzed on the basis of the conventional rigid procedure or if as a beam on an elastic foundation.

Rigid members: $\lambda L < \dfrac{\pi}{4}$ (bending not influenced much by k_s)

Flexible members: $\lambda L > \pi$ (bending heavily localized)

The author has found the above criteria of limited application because of the influence of load position on the member.

The classical solution presented here has several distinct disadvantages over the finite-element solution presented in the next section, such as:

1. Assumes weightless beam and weight is a factor in footing separation.
2. Difficult to remove soil effect when footing tends to separate from soil.
3. Difficult to program boundary conditions of no rotation or deflection at selected points.
4. Difficult to apply multiple types of loads to a footing.
5. Difficult to change footing properties of I, D, and B.
6. Difficult to allow for change in subgrade reaction along footing.

9-9. FINITE-ELEMENT SOLUTION OF BEAM ON ELASTIC FOUNDATION

This solution is the most general solution available to solve the beam on an elastic-foundation problem. It can account for any boundary or loading condition and can be solved on computers of modest core capacity. A program listing is in the Appendix so the reader will not have to reproduce one from the beginning. Only the basic elements of the theory are given here. For a more extensive and easily readable coverage, the reader is referred to Wang (1970).

General Equations in Solution

For the following development refer to Fig. 9-15. At any node (junction of two or more members) on the structure we may write

$$P_i = A_i F_i$$

which states that the external node force P is equated to the internal member forces F using a bridging constant A. It is understood that P and F are used for either forces or moments and that this equation is shorthand notation for several values of $A_i F_i$ summed to equal the ith nodal force.

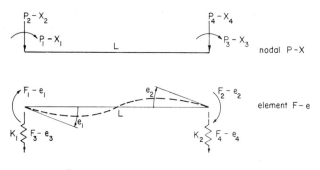

Figure 9-15. External (nodal) and internal (member) finite-element forces.

For the full set of nodes on any structure and deleting subscripts and braces or brackets, this becomes

$$P = AF \tag{a}$$

An equation relating internal-member deformation e at any node to the external nodal displacements is

$$e = \mathbf{B}X$$

where both e and X may be rotations (radians) or translations. From the reciprocal theorem in structural mechanics it can be shown that the B matrix is exactly the transpose of the A matrix, which is a convenience indeed; thus

$$e = \mathbf{A}^T X \tag{b}$$

The internal-member forces F are related to the internal-member displacements as

$$F = \mathbf{S}e \tag{c}$$

These three equations are the fundamental equations in the finite-element method of analysis:

Substituting (b) into (c)

$$F = \mathbf{S}e = \mathbf{SA}^T X \tag{d}$$

Substituting (d) into (a),

$$P = \mathbf{A}F = \mathbf{ASA}^T X \tag{e}$$

Note the order of terms used in developing Eqs. (d) and (e). Now the only unknowns in this system of equations are the X's; so the ASA^T is inverted to obtain

$$X = (\mathbf{ASA}^T)^{-1}P \tag{f}$$

and with the X's we can back-substitute into Eq. (d) to obtain the internal-member forces which are necessary for design. This method gives two significant pieces of information: (1) design data and (2) deformation data.

The ASA^T matrix above is often called a global matrix, since it represents the system of equations for each P or X nodal entry. It is convenient to build it using one finite element of the structure at a time and use superposition to build the global ASA^T from the element $EASA^T$. This is easily accomplished, since every entry in both the global and element ASA^T matrix is defined by two subscripts (I, J). Every entry from the $EASA^T$ with a unique set of subscripts is placed into that subscript location in the ASA^T; i.e., for $I = 2$, $J = 5$ all $(2, 5)$ subscripts in $EASA^T$ are added into the $(2, 5)$ coordinate location of the global ASA^T.

Developing the Element A Matrix

Consider the single element shown in Fig. 9-16b coded with four values of P-X (note that two of these P-X values will be common to the next member) and the forces on the element (Fig. 9-16c). The forces on the element include two internal bending

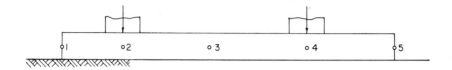

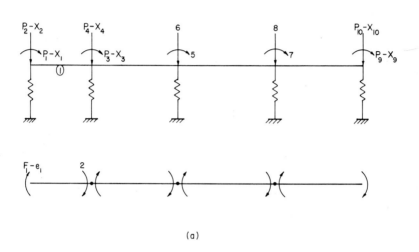

(a)

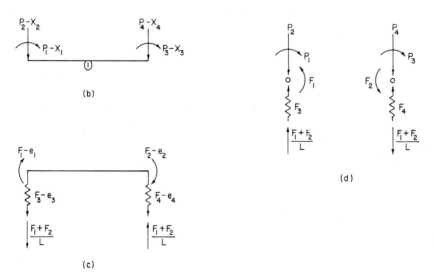

(b)

(c)

(d)

Figure 9-16. (a) Structure and broken into finite elements with global $P - X$; (b) $P - X$ of first element; (c) element forces of any (including first) element; (d) nodal forces: external = P; internal = F.

moments and two element soil "springs" and the shear effect of the bending moments. The sign convention used is consistent with the computer program in the Appendix. Note, too, that the soil "spring" on the node is opposite in direction to that on the element.

Now on node 1 summing moments (Fig. 9-16d)

$$P_1 = F_1 + OF_2 + OF_3 + OF_4$$

similarly, summing forces

$$P_2 = \frac{F_1}{L} + \frac{F_2}{L} + F_3 + OF_4$$

$$P_3 = OF_1 + F_2 + OF_3 + OF_4$$

and

$$P_4 = -\frac{F_1}{L} - \frac{F_2}{L} + OF_3 + F_4$$

Placing into matrix form, the element A is

$$EA = \begin{array}{c|c|c|c|c}
\diagdown \!\!\begin{array}{c} F \\ P \end{array} & 1 & 2 & 3 & 4 \\
\hline
1 & 1 & 0 & 0 & 0 \\
\hline
2 & 1/L & 1/L & 1 & 0 \\
\hline
3 & 0 & 1 & 0 & 0 \\
\hline
4 & -1/L & -1/L & 0 & 1
\end{array}$$

The EA matrix for member 2 would contain P_3 to P_6; it is not necessary to resubscript the F values.

Developing the S Matrix

Referring to Fig. 9-17 and using conjugate-beam (moment-area) principles, the end slopes e_1 and e_2 are

$$\frac{F_1 L}{3EI} - \frac{F_2 L}{6EI} = e_1 \qquad (g)$$

$$-\frac{F_1 L}{6EI} + \frac{F_2 L}{3EI} = e_2 \qquad (h)$$

Solving Eqs. (g) and (h) for F, we obtain

$$F_1 = \frac{4EI}{L} e_1 + \frac{2EI}{L} e_2$$

$$F_2 = \frac{2EI}{L} e_1 + \frac{4EI}{L} e_2$$

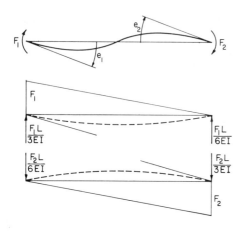

Figure 9-17. Congugate-beam relationships between end moments and beam rotations.

The forces F_3 and F_4 are obtained from the spring equation for force deflection as

$$F_3 = K_1 e_3 \qquad F_4 = K_2 e_4$$

The soil "spring" will have units of FL^{-1} obtained from the modulus of subgrade reaction as

$$K_1 = \frac{L}{2} Bk_s \qquad \text{and} \qquad K_2 = \frac{L}{2} Bk_s$$

The effect of the soil accumulates for interior nodes because of superposition, so that the problem "sees" the sum of two springs from the adjacent elements. Bowles (1974a) shows that best results are obtained by doubling the end springs. This was done to make a best fit of the measured data of Vesic and Johnson (1963) with computations. This is incorporated into the computer program in the Appendix. There is some logic in this in that if higher edge pressures are obtained for footings, then this translates into "stiffer" end soil springs.

The element S matrix is

	F e	1	2	3	4
	1	$\dfrac{4EI}{L}$	$\dfrac{2EI}{L}$	0	0
$ES =$	2	$\dfrac{2EI}{L}$	$\dfrac{4EI}{L}$	0	0
	3	0	0	K_1	0
	4	0	0	0	K_2

Developing the SA^T Element and element ASA^T Matrices

The ESA^T matrix is formed by multiplying the ES and the transpose of the EA matrix (in the computer program this is done in place by proper use of subscripting) as follows and noting that A^T goes with e and X. The $EASA^T$ is obtained in a similar manner as shown below:

$$EA^T =$$

e \\ X	1	2	3	4
1	1	$1/L$	0	$-1/L$
2	0	$1/L$	1	$-1/L$
3	0	1	0	0
4	0	0	0	1

$$ES =$$

F \\ e	1	2	3	4
1	$\dfrac{4EI}{L}$	$\dfrac{2EI}{L}$	0	0
2	$\dfrac{2EI}{L}$	$\dfrac{4EI}{L}$	0	0
3	0	0	K_1	0
4	0	0	0	K_2

$$\dashrightarrow \quad ESA^T =$$

$\dfrac{4EI}{L}$	$\dfrac{6EI}{L^2}$	$\dfrac{2EI}{L}$	$\dfrac{-6EI}{L^2}$
$\dfrac{2EI}{L}$	$\dfrac{6EI}{L^2}$	$\dfrac{4EI}{L}$	$\dfrac{-6EI}{L^2}$
0	K_1	0	0
0	0	0	K_2

$$EA =$$

1	0	0	0
$1/L$	$1/L$	1	0
0	1	0	0
$-1/L$	$-1/L$	0	0

$$\dashrightarrow \quad EASA^T =$$

$\dfrac{4EI}{L}$	$\dfrac{6EI}{L^2}$	$\dfrac{2EI}{L}$	$\dfrac{-6EI}{L^2}$
$\dfrac{6EI}{L^2}$	$\dfrac{12EI}{L^3} + K_1$	$\dfrac{6EI}{L^2}$	$\dfrac{-12EI}{L^3}$
$\dfrac{2EI}{L}$	$\dfrac{6EI}{L^2}$	$\dfrac{4EI}{L}$	$\dfrac{-6EI}{L^2}$
$\dfrac{-6EI}{L^2}$	$\dfrac{-12EI}{L^3}$	$\dfrac{-6EI}{L^2}$	$\dfrac{+12EI}{L^3} + K_2$

A check on the correct formation of the $EASA^T$ and the global ASA^T is that it is always symmetrical and there cannot be a zero on the diagonal. Note that the soil spring is an additive term to only the appropriate diagonal term. This allows easy removal of a spring for tension effect while still being able to obtain a solution, since there is still the shear effect at the point (not having a zero on the diagonal).

The computer program develops the EA and ES for each finite element in turn from input data describing the member so that e, I, L, and computations (or read in) for K_1 and K_2 can be made. The program performs matrix operations to form the ESA^T and $EASA^T$ and with proper instructions identifies the P-X coding to take the $EASA^T$ entries and put them in the proper global ASA^T locations.

When this has been done for all the finite elements (number of members NM), a global ASA^T of size $NP \times NP$ will have been developed as follows:

$$P_{NP} = A_{NP \times NF} S_{NF \times NF} A^T_{NF \times NP} \times X_{NP}$$

and cancelling interior terms as shown gives

$$P_{NP} = ASA^T_{NP \times NP} X_{NP}$$

which indicates that the system of equations is just sufficient (which yields a square coefficient matrix, the only type which can be inverted). It also gives a quick estimate of computer needs, as the matrix is always the size of a number of P's $(NP \times NP)$.

The ASA^T is inverted and multiplied by the external load matrix P to obtain the unknown X's. The computer program then rebuilds the EA, ES, and ESA^T and computes the F values for each finite element in turn, and the values are immediately printed to save storage.

If the footing separates or the deflections are larger than X_{max}(XMAX) and the switch is activated by input control, the program zeroes the appropriate element springs where tension is developed by inspection for negative X values, and computes a negative P-matrix entry as

$$-P_i = K_i(\text{XMAX})$$

with proper account made for the fact that two element soil springs are involved from the two adjacent finite elements. Once this nonlinear P-matrix entry is computed, the corresponding element springs are made zero. This operation, if specified, is done until convergence, the problem becomes unstable from zeroing too many soil springs, or a specified number of cycles is made. The reader should consult the program documentation in the Appendix with this program and closely inspect the "comment" cards interspersed through the program as additional documentation.

This computer program, in addition to solving beams on elastic foundations, will solve ring foundations, sheet-pile walls, braced excavation problems, and laterally loaded piles as well as many structure problems by inputing the subgrade modulus as zero.

Reworking Example 9-1 using $L = 19.75$ and $B = 8.17$ ft gave the approximate maximum bending moment 471.44 ft-kips at 9 ft from the left end using 11 finite elements with lengths of 1, 2, 2, 2, 2, 2, 2, 1.875, 1.25, 1.55, and 2.075 ft. This allowed placing half of each column load at the adjacent nodes since the columns fell on elements 1 and 9. Reworking the problem using point loads for the column, 13 finite elements with lengths of 0.5, 0.5, 2, 2, 2, 2, 2, 2, 1.875, 0.625, 0.625, 1.55, and 2.075 ft gave the maximum bending moment of 494.19 at 7 ft from the left end. The value from Example 9-1 was 597.5 ft-kips, about 27 percent larger than the finite-element solution—the order of magnitude usually found when working combined footings as beams on elastic foundations vs. the rigid (conventional) method. It should be noted that how the column is programmed to the footing also affects the problem. In this case (the placing of column loads at two nodes rather than the less correct point-load solution) the bending moment is reduced about 5 percent along with a location change.

Example 9-8. Use the computer program in the Appendix and solve a beam-on-elastic-foundation problem. Given data: $L = 15$ ft (see Fig. E9-8a). Column loads: $P_1 = 300$ kips (factored $D + L$),

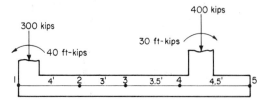

Figure E9-8a

$M = -40$ ft-kips

$P_2 = 400$ kips, $M = 30$ ft-kips

$B = 8.0$ ft, $D = 24$ in

$f'_c = 3,000$ psi $(E_c = 452,700$ ksf$)$

$\gamma_c = 0.145$ kcf

$K_s = AS + BSZ*EXPO$ (computer identification)

$= 144 + OZ^1$

SOLUTION. We will put loads of column 1 at node 1 as a convenience. We will use $NM = 4$, $NP = 10$ in order to show the global ASA^T matrix (with 1,000 factored).
Card input data are as follows:

CARD
1 Title (see first line of output sheet)
2 UT1   UT3, UT4, UT5, UT6
   FT   KIPS FT-K KSF K/CU FT
   this card allows identification of input data as in fps or SI units
3 10 4 4 1 0 1 0 0 0 0   (16I5)
   $NP = 10$, $NM = 4$, No of P-matrix entries, $NNZP = 4$, No of load conditions, $NLC = 1$, $IPILE = 0$,
   $IBEAM = 1$, $ISPILE = 0$, $IRING = 0$, $IBRAC = 0$, $NCYC = 1$
4 0 1 0 0 0 1 0 0 0   (16I5)
   $JJS = 0$ (no soil springs read in), $JTSOIL = 0$, $NONLIN$ = switch to activate $XMAX = 0$, $IAR = 0$,
   $NZEROX = 0$, $LIST = 1$, $INERP = 0$, $INERB = 0$, $IPRESS = 0$
5 452700 0.145   1.0   1.   1.   1.   1.   (8F10.4)
        E γ XMAX FACP DEMB CONV FAC2
6   144.   0.   1.   (8F10.4)
7   4.   3.   3.5   4.5   H(I)   (8F10.4)
8   8.   8.   8.   8.   B(I)   (8F10.4)
9   2.0   2.0   2.0   2.0   T(I)   (8F10.4)
10 1 1 -40.   K, J PR(K, J) $K = NP$
11 2 1   300. $J = NLC$
12 7 1   30. $PR(K, J) = $ load
13 8 1   400.

The output is displayed in Fig. E9-8b.

COMMENTS:
1. Note UT data are written back to identify input/output in units.
2. Note member data show $NPE(I)$, $I = 1,4$ as well as length, width, thickness, and computed I.
3. Note soil springs $SSK(I,J)$ for each end of each member
   $SSK(I,1) = $ near end spring
   $SSK(I,2) = $ far end spring
4. The ASA^T is symmetrical (note that it is divided by 1,000).

```
SOLUTION FOR BEAM ON ELASTIC FOUNDATION *******

MODULUS OF ELAST =      452700.0 KSF     UNIT WT =  0.1450 K/CU FT
NO NODES REQ CORRECT =     0    NODE SOIL STARTS FOR PILES =   1
MAX NON-LIN SOIL DEFORM =   1.000 FT

        SOIL MODULUS =      144.000+      0.0  *Z**1.000 K/CU FT
NO OF NON-ZERO P-MATRIX ENTRIES =     4   NO OF LOAD CONDITIONS =    1
     GROUND LINE REDUCTION FACTOR = 1.000     NO OF NODAL PRESSURE ENTRIES =    0
```

MEMNO	NP1	NP2	NP3	NP4	LENGTH	WIDTH	THICK	INERTIA,FT**4
1	1	2	3	4	4.000	8.000	2.000	5.33333
2	3	4	5	6	3.000	8.000	2.000	5.33333
3	5	6	7	8	3.500	8.000	2.000	5.33333
4	7	8	9	10	4.500	8.000	2.000	5.33333

MEMNO	SOIL MODULUS	SPRINGS-SOIL, A.R., OR STRUT	
1	144.000	4608.000	2304.000
2	144.000	1728.000	1728.000
3	144.000	2016.000	2016.000
4	144.000	2592.000	5184.000

```
    THE ASAT MATRIX CORRECTED FOR ANY BOUNDARY CONDITIONS AND 1000 FACTORED

 1   2414.40    905.40   1207.20   -905.40      0.0       0.0       0.0       0.0       0.0       0.0
 2    905.40    457.31    905.40   -452.70      0.0       0.0       0.0       0.0       0.0       0.0
 3   1207.20    905.40   5633.60    704.20   1609.60  -1609.60      0.0       0.0       0.0       0.0
 4   -905.40   -452.70    704.20   1529.80   1609.60  -1073.07      0.0       0.0       0.0       0.0
 5      0.0       0.0    1609.60   1609.60   5978.51   -427.04   1379.66  -1182.56      0.0       0.0
 6      0.0       0.0   -1609.60  -1073.07   -427.04   1752.56   1182.56   -675.75      0.0       0.0
 7      0.0       0.0       0.0       0.0    1379.66   1182.56   4905.45   -467.19   1073.07   -715.38
 8      0.0       0.0       0.0       0.0   -1182.56   -675.75   -467.19    998.30    715.38   -317.95
 9      0.0       0.0       0.0       0.0      0.0       0.0    1073.07    715.38   2146.13   -715.38
10      0.0       0.0       0.0       0.0      0.0       0.0    -715.38   -317.95   -715.38    323.13

LOAD MATRIX FT-K OR KIPS FOR NLC =   1

  1    -40.0000      2     304.6399
  3      0.0         4       8.1200
  5      0.0         6       7.5400
  7     30.0000      8     409.2798
  9      0.0        10       5.2200

FORCE MATRIX FOR NLC =   1
         MOMENTS, FT-K                    SOIL REACTIONS, KIPS

  1    -40.008     391.691              216.713      89.179
  2   -391.668     211.848               66.884      58.344
  3   -211.848    -414.090               68.068      57.652
  4    444.043      -0.020               74.124     103.901

SUM SOIL REACTIONS =  734.865 KIPS   SUM APPLIED FORCES =  734.799 KIPS
```

NODE NO	NODE ROTAT,RAD	NODE DEFL, FT	SOIL PRESS, KSF
1	-0.0022111	0.0470297	6.7723
2	-0.0018535	0.0387062	5.5737
3	-0.0014786	0.0337640	4.8620
4	-0.0016251	0.0285972	4.1180
5	-0.0020390	0.0200426	2.8861

Figure E9-8b

The first entry $ASA^T(1,1) = \dfrac{4EI}{L}$ (see $EASA^T$) = 2,414.398 × 1,000.

$$ASA^T(2,2) = \frac{12EI}{L^3} + K_1 = \frac{452,699.7 + 4,608}{1,000} = 457.307$$

$$ASA^T(4,4) = EASA_1^T(4,4) + EASA_2^T(4,4)$$

$$= +\frac{12EI}{L^3} + K_2 \text{ (element 1)} + \frac{12EI}{L^3} + K_2 \text{ (element 2)}$$

$$= 452,697 + 2,304 + 1,073,060 + 1,728 = 1,529.789$$

(slight error due to 5.333 round-off)

5. $\sum F_v$ is nearly satisfied, 734.9 vs. 734.8 kips
6. $\sum M$ (the F's) at each node is nearly zero.
   Node 1: -40.008, but adding 40 from load matrix with attention to signs gives zero
   Node 2: 391.691 vs. -391.668 from far end of member 1 and near end of member 2
   Node 3: 211.848 vs. 211.848
   Node 4: -414.09 vs. 444.04, but adding in 40 ft-kips applied load gives nearly zero
   Node 5: 0.020 vs. should be 0.00; this is computer round-off and due to insufficient NM

9-10. BRIDGE PIERS

Bridge piers are usually designed as conventional rigid footings but may be analyzed as beams on elastic foundations. Figure 9-18 illustrates several bridge-pier configurations. It is necessary that the designer evaluate the pier cap loads. Commonly this is analyzed as a rigid frame with pinned-base columns. The columns are assumed pinned to the footing as shown in Fig. 9-19. This is a simplification which makes for easy footing analysis; however, for the finite-element method it is not necessary to make this simplification. This simplification introduces some error, since the columns are usually rigidly attached to the pier foundation.

The foundation is usually so rigid that

$$q = \frac{P}{A} \pm \frac{Mc}{I}$$

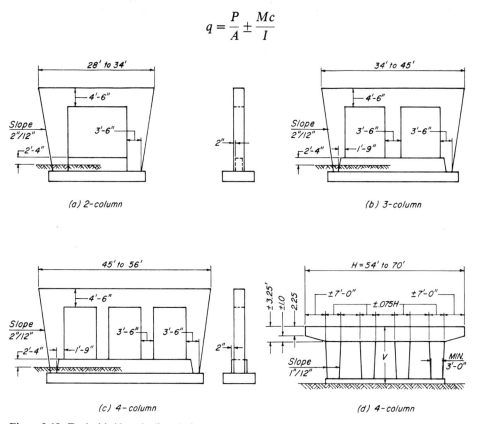

(a) 2-column

(b) 3-column

(c) 4-column

(d) 4-column

Figure 9-18. Typical bridge pier/foundation geometry.

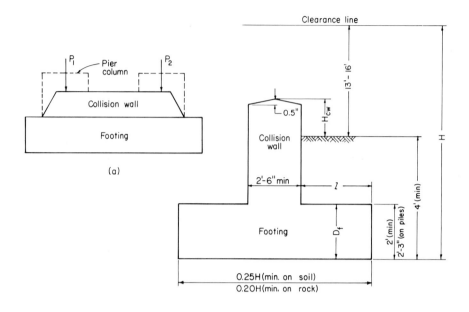

If: $l \leq 0.67 D_f$ check shear only and use shrinkage steel in bottom.

$H_{cw} = 2'-4''$ for grade separation structures; $2'-0''$ in streams; $6'-0''$ adjacent to railroad tracks.

$P_u = 1.3 [D + 1.67(L + Impact)]$

HS20-44 loading, AASHO Art. 1.5.17B

Shrinkage and temp. steel $= 0.125$ sq.in. at $s \leq 18''$
Shrinkage and temp. steel not req'd. if covered by at least $1.5'$ of earth

(b)

Figure 9-19. General bridge-pier-footing details.

is valid—to being overly conservative. Longitudinal moments may be such that reinforcing-steel requirements control. Transverse steel will seldom be required for bending unless $l > \frac{2}{3}D_c$ [AASHO (1973, Art. 1.4.6)] as shown in Fig. 9-19b.

The latest AASHO specifications contain ultimate-strength design provisions applicable to foundation design in:

Art. 1.5.17   (loads and load factors)
Art. 1.5.18   Concrete strength requirements
Art. 1.5.19 ϕ-factors*
Art. 1.5.20   Flexure*
Art. 1.5.21   Wide-beam shear*
Art. 1.5.21E   Footing applications*
Art. 1.5.23   Bearing*
Art. 1.5.29E   Development of reinforcement*

* Same as ACI requirements for foundations.

Minimum reinforcement for temperature and shrinkage is $\frac{1}{8}$ sq in/ft in both directions (Art. 1.5.6D) except that it is not necessary if the member is covered by at least $1\frac{1}{2}$ ft of earth.

Bridge piers on piles will be considered in later chapters.

9-11. RING FOUNDATIONS

Ring foundations can be used for water-tower structures, transmission towers, TV antennas, and various other possible superstructures. The ring foundation considered here is a relatively narrow circular beam as opposed to a circular mat to be considered in the next chapter. It would be presumed that the designer has developed a depth to satisfy diagonal tension, although several computer runs can be easily done to obtain sufficient data to establish a depth, then check it to see if the internal forces have changed as the cross section is modified.

The finite-element method (and computer program) of the preceding section can be used to obtain the bending moments of ring foundations. Bowles (1974a) has shown that the results compare very well with closed-form solutions. The finite-element method, however, is more adaptable to boundary conditions for which the mathematical solution is difficult to obtain. The program switch (IRING = 1) is used so that the program properly executes a ring-foundation solution.

The ring solution is preprogrammed to divide a ring into 20 finite elements which closes on itself as illustrated in Fig. 9-20 and the soil springs are internally computed from reading in a single value of k_s; however, as usual it is possible to override the computed spring values with read-in values by setting JJS to the desired number of values to read (number of members, since the program expects the user to read spring values for each end of the member). Loads which fall between nodes can be prorated to adjacent nodes using simple beam theory with negligible error. Alternatively, one could treat the finite element as a fixed-end beam and compute both shear and fixed-end moments to use in prorating the load to adjacent nodes, but this is considerable extra work for very slight computational improvement.

The loads being applied to the ring should be placed on the radius defining the center of area and not on the mean radius to reduce twisting, since one of the basic assumptions in beam-on-elastic-foundation theory is a uniform deflection across the width B. Next consideration should be given to the tangential twist, which is developed from the fact that the inner radius is smaller and the outer radius larger than the radius defining the center of area. The finite-element length is computed as

$$R_m = \sqrt{\frac{\text{ID}^2 + \text{OD}^2}{8}}$$

$$L = R_m(0.31416)$$

$$K = k_s\left(\frac{\text{Area}}{40}\right)$$

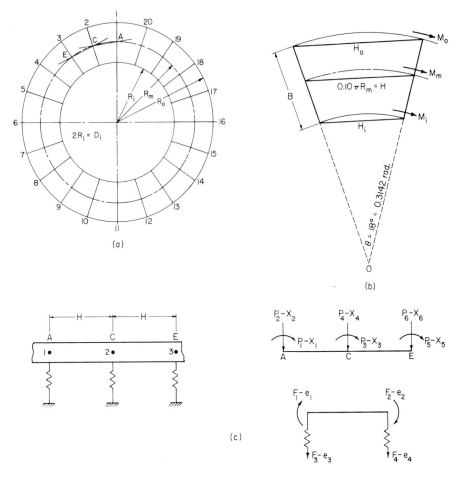

Figure 9-20. Matrix solution of ring foundation. (a) Ring foundation with 20 segments; (b) tangential moments on one side of the segment; (c) coding for matrix solution, considering segments *ACE* of part (a).

From the central finite difference, expressions for moment are as follows:

$$M = \frac{EI}{\Delta x^2}(y_{n+1} - 2y_n + y_{n-1})$$

it can be seen that the bending moment is inversely proportional to L^2, since by definition the deflections are constant; thus the bending moment at ID is larger than the moment computed based on R_m and the OD moment will be smaller, resulting in

$$M_i = \left(\frac{2R_m}{ID}\right)^2 M_m \quad \text{and} \quad M_0 = \left(\frac{2R_m}{OD}\right)^2 M_m$$

where M_i, M_m, and M_o are interior, as computed from computer program and outside bending moments, respectively. The difference between M_i and M_o represents ring twisting, which may require some torsional reinforcing steel, although most of the time the radial-shrinkage steel requirements will satisfy both shrinkage and torsion. The following example illustrates three equally spaced loads. Water towers, for example, will have overturning moments causing uplift forces in one or more columns and increased vertical loads in the other columns. The moment effects can be computed analogous to $P/A \pm Mc/I$.

Example 9-9. Find bending moments at selected points for the ring foundation shown in Fig. 9-20. Given data: ID = 14.5 m, OD = 16.0 m, DC = 0.76 m, $k_s = A + BZ^1$, $E_c = 224{,}100 = 22{,}407{,}000$ kPa, $k_s = 13{,}600 + 0Z^1$ kN/m³
Consider beam weightless
Three equal spaced column loads at 120° of 675 kN.
We will put 1 at node 1, and the other two will be on members 7 and 14.

SOLUTION. *Step 1.* Find load positions and prorate load to adjacent nodes using simple beam theory. Load 2 is $120 - 6 \times 18 = 12°$ into member 7 or $\frac{2}{3}$ of length:

$$R_m = \sqrt{\frac{14.5^2 + 16^2}{8}} = 7.63 \text{ m}$$

$$L = 7.63(0.31416) = 2.398 \text{ m}$$

$$\tfrac{2}{3}L = 1.60 \text{ m} \qquad L/3 = 0.798 \text{ m}$$

At node 7

$$P_7 = \frac{0.798}{2.398}(675) = 224.6 \text{ kN}$$

At node 8

$$P_8 = \frac{1.6}{2.398}(675) = 450.4 \text{ kN}$$

Load 2 is located between nodes 14 and 15 and at 6° from 14

$$L/3 = 0.798 \qquad \text{and} \qquad \tfrac{2}{3}L = 1.60 \text{ as before}$$

Also at node 14

$$P_{14} = P_8 = 450.4 \text{ kN}$$

$$P_{15} = P_7 = 224.6 \text{ kN}$$

Step 2. Set up input cards as follows:

Card No
1 Title (any 80 alphanumeric characters)
2 M kN kN-M kPa kN/CU M
Start in Col. 1, 11, 21, etc.
3 40 20 5 1 0 0 0 1 0 0
4 0 1 0 0 0 0 0 0 0
5 22407000. 0. 1. 1. 1. 1. 1.
6 13600. 0. 1.
7 14.5 16. 0.76
8 2 1 675
9 14 1 224.6
10 16 1 450.4
11 28 1 450.4
12 30 1 224.6

Step 3. Typical output is as follows:

Member No.	Moment kN-m (near end)	
	Near end	Far end
1	625.3	−49.9
2	49.9	176.1
3	−176.1	230.9
6	−95.9	−246.3
7	246.3	−437.9
8	437.9	22.4
9	−22.4	194.1
20	49.9	−625.3

The sum of vertical forces computed = 2,025 kN

Read-in = 2,025 kN

From this two checks are possible:

    (1) $\sum F_v = 0$

    (2) Far end M of member i = near end M of member $i + 1$

These are shown in the partial-output moment table above.

9-12. GENERAL COMMENTS ON THE FINITE-ELEMENT PROCEDURE

The following comments are observations made from solving a large number of different problems using the finite-element method.

1. One can only check a problem by
   a. Carefully checking the input data for correct dimensions and elastic properties
   b. Checking the sum of moments = 0 at nodes and the sum of soil reactions equal to applied loads. Note how applied moments were treated in Example 9-8.
2. One should use at least eight finite elements, but it is not usually necessary to use more than 20. The number of finite elements used (NM) depends some on the length of the member.
3. One should not use a very short member next to a long member. Use more finite elements and effect a transition between short and long members.
4. The value of k_s directly affects the deflection (but the values of deflection may not be critical) but has very little effect on the computed bending moments—at least for reasonable values of k_s as one might obtain from $k_s = 36q_a$.

PROBLEMS

9-1. Design a continuous rectangular footing for the conditions shown in Fig. P9-1 using the assigned values given in Table P9-1a (fps units) and the conventional rigid method.

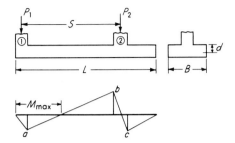

Figure P9-1

Table P9-1a. Units: Col. size, in or cm; D, L = kips or kN; f_c', f_y = ksi or kg/cm²; q_a = ksf or kPa; S = ft or m

Prob.	Col.	Size	Rebars	D	L	f_c'	f_y	q_a	S
9-1a	1	11	4 No. 9	90	50				
	2	13	4 No. 10	120	70	3.0	50	2.0	15
9-1b*	1	14	8 No. 8	200	100				
	2	17	8 No. 10	250	200	4.0	60	4.0	18
9-1c	1	13	8 No. 8	125	100				
	2	16	4 No. 8	130	200	3.5	50	2.5	20
9-1d*	1	13	4 No. 10	120	70				
	2	15	8 No. 9	175	100	3.0	60	2.5	16
9-2a*	1	33	8, 25 mm	580	312				
	2	38	8, 25 mm	670	423	211	4,219	175	4.85
9-2b	1	33	8, 25 mm	712	400				
	2	38	8, 25 mm	780	440	211	3,515	145	5.50
9-2c*	1	40	8, 20 mm	712	890				
	2	44	8, 30 mm	1,112	890	281	4,219	145	6.10
9-2d	1	40	8, 25 mm	712	890				
	2	44	8, 30 mm	1,112	900	281	4,219	190	5.50

* Partial answers in Table P9-1b.

Table 9-1b. For Probs. 9-1 and 9-2. Units: L, B = ft or m, V = kips or kN, M = ft-kips or kN-m, d = in or cm

Prob.	Length	Width	V_a	V_b	V_c	M_{max} at	M at ext. face of col. 2	d
a	5.689	1.99	−1,168.4	1,201.6	−255.1	−1,487.4 at 2.55	61.72	66.5
b	22.96	8.17	−391.91	437.64	−182.05	−1,755.91 at 9.06	333.74	29.82
c	7.112	3.50	−2,196.0	2,260.0	−464.6	−3,513 at 3.20	137.6	67.0
d	20.00	9.30	−248.95	271.65	−99.48	−1,017.9 at 8.18	140.98	20.40

9-2. Design a continuous rectangular footing for the conditions shown in Fig. P9-1 using the assigned values given in Table P9-1a (metric units) and the conventional rigid method.

9-3. Proportion a trapezoidal-shaped footing using the assigned problem in Table P9-3 and as shown in Fig. P9-3. Draw the shear and moment diagrams. Note problems are both fps and metric. If dowels are needed to transfer column load, indicate the load to be carried; otherwise use 0.5 percent A_{col}.

Table P9-3. Use units same as Prob. 9-1

Prob.	Col.	w	D	L	q_a	S
a*	1	24	250	200	4.0	20.0
	2	16	180	140		
b	1	16	180	170	3.0	15.0
	2	16	150	100		
c	1	0.50	1,400	1,200	120	5.20
	2	0.48	1,150	650		
d*	1	0.50	2,025	1,100	195	4.87
	2	0.48	1,125	1,150		

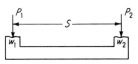

Figure P9-3

* Partial answer.

Prob.	a, ft or m	b, ft or m	M_{max}	d
a	5.12	12.64	2,693.15 at 10.38 ft	41.5 wide-beam
d	3.12	7.25	4,536.3 at 2.54 m	97.45 diagonal tension b

9-4. Design the trapezoidal-shaped footing for which the shear and moment diagrams were drawn in Prob. 9-3, using $f'_c = 3$ ksi and $f_y = 60$ ksi and USD.

9-5. Proportion a strap footing for the following conditions:

$w_1 = 16$ in + 6-in edge distance $D = 65$ $L = 40$ kips
$w_2 = 16$ in $D = 85$ $L = 60$ kips
$S = 18$ ft $q_a = 3$ ksf

9-6. Design d, steel requirements, and strap for Prob. 9-5. Use $f'_c = 3$ and $f_y = 60$ ksi.

9-7. Proportion a strap footing for the following conditions:

$w_1 = 40$ cm + 15-cm edge distance $D = 300$ $L = 180$ kN
$w_2 = 42$ cm $D = 385$ $L = 270$ kN
$S = 5.5$ m $q_a = 150$ kPa

9-8. Design d, steel requirements, and strap for Prob. 9-7. Use $f'_c = 211$ and $f_y = 4,219$ kg/cm².

9-9. Proportion a footing for $P = 120$ and $M = 180$ ft-kips, $q_a = 5$ ksf. Keep R in middle 1/3.

9-10. Proportion a footing for $P = 450$ kN and $M = 250$ kN-m, $q_a = 240$ kPa. Keep R in middle 1/3.

9-11. Use the footing dimensions from the assigned problem of Prob. 9-1 as a beam on an elastic foundation. Compare answers of using part of the column load on two adjacent nodes vs. using as point load. Use $k_s = 36q_a$, $E_c = 57,400\sqrt{f'_c}$.

9-12. Use the footing dimensions from the assigned problem of Prob. 9-2 as a beam on an elastic foundation. Compare answers of using part of the column load on two adjacent nodes vs. using as point load. Use $k_s = 120q_a$, $E_c = 15,100\sqrt{f'_c}$ (metric).

9-13. Use proportions from the assigned problem of Probs. 9-3 and 9-4 or Probs. 9-5 and 9-6 as a beam on elastic foundation. Use k_s and E_c as given in Prob. 9-11 or above. Use the average width of the finite element as $B(I)$.

9-14. Redo the notched footing of Example 9-6 if the notch is made 3 × 2 ft. What is the percent increase in pressure over the unnotched case?

Answer: $I_{xy} = 89.4$ ft^4 % increase $\cong 39\%$.

9-15. Design a ring foundation similar to Example 9-9 assuming a water tower with four equally spaced columns on a diam = 6.5 m. The tank holds 285 m^3 of water, and the empty structure weighs 2,200 kN. The wind moment is 2,250 kN-m. Take the maximum allowable soil pressure ($F = 2$) as 200 kPa. Use concrete properties of Example 9-9. Find D, ID, and OD and circumferential rebars for bending.

9-16. Design a bridge pier assuming three equally spaced columns as shown in Fig. P9-16 carrying:

Col	Case 1	Case 2
1	95	95
2	110	90
3	95	55

Take $f'_c = 3{,}000$, $f_y = 60{,}000$ psi, and $q_a = 6$ ksf.

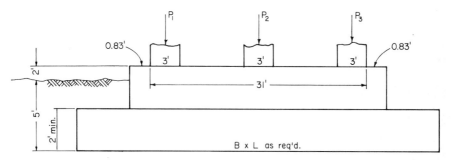

Figure P9-16

TEN

MAT FOUNDATIONS

10-1. INTRODUCTION

A mat foundation is a large concrete slab which transmits the loadings from several columns in a building or the entire building loads to the ground. It differs from a combined footing, which may also support several columns in that the columns are in a single line on a combined footing. A mat is often used when the soil is of such poor quality, or the column loads are so large, that more than 50 percent of the building-plan area is covered by footings. Considering current building costs, however, one should also investigate the costs of individual footings even if they encompass the entire area and using fiberboard (or other material) as separators (Fig. 10-1), since individual footings will not require negative reinforcing steel.

Mats may be supported by piles under certain circumstances such as high groundwater table, very poor soil, or where it is particularly important to control settlement (and the designer is uncertain of the soil response).

Mats are also used beneath silo clusters, storage tanks, chimneys, and other structures where a single foundation element beneath all parts of the structure is required.

10-2. TYPES OF MAT FOUNDATIONS

Figure 10-2 illustrates several possible mat-foundation configurations. Probably the most common mat design consists in a flat concrete slab 4 to 5 ft ($1\frac{1}{2}$ to 2 m) thick

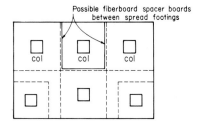

Figure 10-1. Mat vs. possible use of spread footings to save labor and forming costs.

and with continuous two-way reinforcing top and bottom. This type of foundation tends to be heavily overdesigned for three major reasons:

1. Additional cost of, and uncertainty in, analysis.
2. The extra cost of overdesign of this element of the structure will generally be quite small for reasonable amounts of overdesign relative to total project cost.
3. The extra safety factor provided for the additional cost.

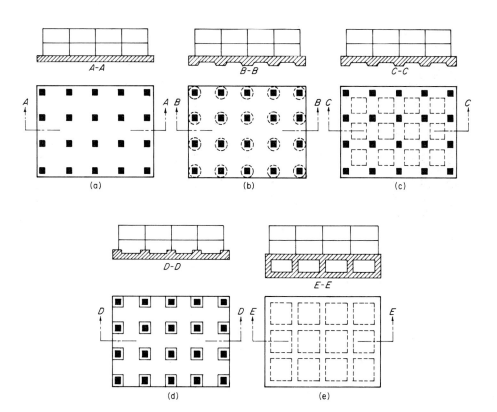

Figure 10-2. Common types of mat foundations. (*a*) Flat plate; (*b*) plate thickened under columns; (*c*) beam-and-slab; (*d*) plate with pedestals; (*e*) basement walls as part of mat.

10-3. BEARING CAPACITY OF MAT FOUNDATIONS

A mat must be stable against excessive settlements, which may be long-term (consolidation) as well as relatively rapidly occurring settlements (elastic or immediate).

A mat must be stable against a deep shear failure which may result in either a rotational failure [for example, the Transcona elevator failure (White, 1953)] or a vertical (or punching) failure. A punching failure alone would, if uniform vertically, not be particularly serious, as the effect would simply be a large settlement which could probably be landscaped; however, as the effect is not likely to be uniform and the end result is not predictable, this mode should be treated with concern equal to that for the deep-seated shear failure.

The bearing-capacity equations of Chap. 4 may be used to evaluate the soil capacity

$$q_{ult} = cN_c s_c i_c d_c + \gamma D_f n_q s_q i_q d_q + \tfrac{1}{2}\gamma BN_\gamma s_\gamma i_\gamma d_\gamma \qquad (4\text{-}6)$$

$$q_{ult} = 5.14c(1 + s'_c + d'_c - i'_c) + \bar{q} \qquad (4\text{-}6a)$$

where all terms have been defined in Chap. 4. The shape, depth, and inclination factors are in Table 4-3. Use B = least mat dimension and D = depth of mat (Fig. 10-3).

The allowable soil pressure from the above equation is

$$q_a = \frac{q_{ult}}{F}$$

where F = safety factor = 3 for cohesive soils
            = 2 for cohesionless soils

$F = 2$ may also be used for certain combinations of $D + L$ loads where the live loads, L, include wind, snow, or earthquake effects.

When the bearing capacity is established on the basis of penetration tests (SPT), one may use the Meyerhof (1956) equation from Chap. 4 and for any allowable settlement, S_a.

$$q_a = \frac{N}{4} S_a K_d \qquad \text{ksf} \qquad (4\text{-}10)$$

With cone-penetration data [Eq. (4-13a) doubled for mats] the bearing capacity is

$$q_a = \frac{q_c}{12}\frac{B+1}{B} K_d \qquad \text{ksf} \qquad (10\text{-}1)$$

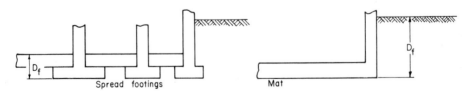

Figure 10-3. Increase in bearing capacity by using a mat foundation.

10-4. MAT SETTLEMENTS

Mat foundations are commonly used where settlements may be a problem as where a site contains erratic deposits or lenses of compressible materials, suspended boulders, etc. The settlement tends to be controlled via:

1. Lower soil contact pressures
2. Displaced volume of soil (flotation effect)
3. Bridging effects due to
   a. Mat rigidity
   b. Superstructure-rigidity contribution to the mat

The flotation effect should enable most mat settlements, even where consolidation is a problem or piles are used, to be limited to 2 to 3 in (5 to 8 cm). A problem of more considerable concern is the differential settlement. Again the mat tends to reduce this value as shown in Fig. 10-4. It can be seen that bending moments ($6EI\Delta/L^2$) and shear forces ($12EI\Delta/L^3$) induced in the superstructure depend on relative movement Δ between beam ends; and owing to continuity of the mat, this tends to be less than for spread footings.

Computer methods can allow one to estimate the upper and lower bound of expected differential-settlement effects. Observations of buildings supported on mats and on spread footings has produced the rule of thumb [Terzaghi and Peck (1967, p. 515)] of

Foundation type	Expected max settlement, in	Expected differential settlement, in
Spread	1	0.75
Mat	2	0.75

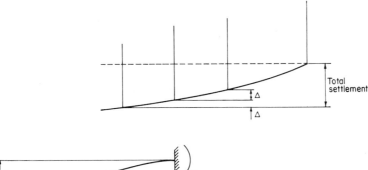

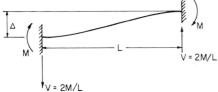

Figure 10-4. Reduction of bending moments in superstructure by using mat foundation. Bending moment M is based on differential settlement between columns and not on total settlement.

Various other studies [Grant et al. (1974)] furnish additional data which may be of use in estimating structural distress due to differential settlements.

Field observations as reported in literature indicate that mat settlements are almost always far less than the predicted values. Elastic or immediate settlement values are usually predicted on the basis of the theory of elasticity equation of Chap. 5:

$$S = qB \frac{1 - \mu^2}{E_s} I_w \qquad (5\text{-}10)$$

or from theory of consolidation. The finite-element method used in this chapter will give as good an estimate of settlements as Eq. (5-10) if E_s is used to compute the modulus of subgrade reaction k_s.

Corrections for the strata thickness and depth of mat into the soil should be made when using this equation, as outlined in Chap. 5. For layered soils a finite-element analysis can be made using a two-dimensional element mesh and a single point load to obtain the settlement as:

1. D_1 = settlement of a homogeneous stratum for a unit load at center of mat using E_s, μ of top layer.
2. D_2 = settlement of the layered strata for unit load with all E_s, μ.
3. S = settlement from Eq. (5-10) or other methods. Then by simple proportion

$$S_t = S \frac{D_2}{D_1} \qquad (10\text{-}2)$$

10-5. DESIGN OF MAT FOUNDATIONS (APPROXIMATE METHOD)

There are three methods which can be used for the design for mat foundations:

1. Conventional—where columns are reasonably regularly spaced in lines in the X and Y directions and column loads do not vary much over 20 percent [an ACI Committee 436 (1966) recommendation]. If the mat loading conditions are different from this, it is necessary to use either the finite-difference or finite-element method for the bending analysis.
2. Finite differences
3. Finite element

These methods will be considered separately in the following sections.

The *conventional* method uses an assumed linear pressure diagram on the base of the mat (refer to Fig. 10-5) computed as

$$q = R\left(\frac{1}{A} \pm \frac{e_x x}{I_y} \pm \frac{e_y y}{I_x}\right) \qquad (10\text{-}3)$$

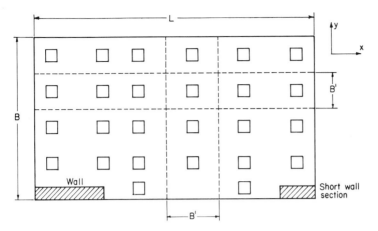

Figure 10-5. Rigid-mat design. Walls may create problems in a "rigid" design. Uneven spacing may also invalidate procedure.

where $R = \Sigma P_i$ = resultant of vertical column and wall loads onto mat
A = plan area of mat
I_x, I_y = moment of inertia of plan area of mat with respect to X and Y axes, respectively ($I_x = LB^3/12$; $I_y = BL^3/12$; Fig. 10-5)
e_x, e_y = eccentricity of the resultant vertical loads with respect to the centroidal axes
x, y = coordinate locations where soil pressures are desired

Since Re_x, Re_y are moments about the Y and X axis, any column moments may be algebraically added to obtain the total moment for computing the soil pressures at selected locations beneath the mat. Column moments should also be included in the computations to locate the eccentricities e_x, e_y of the resultant of the column loads.

The depth of the mat is selected to satisfy strength design of diagonal-tension (or punching) stresses of

$$v_c = 4\phi\sqrt{f'_c} \quad \text{psi} \quad (1.06\phi\sqrt{f'_c} \quad \text{kg/cm}^2)$$

where f'_c is in units of either psi or kg/cm². The ϕ factor is taken as 0.85. Diagonal tension is investigated at a distance of $d/2$ from the column faces for either round or square columns. Do not convert a round column into an equivalent square column for this investigation.

Exterior wall columns may require investigation of a three-sided failure zone, interior columns will have a four-sided zone, and corner columns may have only a two-sided failure zone.

Perimeter load-bearing walls (and possible interior) should be investigated for wide-beam shear at a distance of d from the wall face, and the depth should satisfy

$$v_c = 2\phi\sqrt{f'_c} \quad \text{psi} \quad (0.53\phi\sqrt{f'_c} \quad \text{kg/cm}^2)$$

where terms have been defined previously.

The depth obtained to satisfy shear requirements above is required in the absence of shear reinforcement. Shear reinforcement to reduce the mat depth is generally not recommended, because the lesser depth reduces the mat rigidity, although the ACI Code Art. 11.11 allows both shear and shear-head reinforcement to reduce overall depth. It is suggested where the potential need exists for shear reinforcement that the Report of the Joint ASCE-ACI Committee (1974) be consulted. Reduced mat rigidity will tend to localize column and wall-load effects instead of spreading them as assumed in the design based on linear soil-pressure distribution.

The overall mat depth must be sufficient to satisfy structural requirements and the 3 in (7.5 cm) of reinforcing-bar protective cover.

Next in *conventional design*, the mat is divided into convenient strips of width B' as in Fig. 10-5 in both the X and Y directions. Since the strips will usually have several columns on them, we may:

1. Treat the strip as a continuous beam and apply slope-deflection or moment-distribution methods to find bending moments at selected locations. It may be necessary to "adjust" the soil pressure or column loads using these more refined methods to satisfy statics of the individual beams (which has been lost by ignoring shear transfer at the edges of the strips to adjacent strips).
2. Simply compute the bending moments as:
   For interior spans [at column locations $(+)$ moment and between columns for $(-)$ moment] $M = wL^2/10$
   For moments in the exterior spans

$$M = wL^2/8 \qquad \text{or} \qquad M = wL^2/9$$

These computations recognize that fully fixed-beam moments are $+wL^2/12$ and $-wL^2/24$ and simply supported beam moments are $-wL^2/8$ for this type of loading configuration. The latter procedure gives rather conservative values of design moment and is generally preferred over the refinements of method 1 owing to the approximations associated with dividing the mat into strips. The effect of having the design moment be conservative is to use somewhat larger amounts of reinforcing steel. Note, however, that by not using shear-reinforcement steel, the mat is already deeper and thus the steel "moment arm" $(d - a/2)$ is larger, reducing the computed steel requirement somewhat for any given bending moment.

Example 10-1. Design a mat foundation by the conventional (rigid) method. Given data: All columns 15×15 in; $f'_c = 3,000$ psi; $f_y = 50,000$ psi, and $q_a = 1.0$ ksf (net). Refer to Fig. E10-1a assuming an average $LF = 1.6$ for given loads.

SOLUTION. *Step 1.* Find location of resultant.

$$F_v = 2(85) + 4(330) + 2(110) + 2(250) + 90 + 97 = 2,397 \text{ kips}$$

Find e_x by taking moments along line AB through the columns.

$$2,397\bar{x} = 2(330)(20) + 2(110)(20) + 2(250)(40) + 40(90 + 97)$$

$$\bar{x} = \frac{45,080}{2,397} = 18.81 \text{ ft} \qquad \text{from which } e_x = -1.19 \text{ ft (west of center)}$$

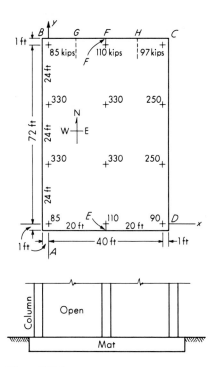

Figure E10-1a

Find e_y by summing moments along line AD through the columns.

$$2,397\bar{y} = 2(330)(24) + 2(330)(48) + 72(85 + 110 + 97)$$

$$+ 24[250 + 2(250)]$$

$$\bar{y} = \frac{86,544}{2,397} = 36.11 \text{ ft} \qquad \text{and } e_y = 0.11 \text{ ft (north of center)}$$

Table E10-1

Point	V/A	x	$0.006x$	q, ksf
A	+0.77	−21	+0.13	+0.90 (compression)
B	0.77	−21	0.13	0.90
C	0.77	21	−0.13	0.64
D	0.77	21	−0.13	0.64
E	0.77		0	0.77
F	0.77		0	0.77
G	0.77	−10	+0.06	0.83
H	0.77	10	0.06	0.71

Step 2. Compute the soil pressure at selected points beneath the mat.

$$I_x = \frac{bh^3}{12} = \frac{42(74)^3}{12} = 14.18 \times 10^5 \text{ ft}^4 \qquad M_x = 0.11(2,397) = 264 \text{ ft-kips}$$

$$I_y = \frac{74(42)^3}{12} = 4.57 \times 10^5 \text{ ft}^4 \qquad M_y = 1.19(2,397) = 2,852 \text{ ft-kips}$$

$$q = \frac{V}{A} \pm \frac{M_x y}{I_x} \pm \frac{M_y x}{I_y}$$

$$= \frac{2,397}{42(74)} \pm \frac{264y}{14.20 \times 10^5} \pm \frac{2,850x}{4.56 \times 10^5}$$

$$= 0.77 \pm 0.00018y \pm 0.006x$$
$$\underset{\text{neglect as } \geq 0}{\uparrow}$$

Table E10-1 summarizes the soil pressures of selected locations beneath mat.

Step 3. Solve for depth to satisfy shear requirements: The maximum shear will occur as diagonal-tension shear at one of the 330-kip columns (exterior), since it is obvious wide-beam shear does not control. When the location of critical shear is not obvious, it may be necessary to check several locations for the critical location (refer to figure). Perimeter in shear $= 2\left(1 + \dfrac{1.25 + d}{2}\right) + 1.25 + d$ or Perimeter $= 4.50 + 2d$.

Allowable shear stress $= 26.81$ (Table 8-2). We will neglect the upward soil pressure in zone *mnop* as a negligible force.

$$d(4.50 + 2d)(26.81) = 330(1.6)$$

$$2d^2 + 4.5d = 19.7$$

$d = 2.21$ ft   Use $d = 27.5$ in, $D = 32$ in (overall depth)

Step 4. Divide mat into 3 beams in N-S direction.

AB with $B' = 11$ ft and $q = (0.90 + 0.83)/2 = 0.865$ ksf

EF with $B' = 20$ ft and $q = (0.83 + 0.71)/2 = 0.77$ ksf

DC with $B' = 11$ ft and $q = (0.71 + 0.64)/2 = 0.68$ ksf

Check statics of strips:

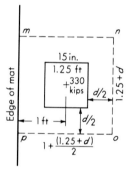

Figure E10-1b

Sum column loads $AB = 830$ kips

Soil reaction $AB = 11(74)(0.865) = 704$ kips

Sum column loads $EF = 880$

Soil reaction $EF = $ 1,139.6

Sum column loads $DC = 687$

Soil reaction $DC = $ 553.4

                        2,397.0   2,397.0

The mat can carry across any strip at least (wide-beam shear) $13.4(d)74 = 2,272$ kips

Step 5. Compute bending moments: With the large difference in column loads (greater than 20 percent as suggested by ACI Committee 436), an approximate analysis may not be completely valid.

Compute moments as $\dfrac{wL^2}{10}$

Strip	M, ft-kips
AB	$(0.1)(0.865)(24)^2 = 49.8$
EF	44.4
DC	39.2

Next let us average the load and reactions on strip AB and recompute moments

$R = 704$

$V = 830$

Av. $= 767$ kips $q = 0.94$ $R = \dfrac{767}{830} = 0.924$

Column loads $= 85(0.924) = 78.5$

$330(0.924) = 30.5$

For strip EF average load $= 1{,}009.8$; $q = 0.68$; $R = 1.15$. Find moments at selected locations (ft-kips/ft)

	AB	EF	CD
13 ft from B	13.34	24.8	39.2
25 ft from B	-115.3	-54.4	-38.6
37 ft from b	46.6	-3.7	20.2

These moments differ considerably from previous values.

Step 6. Compute required steel N-S.

$$a = \frac{50A_s}{0.85(3)(12)} = 1.634A_s$$

$$A_s(27.5 - 0.82A_s) = \frac{M_u(12)}{(0.9)f_y}$$

For strip AB check $M = 115.3$ ft-kips/ft

$$A_s^2 - 33.54A_s = -37.5$$

$$A_s = 1.16 \text{ sq in/ft}$$

$$p = \frac{1.16}{(28.5)(12)} = 0.0033$$

Since we are analyzing as a beam we will use (Art. 10.5)

$$p = \frac{200}{f_y} = 0.004 > 0.0033$$

$$A_s = 28.5(12)(0.004) = 1.37 \text{ sq in/ft}$$

other two strips will be governed by

$$p = 0.004 \text{ also}$$

$$A_s(\text{total}) = 1.37(42) = 57.46 \text{ sq in}$$

Use 73 No. 8 bars at 6.9-in spacing across bottom of mat in negative zone and run 37 of the bars full length.

Run 37 No. 8 bars full length across top of mat for (+) moments.

The E-W reinforcement would be designed similarly taking 4 equivalent beams of say 13, 24, 24, and 13 ft.

10-6. FINITE-DIFFERENCE METHOD FOR MATS

The finite-difference method uses the fourth-order differential equation found in any text on the theory of plates and shells [Timoshenko and Woinowsky-Krieger (1959)].

$$\frac{\partial^4 w}{\partial x^4} + \frac{2\partial^4 w}{\partial x^2\,\partial y^2} + \frac{\partial^4 w}{\partial y^4} = \frac{q}{D} + \frac{P}{D(\partial x\,\partial y)} \tag{10-4}$$

which can be transposed into a finite-difference equation when $r = 1$ (Fig. 10-6):

$$20w_0 - 8(w_T + w_B + w_R + w_L) + 2(w_{TL} + w_{TR} + w_{BL} + w_{BR})$$

$$+ (w_{TT} + w_{BB} + w_{LL} + w_{RR}) = \frac{qh^4}{D} + \frac{Ph^2}{D} \tag{10-5}$$

When $r \neq 1$, this becomes [see Bowles (1974a, Chap. 7) for a computer program]

$$\left(\frac{6}{r^4} + \frac{8}{r^2} + 6\right)w_0 + \left(-\frac{4}{r^4} - \frac{4}{r^2}\right)(w_L + w_R) + \left(-\frac{4}{r^2} - 4\right)(w_T + w_B)$$

$$+ \frac{2}{r^2}(w_{TL} + w_{TR} + w_{BL} + w_{BR}) + w_{TT} + w_{BB} + \frac{1}{r^4}(w_{LL} + w_{RR})$$

$$= \frac{qrh^2}{D} + \frac{Ph^2}{rD} \tag{10-6}$$

It is usual to use the concept of modulus of subgrade reaction in the solution of this type of problem. Christian (1975) has discussed this concept, and it is widely used. A few authors tend to downgrade the method [e.g., Lee (1974), Lee and Brown (1972)] with very little computational/performance evidence to indicate any alternative method (elastic method using E_s, μ) is superior. Gibson (1967) indicates that for a

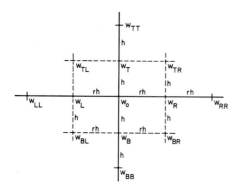

Figure 10-6. Finite-difference grid of elements of $rh \times h$ dimension.

homogeneous elastic, half-space with E_s increasing linearly with depth the elastic solution is the same as for the subgrade-reaction concept.

The analytical methods for mat foundations using the concept of subgrade reaction are considerably simplified over other methods of computation. This is because the subgrade reaction becomes converted to an equivalent spring applied at the node of interest. The node of interest in the matrix of simultaneous equations resulting from the finite-difference or finite-element method of the next section is the diagonal term.

Depending on the procedure, and identifying the subgrade-reaction contribution as K_i, we simply augment the deflection-coefficient matrix and, as illustrated in Sec. 9-9,

$$(A_{ii} + K_i)w = C \qquad (a)$$

This bookkeeping is quite simple. In the particular case of the finite-difference method we have

$$\left(\frac{6}{r^4} + \frac{8}{r^2} + 6\right)w_0 + k_s\frac{rh^2}{D}w_0 \qquad (b)$$

but $rh \times h =$ the contributing area of soil to the interior node o; thus $k_s rh^2 = FL^{-1} =$ units of a spring or we have

$$\left(\frac{6}{r^4} + \frac{8}{r^2} + 6 + \frac{K}{D}\right)w_0 \qquad (c)$$

when $r = 1$, we have, of course,

$$\left(20 + \frac{K}{D}\right)w_0 \qquad (d)$$

modifications are as follows:

$$\text{Corners} \quad K_i = \frac{k}{4}$$

$$\text{Sides} \quad K_i = \frac{k}{2}$$

$$\text{Reentrant corners} \quad K_i = 0.75K$$

The finite-difference method is particularly attractive because the matrix is of size

$$NP = \text{number of nodes}$$

This matrix will be symmetrical $(A = A^T)$ and of the general form

$$Aw = C$$

and inverting one obtains the deflection matrix as

$$w = A^{-1}C$$

From symmetry one may utilize a band-reduction method of reducing the matrix, rather than inverting it.

The finite-difference method has several major disadvantages including:

1. It is necessary to use modifications of Eq. (10-6) for boundary conditions; thus the program requires many subroutines.
2. It is difficult to allow for holes, notches, reentrant corners, etc.
3. It is difficult to program boundary conditions of zero rotation and/or deflections.
4. It is difficult to account for moments applied at nodes. Here, the author has tried using equivalent couples at adjacent nodes to account for the bending moment, but the solution has never been satisfactory. A problem is that the finite-difference method is width-dependent, whereas loads are not. A listing of the necessary equations for various nodes and a computer program for their solution are readily available in Bowles (1974a) and will not be repeated here.

10-7. FINITE-ELEMENT METHOD FOR MAT FOUNDATIONS

There are two basic approaches to the generalized finite-element method (discounting that the finite-difference method is a "finite element"). They are:

1. Grid analysis (or line element)
2. Finite element

In the *finite-element* analysis, element continuity is maintained through use of displacement functions. The displacement function is of the form

$$u = a_1 + a_2 X + a_3 Y + a_4 X^2 + a_5 XY + a_6 Y^2 + a_7 X^3 + a_8 X^2 Y + a_9 XY^2$$
$$+ a_{10} Y^3 + a_{11} X^4 + a_{12} X^3 Y + a_{13} X^2 Y^2 + a_{14} XY^3 + a_{15} Y^4 \qquad (10-7)$$

With a rectangular plate and three general displacements at each corner (node) only 12 unknowns of Eq. (10-7) are necessary. This results in reducing the general displacement equation to one with 12 coefficients instead of 15. Which three are best to discard becomes a considerable exercise in both engineering judgment and computational ability/tenacity. Various procedures have been and are being periodically

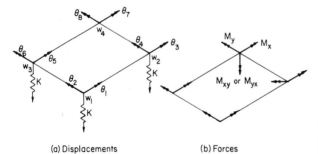

(a) Displacements   (b) Forces

Figure 10-7. Finite-element method using a rectangular plate element.

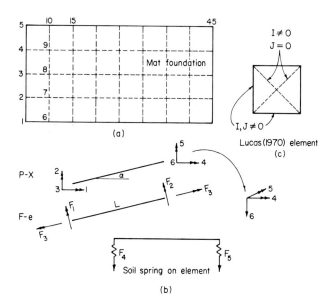

Lucas (1970) element
(c)

Figure 10-8. Method of finite-element (grid) analysis. Note that orientation of node numbers above results in a banded stiffness matrix of minimum width (18 vs. 30 by numbering left to right). A finite-element grid proposed by Lucas (1970).

proposed to reduce and solve the resulting matrix such as the finite element confer-ences at McGill University (1972), Wright Patterson AFB (1965, 1968, 1971), and regular papers in the *International Journal of Solids and Mechanics* and the *Interna-tional Journal for Numerical Methods in Engineering*. The end result in any case is difficult to interpret. For this reason, for computational ease, and since a comparison of solutions displays little difference, the author uses the grid-element approach [Bowles (1974a)]. Figures 10-7 and 10-8 illustrate the essential differences between the two procedures. Figure 10-8 also shows an alternative grid approach proposed by Lucas (1970) which appears to have insignificant advantage over the author's method as far as comparing deflections is concerned.

The finite-element method (grid) uses the same equations as in Sec. 9-9:

$$P = AF \qquad e = A^T X \qquad F = Se = SA^T X$$

$$P = ASA^T X \qquad \text{and} \qquad X = (ASA^T)^{-1} P$$

As before, it is necessary to develop the element EA and ES matrices, with the computer taking care of the remainder of the problem including the global ASA^T-matrix.

Referring to Fig. 10-8, the element EA matrix is built by summing ΣF at each node. For example, at node 1

$$P_1 = F_1 \sin \alpha + OF_2 - F_3 \cos \alpha + OF_4 + OF_5$$

$$P_2 = F_1 \cos \alpha + OF_2 - F_3 \sin \alpha + OF_4 + OF_5$$

$$P_3 = \frac{F_1}{L} \qquad + \frac{F_2}{L} + OF_3 \qquad - F_4 + OF_5$$

and the resulting matrix is

$$EA =$$

P \ F	1	2	3	4	5
1	$-\sin\alpha$	0	$-\cos\alpha$	0	0
2	$\cos\alpha$	0	$-\sin\alpha$	0	0
3	$\dfrac{1}{L}$	$\dfrac{1}{L}$	0	$-1.$	0
4	0	$-\sin\alpha$	$\cos\alpha$	0	0
5	0	$\cos\alpha$	$\sin\alpha$	0	0
6	$-\dfrac{1}{L}$	$-\dfrac{1}{L}$	0	0	$-1.$

Similar to the *ES* of Sec. 9-9 but including a torsion factor for F_3, the mat *ES* matrix is

$$ES =$$

F \ e	1	2	3	4	5
1	$\dfrac{4EI}{L}$	$\dfrac{2EI}{L}$	0	0	0
2	$\dfrac{2EI}{L}$	$\dfrac{4EI}{L}$	0	0	0
3	0	0	$\dfrac{\Omega GJ}{L}$		
4	0	0	0	K_1	0
5	0	0	0	0	K_2

The torsion constant J for rectangular sections is computed from Seely and Smith's (1952) procedure using a curve-fitting subroutine. Both I and J for ribbed sections would have to be separately entered. The major disadvantages of this method of analysis are:

1. Large matrix (even using a band-matrix method of solution) to solve ($NP = 3 \times$ No. of nodes)
2. Large number of data cards

The principal advantages are:

1. Ease of programming for reentrant corners, slots, or holes
2. Use any boundary conditions
3. Used for circular plates with slight modifications
4. Used for counterfort retaining walls (plate fixed on three edges)
5. Used for pile-cap solutions

A comparison of the grid and rectangle FEM models by the author with values from Timoshenko and Woinowsky-Krieger (1959) for plates simply supported and with clamped edges (not on soil) yielded almost identical deflections for all three methods if the torsion constant in the grid analysis is multiplied by

$$\Omega = 0.75 \text{ for } b/a = 1 \text{ (square)}$$

$$\Omega = 1.00 \text{ for } b/a > 1 \text{ (rectangle)}$$

$$\Omega = 1.00 \text{ for clamped edges}$$

For plates on soil, one can use $\Omega = 1.00$ for all b/a, since the soil contribution alters the deflection pattern. Moments of the author's grid method and the values obtained from Timoshenko and Goodier (1951) are in good agreement.

The grid-analysis method is recommended because of its greater computational simplicity and with the modulus-of-subgrade-reaction concept. The finite-element methods allow greater ease of programming boundary conditions such as zero rotations at wall intersections or for rigidly attached columns. Column loads and moments can be easily included.

The concept of subgrade reaction with spring contributions at nodes is easy to modify for soil separation, since the diagonal term is the only coefficient in the stiffness matrix with the soil spring K_i

$$(A_{ii} + K_i)X_i = P_i$$

thus, for footing separation we simply make $K_i = 0$, rebuild the stiffness matrix, and resolve for the displacements X_i.

Generally one should include the mat weight in the analysis. The mat does not cause internal bending moments due to self-weight since the concrete is poured directly on the subbase and in the fluid condition conforms to any surface irregularities prior to hardening. Should the loads cause separation, however, the mat weight tends to counter this. Deflections will be larger when including mat weight, since the soil springs react to all vertical loads.

Preliminary Work

Generally, the depth of the mat is established from shear requirements as indicated earlier. This depth + steel cover is used to compute the moment of inertia or D:

$$I = \frac{BT^3}{12} \qquad D = \frac{ET^3}{12(1 - \mu^2)}$$

The bending moments obtained from the latter analysis are used to design the reinforcement in both directions for the mat.

Total deflections are sensitive to the value of k_s used. Bending moments are much less so, but the designer should try to use a realistic minimum and a probable maximum value of k_s and obtain at least two solutions. The design would be based on the best information available or the worst conditions obtained from either of the two solutions (generally when k_s is minimum).

The author does not particularly recommend the finite-difference method of analysis for mats, for the reasons cited earlier. In passing, however, it should be noted that the "strip" or conventional method of mat analysis could be carried out as a beam on an elastic foundation of Sec. 9-9.

Establishing Finite-Element Grid (Variables in Brackets Refer to Computer Program in Appendix)

The problem is begun by laying the mat out in plan with all column and wall locations. Next lay a grid on this plan such that grid intersections (nodes) occur at all column faces (if column fixity is established), also at all wall faces or wall lines depending on wall dimensions. Grid elements do not have to be the same size, but best results are obtained if very small members are not adjacent to large ones (e.g., a member 0.2 ft long connecting to a 2-ft-long member is not as satisfactory as a 1-ft connecting to a 2-ft member). For pinned columns the grid can be at convenient divisions. The load matrix can be developed using column locations.

Develop a card generator to punch data cards for the member number ($MEMNO$) and NP values for each element $[NPE(I)]$ for the six values H, V, and B (refer to Fig. 10-9). A card generator is a necessity, since the input data are enormous.

Develop the nonzero P-matrix entries for each load condition (also NLC), (K), $(NNZP)$, and $[I, P(I, K)]$.

Establish if any changes in soil modulus are required. This may be accounted for in the card generator; however, for local soft spots, holes in the ground, hard spots, etc., it may be more convenient to hand-compute the effect and insert it into the problem (via use of $JJS > 0$ in program).

Establish any zero boundary conditions ($NOZX$).

Compute the number of NP's in the matrix: $NP = 3 \times$ number of nodes; also count the number of members (NM) to be used in the grid.

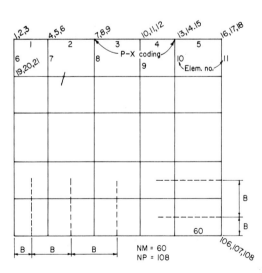

Figure 10-9. Typical coding for a mat. Note elements are not all the same size. Program "sees" element widths B as shown above. For horizontal members $L = H$ and $V = 0$; vertical members $L = -V$ and $H = 0$.

Compute the bandwidth of the matrix as follows:

1. Find the minimum NP value at various nodes
2. Find the maximum NP value at adjacent nodes
3. Compute bandwidth as:

$$NBAND = NP_{max} - NP_{min} + 1$$

the size of the band matrix in the core (computer core requirement) is

$$STIFF = NBAND \times NP$$

The Solution Procedure

With the displacements from $X = (ASA^T)^{-1}P$ we can solve $F = SA^TX$ for each element in turn to find the element forces.

The computer program performs the necessary matrix multiplication to form the element SA^T ($ESAT$) and the element ASA^T ($EASAT$). The element ASA^T will be of size 6×6. The element ASA^T is sorted for values to be placed at the appropriate locations in the global ASA^T ($ASAT$) for later banding and solution. Normally the ASA^T will have to be put on a disk or tape file capable of *random access*.

The computer routine next recalls the ASA^T from disk (or tape) and stores the band in core (refer to Fig. 10-10), filling the lower right corner with zeros. Zero boundary conditions are applied if specified, which results in zeroing the appropriate *row* and *upper diagonal* of the band matrix and placing a 1.0 in the first column as shown below (typical):

$$
\begin{matrix}
& & & & & 0. \\
& & & & 0. & \\
& & & 0. & & \\
& & 0. & & & \\
& 0. & & & & \\
1.0 & 0. & 0. & 0. & 0. & 0.
\end{matrix}
$$

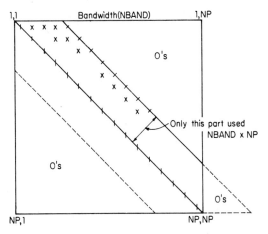

Figure 10-10. Symmetrical ASA^T matrix. Only part used in reduction is as shown.

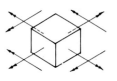

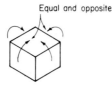

Equal and opposite

(a) Node and moments

(b) Node with moments summed for checking

Figure 10-11. Checking moments in output for statics at a node. (*a*) Node and moments; (*b*) node with moments summed for checking.

With the band-reduction method, the displacements are exchanged with the P matrix at the end of the reduction. If it is desired to save the original P matrix for any reason, it must be stored in some alternative location (in the computer program it is put on disk at the end of the ASA^T). The X's (or redefined P's) are used to compute the F's. Also they should be scanned to see if mat-soil separation has occurred at any nodes. If negative deflections occur (tension soil springs), the stiffness matrix is rebuilt with no springs ($K = 0$) at those nodes, and the problem is cycled until the solution converges. Converging is understood to be when

$$N_i \text{ of } K = 0 \text{ is equal to } N_{i-1} \text{ of } K = 0$$

It is useful to obtain the F's for each cycle to observe the effect of the mat separation on the problem. For large problems it should be evident that this cycling can use a rather large amount of computation time.

When convergence is achieved, the program should compute the moments on each side of all nodes (for comparing $\Sigma M = 0$ as a statics check), noting that corner and side nodes will be set to zero on the appropriate sides (Fig. 10-11). This operation is extremely laborious unless done on the computer. It may have to be done by hand for circular mats, as special considerations are involved.

It is useful to sum the element spring forces for a statics check of vertical mat equilibrium and to compute the soil pressure at each node as

$$q = k_s X$$

where X is the translation displacement at that node. It can be observed that the soil pressure as computed above is nearly independent of k_s but the deflection X is almost directly dependent upon k_s. This method will be illustrated by the following examples.

Example 10-2. Redo the spread footing (Example 8-2) and compare maximum bending moments from a pinned and fixed column boundary condition with the ACI method used in Example 8-2. Given data:

$$D = 71 \qquad L = 100$$

$$q_a = 4 \qquad f'_c = 3,000$$

$$\text{Column} = 14 \times 14 \text{ in}$$

$$D_c = 17 \text{ in} \qquad B = 6.75 \text{ ft}$$

Figure E10-2 illustrates the grid.

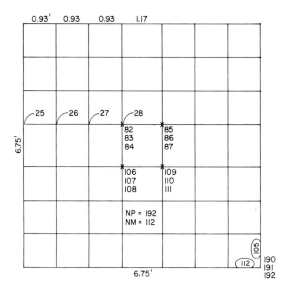

Figure E10-2

SOLUTION. We will use the computer program in the Appendix.

$$P_u = 1.4(71) + 1.7(100) = 269.4$$

for $\gamma_c = 145$ pcf $E_c = 57,400\sqrt{f'_c}(0.144) = 453,000$ ksf

$$G = \frac{E}{2(1 + \mu)} = \frac{453,000}{2(1 + 0.15)} = 196,800 \text{ ksf}$$

assume $k_s = 50$ kcf [could use $36(4) = 144$]

From counting on the grid or card-generator output

$$\text{Number of } P\text{'s, } NP = 192$$

$$\text{Number of grid members, } NM = 112$$

$$NBAND = 27 - 1 + 1 = 27$$

Data cards are as follows:

Card No.

1	Title card
2	Units card
3	192 112 4 1 1 27 8 8 64 0 0 1
	No list of matrices but repeat problem
4	0 2 0 0 0 max. repeats = 2
5	453000. 196800. 50. 1.417 0.0 1.
	take unit weight = 0
6	to include 118 member card data
119	113 = NM + 1 and is sole entry
120	84 67.35 Read I, $P(I, K)$ $K = 1$ for NLC
121	87 67.35
122	108 67.35
123	111 67.35
124	8 2 $NOZX$ and $IREPT$ (sta 9880)
125–128	repeat cards 120–123
129	82 83 85 86 106 107 109 110

these NP's are being given a zero rotations i.e., $X(82) = 0.0$; $X(83) = 0.0$; etc.

Because of symmetry, only one-fourth of output is shown.

Node	Distance	Moments		
		Pinned	Fixed	ACI
25	0	5.3	4.3	23.0
26	0.93	16.2	12.2	23.0
27	0.93	23.4	18.8	23.0
28	0.93	34.6	50.7	23.0

The average "fixed" moment is

$$\frac{6.75M}{2} = 0.93(4.3 + 12.2 + 18.8) + \frac{0.93 + 1.17}{2}(50.7)$$

$$M = \frac{86.064(2)}{6.75} = 25.5 \cong 23.0$$

The value of 50.7 ft-kips in the column zone is too high. With working stress design the nominal $F = 1/0.45 = 2.22$; with USD $LF \cong 1.55$

actual $F = \dfrac{50.7}{23} = 2.20 \cong 2.25$

$$> 1.55$$

Example 10-3. Redo Example 9-6 and compare soil pressure with Example 9-6. Given data:
$P = 540$ kips

$$f'_c = 3,000 \text{ psi} \qquad E_c = 452,700 \text{ ksf}$$

$$\text{Column} = 2 \times 2 \text{ ft}$$

$$D = 1.67 \text{ ft}$$

SOLUTION. We will lay a grid as shown and use a card generator to develop the cards, then correct the data for members 96, 97, 103 and remove and correct the *MEMNO* as required to obtain the shape shown.

$$NP = 186 \qquad NM = 108$$

$$NBAND = 27 - 1 + 1 = 27$$

$$k_s = 36q_a = 36(6) = 216$$

The cards are as follows:

Card No	
1	Title
2	FT KIPS KSF K/CU FT
3	186 108 4 1 1 27 8 8 62 0 0 1
4	0 2 0 0 0
5	452700. 196800. 216. 1.67 0.0 1.0
6–114	Member data cards
115	109 (NM + 1)
116	84 135.
117	87 135.
118	108 135.
119	111 135.
120	8 2
121–124	repeat 116–119
125	

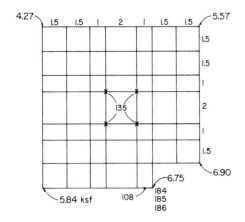

Figure E10-3

The output is:

Point	Example 9-6	Pinned col.	Fixed col.
A	5.84	5.309	5.292
B	6.75	6.583	5.655
C	6.51	6.528	5.755
D	6.90	6.629	5.602
E	5.57	5.00	5.287
F	4.27	4.069	5.308

From this it is evident (and intuitively correct) that column fixity will cause the soil pressure to remain relatively uniform beneath a notched footing.

Example 10-4. Redo Example 10-1 using the mat program in the Appendix.

Given data: Refer to Example 10-1 for loading

Refer to accompanying sketch for grid

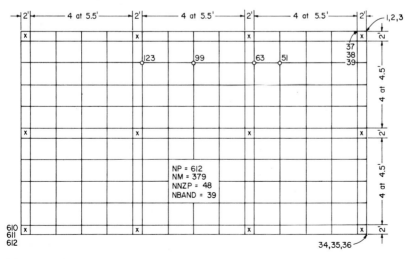

Figure E10-4

SOLUTION. *Step 1.* Compute necessary properties.

$$E_c = 57,400\sqrt{f'_c} = 452,700 \text{ ksf} \qquad \text{take } \mu = 0.15$$

$$G = \frac{E}{2(1 + \mu)} = 452,700/(2.3) = 196,800 \text{ ksf}$$

$$k_s = 36q_a = 36 \text{ kcf} \qquad D = 2.667 \text{ ft (32 in)}$$

Take unit weight = 0.0 kcf since footing separation does not occur.

Step 2. Use card generator to develop member data cards:

$$\text{Number of members } NM = 379$$

$$NP = 3 \times 204 = 612$$

$$\text{No boundary conditions } NOZX = 0$$

The grid was selected to put one-fourth of column load at the four adjacent nodes, therefore, $NNZP = 4 \times 12 = 48$.

$$\text{The bandwidth } NBAND = 39$$

$$STIFF(I) = 39 \times 612 = 23,868 \text{ entries}$$

The first 12 *P*-matrix entries $[K, P(I, K)]$ are:

3	21.25	39	21.25
6	21.25	42	21.25
18	27.5	54	27.50
21	27.5	57	27.50
33	24.25	69	24.25
36	24.25	72	24.25

Because the band-reduction routine sweeps forward, obtaining $X(612)$ when the last step is done, then uses $X(612)$ to back-substitute into equation 611 to find $X(611)$, etc., errors accumulate. It is necessary when doing this large a problem on IBM 360/370 or other 7-digit machines to use DOUBLE PRECISION for *ASAT*, *STIFF*, and *P* to obtain a solution.

Step 3. Selected output: Use nodes close to values computed in Example 10-1 (computer values divided by element width to obtain ft-kips/ft).

Node	$M_y(B = 4.5)$	$M_x(B = 5.5)$	$M_x(\text{used})$
51	− 36.6	52.1	− 13.34 (ft-kips/ft)
63	− 48.9	− 74.92	− 115.3
99	− 9.9	35.1	46.6
123	− 42.3	− 66.5	− 115.3

From this it is evident that the approximate design of Example 10-1 for the N-S direction is adequate, primarily because minimum reinforcing-steel requirements control.

Maximum differential settlement (nearly independent of k_s)

$$\text{Node } 121 = 0.03779 \text{ ft } (NP = 363)$$

$$\text{Node } 200 = 0.01481 \text{ ft } (NP = 600)$$

$$\Delta S = 0.02298 \text{ ft (approx. } \tfrac{1}{4} \text{ in)}$$

10-8. MAT–SUPERSTRUCTURE INTERACTION

It is easy to interact the superstructure rigidity into the problem. It is necessary to have a computer program to solve a superstructure frame of the type in Fig. 10-12 (such a program can be found in Wang (1970, program k). The second program is either a mat or the beam-on-elastic-foundation program. One then does the following:

1. Solve the frame as if it is rigidly attached to the foundation (rotation = displacement = 0.0).
2. From this program obtain the column bending and axial forces of the columns attached to the foundation.
3. Put values from step 2 into the foundation (mat or elastic-beam) program and solve for resulting deflections (rotation and displacements).
4. From step 3 compute differential column elongations. Scan all Δ, and the smallest becomes reference value Δ_1. Next compute for all other columns along any row

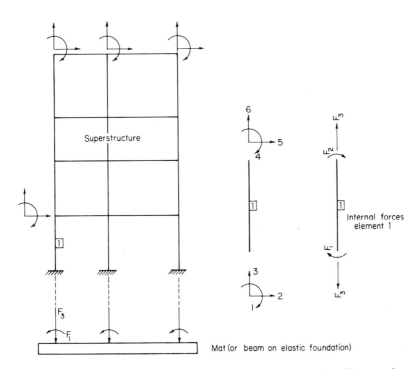

Mat (or beam on elastic foundation)

Figure 10-12. Interaction of superstructure and foundation via iteration. The use of a program such as Wang (1970) for structure coding shown allows direct output of the necessary column forces reacting with foundation.

$P_i = (\Delta_i - \Delta_1)(AE/L)$. Also compute moments and shear due to footing rotation as

$$M_1 = \frac{4EI}{L}\theta \qquad M_2 = \frac{2EI}{L}\theta$$

$$V = \frac{M_1 + M_2}{L}$$

5. These values are used to modify the P matrix of step 1.
6. Repeat steps 1 to 5 as required to obtain the necessary convergence (i.e., values obtained on current cycle vs. values of previous cycle).

Generally satisfactory results will be obtained using a single bay and the beam on elastic foundation rather than the entire structure on a mat.

10-9. CIRCULAR MATS

Circular mats or plates on elastic foundations are widely used for chimneys and various tower structures, silos, etc. The mat, analyzed as an equivalent plate, may be octagon- or decagon-shaped to reduce forming. Little analytical work has been published except a few closed-form solutions [Egorov (1965), Egorov and Serebry-anyi (1963)], a finite-element method by Hooper (1974), and the practical applications of Chu and Afandi (1966) and Smith and Zar (1964).

The mat program in the Appendix can be used after making suitable adjustments to solve this problem. Referring to Fig. 10-13, the circular plate is divided into the grid shown. With three NP's at a node the matrix will be quite large and the bandwidth (calculated as for a rectangular grid, $NBAND$) will be much larger than the rectangular grid. It may be possible, however, from symmetry—even with over-turning moments—to use only one-fourth or one-half the plate [the appropriate NP's (rotation X's) are made zero at the line(s) of symmetry].

It will be necessary to use a card generator to develop the member data cards as one-fourth of the plate of Fig. 10-13 requires:

24 radial members

20 tangential members

$44 = NM$

Also

$NP = 25$ nodes $\times$ 3 = 75

Boundary conditions are 10 zero rotations

At $NX = 1, 2, 4, 20, 22, 38, 40, 56, 58, 74$

$NBAND = 21$

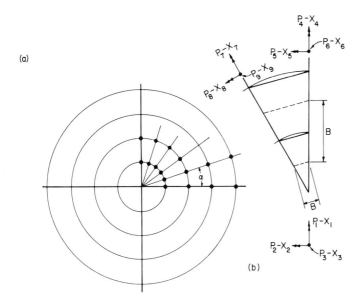

Figure 10-13. A finite-element (grid) method to solve for displacements and bending moments in a circular-mat foundation. Matrix is of size 3 × No. of nodes. (*a*) Circular mat. Use one-fourth of mat if symmetrical, as a large number of nodes results for even medium-sized alpha angles. (*b*) A sector showing method of obtaining grid-element cross section and coding.

It will be necessary to make several approximations in this solution:

1. Average the widths of the triangular- and trapezoidal-shaped elements for *B*. Use lengths as radial distance between concentric rings and the chord distance tangentially (do this in the card generator).
2. Use straight segments between nodes.
3. Orient the reference axis so that all *P-X*'s are radial and tangential except the first set at the center of the plate.

This adjustment will require *A*-matrix subroutines to take into account the first radial members (using the center as reference), then a routine for the tangential elements, and a third routine for the remaining radial elements. Note that the general mat program used *H*, *V*, and *L* which for rectangular grids could have been simplified; now, however, *H* and *V* must be used, as they define the location of the member and its $EASA^T$ contribution. This allows easy and accurate boundary-condition adjustments.

Work by the author shows that the computed deflections using this procedure and compared with theory give results good to 0.001. Moments (properly interpreted) are "exact" to not more than +20 percent larger. Thus these results are quite adequate for design.

Circular mats are designed similarly to any other footing. The plan dimension should be such that the allowable soil pressure is not exceeded. Overturning wind (or earthquake) moments may be a significant design parameter for chimney structures.

The allowable soil pressure is based on Eq. (4-6) with allowance for footing depth. If the circular mat contains an important structure such as a chimney, the soil parameters should be accurately determined.

Elastic settlements can be computed using Eq. (5-10) or Eq. (5-13) and taking diameter = B. Consolidation settlements will require checking. Differential settlement will have to be carefully controlled, as tall structures will topple when the line of action of the gravity forces falls out of the base. On poor soils, piles may be required or the soil may have to be stabilized to support the structure adequately.

The plate depth is designed for wide-beam or diagonal-tension shear as appropriate. The computer program would then be used to find soil pressures (which may be critical when the eccentricity is out of the middle one-third) and the bending moments for the various loading conditions. Revision as required is made. Chimney walls or other superstructure attachments to the mat may require adjusting boundary conditions for rotational restraint.

PROBLEMS

10-1. Design the E-W steel for the mat of Example 10-1.

10-2. Verify the output of Example 10-2; redo with different k_s.

10-3. Verify the output of Example 10-3; redo with a different k_s.

10-4. Redo Example 9-5 using the mat program and (a) pin the column at node 6,6 (the load); (b) apply rotational restraint (x and y) at the node and compare results.

10-5. As a group project analyze Example 10-1 using the mat program in the Appendix. Make changes in the DIMENSION statements. On IBM 360/370 computers use DOUBLE PRECISION for *ASAT*, *STIFF*, and *P* matrices.

10-6. Write a card generator for a circular plate based on Fig. 10-13.

10-7. Modify the *A* matrix of the mat program to solve a circular plate.

10-8. Design a chimney foundation for the following data:

Chimney height above ground	825 ft
Bottom OD	64 ft
Top ID	22 ft
Bottom wall thickness	30 in
Top wall thickness	8 in
Weight above foundation including liner	23,200 kips

Design for wind in your area and for dead load and dead load + wind. Take $q_a = 8$ ksf, and place base of foundation 12 ft below ground surface.

ELEVEN

LATERAL EARTH PRESSURE

11-1. THE LATERAL-EARTH-PRESSURE PROBLEM

The solution of many foundation-engineering problems requires a knowledge of the lateral pressure which may be exerted by the earth. Retaining walls, sheet-pile walls, trench excavations, silo walls withstanding grain pressure, tunnel walls, and other types of underground structures require an estimate of the lateral pressure on the member for structural design. The method of plastic equilibrium as defined by the Mohr rupture envelope (Figs. 2-4 and 11-1a) is most generally used for the analysis of lateral-pressure problems. In a few situations such as tunnel and buried conduits the finite element of the elastic continuum can be used to particular advantage.

The magnitude and distribution of earth pressure is a function of displacement and strain and is generally indeterminate. As a consequence it is convenient to study the soil in a state of *plastic* equilibrium, as indicated in Fig. 11-1. In this figure, every circle drawn through point A (of which an infinite number can be formed) represents a state of elastic equilibrium and satisfies the requirements for elastic equilibrium as long as one of the principal stresses (σ_1 or σ_3) is equal to OA. There are only two circles that can be drawn through point A which just touch the rupture line OB. These two circles represent the plastic-equilibrium conditions which will be studied in the remainder of this chapter.

Plastic-equilibrium conditions acting on an element of soil are shown in Fig. 11-1b where the element is initially subject to principal stresses (no shear on the stress planes) of $\sigma_1 = OA$ and $\sigma_3 = OE$. Now if the vertical pressure OA is held constant and the lateral pressure is increased to failure at a stress of OD as shown in

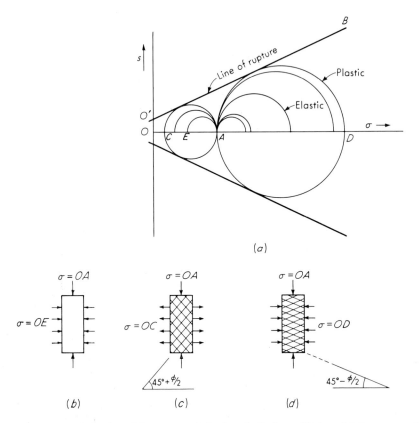

Figure 11-1. Illustration of the concept of elastic and plastic equilibrium. (*a*) Stress states before rupture (elastic) and at rupture (plastic); (*b*) initial condition of constant confining stress OA; (*c*) shear-rupture surfaces for active-pressure theory; (*d*) shear-rupture surfaces for passive-pressure theory.

Fig. 11-1*d*, the principal stresses become rotated, with the major principal stress becoming OD. The resulting Mohr's circle can be plotted as shown in Fig. 11-1*a* with a diameter of AD. The failure planes in Fig. 11-1*d* make angles of intersection with the major principal plane of $45 + \phi/2$ or with the horizontal plane of $45 - \phi/2$.

For the initial conditions of Fig. 11-1*b*, if the lateral pressure is reduced to OC, rupture again occurs when the circle of diameter CA just touches the rupture line $O'B$. For this condition the stress OA is the major principal stress and the rupture planes make angles of $45 + \phi/2$ with the major principal plane as shown in Fig. 11-1*c*.

Figure 11-2 illustrates the concept of earth pressure at rest for both normally consolidated and overconsolidated materials ($OCR > 1$). Here the soil is in a condition of elastic equilibrium and failure cannot take place until stress conditions change sufficiently to increase the stress circle until it just touches the rupture envelope defined by the Mohr-Coulomb strength equation of Chap. 2:

$$s = c + \bar{\sigma} \tan \phi_e \qquad (2\text{-}13)$$

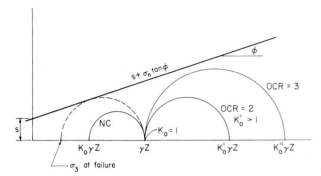

Figure 11-2. Qualitative relationships between normally and overconsolidated soils for at-rest (K_0) conditions.

In normally consolidated material the vertical stress on a horizontal plane will be γh and the lateral stress (on a vertical plane) will be $K_0\gamma h$. The at-rest earth-pressure coefficient K_0 was given earlier as

$$K_0 = 1 - \sin\phi \qquad (2\text{-}22)$$

This stress state is represented by the Mohr's circle labeled *NC* in Fig. 11-2. When K_0 is greater than 1, the resulting stress states are represented by the Mohr's circles appropriately labeled with the OCR ratio. Carefully note that when the OCR is greater than 1, the major principal stress plane is vertical and when $K_0 \le 1$, the major principal stress plane is horizontal. These stress states are the result of geologic processes in forming the soil mass under consideration. These at-rest stress states are altered by any construction conditions resulting in stress changes.

Active Earth Pressure

When the conditions of Fig. 11-1b represent the initial at-rest earth-pressure conditions, the soil is in a state of elastic equilibrium. If the minor principal stress is further decreased while σ_1 remains constant as in Fig. 11-1c, the Mohr's circle increases in size until it reaches a point of tangency with the rupture line. This represents the limiting state of plastic equilibrium, and the lateral pressure is a minimum. This minimum value of lateral earth pressure is termed *active earth pressure*. This value can be obtained by solving for the limiting shear stress as done with Eqs. (a), (b), and (c) of Sec. 2-12 to obtain

$$\sigma_3 = \sigma_1 \tan^2\left(45° - \frac{\phi}{2}\right) - 2c\tan\left(45° - \frac{\phi}{2}\right) \qquad (2\text{-}15)$$

This equation was developed by Coulomb in a considerably different form about 1776, and it appears that Bell (1915) first presented Coulomb's equation in the above form. The above equation is often written using the following trigonometric relationships:

$$\tan^2\left(45 - \frac{\phi}{2}\right) = \frac{1 - \sin\phi}{1 + \sin\phi} \qquad \tan\left(45 - \frac{\phi}{2}\right) = \frac{1 - \sin\phi}{\cos\phi}$$

For relationships involving $45 + \phi/2$ reverse signs.

Stress is always accompanied by deformation; therefore, for a reduction in the lateral stress of Fig. 11-1c there must be a soil expansion. Where walls are involved, it follows that the wall must move away from the soil mass in order to develop the active stress state. The movement must be of sufficient magnitude to develop the minimum value of σ_3 of Eq. (2-15); otherwise the lateral pressure is a larger than minimum value (but less than the initial $K_0 \gamma h$ value). The following values may be considered as typical amounts of wall translation (top movement) to develop the active earth pressure:

Soil and condition	Amount of translation
Cohesionless, dense	0.001 to 0.002H
Cohesionless, loose	0.002 to 0.004H
Cohesive, firm	0.01 to 0.02H
Cohesive, soft	0.02 to 0.05H

Passive Earth Pressure

If instead of reducing the lateral pressure as Fig. 11-1c, the pressure is increased as Fig. 11-1d, the Mohr's circle will decrease to a point when $\sigma_3 = \sigma_1$. The principal planes will then rotate 90° and the circle will increase in size until failure occurs at the maximum major principal stress of OD. This maximum (and limiting) value of lateral stress is termed *passive earth pressure*. It can be computed from Mohr's-circle geometry to obtain Eq. (2-14), repeated here for convenience:

$$\sigma_1 = \sigma_3 \tan^2 \left(45° + \frac{\phi}{2}\right) + 2c \tan \left(45° + \frac{\phi}{2}\right) \qquad (2\text{-}14)$$

Figure 11-3 illustrates relative movements and order of magnitude of the lateral-earth-pressure coefficients for various soils.

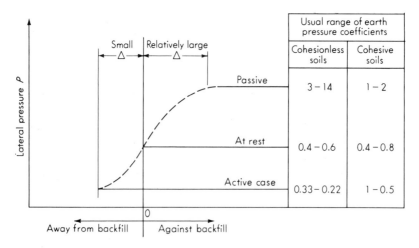

Figure 11-3. Illustration of active and passive pressures with usual range of values for cohesionless and cohesive soil.

11-2. COULOMB EARTH-PRESSURE THEORY

The basic assumptions for the earth-pressure theory proposed by C. A. Coulomb in 1776 (*Coulomb's earth-pressure theory*) are as follows:

1. The soil is isotropic and homogeneous and possesses both internal friction and cohesion.
2. The rupture surface is a plane surface. Coulomb recognized that this was not so, but it greatly simplifies the computations. The backfill surface is planar.
3. The friction forces are distributed uniformly along the plane rupture surface and $f = \tan \phi$. (Symbol for friction coefficient $= f$.)
4. The failure wedge is a rigid body.
5. There is wall friction; i.e., the failure wedge moves along the back of the wall, developing friction forces along the wall boundary.
6. Failure is a two-dimensional problem; consider a unit length of an infinitely long body.

The principal deficiencies in the Coulomb theory are in the assumption of an ideal soil and that a plane defines the rupture surface.

The equations based on the Coulomb theory for a cohesionless soil can be derived from Figs. 11-4 and 11-5. The weight of the soil wedge ABE is

$$W = \gamma A(1) = \frac{\gamma H^2}{2 \sin^2 \alpha} \left[\sin (\alpha + \rho) \frac{\sin (\alpha + \beta)}{\sin (\rho - \beta)} \right] \qquad (a)$$

The active force P_a is a component of the weight vector as illustrated in Fig. 11-5c. Applying the law of sines to Fig. 11-5c,

$$\frac{P_a}{\sin (\rho - \phi)} = \frac{W}{\sin (180 - \alpha - \rho + \phi + \delta)}$$

or

$$P_a = \frac{W \sin (\rho - \phi)}{\sin (180 - \alpha - \rho + \phi + \delta)} \qquad (b)$$

From Eq. (b) it can be seen that the value of $P_a = f(\rho)$; that is, all other terms for a given problem are constant, and the value of P_a of primary interest is the largest possible value. Combining Eqs. (a) and (b), we obtain

$$P_a = \frac{\gamma H^2}{2 \sin^2 \alpha} \left[\sin (\alpha + \rho) \frac{\sin (\alpha + \beta)}{\sin (\rho - \beta)} \right] \frac{\sin (\rho - \phi)}{\sin (180 - \alpha - \rho + \phi + \delta)} \qquad (c)$$

Equating the first derivative to zero,

$$\frac{dP_a}{d\rho} = 0$$

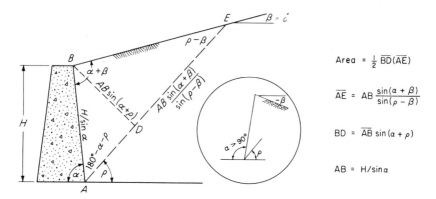

Figure 11-4. Failure wedge used in deriving the Coulomb equation for active pressure. Note β may be $\pm$ and $\alpha > 90°$.

the maximum value of active wall force P_a is found to be

$$P_a = \frac{\gamma H^2}{2} \frac{\sin^2 (\alpha + \phi)}{\sin^2 \alpha \sin (\alpha - \delta)\left[1 + \sqrt{\dfrac{\sin (\phi + \delta) \sin (\phi - \beta)}{\sin (\alpha - \delta) \sin (\alpha + \beta)}}\right]^2} \qquad (11\text{-}1)$$

If $\beta = \delta = 0$ and $\alpha = 90°$ (a smooth vertical wall with horizontal backfill), Eq. (11-1) simplifies to

$$P_a = \frac{\gamma H^2}{2} \frac{(1 - \sin \phi)}{(1 + \sin \phi)} = \frac{\gamma H^2}{2} \tan^2 \left(45 - \frac{\phi}{2}\right) \qquad (11\text{-}2)$$

which is also the Rankine equation for active earth pressure considered in the next

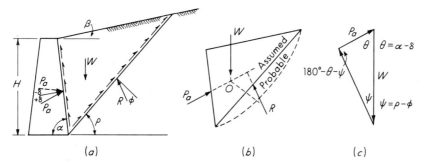

Figure 11-5. (a) Assumed conditions for failure; (b) indicates resultant of forces is not through point 0; hence static equilibrium is not satisfied; (c) force triangle to establish P_a.

Table 11-1. Active-earth-pressure coefficients K_a based on the Coulomb equation (11-3)

ALPHA = 90.   BETA = 0.

δ	$\phi=26$	28	30	32	34	36	38	40
0	0.390	0.361	0.333	0.307	0.283	0.260	0.238	0.217
16	0.349	0.324	0.300	0.278	0.257	0.237	0.218	0.201
17	0.348	0.323	0.299	0.277	0.256	0.237	0.218	0.200
20	0.345	0.320	0.297	0.276	0.255	0.235	0.217	0.199
22	0.343	0.319	0.296	0.275	0.254	0.235	0.217	0.199

ALPHA = 90.   BETA = 5.

δ	$\phi=26$	28	30	32	34	36	38	40
0	0.414	0.382	0.352	0.323	0.297	0.272	0.249	0.227
16	0.373	0.345	0.319	0.294	0.271	0.250	0.229	0.210
17	0.372	0.344	0.318	0.294	0.271	0.249	0.229	0.210
20	0.370	0.342	0.316	0.292	0.270	0.248	0.228	0.209
22	0.369	0.341	0.316	0.292	0.269	0.248	0.228	0.209

ALPHA = 90.   BETA = 10.

δ	$\phi=26$	28	30	32	34	36	38	40
0	0.443	0.407	0.374	0.343	0.314	0.286	0.261	0.238
16	0.404	0.372	0.342	0.315	0.289	0.265	0.242	0.221
17	0.404	0.371	0.342	0.315	0.288	0.264	0.242	0.221
20	0.402	0.370	0.340	0.313	0.287	0.263	0.241	0.220
22	0.401	0.369	0.340	0.312	0.287	0.263	0.241	0.220

Table 11-2. Passive-earth-pressure coefficients K_p based on the Coulomb equation (11-6)

ALPHA = 90.   BETA = 0.

δ	$\phi=26$	28	30	32	34	36	38	40
0	2.561	2.770	3.000	3.255	3.537	3.852	4.204	4.599
16	4.195	4.652	5.174	5.775	6.469	7.279	8.230	9.356
17	4.346	4.830	5.385	6.025	6.767	7.636	8.662	9.882
20	4.857	5.436	6.105	6.886	7.804	8.892	10.194	11.771
22	5.253	5.910	6.675	7.574	8.641	9.919	11.466	13.364

ALPHA = 90.   BETA = 5.

δ	$\phi=26$	28	30	32	34	36	38	40
0	2.943	3.203	3.492	3.815	4.177	4.585	5.046	5.572
16	5.475	5.878	6.609	7.464	8.474	9.678	11.128	12.894
17	5.746	6.146	6.929	7.850	8.942	10.251	11.436	13.781
20	6.249	7.074	8.049	9.212	10.613	12.321	14.433	17.083
22	6.864	7.820	8.960	10.334	12.011	14.083	16.685	20.011

ALPHA = 90.   BETA = 10.

δ	$\phi=26$	28	30	32	34	36	38	40
0	3.385	3.713	4.080	4.496	4.968	5.507	6.125	6.841
16	6.652	7.545	8.605	9.876	11.417	13.309	15.665	18.647
17	6.992	7.956	9.105	10.492	12.183	14.274	16.899	20.254
20	8.186	9.414	10.903	12.733	15.014	17.903	21.636	26.569
22	9.164	10.625	12.421	14.659	17.497	21.164	26.013	32.602

section. Equation (11-2) takes the general form

$$P_a = \frac{\gamma H^2}{2} K_a$$

where

$$K_a = \frac{\sin^2 (\alpha + \phi)}{\sin^2 \alpha \sin (\alpha - \delta)\left[1 + \sqrt{\frac{\sin (\phi + \delta) \sin (\phi - \beta)}{\sin (\alpha - \delta) \sin (\alpha + \beta)}}\right]^2} \qquad (11\text{-}3)$$

and is a coefficient which considers α, β, δ, and ϕ, but is independent of γ and H. Table 11-1 gives values of K_a for selected angular values, and a computer program can easily be written to solve for values of K_a for other angle combinations.

Passive earth pressure is derived similarly except that the inclination at the wall and the force triangle will be shown as in Fig. 11-6.

From Fig. 11-6 the weight of the assumed failure mass is

$$W = \frac{\gamma H^2}{2} \sin (\alpha + \rho) \frac{\sin (\alpha + \beta)}{\sin (\rho - \beta)} \qquad (d)$$

and from the force triangle, using the law of sines

$$P_p = W \frac{\sin (\rho + \phi)}{\sin (180 - \rho - \phi - \delta - \alpha)} \qquad (e)$$

Setting the derivative $dP_p/d\rho = 0$ gives the minimum value of P_p as

$$P_p = \frac{\gamma H^2}{2} \frac{\sin^2 (\alpha - \phi)}{\sin^2 \alpha \sin (\alpha + \delta)\left[1 - \sqrt{\frac{\sin (\phi + \delta) \sin (\phi + \beta)}{\sin (\alpha + \delta) \sin (\alpha + \beta)}}\right]^2} \qquad (11\text{-}4)$$

For a smooth vertical wall with horizontal backfill ($\delta = \beta = 0$ and $\alpha = 90°$), Eq. (11-4) simplifies to

$$P_p = \frac{\gamma H^2}{2} \frac{1 + \sin \phi}{1 - \sin \phi} = \frac{\gamma H^2}{2} \tan^2 \left(45 + \frac{\phi}{2}\right) \qquad (11\text{-}5)$$

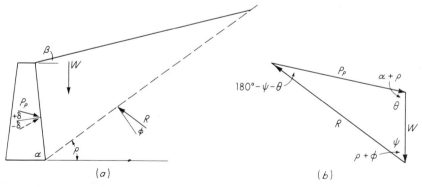

Figure 11-6. (a) Failure wedge and acting forces for passive pressure; (b) force polygon for computation of passive pressure.

Equation (11-4) can also be written

$$P_p = \frac{\gamma H^2}{2} K_p$$

where

$$K_p = \frac{\sin^2 (\alpha - \phi)}{\sin^2 \alpha \sin (\alpha + \delta) \left[1 - \sqrt{\dfrac{\sin (\phi + \delta) \sin (\phi + \beta)}{\sin (\alpha + \delta) \sin (\alpha + \beta)}} \right]^2}$$ (11-6)

Table 11-2 gives values for K_p for selected angular values of ϕ, α, δ, and β.

Example 11-1. What is the total force per foot of wall for the soil-wall system shown, using the Coulomb equations?

SOLUTION. Take

$\delta = 2\phi/3 = 20°$

$p = \gamma h K_a \qquad K_a = 0.34 \qquad$ Table 11-1

$P = \displaystyle\int_0^H \gamma h K_a \ dh = \tfrac{1}{2}\gamma H^2 K_a$

$\qquad = \tfrac{1}{2}(17.52)(5)^2(0.34) = 74.46 \ \text{kN/m}$

$P\bar{y} = \displaystyle\int_0^H \gamma h K_a(h \ dh) = \frac{\gamma H^3}{3} K_a$

$\bar{y} = \tfrac{2}{3}H \qquad$ from top of wall

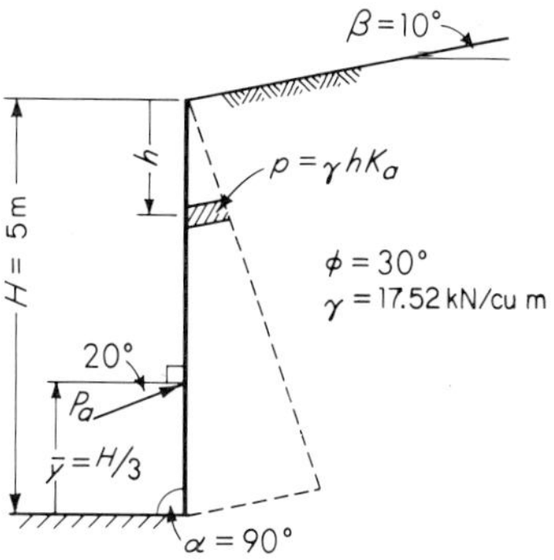

Figure E11-1

with $\delta = 20°$, P_a will act as shown.

Example 11-2. What is the total force and location of the resultant for the wall-soil system shown, using the Coulomb solution and taking $\delta = 0$?

SOLUTION. With $\beta = \delta = 0$ the solution is the same as the Rankine solution of the next section:

$K_{a1} = 0.307 \qquad K_{a2} = 0.333 \qquad$ Table 11-1

$p_1 = (\gamma h° + q)K_a = 2(0.307) = 0.614 \ \text{ksf}$

$p_2 = \gamma h k_a = 0.614 + 0.105(0.307)(10)$

$\qquad = 0.614 + 0.322 \ \text{ksf}$

$p_2' = (2.0 + 1.05)0.333 = 1.02 \ \text{ksf}$

$p_3 = (0.1225 - 0.0625)(0.333)(10) = 0.20 \ \text{ksf}$

$p_4 = \gamma_w h = 0.0625 \times 10 = 0.625 \ \text{ksf}$

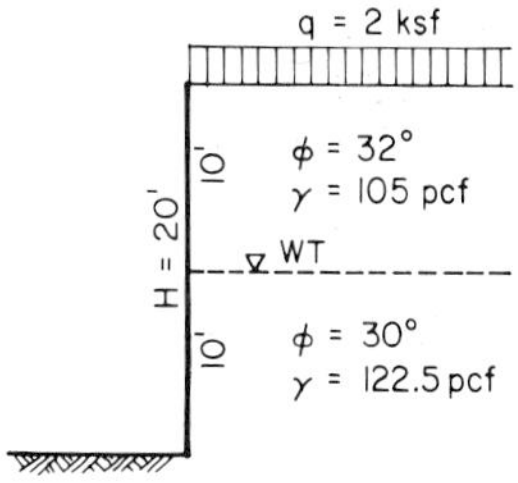

Figure E11-2a

These values are shown on the pressure diagram from which areas can be numerically integrated to obtain forces as follows:

$P_1 = 0.614(10) = 6.14 \ \text{kips} \qquad P_2 = 0.322(\tfrac{10}{2}) = 1.61 \ \text{kips}$

$P_3 = 1.02(10) = 10.2 \ \text{kips} \qquad P_4 = \tfrac{10}{2}(0.825) = 4.12 \ \text{kips}$

$R = \displaystyle\sum_1^4 P_i = 6.14 + 1.61 + 10.2 + 4.12 = 22.07 \ \text{kips}$

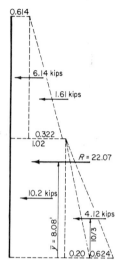

Figure E11-2b

Find $\bar{y}$ above base.

$$22.07\bar{y} = 4.12(\tfrac{10}{3}) + 10.2(\tfrac{10}{2}) + 1.61(10 + \tfrac{10}{3})$$

$$+ 6.14(10 + \tfrac{10}{2})$$

$$= 13.7 + 51.0 + 21.5 + 92.1 \text{ and } \bar{y} = \frac{178.3}{22.07} = 8.08 \text{ ft}$$

11-3. RANKINE EARTH PRESSURES

Rankine (1857) considered the soil in a state of plastic equilibrium and used essentially the same assumptions as Coulomb, except that he assumed no wall cohesion or wall friction, which greatly simplifies the problem. The Rankine case is illustrated in Fig. 11-7.

Substituting Eq. (1) into Eq. (2) of Fig. 11-7 and setting $dP_a/d\rho = 0$, one obtains for P_a,

$$P_a = \frac{\gamma H^2}{2} \cos \beta \, \frac{\cos \beta - \sqrt{\cos^2 \beta - \cos^2 \phi}}{\cos \beta + \sqrt{\cos^2 \beta - \cos^2 \phi}} = \frac{1}{2} \gamma H^2 K_a \qquad (11\text{-}7)$$

where

$$K_a = \cos \beta \frac{\cos \beta - \sqrt{\cos^2 \beta - \cos^2 \phi}}{\cos \beta + \sqrt{\cos^2 \beta - \cos^2 \phi}} \qquad (11\text{-}8)$$

Table 11-3 gives values for K_a for various $\phi - \beta$ combinations. When $\delta = 0$ (smooth wall), no wall shearing stresses develop, and the active stress p_a becomes a principal stress. When this occurs, it can be shown by an analysis of the stress on the soil element that the active force P_a acts parallel to the slope of the backfill; i.e., the force acts on the face of the retaining wall at the angle β with a perpendicular to the vertical axis. When the ground surface is level ($\beta = 0$), Eq. (11-7) simplifies to Eq. (11-2).

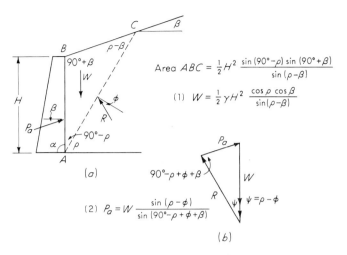

$$\text{Area } ABC = \frac{1}{2}H^2 \frac{\sin(90°-\rho)\sin(90°+\beta)}{\sin(\rho-\beta)}$$

$$(1) \quad W = \frac{1}{2}\gamma H^2 \frac{\cos\rho \cos\beta}{\sin(\rho-\beta)}$$

$$(2) \quad P_a = W \frac{\sin(\rho-\phi)}{\sin(90°-\rho+\phi+\beta)}$$

Figure 11-7. (a) Soil-structure system for the Rankine solution for $\alpha = 90°$; (b) force triangle in the Rankine solution.

By analogy, the passive pressure for the Rankine solution can be derived as

$$P_p = \frac{\gamma H^2}{2}\cos\beta \frac{\cos\beta + \sqrt{\cos^2\beta - \cos^2\phi}}{\cos\beta - \sqrt{\cos^2\beta - \cos^2\phi}} = \frac{1}{2}\gamma H^2 K_p \qquad (11\text{-}9)$$

Values of K_p are given in Table 11-4 for selected values of β and ϕ.

Example 11-3. What is the total force per foot of wall for the soil-wall system shown in Example 11-1, using the Rankine equation?

SOLUTION. $p = \gamma h K_a$ $K_a = 0.35$ Table 11-3

$$P = \int_0^H p \, dh = \frac{1}{2}\gamma H^2 K_a = \frac{1}{2}(17.52)(5)^2(0.35) = 76.65 \text{ kN/m}$$

This compares with 74.46 kN/m by the Coulomb equation, or a difference of about 3 percent.

Table 11-3. Active-earth-pressure coefficients K_a for the Rankine equation (11-8), values not given for $\beta > \phi$

β	$\phi=$ 26	28	30	32	34	36	38	40	42
0	0.3905	0.3610	0.3333	0.3073	0.2827	0.2596	0.2379	0.2174	0.1982
5	0.3959	0.3656	0.3372	0.3105	0.2855	0.2620	0.2399	0.2192	0.1997
10	0.4134	0.3802	0.3495	0.3210	0.2944	0.2696	0.2464	0.2247	0.2044
15	0.4480	0.4086	0.3730	0.3405	0.3108	0.2834	0.2581	0.2346	0.2129
20	0.5152	0.4605	0.4142	0.3739	0.3381	0.3060	0.2769	0.2504	0.2262
25	0.6999	0.5727	0.4936	0.4336	0.3847	0.3431	0.3070	0.2750	0.2465
30	0.0000	0.0000	0.8660	0.5741	0.4776	C.4105	0.3582	0.3151	0.2784
35	0.0000	0.0000	0.0000	0.0000	0.0000	0.5971	0.4677	0.3906	0.3340
40	0.0000	0.0000	0.0000	0.0000	C.0000	0.00C0	0.0000	0.7660	0.4668

Table 11-4. Passive-earth-pressure coefficients K_p for the Rankine equation (11-9)

β	$\phi=$ 26	28	30	32	34	36	38	40	42
0	2.5611	2.7698	3.0000	3.2546	3.5371	3.8518	4.2037	4.5989	5.0447
5	2.5070	2.7145	2.9431	3.1957	3.4757	3.7875	4.1360	4.5272	4.9684
10	2.3463	2.5507	2.7748	3.0216	3.2946	3.5979	3.9365	4.3161	4.7437
15	2.C826	2.2836	2.5017	2.7401	3.0024	2.2925	3.6154	3.9766	4.3827
20	1.7141	1.9175	2.1318	2.3618	2.6116	2.8857	3.1888	3.5262	3.9044
25	1.1736	1.4343	1.6641	1.8942	2.1352	2.3938	2.6758	2.9867	3.3328
30	0.0000	0.0000	0.8660	1.3064	1.5705	1.8269	2.0937	2.3802	2.6940
35	0.0000	0.0000	0.0000	0.0000	0.0000	1.1238	1.4347	1.7177	2.0088
40	0.0000	0.0000	0.0000	0.0000	0.0000	0.0000	0.0000	0.7660	1.2570

Example 11-4. Draw the pressure diagram for the wall system shown.

SOLUTION. Compute K_a, using the Rankine solution.

$$K_a = \tan^2\left(45° - \frac{\phi}{2}\right) = \tan^2 40° = 0.704$$

$$\sqrt{K_a} = 0.84$$

At top: $h = 0$, and

$$p = \gamma h K_a - 2c\sqrt{K_a} = -2(10.5)(0.84) = -17.64 \text{ kPa}$$

At $p = 0$: $\gamma h K_a - 2c\sqrt{K_a} = 0$

$$h = \frac{2c\sqrt{K_a}}{\gamma K_a} = \frac{2c}{\gamma\sqrt{K_a}} = \frac{2(10.5)}{17.52(0.84)} = 1.43 \text{ m}$$

Note: This value of h is the depth of a potential tension crack. At base, the lateral pressure is

$$p = 17.52(6.5)(0.704) - 2(10.5)(0.84) = 62.53 \text{ kPa}$$

The resultant force is found as $\sum F_h = R$. The location of the resultant may be found by summing moments at the base or by inspection, depending on the complexity of the pressure diagram. The tension zone $\overline{ab}$ is usually neglected for finding the magnitude and location of the resultant.

Neglecting the tension zone   *Using alternative pressure diagram acd*

$$R = 62.53\left(\frac{5.07}{2}\right) = 158.52 \text{ kN/m} \qquad R = 62.53\left(\frac{6.5}{2}\right) = 203.22 \text{ kN/m}$$

$$\bar{y} = \frac{5.07}{3} = 1.69 \text{ m above } c \qquad\qquad \bar{y} = \frac{6.5}{3} = 2.17 \text{ m}$$

(by inspection)

With water in the tension crack:

$$R = 158.52 + 9.807(1.43)^2/2 = 168.55 \text{ kN/m}$$

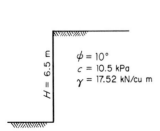

$\phi = 10°$
$c = 10.5$ kPa
$\gamma = 17.52$ kN/cu m

$H = 6.5$ m

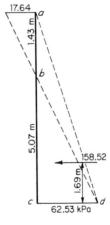

17.64

a

1.43 m

b

5.07 m

158.52

1.69 m

c

d

62.53 kPa

Figure E11-4a   **Figure E11-4b**

the overturning moment is

$$M_0 = 158.52(1.69) + 10.03(5.07 + 1.43/3) = 323.52 \text{ kN-m/m}$$

$$\bar{y} = 323.52/168.55 = 1.92 \text{ m above } c$$

In this case the water-in-crack solution is between the two previous solutions, from which it appears that the alternative pressure diagram *acd* provides a conservative solution.

Example 11-5. Plot the pressure diagram for the wall system shown.

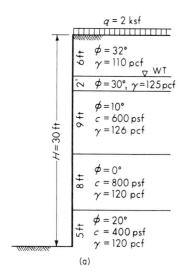

Figure E11-5*a*

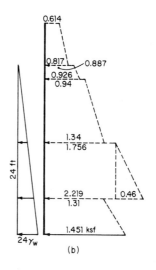

Figure E11-5*b*

SOLUTION. Using Rankine values for K_a,

$K_1 = 0.307 \qquad K_2 = 0.333$

$K_3 = 0.704 \qquad \sqrt{K_3} = 0.84$

$K_4 = 1 \qquad \sqrt{K_4} = 1$

$K_5 = 0.49 \qquad \sqrt{K_5} = 0.70$

at $h = 0$: $p = 2(0.307) = 0.614$ ksf

at $h = 6$ ft: $\Delta p = 6(0.307)(0.11) = 0.203$ ksf; $6 + dh$: $\Delta p = 0.887$

at $h = 8$ ft: $\Delta p = 2(0.33)(0.125 - 0.0625) = 0.0413$ ksf

at $h = 8 + dh$: $p = [2 + 6(0.11) + 2(0.0625)](0.70) - 2(0.6)(0.84)$

$\qquad\qquad\qquad = 2.785(0.705) - 1.2(0.84) = 0.94$ ksf

at 17 ft: $\Delta p = 9(0.126 - 0.0625)(0.70) = 0.40$ ksf

at $17 + dh$: $p = 3.356(1.0) - 2(0.8)(1) = 1.756$ ksf

at 25 ft: $\Delta p = 8(0.120 - 0.0625)(1) = 0.46$ ksf

at $25 + dh$: $p = 3.816(0.49) - 2(0.4)(0.7) = 1.31$ ksf

at 30 ft: $\Delta p = 5(0.120 - 0.0625)(0.49) = 0.141$ ksf

Note that the total pressure profile against the wall is the sum of the soil and water pressures.

11-4. ACTIVE AND PASSIVE EARTH PRESSURE USING THEORY OF PLASTICITY

The passive-earth-pressure theory consistently overestimates the passive pressure developed in field and model tests. This may or may not be conservative, depending on the need for the passive-pressure value. Because of the problem of overestimation, Caquot and Kerisel (1948) produced tables of earth pressure based on nonplane-failure surfaces; later Janbu (1957) and more recently Shields and Tolunay (1973) proposed an approach to the earth-pressure problem similar to the method of slices used in slope-stability analyses. Sokolovski (1960) presented a finite-difference solution using a highly mathematical method. All these methods give smaller values for the passive-earth-pressure coefficient. None of these methods significantly improves on the Coulomb or Rankine active-earth-pressure coefficients.

Rosenfarb and Chen (1972) developed a closed-form solution using plasticity theory which also solves the earth-pressure problem for active and passive pressure. The closed-form solution requires a computer program and iteration on the computer, which is not particularly difficult. This method is included here because of the greater clarity over the alternative methods. A computer program is included in the Appendix to obtain passive-earth-pressure coefficients for cohesionless backfill.

Rosenfarb and Chen considered several failure surfaces, and the combination of a so-called "log-sandwich" mechanism gave results which compared most favorably with the Sokolovski solution, which has been accepted as correct by many persons. Figure 11-8 illustrates the passive log-sandwich mechanism. From this and appropriate consideration of velocity components the following equations are obtained:

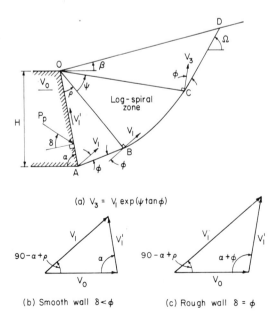

(a) $V_3 = V_1 \exp(\psi \tan \phi)$

(b) Smooth wall $\delta < \phi$

(c) Rough wall $\delta = \phi$

Figure 11-8. (a) Passive log-sandwich mechanism; (b, c) velocity diagrams. [*Rosenfarb and Chen (1972)*.]

Cohesionless Soil

For a smooth wall:

$$\begin{Bmatrix} K_{a\gamma} \\ K_{p\gamma} \end{Bmatrix} = \frac{\mp \sec \delta}{\mp \sin \alpha + \tan \delta \cos \alpha - [\tan \delta \cos (\alpha - \rho)/\cos \rho]}$$

$$\times \left(\frac{\tan \rho \cos (\rho \pm \varphi) \cos (\alpha - \rho)}{\sin \alpha \cos \phi} + \frac{\cos^2 (\rho \pm \phi)}{\cos \rho \sin \alpha \cos^2 \phi (1 + 9 \tan^2 \phi)} \right.$$

$$\times \left\{ \cos (\alpha - \rho)[\pm 3 \tan \phi + (\mp 3 \tan \phi \cos \Psi + \sin \Psi) \exp (\mp 3\Psi \tan \phi)] \right.$$

$$\left. + \sin (\alpha - \rho)[1 + (\mp 3 \tan \phi \sin \Psi - \cos \Psi) \exp (\mp 3\Psi \tan \phi)] \right\}$$

$$\left. + \frac{\cos^2 (\rho \pm \phi) \sin (\alpha - \rho - \Psi + \beta) \cos (\alpha - \rho - \Psi) \exp (\mp 3\Psi \tan \phi)}{\cos \phi \sin \alpha \cos (\alpha - \rho - \Psi \mp \phi + \beta) \cos \rho} \right) \qquad (11\text{-}10)$$

For a rough wall:

$$\begin{Bmatrix} K_{a\gamma} \\ K_{p\gamma} \end{Bmatrix} = \frac{\mp \sec \delta}{\mp \sin \alpha + \tan \delta \cos \alpha} \left(\frac{\sin^2 \rho \cos (\rho \pm \phi) \cos (\alpha - \rho) \sin (\alpha \mp \phi)}{\sin^2 \alpha \cos \phi \cos (\rho \mp \phi)} \right.$$

$$\mp \frac{\cos^2 (\rho \pm \phi) \sin (\alpha \mp \phi)}{\sin^2 \alpha \cos^2 \phi (1 + 9 \tan^2 \phi) \cos (\rho \mp \phi)}$$

$$\times \left\{ \cos (\alpha - \rho)[\pm 3 \tan \phi + (\mp 3 \tan \phi \cos \Psi + \sin \Psi) \exp (\mp 3\Psi \tan \phi)] \right.$$

$$\left. + \sin (\alpha - \rho)[1 + (\mp 3 \tan \phi \sin \Psi - \cos \Psi) \exp (\mp 3\Psi \tan \phi)] \right\}$$

$$\left. + \frac{\cos^2 (\rho \pm \phi) \sin (\alpha - \rho - \Psi + \beta) \cos (\alpha - \rho - \Psi) \sin (\alpha \mp \phi) \exp (\mp 3\Psi \tan \phi)}{\sin^2 \alpha \cos \phi \cos (\alpha - \rho - \Psi + \beta \mp \phi) \cos (\rho \mp \phi)} \right)$$
$$(11\text{-}11)$$

Cohesive Soil

For a smooth wall:

$$\begin{Bmatrix} K_{ac} \\ K_{pc} \end{Bmatrix} = \frac{\sec \delta}{\mp \sin \alpha + \tan \delta \cos \alpha - [\tan \delta \cos (\alpha - \rho)/\cos \rho]}$$

$$\times \left\{ \tan \rho + \frac{\cos (\rho \pm \phi) \sin (\alpha - \rho - \Psi + \beta) \exp (\mp \Psi \tan \phi)}{\cos \rho \cos (\alpha - \rho - \Psi \mp \phi + \beta)} \right.$$

$$\left. \mp \frac{\cos (\rho \pm \phi)[\exp (\mp 2\Psi \tan \phi) - 1]}{\sin \phi \cos \rho} \right\} \qquad (11\text{-}12)$$

For a rough wall:

$$\left|\frac{K_{ac}}{K_{pc}}\right| = \frac{\sec \delta}{\mp \sin \alpha + \tan \delta \cos \alpha} \left\{ \frac{\cos \phi \cos (\alpha - \rho)}{\sin \alpha \cos (\rho \mp \phi)} + \frac{\sin \rho \sin (\alpha \mp \phi)}{\sin \alpha \cos (\rho \mp \phi)} \right.$$

$$+ \frac{\cos (\rho \pm \phi) \sin (\alpha - \rho - \Psi + \beta) \sin (\alpha \mp \phi) \exp (\mp \Psi \tan \phi)}{\sin \alpha \cos (\alpha - \rho - \Psi \mp \phi + \beta) \cos (\rho \mp \phi)}$$

$$\left. \mp \frac{\cos (\rho \pm \phi) \sin (\alpha \mp \phi)[\exp (\mp 2\Psi \tan \phi) - 1]}{\sin \phi \sin \alpha \cos (\rho \mp \phi)} \right\} \tag{11-13}$$

In solving Eqs. (11-10) through (11-13), it is necessary to solve for the maximum value of K_p or K_a. The maximizing of these equations depends on the two variables ρ and ψ. This requires a search routine, which is included with the computer program in the Appendix. The values of the two dependent variables are initialized to approximately

$$\rho \cong 0.5(\alpha + \beta)$$

$$\psi \cong 0.2(\alpha + \beta)$$

With these initial values, the search routine is used to revise the values until convergence is obtained. In most cases values from which K_p is computed are found after

Table 11-5. Selected values of K_p using limit analysis and computer program in Appendix for $\alpha = 90°$ (vertical wall). $\beta = 0°$ is not shown because same as Coulomb value. Intermediate values can be obtained by plotting K_p. Values shown are for granular soil.

β	$\phi = 30$	35	40	45°
$\delta = 0$				
10	4.01	5.20	6.68	8.93
20	5.25	7.03	9.68	13.8
30	6.74	9.50	14.0	21.5
$\delta = 10$				
10	5.70	7.61	10.4	14.9
20	7.79	10.9	15.9	24.4
30	10.3	14.7	23.6	39.6
$\delta = 20$				
10	7.94	11.2	16.3	24.9
20	11.2	16.5	25.6	42.4
30	15.1	23.2	41.0	70.2
$\delta = 30$				
10	10.6	15.8	24.6	40.7
20	15.2	23.2	39.5	70.3
30	20.8	34.8	62.0	0*

* No solution after 46 iterations.

not more than 20 iterations. The program shuts off after 46 iterations. In a few cases, the program will be unable to find a solution, but this is a program problem and not a deficiency in the plasticity theory used to develop the concept. Table 11-5 gives selected values of K_p for cohesionless soils. Values of $\beta = \delta = 0$ are not shown, as they are identical to the Coulomb or Rankine solution.

The "smooth" wall solution is used for wall friction $\delta < \phi$; when $\delta = \phi$ the "rough" wall equation is used. Equations (11-12) and (11-13) can readily be programmed, using the same routines to solve an equation for minimum or maximum with two dependent variables, to obtain passive-pressure coefficients for cohesive soil. This solution does not give greatly different values from the Coulomb passive-pressure theory until the ϕ angle becomes large and with δ on the order of $\phi/2$ or more and $\beta > 0°$.

11-5. EARTH PRESSURE ON WALLS, SOIL-TENSION EFFECTS, RUPTURE ZONE

The Rankine or Coulomb earth-pressure equations can be used to obtain the force and its point of application acting on the wall for design. Soil-tension concepts can also be investigated. These will be taken up in the following discussion.

Earth Forces on Walls

From Eq. (2-15) and temporarily considering a $c = 0$ soil and referring to Fig. 11-9a, the wall force can be computed as

$$P_a = \int_0^H \sigma_3 K_a \, dh = \int_0^H \gamma h K_a = \frac{\gamma h^2 K_a}{2}\bigg]_0^H = \frac{\gamma H^2}{2} K_a \qquad (f)$$

from which it is evident that the soil-pressure diagram is hydrostatic (linearly increases with depth) as shown in the figure. If there is a surcharge q on the backfill as shown in Fig. 11-9c (other surcharges will be considered in Sec. 11-11), the wall force

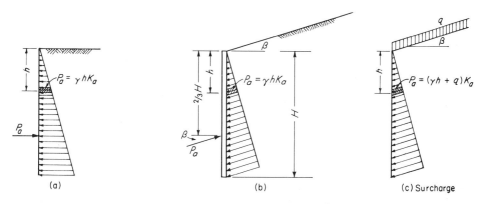

(a) (b) (c) Surcharge

Figure 11-9. Pressure diagrams for active Rankine earth pressure.

can be computed as

$$P_a = \int_0^H (\gamma h + q)K_a \ dz = \left(\frac{\gamma H^2}{2} + qH\right)K_a \qquad (g)$$

The point of application requires taking moments about a convenient point, and for the case with surcharge and the top of the wall we have

$$P_a y = \int_0^H (\gamma h + q)K_a h \ dz = \left(\frac{\gamma H^3}{3} + \frac{qH^2}{2}\right)K_a \qquad (h)$$

and inserting the value of P_a from Eq. (g), the distance from the top of the wall is

$$y = \frac{1}{3} \frac{2\gamma H^2 + 3qH}{\gamma H + 2q}$$

and from the bottom of the wall $\bar{y} = H - y$, which simplifies to

$$\bar{y} = \frac{H}{3}$$

when the surcharge $q = 0$. It is not correct to convert the surcharge to an equivalent additional wall height and use $\bar{y}$ to the centroid of a triangle, because the surcharge effect is rectangular against the wall.

Soil-Tension Effects on Backfill and Open Trenches

When cohesive soil is used for backfill, one may expect a tension zone to develop as given by Eq. (2-15). With cohesion not zero, Eq. (f) becomes

$$P_a = \int_0^H (\gamma h K_a - 2c\sqrt{K_a}) \ dz \qquad (i)$$

Now let us find the depth h_t for which $\sigma_3 = 0$. Solving Eq. (2-15), we have

$$h_t = \frac{2c}{\gamma\sqrt{K_a}} \qquad (11\text{-}14)$$

This is the depth of a theoretical tension crack in the soil behind the wall. This crack can form at the interface of the wall-soil mass and at some distance back from the wall as shown in Fig. 11-10.

The other value of interest is the theoretical depth H_c a vertical excavation may stand. This can be found by equating P_a from Eq. (i) to zero, obtaining

$$H_c = \frac{4c}{\gamma\sqrt{K_a}} \qquad (11\text{-}15)$$

There may be some question of what to use for K_a in Eqs. (11-14) and (11-15) when $B > 0$, since the use of Eq. (2-15) as developed was restricted to horizontal

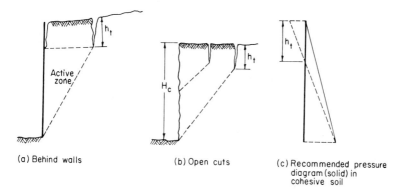

(a) Behind walls   (b) Open cuts   (c) Recommended pressure
                                              diagram (solid) in
                                              cohesive soil

Figure 11-10. Tension cracks.

backfill. In the absence of any better information, use K_a from Table 11-3 (Rankine values).

The tension zone (Fig. 11-10c) should not be relied on to reduce lateral pressures. Instead one should assume that it can form and will possibly fill with water. The depth of water (not the quantity) can increase the overturning pressure against the wall considerably owing to both the hydrostatic force of $\gamma_w h_t$ and the larger moment arm caused by combining the hydrostatic force with the already existing lateral pressure. It is recommended for these problems that, at the very least, the wall pressure be taken as shown in Fig. 11-10c (see Example 11-4).

One cannot rely on Eq. (11-15) to predict the critical embankment height accurately, for several reasons:

1. Once the tension crack forms, Eq. (2-25) is not valid for the full depth of excavation.
2. Cohesive soils tend to lose cohesion when exposed in excavations owing to moisture adsorption and/or drying.
3. Equipment loads adjacent to the excavation.

Because of these factors, Eq. (11-15) should be modified for design to

$$H_c = \frac{Gc}{\gamma\sqrt{K_a}}$$

where the value G is on the order of 2.67 to 3.

One may observe tension cracks in the ground surface adjacent to open cuts as cracks paralleling the excavation. They can sometimes be seen in pavements adjacent to excavations. Because of the tendency for cracks to form and cause cave-ins, OSHA (Occupational Safety and Health Act) requires all excavations over 5 ft in depth in unstable or soft material to be sheeted, braced, shored, or sloped (Arts. 1926.651c and 1926.652b).

Rupture Zone

The solution of Eq. (11-1) or Eq. (1) in Fig. 11-7 for ρ with a horizontal backfill ($\beta = 0$) gives

$$\rho = 45 + \frac{\phi}{2} \text{ (active)} \quad \text{or} \quad \rho = 45 - \frac{\phi}{2} \text{ (passive)}$$

In general for sloping backfills ρ is not equal to $45 + \phi/2$ but is dependent on α, β, and δ. After some trigonometric manipulations of Eq. (11-1), one can obtain an expression for ρ as

$$\sin 2\rho = \frac{FN \pm M\sqrt{M^2 + N^2 - F^2}}{M^2 + N^2} \tag{11-16}$$

where $M = \cos(2\alpha - \phi - \delta) - \dfrac{\sin(\phi + \delta)}{\sin(\phi - \beta)} \cos(\phi + \beta)$

$\quad N = \sin(2\alpha - \phi - \delta) + \dfrac{\sin(\phi + \delta)}{\sin(\phi - \delta)} \sin(\phi + \beta)$

$\quad F = \dfrac{\sin(\delta + \beta)}{\sin(\phi - \beta)}$

11-6. RELIABILITY OF LATERAL EARTH PRESSURES

Several sets of wall tests have been performed to check the validity of the active- and passive-earth-pressure concepts. These include the tests of Terzaghi (1934), Peck and Ireland (1961), Rowe and Peaker (1965), Mackey and Kirk (1967), James and Bransby (1970), and Rehnman and Broms (1972). Field and model tests tend to confirm the active-earth-pressure concept reasonably well if the backfill is carefully placed so that compaction effects do not create excessive stresses and if the wall rotates (or translates) sufficiently to mobilize fully the "active" zone and the shear stresses along the rupture surface. For cohesionless soil, the "active" rupture surface is approximately a plane as assumed in the development of the equations.

11-7. SOIL PROPERTIES AND LATERAL EARTH PRESSURE

It is evident from the use of the Mohr-Coulomb strength equation and as used in the previous examples, that effective stresses are used (together with hydrostatic pressure) to obtain the lateral pressure on a wall. The usual condition of soil behind walls will be as shown in Fig. 11-11, where in the case of permanent walls, the retained material will be of one type. The case of Fig. 11-11c, where temporary bracing may be employed, can and often does retain layered soils. This condition will be con-

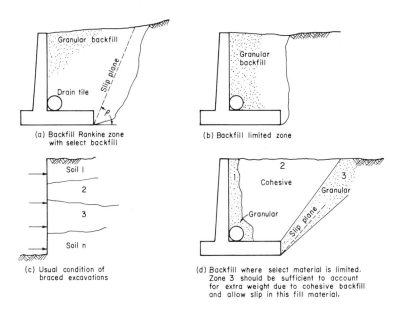

Figure 11-11. Various backfill conditions.

sidered in Chap. 14. For the other cases, which could include retaining walls, foundation (basement) walls, or bridge abutments, the lateral earth pressure will vary from active earth pressure, where the wall can move adequately to allow the shear strength to mobilize, to at-rest or somewhat above at-rest values depending on soil placement and whether all movement is restrained.

Soil Parameters

Soil parameters for use in computing lateral earth pressure will be:

1. Drained values for sand for reasons cited in Chap. 2. Plane-strain ϕ values should generally be used, as this condition is more likely than triaxial conditions for long walls.
2. CU or UU most likely for cohesive soils. These values will be adequate for normally consolidated materials. Note that this gives $\phi \cong 0°$.
3. One of the following for overconsolidated materials:
   a. CD and neglecting any cohesion
   b. UU at the creep threshold
   c. ϕ angle between peak and residual strengths

   The strength parameters for overconsolidated materials should be obtained from triaxial tests with chamber pressure decreasing (hold σ_1 constant and decrease σ_3).

   When the active-earth-pressure zone cannot form as in Fig. 11-11b because of interference from the natural soil embankment, the lateral earth pressure will be on the order of at-rest values. In fact, Fig. 11-11a will reproduce active earth pressures reliably only if backfilled with cohesionless material. Cohesive material will tend to form tension cracks, which considerably alter the assumptions involved in either the Rankine or Coulomb theories.

Water in Backfill

Water in the backfill is particularly undesirable. It increases the unit weight and lateral pressure if the soil is wet. If a water table is allowed to stabilize, the effect is considerably worse, since the ϕ angle of water is zero and the resulting pressure coefficient is 1.0 (as illustrated in Examples 11-2 and 11-5). Additionally the water may freeze, creating ice pressure for which no theory is currently available to estimate its magnitude.

The water problem can be avoided by constructing drainage holes through the wall and providing free-draining backfill.

Angle of Wall Friction

It appears that wall friction depends not only on the soil properties but on the amount and direction of wall movement [Morgenstern and Eisenstein (1970)]. Indications are that maximum wall friction may not occur simultaneously with maximum shearing resistance and that wall friction is not a constant value across the wall. Also to be taken into consideration will be the wall material and the surface condition of the wall interfacing with the soil. Considerable engineering judgment must be applied to obtain realistic values of wall friction. Values of wall friction can range from $0°$ to ϕ, and values of 0.4 to 0.6ϕ do not seem unrealistic; values of $\delta = 0°$ or $\delta = \phi$ are generally too low and high, respectively, for most practical cases. Table 11-6 presents wall-friction-angle values for several structural materials.

Wall Adhesion

It is expected that in the upper region of the wall a tension crack will form—especially if the soil dries out. The value of adhesion for the wall zone below the tension crack is usually taken as 0.5 to $0.7c$ with a maximum value of not much over 1,000 psf (50 kPa).

11-8. EARTH-PRESSURE THEORIES IN RETAINING-WALL PROBLEMS

Both the Rankine and Coulomb earth-pressure solutions satisfy $\sum F_H$ and $\sum F_v = 0$. As can be seen from the failure wedges, neither theory will satisfy the statical requirement that the sum of the moments at any point equals zero. The moment requirements are not satisfied, because the rupture surface will be somewhat curved in the actual problem. There is also an angle of wall friction which may not be correctly evaluated or considered, and the distribution of stress on the rupture surface will not

Table 11-6. Friction angles δ between various foundation materials and soil or rock*

Interface materials	Friction angle, δ, degrees†
Mass concrete or masonry on the following:	
Clean sound rock	35
Clean gravel, gravel-sand mixtures, coarse sand	29–31
Clean fine to medium sand, silty medium to coarse sand, silty or clayey gravel	24–29
Clean fine sand, silty or clayey fine to medium sand	19–24
Fine sandy silt, nonplastic silt	17–19
Very stiff and hard residual or preconsolidated clay	22–26
Medium stiff and stiff clay and silty clay	17–19
Steel sheet piles against:	
Clean gravel, gravel-sand mixture, well-graded rock fill with spalls	22
Clean sand, silty sand-gravel mixture, single-size hard-rock fill	17
Silty sand, gravel or sand mixed with silt or clay	14
Fine sandy silt, nonplastic silt	11
Formed concrete or concrete sheetpiling against:	
Clean gravel, gravel-sand mixtures, well-graded rock fill with spalls	22–26
Clean sand, silty sand-gravel mixture, single size hard rock fill	17–22
Silty sand, gravel or sand mixed with silt or clay	17
Fine sandy silt, nonplastic silt	14
Various structural materials:	
Masonry on masonry, igneous and metamorphic rocks:	
Dressed soft rock on dressed soft rock	35
Dressed hard rock on dressed soft rock	33
Dressed hard rock on dressed hard rock	29
Masonry on wood (cross grain)	26
Steel on steel at sheet-pile interlocks	17
Wood on soil	14–16‡

* Based in part on NAFAC (1971).
† Single values ±2°. Alternate for concrete on soil is $\delta = \phi$.
‡ May be higher in dense sand or if sand penetrates wood.

be uniform. The actual physical system does, however, satisfy statics, but this system is less simple than that idealized in developing the equations.

The Rankine solution is often used because the equations are simple, especially for no cohesion and a horizontal backfill, and because of the uncertainty of evaluating the wall friction δ. The Rankine solution gives slightly larger values of P_a than does the Coulomb equation. Since the Rankine equation for cohesionless soil has the same form as the hydrostatic pressure equation,

$$P_a = \tfrac{1}{2}H^2(\gamma K_a)$$

where the γK_a term is synonymous with the unit weight of a fluid, arbitrary (handbook) values such as 30 to 50 pcf are sometimes used. This method is also termed the

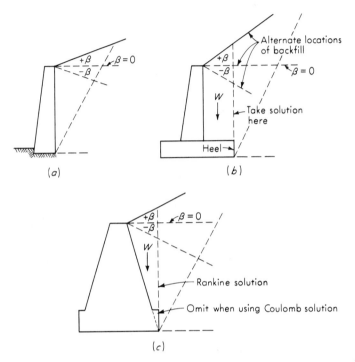

Figure 11-12. Location for solutions using the Rankine procedure. (*a*) Wall and base do not interfere with failure wedge; (*b*) footing will interfere with formation of failure wedge unless located as shown; (*c*) backslope and footing will interfere with failure wedge unless located as indicated. In (*b*) and (*c*) include weight *W* in stability computations.

equivalent-fluid method. Note that one should investigate γK_a using K_a based on backfill geometry or considerable error may result from using "equivalent fluid."

Three important limitations on the Rankine solution are (1) the backfill must be a plane surface; (2) the wall must not interfere with the failure wedge; and (3) there is no wall friction or wall cohesion. If these criteria are met, solutions by the Rankine equations can be taken as indicated in Fig. 11-12. The Coulomb solution may also be applied to Fig. 11-12c by neglecting the heel projection, as indicated by the dashed-line projection of the back face of the wall through the heel.

The Coulomb solution also was not designed for use with nonplanar backfills. If the wall interferes with the failure wedge, as it does in cantilever and counterfort walls, both methods must be modified by computing the resultant pressure through the heel on a vertical plane.

If the heel is long enough for a ρ value as in Fig. 11-13, then P_a will act at β, since the soil will consist in two plastic equilibrium zones symmetrical about AB'.

Retaining walls are not designed to withstand passive pressure; however, certain problems do require consideration of passive pressures. Examples of problems requiring passive-pressure analysis are the anchors, or "deadmen," of Sec. 13-9 and the possible beneficial effect of the soil in front of a retaining wall on sliding stability, considered in the next chapter.

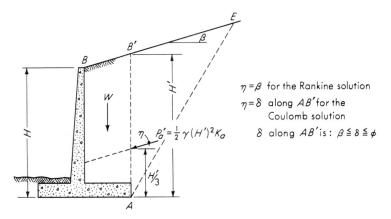

$\eta = \beta$ for the Rankine solution

$\eta = \delta$ along AB' for the Coulomb solution

δ along AB' is: $\beta \leqq \delta \leqq \phi$

$P'_a = \frac{1}{2}\gamma(H')^2 K_a$

Figure 11-13. Equivalent Rankine and Coulomb solutions for cantilever and counterfort retaining walls.

11-9. GRAPHICAL AND COMPUTER SOLUTIONS FOR LATERAL EARTH PRESSURE

There are several graphical solutions for obtaining lateral forces when the backfill is irregular-shaped or there are concentrated loads, both conditions that do not fit the Coulomb or Rankine theories. Among the many solutions are Culmann's (ca. 1866), the trial-wedge method (ca. 1877), and the logarithmic spiral. An analytical solution based on the theory of elasticity of Sec. 11-10 can also be used.

The Culmann Solution

The Culmann solution considers wall friction δ, irregularity of the backfill, any surcharges (either concentrated or distributed loads), and the angle of internal friction of the soil. In this discussion the solution is applicable only to cohesionless soils, although with modifications it can be used for soils with cohesion. This method can be adapted to stratified deposits of varying densities, but the angle of internal friction must be the same throughout the soil mass. A rigid, plane rupture surface is assumed. Essentially, the solution is a graphical determination of the maximum value of soil pressure, and a given problem may have several graphical maximum points, of which the largest value is chosen as the design value. A solution can be made for both active and passive pressure.

Steps in the Culmann solution for active pressure (Fig. 11-14) are as follows:

1. Draw the retaining wall to any convenient scale, together with the ground line, location of surface irregularities, point loads, surcharges, and the base of the wall when the retaining wall is a cantilever type.
2. From the point A lay off the angle ϕ with the horizontal plane, locating the line AC. Note that in the case of a cantilever wall, the point A is at the base of the heel, as shown in Fig. 11-14b.

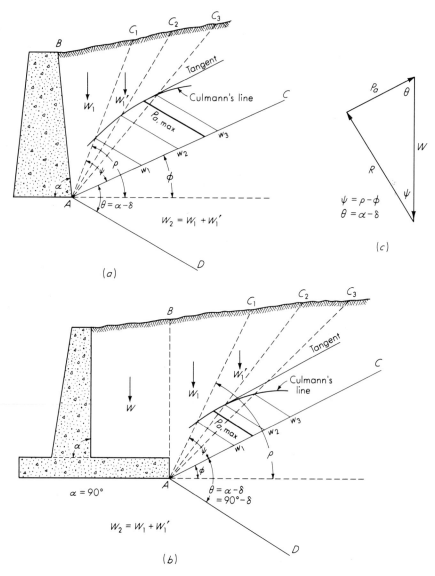

Figure 11-14. Culmann's solution for active earth pressure. (*a*) No interference with wall or footing; (*b*) cantilever retaining wall; (*c*) force polygon used in the Culmann graphical solution.

3. Lay off the line AD at an angle of θ with line AC. The angle θ is computed as

$$\theta = \alpha - \delta$$

where α = angle back of wall makes with the horizontal
δ = angle of wall friction
4. Draw assumed failure wedges as ABC_1, ABC_2, ..., ABC_n. These should be made utilizing the backfill surface as a guide, so that geometrical shapes such as triangles and rectangles are formed.

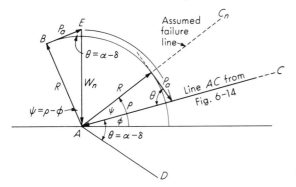

Figure 11-15. Force triangle used in Culmann's graphical solution. The θ angle is constant; ρ is defined as the angle from the horizontal to the failure surface; therefore, with the orientation shown, ψ can be found graphically.

5. Find the weight w_n of each of the wedges by treating as triangles, trapezoids, or rectangles, depending on the soil stratification, water in the soil, and other conditions of geometry.
6. Along the line AC, plot to a convenient weight scale the wedge weights locating the points $w_1, w_2, \ldots, w_n$.
7. Through the points just established (step 6), draw lines parallel to AD to intersect the corresponding side of the triangle, as w_1 to side AC_1, w_2 to side $AC_2, \ldots, w_n$ to side AC_n.
8. Through the locus of points established on the assumed failure wedges, draw a smooth curve (the Culmann line). Tangent to this curve and parallel to the line AC draw a tangent line. It may be possible to draw tangents to the curve at several points; if so, draw all possible tangents.
9. Through the tangent point established in step 8, project a line back to the AC line, which is also parallel to AD. The value of this to the weight scale is P_a, and a line through the tangent point from A is the failure surface. When several tangents are drawn, choose the largest value of P_a.

The basis for Culmann's procedure is the solution of the force triangle shown in Fig. 11-15. The triangle is rotated so that the location of the trial failure wedges automatically yields the angle ψ without recourse to measuring ψ each time. The line AD is laid off for use in projecting the instant value of P_a at the proper slope since θ is constant for a particular problem. The slope of R is automatically established from the slope of the weight line AC; thus, with all slopes and one side w_n known, the force triangle is readily solved.

To find the point of application of P_a, the following procedure [Terzaghi (1943)] is recommended:

Case 1. No concentrated loads (Fig. 11-16a), but may have other surcharges.

a. Find the center of gravity of the failure wedge graphically or by trimming a cardboard model and hanging by a thread at two or three points.
b. Through the center of gravity and parallel to the failure surface draw a line of action of P_a until it intercepts AB (wall or plane through the heel of the cantilever wall). P_a acts at an angle of δ or β to a perpendicular to AB.

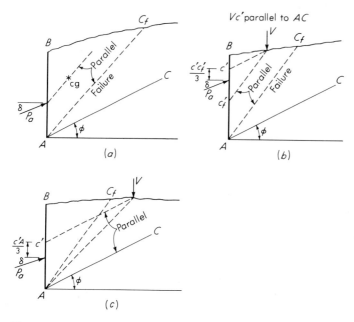

Figure 11-16. Procedures for location of point of application of P_a for (a) irregular backfill; (b) concentrated or line load inside failure zone; (c) concentrated or line load outside failure zone (but inside zone ABC).

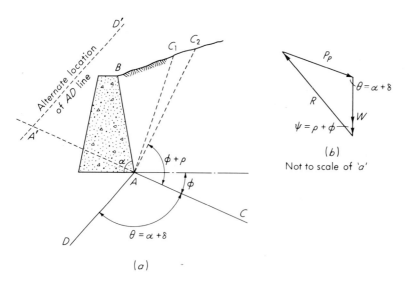

Figure 11-17. (a) Culmann's solution for passive pressure; (b) force polygon, which is graphically solved in the method.

Case 2. Concentrated load or line load within the failure wedge (Fig. 11-16*b*).

a. Parallel to *AC* draw line *Vc'*, and parallel to *AC_f* draw *Vc'_f*.
b. Take one-third of distance *c'c'_f* from *c'* for the point of application of P_a.

Case 3. Concentrated load or line load outside the failure wedge (Fig. 11-16*c*).

a. Draw a line from the concentrated load to *A(VA)*.
b. Draw *Vc'* parallel to *AC*.
c. Take one-third of *c'A* from *c'* as point of application of P_a. If the surcharge falls out of zone *ABC*, the problem should be treated as if no surcharge were present.

For passive earth pressure, Culmann's procedure is modified as indicated in Fig. 11-17 (see also Example 11-7).

Example 11-6. For the given data and wall-soil system shown, find the total active force P_a and the location of the resultant, using Culmann's graphical solution.

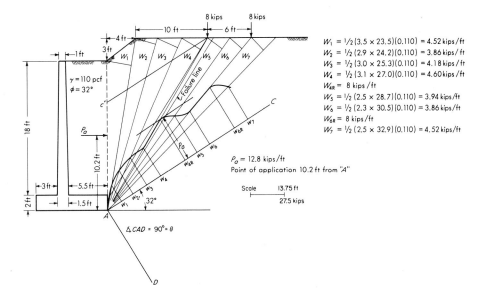

$W_1 = 1/2 (3.5 \times 23.5)(0.110) = 4.52$ kips/ft
$W_2 = 1/2 (2.9 \times 24.2)(0.110) = 3.86$ kips/ft
$W_3 = 1/2 (3.0 \times 25.3)(0.110) = 4.18$ kips/ft
$W_4 = 1/2 (3.1 \times 27.0)(0.110) = 4.60$ kips/ft
$W_{4R} = 8$ kips/ft
$W_5 = 1/2 (2.5 \times 28.7)(0.110) = 3.94$ kips/ft
$W_6 = 1/2 (2.3 \times 30.5)(0.110) = 3.86$ kips/ft
$W_{6R} = 8$ kips/ft
$W_7 = 1/2 (2.5 \times 32.9)(0.110) = 4.52$ kips/ft

$P_a = 12.8$ kips/ft
Point of application 10.2 ft from "*A*"

Figure E11-6

Example 11-7. Redo Example 11-6 for the passive pressure by the Culmann graphical procedure. Omit finding the location of the point of application of the resultant.

SOLUTION. See Fig. E11-7 for geometry.

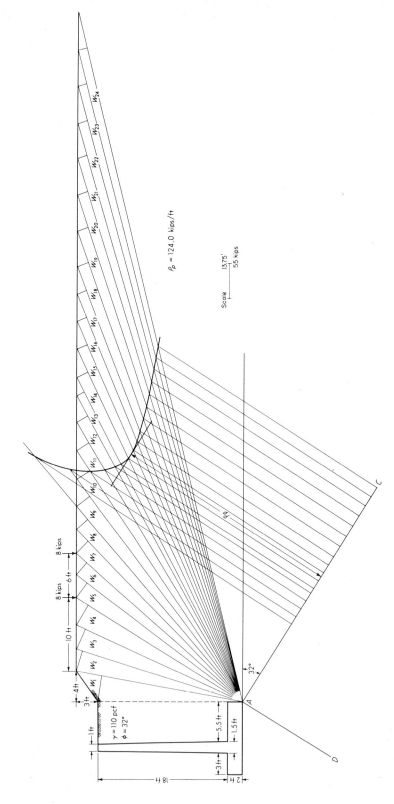

Figure E11-7

The Trial-Wedge Solution

The trial-wedge solution is a graphical solution, very similar to the Culmann solution, which is advantageous for soils with cohesion. There are two approaches to this problem, one using plane failure surfaces and the other using the logarithmic spiral.

The procedure is based on a force polygon for the forces which act on any failure wedge, including any or all of the following forces: wall friction, wall adhesion (or cohesion), friction and cohesion forces on the failure surface, and the weight of the failure wedge. The idealized conditions are illustrated with the corresponding force polygon in Fig. 11-18a and b. As long as all the forces of the system are included, the force polygon can be drawn using the forces in any order, but since C_w is a constant value in a problem, the order of forces in Fig. 11-18b is recommended.

For simplicity, the method will be described in a step-by-step procedure:

1. Draw the wall and ground surface to an appropriate scale and compute the depth of the tension crack as

$$h_t = \frac{2c}{\gamma\sqrt{K_a}}$$

This value of h_t should be plotted at sufficient points to establish the tension-crack profile.

2. Lay off trial wedges as $ABE_1 D_1$, $ABE_2 D_2$, ..., and compute the weight of the corresponding wedges as w_1, w_2, ..., w_n.

3. Compute C_w and C_s (note that C_w is a constant) and lay off C_w as indicated in Fig. 11-18b to the wall slope and to the appropriate force scale. As a tension crack

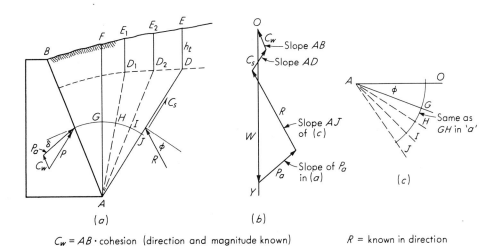

$C_w = AB \cdot \text{cohesion (direction and magnitude known)}$ $R = \text{known in direction}$
$C_s = AD \cdot \text{cohesion (direction and magnitude known)}$ $P_a = \text{known in direction}$
$W = \text{weight of trial wedge (direction and magnitude known)}$

Figure 11-18. The trial-wedge solution. (a) Forces acting on a trial wedge $ABED$; (b) forces acting on $ABED$ formed into the force polygon; (c) rapid method of establishing the slope of R.

can form along the wall, the length AB should also be adjusted. Also draw the weight vectors $w_1, w_2, \ldots, w_n$ along the line OY.

4. From the terminus of C_w lay off C_s at the slope of the assumed trial failure wedges.

5. Through points $w_1, w_2, \ldots, w_n$ established in step 3, lay off a vector P_a to the correct slope. Note that the slope of P_a is constant.

6. Through the terminus of C_s lay off the vector R to the appropriate slope. The slope is at the angle ϕ to a perpendicular to the assumed failure surfaces $AD_1, AD_2, AD_3, \ldots$.

7. The intersection of the R and P_a vectors establishes a locus of points, through which a smooth curve is drawn.

8. Draw a tangent to the curve obtained in step 7, parallel to the weight vector, and draw the vector P_a through the point of tangency. As with the Culmann solution, several maximum values may be obtained. The largest possible value of P_a is the design value.

The slope of the R vector can be established conveniently (Fig. 11-18c) as follows:

1. To some radius r draw the arc GJ from the vertical line AF in Fig. 11-18a.

Example 11-8. Redo Example 11-6, using a ϕ-c soil, by the trial-wedge method.

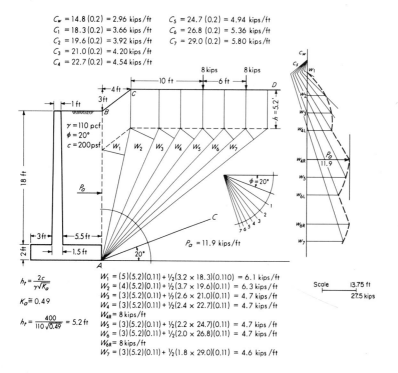

$C_w = 14.8\,(0.2) = 2.96$ kips/ft $\quad C_5 = 24.7\,(0.2) = 4.94$ kips/ft
$C_1 = 18.3\,(0.2) = 3.66$ kips/ft $\quad C_6 = 26.8\,(0.2) = 5.36$ kips/ft
$C_2 = 19.6\,(0.2) = 3.92$ kips/ft $\quad C_7 = 29.0\,(0.2) = 5.80$ kips/ft
$C_3 = 21.0\,(0.2) = 4.20$ kips/ft
$C_4 = 22.7\,(0.2) = 4.54$ kips/ft

$\gamma = 110$ pcf
$\phi = 20°$
$c = 200$ psf

$\phi = 20°$

$P_a = 11.9$ kips/ft

Scale   13.75 ft
        27.5 kips

$h_r = \dfrac{2c}{\gamma\sqrt{K_o}}$

$K_o \cong 0.49$

$h_r = \dfrac{400}{110\sqrt{0.49}} = 5.2$ ft

$W_1 = (5)(5.2)(0.11) + \frac{1}{2}(3.2 \times 18.3)(0.110) = 6.1$ kips/ft
$W_2 = (4)(5.2)(0.11) + \frac{1}{2}(3.7 \times 19.6)(0.11) = 6.3$ kips/ft
$W_3 = (3)(5.2)(0.11) + \frac{1}{2}(2.6 \times 21.0)(0.11) = 4.7$ kips/ft
$W_4 = (3)(5.2)(0.11) + \frac{1}{2}(2.4 \times 22.7)(0.11) = 4.7$ kips/ft
$W_{4R} = 8$ kips/ft
$W_5 = (3)(5.2)(0.11) + \frac{1}{2}(2.2 \times 24.7)(0.11) = 4.7$ kips/ft
$W_6 = (3)(5.2)(0.11) + \frac{1}{2}(2.0 \times 26.8)(0.11) = 4.7$ kips/ft
$W_{6R} = 8$ kips/ft
$W_7 = (3)(5.2)(0.11) + \frac{1}{2}(1.8 \times 29.0)(0.11) = 4.6$ kips/ft

Figure E11-8

2. Draw a horizontal line AO and lay off the angle ϕ as shown. With the same r used in step 1, draw arc OJ.
3. Then AG is the slope of the vector R to failure plane AF.
4. Now lay off arcs GH, HI, IJ in Fig. 11-18c to the same arc length used in step 1.
5. The slopes of lines AH, AI, and AJ of Fig. 11-18c are the corresponding slopes of the vector R to failure surface AD_1, AD_2, ...

In cohesionless materials the values C_w and C_s are zero, and the trial-wedge solution is the same as the Culmann solution, except for the orientation of the force polygon.

Computer Solution of Trial Wedge

The simplest method of solving the trial wedge (or Culmann method) is using a computer solution (program in Appendix). In a computer solution, it is necessary to do the following:

1. Draw the wall-soil geometry to scale and obtain coordinates for points A, B (and F of Fig. 11-18).
2. Obtain sufficient additional coordinates to define the top of backfill and the coordinates of all concentrated loads.
3. Program a solution of the force triangle of Fig. 11-18b for P_a. This involves two unknown variables P_a and R, but all angles and slopes are known and the force polygon must close; therefore, a direct analytical solution for P_a can be made.
4. Start with a soil wedge, say AFD_1E_1, with angle FAD_1 of about $5°$ (depending on whether AF is vertical) and increment the wedge angle in $1°$ increments. Solve the soil wedge twice at all point loads (dx to left and dx to right).
5. Sort the P_a values computed from steps 3 and 4 until all concentrated loads have been accounted for and P_a has decreased at least two consecutive times, stop the computations, and print out the maximum P_a and the corresponding ρ angle. This procedure allows for study of parametric effects (ϕ, δ, c, γ) much more easily than with the graphical procedures previously discussed. The point of application of P_a is obtained similar to the Culmann method.

Example 11-9. Redo Example 11-8 using the computer program in the Appendix.

SOLUTION. *Step 1.* Establish coordinates for points A, B, C, D and the two concentrated loads.

Coordinates are:	A	B	C	D	Load 1	Load 2
$X =$	100.0	100.0	104.0	180.0	114.0	120.0
$Y =$	100.0	120.0	123.0	123.0	123.0	123.0

Coordinates of D are arbitrarily taken such that line is always defined.

Step 2. Card input data.

Card No.

1 TITLE
2 Units (FT KIPS KSF K/CU FT)
3 0.110 20.0 90.0 0.0 0.200 1.00
  Unit Weight, ϕ, α, δ, cohesion, and wall adhesion factor
4 1 100.0 120.0 104.0 120.0 (line No. and end coordinates)
5 2 104.0 123.0 180.0 123.0 (2nd line)
6 100.0 100.0 100.0 120.0 (XSTART, YSTART, XTOP, YTOP)
7 1 8.0 114.0 123.0 (1st load and coordinates)
8 2 8.0 120.0 123.0 (2nd load and coordinates)

The output gives $P_a = 12.583$ kips at $\rho = 58.671°$. With $IP > 1$, these data will solve Example 11-10 to obtain $P_p = 119.625$ kips at $\rho = 23.991°$.

Logarithmic-Spiral Trial Wedge

The logarithmic spiral may be used to develop the trial wedge. The log spiral uses the principle that any force vector makes an angle of ϕ with the tangent to the spiral and the line of action passes through O, the center of the spiral. The method is basically as follows [Terzaghi (1943, p. 110), Hijab (1956)]:

1. On a piece of tracing paper lay out a portion of a logarithmic spiral using the equation

$$r = r_0 \exp (\theta \tan \phi) \qquad (11\text{-}17)$$

where r = radius of spiral at angle θ from r_0
  r_0 = starting radius (use any arbitrary value which will allow fitting onto wall-soil geometry)
  ϕ = angle of internal friction of soil
  θ = angle between r and r_0 , rad

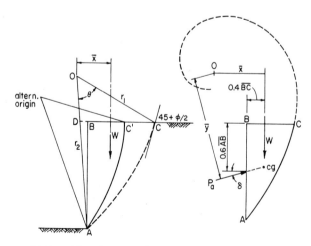

(a) Curve AC preferred to AC'

(b) Approximate location of spiral centroid

Figure 11-19. Logarithmic spiral used to obtain trial wedge. Use several trial centers O and exit C; P_a is minimum obtained from the several trials.

2. Draw wall-soil geometry to scale.
3. Place the logarithmic spiral on the wall-soil system by eye (making an approxi-mate 90° exit angle as shown in Fig. 11-19a). Scale the perpendicular distance from the spiral origin (point O) to B. Note that the exit angle should be $45 \pm \phi/2$, since the ground surface is a principal plane (no shear stress), and if 90° represents a significant variation use $45 \pm \phi/2$.
4. Compute the weight vector of the failure wedge as

$$\text{Area } OAC = \frac{r_1^2 - r_2^2}{4 \tan \phi}$$

and the areas of the two triangles ODC and ADB. Exclude appropriate areas to obtain the area of the wedge ABC, which is multiplied by the soil weight. Alterna-tively use a planimeter to obtain the area of ABC directly (use the average of at least two tracings).
5. Obtain the perpendicular distance to the wedge centroid to obtain

$$M_0 = W\bar{X}$$

Also obtain the perpendicular to the line of action of P_a as $\bar{Y}$

$$M_r = P_a \bar{Y}$$

Hijab gives some equations for obtaining the distances $\bar{X}$, $\bar{Y}$; however, for all practical purposes the distances can be obtained as shown in Fig. 11-19b.
6. Compute P_a as

$$P_a = \frac{M_0}{\bar{Y}}$$

7. Repeat several times to obtain the largest value of P_a.

11-10. LATERAL PRESSURES BY THEORY OF ELASTICITY FOR SURCHARGES

In the preceding discussions, methods for finding the lateral pressure have been presented, using simplified assumptions for the location of the point of application of the total pressure and the magnitude of the resultant total pressure.

Tests by Spangler (1936), Spangler and Mickle (1956), and others indicate that lateral pressures can be computed for various types of surcharges by using modified forms of the theory-of-elasticity equations. From Chap. 5 an equation for lateral pressure, using the theory of elasticity, was presented as

$$\sigma_h = \frac{q}{2\pi z^2} \left[3 \sin^2 \theta \cos^3 \theta - \frac{(1 - 2\mu) \cos^2 \theta}{1 + \cos \theta} \right] \qquad (5\text{-}7)$$

If we refer to Fig. 5-2, and let $r = x$, and redefine the terms slightly as

$$x = mH \qquad \text{and} \qquad z = nH$$

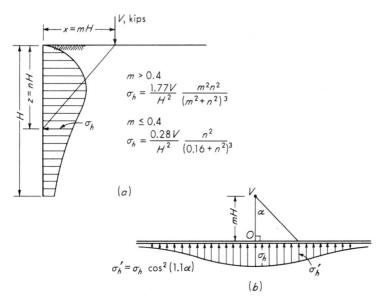

Figure 11-20. (a) Lateral pressure against rigid wall due to a point load and $\mu = 0.5$; (b) lateral pressure at points along the wall on each side of a perpendicular from the concentrated load V to the wall.

and take a value for Poisson's ratio of $\mu = 0.5$, we may rewrite the equation as

$$\sigma_h = \frac{3q}{2\pi H^2} \frac{m^2 n}{(m^2 + n^2)^{5/2}} \qquad (11\text{-}18)$$

The theoretical form of this equation requires adjustment [Mindlin (1936a)] when computing the lateral pressure against a rigid wall to make the equation values compare with the measured test values.

Case 1: Point load. Equations in Fig. 11-20 can be used for this case, which was investigated by Spangler and others. These equations are based on mH being the perpendicular distance to the wall, as shown in Fig. 11-20. In the equations shown in the figure the given coefficients have been adjusted to make the theoretical agree with the measured pressure.

Case 2: Line load. The engineer should inspect the relative dimensions of the retaining wall and of the structure to decide if the loading may be considered a line load or a strip load (case 3). A concrete-block wall or fence could be considered a line load; a conduit laid on the ground is another example; wide strip loads may be considered as a series of parallel-line loads.

For line loads (Fig. 11-21), from the Boussinesq equations and using the ratios m, n as before

$$\sigma_h = \frac{2q}{\pi H} \frac{m^2 n}{(m^2 + n^2)^2} \qquad (11\text{-}19)$$

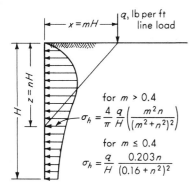

for $m > 0.4$

$$\sigma_h = \frac{4}{\pi} \frac{q}{H} \left(\frac{m^2 n}{(m^2 + n^2)^2} \right)$$

for $m \le 0.4$

$$\sigma_h = \frac{q}{H} \frac{0.203 n}{(0.16 + n^2)^2}$$

Figure 11-21. Lateral pressure against rigid wall due to a line load and $\mu = 0.5$.

However, the measured values from tests were found to be approximately twice [Terzaghi (1954, p. 1252)] this value; therefore, modifying Eq. (11-19), we obtain the value shown in Fig. 11-21.

Case 3: Strip load. A strip load is a loading intensity with a finite width, such as a highway, railroad, or earth embankment, which is parallel to the retaining structure.

Terzaghi (1943, p. 376) presented an equation which has been doubled for this case:

$$\sigma_h = \frac{2q}{\pi} (\beta - \sin \beta \cos 2\alpha) \tag{11-20}$$

where β is in radians and the other terms are as identified in Fig. 11-22.

Newmark (1942) used the theory-of-elasticity equation to construct an influence chart for lateral pressure. The influence chart in Fig. 11-23 uses a Poisson's ratio value of 0.5, which reduces Eq. (5-7) to the simplest case. This chart can be used similarly to the Boussinesq or Westergaard influence charts of Chap. 5, by plotting the shape of the load to the scale of the chart for the depth desired and counting the squares enclosed in the area.

As with the vertical-stress influence charts [Eq. (5-6)],

$$\sigma_h = IMq$$

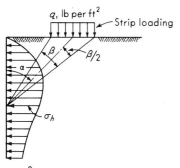

$$\sigma_h = \frac{2q}{\pi} (\beta - \sin \beta \cos 2\alpha)$$

Figure 11-22. Lateral pressure against rigid wall due to a strip load.

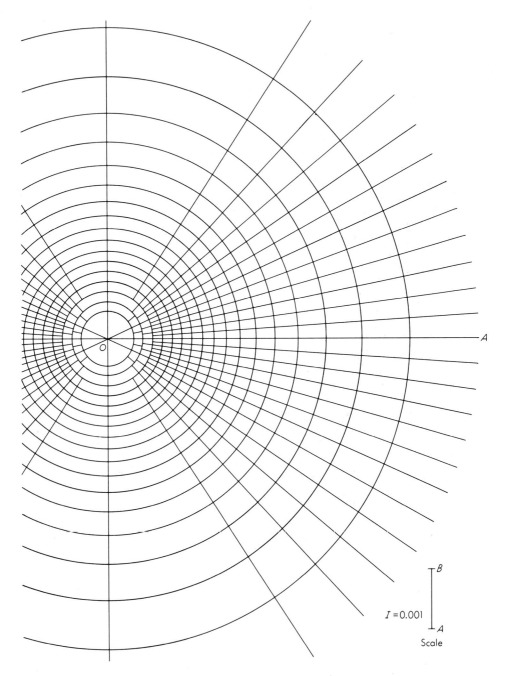

Figure 11-23. Influene chart for computing lateral pressure at point O for any type of loading in the influence field ($\mu = 0.5$). [*After Newmark (1942)*.]

This method provides a rapid solution for irregularly shaped areas. The computed lateral pressure by this method should be doubled for rigid walls, as before. From the use of the charts one may plot a wall-pressure diagram which will indicate the location of the resultant pressure of the surcharge.

This may be combined with the pressure from the conventional Coulomb or Rankine analysis for the backfill to find the value of total pressure, as well as the location of the resultant pressure, which is more realistic than the reliance on a graphical solution, with the approximations for point of application of pressure given in cases 2 and 3 of Sec. 11-9.

The most rapid and reliable method of obtaining the lateral pressure, wall force, and location of resultant is to program Eq. (5-7) on the computer as indicated could be done for the vertical pressure in Chap. 5 (program is in Appendix). Load configurations are treated as follows:

1. Point load—use a single unit area with $q = V$.
2. Line load—use a strip one unit wide with $q = V(FL^{-1})$.
3. Strip load—same as line load except divide strip into appropriate widths.
4. Area loads—divide area into unit areas and sum the area. For irregular shapes it may require integrating more than one area and adding the effects. For round tanks use an equivalent square area.

Example 11-10. Plot the lateral pressure-distribution curve against a rigid retaining wall for a 10-ft-diameter tank located 15 ft from the wall and loading the soil with a surface contact pressure of 2.0 ksf. After finding the pressure intensity, find the total force acting against a 1-ft strip, and find the

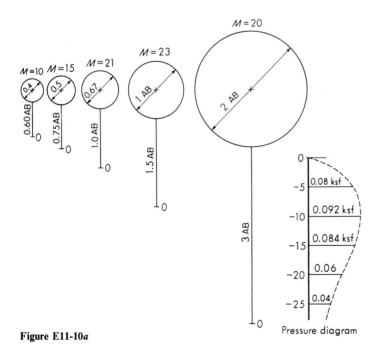

Figure E11-10a

location of the resultant $\bar{y}$ above the bottom of the wall. Take the wall as 25 ft high, and use Fig. 11-23 for the solution.

SOLUTION. Five points will be taken, as follows:

Elevation, ft	Scale	Dimension used for plotting, in
-5	1 in = 5 ft	$D = 2, OA = 3$
-10	1 in = 10 ft	$D = 1, OA = 1.5$
-15	1 in = 15 ft	$D = 0.67, OA = 1$
-20	1 in = 20 ft	$D = 0.5, OA = 0.75$
-25	1 in = 25 ft	$D = 0.4, OA = 0.6$
		$q = q_0(MI)$

The pressure will be doubled for rigid wall. The total force against a 1-ft strip, using the trapezoidal rule, is

$$P = H\left(\frac{h_0 + h_n}{2} + h_1 + h_2 + \cdots + h_{n-1}\right)$$

$$= 5\left(\frac{0 + 0.04}{2} + 0.08 + 0.92 + 0.084 + 0.06\right)$$

$$= 5(0.336) = 1.68 \text{ kips}$$

To find the location of the resultant, assume that each force on a segment is at midpoint, except the first, which is taken as a triangle.

$$P_1 = \frac{0.08}{2}(5) \qquad = 0.2 \text{ kip}$$

$$
\begin{aligned}
P_2 &= 0.172(2.5) &= 0.43 \text{ kip} \\
P_3 &= 0.176(2.5) &= 0.44 \text{ kip} \\
P_4 &= 0.144(2.5) &= 0.36 \text{ kip} \\
P_5 &= 0.10(2.5) &= 0.25 \text{ kip} \\
\end{aligned}
$$

$$1.68 \text{ kips}$$

$$\sum M_{\text{base}} = P\bar{y}$$

$$1.68\bar{y} = 0.25(2.5) + 0.36(7.5) + 0.44(12.5) + 0.43(17.5) + 0.2(21.7)$$

$$= 0.625 + 2.7 + 5.5 + 7.53 + 4.34$$

$$\bar{y} = \frac{20.70}{1.68} = 12.35 \text{ ft} \qquad \text{above base}$$

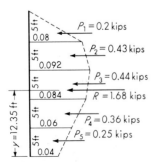

Figure E11-10b

Example 10-10a. Redo Example 11-10 using the computer program.

SOLUTION. *Step 1.* Compute equivalent square $B = \sqrt{78.54} = 8.86$ ft; $DOP = 4.43$ ft using 9 spaces; $WSQ = 8.86/9 = 0.984$ ft

The total load is $qA = 2(78.54) = 157$ kips

Poisson's ratio $= 0.5(AMU)$

Start at ground surface, $DY = 0.0$; wall height, $HTWALL = 25.0$ ft

Step 2. Set up card input.

Card No.

1	TITLE
2	Units (FT KIPS FT-K KSF)
3	6 9 9 0 (6 vertical increments, 9 × 9 grid of equivalent area)
4	0.984 157.0 0.50 4.43 15.0 0.0 25.00

(Check comment cards and computer program for variable identity.)

The output gives $Q_h = 0.975$ kip (not doubled for rigid wall)

$$\bar{Y} = 12.581 \text{ ft}$$

11-11. OTHER CAUSES OF LATERAL PRESSURE

Ice Formation

Lateral pressures may be developed when pore water freezes. This would be a minor problem in an unsaturated soil unless ice lenses form. The problem can be eliminated by using granular (free-draining) backfill and providing drainage outlets to carry the water away.

Earth Pressures from Earthquakes

The lateral pressure may be increased against a retaining wall because of the vibration of the ground. The increase should be taken as at least 10 percent of the design pressure for walls of normal height [NBC (1967, p. 344)]. Local building codes may specify the increase factor to use. Longer-duration vibrations such as, for example, reciprocating machinery, may increase the lateral pressure much more than 10 percent. Barkan (1962) indicates that the friction coefficient (tan ϕ) with vibrations may be 25 to 30 percent smaller than without vibrations. This would increase the lateral pressure against a wall for a soil of $\phi = 30°$ about 30 percent.

Seed and Whitman (1970) indicate the following additional factors to consider in designing retaining walls to resist earthquakes:

1. Model wall tests may reasonably predict performance.
2. The location of the wall resultant is at 0.5 to 0.67H above the base.
3. The failure surface will be at a flatter slope than the active-earth-pressure value.

4. Smaller safety factors (order of 1.1 to 1.2) can be used for resisting earthquake forces. The latter condition results in little—if any—increase in design due to earthquake forces.

Swelling Pressure

If an expansive clay is placed behind a retaining wall and becomes wet, large pressures may be developed. The problem can be somewhat alleviated by placing the clay under carefully controlled conditions of no lumps and at a water content considerably above optimum. The problem can be considerably alleviated by using granular backfill; however, this is not always possible. Lateral pressure is not likely to be developed when building against overconsolidated clay, as the high initial K_0 stresses will be lost as soon as the excavation is opened. Vertical rather than lateral expansion is more likely to be a problem in overconsolidated clay.

Thrust Due to Temperature

Walls providing restraint to members which may undergo thermal expansion and contraction may develop unwanted stresses. This problem can be solved by backing the restraint with rollers, hinges, or expansion joints.

11-12. PRESSURES IN SILOS, GRAIN ELEVATORS, AND COAL BUNKERS

Lateral pressure of agriculture products against the walls of grain-storage containers is similar to lateral-earth-pressure problems earlier in this chapter. It is necessary to obtain the internal and wall friction angles of the material. These values depend on the material being contained, its water content and density. Wall friction depends on the wall material being used and the factors cited earlier for soil. Table 11-7 gives representative values for several agricultural grains for which containment structures may be required. Grain is often measured in terms of bushels; a bushel may be taken as 1.24 cu ft or 0.0352 m^3.

The grain pressure (or coal pressure) for relatively shallow containment structures, say under 7 m in height, and with a small height/width ratio (not over 2; see also Fig. 11-24c) can be computed using the Rankine [Eq. (11-7)] earth-pressure equation with β taken as the angle of repose of the grain or coal. This will be conservative, as the wall friction is neglected. Plane-strain ϕ angles are recommended for this condition.

Coal bunkers often have sloping hopper bottoms (Fig. 11-24c), which requires obtaining the normal and tangential components of pressure. These values can be obtained from the geometry of the problem and an ellipse of stress analysis to obtain:

$$P_t = h(\gamma - \gamma K_a) \sin \alpha \cos \alpha \qquad \text{(psf or kPa)}$$

$$P_n = \gamma h \cos^2 \alpha + \gamma h K_a \sin^2 \alpha$$

Table 11-7. Angle of internal friction and other data for selected grain and other bulk storage materials. These values are representative; actual values should be obtained from tests

Grain	ϕ*	ϕ_r†	Concrete	Wood	Brick	γ, g/cm³§
Wheat	28	25	28	25	26	0.75–0.85
Rye	29	24	25	25	27	0.72–0.82
Barley	32	30	29	26	27	0.65–0.75
Oats	33	29	28	26	28	0.42–0.55
Corn	35	32	28	25	28	0.65–0.79
Beans	33	27	28	25	27	0.83–0.88
Peas	34	30	27	24	27	0.70–0.80
Flour		40	17	17		0.60–0.70
Sugar		35	23	22		0.95–1.05
Coal	35	35	30		35	0.75–1.10
Cement	38	42	22			1.01–1.60
Iron ore		40	26	26		2.55–2.75
Lime		35	26	26		0.70–0.96

* $\pm 2°$ based on w percent.
† Angle of repose also $\pm 2°$.
‡ For metal walls use 16–18°; use 17–20° for coal.
§ pcf = 62.5 × g/cm³; kN/m³ = 9.807 × g/cm³.

where h = depth from free surface to point under consideration
  α = angle of bunker side from horizontal
  γ = unit weight of material
  K_a = Rankine or Coulomb (with/side friction) lateral-earth-pressure coefficient

Grain elevators, silos, and deep coal bunkers (Fig. 11-24) are deep bins and require a modified analysis for the lateral and vertical wall pressure for wall design. When grain elevators are emptied, dynamic-pressure forces can develop which have caused walls to split. The dynamic pressure is not fully understood [Mackey and Mason (1972), Deutsch and Clyde (1967), Turitzin (1963, 1969a, 1969b)], but measurements indicate that it is quite large—as much as 3.3 times the usual design pressure. It appears that not all silo walls, even using the same design procedures, are distressed by dynamic pressures, but it is not known whether this is due to the erratic nature of dynamic-pressure formation or whether the walls were initially "overdesigned." Dynamic-pressure formation appears to be caused by the "plug" of Fig. 11-25 flowing into the "pipe." Dynamic pressure is a minimum if the drawoff orifice is centered in the middle third of the container [ACI (1975)] and a maximum when to one side. Side-location valves or orifices are not generally recommended for use, and the dynamic coefficients suggested here are for centrally located orifices. If some means is provided to avoid the pipe and plug formation, dynamic pressures do not form.

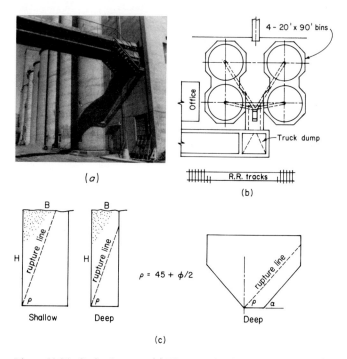

Figure 11-24. Grain elevators. (a) Photograph of typical elevator; (b) general layout of small four-silo group; (c) condition for shallow or deep silo analysis—if potential material-rupture line intersects wall, silo is "deep."

Grain pressures are generally determined by Janssen's or Reimbert's pressure method [Turitzin (1963) is a convenient reference on both methods]. The Janssen method computes the lateral and vertical pressures as

$$\sigma_h = \frac{\gamma R E_1}{\tan \phi'} \tag{11-21}$$

$$E_1 = 1 - \exp\left(-K_a \tan \phi' \frac{h}{R}\right)$$

$$\sigma_v = \frac{\sigma_h}{K_a} \tag{11-22}$$

$$P_v = \gamma R\left(h - \frac{R}{K_a \tan \phi'}\right) E_1 \tag{11-23}$$

The Reimbert equations for calculating the lateral and vertical pressure are

$$\sigma_h = \frac{\gamma R}{\tan \phi'}\left[1 - \left(\frac{h}{C} + 1\right)^{-2}\right] \tag{11-24}$$

$$\sigma_v = \gamma\left(\frac{Ch}{C + h} + \frac{y}{3}\right) \tag{11-25}$$

$$C = \frac{D}{4K_a \tan \phi'} - \frac{y}{3} \tag{11-26}$$

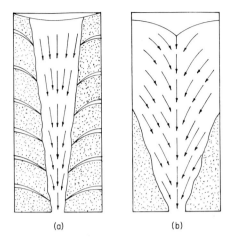

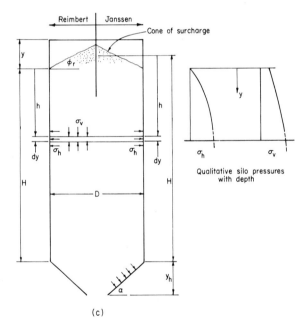

Figure 11-25 (a) Nondynamic silo flow conditions; (b) dynamic flow condition; (c) identification of terms in the Reimbert and Janssen pressure equations.

where [Eqs. (11-21) to (11-26)]

R = hydraulic radius = area/perimeter

K_a = active-earth-pressure coefficient (usually Rankine value and triaxial ϕ angles)

ϕ' = angle of wall friction

h = depth from free grain surface

y = cone of surcharge (Fig. 11-25c); $y = (D/2) \tan \phi_r$ where ϕ_r = angle of repose

D = diameter (for square bins use equivalent diameter)

P = load carried by silo wall per foot of perimeter

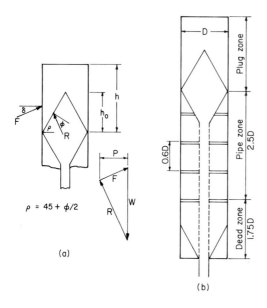

$\rho = 45 + \phi/2$

(a)

(b)

Plug zone

Pipe zone $2.5D$

Dead zone $1.75D$

$0.6D$

Figure 11-26. Alternative method of computing bursting pressure in silos. (*a*) Force polygon in "plug" zone; (*b*) zones for analysis in silo. [*After Mackey and Mason (1972).*]

Mackey and Mason (1972) proposed an analysis based on Fig. 11-26. The bottom "dead" zone of height $1.75D$ is designed based on the Janssen equations. The "pipe" zone is designed based on lateral pressures to hold an arch ring in place by friction where the ring is $0.6D$ thick. The remainder of the wall height is computed using a "wedge" theory as shown in the figure. The angle ρ shown in the figure is computed as $45° + \phi/2$.

The Janssen or Reimbert methods can be used for deep coal bunkers (and cement silos). Dynamic pressures may develop cracks; however, for coal bunkers cracks may be tolerable, since the coal is not likely to be damaged by water as is grain or cement. Safety should be considered, and depending on the effect of safety the lateral pressure may be increased in the middle height of the bin wall by a factor of 1 to 3.5.

When square bins are used, the analysis may proceed based on a hydraulic radius of a square. When rectangular bins ($a \times b$) are used, the lateral pressures are different, being larger on the long side. The pressure may be computed on the short side using the hydraulic radius R for a square of sides $a \times a$. For the long dimension b, use an equivalent square with sides [Rogers (1952)] equal to

$$a' = \frac{2ab - a^2}{b}$$

to obtain R.

Example 11-11. Compute the pressures acting on a wheat grain elevator which is 5 m diameter by 28.6 m high. Use the three methods presented.

Other data: $\gamma(\text{wheat}) = 0.8 \text{ g/cm}^3$ (50 pcf)
$\phi = 28°$ $\phi' = 24°$
Cone of surcharge $y = 0.0$

SOLUTION. We will make a table of σ_h and σ_v for each 5 m of depth.

Table E11-11.

h		Janssen σ_h	Janssen σ_v	Reimbert σ_h	Reimbert σ_v
0	top	0	0	0	0
5		10.45	28.94	13.87	23.88
10		15.94	44.16	17.81	34.33
15		18.83	52.17	19.47	40.20
20		20.35	56.38	20.31	43.95
25		21.15	58.59	20.80	46.56
28.6	bottom	21.48	59.50	21.03	47.99

Janssen's method

$$R = \frac{A}{P} = \frac{D}{4} = 1.25$$

$$K_a = \tan^2 (45 - 28/2) = 0.361$$

$$\tan \phi' = \tan 24° = 0.445$$

$$\gamma = 0.8(9.807) = 7.846 \text{ kN/m}^3$$

and $\qquad \sigma_h = \dfrac{7.846(1.25)}{0.445} E_1 \qquad \sigma_v = \sigma_h/0.361$

results are shown in Table E11-11.

Checking pressure at base of silo:

$$P_v = 7.846(1.25)\left[28.6 - \frac{1.25}{(0.445)(0.361)}\right]0.975 = 199.0 \text{ kN/m}$$

$$P_{total} = 199.0(\pi)(5) = 3,126.0 \text{ kN for perimeter}$$

$$P_{wheat} = 0.7854(5)^2(28.6)(7.846) = 4,406.0 \text{ kN}$$

$$\Delta P = 4,406.0 - 3,126 = 1,280$$

The vertical wheat pressure on the base of the silo is

$$\sigma_v = \frac{\Delta P}{A} = \frac{1,280.0}{(0.7854)(5)^2} = 65.19 \text{ vs. } 59.5 \text{ kPa} \qquad \text{from Table E11.11} \qquad O.K.$$

The wall would be designed for hoop-tension stresses based on σ_h and vertical-compression stresses based on P plus overturning due to wind or earthquake plus the weight of wall material. In the zone -12 to -24 m one probably should use a dynamic-load factor of 2.5 to 3.5 if a central hopper is used to empty the bin. Alternatively, provide a means to avoid the pipe and plug formation and make the design using the pressures in Table E11-11.

Reimbert method

$$C = \frac{D}{4K_a \tan \phi'} = \frac{5}{4(0.361)(0.445)} = 7.781$$

$$\sigma_h = \frac{7.846(1.25)}{0.445}\left[1 - \left(\frac{h}{7.781} + 1\right)^{-2}\right]$$

$$\sigma_v = 7.846\left(\frac{7.781h}{7.781 + h}\right)$$

Check:

$$P/m \text{ of wall} = \text{area lateral-pressure diagram} \times \tan \phi' \text{ using}$$
$$\text{average end areas}$$

$$P = \frac{\sigma_1 + \sigma_2}{2} h_1 + \frac{\sigma_2 + \sigma_3}{2} h_2 + \cdots$$
$$= 629.4$$

$$P_v = 629.4 \tan \phi'(\pi)(5) = 4{,}399.4 \cong 4{,}406 \text{ kN}$$

Note very little difference in lateral pressures except in top areas; considerable difference in vertical pressures. Use pressures in same manner including dynamic-load factors (if required) to design walls.

Mackey and Mason method

In bottom $1.75(5) = 8.75$ m use Janssen pressure distribution
In next (pipe) zone $2.5(5) = 12.5$ m   use arching
In top $28.6 - 12.5 - 8.75 = 7.35$ m   use wedge
   Considering any ring arch in the pipe zone, the weight of an arch ring of height $0.6D$ is

$$W = \gamma A h$$
$$= 7.846(0.7854)(5)^2(0.6)(5) = 462.17 \text{ kN}$$

Friction resistance $= 0.5\gamma h^2 K(\tan \phi')\pi D = 246.8K = W$

$$\text{Solving, } K = \frac{462.17}{246.8} = 1.87$$

The lateral pressure for each ring arch (varying from 0 at top to maximum at $0.6D$) is

$$\sigma_h = \gamma h K = 7.846(0.6)(5)(1.87) = 44.01 \text{ kPa}$$

This value compares with values of 15.94 and 17.81 kPa of previous methods.
   In top "plug" zone the weight of plug is

$$W = 0.7854D^2\gamma\left(h - \frac{h_0}{3}\right)$$

$$\theta = 45 + \phi/2 = 59° \qquad h_0 = 2.5 \tan 59° = 4.16 \text{ m}$$

$$W = 0.7854(5)^2(7.846)\left(7.35 - \frac{4.16}{3}\right) = 918.65 \text{ kN}$$

The active earth wedge can be solved directly for P_a to give

$$P_a = \frac{W}{\sin \phi' + \cos \phi' \tan (45 + \phi/2)}$$

$$= \frac{918.65}{0.407 + 0.914(1.664)} = 476.44 \text{ kN}$$

If we assume average lateral pressure on plug height

$$P_a = \sigma_h A = \sigma_h \pi D h$$

$$\sigma_h = \frac{476.44}{\pi(5)(7.35)} = 4.13 \text{ kPa}$$

PROBLEMS

11-1. A retaining wall contains a backfill with the following characteristics:

$$\gamma = 105 \text{ pcf} \qquad \phi = 28° \qquad c = 0 \text{ psf} \qquad H = 15 \text{ ft}$$

Required: Find the lateral pressure against the wall (total force per foot of width).

(a) Using Coulomb's equation and a value of wall friction of $\frac{2}{3}\phi$, plot the wall-pressure profile and show the value at midheight and at the wall base.

Answer: $R = 3.82$ kips/ft at 5 ft above base.

(b) Redo part a if $\beta = 15°$.

Answer: 4.96 kips/ft at 5 ft above base.

11-2. Redo Prob. 11-1 using the Rankine equation and taking the wall friction as zero.

Answer: (a) $R = 4.26$ kips/ft; (b) $R = 4.83$ kips/ft.

11-3. Compute the lateral force per foot of width and show the location of the resultant for a wall under the following conditions. Use Coulomb's equation and take $\delta = 2\phi/3$.

$$\gamma = 17.58 \text{ kN/m}^3 \qquad c = 19.2 \text{ kPa} \qquad H = 6.5 \text{ m} \qquad \beta = 0° \qquad \phi = 25°$$

11-4. Redo Prob. 11-3 if $\beta = 10°$.

11-5. Redo Prob. 11-3 using the Rankine equation and neglect the tension zone.

Answer: $R = 33.67$ kN/m; $\bar{y} = 1.02$ m above base.

11-6. Redo Prob. 11-5 using water in the tension crack.

11-7. Redo Prob. 11-3 using a triangular pressure diagram to top of wall.

Answer: $R = 71.27$ kN/m; $\bar{y} = 2.17$ m above base.

11-8. Redo Prob. 11-3 and neglect the zone of apparent soil adhesion (tension) to the wall and with a surcharge of 92 kPa.

11-9. Compute the lateral force and show the location of the resultant, using the Rankine equations for the wall-soil system shown in Fig. P11-9.

Answer: $R = 26.6$ kips; $\bar{y} = 8.02$ ft from base.

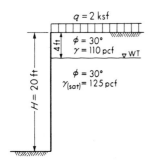

Figure P11-9

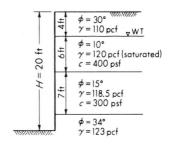

Figure P11-10

11-10. Compute the lateral force and show the location of the resultant for the wall system shown in Fig. P11-10. Use the modified Rankine equations.

11-11. For the conditions given in Fig. P11-11, find the active earth pressure and point of application.

Answer: $P_a = 166$ kN/m; $\rho = 63.3°$.

11-12. For the conditions given in Fig. P11-12, find the active pressure and the point of application.

11-13. For the cantilever retaining wall shown in Fig. P11-13, find the active earth pressure and the point of application, using Culmann's graphical solution.

Answer: $P_a = 93.3$ kN/m; $\rho = 62.7°$.

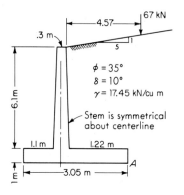

$\phi = 35°$
$\delta = 10°$
$\gamma = 17.45$ kN/cu m

Stem is symmetrical about centerline

Figure P11-11

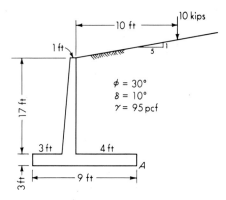

$\phi = 30°$
$\delta = 10°$
$\gamma = 95$ pcf

Figure P11-12

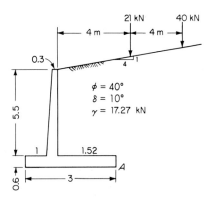

$\phi = 40°$
$\delta = 10°$
$\gamma = 17.27$ kN

Figure P11-13

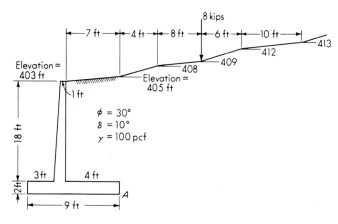

$\phi = 30°$
$\delta = 10°$
$\gamma = 100$ pcf

Figure P11-14

11-14. Determine P_a and the point of application due to the conditions shown in Fig. P11-14. Use Culmann's graphical solution.

Answer: P_a = 13.5 kips at 11.6 ft vertically above point A.

For Probs. 11-15 to 11-18 use $\delta = 2\phi/3$.

11-15. Find the active earth pressure of Prob. 11-11 by the trial-wedge method if the soil has cohesion. Soil parameters are $\delta = 23.3$; $c = 5$ kPa; adhesion $= 0.75c$.

Answer: P_a = 113.4 kN; $\rho = 61.1°$.

11-16. Find the active earth pressure of Prob. 11-12 by the trial-wedge method if the soil has cohesion. Soil parameters are $\phi = 25°$, $h_t = 4.0$ ft, $\gamma = 95$ pcf.

11-17. Find the active earth pressure of Prob. 11-13 by the trial-wedge method if the soil has cohesion. Soil parameters are $\phi = 20°$, $c = 7.17$ kPa, $\delta = 13.3$, $\gamma = 17.27$ kN/m³.

Answer: P_a = 144.4 kN/m; $\rho = 46.9°$.

11-18. Find the active earth pressure of Prob. 11-14 by the trial-wedge method if the soil has cohesion. Soil parameters are $\phi = 23°$, $h_t = 4.0$ ft, $\gamma = 100$ pcf.

Answer: P_a = 11.4 kips.

11-19. Write a computer program to solve for the Rankine active-pressure coefficients for each 2° of angle from 0 to 28°; each degree from 28 to 38°; then each 2° to $\phi = 46°$. Take β from 0 to 25° by increments of 5°.

11-20. Write a computer program to solve the Rankine passive-pressure coefficients for each 2° of angle from $\phi = 0°$ to $\phi = 46°$. Take β as in Prob. 11-19.

11-21. A 6-m-square storage bin with a contact pressure of 150 kPa is located 5 m from a rigid 9-m-high retaining wall. Refer to Example 11-10 and take $\mu = 0.3$. Plot the pressure increase due to this load; find the total force and the location of the resultant.

Answer: 53.6 kN (not doubled); $\bar{y}$ = 4.07 m from bottom of wall.

11-22. Redo Example 11-9 if the surface contact pressure is 3 ksf instead of 2 ksf as in the example.

11-23. Redo Example 11-9 if the diameter of the tank is 15 ft.

11-24. Establish dimensions and plot the pressure profile for a 20,000-bushel wheat-storage silo. Use an H/D ratio of at least 6.

11-25. A steel-plate coal bunker is 50 ft deep; the slope of the bottom part of the hopper $\alpha = 50°$ as shown in Fig. P11-25. Plot the normal pressure profile along ABC, when full of coal of $\gamma = 50$ pcf.

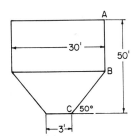

Figure P11-25

TWELVE

RETAINING WALLS

12-1. INTRODUCTION

Retaining walls are structures used to provide stability for earth or other material where conditions disallow the mass to assume its natural slope, and are commonly used to hold back or support soil banks, coal or ore piles, and water.

Retaining walls are classified, based on the method of achieving stability, into six principal types (Fig. 12-1). The *gravity* wall depends upon its weight, as the name implies, for stability. The *cantilever* wall is a reinforced-concrete wall that utilizes cantilever action to retain the mass behind the wall from assuming a natural slope. Stability of this wall is partially achieved from the weight of soil on the heel portion of the base slab. A *counterfort* retaining wall is similar to a cantilever retaining wall, except that it is used where the cantilever is long or for very high pressures behind the wall, and has counterforts, which tie the wall and base together, built at intervals along the wall to reduce the bending moments and shears. As indicated in Fig. 12-1c, the counterfort is behind the wall and subjected to tensile forces. A *buttressed* retaining wall is similar to a counterfort wall, except that the bracing is in front of the wall and is in compression instead of tension. Two other types of walls not considered further are *crib* walls, which are built-up members of pieces of precast concrete, metal, or timber and are supported by anchor pieces embedded in the soil for stability, and *semigravity* walls, which are walls intermediate between a true gravity and a cantilever wall.

Bridge abutments (Fig. 12-1f) are often retaining walls with wing wall extensions to retain the approach fill and provide protection against erosion. They differ in two major respects from the usual retaining wall in:

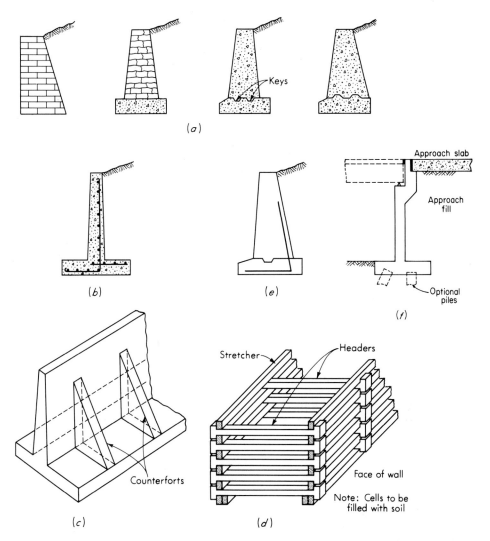

Figure 12-1. Types of retaining walls. (*a*) Gravity walls of stone masonry, brick, or plain concrete. Weight provides overturning and sliding stability; (*b*) cantilever wall; (*c*) counterfort, or buttressed wall. If backfill covers counterforts, the wall is termed a counterfort; (*d*) crib wall; (*e*) semigravity wall (small amount of steel reinforcement is used); (*f*) bridge abutment.

1. They carry end reactions from the bridge span
2. They are restrained at the top so that an active earth pressure is unlikely to develop.

Foundation walls of buildings including residential construction are retaining walls whose function is to contain the earth out of the basements.

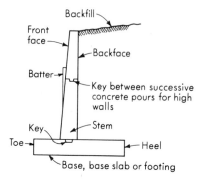

Figure 12-2. Principal terms used with retaining walls.

Retaining walls must be of adequate proportions to resist overturning (or excessive tilting) and sliding as well as being structurally adequate.

Terms used in retaining-wall design are shown in Fig. 12-2. Note that the front base projection is also the "toe"; the back slab projection is also the "heel."

12-2. COMMON PROPORTIONS OF RETAINING WALLS

Retaining-wall design proceeds with the selection of tentative dimensions, which are then analyzed for stability and structural requirements and are revised as required. Since this is a trial process, several solutions to the problem may be obtained, all of which are satisfactory. A computer solution greatly simplifies the work in retaining-wall design and provides the only practical means to optimize the design.

Cantilever Wall

Dimensions for a retaining wall should be adequate for structural stability and to satisfy local building-code requirements. The tentative dimensions shown in Fig. 12-3 are based in part on the history of satisfactorily constructed walls, and may be used in the absence of other data, but may result in an overly conservative design. The top of the stem should generally be not less than 8 in, ACI Art. 14.2(k), and

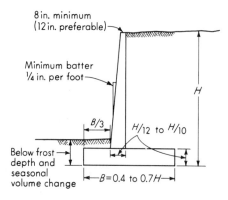

Figure 12-3. Tentative design dimensions for a cantilever retaining wall.

preferably not less than 12 in, so that proper placement of concrete may be effected and, if part of the end spalls off or is broken off, a sufficient amount will remain to satisfy structural and aesthetic requirements. The base of the stem should be at least thick enough to satisfy the shear requirements without the use of shear reinforcing steel (see Example 12-3 for further comment on this point).

The base-slab dimensions should be such that the resultant of the vertical loads falls within the middle one-third. If the resultant falls outside the middle one-third, the toe pressures will be excessively large, only a part of the footing will be effective, and computations should be made as indicated in Sec. 9-5.

A batter is usually used for cantilever retaining walls to effect some savings in material. A front batter is preferable so that. wall movement to develop active pressure is not noticeable. Some organizations place the batter on the backface. Most highway retaining walls and nearly all bridge abutments have the exposed face vertical. A slight increase in wall stability is usually obtained when the batter is on the back face. The included computer program allows batter on either face, but a slight approximation is made when the batter is on the back face by averaging the material weights in the two triangular zones and for the moment arm to the equivalent force. Low walls of under 3 m in height and all foundation walls are often built constant thickness to save on forming costs.

Counterfort Retaining Walls

Typical proportions for counterfort retaining walls are as shown in Fig. 12-4. These dimensions are only a guide, and thinner wall sections may be used if structural stability is satisfied. Walls 4 to 6 in thick have been reportedly constructed in Great Britain. The use of a counterfort will be determined by the relative costs of forms, concrete, reinforcing, and labor. It is doubtful if a counterfort wall will provide any relative construction economy unless it is over 20 ft in height.

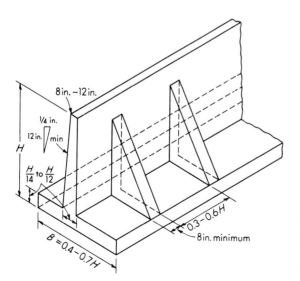

Figure 12-4. Tentative design dimensions for a counterfort retaining wall. Depth of base should be adequate for frost and below soils which undergo volume change. This wall may not be economical unless $H \geq 20$ to 25 ft.

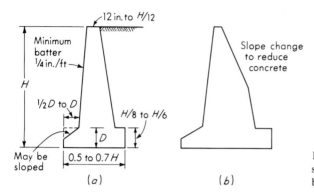

Figure 12-5. (a) Tentative dimensions for a gravity retaining wall; (b) broken-back retaining wall.

The spacing of the counterforts is a trial process to give a minimum cost. The most economical spacing appears to be from one-third to one-half the height of the wall. A counterfort may be built into the beginning of the wall or by allowing a part of the wall to overhang (Fig. 12-13b). The overhanging configuration may prove to be more economical, since this saves the concrete and formwork on the two counterforts at the joint. By conventional beam theory, bending moments in the face slab will be the same at the counterfort of the cantilevered part of the wall as at the interior counterforts if the length of overhang is made 0.41l and a spacing between counterforts of l. The counterfort wall may be constructed without a toe if additional front clearance is needed and the sliding and overturning stability requirements are met.

Counterfort walls are most economically designed considering the wall and heel as plates fixed on three edges and using the finite-element program in the Appendix (mat program). The input data are fairly extensive, but the output will be more reliable than conventional methods.

Gravity Walls

Gravity-wall dimensions may be taken as shown in Fig. 12-5. Gravity walls, generally, are trapezoidal-shaped but also may be built with broken backs. The base and other dimensions should be such that the resultant falls within the middle one-third of the base. The top width of the stem should be 9 to 12 in, with 12 in minimum preferred. If the heel projection is only 4 to 6 in, the Coulomb equation may be used for evaluating the lateral earth pressure, with the surface of sliding taken along the back face of the wall. The Rankine solution may also be used on a section taken through the heel. Because of the massive proportions and resulting low concrete stresses, low-strength concrete can generally be used for the wall construction.

A critical section for analysis of tensile flexure stresses will occur through the junction of the toe portion at the front face of the wall.

12-3. SOIL PROPERTIES FOR RETAINING WALLS

Section 11-7 indicated the method of obtaining the soil parameters c and ϕ for computing lateral pressure. In general, retaining walls will be constructed after an earth embankment is open so that overconsolidated soil properties are not of consequence. The plane-strain, drained ϕ angle for sand as obtained in a direct shear test is directly applicable as retaining walls are sufficiently long that plane-strain failure conditions would be considered. An extensive literature survey and some testing by Lee (1970) indicate that for both cohesionless and ϕ-c soils the ϕ angle from triaxial tests is too low and may be increased by about 10 percent to use for earth-pressure coefficients.

Plane-strain tests (CD and UU) are appropriate for cohesive soil parameters for wall design as indicated in Sec. 11-7. The backfill material will be remolded during placing and compaction (if any). Soil properties should be obtained for the remolded state at the water content and compaction density. Even so, the prediction of the resulting lateral pressure will be something of an estimate for both cohesionless and cohesive backfill, and more so for the cohesive materials. This estimate is also due to the fact that the theoretical active earth pressures are computed from a "backfill-in-place" condition from which the wall moves laterally and rotates sufficiently to allow active pressure to develop. Practically, the backfill is placed in stages to the desired depth with wall-soil interaction throughout the placing process.

The maximum lateral pressure for cohesionless materials is likely to be on the order of K_0 conditions near the bottom of the backfill to approximately K_a conditions in the upper one-half of the wall, since most concrete walls are relatively rigid. Studies by Coyle et al. (1972) and Prescott et al. (1973) on full-scale instrumented walls using compacted granular backfill have consistently measured higher-than-theory (Rankine or Coulomb) lateral pressures. Pressures in the bottom one-half of the walls have been as much as $2\frac{1}{2}$ times theoretical values, although in the upper one-half the pressures tended to be quite close to theoretical values.

The maximum lateral pressure for cohesive materials may approach hydrostatic conditions ($K = 1.0$) depending on the soil, water content at placement, size of the backfill zone, and wall rigidity. Narrow backfill zones or backfill zones of a size that permits use of massive compaction equipment and rigid walls may cause lateral pressures to develop in excess of hydrostatic conditions, according to Broms (1971). The high lateral pressure will reduce with time through creep; however, the wall must be of sufficient proportions to sustain the initially high lateral pressures without unsightly deformations or structural failure.

Retaining walls seldom are required to contain a water table (but are often used on hydraulic structures) and are usually constructed with drainage facilities ranging from weep holes through the wall to lateral drainage pipes laid parallel to the wall on the heel slab. This limits the lateral-pressure computation to, at most, the saturated unit weight of the backfill.

A study by Peck et al. (1948) indicated that most retaining-wall failures are caused by a bearing-capacity failure of wall footings founded on clay; very few structural failures were recorded. This may be partly attributed to the fact that during this time span concrete-wall design used an apparent load factor (based on $0.45f'_c$) of 2.22, which is considerably higher than the current USD load factors recommended by ACI.

A minimum load factor is recommended for cohesionless backfill of 1.7 obtained somewhat as follows [and using the test data of Coyle et al. (1972)]:

Let σ = average Rankine or Coulomb stress at base of wall of height H

2σ = pressure increase in bottom half of wall (total = $\sigma 3$)

$$P_t = \sigma \frac{H}{2} + 2\sigma \frac{H}{4} = \sigma H$$

also P_t = Rankine (or Coulomb value) × load factor

$$\sigma_R \frac{H}{2} \text{(load factor)} = \sigma H$$

from which the load factor = 2.00. This value is satisfactory for very rigid walls; however, since any wall should move somewhat prior to failure, resulting in a reduced pressure and less likelihood of failure, the factor should be reduced to, say, 1.7. A factor of 1.4 as in the current ACI Code for dead loads is considered by the author to be too low.

Compaction-induced lateral pressure against massive walls which may be un-yielding should use a factor of at least 1.5 and the modified pressure diagram of Fig. 12-6 based somewhat on a compromise of Brom's (1971) and Rowe's (1954) work and practical considerations taking into account safety and the history of wall performance.

The load factors considered here are for use in concrete design of stem and base and are not to be used for total wall stability. Total wall stability should be based on active earth pressures, since the wall must translate and/or rotate prior to failure, which will reduce the lateral pressure to active values regardless of what the pressure is prior to wall movement.

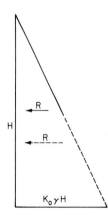

Figure 12-6. Pressure diagram for very rigid retaining walls. If some lateral movement can take place the resultant R can be placed at 1/3 point; with no movement place R at 1/2 point. Note use of K_0, not K_a.

12-4. STABILITY OF WALLS

Retaining walls must provide adequate stability against sliding, as shown in Fig. 12-7. The soil in front of the wall provides a passive-earth-pressure resistance as the wall tends to slide into it. If the soil is excavated or eroded after the wall is built, the passive-pressure component is not available and sliding instability may occur. If there is certainty of no loss of toe soil the designer may use the passive pressure in this zone as part of the sliding resistance.

Additional sliding stability may be derived from the use of a key beneath the base. Unless the key is quite deep, however, the sliding zone (Fig. 12-8) may bridge over the key in taking that path of least resistance. A key into firm soil or rock may be quite advantageous, since the resistance is now the force necessary to shear the key from the base slab.

The best key location is at the heel as indicated in Fig. 12-8. This location creates a slightly larger sliding-resistance distance L, as well as an additional component of force from the upward-sloping plane. The lesser of the two values

1. Passive pressure developed to the bottom of the key
2. Sliding resistance up plane $\overline{ab}$

is used in computing sliding stability.

The sliding resistance along the base is taken as fR, where R includes all the vertical forces, including the vertical component of P_a, acting on the base.

The coefficient of friction between the base and the soil may be taken as

$$f = \tan \phi \qquad \text{to} \qquad 0.67 \tan \phi$$

and base cohesion c' as

$$c = 0.5c \qquad \text{to} \qquad 0.75c$$

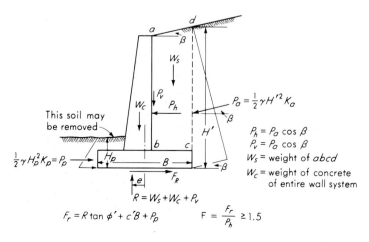

Figure 12-7. Forces involved in the sliding stability of a retaining wall.

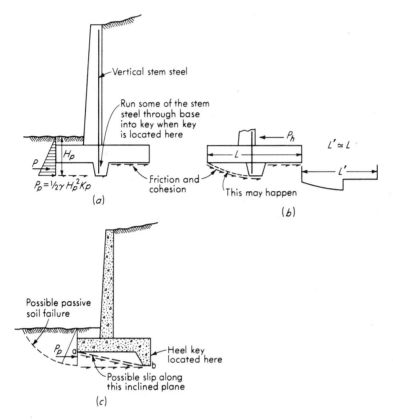

Figure 12-8. Stability against sliding using a base key. (a) Base key near stem so that stem steel may be run into the key; but (b) the sliding surface may develop as shown here where little aid is gained from using the key; (c) heel key which presents two possible modes of failure (passive and slip along the plane).

The base soil is usually compacted prior to pouring the base slab. The wet concrete will always attach to the ground such that $f = \tan \phi$ is not unrealistic. The cohesion may be considerably destroyed; thus values of 0.5 to $0.75c$ are more appropriate.

The safety factor against sliding should be at least 1.5 for cohesionless backfill and about 2.0 for cohesive backfill computed as follows:

$$F_{\text{sliding}} = \frac{\text{sum resisting forces}}{\text{sum driving forces}} \qquad (a)$$

The usual safety factor against overturning with respect to the toe is 1.5, with a value of 2.0 suggested for cohesive soil.

$$F_{\text{overturning}} = \frac{\text{sum of moments to resist overturning}}{\text{sum of overturning moments}} \qquad (b)$$

The safety factor can be computed in several ways depending on the interpretation of what goes in the numerator or denominator of Eqs. (*a*) and (*b*) as illustrated in Example 12-1.

Measured pressures against the stem of retaining walls are generally higher than active-pressure values. However, when considering total wall stability, the wall will slide or the stem will rotate when the pressures are sufficiently large, resulting in wall movements sufficient to develop the active-pressure state in the backfill and a corresponding pressure reduction. For this reason the use of active earth pressure in Eqs. (*a*) and (*b*) is appropriate.

Safety factors for basement walls and bridge abutments are computed similarly except that the wall is usually rigid and movements which are necessary to achieve active pressure are restrained, resulting in larger wall pressures. These larger pressures, generally on the order of K_0 conditions, are used for both structural-design (and with the appropriate load factor) and for wall-stability computations.

12-5. RETAINING-WALL FORCES

The forces acting on a retaining wall are customarily taken per unit of width for both gravity and cantilever walls. Counterforted walls may be considered as a unit between joints, or as a unit centered on two buttresses.

Gravity Walls

The forces on a gravity wall are as indicated in Fig. 12-9. The active earth pressure is computed by either the Rankine or Coulomb methods, presented in Chap. 11. If the Coulomb method is used, it is assumed that there is incipient sliding on the back face of the wall, and the earth pressure acts at the angle of wall friction δ to a normal with the wall. The Rankine solution applies to P_a acting at the angle β on a vertical plane through the heel. The vector can then be added to the weight vector of the wedge of soil W between the vertical plane and the back of the wall to get the direction and magnitude of the resultant P_a on the wall. The vertical resultant R acting on the base is equal to the sum of the forces acting downward, and will have an eccentricity e with respect to the geometrical center of the base. Taking moments about the toe,

$$\bar{x} = \frac{\text{sum of overturning moments (net)}}{R} \tag{c}$$

If the width of the base is B, the eccentricity of the base can then be computed as

$$e = \frac{B}{2} - \bar{x} \tag{d}$$

Cantilever Walls

Forces on a cantilever wall are shown in Fig. 12-10. Because of difficulty in evaluating wall friction, it is usual practice up to about $H = 20$ ft to use the Rankine solution for active pressure (i.e., taking $\delta = 0$).

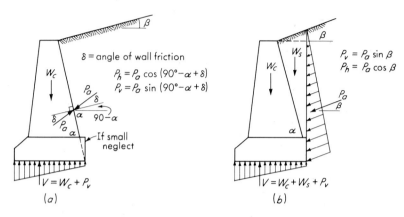

Figure 12-9. Forces on a gravity wall. (*a*) Coulomb analysis; (*b*) Rankine analysis.

For walls over 20 ft in height the Coulomb equation is more economical. It should be noted that the wall height H for use in determining the maximum shear and bending moment acting on the stem is not the same as that used to find the driving forces for sliding computations. The designer must also make a decision on whether to use passive pressure from the soil in front of the toe and whether the soil

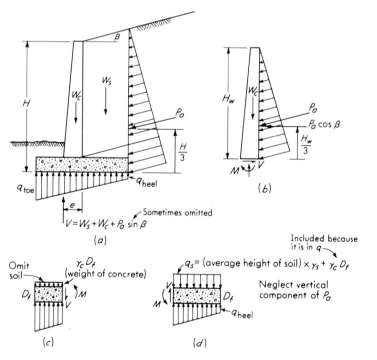

Figure 12-10. Forces on cantilever wall. (*a*) Entire unit; (*b*) stem; (*c*) toe; (*d*) heel. Note that $M_1 + M_2 + M_3 \cong 0.0$.

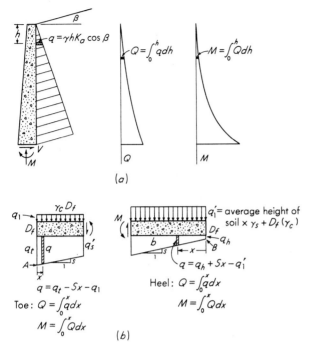

β

$q = \gamma h K_o \cos \beta$

$Q = \int_0^h q\,dh$

$M = \int_0^h Q\,dh$

Q

M

(a)

$\gamma_c D_f$

q_1

D_f

q_t | q | q_s'

A

$q = q_t - Sx - q_1$

Toe: $Q = \int_0^x q\,dx$

$M = \int_0^x Q\,dx$

q_1' = average height of
soil $\times \gamma_s + D_f(\gamma_c)$

M

D_f

b

q_h

$q = q_h + Sx - q_1'$

Heel: $Q = \int_0^x q\,dx$

$M = \int_0^x Q\,dx$

(b)

Figure 12-11. Cantilever retaining wall. (a) Stem shear and moments; (b) toe and heel shears and moments.

covering the top portion will be available for resisting overturning moments and sliding.

Forces on the stem and base slab are shown in Fig. 12-11. The triangular pressure diagram on the stem will yield a shear diagram that is a second-degree curve and a moment diagram that is a third-degree curve. It will be found to be easier to construct the shear and moment diagram by use of the differential equations for shear and moment as shown in the figure.

The use of the differential equation enables rapid computation of the cutoff points for the reinforcing steel, since it is uneconomical to use a constant amount of reinforcing for the entire wall height.

From Fig. 12-11b the weight of the base slab is deducted from the gross-soil-pressure diagram before computing the shear and moment diagrams. The eccentricity is computed by Eqs. (c) and (d). It is also convenient to use the differential equations for computing the shears and moments of the base slab if the shape of the curve is desired or if numerical values are to be evaluated at points other than the ends.

Counterfort Walls

Counterfort walls are indeterminate problems which can be solved using plate theory, at the expense of considerable labor. Simplified methods are commonly used to solve the problem, which generally will result in a wall somewhat overdesigned. The weight of the counterfort is usually neglected in the computations.

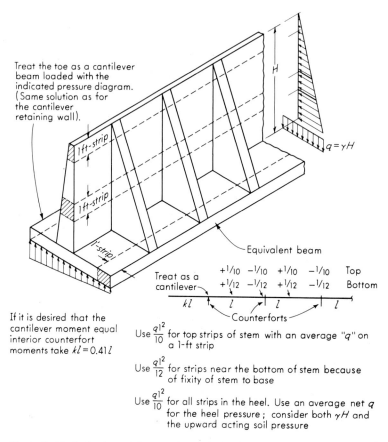

Treat the toe as a cantilever
beam loaded with the
indicated pressure diagram.
(Same solution as for
the cantilever
retaining wall).

1 ft-strip

1 ft-strip

1-strip

$q = \gamma H$

H

Equivalent beam

Treat as a
cantilever

+¹/₁₀ −¹/₁₀ +¹/₁₀ −¹/₁₀   Top

+¹/₁₂ −¹/₁₂ +¹/₁₂ −¹/₁₂   Bottom

$k l$ l l l

Counterforts

If it is desired that the
cantilever moment equal
interior counterfort
moments take $k l = 0.41 l$

Use $\dfrac{q l^2}{10}$ for top strips of stem with an average "q" on
a 1-ft strip

Use $\dfrac{q l^2}{12}$ for strips near the bottom of stem because
of fixity of stem to base

Use $\dfrac{q l^2}{10}$ for all strips in the heel. Use an average net q
for the heel pressure; consider both γH and
the upward acting soil pressure

Figure 12-12. Reduction of the complex analysis of a counterfort retaining wall to a system of simple
beams for rapid design.

Figure 12-12 indicates a simplified and conservative solution to a counterfort-
wall problem. The face slab of the wall is taken as a continuous slab made up of a
series of equivalent-unit-width beams. Since the pressure distribution is triangular,
the equivalent beams should be analyzed for strips at the junction of the wall and
base, and at two or three intermediate locations between the top of the wall and the
base, so that adjustments in reinforcing-steel requirements can be made as the pressure decreases. Moment distribution may be employed for finding the bending
moments; however, because of approximations being made, continuous-beam
coefficients may be used. Values of $w l^2 / 12$ or $w l^2 / 14$ may be used for the lower strips,
because of the lower edge being fastened to the base, and $w l^2 / 9$ or $w l^2 / 10$ for the
upper strips. For a conservative solution the same coefficients for both positive and
negative moments may be used, although smaller values for the positive moments
may be used, as the designer considers appropriate. The toe of the base is treated as a
cantilever beam and the heel as a continuous beam, similar to the treatment of the
face slab. The ACI Code requirements for shrinkage steel should be satisfied in the
directions not analyzed in this manner. The counterfort member can be treated as a

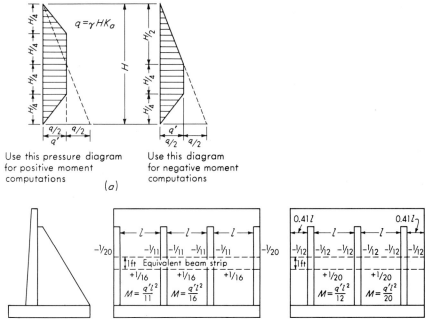

Use this pressure diagram for positive moment computations

Use this diagram for negative moment computations

(a)

Use q' from the shaded portions of the pressure diagrams in (a). Moment coefficients are shown. Compute moments for several strips near top, midheight and near bottom.

(b)

Figure 12-13. Computation of bending moments in the horizontal direction for the counterfort stem. [*After Huntington (1957).*]

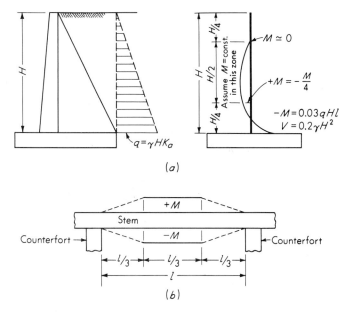

(a)

(b)

Figure 12-14. Distribution of vertical moments in a counterfort wall stem for Huntington's procedure. (a) Distribution of moment vertically in stem; (b) distribution of moment horizontally in stem. Assume that both positive and negative moments vary linearly as shown.

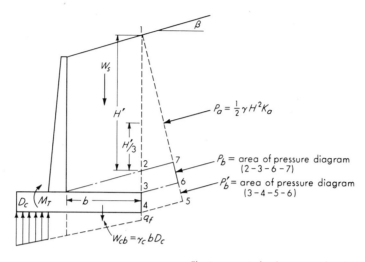

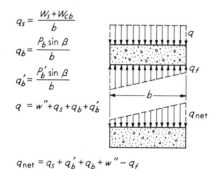

$$q_s = \frac{W_s + W_{cb}}{b}$$

$$q_b = \frac{P_b \sin \beta}{b}$$

$$q_b' = \frac{P_b' \sin \beta}{b}$$

$$q = w'' + q_s + q_b + q_b'$$

$$q_{net} = q_s + q_b' + q_b + w'' - q_f$$

The increase in heel pressure due to the toe moment is:

$$w' = \frac{2.4 M_t}{b^2} \qquad W' = \frac{2}{3} w' b$$

M_t = toe moment value at front face of wall

Note that w' is parabolic but may be approximated as a uniform pressure w''
$$w'' = W'/b$$

Assume pressures q_b', q_b, and q are constant and uniformly distributed across b.

If $\beta = 0$, there is only q and w'' to consider.

Since w'', q_b, and q_b' are small the design will usually be sufficiently accurate to neglect these pressures.

Figure 12-15. Forces on the heel slab of a counterfort wall as proposed by Huntington (1957).

wedge-shaped T beam, which includes the applicable portion of the wall stem as the flange, but these beams are so massive that the concrete stresses will be so low that an analysis is usually not required. Tensile steel will be required at the junction of the base (heel) and the counterfort to resist the moment, tending to tip the wall over, and the quantity can be conservatively computed, treating the counterfort alone as a beam. Tensile steel will also be required running horizontally from the counterfort into the stem to tie the wall and counterfort together. The slope of the counterfort member may be controlled by the bond-stress requirements of this reinforcement.

Huntington (1957) presented a solution for the counterfort-wall problem, as shown in Figs. 12-13 to 12-15. Huntington also recommended a value for Q_{xz} in the middle half of the wall, at the base of $0.2\gamma H^2$, to be used as long as the ratio of counterfort spacing to the wall height l/H is not greater than 0.5.

Example 12-1. Analyze the retaining wall shown in Fig. E12-1*a* for overturning and sliding stability. Note that this wall is backfilled in a limited zone with a cohesive soil. The batter is on the back face of the wall to illustrate the method used in the computer program to approximate this solution. Most of the output is obtained from running this wall with the computer program in the Appendix. Dimensions shown were optimized by the computer to satisfy stability and strength requirements. The load factor for USD was taken as 1.8 (1) to allow somewhat for cohesive soil and water in tension crack and (2) because of the uncertainty of the earth pressure, since any "Rankine" zone will be partly in fill and partly in original soil.

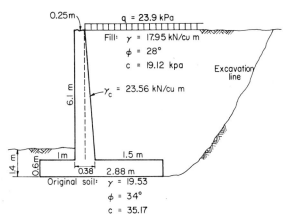

Figure E12-1a

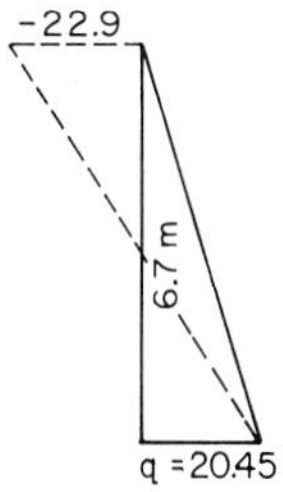

$\bar{\sigma}_v K_a \mu$

Figure E12-1b

SOLUTION. *Step 1.* Find the ϕ angle for an equivalent cohesionless soil (see Fig. E12-1*b*)

$$q = \gamma H K_a - 2c\sqrt{K_a} = \gamma H K_a'$$

Rankine $K_a = 0.361$ for $\phi = 28°$

$$q = 17.95(6.7)(0.361) - 2(19.12)(0.60)$$

$$= 43.42 - 22.94 = 20.48$$

$$K_a' = \frac{20.48}{(6.7)(17.95)} = 0.170$$

$$45 - \phi/2 = 22.42°$$

$$\phi = 45.16° \qquad \text{Use } \phi' = 45° \qquad K_a' = 0.172$$

This value of ϕ' and $\gamma = 17.95$ is used in the computer program.

Step 2. Compute P_a

$$P_a = (0.5\gamma H^2 + qH)K_a'$$

$$= [0.5(17.95)(6.7)^2 + 23.90(6.7)](0.171)$$

$$= 68.89 + 27.38 = 96.28$$

Step 3. Compute overturning stability

Set up table and refer to Fig. E12-1*c*.

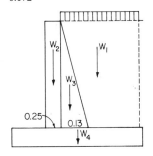

Figure E12-1c

Part	Wt of part, kN		Arm, m	Moment, kN-m
1	$1.5(23.9) + 6.1(1.5)(17.95)$	$= 200.09$	2.130	426.19
2	$23.56(0.25)(6.1)$	$= 35.94$	1.125	40.43
3	$0.13[(6.1)(23.56 + 17.95)0.5 + 23.90]$	$= 19.57$	1.315	25.73
4	$23.56(0.6)(2.88)$	$= 40.71$	1.440	58.62
$P_{a(v)}$		$= 0.0$	2.880	0.0
		$\sum F_v = 296.31$ kN		$\sum M_r = 550.98$

The location of P_a is at $\bar{y}$:

$$\bar{y} = \frac{68.89(6.7/3) + 27.38(6.7/2)}{96.28} = \frac{245.58}{96.28} = 2.55 \text{ m}$$

The overturning safety factor is

$$F = \frac{M_r}{M_0} = \frac{550.98}{245.58} = 2.24 < 1.5 \qquad O.K.$$

Step 4. Compute sliding F.
Use base soil parameters and

$$c' = 0.67c = 0.67(35.17) = 23.56$$

$$\tan \phi' = \tan \phi = \tan 34° = 0.675$$

$$F_r = c'B + F_v \tan \phi' = 23.56(2.88)$$

$$+ 296.31(0.675) = 267.86$$

$$F = \frac{F_r}{F_d} = \frac{267.86}{96.28} = 2.78$$

Step 5. Locate the resultant on the base of the footing. From rigid-body statics a moment summation can be taken at any location. Using the toe, as we already have most of the moments computed:

$$\sum M = M_r - M_0 = 550.98 - 245.58 = 305.40$$

$$X = \frac{\sum M}{\sum F_v} = \frac{305.40}{296.31} = 1.03 \text{ m from toe}$$

$$e = B/2 - X = 1.44 - 1.03 = 0.41 \text{ m}$$

$$L/6 = 2.88/6 = 0.48 > 0.41;$$

therefore, the resultant is in the middle one-third of base.

Step 6. Compute passive pressure in front of wall and recompute F.

$$K_p = \tan^2 (45 + \phi/2) = \tan^2 62° = 3.537$$

$$P_p = 0.5(19.53)(1.4)^2(3.537) = 67.7 \text{ kN}$$

Now how do we apply P_p?

(1) $P_p = -$driving force   (2) $P_p = $ resisting force

$$F = \frac{267.86}{96.28 - 67.7} = 9.37 \qquad\qquad F = \frac{267.86 + 67.7}{96.28} = 3.48$$

At least two other ways of computing F exist, including 267.86 taken as a $(-)$ driving force and 96.28 considered as a $(-)$ resisting force.

12-6. ALLOWABLE BEARING CAPACITY

Stability of the base against a bearing-capacity failure is achieved by using a suitable safety factor with the ultimate bearing pressure where the safety factor is usually taken as 2.0 for granular soils and 3.0 for cohesive soil. The allowable soil pressure can be computed from Eq. (4-6) with shape factors deleted, since this type of footing is a strip:

$$q_{ult} = cN_c d_c i_c + qN_q d_q i_q + \tfrac{1}{2}\gamma BN_\gamma d_\gamma i_\gamma$$

where i is the inclination factor, considering horizontal forces (in this case the horizontal component of P_a) and the vertical forces V, and d is a depth factor. Use $B' = B - 2e$ for A_f and for the D/B ratio.

The depth of the base D must be sufficient to place the footing below the zone of seasonal moisture variation, frost depth, or scour and deep enough to provide adequate bearing capacity and sliding resistance.

The intensity of soil pressure is computed for a rigid footing and a linear pressure distribution as

$$q = \frac{V}{A} \pm \frac{Vec}{I} \le q_a$$

This equation is valid as long as $e \le L/6$. The base width should be adjusted until $e \le L/6$ for maximum efficiency of footing and minimum difference in soil pressure beneath the footing.

When the soil is of low bearing capacity and/or it is not practical to use a larger base slab, it will be necessary to use a pile foundation.

12-7. SETTLEMENTS

Walls with base on granular soils should undergo most of the expected settlement by the time the wall construction and backfill are completed. Walls on cohesive soils for which consolidation theory is applicable will continue to settle for some time after completion of construction. The resultant force should be kept near the middle of the base for these soils to keep the settlement relatively uniform and reduce tilting. The intensity of the soil pressure at the toe is twice as great when the eccentricity of the resultant is $L/6$ as when the eccentricity is zero.

When the footing is on rock, two items must be considered. First, there must be sufficient rotation of the base and wall so that active pressure is developed. This can be accomplished by placing an earth *pad* beneath the base 15 to 30 cm thick, or constructing the stem with sufficient flexibility to yield with the soil pressure. The second problem is to avoid high toe pressures, which may break the toe away from the remainder of the base. This can be avoided by proportioning the footing so that the resultant falls near its center.

Differential settlements may be a problem in the longitudinal direction if the wall is long, although there will be a tendency to bridge local deposits of poor material.

However, if the material is different in bearing quality for a considerable distance, this will have to be taken into account, or else the wall may crack vertically. The soil may be replaced, compacted, or stabilized, or the soil bearing pressure may be reduced by increasing the footing width. When the computed settlements, either elastic or consolidation, are too large, pile foundations are necessary.

12-8. TILTING

A certain amount of wall tilt is desirable so that the active-pressure concept is valid as given in Sec. 11-1.

Since the stem is essentially a flexible cantilever beam, the yield necessary to achieve the active state is easy to attain ($0.002H$ or less) if the backfill is granular material, since the necessary wall yield is so small. The yield necessary to provide the active condition for clays is considerable—on the order of ten times or more as much yield as when the backfill is granular. This provides an additional reason for the preference of granular backfill.

Undesirable wall tilting is indicated in Fig. 12-16a and b, where the tilt is associated with a foundation failure. In Fig. 12-16a the toe pressure is excessive, and in Fig. 12-16b the substratum is weak and compressible, resulting in an excessive settlement due to the weight of the backfill. This latter problem may possibly be alleviated by using cinders or other lightweight material for backfill. It may be necessary to use piles to transmit the wall loads to a firmer stratum, or spread the load more uniformly through more strata. This problem must be carefully analyzed, as using piles may not provide a solution. If the forces causing the back rotation are large enough, they may simply bend or push the piles forward and/or downward with additional back rotation. Tschebotarioff (1970) describes several abutment failures attributed to inadequate assessment of the forces and direction of application in this problem.

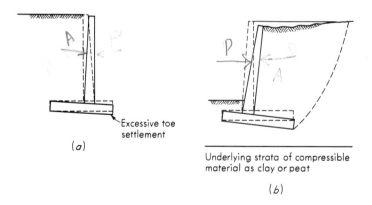

Figure 12-16. Settlement failures. (a) Excessive forward tilt due to a high toe pressure; (b) excessive settlement and tilt due to backfill. This is a common potential problem at bridge abutments caused by the approach fill.

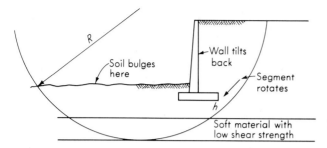

Figure 12-17. Soil shear failure. May be analyzed by the Swedish-circle method. A "shallow" failure occurs when base soil fails. A "deep" failure occurs if the poor soil stratum is underlying a better soil as in the figure.

If the soil underlying the base is stratified with poor material, as indicated in Fig. 12-17, resulting in the failure circles (shear failure) shown, excessive tilting may also occur. This analysis should be made if the poor stratum is within the depth 1.5 to 2H of the base of the wall. The analysis of the circle of failure (also called the Swedish-circle method) proceeds as follows:

1. Draw the wall-soil system to a convenient scale.
2. Through the heel point h, draw a circle with the radius sufficient to penetrate the soft stratum. This applies for $\beta = 0$ to perhaps $\beta = 10°$. For larger values of β, irregular surfaces, or surcharges, the assumed failure circle may not pass through the heel, and other circle locations should be investigated.
3. Compute all the forces but the weight of the segment itself which tend to act on this segment of soil and the moment arms with respect to point O, the center of the trial circle.
4. For the segment: (a) If the soil is cohesive and all of one type or if the failure surface is all cohesive ($\phi = 0$), the resisting force is simply the cohesion multiplied by the arc length, and the driving moments will be all forces acting on the system tending to cause the indicated rotation, including the applicable portion of the weight of the segment.

    (b) If there is internal friction ($\phi \neq 0$), divide the circular segment into several pieces so that the area of each may be treated as a rectangle or triangle, and compute the weight. Assume that the weight W acts through the midpoint of the arc of the part under consideration, and draw a ray from O through this point. Plot W vertically to scale, and resolve along the radius vector into a normal force vector N through the center of the circle and a tangential force. The frictional resistance is fN. Since this slip is soil-to-soil, we may take the coefficient of friction as $f = \tan \phi$.

    (c) When the soil has both cohesion and internal friction, it is treated as a combination of conditions a and b.
5. Summing moments about point O, the safety factor is

$$F = \frac{\Sigma M_r}{\Sigma M_0}$$

where the moments are computed as forces multiplied by the moment arms with respect to point O. The safety factor should be at least 2.0 for this analysis. Several trial circles should be drawn, the safety factor computed, and the minimum value taken. It may be convenient to plot the safety factors as indicated in Example 12-2, and graphically determine the minimum value and its approximate location.

Example 12-2. For the retaining wall shown overlying a soft-clay deposit, determine F against a deep-foundation failure for a trial circle through the heel.

Table E12-2

Segment	Weight of segment, kips	N, kips	T, kips
1	$3(21.5)(0.110) + 2(3)(0.15) = 8.0$	5.7	5.5
2	$3(24)(0.110) + 2(3)(0.15) = 8.8$	7.0	5.3
3	$1(20)(0.110) + 1(2)(0.15) = 2.5$	2.2	1.4
4	$1.5(20)(0.15) = 4.5$	4.0	2.1
5	$6.8(4)(0.110) = 3.0$	2.4	-1.8
6	$(2.6)(4.8)(0.110) = 1.4$	1.1	-1.2
		$\sum N = 22.4$	$\sum T = 11.3$

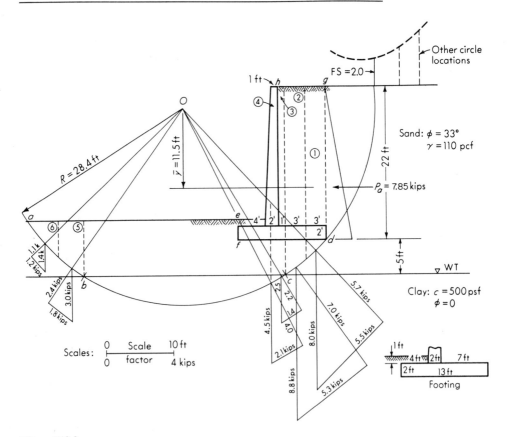

Figure E12-2

Computing the safety factor

$$F = \frac{(f \sum N + c \, \widehat{bc})R}{R \sum T + P_a \bar{y}}$$

$$f = \tan \phi = \tan 33° = 0.65$$

Since O is centered with respect to $\widehat{bc}$, it is not necessary to break into slices and find T. Some correction was applied to include the concrete wall in segments 3 and 4. Arc $\widehat{bOc} = 65°$.

$$f \sum N = 0.65(22.4) = 14.6 \text{ kips}$$

$$\widehat{bc} = R\theta = 28.4(1.134) = 32.2$$

$$1(\widehat{bc})c = 1(32.2)(0.5) = 16.1 \text{ kips}$$

$$R \sum T = 28.4(11.3) = 321 \text{ ft-kips}$$

$$P_a \bar{y} = 11.5(7.85) = 90.2 \text{ ft-kips}$$

$$F = \frac{(14.6 + 16.1)(28.4)}{321 + 90.2} = \frac{871.9}{411.2}$$

$$= 2.12(2.37 \text{ using a computer program})$$

Other trial-circle locations should be examined to see if a smaller F can be found. The minimum $F = 1.89$ (computer) was found at 4 ft to left and 4 ft above point O.

The slope computer program in Chap. 14 of Bowles (1974a) can be used to solve this problem. The wall *efdgh* is removed and replaced with an equivalent vertical force. The active earth pressure is computed and its point of application and slope established, and a series of trial circles with varying x, y coordinates for point O and constant entrance x, y coordinates of point d are used to obtain the minimum F.

12-9. DESIGN OF GRAVITY AND SEMIGRAVITY WALLS

The design of walls of the gravity type will be illustrated by an example. The semigravity wall is intermediate between a cantilever and a gravity wall, and will be left for the reader to analyze.

The first step in the design of a gravity wall is to select proportions. Figure 12-5 is used as a guide for selecting initial wall dimensions. Figure 12-18 indicates critical sections and the method of computing concrete stresses. Also shown are the allowable ACI (USD) Code concrete stresses in fps and metric units.

Example 12-3. Design a solid gravity retaining wall which is to retain an 18-ft embankment. The wall is on a soil of $\phi = 36°$, $\gamma = 120$ pcf. The backfill slopes 10° to the horizontal, and $\phi = 32°$; $\gamma = 110$ pcf. The allowable bearing capacity of the soil is 6 ksf. The base will be 4 ft in the ground. The ϕ angles were determined by direct shear testing.

$$f'_c = 3,000 \text{ psi} \qquad \text{refer to Fig. E12-3a}$$

SOLUTION. *Step 1.* Use graphic construction and find dimensions shown on Fig. E12-3a,b,c.

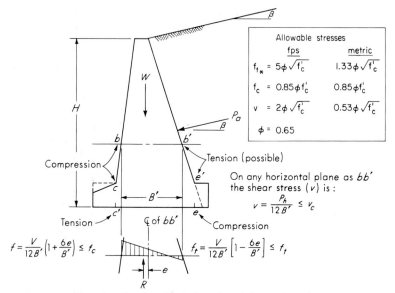

Allowable stresses

	fps	metric
f_{t_*} =	$5\phi\sqrt{f'_c}$	$1.33\phi\sqrt{f'_c}$
f_c =	$0.85\phi f'_c$	$0.85\phi f'_c$
v =	$2\phi\sqrt{f'_c}$	$0.53\phi\sqrt{f'_c}$
ϕ = 0.65		

On any horizontal plane as bb' the shear stress (v) is:

$$v = \frac{P_h}{12B'} \le v_c$$

$$f = \frac{V}{12B'}\left(1+\frac{6e}{B'}\right) \le f_c \qquad f_t = \frac{V}{12B'}\left[1-\frac{6e}{B'}\right] \le f_t$$

Note that R varies with the location of the section taken.

Figure 12-18. Design of a gravity retaining wall with critical points indicated.

Step 2. Use Rankine solution.

$$K_a = 0.321$$

$$P_a = 0.5\gamma H^2 K_a = 0.5(0.110)(22.9)^2(0.321) = 9.3 \text{ kips/ft}$$

$$P_{ah} = 9.3\cos 10° = 9.1 \text{ kips} \qquad P_{av} = 9.3\sin 10° = 1.6 \text{ kips}$$

$$W = 0.5(19.9)(0.110)(5.09) = 5.57 \text{ kips}$$

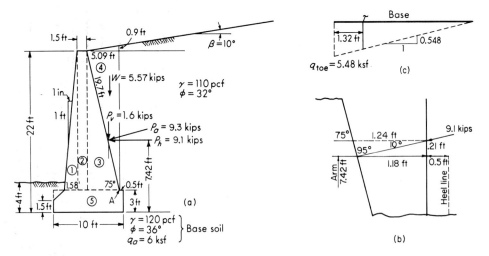

Figure E12-3

Step 3. Compute wall stability. Assume rectangular toe and neglect 1 ft of soil cover over toe. Use the following Table E12-3. Arm is with respect to toe of base slab.

Table E12-3

Part	Weight, kips	Arm, ft	Moment, kips/ft
1	$0.5(1.58)(19)(0.15) = 2.25$	2.39	5.4
2	$1.5(19)(0.15) = 4.27$	3.66	15.6
3	$0.5(5.10)(19)(0.15) = 7.26$	6.10	44.3
4	Soil $= 5.57$	7.80	43.4
5 (approx.)	$10(3)(0.15) = 4.50$	5.0	22.5
	$P_v = 1.6$	8.32	13.3
	$\sum V = 25.45$ kips		144.5 kips-ft

$$M \text{ (overturning)} = 9.1(7.42) = 67.5 \text{ kips}$$

$$F \text{ (overturning)} = \frac{144.4}{67.5} = 2.14 > 1.5 \qquad O.K.$$

Compute safety factor against sliding.

$$f = 0.9 \tan \phi = 0.9 \tan 36 = 0.65 \qquad \text{(estimate 0.9 value)}$$

$$F_R = 25.45(0.65) = 16.5 \text{ kips}$$

$$F = \frac{16.5}{9.1} = 1.82 > 1.5 \qquad O.K.$$

Step 4. Locate resultant on base of footing.

$$\bar{x} = \frac{144.5 - 67.5}{25.45} = 3.03 \text{ ft}$$

$$e = 5 - 3.03 = 1.97 \text{ ft} \qquad \text{slightly out of middle 1/3}$$

Step 5. Compute soil pressure.

$$q = \frac{V}{L}\left(1 \pm \frac{6e}{L}\right) = \frac{25.45}{10}\left[1 \pm \frac{6(1.97)}{10}\right]$$

$$q_{max} = 2.54(1 + 1.18) = 5.5 \text{ ksf} < 6 \qquad O.K.$$

$$q_{min} = 2.54(-0.18) = -0.46 \text{ ksf} \qquad \text{(say 0)}$$

Step 6. Check shear and tensile stresses in toe. Refer to pressure Fig. E12-3c and use $LF = 2.0$;

$$q = 5.5 - 0.55x$$

$$V = 5.5x - 0.55x^2/2$$

and

$$M = 5.5x^2/2 - 0.55x^3/6$$

with limits 0 to 1.32

$$V = 6.78 \text{ kips/ft} \qquad M = 4.58 \text{ ft-kips/ft}$$

$$v_c = 2\phi\sqrt{f_c'} = 93 \text{ psi} \qquad f_t = 5\phi\sqrt{f_c'} = 178 \text{ psi}$$

$$v = \frac{6.78(2)}{12(36)} = 0.031 \ll 0.093 \text{ ksi}$$

$$f = \frac{Mc}{I} = \frac{6M}{bh^2} = \frac{6(4.58)(12)(2)}{12(36)^2} = 0.42 \ll 0.178 \text{ ksi}$$

Step 7. Check stresses in stem at intersection with base. Neglect P/A due to P_{av} and weight of stem. Include moment due to P_{ah}, P_{av}, and W and sum moments at A. Use $LF = 2.0$;

$$M = 9.1(7.42 - 3) + 1.6(1.18 + 0.5) + 5.56(1.70) = 52.4 \text{ ft-kips/ft}$$

$$f = \frac{6(52.4)(2)(12)}{12(8.18 \times 12)^2} = 0.065 \text{ ksi} \ll 0.178 \qquad O.K.$$

The compressive stresses would be approximately

$$f_c = +\frac{P}{A} + \frac{6M}{bt^2}$$

which are by inspection negligible. Although this design may not be the most economical, it is adequate. For mass concrete members of this type, it is possible in some cases to use boulders as filler to reduce the quantity of cement paste.

12-10. WALL JOINTS

Joints may be constructed into a retaining wall between successive pours of concrete both horizontally and vertically. This type of joint, termed a *construction joint*, may be keyed as indicated in Fig. 12-19, although the surface can simply be cleaned and roughened. *Contraction* joints may be placed at intervals vertically, so that when

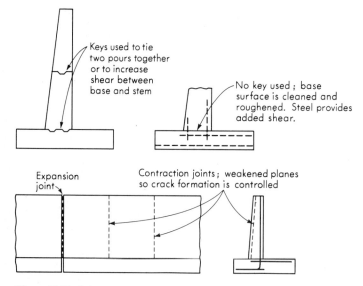

Figure 12-19. Joints.

shrinkage tensile stresses become great enough to tend to pull the concrete apart, the failure will occur across the joint, and the crack will appear as a neat line rather than jagged or haphazardly formed. The contraction joint is a weakened plane built into the concrete, so that the location of the failure is predetermined. Expansion joints are vertical joints which completely separate two portions of the wall. The reinforcing steel is generally continuous through all joints to maintain horizontal and vertical alignment. When the steel runs across the joint, one end of all the bars on one side of the joint are greased or sheathed so that, being unbonded, the design expansions can take place.

To take care of both shrinkage effects and differential settlements, long walls should have contraction joints at intervals of 25 to 40 ft (8 to 12 m). Vertical-expansion joints may also be placed along the wall at spacings of 60 to 100 ft (18 to 30 m). The actual space to allow at the joint for expansion will be difficult to evaluate, although the maximum theoretical space S_e can be computed as

$$S_e = \alpha T(L)$$

where α = thermal coefficient for concrete (about 0.000005 per degree Fahrenheit)

T = change in temperature from construction temperature

L = joint spacing

The wall must slide on the soil underlying the base, and simultaneously shear the soil behind the wall in order for any wall expansion (or contraction) to take place. These resisting forces may be large enough to cancel the expansion forces, and no expansion joint at all may actually be required.

12-11. DRAINAGE

It has been pointed out, in Chap. 11, that it is much more desirable to provide soil drainage than to design a retaining wall for the larger lateral pressure which will be induced if the backfill does not readily drain.

Drainage can be accomplished as in Fig. 12-20 by providing *weep holes* (pipes which permit passage of water from the backfill to the front) and/or *longitudinal*

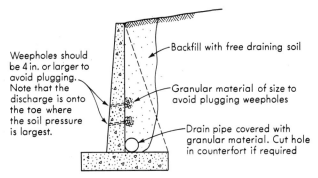

Weepholes should be 4 in. or larger to avoid plugging. Note that the discharge is onto the toe where the soil pressure is largest.

Backfill with free draining soil

Granular material of size to avoid plugging weepholes

Drain pipe covered with granular material. Cut hole in counterfort if required

If weepholes are used with a counterfort wall at least one weephole should be located between counterforts

Figure 12-20. Drainage of retaining walls.

drains along the back face. In either case a layer of free-draining granular material should be placed along the back of the wall and surrounding the entrance to the weep holes or drain-pipe openings. This material must possess adequate gradation so that the pipes are not clogged with leached-out material. Total reliance on weep holes is somewhat risky unless they are of sufficient size and are carefully protected with a filter on the backfill side so that they will not clog.

12-12. ABUTMENT WING AND RETAINING WALLS OF VARYING HEIGHT

Seldom is a long retaining wall of constant height except possibly when used in hydraulic structures. The conventional analysis considers a constant height, and it would appear that when the wall varies in height both nonplane-strain conditions and a certain amount of wall twist will develop. For the usual conditions of a change in wall height developed by a uniform slope, the wall is overdesigned sufficiently to absorb the twist both because the wall thickness is held constant over enough of the length to be adequate and because longitudinal steel for temperature and shrinkage requirements will normally be sufficient to carry the twisting moments. Where abrupt changes in height occur, one may use the mat program in the Appendix and treat the wall as a cantilever plate restrained against rotation and translation along the bottom and subjected to a triangular pressure diagram. The base-slab thickness will normally be of constant thickness; however, for retaining walls the width may be reduced to account for the smaller overturning moments.

Abutments are relatively short in plan; thus it would be more appropriate to use triaxial soil parameters ($0.9 \times$ plane strain ϕ). There are two approaches to the wing-wall design, as indicated in Fig. 12-21. One method is to separate the wing wall and the abutment proper by a joint which is watertight (to avoid loss of fines behind the wall) and self-supporting. The other method is to construct the wing wall mono-

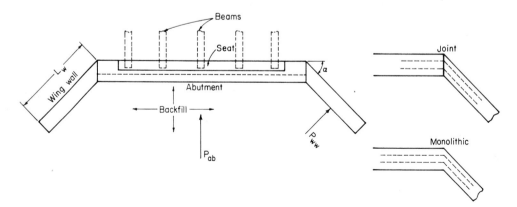

Figure 12-21. Bridge abutment and wing-wall earth pressure and methods of construction.

lithically with the abutment. If the wing wall is monolithic with the abutment, the joint must be designed for shear, tension, and twisting moment. The differential shear can be obtained by obtaining the total wing-wall force P_{ww} and direction, and the total abutment force P_{ab}. The shear S to be resisted is approximately obtained as

$$S = P_{ww} \cos \alpha - \frac{P_{ab}}{2}$$

The tensile force to be resisted is approximately

$$T = P_{ww} \sin \alpha$$

The approximate joint moment is

$$M = \frac{P_{ww} L_w}{2}$$

The wing-wall force and point of application will require computing forces for several unit-width sections for wing walls which vary in height and using statics to find the resultant P_{ww} and the point of application.

12-13. DESIGN OF A CANTILEVER RETAINING WALL

The design of a cantilever retaining wall consists in the following steps:

1. Obtain soil parameters ϕ and c for both backfill and base soil. Use cohesionless backfill in a zone slightly larger than the Rankine zone if possible.
2. Estimate the load factor (in the computer program this is FAC). Use at least 1.7 for cohesionless backfill in the full Rankine zone. Use more than 1.7 for cohesive backfill, or limited backfill zone. Limit the load factor to an equivalent $K_a = 1.10$ regardless of the backfill.
3. Select trial dimensions using Fig. 12-3 as a guide. If a computer program is used, reduce these dimensions by about 20 percent for input and let the computer optimize the design. Note that it is not necessary or always desirable to have a toe dimension.
4. Compute the stem thickness t based on wide-beam shear and take the critical section at the base-slab junction. Some persons have advocated taking the section at t from the base slab, but this is not conservative because the base and stem are almost always poured separately and regardless of how well the shear key is constructed, a weakened shear plane exists at the joint. Use the stem height and apply the load factor to the resulting shear force.
5. Compute the stem moments and reinforcing-steel requirements at several points so that the full amount of steel is not used throughout the wall height, since the moment varies as the third power of h measured down the wall. Use the load factor in this computation.
6. Compute overturning, sliding, and (if applicable) deep-seated shear failure as in

Example 12-2. Revise the base-slab dimensions as necessary. Find $\bar{x}$ and the eccentricity of the resultant on the base of the footing. The eccentricity should be $e \leq L/6$. Use the full wall height (stem + base slab + heel $\times \tan \beta$) for this computation. Include passive pressure in front of wall to increase sliding resistance if applicable.

7. Compute the bearing capacity of the footing considering that it is a strip and thus all shape factors = 1.0. Use Eq. (4-6) and depth and inclination factors from Table 4-3. For inclination factors use H = computed horizontal component of earth pressure; V = sum of the vertical forces of the retaining-wall system. Use $B' = B - 2e$ in Eq. (4-6) and where B appears in Table 4-3. The allowable bearing capacity is

$$q_a = q_{ult}/2 \text{ (cohesionless base soil)}$$

$$q_a = q_{ult}/3 \text{ (ϕ-c base soil)}$$

Increase the base-slab width if the actual soil pressure as computed by

$$q = \frac{V}{B}\left(1 \pm \frac{6e}{B}\right)$$

is larger than the allowable bearing capacity.

8. Establish the pressure profile along the footing base. Normally neglect the soil over the toe of the footing, and include the soil over the heel part. Include the weight of the base slab in the pressure profile, since it has been used to obtain V.

9. Using the pressure profile and the differential equations for shear and moment, compute the shear and bending moment at the intersection of the toe and stem and find the depth to satisfy wide-beam shear *at the intersection*. Apply the load factor from step 2 to both the shear and bending moment.

10. Using the pressure profile and the differential equations for shear and moment, compute the shear and bending moment at the center of the stem tension-reinforcing steel. This can be taken as 3.5 in (or 9.0 cm) from the back face of the stem. Check the depth for wide-beam shear *at the back face* and find the reinforcing steel necessary for bending moment. This steel goes near the top surface of the heel slab. Apply the load factor from step 2 to both the shear and bending moment.

11. Dowels will be required to join the stem and base slab; however, using bent bars placed on line with the toe steel and of A_s = required value at stem base will be satisfactory.

AASHO (1973) in Art. 1.4.8B states: "The rear projection or heel of base slabs shall be designed to support the entire weight of the superimposed materials, unless a more exact method is used." It would appear that step 8 would satisfy the requirement of a more exact method. The computer program in the Appendix considers this (HLOSS > 0) but does not use the current ACI shear and bending stresses. Because of the unlikely occurrence of this mode of failure, the shear stresses are increased approximately 55 percent and the bending moment is divided by the load factor and compared with the moments from steps 9 and 10. With this adjustment, the most critical values either from the conventional procedure or from this alternative are used for design.

Example 12-4. Design a cantilever retaining wall for the conditions of Fig. E12-4a.

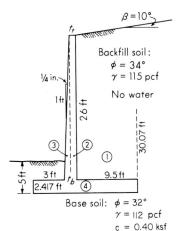

Base soil: $\phi = 32°$
$\gamma = 112$ pcf
$c = 0.40$ ksf   **Figure E12-4a**

Other data: $f'_c = 3$ ksi $f_y = 60$ ksi

$\gamma = 0.15$ kcf $LF = 1.8$

Batter on front face of wall $= 1:48$

Top $= 16$ in

SOLUTION. (Values shown in sketch from optimizing using computer program.) Estimate 3.5 in from cgs to soil interface to allow approximately 3.0 in of clear steel cover.

Step 1. Establish stem dimensions; round to even values.

$$K_a = 0.294 \quad \text{[Eq. (11-7)]}$$

Stem uses $H = 26$ ft

$$P_a = 0.5(0.115)(26)^2(0.294) = 11.43 \text{ kips/ft}$$

$$P_{ah} = 11.43 \cos 10° = 11.25 \text{ kips/ft}$$

$$v_c = 2\phi\sqrt{f'_c} = 0.09311 \text{ ksi}$$

$$t = \frac{11.25(1.8)}{(0.093)(12)} = 18.14 + 3.5 = 21.6$$

$$\text{Top} = 21.6 - 26(0.25) = 15.1 \quad \text{Use 16 in}$$

To maintain even dimensions

$$t = 16 + 26(0.25) = 22.5 \quad \text{Use 23 in}$$

The computer program next computes the bending moment at the 0.1 points and steel requirements. Here this is done in step 7.

Step 2. Compute stability of wall.

$$H' = 26 + 2.417 + 9.5 \sin 10° = 30.07 \text{ ft}$$

$$P_{ah} = \left(\frac{30.07}{26}\right)^2 11.25 = 15.05 \text{ kips/ft}$$

$$P_{av} = 15.05 \tan 10° = 2.65 \text{ kips}$$

From which we can set up a table to obtain $\sum F_v$ and $\sum M$ (refer to Fig. E12-4a).

Part	Weight	Arm	Moment
1	$0.5(26 + 27.65)(0.115)9.5 = 29.31$	9.67	283.3
2	$1.33(26)0.15 = 5.19$	4.25	22.1
3	$0.59(26)(0.15)0.5 = 1.15$	3.39	3.9
4	$2.417(14.417)0.15 = 5.23$	7.21	37.7
P_{av}		2.65 14.42	38.2

$$\sum F = 43.53 \quad \sum M_r = 385.2$$

$$M_0 = 15.05(30.07/3) = 150.9$$

$$F_0 = \frac{385.2}{150.9} = 2.55 \quad \text{overturning} > 1.5 \quad O.K.$$

The safety factor against sliding will use at least 3 ft or 1 m of depth if $D >$ than this or the full depth if specified (in computer program). Computer program does not increase footing width for F (sliding) < 1.5, as one may use a heel key as an alternative. Use $\tan \phi$ for sliding but $0.67c$.

$$F_r = 43.53 \tan 32° + 0.67(14.42)(0.4) + 0.5(0.112)(3)^2 K_p$$
$$+ 2.0(0.4)(3)\sqrt{K_p}$$
$$= 27.2 + 3.85 + 5.97 = 37.02 \text{ kips}$$

$$F = \frac{37.02}{15.05} = 2.46 > 1.5 \quad O.K. \text{ for sliding}$$

$$\bar{x} = \frac{385.2 - 150.9}{43.53} = 5.38 \text{ ft}$$

$$e = 14.42/2 - 5.38 = 1.83 \text{ ft} < \frac{L}{6}$$

Step 3. Compute bearing capacity. Use equations in Table 4-3 and actual soil pressure.

$$q_{ult} = cN_c d_c i_c + qN_q d_q i_q + \tfrac{1}{2}\gamma BN_\gamma d_\gamma i_\gamma \qquad \text{[Eq. (4-6)]}$$

$$B' = 14.417 - 2(1.83) = 10.76$$

$$N_c = 35.5 \qquad N_q = 23.2 \qquad N_\gamma = 20.8$$

$$i_c = 0.42 \qquad i_q = 0.44 \qquad i_\gamma = 0.309$$

$$d_c = 1.19 \qquad d_q = 1.13 \qquad d_\gamma = 1.0$$

$$q_{ult} = 0.4(35.5)(0.42)1.19 + 5(0.112)(23.2)1.13(0.44) + 0.5(0.112)(10.8)20.8(0.309)$$

$$= 7.1 + 6.5 + 3.9 = 17.5$$

$$q_a = 17.5/3 = 5.8 \text{ ksf}$$

Actual soil pressure

$$q = \frac{V}{L}\left(1 \pm \frac{6e}{L}\right) = \frac{43.5}{14.42}\left[1 \pm \frac{6(1.83)}{14.42}\right] = 3.02(1 \pm 0.76)$$

$$= 5.3 \text{ ksf max at toe}$$

$$= 0.7 \text{ ksf min at heel}$$

Step 4. Compute base-slab shear and bending moments. Refer to Fig. E12-4b.

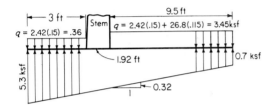

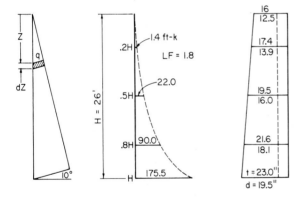

Figure E12-4b

For toe at stem face $x = 3.00$

$$q = 5.3 - 0.36 - 0.32x$$

$$V = 4.94x - \frac{0.32x^2}{2} = 13.4 \text{ kips}$$

$$M = \frac{4.94x^2}{2} - \frac{0.32x^3}{6} = 20.8 \text{ ft-kips}$$

For heel at approx. cg of tension steel

$$x = 9.5 + 3.5/12 = 9.79 \text{ ft for moment} \qquad \text{Use 9.5 for shear}$$

Use average height of soil on heel for downward pressure; include $P_{av} = 2.65$ kips

$$q = -0.7 + 3.45 - 0.32x$$

$$V = 2.75x - \frac{0.32x^2}{2} + P_{av} = 14.3 \text{ kips}$$

$$M = \frac{2.75x^2}{2} - \frac{0.32x^3}{6} + P_{av}x = 107.7 \text{ ft-kips}$$

Step 5. Check base-slab shear stress using largest base V, $LF = 1.8$, $d = 2.417 - 0.29 = 2.13$ ft

$$v = \frac{14.3(1.8)}{12(25.5)} = 0.084 < 0.093 \qquad O.K.$$

Note we could reduce the base slab by about 1 to 1.5 in. Leave at 2 ft 5 in.

Step 6. Compute toe (bottom of footing) and heel (top of footing) reinforcing-steel requirements

$$p_{max} = 0.016 \qquad \text{Table 8-1}$$

$$p_{min} = 200/f_y = 0.0033$$

$$0.5a = \frac{A_s f_y}{0.85 f'_c b} = 0.98 A_s$$

For heel

$$A_s(25.5 - 0.98 A_s) = \frac{107.7(12)(1.8)}{0.9(60)}$$

$$A_s^2 - 26.02 A_s = -43.96$$

$$A_s = 1.81 \text{ sq in/ft} \qquad p = 0.006$$

For toe

$$A_s^2 - 26.02 A_s = -8.49$$

$$A_s = 0.34 \text{ sq in} \qquad p = 0.0011 < 0.0033$$

$$A_s = 0.0033(12)(25.5) = 1.02 \text{ sq in/ft}$$

Step 7. Compute stem steel at top, $0.5H$, $0.8H$, and H using $LF = 1.8$.

Point	M, ft-kips	Wall thickness	d	A_s*	p
0	0	16.0	12.5	0.5	0.0033
0.5H	21.97	19.5	16.0	0.64	0.0033
0.8H	90.00	21.60	18.1	1.18	0.005
H	175.8	23.0	19.5	2.26	0.010

* sq in/ft

From this it is evident minimum requirements control top half of wall.

Use 3 No. 8 bars/ft ($A_s = 2.37$ sq in) from the base slab to $26 - 0.8(26) + d$ (per ACI 12.1.3)

$$y = 5.2 + 1.5 = 6.7 \text{ ft}$$

Run 1 No. 8 bar full wall height

Run 2 No. 4 bars/ft full wall height on front face to carry part of horizontal shear $0.4(60)(2) \times (0.2) = 9.6$ kips $> 15.05/2$, also to hold longitudinal shrinkage and temperature steel in place.

For longitudinal shrinkage and temperature steel

Use $\dfrac{16 + 19.5}{2}(12)(0.002) = 0.43$ sq in/ft in top half

Use 2 No. 5 at one on each face

Use $\dfrac{19.5 + 23}{2}12(0.002) = 0.51$ sq in/ft in bottom half

Use 2 No. 5 at one on each face.

Step 8. Design sketch (see Fig. E12-4c).

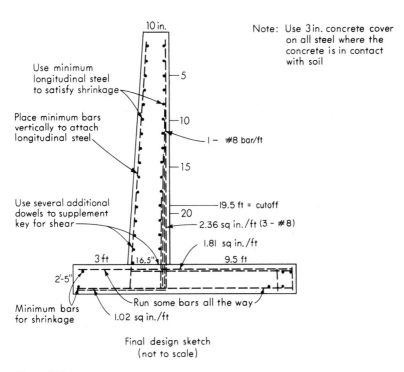

10 in.

Note: Use 3 in. concrete cover
on all steel where the
concrete is in contact
with soil

Use minimum
longitudinal steel
to satisfy shrinkage

5

Place minimum bars
vertically to attach
longitudinal steel

10

1 — #8 bar/ft

15

Use several additional
dowels to supplement
key for shear

19.5 ft = cutoff

20

2.36 sq in./ft (3 - #8)

1.81 sq in./ft

3 ft   16.5"   9.5 ft

2'-5"

Minimum bars
for shrinkage

Run some bars all the way

1.02 sq in./ft

Final design sketch
(not to scale)

Figure E12-4c

12-14. DESIGN OF A COUNTERFORT RETAINING WALL

The design of a counterfort retaining wall is similar in many respects to that of a cantilever retaining wall as far as computing overturning and sliding stability, bearing capacity, and soil pressure beneath the base slab are concerned. The retaining-wall computer program in the Appendix can be used to obtain the wall stability, bearing capacity, soil pressure, and toe shear and moments for a counterfort wall.

From this point an approximate design would proceed as follows:

1. Divide the stem into several horizontal zones and obtain longitudinal bending

moments using Fig. 12-13. About three strips—top, midheight, and at base—should be sufficient. Use these moments to find required horizontal reinforcing steel and run the required amount the full length on front and back face.

2. Divide the stem into several vertical strips, compute the vertical bending moments and shear at the base of the stem (junction with base slab), and check stem thickness for shear adequacy. Use Fig. 12-14 for this analysis. Consider using cutoff points for the vertical steel.

3. Divide the heel slab into several longitudinal strips, and using the pressure diagrams of Fig. 12-15 and the moment equations of Fig. 12-13 to obtain the longitudinal bending moments. Use these moments to find the longitudinal reinforcing steel in the base slab.

4. Treat the base slab as a cantilever similar to the cantilever retaining wall, and find the shear at the back face of the stem and bending moment at this location. Revise the base thickness if necessary for shear requirements. Use the bending moment to compute required perpendicular heel-slab reinforcing steel.

5. Treat the toe of the base slab identical to a cantilever retaining wall.

6. Analyze the counterforts. They carry (refer to Fig. 12-14) a shear of Q_c computed as follows:

$$Q_{total} = qLH \text{ for each full counterfort spacing}$$

$$Q' = 0.2LH = \text{shear carried along base of wall}$$

$$Q_c = qLH - 0.2qLH = 0.8qLH$$

$$= \text{lateral wall shear carried by counterfort}$$

Assume a linear increase of shear (based on shape of pressure diagram) with depth to obtain tension steel to tie the counterfort to the wall. Using this same linear increase find the location of the shear resultant and compute the moment the counterfort must carry into the base slab as shown in Fig. 12-22. Use bent reinforcing bars so that embedment depth is not critical.

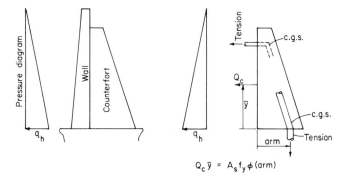

$$Q_c \bar{y} = A_s f_y \phi \text{ (arm)}$$

Figure 12-22. Structural design of counterfort wall. Make thickness to contain reinforcing steel with adequate cover.

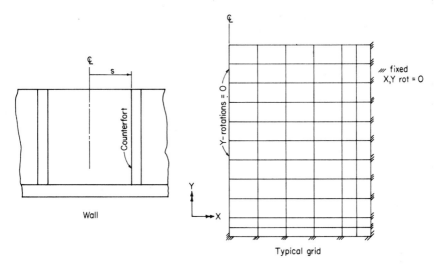

Figure 12-23. Typical layout for using mat program to solve a plate fixed on three edges. Note use of closer grid spacing at fixed edges to develop plate curvature better.

The alternative method of designing a counterfort retaining wall is to treat the system (wall and heel slab) as plates fixed on three edges with the appropriate pressure diagrams—hydrostatic in form for the wall and the combination of soil pressure and soil overlying the heel slab as in cantilever-retaining-wall analysis (e.g., see Fig. 12-10). Use a card generator to develop appropriate load matrices (P matrix) and member data cards for the two members. Use the mat program in the Appendix with I3EDGE = 1. It will be necessary to establish boundary conditions for this problem. It will be necessary also to solve half the plate as in Fig. 12-23 to minimize the size of the matrix to be reduced. Since the plate is approximately symmetrical with respect to the vertical axis, one would set the Y rotations = 0.0 at the center line; at the fixed edges all three displacements = 0.0 (X and Y rotations and Z translation). A similar situation is used to describe the heel slab. Typical gridding and partial coding for this problem is shown in Fig. 12-23. Note the use of closer grid spacing near the fixed edges. It is possible to account for varying thickness of the wall by dividing the wall height into several zones and using a constant (but different for each zone) thickness in the zone (switch in computer program ITHICK = 1). The use of a varying thickness instead of a constant thickness may vary the computed bending moments 20 percent or more [see also Jofriet (1975)].

12-15. BASEMENT OR FOUNDATION WALLS; WALLS FOR RESIDENTIAL CONSTRUCTION

Walls for building foundations, and basement walls for both residential construction and larger structures require the same design considerations. Normally these walls are backfilled with any material available at the site and in a very limited backfill zone. The tops of these walls are usually restrained from lateral movement so that active-pressure conditions are not very likely to be obtained. If active-pressure conditions are obtained and especially if the backfill is cohesive, the lateral-wall deformations would be likely to be noticeable. For this reason the lateral-pressure coefficient should be taken for a K_0 condition. The structural design would proceed as for other types of retaining walls.

Backfill for residential basement walls should be carefully placed and of good quality and preferably granular, with the wall provided with a perimeter drain placed on the wall footing. This type of construction will ensure a dry basement and is more economical than later having to dry the basement by digging out and replacing the fill and/or installing a perimeter drain. For much residential construction, however, the backfill for walls consists in any material available—usually material excavated from the basement and including large quantities of wood fragments from cutting the framing and other materials to proper size. The backfill is seldom compacted; however, since many basement walls are not designed and consist in concrete blocks and mortar, the lateral earth pressure they can sustain is rather low. It is not uncommon to observe walls propped into place in subdivisions after heavy rains have densified and saturated the cohesive backfill. It is never possible to push the walls back into place; thus many basement walls remain permanently bulged.

12-16. REINFORCED-EARTH RETAINING STRUCTURES

A relatively recent method of providing a retaining structure is to use the concept of reinforced earth [Vidal (1969), Lee et al. (1973)]. Figure 12-24 illustrates the general concept. The earth is reinforced with strips of dissimilar material, metal, fiber, wood, etc., which effectively imparts "cohesion" to the retained mass. The design of a reinforced-earth wall consists in evaluating the active lateral earth pressure at the various levels and providing a reinforcing strip which can carry that amount of tension as developed by the length of the strip protruding behind the "active Rankine" earth zone. It is evident that the strips near the top protrude beyond the active zone much less than the lower strips; thus for equal-length reinforcing strips there is compensation.

The wall surface, or "skin," can be almost any material which is weather/corrosion-resistant and reasonably flexible so that the skin deforms to fit the soil deformation. Thin metal or thin precast concrete plates can be used. Metal plates may be treated on the back with asphalt compounds to increase the service life. The skin must be fitted by bolting or welding in the case of metal plates. Tongue-and-groove or other means are used with concrete so that the backfill does not wash through and fail the wall. The reinforcing elements may be wire mesh, steel cables, steel strips, aluminum-alloy strips, or plastic fabrics.

The soil backfill should be a granular material for free-draining and angle-of-internal-friction characteristics. The former reduces the lateral pressures and the latter enables sufficient friction to be developed on the reinforcement (analogous to bond stress for reinforced concrete) that the system is held in place.

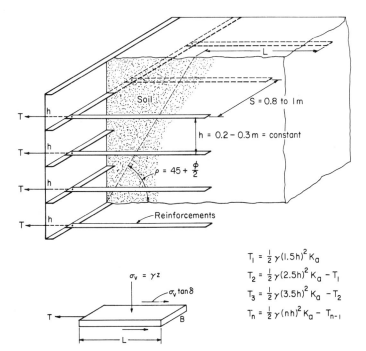

$$T_1 = \tfrac{1}{2}\gamma(1.5h)^2 K_a$$

$$T_2 = \tfrac{1}{2}\gamma(2.5h)^2 K_a - T_1$$

$$T_3 = \tfrac{1}{2}\gamma(3.5h)^2 K_a - T_2$$

$$T_n = \tfrac{1}{2}\gamma(nh)^2 K_a - T_{n-1}$$

Figure 12-24. Reinforced-earth concept.

Reinforced-earth walls can be used with surcharges on the backfill as would be obtained from construction of a highway across the backfill.

Example 12-5. Design the reinforcement and skin for the wall of Fig. 12-24 if $H = 1.2$ m and $h = 0.3$ m (to illustrate method). Other data: $\gamma = 17.1$ kN/m^3; $\phi = 32°$; take $\delta = 25°$; $s = 1$ m.

SOLUTION. *Step 1.* Compute tensions in reinforcements

$$T_1 = 0.5(17.1)(0.45)^2 0.307 = 0.53 \text{ kN/m}$$

$$T_2 = 0.5(17.1)(0.75)^2 0.307 - 0.53 = 0.95 \text{ kN/m}$$

$$T_3 = 0.5(17.1)(1.05)^2 0.307 - 0.95 = 1.94 \text{ kN/m}$$

$$T_4 = 0.5(17.1)(1.2)^2 0.307 - 1.94 = 1.84 \text{ kN/m}$$

Step 2. Find A and L of reinforcements. Use allowable steel stress $= 20$ ksi or $1,406.2$ kg/cm^2

$$A, \text{cm}^2 = \frac{1.84}{1,406.2(98.07)} = 0.000013 \text{ cm}^2 \qquad \text{area required}$$

The usual situation is one where the reinforcement is arbitrarily selected. Use strips 7.5 cm wide × 0.31 cm thick (3 in × 1/8) with $A = 2.32$ cm^2.

The reinforcements must be at least long enough to go past the Rankine active zone.

$$L_R = 1.2 \tan(45 - \phi/2) = 0.67 \text{ m at top}$$

Check T_1 (use friction both sides of strip and only the length beyond the Rankine zone)

$$T_s = \gamma h \tan \delta \, (2)(L - L_R)w$$

$$T_s = 17.1(0.3)(\tan 25)(2)(L - 0.67)0.075$$

$$= 0.359(L - 0.67)$$

Solving for L using $F = 1.2$ for upper layer of reinforcement

$$0.359(L - 0.67) = 0.53(1.2)$$

$$L = 1.77 + 0.67 = 2.44 \text{ m}$$

Check T_4 for $L = 2.44$ m and $L_R = 0.0$

$$T_s = 17.1(1.2)(\tan 25)(2)(0.075)(2.44)$$

$$= 3.50 > 1.84(1.2) \qquad O.K.$$

Use reinforcement strips 7.5×0.31 cm $\times L = 2.5$ m.

Step 3. Skin design

For the maximum forces involved (1.84 kN/m) any metal thickness of, say, 0.31 cm and above can be used. This thickness is to alleviate corrosion.

Note that the method of analysis used here is not valid for cables or wire mesh. For these materials it would be necessary to evaluate the pullout resistance in the laboratory to effect a wall design.

PROBLEMS

12-1. For the assigned retaining wall problems listed in Table P12-1 (Fig. P12-1) analyze the following as required:

(*a*) Draw shear and moment diagrams for the stem.

(*b*) Analyze the wall for overturning and sliding stability and adjust the base and toe dimensions until the wall is stable.

(*c*) For the base dimensions which provide a stable wall, draw shear and moment diagrams for the toe and heel. Use three points ($x = 0$ is one) to plot the diagrams.

(*d*) Compute the required steel reinforcement for the stem and base. Make a final sketch of your design. Also check and adjust concrete thickness to satisfy shear. Use $f'_c = 3$ ksi; $f = 60$ ksi, except as noted.

Table P12-1

Prob.	H, ft or m	γ_1*	γ_2	β, deg	ϕ_1, deg	ϕ_2, deg	c†	D, ft or m
1a	15 ft	115	125	15	34	15	900	3.5 ft
1b	6 m	18.0	20.1	0	34	20	24.0	1.2 m
1c	10 ft	100	125	0	26	20	0	3.0 ft
1d	4 m	18.1	17.3	10	28	40	0	1.3 m
1e	30 ft	105	130	20	32	20	750	3.5 ft
1f	5 m	14.9	20.4	25	33	30	28.7	1.5 m
1g	27 ft	112	122	0	26	0	1,200	5.0 ft
1h	9 m	18.1	19.6	15	34	20	31.5	1.1 m

* pcf or kN/m³ by inspection of data
† psf or kPa by inspection of problem data.

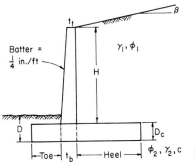

Figure P12-1

12-2. Design a counterfort wall for the following conditions:

$$f'_c = 3.5 \text{ ksi}, \quad f_y = 60 \text{ ksi}, \quad D = 4.5 \text{ ft}, \quad H = 39.0 \text{ ft}$$

Backfill: $\gamma = 118$ pcf, $\phi = 36°$, $c = 0.40$
zone limited to heel slab + 3 ft backfill placed
against ϕ-c soil, which is at a cut about 10°
from vertical.
Base soil: $\gamma = 115$ pcf, $\phi = 34°$, $c = 600$ psf
Surcharge = 500 psf.
Assume you can use 3.0 ft of toe soil for passive pressure.

12-3. Design a counterfort wall for the following conditions: $f'_c = 211$ kg/cm², $f_y = 4,219$ kg/cm²,
$D = 1.4$ m, $H = 11.5$ m
Backfill: $\gamma = 18.8$ kN/m³, $\phi = 37°$, $c = 26.2$ kPa
Assume fill in about a half Rankine zone.
Base soil: $\gamma = 18.2$ kN/m³, $\phi = 31°$, $c = 32.4$ kPa
Surcharge: 24 kPa
Assume that you can use passive pressure in front of the wall for sliding stability.

12-4. Check Example 12-2 for other circle centers, say, 2, 4, 6 ft to left of O and $y = O \pm 2$ ft.

12-5. Redo Example 12-2 using SI units for several trial-circle centers.

12-6. What is F for the other two alternatives of Example 12-1?

12-7. What size reinforcing strips would be required in Example 12-5 if $H = 8$ m and $h = 0.5$ m?

THIRTEEN

SHEET-PILE WALLS—CANTILEVERED AND ANCHORED

13-1. INTRODUCTION

This chapter presents methods to analyze and design cantilever and anchored sheet-pile structures (Fig. 13-1). Types of materials used for sheet piles are also given as background for this chapter and the following chapters, which consider other foundation structures built with sheetpiling.

This chapter will present briefly the classical methods of analyzing cantilever and anchored sheet piles, then proceed to the finite-element method, which has been shown by the author [Bowles (1974a)] to be the most rational method of analysis of sheet-pile walls. This method can be extended to obtain the best currently available method of analyzing braced excavations, as will be shown in Chap. 14. The finite-element is quite similar to the finite-difference (sometimes called beam-column) method used by some organizations. The finite-element method has the advantage over the classical methods of directly applying "moment reduction" during the analysis and of obtaining the deflections and bending moments at the nodal points of the wall. The classical methods made simplifying assumptions, and implicitly assumed dredge-line deflections were zero and the moments computed were, generally, too large and required moment reduction for design. The finite-element procedure also gives directly the soil pressure required for wall stability for comparison with the bearing ability of the soil.

Sheet-pile walls are widely used for both large and small waterfront structures ranging from small pleasure-boat launching facilities to large oceangoing-ship dock

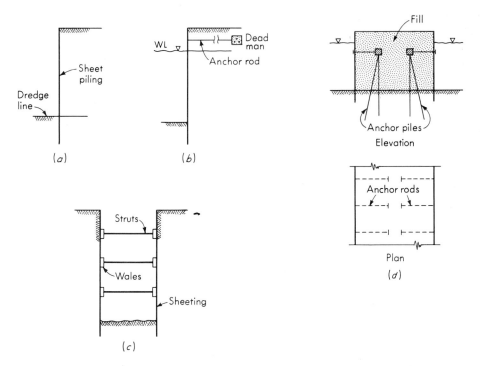

Figure 13-1. Sheetpiling structures. (*a*) Cantilever sheetpiling; (*b*) anchored sheetpiling; (*c*) braced sheeting; (*d*) anchored bulkheads.

facilities. Piers jutting into the harbor consisting in two rows of sheetpiling as in Fig. 13-1*d* are widely used. Sheetpiling is also used for slope stability and erosion protection.

13-2. SOIL PROPERTIES FOR SHEET-PILE WALLS

The sheet-pile problem is again one of lateral pressure. The concept used in Chap. 11 can be directly applied to this problem. Plane-strain conditions are appropriate for long walls. If the wall is relatively short or a cellular structure is considered, triaxial strain conditions are more appropriate.

Consolidated-drained (CD) conditions will be obtained from cohesionless soil tests. Where the embedment soil and/or backfill is cohesive, consolidated-undrained (CU) or unconsolidated-undrained (UU) tests are required. As before, if triaxial tests are used for plane-strain conditions, one may increase the ϕ angle by about 10 percent for use in the lateral-pressure equations. Tests on the appropriate state of the backfill are necessary (remolded where fill is needed, undisturbed if wall retains in situ material).

Either the Rankine or Coulomb active-lateral-pressure equations may be used to compute the lateral pressure of the backfill against the wall. The Coulomb pressure equations are recommended because the sheet-pile wall is a flexible structure with large lateral deformations relative to, say, the retaining walls of Chap. 12. With these large lateral deformations there is certain to be slip between soil and wall with wall friction and/or wall adhesion being developed. In the classical solution, passive pressure is computed for the soil below the dredge line (Fig. 13-1), which resists horizontal wall movement. The passive pressure may be computed by either the Rankine or Coulomb pressure equations. The alternative passive-earth-pressure values using the theory of plasticity are not recommended for this problem, since with the slope β of the soil being zero, the values are little different from the Coulomb values. In all earth-pressure computations use effective stresses.

The angle of wall friction may be estimated from Table 11-6; however, for Z or deep web shapes part of the friction is soil to soil and about half is soil to sheet pile. Where this is the case use

$$\tan \delta' = \frac{\tan \phi + \tan \delta}{2}$$

The δ value is from Table 11-6, and ϕ is the angle of internal friction of backfill.

The finite-element procedure uses the active-lateral-earth-pressure zone in the backfill as in the classical methods. In the passive zone, the concept of modulus of subgrade reaction is used. The author has shown [Bowles (1974a)] that this model is reasonably correct by using it to analyze full-scale field walls and for reanalyzing model sheet-pile wall studies of Tschebotarioff (1949) and of Rowe (1952). Rowe's model walls were constructed by filling both sides of the wall simultaneously, then excavating the front side. This creates an "ideal" wall which is easy to model by the finite-element procedure in spite of the fact that the subgrade modulus must be estimated. Tschebotarioff's walls were built and backfilled as one would in the field, i.e., constructing the wall, then backfilling in stages. These walls could be modeled but with more difficulty, as one would expect. The most important advantages of the finite-element method of solution are:

1. Estimates of deflection
2. Rational evaluation of changed wall geometry (loss of dredge line, etc.)
3. Estimate of actual lateral passive pressure to see if the solution is possible
4. Effects of large increases in embedment depth
5. Ease of handling ϕ-c soils. Most texts consider either ϕ or c soils. ϕ-c soils are extremely difficult to use in classical sheet-pile analysis, as are stratified soils.

Estimates of the modulus of subgrade reaction can be obtained from equations given in Sec. 9-7. In general, the value will increase in depth according to Eq. (9-16):

$$k_s = A + BZ^n \tag{9-16}$$

Lateral pressure due to surcharges can be analyzed using Eq. (5-7) and the computer program in the Appendix in the method outlined at the end of Sec. 11-10.

13-3. TYPES OF SHEETPILING

Sheetpiling materials may be of timber, reinforced concrete, or steel. Design stresses are usually higher than in building contruction and may be from 0.6 to $0.9F_y$ for steel and wood and as much as $0.75f'_c$ for unfactored loads. Actual design stresses depend on engineering judgment, effect of a wall failure, and local building codes.

Timber Sheetpiling

Timber sheetpiling is used for short spans, light lateral loads, and commonly, for temporary structures in the form of braced sheeting. If it is used in permanent structures above water level, it requires preservative treatment, and even so, the useful life is relatively short.

Timber piling probably finds its greatest use as braced sheeting for temporary retaining structures in excavations.

Driving of wood sheeting is somewhat troublesome, since a driving cap is required, and driving in hard soil with large gravel tends to split the piling. The sheeting may be pointed, generally, as shown in Fig. 13-2, and placed so that the piling tends to wedge against the previously driven pile. The tongue-and-groove joints indicated will provide a reasonably well-jointed wall only if small stones or soil does not get wedged into the grooves.

Reinforced-Concrete Sheet Piles

These sheet piles are precast concrete members, usually with a tongue-and-groove joint. They are designed for the computed service stresses, but the handling and driving stresses, which may be considerable because of their weight, must also be taken into account. The points are usually cast with a bevel, which tends to wedge the pile being driven against the previously driven pile. Typical dimensions are shown in Fig. 13-3, from which it is seen that the piles are relatively bulky, and thus displace a large relative volume of soil. This large volume displacement of soil tends to increase the driving resistance. The relatively large sizes, coupled with the high unit weight of

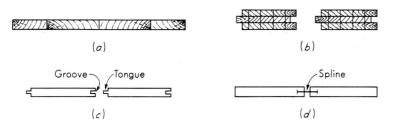

Figure 13-2. Wood sheet piles. (*a*) Ends butted together; (*b*) fabricated tongue and groove (Wakefield); (*c*) milled tongue and groove; (*d*) metal spline to fasten adjacent sheeting together.

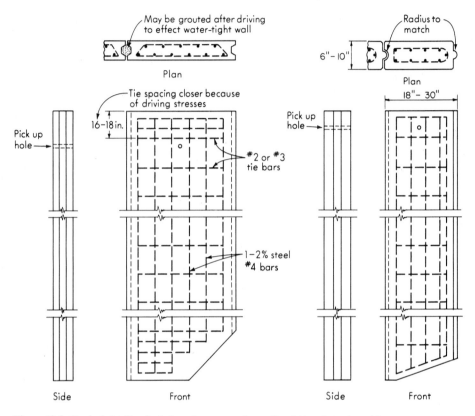

Figure 13-3. Typical details of reinforced-concrete sheet piles. [*After PCA (1951).*]

concrete, means that the piles are quite heavy and may not be competitive with other types of piles unless they are cast in close proximity to the job site.

By cleaning and grouting the joints after driving, a reasonably watertight wall may be achieved. If the wall is grouted, however, it may be necessary to provide expansion joints at intervals along the wall.

Steel Sheetpiling

Steel sheetpiling is the most common type used because of several advantages over other materials. Principal advantages are as follows:

1. It is resistant to high driving stresses as developed in hard or rocky material.
2. It is of relatively light weight.
3. It may be reused several times.
4. It has a long service life either above or below water with modest protection [NBS (1962)].
5. It is easy to increase the pile length by either welding or bolting.
6. Joints are less apt to deform when wedged full with soil and stones during driving.

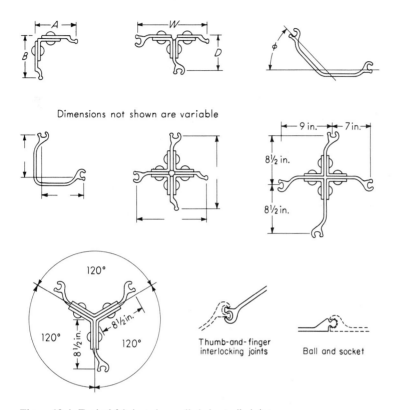

Dimensions not shown are variable

Thumb-and-finger interlocking joints

Ball and socket

Figure 13-4. Typical fabricated or rolled sheet-pile joints.

Steel sheetpiling is available in several shapes and joints (see Table A-3 in the Appendix).

The arch web and Z piles are used to resist large bending moments, as in anchored or cantilever walls. The deep-arch web and Z piles are used in cases where the larger bending moments are to be resisted. Where the bending moments are somewhat less, the shallow-arch piles with corresponding smaller section moduli can be used. Straight-web sheet piles are used where the web will be subjected to tension, as in cellular cofferdams.

To form cofferdams, the sections may be made up or formed into certain standard joints, such as Ts, Ys, and crosses, to effect the joining of several cells or to change directions. Figure 13-4 indicates several joints which are available.

13-4. SAFETY FACTORS FOR SHEET-PILE WALLS

The concept of safety factor with sheet-pile walls is somewhat of a misnomer, since it is not clear just what it means. If we increase the computed embedment depth by 50 percent, this does not apply a safety factor of 1.5 to the problem. That is, the wall is not 50 percent safer, nor does it mean we can increase the lateral pressure 50

percent—we may even be able to increase the lateral pressure 100 percent. We would
have to analyze the new conditions to see just what the "safety" is. Consider also an
example where the computed embedment depth is 1 m. Suppose we increase this to
1.5 m (this is considered as "$F = 1.5$" by many persons). Now suppose the dredge
line is lowered by 0.5 m. In all likelihood the wall will fail, since the active lateral
force increases by the square of the depth above the dredge line. What we should do
here is analyze a change in dredge line of 0.5 m and see what the embedment should
be for this new geometry. The problem of arbitrarily assigning a number to increase
the embedment depth (or divide the passive-pressure coefficient with) as done in the
past is not recommended. The embedment depth must be increased somewhat from
the value required for stability from the current loads and geometry (which is the
practical application of "safety factor" for sheet-pile walls). It is suggested, however,
that the increase in embedment depth be done by analyzing:

1. Some increase in lateral pressure either as an increase in soil unit weight, application of a surcharge, or a decrease in backfill soil parameters based on an engineering judgment of the total problem
2. Some loss of dredge line

and using the most critical embedment depth from these several analyses. With the
finite-element approach, this is very simple to do—with the older classical methods
this is quite difficult, and it is expected that the conventional method of an arbitrary
depth increase will be the extent of applying a safety factor for those using the latter
methods.

The conventional methods use

$$F = 1.2 \text{ to } 1.5 \ (\times \text{ embedment})$$
$$F = 1.5 \text{ to } 2.0 \ (K_p/F \text{ and cohesion}/F)$$

It is not likely that ϕ and c would be this far in error, but the application of F on K_p
and c will certainly compute a larger embedment depth.

13-5. CANTILEVER SHEETPILING

Sheet piles classified in this group depend on an adequate embedment into the soil
below the dredge line so that a driven line of sheeting acts as a wide cantilever beam
(a unit width of wall is treated as an individual cantilever beam) in resisting the
lateral earth pressures developed above the dredge line. These piles are economical
only for moderate wall heights, since the required section modulus increases rapidly
with increase in wall height; the bending moment increases with the cube of the
cantilevered height of the wall. The lateral deflection of this type of wall, because of
the cantilever action, will be relatively large. Erosion and scour in front of the wall,
i.e., lowering of the dredge line, should be controlled, since stability of the wall
depends primarily on the developed pressure in front of the wall.

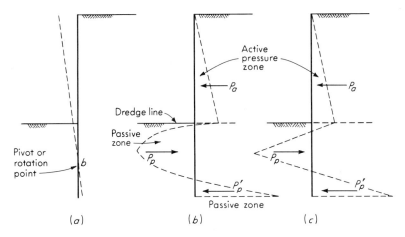

Figure 13-5. (a) Assumed elastic line of the sheetpiling; (b) probable and as obtained in finite-element solution qualitative soil-pressure distribution; (c) simplified pressure diagram for computational purposes (granular soil and no water as shown).

In the usual design of cantilever sheetpiling, simplifying assumptions as indicated in Fig. 13-5c are made. Design procedures will be developed for both granular soils $(c = 0)$ and cohesive soils $(\phi = 0)$. For ϕ-c soils the principles considered may be combined, but with considerable difficulty; this will be left as an exercise for the reader. The Coulomb approach considering wall friction is recommended, although the Rankine approach (with wall friction $= 0$) is widely used.

Cantilever Sheetpiling in Granular Soils

The solution of this problem is to assume that the pile is subjected, on the backfill side, to active pressure to the dredge line. Under the influence of the active pressure the wall tends to rotate, developing passive pressure in front of the wall and active pressure behind the wall. At the pivot point b of Fig. 13-5a, the soil behind the wall goes from active to passive pressure, with active pressure in front of the wall for the remainder of the distance to the bottom of the pile. For computing active and passive pressures, it is convenient to treat the soil mass above the point under consideration as a surcharge.

Since the design of this type of wall requires the solution of a fourth-degree equation for the depth of embedment, it is convenient to derive the equation once for all. This is not too difficult for conditions shown in Figs. 13-5c and 13-6 when it is assumed soil below the dredge line has the same ϕ angle as the backfill above the dredge line. Figure 13-6 identifies the needed terms.

If the method of dividing the passive earth-pressure coefficients by the safety factor is to be followed, the K_p and K'_p terms should be modified before any computations are made.

With the terms defined and shown in Fig. 13-6, a general solution can now be obtained for the sheet-pile wall in a cohesionless soil. First, all the forces above point O should be replaced by a single force resultant R_a located a distance $\bar{y}$ above this

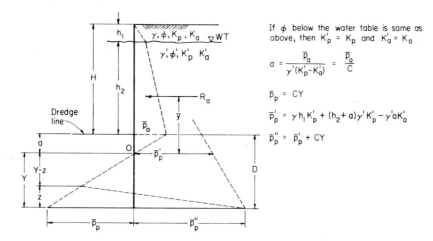

Figure 13-6. Cantilever sheetpiling pressure diagram for a granular soil. The diagram illustrates the possibility of different soil properties below the water table. If other stratification exists, the pressure diagram should be appropriately modified.

point. The point O is located a distance a below the dredge line, where the wall pressure is zero

$$a = \frac{\bar{p}_a}{\gamma'(K_p' - K_a')} = \frac{\bar{p}_a}{\gamma' K'} = \frac{\bar{p}_a}{C} \tag{a}$$

The distance z can be found in terms of Y by statics ($\sum F_H = 0$), or

$$R_a + (\bar{p}_p + \bar{p}_p'')\frac{z}{2} - \bar{p}_p\frac{Y}{2} = 0 \tag{b}$$

and solving for z, we obtain

$$z = \frac{\bar{p}_p Y - 2R_a}{\bar{p}_p + \bar{p}_p''} \tag{c}$$

An additional equation in Y and z can be obtained by summing moments at a convenient location, in this case, the bottom of the pile.

$$R_a(Y + \bar{y}) + \frac{z}{3}(\bar{p}_p + \bar{p}_p'')\frac{z}{2} - \bar{p}_p\left(\frac{Y}{2}\right)\left(\frac{Y}{3}\right) = 0$$

simplifying

$$6R_a(Y + \bar{y}) + z^2(\bar{p}_p + \bar{p}_p'') - \bar{p}_p Y^2 = 0 \tag{d}$$

Substituting Eq. (c) into Eq. (d) and solving for Y, the following fourth-degree equation is formed, which is applicable with or without soil water:

$$Y^4 + Y^3\left(\frac{\bar{p}_p'}{C}\right) - Y^2\left(\frac{8R_a}{C}\right) - Y\left[\frac{6R_a}{C^2}(2\bar{y}C + \bar{p}_p')\right] - \frac{6R_a\bar{y}\bar{p}_p' + 4R_a^2}{C^2} = 0 \quad (13\text{-}1)$$

where all terms are as shown in Fig. 13-6. If there is water, it is considered to be to the same elevation on both sides of the piling, since this equation does not consider hydrostatic pressure.

The "flagpole" problem is solved by making the following adjustments:

$a = 0.0$; $R_a = $ force against pole as from sign projection, etc.

$\bar{y} = $ distance from ground surface to R_a; $\bar{p}'_p = 0.0$

with these adjustments Eq. (13-1) becomes

$$Y^4 - Y^2\left(\frac{8R_a}{C}\right) - Y\left(\frac{12R_a\bar{y}}{C}\right) - \frac{4R_a^2}{C^2} = 0 \qquad (13\text{-}1a)$$

Inspection of the geometry, the terms involved, and the given conditions may simplify the problem. Also, if an arbitrary increase in depth of 20 to 40 percent is preferred to using a safety factor with the K_p term, the appropriate modifications should be made. If the water is at different elevations on each side of the wall, the effect of unbalanced hydrostatic pressure will need to be taken into account. When the difference is not great, or the backfill is a free-draining granular material, the safety factor will generally be adequate, with the assumption of equal hydrostatic pressure on both sides of the wall. On the other hand, if the elevation head is large, say, more than 3 or 4 ft, and the soil has a low coefficient of permeability, a flow net (Sec. 2-18) may be constructed to evaluate the hydrostatic pressure at several points along the wall.

Steps in the solution of a cantilever wall in a granular soil are as follows:

1. Sketch the given conditions.
2. Evaluate the active- and passive-earth-pressure coefficients.
3. Compute the pressures $\bar{p}_p$, $\bar{p}'_p$, $\bar{p}''_p$, the distance a, and the resultant pressure R_a and its location $\bar{y}$. The location of the resultant $\bar{y}$ can be found from the following equation when the pressure diagram is a triangle of base $H + a$ and height $\bar{p}_a$ as

$$\bar{y} = \frac{H + 2a}{3}$$

4. Insert values from step 3 into Eq. (13-1) and evaluate Y. The method of trial and error (assume values of Y and solve) will provide a rapid solution, since answers within about 0.15 m are sufficiently precise. Start Y at about $0.75H$ for first computation, as most cantilever walls will require embedment depths of $0.75H$ to H.
5. Total required length of the pile

$$L = H + D$$

$$D = Y + a$$

Example 13-1. Find the length of embedment D for the sheet-pile conditions shown in the accompanying sketch using clean sand backfill and steel sheet piles.

SOLUTION. This solution will be made by applying an arbitrary percentage increase on the computed depth of 30 percent.

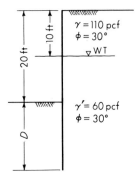

$\gamma = 110$ pcf
$\phi = 30°$

∇ W T

$\gamma' = 60$ pcf
$\phi = 30°$

Figure E13-1

Step 1. Find active- and passive-earth-pressure coefficients. From Table 11-6, $\delta = 17°$. Using the Coulomb equations (11-3) and (11-6) with an electronic calculator

$$K_a = K_a' = 0.299 \qquad \text{also in Tables 11-1 and 11-2}$$

$$K_p = K_p' = 5.385$$

also

$$K' = K_p' - K_a' = 5.385 - 0.299 = 5.086$$

$$C = 0.06(5.086) = 0.31$$

Step 2. Compute a, R_a, and $\bar{y}$ of Fig. 13-6

$$\bar{p}_a = [10(0.110) + 10(0.060)]0.299 = 0.51 \text{ ksf}$$

$$a = \frac{0.51}{C} = \frac{0.51}{0.31} = 1.65 \text{ ft}$$

From pressure diagram (not shown)

$$R_a = 0.33(5) + (0.33 + 0.51)(5) + 0.51(0.5)(1.65) = 6.28 \text{ kips}$$

and R_a acts at $\bar{y}$ above point O.

$$6.28\bar{y} = 0.43(1.65/3) + 0.9(5.0) + 3.3(6.65) + 1.65(15.0)$$

$$\bar{y} = 51.50/6.28 = 8.20 \text{ ft}$$

Step 3. Find $\bar{p}_p'$ of Fig. 13-6

$$\bar{p}_p' = [0.110(10) + 11.65(0.06)]5.385$$

$$- 1.65(0.06)0.299 = 9.66 \text{ ksf}$$

Step 4. Substitute into Eq. (13-1) and find Y. Compute coefficients and constants.

$$\frac{\bar{p}_p'}{C} = \frac{9.66}{0.31} = 31.16 \qquad \frac{8R_a}{C} = 162.06$$

$$\frac{6R_a}{(0.31)^2} = \frac{6(6.28)}{(0.31)^2} = 392.09 \qquad 2\bar{y}(0.31) + \bar{p}_p' = 14.74$$

$$392.09(14.74) = 5,780.97$$

The constant term is

$$\frac{6(6.28)(8.20)(9.66) + 4(6.28)^2}{(0.31)^2} = 32,700$$

$$Y^4 + 31.2Y^3 - 162Y^2 - 5,781Y = 32,700$$

Step 5. Solve by trial and error.

Y	Y^4	$31.2\,Y^3$	$-162\,Y^2$	$-5{,}781\,Y$	$= 32{,}700$
14	38,416	85,613	$-31{,}752$	$-80{,}934$	$= 11{,}342$
15	50,625	105,300	$-36{,}450$	$-86{,}715$	$= 32{,}760$

from which $Y \cong 15$ ft.

$$D = 15 + 1.65 = 16.65$$

$$D' = 16.65(1.3) = 21.6 \ (30 \text{ percent increase})$$

$$L = 21.6 + 20 = 41.6 \text{ ft} = \text{total length of sheet pile}$$

Note that we do not know how large the lateral pressure is or if that pressure is possible. We have no idea at all of the lateral deflection or the bending moment and further note this solution is independent of the sheetpiling used. All we have solved is a rigid-body statics problem.

Cantilever Sheet Pile in Cohesive Soils ($\phi = 0$)

Sheetpiling in cohesive soils is treated somewhat similarly to granular soils. There are, however, certain phenomena associated with cohesive soils which require additional consideration. For example, consolidation may occur in the passive-pressure zones. Tension cracks may form and become filled with water, thus increasing the lateral pressure considerably, as well as changing the location of the resultant. The clay may shrink and pull away from the wall, which also increases the lateral pressure. This latter problem may be allowed for in design by neglecting the theoretical benefits of the wall adhesion, as shown in Fig. 13-7 and as discussed in Chap. 11.

Because of the uncertainty of clay, and since many of the sheet-pile walls are for the purpose of containing a fill on which a structure may be constructed, the piling may be driven in a clay or silty soil, then backfilled with a free-draining granular

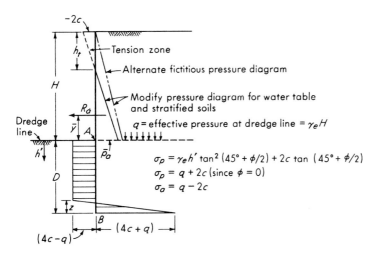

Figure 13-7. Sheetpiling in cohesive soil. The undrained shear strength ($\phi = 0$) case shown is conservative.

material. Of course, it is also possible that the pile will contain, as well as be embedded in, a cohesive material. Both cases will be considered in the following material.

Referring to Fig. 13-7 and from Eqs. (2-14) and (2-15)

$$\sigma_a = \bar{q} \tan^2 \left(45° - \frac{\phi}{2}\right) - 2c \tan \left(45° - \frac{\phi}{2}\right)$$

$$= \bar{q}K_a - 2c\sqrt{K_a}$$

$$\sigma_p = \bar{q} \tan^2 \left(45° + \frac{\phi}{2}\right) + 2c \tan \left(45° + \frac{\phi}{2}\right)$$

$$= \bar{q}K_p + 2c\sqrt{K_p}$$

At point A on the left side of the sheetpiling at the dredge line, $\bar{q} = 0.0$ and the net pressure at A is (note that $\gamma_e hK_a$ or $\gamma_e hK_p$ should be used instead of $\bar{q}$ if $\phi \neq 0$)

$$\sigma_p - \sigma_a = 2c - (\bar{q} - 2c) = 4c - \bar{q}$$

since $K_a = K_p = 1.0$.

At point B the forces are

$$\sigma_p = \bar{q} + \gamma_e D + 2c \text{ acting to the left} \qquad \sigma_a = \gamma_e D - 2c \text{ acting to the right}$$

summing pressures $(\sigma_p - \sigma_a)$, we obtain

$$\gamma_e D + \bar{q} + 2c - (\gamma_e D - 2c) = 4c + \bar{q} \text{ acting to the left}$$

The $\sum F_H = 0$ for wall stability, and summing pressure areas

$$R_a = \frac{z}{2}(4c - \bar{q} + 4c + \bar{q}) - D(4c - \bar{q}) = 0 \qquad (e)$$

solving for z,

$$z = \frac{D(4c - \bar{q}) - R_a}{4c} \qquad (f)$$

Also, for equilibrium, the sum of the moments at any point should be zero, and summing about the base for convenience,

$$R_a(\bar{y} + D) - \frac{D^2}{2}(4c - \bar{q}) + \frac{z^2}{3}(4c) = 0 \qquad (g)$$

Substituting Eq. (f) into Eq. (g), collecting like terms, and simplifying,

$$D^2(4c - \bar{q}) - 2DR_a - \frac{R_a(12c\bar{y} + R_a)}{2c + \bar{q}} = 0 \qquad (13-2)$$

where all terms are identified in Fig. 13-7 but using effective pressure at dredge line $= \bar{q}$.

The depth computed from Eq. (13-2) may be increased 20 to 40 percent or, alternatively, the cohesion c used in this equation may be divided by a safety factor of 1.5 to 2.0, thus directly increasing the computed depth. When applying the safety factor in this manner, the concept of the stability factor of the next section should be considered; otherwise one may arrive at the erroneous conclusion that no wall can be built if the cohesion of a soil is such that

$$\frac{4c}{F} \le \bar{q}$$

Example 13-2a. For the sheet-pile system shown, compute the depth of embedment of the sheetpiling. Use a 30 percent increase applied to the computed embedment depth.

SOLUTION. *Step 1.* Obtain K_a, K_p. Use $\delta = 17°$ Table 11-6

$$K_a = K'_a = 0.299$$

$$K_p = K'_p = 5.385$$

The effective pressure $\bar{q}$ at the dredge line is

$$\bar{q} = 17.3(3) + 9.5(3) = 80.4 \text{ kPa}$$

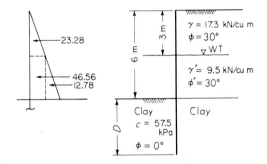

Figure E13-2

Step 2. The lateral active-pressure force is computed and $\bar{y}$ found

$$R_a = 3(17.3)(1.5)0.299 + 3(17.3)(3)0.299$$

$$+ 3(9.5)(1.5)0.299 = 82.6 \text{ kN/m}$$

$$82.6\bar{y} = 12.78(1) + 46.56(1.5) + 23.28(4)$$

$$\bar{y} = \frac{175.74}{82.6} = 2.13 \text{ m}$$

Step 3. Compute coefficients of Eq. (13-2)

$$4c - \bar{q} = 230 - 80.4 = 149.6 \qquad 2R_a = 2(82.6) = 165.2$$

$$12c\bar{y} + R_a = 12(57.5)(2.13) + 82.6 = 1,552.3 \qquad 2c + \bar{q} = 195.4$$

Step 4. Substituting the coefficients into Eq. (13-2), we obtain

$$149.6D^2 - 165.2D - 82.6(1,552.3/195.4) = 0$$

$$D^2 - 1.1D = 4.39$$

solving $D = 2.72$ m

Step 5. Check the solution just obtained to see if statics is satisfied,

$$z = \frac{2.72(149.6) - 82.6}{4(57.5)} = 1.41$$

$$\Sigma F_H = 0 \qquad \text{Eq. } (e) \qquad 82.6 + \frac{1.41}{2} 8(57.5) - 2.72(149.6)$$

$$= 0.012 \quad O.K.$$

Step 6. Apply 30 percent increase on D for embedment depth

$$\text{Total } D = 2.72(1.3) = 3.54 \text{ m}$$

Example 13-2b. Redo Example 13-2a applying $F = 1.5$ to cohesion,

SOLUTION. Start with step 3.

Step 3. Compute revised coefficients of Eq. (13-2)

$$c = \frac{57.5}{1.5} = 38.3$$

$$4c - \bar{q} = 72.9\dagger \qquad 2R_a = 165.2$$

$$12c\bar{y} + R_a = 12(38.3)(2.13) + 82.6 = 896.3 \qquad 2c + \bar{q} = 157.1$$

Step 4. Substituting these coefficients into Eq. (13-2), we obtain

$$72.9D^2 - 165.2D - 82.6\frac{896.3}{157.1} = 0$$

$$D^2 - 2.27D = 6.46$$

solving $D = 3.92$ m

$$\text{Percent difference } \frac{3.92}{2.72} = 144 \text{ percent}$$

$$\text{Percent difference of } F = \frac{1.5}{1.3} = 115 \text{ percent}$$

A logical question is just what the safety factor is here.

13-6. ANCHORED SHEETPILING: FREE-EARTH SUPPORT

Anchored sheetpiling, also termed *anchored bulkheads*, is widely used for dock and harbor structures. This construction provides a vertical wall so that ships may tie up alongside, or to serve as a pier structure, which may jut out into the water. In these

---

† Note that a sufficiently large F applied to smaller values of soil cohesion will yield a $4c - \bar{q}$ value of zero or a negative quantity, indicating a computed wall instability, when, in fact, a stable wall system could be constructed.

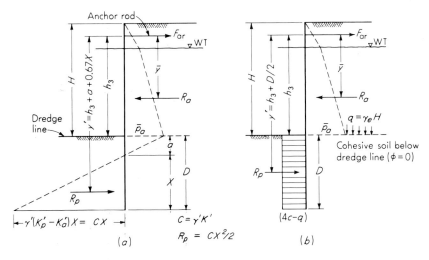

Figure 13-8. Anchored sheetpiling, *free-earth* method. (*a*) All granular soil; (*b*) cohesive soil below dredge line with granular-soil backfill.

cases the sheeting may be required to support laterally a fill on which railroad spurs, roads, or warehouses may be constructed, so that ship cargoes may be transferred to other areas or conveniently stored. The use of an anchor member tends to reduce the lateral deflection, the bending moment, and the depth of penetration of the pile. The use of more than one anchor may be desirable when walls are rather high or it is desirable to control lateral deflections.

There are a number of so-called classical methods currently used to design anchored sheet piles, among which are the *free-earth* method and the *fixed-earth* method.

The *free-earth* method assumes that the piling is rigid and may rotate at the anchor-rod level, with failure occurring by a rotation about the anchor rod. Passive pressure develops in the soil in front of the piling, and active pressure develops behind the wall. After the theoretical embedment is computed, the value may be increased 20 to 40 percent, or K_p may be divided by an appropriate safety factor prior to computation of the embedment length. The assumed pressure diagrams and identification of terms are illustrated in Fig. 13-8.

From Fig. 13-8a the distance a to the point of zero pressure is

$$a = \frac{\bar{p}_a}{\gamma' K'} = \frac{\bar{p}_a}{C}$$

Next, summing moments about the anchor rod to satisfy statics,

$$y' R_p = \bar{y} R_a$$

From substitution of values shown in Fig. 13-8a

$$\bar{y} R_a = C \frac{X^2}{2}\left(h_3 + a + \frac{2}{3}X\right)$$

Combining terms in descending powers of X, we obtain

$$2X^3 + 3X^2(h_3 + a) - \frac{6R_a\bar{y}}{C} = 0 \qquad (13\text{-}3)$$

The force in the anchor rod F_a is found by summing horizontal forces to obtain

$$F_a = R_a - R_p \qquad (h)$$

The embedment depth is $D = X + a$.

For Fig. 13-8b, where the soil below the dredge line is cohesive ($\phi = 0$ or undrained conditions), we may again sum moments about the anchor rod as

$$R_a\bar{y} - D(4c - \bar{q})\left(h_3 + \frac{D}{2}\right) = 0 \qquad (i)$$

which may be rearranged in descending powers of D to give

$$D^2 + 2Dh_3 - \frac{2\bar{y}R_a}{4c - \bar{q}} = 0 \qquad (13\text{-}4)$$

The forces in the anchor rod can be computed by Eq. (h). From inspection of Fig. 13-8b it can be seen that if the passive pressure $\sigma_p \le 0$ the wall is unstable. This occurs at

$$\frac{c}{\bar{q}} = \frac{c}{\gamma H} \le 0.25$$

For a given backfill material, there is a critical value of H beyond which a stable wall cannot be constructed in clay. The ratio of $c/\bar{q}$ is termed the stability number in soil-mechanics literature

$$S_n = \frac{c}{\bar{q}} \qquad (j)$$

If wall adhesion c_a is taken into account, the stability number according to Rowe (1957) can be computed approximately as

$$S_n = \frac{c}{\bar{q}}\sqrt{1 + \frac{c_a}{c}} \qquad (k)$$

For $c_a = 0.56c$, the square-root term is

$$\sqrt{1 + \frac{c_a}{c}} = 1.25 \qquad (l)$$

and the stability number becomes

$$S_n = \frac{1.25c}{\bar{q}} \qquad (m)$$

For a safety factor of 1 and $c/\bar{q} = 0.25$, the stability number becomes approximately 0.31 with wall adhesion considered. In sheet pile design in clay the wall should

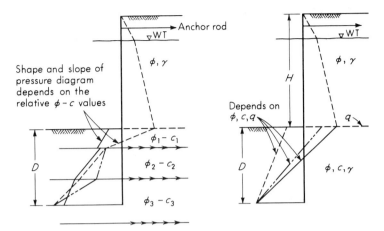

Figure 13-9. Sheetpiling pressure diagrams. (*a*) Erratic deposits; (*b*) sand backfill overlying a cohesive ϕ-*c* soil.

have a stability number of about

$$S_n = 0.3 \times F$$

The free-earth-support method is preferred over the other classical methods because of its greater simplicity. This method can also be used for ϕ-*c* soils. Figure 13-9 indicates qualitative pressure diagrams in erratic deposits or ϕ-*c* soils.

Example 13-3. Compute the embedment depth of the sheetpiling shown in Fig. E13-3*a*. Use the *free-earth* method.

Given data: $\phi = \phi' = 30°$ $\delta = 20°$ $c = 0$
           $\gamma = 0.105$ kcf $\gamma' = 0.066$ kcf

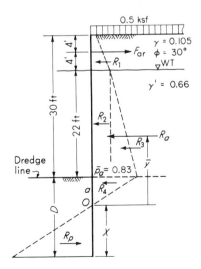

Figure E13-3*a*

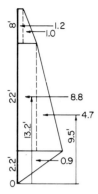

Figure E13-3*b*

SOLUTION. *Step 1.* Compute earth-pressure coefficients.

$$K_a = K'_a = 0.297 \qquad \text{Table 11-1}$$
$$K_p = K'_p = 6.10 \qquad \text{Table 11-2}$$
$$C = \gamma' K' = 0.066(6.10 - 0.297) = 0.38$$

Step 2. Find R_a and $\bar{y}$; refer to Fig. E13-3*b*.

$$P_1 = 8(0.5)(0.297) = 1.2 \text{ kips}$$
$$P_2 = 8(0.105)(0.297)(4) = 1.0 \text{ kips}$$
$$P_3 = 22(0.40) = 8.8 \text{ kips}$$
$$P_4 = 22(0.066)(0.297)(11) = 4.7 \text{ kips}$$
$$\bar{p}_a = 0.83 \text{ ksf}$$
$$a = \frac{0.83}{C} = \frac{0.83}{0.38} = 2.2 \text{ ft}$$
$$P_5 = 0.83(2.2)(0.5) = 0.9 \text{ kip}$$
$$\sum P = R_a$$
$$R_a = 1.2 + 1.0 + 8.8 + 4.7 + 0.9 = 16.6 \text{ kips}$$

Find $\bar{y}$

$$16.6\bar{y} = 1.2(28.2) + 1.0(26.9) + 8.8(13.2) + 4.7(9.5) + 0.9(1.5)$$

$$\bar{y} = \frac{222.9}{16.6} = 13.4 \text{ ft from point } O$$

$$\bar{y} = 32.2 - 13.4 - 4 = 14.8 \text{ ft}$$

Step 3. Compute coefficients for Eq. (13-3).

$$3(h_3 + a) = 3(26 + 2.2) = 84.6$$

$$\frac{6R_a\bar{y}}{C} = \frac{6(16.6)(14.8)}{0.38} = 3,879.2$$

Step 4. Compute X and D.

$$2X^3 + 84.6X^2 - 3,879.2 = 0$$

$$X^2 + 42.3X^2 = 1,939.6$$

Trial

X	X^3	$42.3X^2$	$= 1,939.6$
6	216	1,522.8	$= 1,738.8$
6.5	275	1,787.2	$= 2,062.2$
6.4	262	1,732.6	$= 1,995$

Use $X = 6.3$

$D = 6.3 + 2.2 = 8.5$ ft

At this point no factor has been applied or $F = 1.0$.

Step 5. Find anchor-rod force.

$$R_p = CX^2/2 = 0.38(6.3)^2/2 = 7.5 \text{ kips}$$

and $\sum F_H = 0$

$$F_{a.r.} + R_p - R_a = 0$$

$$F_{a.r.} = 16.6 - 7.5 = 9.1 \text{ kips}$$

Step 6. Sum moments about anchor rod to check statics.

$$\bar{y}R_a - y'R_p = 0$$

$$y' = h_3 + a + 0.67x = 26 + 2.2 + 6.3(0.67) = 32.4 \text{ ft}$$

$$16.6(14.8) - 32.4(7.5) = 245.7 - 243 \cong 0 \qquad O.K.$$

There is some round-off error from using values to 0.1 and not obtaining the "exact" value of X in step 4.

Example 13-4. An anchored retaining wall with water on both sides of the wall and cohesive soil below the dredge line is as shown in the figure. It is required to find the embedment depth and anchor-rod force. Use $F = 1.5$ on cohesion

SOLUTION. Take $\delta = 20°$ and from Example 13-3 obtain

$$K_a = K'_a = 0.297 \qquad K_p = K'_p = 6.10$$

Step 1. Compute R_a and $\bar{y}$. $\quad \gamma H K_a \, \tfrac{H}{2} =$

$$P_1 = 16.5(2.4)(0.297)1.2 = 14.1 \text{ kN}$$

$$P_2 = 11.76(6.7) = 78.8 \text{ kN}$$

$$P_3 = 20.7(6.7)0.5 = 69.3 \text{ kN}$$

$$R_a = 14.1 + 78.8 + 69.3 = 162.2 \text{ kN}$$

$$\bar{q} = 2.4(16.5) + 6.7(10.4) = 109.3 \text{ kPa}$$

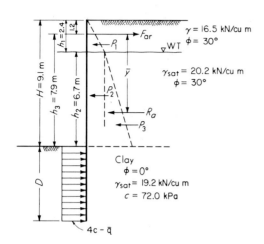

$$P_1 = \gamma H K_a' \frac{H}{2}$$

$$P_2 = 86.7 \, K_a' \, 6.2$$

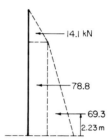

Figure E13-4a   Figure E13-4b

Find $\bar{y}$; y' above the dredge line is

$$162.2y' = 14.1(7.5) + 78.8(3.35) + 69.3(2.23)$$

$$y' = 524.3/162.2 = 3.23 \text{ m}$$

$$\bar{y} = 9.1 - 1.2 - 3.23 = 4.67 \text{ m below anchor rod}$$

Step 2. Compute coefficients of Eq. (13-4).

$$2h_3 = 2(7.9) = 15.8$$

$$\frac{4c}{F} - \bar{q} = 4\frac{72.0}{1.5} - 109.3 = 82.7$$

$$2\bar{y}R_a = 2(4.67)(162.2) = 1,514.9$$

Step 3. Solve for D (Eq. 13-4)) by using the method of "completing the square."

$$D^2 + 15.8D - 1,514.9/82.7 = 0$$

$$D^2 + 15.8D = 18.3$$

$$D = \pm 9.0 - 7.9 = 1.1 \text{ m}$$

Step 4. Find anchor-rod force and check statics.

$$P_p = D\left(\frac{4c}{F} - \bar{q}\right) = 1.1(82.7) = 91.0 \text{ kN}$$

$$\sum F_H = 0$$

$$F_{a.r.} + P_p - R_a = 0$$

$$F_{a.r.} = 162.2 - 91 = 71.2 \text{ kN/m}$$

Check $\sum M = 0$ at anchor rod

$$P_p y' - R_a \bar{y} = 0$$

$$y' = h_3 + D/2 = 7.9 + 1.1/2 = 8.45$$

$$91(8.45) - 162.2(4.67) = 769 - 757.5 = 11.5 \qquad \text{O.K.}$$

round-off errors have accumulated sufficient to affect closure. In SI the numbers are so large that small errors accumulate rapidly.

13-7. ROWE'S MOMENT REDUCTION APPLIED TO THE FREE-EARTH-SUPPORT METHOD

Rowe (1952) has proposed moment reductions for sheetpiling designs based on the free-earth method. This technique may be used for uniform medium-dense to dense silty sand or sand deposits. The soil in front of piles driven in loose silty-sand deposits may undergo excessive compressibility, and should be dimensioned for free-earth support without moment reduction. Piles embedded in clay deposits will undergo consolidation deformation in the passive-pressure zones under lateral pressure, and should also be proportioned based on free-earth support but without a moment reduction. Curves are presented here, however, for designers who may wish to apply moment reduction to sheet piles in clay.

The Rowe moment-reduction theory is based on the following factors:

1. The relative density of granular soils
2. The stability number of cohesive soils, which has been defined in Eq. (m) as

$$S_n = \frac{1.25c}{q}$$

3. A flexibility number (derived for fps units)

$$\rho = \frac{H^4}{EI}$$

where H = total length of piling, ft
EI = modulus of elasticity, psi, and moment of inertia, for a unit width of wall, in^4
4. The relative height of piling α and the relative freeboard of the piling β as shown in Fig. 13-10.

Curves derived from experimental data [Rowe (1952, 1957)] have been published for selected values of S_n, α, log ρ, and moment ratios. These curves are necessary for a design using this method, and are presented in Fig. 13-10a and b for both sand and clay soils.

Design by the moment-reduction method proceeds by first performing a free-earth analysis for the maximum bending moment M_0 and the length of pile. Next the appropriate moment-reduction curve from Fig. 13-10, depending on the anchor-rod location, length of pile, and soil type, is selected, using interpolation if necessary. This curve should be replotted (or traced) on a separate sheet of paper. From a table of sheetpiling sections, the actual bending moments M of the piles can be computed as M = allowable steel stress × section modulus/unit width. The flexibility coefficient $(\rho = H^4/EI)$ is also computed for the corresponding pile section. Using the M/M_0 ratio and ρ, a second curve is plotted for the various sections and superimposed on the standard curve. The intersection of the two curves may not coincide with a pile section, but any pile section lying above the intersection is satisfactory to use, the pile closest to the intersection being the most economical (see Example 13-4).

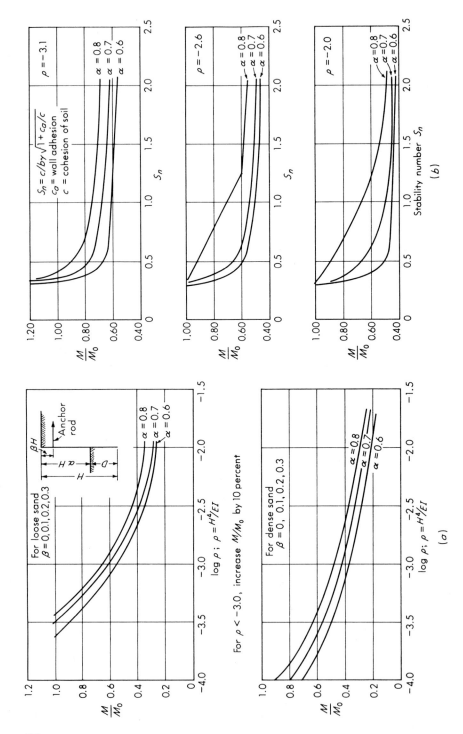

Figure 13-10. Rowe's moment-reduction curves. (*a*) Sheet piles in sand. [*After Rowe (1952).*] (*b*) Sheet piles in clay. [*After Rowe (1957).*]

Example 13-5. Design the sheetpiling necessary to resist the loads and forces found in Example 13-3, using Rowe's moment-reduction technique. Take the allowable bending stress for steel as $f_s = 24$ ksi ($f_y = 36$ ksi). The modulus of elasticity of steel is 29×10^3 ksi. Use an arbitrary depth increase factor of 1.83.

Other data. $\phi = \phi' = 30°$

Surcharge = 0.5 ksf = q_s

$\gamma = 105$ pcf

$\gamma_{sat} = 128.5$ pcf

$D = 8.5(1.83) = 15.55$ ft

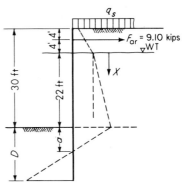

Figure E13-5a

SOLUTION. *Step 1.* The maximum moment will occur at a point of zero shear. We will assume maximum moment occurs between water table and the dredge line; therefore, the following equation for shear may be written, where X is the vertical distance to the point of zero shear as measured from the water table. Summing horizontal forces (shear) to zero and referring to Fig. E13-3b, we obtain

$$1.2 + 1.0 + (q_s + \gamma h_1)K_a X + \gamma K_a' X\left(\frac{X}{2}\right) - F_{a.r.} = 0.0$$

$$2.2 + (0.5 + 0.105(8))(0.297)X + 0.066(0.297)(0.5)X^2 = 9.10$$

$$0.0098X^2 + 0.4X = 6.9$$

$$X^2 + 40.8X = 704.0$$

Completing the square

$$X = \pm 33.5 - 20.4 = 13.1 \text{ ft below water table } < 22 \text{ ft}$$

A correct assumption was made for the location of M_{max}; otherwise a new assumption would be required and a new shear equation written, etc.

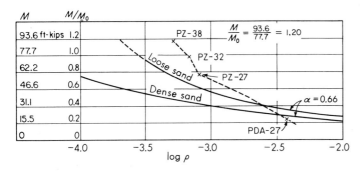

M	M/M₀
93.6 ft-kips	1.2
77.7	1.0
62.2	0.8
46.6	0.6
31.1	0.4
15.5	0.2
0	0

$$\frac{M}{M_0} = \frac{93.6}{77.7} = 1.20$$

PZ-38

PZ-32

PZ-27

Loose sand

Dense sand

$\alpha = 0.66$

PDA-27

−4.0   −3.5   −3.0   −2.5   −2.0

log p

Figure E13-5b

Table E13-5

Pile section	S_x, cu in/ft	I, in^4/ft	$\rho = \dfrac{H^4}{EI} = \dfrac{0.148}{I}$	$\log \rho$	Moment capacity $\dfrac{f_s(S_x)}{12}$, ft-kips	$\dfrac{M}{M_0}$
PZ38	46.8	$\dfrac{421.2}{1.5} = 281$	0.000527	-3.280	93.6	1.20
PZ32	38.3	$\dfrac{385.7}{1.75} = 220$	0.000673	-3.172	76.6	0.99
PZ27	30.2	$\dfrac{276.3}{1.5} = 184$	0.000804	-3.094	60.4	0.78
PDA27	10.7	$\dfrac{53}{1.33} = 39.8$	0.00372	-2.430	21.4	0.28

Sum moments about X just found to obtain $M_{\max}$

$$-0.4X\left(\frac{X}{2}\right) = -0.4(0.5)(13.1)^2 = -34.32$$

$$-0.0098X^2\left(\frac{X}{3}\right) = -0.0098\frac{(13.1)^3}{3} = -7.34$$

$$-1.0\left(13.1 + \frac{8}{3}\right) = -15.77$$

$$(9.1 - 1.2)(13.1 + 4) = 135.09$$

$$\sum M = \text{maximum moment} = \overline{77.66}\ \text{ft-k/ft of wall}$$

Step 2. Compute ρ

$$H = 30 + 15.55 = 45.55\ \text{ft} \qquad \text{total length of pile}$$

$$\rho = \frac{H^4}{EI} = \frac{45.55^4}{29 \times 10^6} = \frac{0.148}{I} \qquad \alpha = \frac{30}{45.55} = 0.66 \qquad \beta = \frac{4}{45.55} = 0.09$$

With this information Table E13-5 can be formed.

Step 3. Construct the M/M_0 vs. $\log \rho$ curve for $\alpha = 0.66$ for the type of sand encountered (i.e., loose or dense) using Fig. 13-10a. Both loose and dense curves are plotted in this example. Note that the values of M shown to the left of the table are obtained by multiplying 77.7 by the M/M_0 ratios of 0.2, 0.4, etc. It would appear that a PZ-27 could be used for this wall system.

13-8. FINITE-ELEMENT ANALYSIS OF SHEET-PILE WALLS

The finite-element method presented in the following material is the most efficient and rational method of sheet-pile-wall analysis currently available. The method gives Rowe's moment reduction directly and the output gives the design moments, and an estimate of the lateral deformation and of the soil pressure developed by passive resistance below the dredge line. The free-earth support and other classical solutions give none of this information directly, although the bending moment can be obtained with some difficulty, as was illustrated in Example 13-5.

The finite-element method is particularly advantageous for sheet-pile walls with multiple anchorage devices (anchor rods, earth anchors, tiebacks, etc.). It is particularly difficult to obtain anchor-rod forces, wall deformations, and bending moments for the multiple-level anchor-rod system using classical methods.

The finite-element method uses the same equations (and the computer program in the Appendix) as given in Chap. 9 and repeated here for convenience:

$$P = AF \qquad e = A^T X \qquad F = Se$$

and substituting to obtain

$$F = SA^T X \qquad P = ASA^T X$$

and finally $X = (ASA^T)^{-1}P$ which are the wall deflections (translation and rotation). With the deflections at each node known, the bending moments and soil nodal reactions (spring forces) can be computed from

$$F = SA^T X$$

Figure 13-11 illustrates the sheet-pile wall and P-X coding as well as the element forces. The problem is actually the beam-on-an-elastic-foundation problem turned $90°$ with the soil springs removed above the dredge line.

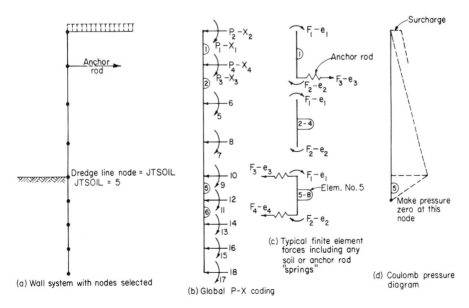

(a) Wall system with nodes selected

(b) Global P-X coding

(c) Typical finite element forces including any soil or anchor rod "springs"

(d) Coulomb pressure diagram

Figure 13-11. Finite-element analysis for sheet-pile wall either cantilever or anchored (including multiple anchors).

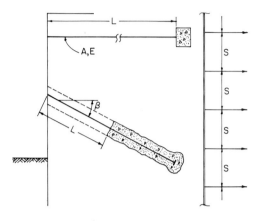

Figure 13-12. Development of anchor-rod "spring" based on spacing, cross section area A, modulus of elasticity E, and length L.

Anchor rods are allowed for by considering that an anchor rod will consist in a member of cross section A, modulus of elasticity E, and some length L. Now the force in a member such as this (similar to the bar of a truss) is from $e = FL/AE$ in mechanics of materials

$$F = \frac{AE}{L} e = \frac{AE}{L} X$$

since e and X are concurrent and parallel and here are equal from geometry because we will always place a node at any anchor-rod location. For earth anchors which may slope from the horizontal, the force component of interest and computed in the analysis is the horizontal component. As we analyze a unit width of wall, the AE/L of the anchor rod is prorated based on the anchor-rod spacing s; and to obtain horizontal effects (β = slope with horizontal as shown in Fig. 13-12), to obtain the equivalent spring, and using computer notation of $SSK(M,1)$ for near-end or $SSK(M,2)$ for far-end element spring, as

$$SSK(M,2) = \frac{AE}{sL} \cos \beta$$

That is, an anchor rod has an effect on the problem as if a soil spring were on the far end of the element just above the anchor-rod location. We could, of course, divide the spring in half and apply the effect as a far-end spring on element M and a near-end spring on element $M + 1$ (the two adjacent elements at the anchor-rod node), but the computed results are identical and dividing the spring represents additional effort and extra card input. Cantilever walls are solved by not reading in any anchor-rod springs.

In this computer program it is possible to read the lateral earth pressure at each node. The pressure is converted to nodal forces (P-matrix entries) by using the average end-area formula and considering the element length. This is done on the first cycle and held for any additional cycles. The force entry at the dredge line is computed based on the element lengths of the element just above the dredge line and the initial length of the first element below the dredge line. This is so that if embedment is increased, the P matrix does not change. It is also possible for the user to read nodal forces directly ($NNZP$ = number of nodal forces or nonzero P's read-in). This may have application for certain tieback wall problems where a known anchor force is being applied.

It is often desirable to reduce the dredge-line soil spring by say, 20 to 50 percent to account for soil disturbance or the fact that the constant AS in the general expression

$$k_s = AS + BS \times Z^n$$

may be near zero at the dredge line but picks up rapidly in a short depth to the value AS. This dredge-line reduction can be accomplished by $FACP$ = value to multiply $SSK(1,L)$ to obtain the reduced spring for computations.

It may be desirable to consider approximate nonlinearity via an $XMAX$ specification ($NONLIN > 0$ and $XMAX$ = maximum deflection for linear elastic soil spring as described in Sec. 9-7). With this specification, the soil springs are made zero where $X > XMAX$ and a $(-)$ P-matrix entry of

$$P(i) = -K_i XMAX$$

is made.

The computer program can be used for both analysis and design. In the design of a sheet-pile wall it will be necessary to analyze several rolled steel sections or the sections arbitrarily selected for concrete-sheeting or concrete-pile walls. It will be desirable to start with a small embedment depth (in design), and by cycling ($NCYC$ = number of cycles to avoid going too long), the depth is incremented by $FAC2$ with recommended values of

1-ft or 0.3-m increments for anchored walls
2-ft or 0.6-m increments for cantilever walls ($IAR = 0$)

The increments are applied until the deflection at the dredge line converges; i.e., the previous computed dredge-line deflection compared with the currently computed dredge-line deflection is within a convergence tolerance ($CONV$) with a value of 0.01 ft (1/8 in) or 0.003 m recommended. With this criterion it is evident that any further increase in depth does not contribute to the problem. It is also evident that if one wishes to apply some kind of safety factor, perhaps a study of the problem with the dredge line lowered some realistic amount should be analyzed with the final embedment depth based on this required value of embedment. Simply to analyze a problem, one would input the given data, set $NONLIN = 0$, $XMAX$ = very large number, and $NCYC = 1$. The output would be for the given input data.

The deflections from the dredge line to the bottom of the sheet pile should be checked to see if a shear slip surface has formed. If all node deflections are positive, the wall has moved forward, and if the movement is sufficient a slip surface will have formed. This will not be detected by the finite-element model but must be interpreted by the user. This can be taken care of by increasing the pile embedment and making a new analysis.

Steps in solving a problem are as follows:

1. Draw soil-wall system and decide on where nodes will go. Put a node at each anchor-rod location, at the dredge line, and at each stratum change. Try to keep the ratio of adjacent member lengths $R = (L_m/L_{m+1}) \leq 5$ and preferably close to 1. Do not exceed the DIMENSION of $ASAT(NP, NP)$ based on $NP = 2 \times$ No. of nodes. Obtain element lengths, look in tables, and obtain I for a unit wall width (1 ft or m). Compute E in ksf or kPa. $JTSOIL =$ node where dredge line occurs.

2. Compute the lateral soil pressure from ground surface to the dredge line, using preferably the Coulomb equation (or the Rankine equation). Where the strata change, use the *average* soil pressure for the nodal value. A small computational error is introduced unless adjacent elements are the same length but will generally be insignificant. Pressure values $PRESS(I)$ are read of $IPRESS = JTSOIL + 1$ with $PRESS(IPRESS) = 0.0$ (node just below dredge line).

3. Compute modulus of subgrade reaction for soil below dredge line. For an estimate use

$$k_s = C_1 q_a \qquad C_1 = 36 \text{ or } 120 \text{ for ksf or kPa}$$

or

$$k_s = C_2(cN_c + \bar{q}N_q + 0.5\gamma BN\gamma)$$

This gives $k_s = AS + BS \times Z^1$

where

$$AS = C_2(N_c + 0.5\gamma BN\gamma)$$

$$BS = C_2 \gamma N_q$$

$$B = \text{unit width}$$

$$C_2 = 12 \text{ or } 40 \text{ for ksf or kPa}$$

These values will give "good" values of bending and other items of interest, but deflections may be substantially in error. Passive soil pressure is nearly independent of k_s.

4. Locate any soil nodes where soil springs are to be read in including anchor-rod nodes for anchored walls. For soil adjustments it is necessary to read values for both elements adjacent to the node.

5. Compute any anchor-rod forces as input if required. Note that anchor-rod input forces must be of correct sign and may result in the wall moving into the backfill, since the final model is a mathematical force system which is forced into equilibrium if possible. If the system is not in static equilibrium, the program usually will not compute.

Example 13-6. Given the sheet-pile wall in Fig. E13-6a, analyze the embedment depth shown and show output checks. For design we would start with the embedment depth 1 to 1.5 m less than shown and increment to the depth needed for stability. We will use two anchor rods to illustrate program capability. Note a cantilever wall is the same solution without anchor-rod springs.

SOLUTION. *Step 1.* Code problem as in Fig. E13-6b and hand-compute pressures shown in Fig. E13-6c using the Coulomb equation. Take $\delta = 17°$ and compute $K_a = 0.277$. At junction of clay at water table

$$\sigma_a = [20 + 17.9(5)]0.277 = 30.33 \text{ kPa in sand}$$

$$\sigma_a = [20 + 17.9(5)]1 - 2c(1) = 48.5 \text{ kPa in clay}$$

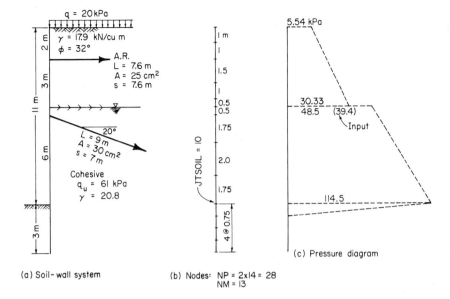

(a) Soil-wall system

(b) Nodes: NP = 2x14 = 28
NM = 13

(c) Pressure diagram

Figure E13-6

Average $\sigma_a = (48.5 + 30.33)0.5 = 39.4$; below water table for rest of clay

$$\sigma_a = 109.5 + 11z - 61 = 48.5 + 11z$$

at dredge line $\sigma_a = 48.5 + 66 = 114.5$ kPa

Step 2. Convert $E = 29,600$ ksi to kPa

$$E = 29,600(6,894.28) = 204,070,688 \text{ say } 204,070,000 \text{ kPa}$$

Find I for *PZ*-38 (table in Appendix) $w = 18$ in, $I = 421.2$ in^4

$$I/m \text{ of width} = \frac{421.2}{1.50}(3.2828)(2.54)^4 \times 10^{-8} = 0.000384 \text{ m}^4/\text{m}$$

Step 3. Compute anchor-rod springs

$$SSK(2,2) = \frac{25(204,070,000)}{7.6(7)10^4} = 9,589.7 \text{ kN/m} \qquad SSK(2,1) = 0.0$$

$$SSK(6,2) = \frac{30(204,070,000)}{9(7)10^4} \cos 20 = 9,131.6 \text{ kN/m} \qquad SSK(6,1) = 0.0$$

Step 4. Miscellaneous input data
1. Do not consider *XMAX* or *NONLIN*
2. Do not reduce dredge-line spring: *FACP* = 1.00
3. Compute

$$k_s = 120q_a = 120q_u = 7,320 \text{ kN/m}^3$$

4. Number of load conditions, $NLC = 1$. From which

$NP = 28$ $NM = 13$ $NNZP = 0$ $NLC = 1$

$IPILE = 0$ $IBEAM = 0$ $ISPILE = 1$ (sheet-pile problem)

$IRING = 0$ $IBRAC = 0$ $NCYC = 1$ (analysis)

$JJS = 2$ (2 anchor rods) $JTSOIL = 10$ $NONLIN = 0$

$IAR = 2$ $NZEROX = 0$ (no specified displacements)

$LIST = 0$ (reduce output) $INERP = 0$ constant moment of inertia

$INERB = 0$ $IPRESS = 11$ $(JTSOIL + 1)$

Step 5. Order of data cards

Card No.

Card No.							
1	TITLE						
2	UNITS	UT1,	UT3,	UT4,	UT5,	UT6	
	M	kN	k	N-M	k	Pa	kN/Cu M
3	28	13	0	1	0	0	1 0 0 1 (16I5)
4	2	10	0	2	0	0	0 0 11 (16I5)
5	204070000.	0.	1.	1.	3.0	0.03	0.3
6	7320.	0.	1.	either BS or EXPO must be 1.			
7	1.	1.	1.5	1.	0.5	0.5	1.75 2.0
8	1.75	0.75	0.75	0.75	0.75	H(I), I = 1, NM	
9	1.0	0.00030	WIDTH, XINER (note, not 0.000384)				
10	3	7					
11	2	0.	9589.7				
12	6	0.	9131.6				
13	5.54	10.5	15.5	22.9	27.9	39.4	54. 73.3
14	95.3	114.5	0.0	PRESS(I), I = 1, IPRESS			

These 14 cards represent input. Output is shown in Fig. E13-6d.

Step 6. Check output

(a) $\sum F_H = 0$ 607.731 vs. applied forces of 607.582 within computer round-off

(b) Moment at top $= 0.015 \cong 0$ O.K.

Moment at bottom $= 0.0$

Moments on each side of each node (far end of member M and near end of $M + 1$) are approximately equal and opposite; e.g., members 2 and 3 are 17.679 and -17.701, respectively, or approximately 0.0.

(c) Looking at deflections, the maximum deflection is 0.047 m at node 9 just above the dredge line

Dredge-line deflection $= 0.0311$ m (3 cm or approx. $1\frac{1}{4}$ in)

(d) The top anchor has 49.62 kN/m and the lower anchor has 304.35 kN/m. Checking lower anchor rod

$$A = 30 \text{ cm}^2 \qquad \beta = 20° \qquad s = 7 \text{ m}$$

$$P = 304.35(7)/\cos 20 = 2{,}267.18 \text{ kN}$$

$$f = \frac{2{,}267.28 \text{ kN}}{0.0030 \text{ m}^2 (98.07)} = 7{,}706 \text{ kg/cm}^2$$

Use lower anchor rod of $F_y = 150$ ksi (high-strength prestress steel wires to provide 30 cm²). Use upper anchor rod of $F_y = 36$ to $F_y = 50$ and nominal diam $= 2.25$ in.

(e) Check bending stress

$$M_{max} = 366.8 \text{ kN-m/m at node 8}$$

```
SOLUTION FOR SHEET PILE WALL ***************

MODULUS OF ELAST = 204070000.0 KPA     UNIT WT =  0.0    KN/CU M
NO NODES REQ CORRECT =   2    NODE SOIL STARTS FOR PILES =  10
MAX NON-LIN SOIL DEFORM =  1.000 M

       SOIL MODULUS =  7320.000+     0.0 *Z**1.000 KN/CU M
NO OF NON-ZERO P-MATRIX ENTRIES =   0   NO OF LOAD CONDITIONS =   1
    GROUND LINE REDUCTION FACTOR = 1.000    NO OF NODAL PRESSURE ENTRIES = 11
```

MEMNO	NP1	NP2	NP3	NP4	LENGTH	WIDTH	THICK	INERTIA,M **4
1	1	2	3	4	1.000	1.000	0.0	0.00030
2	3	4	5	6	1.000	1.000	0.0	0.00030
3	5	6	7	8	1.500	1.000	0.0	0.00030
4	7	8	9	10	1.000	1.000	0.0	0.00030
5	9	10	11	12	0.500	1.000	0.0	0.00030
6	11	12	13	14	0.500	1.000	0.0	0.00030
7	13	14	15	16	1.750	1.000	0.0	0.00030
8	15	16	17	18	2.000	1.000	0.0	0.00030
9	17	18	19	20	1.750	1.000	0.0	0.00030
10	19	20	21	22	0.750	1.000	0.0	0.00030
11	21	22	23	24	0.750	1.000	0.0	0.00030
12	23	24	25	26	0.750	1.000	0.C	0.00030
13	25	26	27	28	0.750	1.000	0.0	0.00030

```
       ANCHOR ROD LOCATED AT NODE NOS =   3   7
```

MEMNO	SOIL MODULUS	SPRINGS-SOIL, A.R., OR STRUT		
1	0.0	0.0	0.0	
2	0.0	0.0	9589.699	
3	0.0	0.0	0.0	
4	0.0	0.0	0.0	
5	0.0	0.0	0.0	
6	0.0	0.0	9131.598	
7	0.0	0.0	0.0	
8	0.0	0.0	0.0	
9	0.0	0.0	0.0	
10	7320.000	2745.000	2745.000	
11	7320.000	2745.000	2745.000	
12	7320.000	2745.000	2745.000	
13	7320.000	2745.000	2745.000	

LOAD MATRIX KN-M OR KN

				WALL PRESSURES, KPA	FOR NLC = 1	
1	0.0	2	3.5967	1	5.5400	
3	0.0	4	10.5067	2	10.5000	
5	0.0	6	20.3917	3	15.5000	
7	0.0	8	27.6083	4	22.9000	
9	0.0	10	21.0500	5	27.9000	
11	0.0	12	19.9583	6	39.4000	
13	0.0	14	65.1625	7	54.0000	
15	0.0	16	139.1416	8	73.3000	
17	0.0	18	176.9541	9	95.3000	
19	0.0	20	123.2124	10	114.5000	
21	0.0	22	0.0	11	0.0	
23	0.0	24	0.0			
25	0.0	26	0.0			
27	0.C	28	0.0			

FORCE MATRIX FOR NLC = 1
        MOMENTS, KN-M   SOIL REACTIONS, KN

1	0.015	3.614	0.0	0.0	
2	-3.620	17.679	0.0	49.619	
3	-17.701	-4.979	0.0	0.0	
4	4.984	7.516	0.0	0.0	
5	-7.457	24.250	0.0	0.0	
6	-24.289	50.941	0.0	304.350	
7	-51.004	-273.840	0.0	0.0	
8	273.840	-366.826	0.0	0.0	
9	366.815	-138.512	0.0	0.0	
10	138.508	-12.328	85.442	58.828	
11	12.336	25.656	58.828	31.534	
12	-25.645	16.332	31.534	4.687	
13	-16.336	0.0	4.687	-21.778	

```
SUM SOIL REACTIONS =  607.731 KN    SUM APPLIED FORCES =  607.582 KN
```

NODE NO	NODE ROTAT,RAD	NODE DEFL, M	SOIL PRESS, KPA
1	0.0076505	-0.0102335	0.0
2	0.0076798	-0.0025732	0.0
3	0.0078532	0.0051742	0.0
4	0.0080085	0.0171398	0.0
5	0.0080291	0.0251416	0.0
6	0.0081582	0.0291828	0.0
7	0.0084644	0.0333293	0.0
8	0.0052900	0.0467141	0.0
9	-0.0051400	0.0473686	0.C
10	-0.0123385	0.0311264	227.8449
11	-0.0132593	0.0214309	156.8741
12	-0.0131780	0.0114879	84.0914
13	-0.0129217	0.0017076	12.4997
14	-0.0128220	-0.0079338	-58.0755

Figure E13-6d

the section modulus is

$$46.8 \frac{in^3}{ft} \times 3.2828 \frac{ft}{m} \times \frac{2.54^3}{(100)^3} = 0.00252 \text{ m}^3/\text{m}$$

$$f_b = \frac{366.8}{0.00252} = 145{,}692.7 \text{ kPa}$$

$$f_b = \frac{145{,}692.7}{98.07} = 1{,}485.6 \text{ kg/cm}^2$$

For A36 steel $F_y = 2{,}531$ kg/cm^2

Therefore, bending approx. $0.6F_y$, PZ-38 section is satisfactory

(f) The maximum soil pressure is 227.8 kPa and is at the dredge line. This is approximately $4 \times q_u$ and the soil response in this zone is nonlinear.

Note the center of pressure is near the dredge line and not 4/3 m from bottom as in free-earth method. This is effectively what Rowe's "moment reduction" allows for.

13-9. WALES AND ANCHORAGES FOR ANCHORED SHEETPILING

Wales are longitudinal members running parallel with and in close contact with the wall as shown in Fig. 13-13. These members are usually made of two rolled channel sections back to back with spacing sufficient to pass the anchor rod (and through a hole in the sheetpiling usually burned in the field with an acetylene torch) but may be W or S shapes. Wales may be attached to the backfill side of the sheetpiling if it is desired to maintain a clear exposed face as for harbor structures. Back-face wales will require weld or bolt tension connections to hold the wall in contact with the wale. The anchor rod will usually consist in either a steel bar upset-threaded so there is no reduction of stress area for the thread root, at least on the wall end, but may be a cable which has been fitted on the wall end with a threaded portion. The threaded portion is a practical necessity in order that the wall may be tightened into align- ment, the wale forced into contact with the wall (although it may be field welded at select locations to the sheeting), and the wall put into contact with the backfill. In addition many tie rods are fitted with turnbuckles (later covered by fill) to take up initial tie-rod slack. The anchor rod is usually treated with a coat of paint or asphal- tic material to give corrosion protection.

The design of the anchor rod proceeds from an analysis of the anchor-rod force developed. In the finite-element analysis, the force developed depends somewhat on the size of the anchor rod selected to start the computations. In any case the deflection at the anchor-rod location should be checked (direct output in the finite- element analysis) to see if the wall movement is sufficient to develop the assumed active pressure. Whether the remainder of the wall moves sufficiently or not, if the movement at the anchor is not sufficient, active pressure may not develop. The wall movements cited in Sec. 11-1 under Active Earth Pressure may be used as a guide. If the movement is not sufficient, the anchor-rod spring constant must be decreased (or increased if too large). The compatibility of the anchor-rod force and the allowable

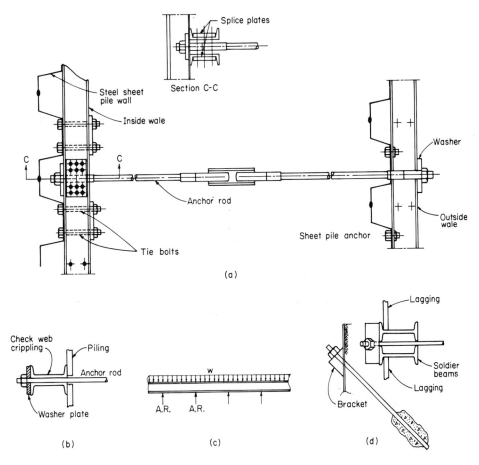

Figure 13-13. Wale location and design. (*a*) Front of wall or back of wall; (*b*) wale design—check web crippling; (*c*) wale design for bending assuming a uniform pressure from anchor rods; (*d*) soldier beams in pairs and using a tieback anchor system.

stresses for the anchor material must be checked. When node movement and anchor-rod stresses are compatible, the anchor-rod design is complete.

The wale is designed for bending and checked for web crippling at the junction of the web and flange. Some control over the web-crippling stresses may be obtained from use of a larger or smaller bearing or washer plate through which the anchor rod passes and against which the holding nut bears (Fig. 13-13*b*). At this location the full anchor-rod force is acting and is depending on the beam action of the wale to distribute the force across the anchor spacing *s* to the piling. The wale may be designed as a simply supported beam spanning between two anchor rods. This is usually too conservative, and preferably, the wale is considered as a continuous member. Finite-element continuous-beam computer programs are recommended for this part of the analysis. If the wale is a pair of channels as a continuous member, the lengths should be staggered so that both channels are not spliced at the same point and it will be necessary to supply a continuous butt weld or a sufficiently adequate bolted splice to effect wale continuity.

Design stresses may be as high as $0.9F_y$ for both the wale and anchor-rod material depending on engineering judgment and effect of wall failure.

Example 13-7. Design wales for lower anchor rod of Example 13-6. Consider a typical interior wale. Use fps units.

Data: $F_{a.r.} = 2{,}267.2$ kN (total)

$$s = 7 \text{ m}$$

SOLUTION. Assume linear distribution of stress. Assume $M = wL^2/10$ is reasonable for end moments; $M = wL^2/40 = $ midspan moment.

Step 1. Find moments

$$w = \frac{2{,}267.2}{7(1)} = 323.9 \text{ kN/m}$$

$$M = \frac{323.9(7)^2}{10} = 1{,}587 \text{ kN-m}$$

$$M = 1{,}170.7 \text{ ft-kips}$$

Step 2. Select sections. Assume anchor bearing plate is 12×18 in long and of adequate thickness.

$$\text{Take } f_b = 0.8F_y = 0.8(36) = 28.8 \text{ ksi}$$

$$S_x = \frac{1{,}170.7(12)}{28.8} = 487.8 \text{ in}^3$$

Use two standard beams (I beams) (AISC handbook)

$$\text{S24} \times 120 \qquad S_x = 2(252) = 503 > 487.8$$

$b_f = 8.05$ in   or about 20 in edge to edge with space between for each rod (or set of rods)

Check web crippling AISC Art. 1.10.10

$$f_b = 0.75f_y \text{ AISC and } k = 2 \text{ in}$$

$$N = 18 \text{ in} = \text{bearing-plate dimension}$$

$$R = \frac{2{,}267.2}{2(4.447)} = 254.9 \text{ kips}$$

$$f = \frac{R}{t_w(N + 2k)} = \frac{254.9}{0.798(18 + 4)} = 14.5 \ll 27 \text{ ksi} \qquad O.K.$$

Use a pair of parallel I beams S24 $\times$ 120. Spot-weld flange to wall at reasonable intervals, say, 2 in of 3/8-in weld at 2-ft intervals, both beams.

Anchorage for sheet piles may be obtained from large cast-concrete blocks or beams (deadmen) embedded in the soil, piles driven with or without a batter to take a horizontal pull, anchor rods extended to an existing structure, or a row of sheetpiling or from use of concrete or grout placed in predrilled holes with a bar or cable embedded in the concrete (Fig. 13-14). The latter method of anchorage, also termed "tieback" construction, is currently quite popular and with the use of high-strength

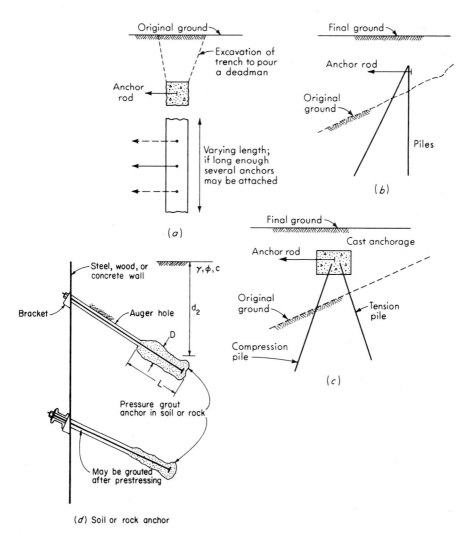

Figure 13-14. Sheet-pile anchorages. (*a*) Cast-in-place deadman; (*b, c*) piles used as anchors; (*d*) tieback consisting in concrete or grout in predrilled hole.

prestressing steel tendons used for the anchor rod (f_y of steel on the order of 180 to 200 ksi). It is not possible to fill the predrilled hole completely with concrete or grout, as this would not allow any stretching action in the prestress cable (anchor rod) to allow the system to work properly. It is possible to fill the void in the drill hole above the grout with additional grout after the anchor rod is pretensioned to the desired or required anchor force.

In all these systems, the anchor will fail under the application of the anchor-rod force during the tightening up of the system after the backfill is in place if the anchor is inadequate. It is necessary, however, to predict in advance the anchor resistance so that the analysis of the system may proceed.

For the tieback system, the anchor resistance can be estimated using Eqs. (4.15) or (4-16) when the base of the anchor is belled. When the base is not belled or the shaft is of variable diameter, the anchor resistance will be (refer to Fig. 13-14d)

$$F_{a.r.} = \pi D \gamma d_2 LK \tan \phi + c_a \pi DL \qquad (13\text{-}5)$$

where c_a is the adhesion of about 0.7 to 0.9c. Use the average shaft diameter for D, $L =$ length of grout and $d_2 =$ average depth of grouted length L. The use of $K = K_0$ can be justified if the concrete (or grout) is placed under pressure, which is often the case; otherwise use $K = K_a$. Values of $K > K_0$ are not recommended because of soil creep. This equation is similar to those of Littlejohn (1970); see also Oosterbaan and Gifford (1972). The simplest method of producing the anchor is to use a hollow-stem auger, drill to the desired depth, connect the grout hose to the hollow stem; gradually withdraw the auger and simultaneously pump the grout under pressure into the void. The soil on the flights above the auger bit retains the grout, thus ensuring a completely filled borehole of the required length. Other methods use a cased hole and grout injected under pressure while the casing is withdrawn. Additional design details for soil and rock anchor systems can be found in PCI (1974).

If the anchor is a deadman and is a long member with a length L considerably greater than the depth and if located so that the depth d_1 (refer to Fig. 13-15 for terms) is less than 0.5 to $0.7d_2$, the allowable anchor pull may be computed as

$$F_{a.r.} = \frac{(P_p - P_a)L}{F} \qquad (13\text{-}6)$$

where F is the factor of safety, say 1.25 to 1.5: P_p and P_a are the passive- and active-earth-pressure forces per unit of anchor width computed as $\frac{1}{2}\gamma d_2^2 K_i$. This equation takes into account that as the member increases in length the principal resistance is passive pressure and that the shear on the two ends of the deadman is negligible. It is self-evident that the anchor must be outside the Rankine active-pressure zone. It should also be far enough out that the passive anchor zone does not intersect the Rankine active zone.

If the deadman consists in essentially a short block of length $L \leq 1.5H$ and embedded in granular soil, the anchor resistance may be computed as

$$F_{a.r.} = \frac{C \gamma d_2^2 LK_p}{F} \qquad (13\text{-}7)$$

with terms also identified in Fig. 13-15. The coefficient C may be taken as 0.5 to 0.65, and K_p is the Rankine value (Table 11-4). Full-scale anchor-block tests by Smith (1957) indicated that $C = 0.65$ is reasonable to account for both passive resistance and end and bottom shear, which on short concrete blocks is considerable. One should use $C = 0.5$ for vertical steel plates. Tests reported by Tschebotarioff (1962) also indicate that $C = 0.65$ is reasonable.

Neely et al. (1973) performed a series of model anchor-plate tests using plates with the H of Fig. 13-15 taken as 2 in and varying the length L. Results were presented in terms of a shape-factor coefficient and a force coefficient. Comparison of that method with Eq. (13-7) by the author for a $\phi = 35°$ and a 2-ft-square plate gives

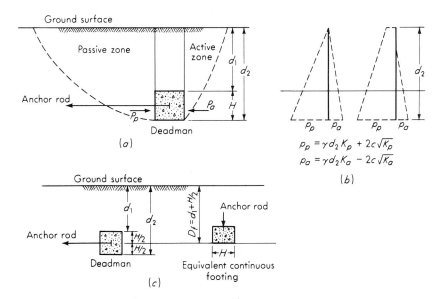

Figure 13-15. Anchor capacity of deadmen. (*a*) Deadman located near surface and $d_1 < 0.5$ to $0.7d_2$; (*b*) earth-pressure diagrams for granular or cohesive soils; (*c*) deadman located at H much smaller than d_2. For this case treat deadman as a continuous footing with D_f as shown. The footing must resist a V = anchor-rod force; use equations given in Chap. 4.

6.1 kips for the Neely method and 6.8 kips from Eq. (13-7). With this small difference and the fact that Neely et al. used models on a single sand of $\phi = f(\gamma)$, Eq. (13-7) is recommended for general use as being simpler.

For cohesive soil ($\phi = 0$) the anchor force can be computed as for a deep footing where the limit of bearing is approximately $9c$ and guided somewhat by tests of Mackenzie (1955) to obtain the following equation:

$$F_{a.r.} = McHL \tag{13-8}$$

where
$$M = 9 \text{ for } d_2/H \geq 3$$

$$M = 9d_2/H \text{ for } 0 \leq d_2/H < 3$$

For ϕ-c soils use Eq. (13-6) with the active and passive earth forces P_a and P_p evaluated to include both friction and cohesion resistance for long anchor blocks and

$$F_{a.r.} = \frac{P_p HL}{F} \tag{13-9}$$

for short anchor blocks.

When the anchor block is very deep, one may compute the resistance by Eq. (13-7) as a reasonable approximation for all values of length L.

There are several other methods of computing the resistance of vertical anchor plates [see, for example, several methods in US Steel (1974)]; however, alternative methods are considerably complicated from the preceding equations with little if any improved reliability of the computed anchor capacity.

PROBLEMS

For Probs. 13-1 to 13-4 refer to Fig. P13-1.

FLAGPOLE PROBLEM

Find D and make a recommendation of the sheet pile to use from the assigned problem. From the table $P = kN$, $r = kN/m^3$, $c = kPa$ when $l = $ meter. Use Rankine earth-pressure coefficients and no water present in embedment soil.

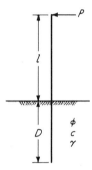

Figure P13-1

Table P13-1

No.	l	P	γ	ϕ	$c*$	Ans.†	Ans.‡
a	25 ft	10.0 kips	110	32°	1.2 ksf	25.0 ft	
b	5 m	50.0 kN	17.3	34°	50		19.78 m
c	6 m	30.0	17.2	35°	60	3.7	
d	18 ft	6.0	115	36°	2.5		13.4

* See Probs. 13-3 and 13-4.
† Ans. is for $F = 1.0$ and $c = 0$.
‡ Ans. is for $F = 1.0$ and $\phi = 0$.

13-1. Do the assigned problem in Table P13-1 for $(c = 0)$ a 20 percent increase of D.

13-2. Do the assigned problem of Table P13-1 for $(c = 0)$ $F = 1.4$ on K_p.

13-3. Do the assigned problem of Table P13-1 for $(\phi = 0)$ a 30 percent increase of D.

13-4. Do the assigned problem of Table P13-1 for $(\phi = 0)$ $F = 1.5$ on K_p.

13-5. Do the assigned problem of Table P13-1 using the finite-element method.

For Probs. 13-6 to 13-14 refer to Fig. P13-2 and Table P13-2.

Find the embedment depth D, draw the moment diagram, and select sheetpiling. Select anchor-rod diameter using either $f_y = 50$ ksi $(3,515$ kg/cm$^2)$ or $f_y = 150$ ksi $(10,546$ kg/cm$^2)$. If $H = $ ft, use fps units.

CANTILEVER WALL

13-6. Find the embedment depth for the assigned problem from Table P13-2 using a reasonable increase on the computed embedment depth D. Use Coulomb earth-pressure coefficients and $\delta = 17°$.

Answer (no increase in depth, $F = 1$): (a) $D = 20$ ft, $R_a = 8.86$ kips; (d) $D = 5.33$ m, $R_a = 119.2$ kN.

13-7. Find the embedment depth for the assigned problem from Table P13-2 using a safety factor of 1.4 on K_p. Use K_a and K_p values from Prob. 13-6.

Table P13-2

	a	b	c	d
H	25 ft	7.5 m	28 ft	7.5 m
h_1	10	3	12	3.5
h_a *	5	1.5	4	1.5
γ_1	105	16.5	110	17.3
ϕ_1	30	30	33	33
γ_2	122.5	19.2	122.5	19.3
ϕ_2	30	30	33	33
γ_3	122.5	19.2	122.5	19.3
ϕ_3	30	30	33	33
c†	1.2 ksf	190	1.0	160
h_b‡	10	3.0	12	3.5
Surcharge‡ q	0.50 ksf	24.0	0.60 ksf	30

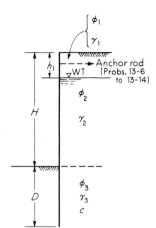

Figure P13-2

* For Prob. 13-11 to 13-14.
† For Probs. 13-8, 13-9, 13-12.
‡ For Prob. 13-14.

13-8. Find the embedment depth for the assigned problem from Table P13-2 if $\phi_3 = 0$ and c is as shown in the table. Use a reasonable percent increase on the computed embedment depth. Use Coulomb earth-pressure coefficients and $\delta = 17°$.

Answer (no increase in depth, $F = 1$): (*a*) $D = 12.91$ ft, $R_a = 8.30$ kips; (*d*) $D = 1.88$ m, $R_a = 113.25$ kN.

13-9. Find the embedment depth for the assigned problem from Table P13-2 if $\phi_3 = 0$ and c is as shown in the table. Use a safety factor of 1.4 (if possible) on c.

13-10. Solve the assigned problem from Table P13-2 using cohesionless, cohesive, or both below the dredge line and the finite-element method. Use $\delta = 17°$.

ANCHORED WALLS
13-11. Solve the assigned problem of Table P13-2 as an anchored wall in cohesionless soil and using the distance to the anchor rod h_a given in the table. Use $\delta = 17°$ and Coulomb earth-pressure coefficients.

Answer (no increase in depth for F): (*b*) $D = 3.3$ m, $F_{a.r.} = 70.9$ kN; (*c*) $D = 10.6$ ft, $F_{a.r.} = 5.6$ kips.

13-12. Solve the assigned problem of Table P13-2 as an anchored wall in cohesive soil below the dredge line using the distance to the anchor rod h_a and cohesion given in the table. Take $\delta = 17°$ and use Coulomb earth-pressure coefficients.

Answer (no increase in depth for F): (*a*) $D = 1.5$ ft, $F_{a.r.} = 3.9$ kips; (*d*) $D = 0.1$ m, $F_{a.r.} = 51.0$ kN.

13-13. Solve the assigned problem of Table P13-2 using the finite-element method both without and with cohesion in the embedment soil.

13-14. Solve the assigned problem of Table P13-2 using a second anchor rod at h_b below the backfill surface and with a surcharge q as given. Use either a hand solution or the finite-element method. Make the solution for both cohesionless and cohesive embedment soil.

FOURTEEN

BRACED COFFERDAMS FOR EXCAVATIONS

14-1. CONSTRUCTION EXCAVATIONS

It is a legal necessity when new construction is begun in a developed area to provide protection to the adjacent existing buildings when excavation in the new site is to any depth which may cause loss of bearing capacity, settlements, or lateral movements to existing property. New construction may include cut-and-cover work when public transportation or public utility systems are installed below ground and the depth is not sufficient to utilize tunneling operations. The new construction may include excavation from depths of 1 to perhaps 15 m or more below existing ground surface for placing a "shallow" foundation or a mat, to the placing of one to three more basements and subbasements.

This type of work requires installation of some kind of system of retaining structure termed a cofferdam or braced sheeting together with a means of holding the retaining structure in position. The retaining structure may be constructed of one of the following:

1. Sheetpiling (steel, concrete, or wood)
2. Soldier beams (or piles) with or without lagging
3. Drilled-in-place concrete piles (or piers)

Systems to hold the retaining wall in place include:

1. Wales and struts (most common) or rakers (Fig. 14-1)
2. Compression rings (when excavation is relatively small in plan)
3. Tieback anchorages

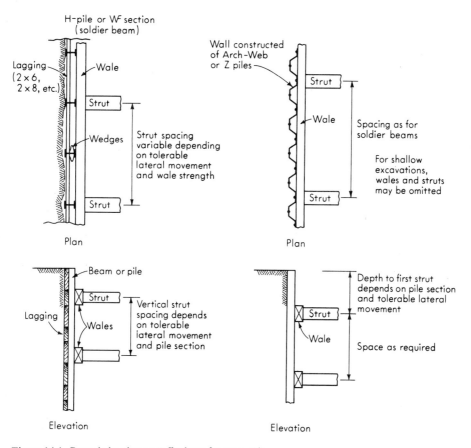

Figure 14-1. Braced sheeting or cofferdams for excavations.

Sheetpiling is commonly used for retaining the excavation because it has the highest strength/weight ratio, and much of the piling is reusable and can generally be easily installed either with sheet-pile hammers or with vibratory driving devices. It is not usable, however, where the subsoil contains many boulders or is dense and the excavation is deep. Where the soil is rocky or quite dense and where sheetpiling will be excessively damaged in driving, a system of soldier beams and lagging is often used (Fig. 14-1). This system consists in a series of H piles (soldier beams) driven on a convenient spacing (often approximately 8 to 10 ft for using standard-length timber). As excavation proceeds, 2-in-thick boards are inserted behind the front flanges, or (as is becoming common because of the accurate excavation required) the boards are placed against the pile and clipped to the front flange using patented fasteners. Where pile-driving vibrations (using both pile hammers and vibratory drivers) may cause damage to adjacent structures or the noise is objectionable, drilled-in-place piles may be used. The piles (or caissons if 76 cm or more in diameter) are drilled on as close centers as practical, and cast-in-place concrete is used.

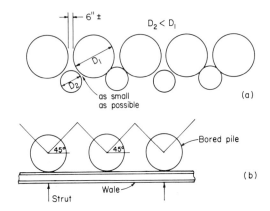

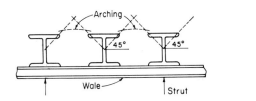

Figure 14-2. Excavation-bracing systems. (*a*) Using two rows of bored piles to effect a watertight wall; (*b*) using a single row of bored piles and relying on arching action to retain the earth; (*c*) using wide-flange shapes instead of bored piles.

Where earth is retained and water is not a factor, the soldier-beam spacing or drilled-in-place piles spacing may be such that lagging or other wall supplement is not required, as "arching" or bridging action of the soil from the lateral pressure developed by the pile will retain the soil across the open space. This zone width may be estimated roughly as the intersection of 45° lines as shown in Fig. 14-2. The piles will, of course, have to be adequately braced to provide the necessary lateral soil resistance.

Where earth and water must be retained, the system will have to be reasonably watertight below the water table and be capable of resisting both soil and hydrostatic pressures. It will seldom be practical to lower the water table, as this will also lower the surrounding soil and/or structures. Sheetpiling joints may allow enough water to pass into the excavation to effectively lower the water table. For these conditions the solutions may be limited to cast-in-place concrete walls or the use of grout around the perimeter of the sheetpiling or cast-in-place walls to reduce the soil permeability.

It is evident that uplift or buoyancy will be a factor for structures whose basements are below the water table. If uplift is approximately equal to the weight of the structure, or larger, it will be necessary to anchor the building to the soil. This can be done using some kind of anchor system such as anchor piles to bedrock, if in close proximity, or perhaps belled piles or vertical "tiebacks."

The sequence of operations is to place the piling (sheet, soldier beams, or cast-in-place) and make any initial waterproofing operations via grouting. During this period a sufficient number of photographs of the surrounding structures should be taken to establish their condition and a select number of ground elevations and control stations established so that ground loss (subsidence accompanying lateral movements into the excavation) can be detected and/or monitored.

The excavation then proceeds, and at selected depths based on both monitoring and prediction of ground loss, wales and struts are installed. The strut system creates obstructions in the excavation area which are undesirable. The alternative to struts is tiebacks, but this involves bringing a drilling machine to drill in the tieback anchorage, obtaining permission to trespass into the adjacent property owner's subsoil, and the problem of encountering public utilities. In spite of this, the tieback is the generally preferable solution where the adjacent property can be trespassed to the extent of installing the tieback anchors. The tieback anchors are left in place after construction is completed, as it would be excessively costly if not physically impossible to remove them.

Ground loss is a very serious problem in excavations in built-up areas. It has not been solved so far with any reliability; where the ground loss has been negligible, it has been a combination of overdesign and luck rather than rational analysis. The finite-element method presented in this chapter is one of the first methods to the author's knowledge of a semirational method of controlling the ground loss.

14-2. SOIL PRESSURES ON BRACED SHEETING OR COFFERDAMS

The braced cofferdam is subjected to the same earth-pressure forces as other retaining structures which may be calculated using the Rankine or Coulomb methods of Chap. 11. The design pressures, however, are different from those computed from the methods of Chap. 11 because of the manner in which the pressures are developed as idealized in Fig. 14-3. In stage 1 of Fig. 14-3 the wall is subjected to an active earth pressure, and wall yield takes place. The lateral deformation depends on cantilever soil-wall interaction as would be obtained by the finite-element program of Chap. 13. Next a strut force is applied to obtain stage 2. No matter how large the strut force

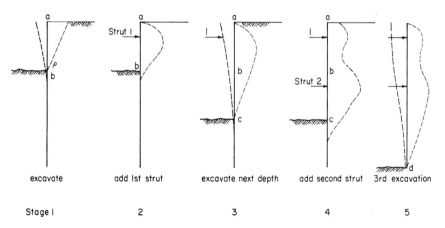

Figure 14-3. Stage development of earth pressure behind an excavation. The actual pressure depends on the strut force more than on the "active" earth pressure. In general, during excavation the strut force will decrease. Lateral movements during excavation stages are not recovered during strut addition stages.

(within practical limitations) the wall and earth is not pushed back to its original position but the strut force, being larger than the active pressure, causes an increase in the wall pressure. The integration of the pressure diagram at the end of stage 2 would be approximately the strut force—not exact because of some uncertainty of how the pressures act at, and below, the excavation line. The excavation as shown for the end of stage 2 causes a new lateral displacement between b and c and probably some loss of strut force as soil moves out of the zone behind the first strut into the displacement between b and c as well as soil creep. The application of the second strut force and/or tightening up of the first strut results in the qualitative diagram of stage 3 beginning and the excavation and additional ground loss due to lateral movement at the end of stage 3 when excavation proceeds from c to d. Thus it is evident that if one measures pressure in back of this wall the pressures measured will be directly related to the strut forces and will have little relation to the actual soil pressures involved in moving the wall into the excavation.

Peck (1943) and later Terzaghi and Peck (1969) proposed empirical pressure diagrams for wall and strut design using measured soil pressures obtained as from the preceding paragraph. Pressures reported by Krey in Berlin for sands were incorporated into the pressure diagrams. These pressure diagrams were obtained as the envelope of the maximum pressures found and plotted for the several projects. The pressure envelope was given a maximum ordinate based on a portion of the active earth pressure using the Coulomb (or Rankine) pressure coefficient. These diagrams with the latest modifications [Terzaghi and Peck (1967), Peck (1969)] are shown in Fig. 14-4 and diagrams by Tschebotarioff (1973) in Fig. 14-5. These diagrams are decidedly conservative, as one would expect. Certainly if one designs a strut force based on this pressure diagram and used simply supported beams for the sheeting as proposed by Terzaghi and Peck, the strut force will produce not more than that pressure diagram owing to creep and ground loss and the sheeting will be certainly overdesigned owing to both pressure-diagram discrepancies and sheeting continuity. This was verified by Lambe et al. (1970) and by Golder et al. (1970) wherein predicted and measured strut loads varied by as much as 100 percent. Swatek et al. (1972), however, found reasonable agreement with the Tschebotarioff pressures in

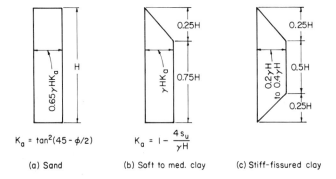

Figure 14-4. Lateral-earth-pressure diagrams as suggested by Peck (1969). Use q_u from unconfined compression tests for the undrained shear strength s_u. Use (b) above for $\dfrac{\gamma H}{c} > 4$; (c) for $\dfrac{\gamma H}{c} \leq 4$.

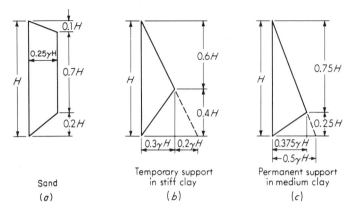

Figure 14-5. Lateral-earth-pressure diagrams against braced sheeting as proposed by Tschebotarioff (1973).

designing the bracing system on a Chicago, Ill., excavation 21.3 m deep. Swatek, however, used a "stage-construction" concept similar to Fig. 14-3 along with the Tschebotarioff pressure diagram. In general, the Tschebotarioff method may be more correct when the excavation depth exceeds about 50 ft (16 m).

Soil Properties

Drained soil parameters for stiff clays and ϕ-c soils in general may be appropriate for lateral pressures behind braced walls where the excavation is open for a considerable length of time, say, more than about 3 months. Bjerrum and Kirkedam (1958) measured pressures in an excavation from September through November where the lateral pressure increased from 20.3 to 63.2 kPa owing to loss of cohesion (based on back computing using initial ϕ and c and final ϕ). Cohesion is often dissipated in cuts because of changes in moisture content, oxidation, tension cracks, and possibly other factors such that on a long-term basis it may not be conservative to rely on cohesion to reduce the lateral pressure. Where the cut is open only 2 to 5 days, the cohesion is relied upon extensively to maintain the excavation sides.

    Soil parameters using triaxial tests with decreasing lateral pressure may be more appropriate than standard triaxial tests. Often, however, the designer must extrapolate lateral pressures and make some estimate of the ϕ angle from penetration-test data. For small projects or rural projects where adjacent property damage is negligible, this may be satisfactory; for other projects this is not good engineering practice, since the net result will be considerable overdesign of the bracing system and also possible large damages to adjacent property. The additional soil engineering will be a relatively small cost when weighed against these alternatives.

    Where the soil is stratified, the soil parameters may be obtained for each layer and treated as in Chap. 11, particularly if stage construction is analyzed. Sometimes it is necessary to substitute the parameters of an equivalent soil for the several layers behind the bracing, as for example, using the pressure diagrams of Fig. 14-4 or 14-5. For these conditions one may use a "weighted" average of the soil mass based on engineering judgment of the contribution (weight) of each layer to the total-equivalent-mass properties.

Plane-strain conditions normally are obtained along the sides (and ends) of sizable excavations. At the corners or ends of narrow excavations, triaxial conditions are likely to be obtained. Plane-strain conditions can be obtained from triaxial tests by increasing ϕ approximately 10 percent. Alternatively, as triaxial values tend to be more conservative than plane-strain values, one may use the triaxial values directly.

14-3. CONVENTIONAL DESIGN OF SINGLE-WALL (BRACED) COFFERDAMS

The conventional design of a braced cofferdam proceeds as follows:

1. Sketch given conditions and indicate all known soil data, stratification, water level, etc.
2. Compute the pressure-distribution diagram by the Tschebotarioff or Terzaghi and Peck method. In the case of a cofferdam in water, the lateral pressure will be only hydrostatic pressure.
3. Design the components (sheeting, wales, and struts).

The design of the components requires some thought. If continuous sheeting is used, the sheeting becomes a continuous beam supported at the wales, cantilevered at the top, and either fixed, partially fixed, hinged, or cantilevered at the bottom, depending on the amount of penetration below the excavation line. This also applies to the soldier beams. The most rapid (and conservative) solution is to assume conditions as in Fig. 14-6. That is, the top is treated as a simple cantilever beam, including

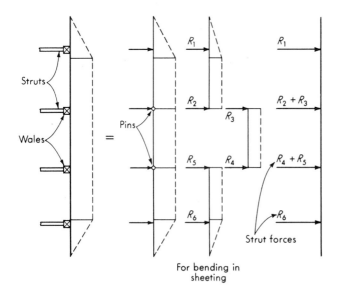

Figure 14-6. Simplified method of analyzing sheting and evaluating strut forces.

the first two struts. The remaining spans between struts are considered as simple beams with a hinge or cantilever at the bottom.

A somewhat more economical solution may be achieved by using moment distribution. Obviously, the span length influences the moment, but it may also require additional struts. It is possible to space the wales (and struts) so that the bending moments are equal at the support points; however, construction considerations may require that the number of struts be a minimum, since these are obstructions in the work area as well as installation problems.

The struts are actually horizontal columns subjected to an axial force (the R_n of Fig. 14-6) and bending due to the dead weight of the member. For narrow excavations the bending moments may be negligible. For wide excavations it may be necessary to splice several members to make a compression member long enough to span between the opposing wales. Since the strut is a column, the carrying capacity is dependent on the l/r ratio. To reduce the l/r ratio for economy or other reasons, intermediate supports (both lateral and vertical) may be used, even though this provides additional work-area obstruction. It may also be necessary to provide a means of maintaining or increasing the strut force against the wale to reduce the potential lateral movement of the supported soil. It should be noted that the strut system should prevent slip, since an attempt to push the slipped soil back into place will be very nearly (if not totally) impossible.

The wales may be treated as continuous members for the length of the wale or as simply supported members pinned at the struts (conservative solution). If the wales are adequately spliced for moment, they may be treated as continuous members for the entire length. Horizontal strut spacing is dependent on the bending (occasionally shear) resistance of the wale.

The first horizontal strut location should not exceed the depth of the potential tension crack,

$$h_t = \frac{2c}{\gamma \sqrt{K_a}}$$

since the formation of this crack will increase the lateral pressure against the sheeting and if the crack fills with water, the pressure will be increased even more. The water may also destroy the cohesion of the soil, as well as soften it. The first strut location

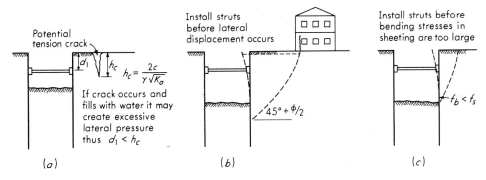

Figure 14-7. Depth of first wale and struts in a braced system.

should also consider the effect of the location of successive potential failure wedges (Fig. 14-7b), whose location depends on the depth of the excavation. Where lateral movement and resulting ground subsidence can be tolerated, the depth to the first strut in sands may be where the allowable bending stress in the sheeting is reached (Fig. 14-7c).

Example 14-1. Partial design of a braced sheeting system.
Given system shown in Fig. 14-1a and the conventional method.
Clearances:

$$V = 10 \text{ ft} \qquad H = 18 \text{ ft}$$

REQUIRED
1. Pressure diagram
2. Select sheeting
3. Design one strut
4. Design one wale

SOLUTION. *Step 1*. Obtain pressure diagram. For convenience use 0.2H instead of 0.25H from Fig. 14-4c. For $\phi = 36°$ the Rankine $K_a = 0.26$

$$\sigma_{max} = 0.8K_a\gamma H$$

$$= 0.8(0.26)(0.120)(31) = 0.773 \text{ ksf}$$

from which Fig. E14.1b can be constructed. Note that $0.2\gamma H = 0.744$ ksf.

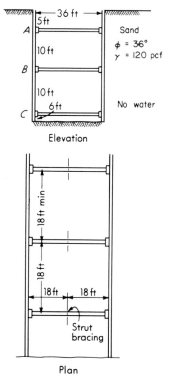

Elevation

Plan

Figure E14-1a.

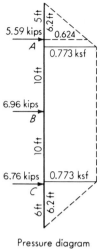

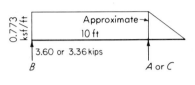

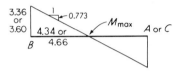

Pressure diagram

Figure E14-1b.   **Figure E14-1c.**

Step 2. Find strut forces (kips/ft of wall)
For struts A and B:

$$\sum M_B = 0$$

$$8.8(4.4)(0.773) + 0.773\left(\frac{6.2}{2}\right)\left(8.8 + \frac{6.2}{3}\right) - 10A = 0$$

$$10A = 29.9 + 26.0$$

$$A = 5.59 \text{ kips/ft of wall}$$

$$\sum M_A = 0$$

$$8.8(0.773)(5.6) - 0.773(3.1)(0.87) - 10B_1 = 0$$

$$10B_1 = 38.09 - 2.08$$

$$B_1 = 3.60 \text{ kips}$$

To find rest of strut force B take $\sum B = 0$ for section BC and round off pressure diagram as shown in Fig. E14-1c

$$10(0.773)(5) + 0.773\left(\frac{6.2}{2}\right)\left(10 + \frac{6.2}{3}\right) - 10C = 0$$

$$10C = 38.7 + 28.9 \qquad C = 6.76 \text{ kips}$$

$$\sum M_C = 0$$

$$38.7 - 0.773\left(\frac{6.2}{2}\right)\left(\frac{6.2}{3}\right) - 10B_2 = 0$$

$$10B_2 = 33.6 \qquad B_2 = 3.36 \text{ kips}$$

$$B_{\text{total}} = B_1 + B_2 = 3.36 + 3.60 = 6.96 \text{ kips/ft}$$

The maximum strut force is strut B with 6.96 kips/ft.

Step 3. Find bending moment/ft of wall

$$M_A = 0.624(5/2)(5/3) = 2.6 \text{ ft-kips}$$

$$M_B = 0 \text{ assumed hinge at } B$$

$$M_C + 0.773\left(\frac{6.0}{2}\right)\left(\frac{6.0}{2}\right) = 4.64 \text{ ft-kips (approx.)}$$

$$M_{max} = \frac{3.36(4.34)}{2} = 7.29, \text{ or } 8.39 \text{ ft-kips}$$

Step 4. Selection of sheeting (use A36 steel and $f_s = 24$ ksi).

$$S_x = \frac{8.39(12)}{24} = 4.19 \text{ cu in}$$

Use PMA-22 $S_x = 5.4$ cu in/ft of wall

Step 5. Design of strut
Largest forces occur at strut B; therefore, for $s = 18$ ft (Fig. E14-1a)

$$P = 6.96s = 6.96(18) = 125.3 \text{ kips}$$

Try a W12 × 45: Assume strut laterally braced at midpoint but not vertically supported; take $C_m = 1.0$, $K = 1$

Unbraced length $l_y = 18$ ft $> L_u$ of 17.5 (approx. 18 ft)

$$l_x = 36 \text{ ft} \qquad r_x/r_y = 2.65; \text{ therefore, } l_y \text{ controls;}$$

$$\text{use } F_b = 22 \text{ ksi}$$

$$P(Kl)^2 = 125.3(36 \times 12)^2 = 23.4 \times 10^6$$

From AISC Handbook tables

$$r_y = 1.94 \text{ in}; r_x = 5.15; B_x = 0.227; a_x = 52.2 \times 10^6, I = 351 \text{ in}^4$$

$$\frac{Kl}{r_y} = \frac{18(12)}{1.94} = 111.3 \qquad F_a = 11.50 \text{ ksi for A36 steel}$$

$$M_b = wl^2/8 = 0.045(36)^2(12/8) = 87.5 \text{ in-kips due to beam weight}$$

$$\Delta = 5wl^4/384EI = \frac{5(0.045)(36)^4(1,728)}{384(29 \times 10^3)351} = 0.167 \text{ in due to beam weight}$$

$$M \text{ due to } P(\Delta) = 125.1(0.167) = 20.9 \text{ in-kips;}$$

$$M_{total} = 20.9 + 8.39(12) = 121.6 \text{ in-kips}$$

$$P + P' = P + C_m B_x M \frac{F_a}{F_{bx}} \frac{a_x}{a_x - P(Kl)^2} \quad \text{AISC Eq. 1.6-1a (modified)}$$

$$= 125.3 + 1(0.227)(121.6)\frac{11.50}{22}\frac{52.2}{52.2 - 23.4}$$

$$= 125.3 + 26.2 = 151.5 < 152 \text{ kips in AISC Column Tables} \qquad O.K.$$

$$P + P' = P \frac{F_a}{0.6F_y} + B_x M_x \frac{F_a}{F_b} \qquad \text{AISC Eq. 1.6 } - 1b \text{ (modified)}$$

$$P + P' = 125.3 \frac{11.50}{22} + 0.227(121.6) \frac{11.50}{22}$$

$$= 65.5 + 14.4 = 79.9 < 152 \text{ kips} \qquad O.K.$$

Therefore, we can use a W12 × 45 rolled section for the struts.

Step 6. Design of first wale. The middle wale is critical. The load per foot is 10(0.773) = 7.73 kips/ft. Neglect bending of the wale due to its dead weight about the Y axis. Since field splices for moment are difficult to make, assume the member is 18 ft long and joined in non-moment-resisting joints at the strut locations. The member should be fastened at selected points along the compression flange to the sheeting to satisfy lateral-support requirements for L_u and to aid in reducing any bending about the Y axis.

$$M = \frac{wl^2}{8} = \frac{7.73(18)^2}{8} = 313.1 \text{ ft-kips}$$

and the required section modulus S is for $f_b = 24$ ksi $= 0.66F_y$

$$S_x = \frac{313(12)}{24} = 156.5 \text{ in}^3$$

A W24 × 68 with an $L_u = 11.7$ ft and $S = 153 < 156.5$ could be used; however, this projects 24 in into excavation. Check section using $F_y = 50$ ksi.

$$S_x = \frac{313(12)}{33} = 113.8 \text{ in}^3 \qquad \text{Use a pair of 12 × 45 sections}$$

S_x furnished $= 2(58.2) = 116.4 \text{ in}^3 > 113.8$

A slightly less conservative section could have been selected using the apparent wale loading of 6.95 kips/ft computed at strut B in step 2 of this example.

14-4. ESTIMATION OF GROUND LOSS AROUND EXCAVATIONS

The estimation of ground loss around excavations is a considerable exercise in engineering judgment. Peck (1969) has given curves as Fig. in 14-8 in nondimensional form which may be used to obtain the order of magnitude. Caspe (1966, but see discussion in November 1966 critical of the method) presented a method of analysis which requires an estimate of the bulkhead deflection and Poisson's ratio. Using these values, Caspe back-computed one of Peck's (1943) excavations in Chicago and obtained reasonable results. A calculation by the author indicates, however, that one could do the following steps and obtain results about equally good:

1. Obtain the lateral wall deflections.
2. Numerically integrate the wall deflections to obtain the volume of deflection V_s. Use average end areas, the trapezoidal formula, or Simpson's 1/3 rule.

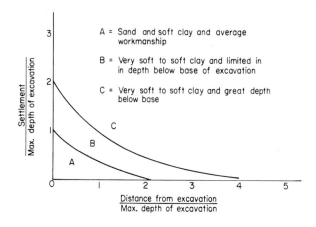

Figure 14-8. Curves for predicting ground loss. [*After Peck (1969).*]

3. Compute or estimate the distance of the settlement influence. The method proposed by Caspe is satisfactory as follows (for clay-base soil):
   a. Compute wall height to dredge line H_w.
   b. Compute a distance below the dredge line $H_p = B$ (for $\phi = 0$) where B = width of excavation. For $\phi > 0$ soil use $H_t \cong 0.5B \tan(45 + \phi/2)$, then

$$H_t = H_w + H_p.$$

   c. Compute distance of settlement influence D as

$$D = H_t \tan\left(45 + \frac{\phi}{2}\right)$$

4. Compute the surface settlement at the wall as

$$S_w = \frac{4V_s}{D}$$

5. Compute remaining settlements assuming a parabolic variation of x from D to wall

$$S_i = S_w\left(\frac{x}{D}\right)^2$$

Example 14-2. Use the values provided by Caspe and verify the above. Figure E14-2 displays data from Caspe and as plotted on Peck's settlement curve.

SOLUTION. *Step 1.* Integrate wall deflections and compute V_s

$$V_s = \frac{4}{12}\left(\frac{1.2 + 0.2}{2} + 1.3 + 1.4 + \cdots + 0.5\right) = 4.4 \text{ cu ft}$$

The settlement-influence distance is taken as 90 ft (given by Caspe) and computed as follows:

$$H_t = 38 \text{ ft} + 52 = 90 \text{ ft} \qquad D = 90 \tan 45 = 90 \text{ ft}$$

Step 2. Compute ground settlements

$$s_w = 4.4(4)/90 = 0.196 \text{ ft} = 2.35 \text{ in (vs. about 2.04 at wall of Peck)}$$

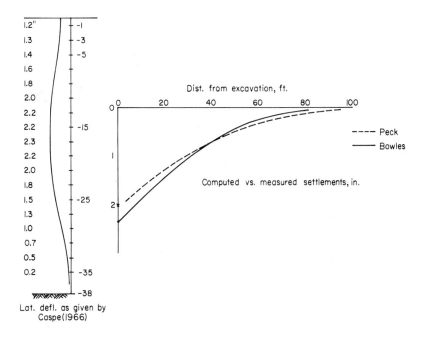

Figure E14-2.

Step 3. Compute settlements at 20, 40, and 60 ft from wall (70, 50, and 30 ft from D).

$$s_{20} = 2.35(70/90)^2 = 1.42 \text{ in (about 1.3 of Peck)}$$
$$s_{40} = 2.35(50/90)^2 = 0.72 \text{ in (about 0.7)}$$
$$`s_{60} = 2.35(30/90)^2 = 0.26 \text{ in (about 0.3)}$$

Several factors complicate the preceding calculations, including possible lack of values for wall movements (the finite-element method of the next section will provide wall movements for initial estimates) and the construction sequence, which may affect the settlement pattern considerably, since the analysis just given considers the completed wall. The preceding calculation may be performed in stages, however, if the finite-element method of wall analysis is used, since the lateral movements and associated movements will be obtained at each stage of construction. The user must make corresponding adjustments in the distance D based on engineering judgment for the several stages.

14-5. FINITE-ELEMENT ANALYSIS FOR BRACED EXCAVATIONS

The finite-element method (and computer program in the Appendix used for sheet-pile walls) can be applied to the solution of braced excavation problems and considering stage construction as follows:

1. Draw proposed wall configuration and soil profile to approximate scale.
2. Select the structural member for the wall sheeting pile(s), sheet pile, or other type of sheeting. Compute the moment of inertia/unit width, and obtain the modulus of elasticity.
3. Compute the active-earth-pressure coefficients for the soils in the various strata (generally use plane-strain, undrained values for long walls). The Coulomb earth-pressure coefficient is generally recommended. Estimate the modulus of subgrade reaction for the various soil strata. Generally it is expected that the modulus of subgrade reaction will increase with depth. Actual values are probably not highly critical, and a lower limiting value can be obtained by observing the output deflections that do not produce wall movements into the backfill.
4. The wall will be analyzed in stages as follows:
   Stage 1: Treat wall as a cantilever for some depth of excavation. At this point analyze the lateral movements for optimum depth of initial excavation vs. lateral movement to attempt to control ground loss. Use active earth pressure to the dredge line as for any cantilever wall. Estimate k_s for soil below dredge line to supply cantilever resistance.
   Stage 2: Apply first strut force (value larger than active earth pressure of stage 1 given on output pages). Apply subgrade reaction behind wall. Do not apply a subgrade-reaction value to the far end of element terminating at the dredge line, as the spring at that point is obtained from the dredge-line soil. With this input, the wall is solved for stage 2. The output is checked and the algebraic sums of the nodal deflections from stage 1 and stage 2 are the nodal deflections at this point. If the net deflections are $(-)$, indicating the wall is moving away from the excavation, the modulus of subgrade reaction used for stage 2 is too small and the values should be increased and the analysis redone. Soil springs are necessary at this stage to absorb the unbalance of horizontal forces; otherwise, the unbalance would be treated by the program as cantilever action below the dredge line.
   Stage 3: With the force system above the initial dredge line from stage 2 as *P*-matrix entries, compute the active pressures from the initial dredge line to the next estimated dredge line using engineering judgment for the pressure at the initial dredge line (zero or treat soil above the initial dredge line as a surcharge, etc.). Read the pressures just computed to the new dredge line. Note that in the computer program one must read values for each node from the top of the wall to the node just below the dredge line. The node just below the dredge line is always 0.0. For stage 3, in the region where the *P* matrix is already developed, the pressures are read in as 0.0. The user should carefully inspect the output to see how the program handles the nodes at the junction of old dredge line and new excavation. The output gives a new set of lateral deformations.
   Stage 4: Repeat stage 2 operation from ground down to the new dredge line with a second strut force at the desired node. Add additional subgrade modulus to additional wall elements. Check output to see if net deflections are reasonable. Note that one may in this stage allow for an increase or decrease in the first strut force to simulate either creep or tightening the strut into the soil.
   Stage 5: Repeat stage 3, and other stages as necessary to stage $N - 1$.
   Stage N-1: Stage *N*-1 is the last excavation stage, with the *N*-1 strut force being

applied and with any necessary adjustments in earlier strut forces. If the wall sheeting extends below the final dredge line, stage *N*-1 is similar to any previous stages where struts are being applied; any unbalanced force will be taken by the soil below the dredge line. If, however, at stage *N*-1, the wall sheeting terminates at or above the dredge line, the last excavation may introduce either lateral or rotational instability into the problem, the result of which is a program cancellation or very large deflections. To avoid this, it is prudent to insert one or more springs behind the wall, say, near the top and base. Any instability will be reflected into the computed force in the springs, and large values will require that the user modify both the "springs" and their location to provide a reasonable answer. If the problem is carefully done, the "springs" can be almost any value at almost any location and the computed spring forces and deflections will be negligible.

Stage N: Same as stage 2 with last strut force applied and remaining wall elements given subgrade-reaction values.

Final wall deflections are the cumulative sums of the deflections from each stage. Instantaneous deflections are obtained as the algebraic and cumulative sum of the deflections at the end of each stage. These deflections can be used with the procedure of Sec. 14-4 to obtain estimates of ground loss, the principal difficulty with using the procedure of Sec. 14-4 now being how to estimate the zone of settlement influence from the wall at the end of the stages.

Note that with the computer program and most of the data input prepared in the initial analysis, it will be relatively simple to modify the solution as field work proceeds and measured deflections are obtained in the initial stages of excavation.

Example 14-3. Make an initial estimate of lateral movements of a braced excavation reported by Lambe (1970) as shown in Fig. E14-3.

SOLUTION. The solution will be illustrated with sketches showing the stages, typical input, and selected output values of deflection.

Step 1. Select all node locations, code wall, estimate soil-modulus values and lateral-earth-pressure coefficient(s).

(a) Values of (measured) strut forces will be taken from Lambe's paper (p. 190) as estimated from a plot of strut forces (kips) at a spacing of $S = 12$ ft.

$$S_1 = 90/12 = 7.5 \text{ kips/ft} \qquad S_2 = 300/12 = 25.0$$

$$S_3 = 215/12 = 17.92 \qquad S_4 = 230/12 = 19.12 \qquad S_5 = 270/12 = 22.5$$

(b) 16 nodes will be taken as shown in Fig. E14-3.

(c) Soil springs behind wall will be as identified in each stage with constant $k_s = 100$ to illustrate method.

(d) Soil springs in front of wall will be computed as follows (take 0.80 for dredge-line reduction factor due to excavation damage):

$$k_s = (\bar{q}N_q + \tfrac{1}{2}\gamma BN_y)12 \qquad \phi = 35°$$

$$= [0.036(33)Z^1 + 0.5(0.036)(40.7)]12$$

$$\cong 9 + 14Z^1$$

use $k_s = 4.5 + 7Z^1$ for second excavation only

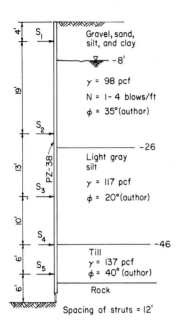

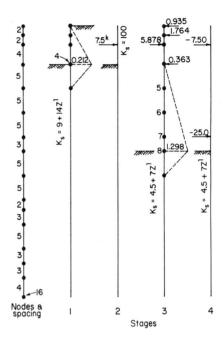

Figure E14-3.

(e) Soil pressures for first excavation are:

Node   (use Rankine coefficients—could have used Coulomb values;
         ϕ angles estimated by author)

1   0

2      $0.098(2)(0.27) = 0.053$

3      $0.098(4)(0.27) = 0.106$

4      $0.098(8)(0.27) = 0.212$

5   (arbitrary) $= 0.0$

(f) $E = 30 \times 10^3 \times 144 = 4{,}320{,}000$ ksf

$$I \text{ for } PZ38 = 280.8 \text{ in}^4/\text{ft} = 0.0135 \text{ ft}^4$$

(g) Card input for first stage is as follows:

Card	Item									
1	Title									
2	Units									
3	32	15	0	1	0	0	1	0	0	1
4	0	4	0	0	0	0	0	0	0	5
5	4320000.	0.	2.	0.8	0.	10.	1.			

Note DEMB, CONV, and FAC2 can be anything

6	9.0	14.	1.0					
7	2.	2.	4.	5.	5.	5.	3.	5.
8	5.	2.	3.	5.	3.	3.	4.	
9	1.0	0.0135						
10	0.0	0.053	0.106	0.212	0.0	(PRESS(I))		

These are data for stage 1

Output: $X(2) = 0.0162$ ft $X(16) = -0.0002$ ft

Step 2. Apply *S*1 (strut 1)

(*a*) Inspect output and see if deflections are realistically satisfactory; if not change depth of excavation and redo.

(*b*) Duplicate cards with changes indicated

1 and 2	same
3	change NNZP = 1 for strut force, IBRAC = 3
5	through 8 same
9	100. 100. 100. 100. 100. 0.0
	(soil modulus behind wall arbitrarily assigned these values, note 0.0 for far end of third element, as this terminates at dredge line and there is already soil there)
10	1. 0.0135 (old card 9)
11	6 1 −7.50 (K, J PR(K, J))

This is all of input stage 2

$$\text{Output: } X(2) = -0.0124 \text{ ft} \qquad X(16) = 0.0002 \text{ ft}$$
$$\text{Output } X(2) \text{ leaves a net } (+) \text{ deflection of } 0.0167 - 0.0124$$
$$= +0.0043 \text{ ft and thus may be satisfactory}$$

Step 3. Stage 3—excavate to node 8 (see Fig. E14-3)

(*a*) Compute pressures include hydrostatic effect:

Node	
1 through 4	= 0.0
5	$5(0.27)(0.0356)(5) + 0.0625(5) = 0.361$
6	$10(0.27)(0.0356) + 0.0625(10) = 0.7201$
7	$15(0.27)(0.0356) + 0.0625(15) = 1.0815$
8	$18(0.27)(0.0356) + 0.0625(18) = 1.2980$
9	arbitrary 0.0

(*b*) Obtain *P* matrix for part supported by first strut from output of soil reactions from stage 2 with attention to signs

$$P(2) = 0.935 \text{ kip}$$

$$P(4) = 0.882 + 0.882 = 1.764 \text{ kips}$$

$$P(6) = 0.811 + 1.622 - 7.5 = -5.878 \text{ kips}$$
$$P(8) = 0.363 \text{ kip (this is at dredge line and could have been}$$
$$\text{neglected; alternatively, continue using values as}$$
$$\text{long as sign does not change)}$$

(*c*) Card input

1 and 2	same
3	NNZP = 4; IBRAC = 0
4	Change JTSOIL = 8; IPRESS = 9
5	Same
6	4.5 7.0 1. (cut k_s by 50 percent for silt stratum)
7–8	Same (H(I))
9	1.0 0.0135
10	2 1 0.935
11	4 1 1.764
12	6 1 −5.878
11	8 1 0.363
12	0. 0. 0. 0. 0.3610 0.7201 1.0815 1.2980
13	0.

Total data

$$\text{Output: } X(2) = 0.3469 \text{ ft}$$

$$X(16) = 0.0973 \text{ ft}$$

Step 4. Stage 4—apply second strut

(a) Data card input

1 and 2	Same
3	NNZP = 2; IBRAC = 7
4	IPRESS = 0
5–8	Same

9   100.   100.   100.   100.   100.   100.   100.   100.

        100.   100.   100.   100.   100.   0.0

10	1.	0.0135	
11	6	1	−7.50
12	14	1	−25.0

$$\text{Output: } X(2) = -0.007 \text{ ft}$$

$$X(16) = 0.0207 < 0.0973 \quad O.K.$$

Step 5. Stage 5—excavate to node 11 with struts 1 and 2 in card input

1 and 2	Same
3	NNZP = 8; IBRAC = 0
4	JJTSOIL = 11; IPRESS = 12
5–6	Same; Change k_s back to 9.0, 14.0, and 1.
7–9	Same

10   0.725   1.348   −5.669   1.992   3.079   7.894   −14.494   0.671

11   0.0   0.0   0.0   0.0   0.0   0.0   0.0   0.0

12   0.4021   0.8042   0.9650   0.0

$$\text{Output: } X(2) = 0.0582 \text{ ft}$$

$$X(16) = 0.00579 \text{ ft}$$

Step 6. Stage 6—add strut 3 (−17.92 kips)

$$\text{Output: } X(2) = -0.00782 \text{ ft}$$

$$X(16) = -0.01392 \text{ ft}$$

Step 7. Stage 7—excavate to node 14

$$\text{NNZP} = 1 \quad \text{IPRESS} = 15$$

$$P(2) = 0.7820 \quad P(12) = 5.8840$$

$$\text{PRESS(I)} = 0.0 \text{ except}$$

$$12 = 0.2172 \quad 13 = 0.5791 \quad 14 = 0.7963$$

$$\text{Output: } X(2) = -0.2123 \text{ ft}$$

$$X(16) = -0.1142$$

Step 8. Stage—8 add strut 4; all k_s behind wall = 100. (NNZP = 4)

$$P(6) = -7.5 \quad P(14) = -25.0$$

$$P(20) = 17.92 \quad P(26) = -19.12 \text{ kips}$$

Step 9. Stage 9—excavate to bottom of wall, since system may be rotationally unstable and "springs" using JJS = 2.

Place springs at:	2	0.0	50
	12	0.0	50

i.e., far-end springs on elements 2 and 12.

These locations are totally arbitrary, and output should be inspected to see if realistic. In this case the top spring carries 1.103 kips tension, the lower spring 3.825 kips compression, indicating that the locations are reasonably satisfactory (the wall was "unbalanced" in this stage by −2.721 kips into the backfill).

$$\text{Output: } X(2) = 0.01226 \text{ ft}$$

$$X(16) = 0.02630 \text{ ft}$$

Step 10. Stage 10—apply fourth strut (−22.5 kips)

$$\text{IPRESS} = 0 \qquad \text{IBRAC} = 16 \qquad \text{NNZP} = 4$$

The soil modulus on far end of member 15 was taken as 0.0, since excavation was below the bottom of the pile. All other k_s values 100.0 as before

$$\text{Output: } X(2) = -0.00733$$

$$X(16) = -0.01737$$

Step 11. Compute final deflections at nodes 1 and 8 (net sum of stage deflections)

At node 1: $X(2) = 0.0162 - 0.0009 + 0.3469 - 0.0072 + 0.0582 - 0.0078$
$- 0.2123 - 0.0073 + 0.1226 - 0.0073 = 0.301$ ft (into excavation)
$X(16) = -0.0002 + 0.0002 + 0.0973 - 0.0207 + 0.0058 - 0.0139 - 0.1142$
$- 0.0177 + 0.0263 - 0.01737 = -0.054$ ft (away from excavation)

General comments and observations concerning this problem:

1. Node 1 appears adequate.
2. Node 8 appears incorrect owing to direction of net deflection; however, it may be possible because of location near the soft silt.
3. This problem is for illustrative purposes, and effort to keep input simple means the best solution was probably not obtained.
   a. Use of rather arbitrary ϕ angles—silt may require an alternative method of obtaining K as $K = 1 - mq_u$.
   b. Use of arbitrary soil modulus of 100 kcf behind wall for full height; it is reasonable that k_s would be different with depth.
   c. Use of only two somewhat arbitrary values of k_s for dredge-line soils.
   d. Use of constant values of strut forces for all stages—a more reasonable method might be to allow some arbitrary strut-force loss of, say, 10 to 20 percent during excavation. These values were taken simply as measured values shown by Lambe (1970). In fact, in this type of analysis the strut force may be arbitrarily selected and an analysis made, those strut forces proving satisfactory being used for design. One might also estimate the strut "spring" as AE/L and input that value instead of a strut force and compute a force necessary for equilibrium. This value could then be increased to obtain the design value.
4. It would probably be more correct in computing the lateral pressure for each stage to treat the soil above the preceding excavation as a surcharge so that the first lateral-pressure value is qK_a instead of 0.0 as used in this example.

14-6. INSTABILITY DUE TO HEAVE OF BOTTOM OF EXCAVATION

When a cofferdam is located over, or in, a soft clay stratum (Fig. 14-9), the clay may flow into the excavation if sufficient soil and/or water is removed from the cell.

Referring to Fig. 14-9, a theoretical analysis can be made using Eq. (2-15), as follows:

$$\sigma_3 = \sigma_1 K_a - 2c\sqrt{K_a}$$

From the figure it is evident that the lateral pressure on an element of soil at point $A(\sigma_3)$ is also the lateral pressure on the element at point B. At point B the element is subjected to a vertical pressure of σ_1' and the same σ_3 as at A. When all the soil over the element at B is removed, σ_1' will reduce to zero. At any time prior to this the soil element will be subjected to a minor principal stress of γh and the major principal stress of σ_3

$$\sigma_3' = \sigma_1' K_a - 2c\sqrt{K_a}$$

or

$$\gamma h = \sigma_3 K_a - 2c\sqrt{K_a}$$

For undrained conditions ($\phi = 0$), where $\gamma h = 0$, clay will flow into the excavation when $\sigma_3 \geq 2c$. An $F = 1.25$ to 1.5 should be allowed for this condition.

An alternative method of computing the stability of the bottom of the excavation in a cofferdam against heave was proposed by Terzaghi (1943, p. 231) for wide and relatively shallow (small D/B ratios) excavations. Skempton (1951) made proposals which can be used for the case of large D/B ratios. Bjerrum and Eide (1956) made comparisons of these methods on actual failures (cases of heave) and generally established the validity of the two methods.

The Terzaghi case for $D/B < 1$ and long excavation (Fig. 14-10) considers the bottom of the excavation as a strip footing of width B. The failure zone is inclined at $45°$ to the base, and the remainder of the zone is circular, with a radius of $r = B\sqrt{2}/2$. In this case the shear resistance in the zone of depth D is also considered.

From Fig. 14-10, for a cohesive soil in undrained conditions ($\phi = 0$) and using

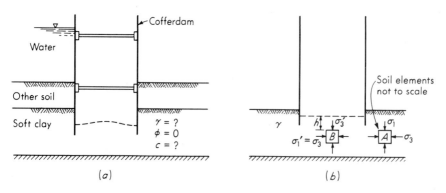

Figure 14-9. (a) Cofferdam on soft clay; (b) theoretical solution.

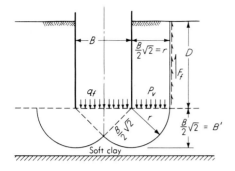

Figure 14-10. Stability of excavation against bottom heave for small D/B ratios.

Terzaghi's N_c factor (others have used the Hanson value of 5.1)

$$q_f = cN_c = 5.7c$$

For a strip 1 ft wide the shear resistance for depth D is cD as in Fig. 14-10. The pressure p_v on a horizontal plane of width $B\sqrt{2}/2 = B'$ at depth D with an additional surcharge, q can be found as

$$B'p_v = \gamma DB' - F_f + qB'$$

or

$$p_v = \gamma D - \frac{F_f}{B'} + q$$

At failure (or heave) $q_f = p_v$ and let $F_f/B' + q = M$ to obtain

$$\gamma D - M = 5.7c$$

and the critical depth D_c occurs at this equality $(F = 1.0)$:

$$D_c = \frac{5.7c + M}{\gamma} \tag{14-1}$$

The safety factor, in general, is

$$F = \frac{5.7c + M}{\gamma D} > 1.25 \tag{14-2}$$

All terms are as defined in Fig. 14-10. This equation is quite reliable according to Bjerrum and Eide (1956) as long as $D/B < 1$.

When $D/B > 1$ and for undrained conditions $(\phi = 0)$ studies by Bjerrum and Eide (1956) indicate the base of the excavation can be treated as a foundation under unloading. The bearing-capacity factor N_c is adjusted for the depth and shape of the foundation (excavation) as shown in Fig. 14-11 and with identification of terms used in the following discussion. The capacity of the equivalent footing is

$$q_f = cN_c + \gamma D + q$$

and the critical depth D_c occurs when $q_f = 0$, resulting in opposite signs of $(\gamma D + q)$ and cN_c to obtain

$$D_c = \frac{cN_c}{\gamma + q}$$

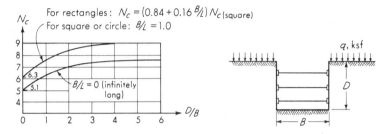

For rectangles: $N_c = (0.84 + 0.16 \,{}^B\!/_L) N_{c \,(\text{square})}$
For square or circle: ${}^B\!/_L = 1.0$

Figure 14-11. Bearing-capacity factors and identification of terms for *Eq.* (14-3). [*After Bjerrum and Eide* (1956).]

and the safety factor is computed as

$$F = \frac{cN_c}{\gamma D + q} > 1.25 \tag{14-3}$$

This equation was used to investigate 14 deep excavations and in nearly all cases predicted failure within ± 16 percent.

14-7. OTHER CAUSES OF COFFERDAM INSTABILITY

A bottom failure may occur because of a piping, or " quick," condition if the hydraulic gradient h/L is too large and if the soil is sandy. In Fig. 14-12*a* a flow-net analysis may be used to estimate when a quick condition may occur. Possible remedies are to drive the piling deeper to increase L or reduce h by less pumping from inside the cell. In a few cases it may be possible to use a surcharge inside the cell.

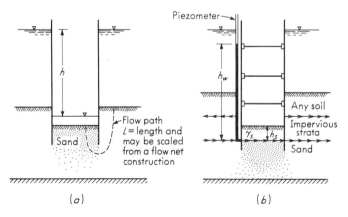

Figure 14-12. (*a*) Conditions for piping, or quick, conditions; (*b*) conditions for a "blow-in."

In Fig. 14-12b, the bottom of the excavation may "blow in" if the pressure head indicated by the piezometer is too great, as follows ($F = 1.0$):

$$\gamma_w h_w = \gamma_s h_s$$

This equation is slightly conservative since the shear, or wall adhesion, along the sides of the cell is neglected, but on the other hand, if there are soil defects in the impervious layer, the blow-in may be local; therefore, in the absence of better data, the equality as given should be used. The safety factor is

$$F = \frac{\gamma_s h_s}{\gamma_w h_w} > 1.25$$

14-8. CONSTRUCTION DEWATERING

Figure 14-12 indicates that water inflow into an excavation can cause a bottom failure. Where it is impractical or impossible to lower the water table owing to possible damage claims from resulting settlements or water-supply curtailment, it is necessary to create a nearly impervious water barrier around the excavation and provide sump holes (small pits below the bottom of the excavation) at selected locations, pump the remaining seepage inflow, and simultaneously monitor the groundwater level outside the excavation to make sure the level is not being substantially altered.

Where it is possible to depress the water table in the vicinity of the excavation, a system of perimeter wells are installed. This system may consist in a single row of closely spaced wellpoints around the site. A wellpoint is simply a section of small-diameter pipe with perforations (or screen) on one end which is inserted in the ground. If the soil is pervious in the area of the pipe screen, the application of a vacuum from a water pump to the top of the pipe will pull water in the vicinity of the pipe into the system. This system will be limited in the height of water raised owing to vacuum to about 6 m. Theoretically water can be raised higher, but this type of system is less than theoretical. More than one set of perimeter wells can be installed as illustrated in Fig. 14-13. This type of system is seldom "designed"; it is contracted by companies who specialize in this work, and while rough computations can be made, the field performance determines the number of wellpoints and amount of pumping required.

Where wellpoints are not satisfactory or practical, one may resort to a system of perimeter wells which may fully penetrate the water-bearing stratum (acquifer) or only partially penetrate if the stratum is quite deep. Again only estimates of the quantity of water can be made as follows:

One may use a plan flow net as in Fig. 14-14 to obtain the seepage quantity. A plan flow net is similar to a section flow net as in Chap. 2. The equipotential drops are now contour lines of equal elevation intersecting the flow paths at the same angle. It is necessary to establish sufficient contour lines to represent the required amount of drawdown to provide a dry work area. Some approximation is required, since it is

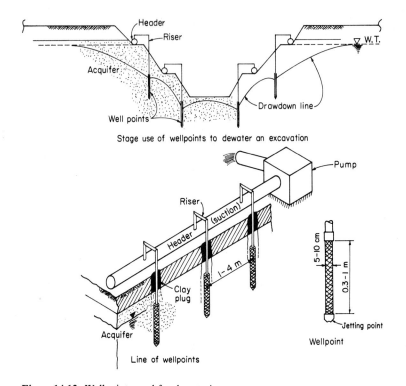

Stage use of wellpoints to dewater an excavation

Figure 14-13. Wellpoints used for dewatering.

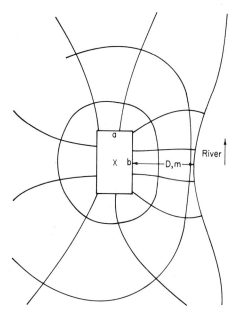

Figure 14-14. Plan flow net. Note only necessary to draw enough flow and equipotential lines to obtain N_f, N_e.

not likely that the piezometric head is constant for a large distance around an excavation. Further approximation is necessary because a system of wells located around the excavation will not draw down the water to a constant contour elevation within the excavation. The water elevation will be a minimum at and higher away from any well.

From a plan flow net the quantity of water can be estimated as

$$Q = \alpha k \, \Delta H \frac{N_f}{N_e} L \tag{14-4}$$

where N_f = number of flow paths including decimal fractions

N_e = number of equipotential drops (always integer)

$\Delta H = H^2 - h_w^2$ for gravity (see Fig. 14-15)

$\quad = H - h_w$ for artesian

$L = 1.0$ for gravity flow

$\quad =$ thickness of acquifer for artesian flow

$k =$ coefficient of permeability in consistent units with H and L

$\alpha = 0.5$ for gravity flow

$\quad = 1.0$ for artesian flow

An estimate of the number of wells and the flow quantity per well is obtained by placing one well in the center of each flow path. The resulting flow quantity per well is then

$$\text{No. of wells} = N_f$$

$$\text{Flow per well} = Q/N_f$$

An estimate of the quantity of water which must be pumped to dewater an excavation can be obtained by treating the excavation as a large well (Fig. 14-15) and using the gravity-flow-well equation

$$Q = \frac{\pi k (H^2 - h_w^2)}{\ln (R/r_w)} \tag{14-5}$$

where terms not previously defined are:

$H =$ surface elevation of water at the maximum drawdown influence a distance R from well

$h_w =$ surface elevation of water in well

$r_w =$ well radius

This equation is for "gravity" wells; that is, the piezometric head and static water level are coincident, which is the likely case for pumping down of the water

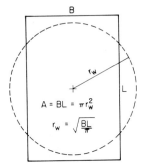

Excavation as large well of radius, r_w

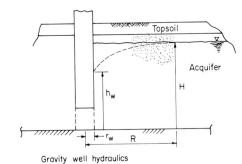

Gravity well hydraulics

Figure 14-15. Approximate computation for flow quantity to dewater an excavation.

table for a large excavation. The maximum radius of drawdown influence R is not likely to be known; however, one may estimate several values of R/r_w and obtain the corresponding Q values to obtain probable pumping quantities. The value of the static groundwater level H is likely to be known, and h_w would normally be estimated as 1 to 2 m below the bottom of the excavation.

This estimate of well pumping to dewater an excavation should be satisfactory for most applications. It is not likely to be correct, primarily because the coefficient of permeability k will be very difficult to evaluate unless field pumping tests are performed. It is usually sufficient to obtain the order of magnitude of the pumping for dewatering and contract to pay for the actual pumping required.

Example 14-4. Estimate the flow quantity to dewater the excavation shown in Fig. 14-14. Other data: $H = 50$ m; $h = 34$ m; $\Delta H = 15$;

$$k = 2 \times 10^{-1} \text{ m/day;} \quad a = 60 \text{ m;} \quad b = 100 \text{ m,} \quad D = 100 \text{ m.}$$

The soil profile is as shown

SOLUTION. We will use a plan flow net (Fig. 14-14 was originally drawn to scale) and check the results using Eq. (14-5).

Step 1. Compute Q for plan flow net (assume gravity flow after drawdown is stabilized). From Fig. 14-14: $n_f = 10$, $n_c = 2.1$

$$H = 50 \qquad H^2 = 2{,}500$$

$$h_w = 34 \qquad h_w^2 = 1{,}156$$

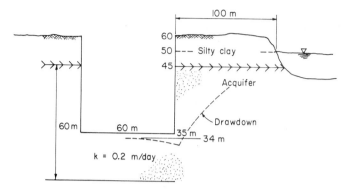

Figure E14-4.

$$H^2 - h_w^2 = 1,344$$

$$Q = 0.5(0.2)(1,344)\frac{10}{2.1} = 640 \text{ m}^3/\text{day}$$

(approx. 120 gpm)

Step 2. Check results using Eq. (14-5):

$$Q = \frac{\pi k (H^2 - h_w^2)}{\ln (R/r_w)}$$

$R = 100$ m (unless we draw down the river)

$$r_w = \sqrt{\frac{A}{\pi}} = \sqrt{\frac{60(100)}{\pi}} = 24.66 \qquad \text{use 25 m}$$

$$Q = \pi(0.2)(2,500 - 1,156)/\ln (100/25) = 609.1 \text{ m}^3/\text{day}$$

14-9. SLURRY-WALL (OR -TRENCH) CONSTRUCTION

The use of a ground cavity filled with a viscous fluid, termed a slurry, is a development from the early 1960s. The basic method has been in use for some time for oil-well and soil-exploration drilling to maintain boreholes in caving soils without casing or in the case of oil wells to retain natural gas using a hydrostatic pressure head. The method is now being applied to maintain the sides of narrow excavations without shoring or other bracing. Walls constructed in excavations where a "slurry" is used to maintain the excavation are termed slurry walls. Open trenches which are later backfilled or filled with clay or lean concrete to affect cutoff walls, as for dams, are termed "slurry trenches."

Basically, slurry construction consists in making an evaluation of the required density and properties of the slurry based on the soil profile, then providing a means to develop large quantities of the water admixture and as the excavation proceeds, keeping the ground cavity filled to the necessary depth with the slurry. When excavation is complete, the slurry-filled cavity is periodically agitated to keep the admixture in suspension. Obviously the agitation must be carefully done to avoid wall collapse. Next the cavity is filled with clay or lean concrete, in the case of cutoff walls, by carefully placing the material under the slurry which is displaced from the top and discarded or saved for use in the next trench section if stage construction is employed. Concrete must be tremie placed in a slurry trench so that only a small portion is exposed to the slurry and to avoid slurry seams where adjacent pours unite. The method is quite expensive; however, where a watertight wall is required as for construction below groundwater table and in conditions where it is difficult to control ground loss, the method offers economic advantages over other procedures. Slurry construction depends upon two factors for successful performance:

1. Formation of a filter skin at the interface of the slurry and excavation via gel action and particulate precipitation
2. Stabilizing lateral pressure due to the density of the slurry pushing against the filter skin and sidewalls of the excavation

The slurry must be of sufficient viscosity that it does not drain easily out through the walls of the excavation including resistance obtained from the filter skin. If the filter skin forms reasonably well, it is expected that the exfiltration loss will be minimal and the filter-skin penetration into the sides of the excavation may be on the order of only a few centimeters where fine-grained soils are supported. A slurry excavation in gravel was reported by La Russo (1963) to have penetrated some 16 m into the surrounding soil, but this may be considered exceptional.

Slurry construction can be used for both caving and cohesive soils and has been for caissons as well as wall and trench construction [O'Neill and Reese (1972), Lorenz (1963)]. Slurry densities up to 1.92 g/cm^3 can be obtained using a mixture of:

$$\text{Barite (barium sulfate) of } G = 4.3 \text{ to } 4.5$$

$$\text{Bentonite (clay for gel action) of } G = 2.13 \text{ to } 2.18$$

Other materials including silt, clay, and fine sand from the excavation may be used to reduce the quantity of commercial admixture. Commonly, slurry densities of 1.15 to 1.25 g/cm^3 are employed using a mixture of bentonite, barite, and a dispersing agent to reduce the tendency of the clay to floc. The gel is a natural by-product of the admixture, and the basic design element consists in determining the required density of the slurry. Referring to Fig. 14-16a, for clay without a slurry, the critical depth is as computed in Chap. 11:

$$H_c = \frac{4c}{\gamma \sqrt{K_a}}$$

With slurry in the trench, a horizontal force summation, and for *undrained conditions*, which is the usual case, gives

$$-0.5\gamma_f H^2 + 0.5\gamma H^2 - 2cH = 0$$

and solving

$$H = \frac{4c}{\gamma - \gamma_f}$$

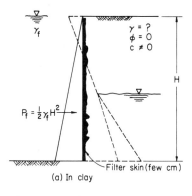
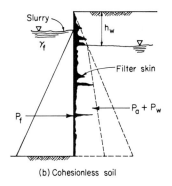

Figure 14-16. Slurry walls.

and for a safety factor F

$$F = \frac{4c}{H(\gamma - \gamma_f)} \qquad (14\text{-}6)$$

The safety factor (or depth of excavation H) can be increased by increasing γ_f. This equation was presented by Nash and Jones (1963) and verified by Meyerhof (1972).

In cohesionless soils (Fig. 14-16b) the slurry density is obtained as

$$0.5\gamma_f H^2 - P_a - P_w = 0$$

Example 14-5. Show the effect of slurry density on depth of excavations in a cohesive soil with $F = 1.5$.

Other data: $c = 35$ kPa; $\qquad \gamma = 18.2$ kN/m^3

SOLUTION. Use several γ_f values

$$H = \frac{4(35)}{1.5(18.2 - \gamma_f)} = \frac{93.33}{18.2 - \gamma_f}$$

γ_f, g/cm^3	γ_f, kN/m^3	H, m
1.1	10.79	12.6
1.15	11.28	13.5
1.2	11.77	14.5
1.25	12.26	15.7

PROBLEMS

14-1. Reanalyze Example 14-1 for $\phi = 30°$.

14-2. Reanalyze Example 14-1 using the finite-element procedure.

14-3. Verify Example 14-3 using the finite-element method.

14-4. Estimate the ground loss of Prob. 14-3.

14-5. Redo Example 14-3 considering the lateral pressure including a surcharge from the soil of the preceding stage.

14-6. Design the wales and struts of Example 14-3 based on forces and bending moments of Prob. 14-5.

14-7. Referring to Example 14-4 and Fig. 14-14 compute the assigned flow quantity if:

    (a) $B \times L = 70 \times 110$ m $D = 100$ m
    (b) $B \times L = 210 \times 300$ ft $D = 325$ ft
    (c) $B \times L = 50 \times 90$ m $D = 80$ m
    (d) $B \times L = 150 \times 270$ ft $D = 250$ ft

Use k_s, H, and h_w, the same as in Example 14-4.

14-8. Check the assigned Prob. 14-7 using a plan-flow-net sketch.

14-9. Design the mix proportions to provide a slurry of $\gamma = 1.2$ g/cm^3. Use water, barite, and bentonite. Use at least 20 percent bentonite in the admixture.

14-10. Design a slurry mixture for the wall of Fig. 14-16b if $h_w = 2$ m, $\gamma = 105$ pcf, and the trench is to be 8 m deep (assume G_s to compute γ_{sat}).

FIFTEEN

CELLULAR COFFERDAMS

15-1. CELLULAR COFFERDAMS: TYPES AND USES

Cellular cofferdams are of three basic types as illustrated in Fig. 15-1, namely, circular, diaphragm, and cloverleaf. These structures are usually constructed of straight-web sheetpiling to take advantage of the high tensile stresses developed in the cells and to avoid bending stresses as would also be developed using Z or deep-web sheetpiling.

Cellular cofferdams are used primarily as retaining structures, with the retained material usually being water. They depend for stability on the interaction of the soil used to fill the cell and the steel sheetpiling. Either material used alone is unsatisfactory; both materials in combination provide a satisfactory means to develop a dry work area in water-covered areas such as ocean- or lakefront structures or dam construction in rivers.

Circular cells are usually used in conjunction with some kind of connecting method to provide the complete cofferdam for allowing dry construction. Figure 15-1 illustrates several methods of joining the several circular cells for this purpose. Occasionally isolated cells may be used for anchors to which an arc of sheetpiling is connected to provide a retaining wall. A fairly recent innovation termed a "sand island" uses a large circular cell filled with sand and capped with concrete. The "sand island" is usually left in place in this type of construction.

The cofferdam cells are constructed by assembling the necessary number of sheet piles around a wooden template consisting of two rings (or other shape) spaced vertically about 10 ft apart which has been anchored into correct position (usually with four or more steel H piles). The sheetpiling is then placed into position with the

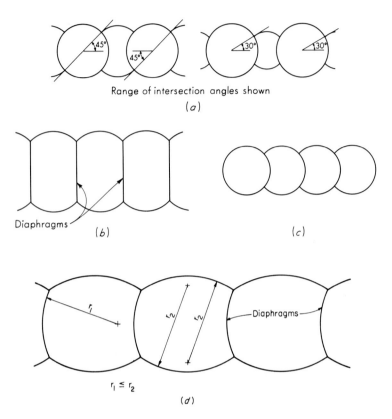

Figure 15-1. Cellular cofferdams. (*a*) Circular, economical for deep cells; (*b*) diaphragm, may be economical in quiet water; (*c*) modified circular (special case of *d*); (*d*) diaphragm with circular crosswalls.

pile sections which have been fabricated to connect the cells (wyes or tees) set first into position and as accurately as possible. These become *key* piles; in deep water it may be necessary to add additional key piles made using regular sheets with light beams or angles bolted on to increase the rigidity. The remainder of the piles are then set both ways from the key piles to close the cell. At this point the piles are resting in the overburden at the bottom of the river. If the closure piles do not slip to the bottom easily in the interlocks, the adjacent piling is picked up in multiples and "shaken out" until all the sheets are free-running in the interlocks. Driving can now commence and the piles are driven—usually in pairs about 1 to 2 m, then the next pair, etc., around the cell. The operation is then repeated, however, either using a new starting pair of piles or going in reverse to avoid distortion of the cell from systematic accumulation of driving effects to one side. Two or more piles are set for the connecting arc before the key (fabricated) piles are driven to grade. Splices are made by cutting the first piles in staggers so that the splice will vary up and down some 1 to 2 m and fall above cell water or ground level.

It is difficult to set sheetpiling in fast-moving water or on windy days, and the operations are greatly slowed when either factor is present.

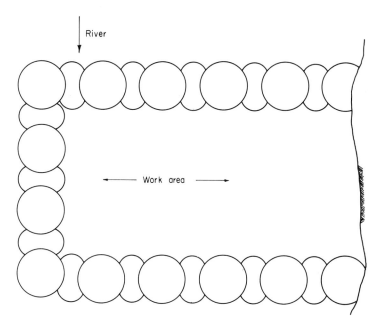

Figure 15-2. Cofferdam work area.

The template should be positioned within about 15 to 30 cm of alignment for circular cells and to less than 15 cm for diaphragm cells. Closer tolerance than this is usually not necessary owing to cell distortion during filling and dewatering operations.

By connecting a series of the cells (Fig. 15-2) around the perimeter of a work area, and filling them with soil, a water barrier is obtained. A series of pumps can then pump out the work area termed a cofferdam; and with a dewatering system to remove the water which percolates through and beneath the cell wall from the differential water head, a reasonably dry work area is made.

Cellular cofferdams may also be used for structures such as breakwaters and retaining walls, or the cells may be built out into the water to function as a pier-type structure. In these cases the cell fill may function as the base for a road, railroad, or warehouse.

The circular cells (Fig. 15-1a) consist of circles of different radii (occasionally of the same radius) intersecting as shown. The cell intersection angle is usually between 30 and 45° (Fig. 15-3). The joint is often a tee, i.e., the intersection angle is 90°, but other angles can be used. A 30° angle on the connector has been used, and may be a better solution for large-diameter cells where high tee stresses will exist.

Sheetpiling interlocks allow a maximum of about 10° deflection between pieces. This results in a minimum cell radius,

$$R = \frac{\text{driving distance, ft or m}}{2 \sin 10°}$$

For a PZ-28 section, $R = 3.6$ ft or the minimum cell diameter $= 7.2$ ft.

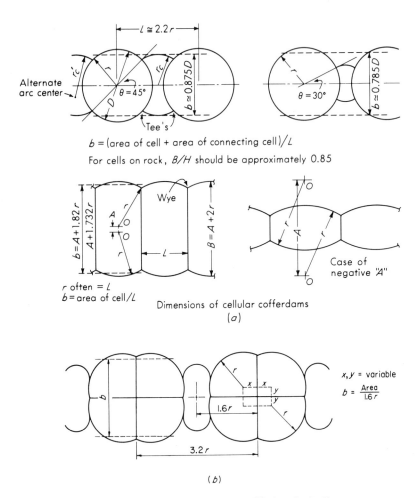

b = (area of cell + area of connecting cell)/L

For cells on rock, B/H should be approximately 0.85

r often = L
b = area of cell/L

Dimensions of cellular cofferdams

(a)

x, y = variable

$b = \dfrac{\text{Area}}{1.6\,r}$

(b)

Figure 15-3. (a) Dimensions of cellular cofferdams; (b) cloverleaf cell.

Diaphragm cells are made up of a series of circular arcs connected by crosswalls (diaphragms) using 120° intersection pieces (Fig. 13-4). The radius of the arc is often made equal to the cell width L (Fig. 15-3a) in order that the interlock tension in the arcs and diaphragms may be equal. The distance A shown in the figure may be either positive for high wide cells or negative for low narrow cells. A wide cell will be necessary for stability when a large head of water is to be resisted.

Other cell types, such as cloverleafs (Fig. 15-3b) and ellipsoidal shapes (not shown), may be fabricated, depending on the purpose, cell height (head of water), type of fill, amount of tolerable distortion, and location. The cloverleaf type has been used considerably as a corner, or anchor, cell in conjunction with circular cells. This type of cell can also be used to reduce the effective diameter of a cell when a large cell width is required for stability against a high head of water.

The circular cell is generally preferable to the other cellular types for the following reasons:

1. It is stable as a single unit, and can be filled as soon as it is constructed.
2. The diaphragm-type cell will distort unless the various units are filled essentially simultaneously with not over 1 or 2 m of differential soil height in adjacent cells; the use of a circular diaphragm cell reduces this requirement if filling is against the concave [Cushing and Moline (1975)].
3. The collapse of a diaphragm cell may fail the cofferdam, whereas the collapse of a circular cell is generally a local failure.
4. The circular cell is easier to form using templates.
5. The circular cell usually requires less sheetpiling, but this will depend somewhat on the diaphragm crosswall spacing.

Increasing the size of a circular cofferdam cell does not necessarily increase the total quantity of sheetpiling for the cofferdam, since the total number of cells will be reduced. This is not true for the diaphragm-type cell. The quantity of cell fill is directly dependent on the cell dimensions for all types of cofferdams.

15-2. CELL FILL

The cell fill provides mass (or weight) for stability and a reduced coefficient of permeability for retaining water without excessive pumping. This must be balanced against the lateral-pressure effects of the soil-water mixture and the resulting stresses which the sheet-pile interlocks must resist before rupture and/or cofferdam failure.

For mass it is apparent that any soil could be used and the higher the density the better. For permeability considerations alone, clay is the best possible fill. The earth-pressure coefficient of sand with a high angle of internal friction gives the minimum lateral pressure which must be resisted by "hoop tension" considerations in the interlocks. Considering all these factors, it develops that the best cell fill:

1. Is free-draining (large coefficient of permeability)
2. Has a high angle of internal friction
3. Contains small amounts of No. 200 sieve material
4. Is resistant to scour (nonsilty or clayey)

Cell fills are sometimes used which do not meet the above criteria, but the closer the fill material approaches these criteria the more economical the design in terms of sheetpiling, which is usually the most expensive portion of the cofferdam.

Cell fill is often placed hydraulically; i.e., the material is obtained from the river bottom if at all possible. The material is dredged up and pumped through a pipe system and discharged into the cells which are in position and full of water to the river level. This operation may substantially reduce the fines which are often present in river-bottom material and which are temporarily suspended in the water and wash overboard. Of course, if material is not available close by, fill may have to be brought in by barge, truck, or rail. In any case it will generally be deposited under water. The effect of this on the angle of internal friction should be carefully evaluated. It appears that this method of soil deposition seldom produces an angle of internal friction over about $30° \pm 2°$. Unless satisfactory drained triaxial tests are performed on the soil and at the expected cell density, it is suggested that the designer limit ϕ to 28 to 30° for design (or preliminary design).

15-3. STABILITY OF CELLULAR COFFERDAMS

There are currently at least three methods of analysis of cellular cofferdams:

1. Former Tennessee Valley Authority (TVA) method, also called Terzaghi's method
2. Cummings method
3. Hansen's method

Of these methods the TVA (1966) and Cummings (1960) are commonly used in the United States and elsewhere. The Hansen method and as modified by Ovesen (1962) is used considerably less. Since more cofferdams have been built by TVA and the U.S. Corps of Engineers than elsewhere, the TVA and Cummings methods have much to commend them, even though none of the methods appear strictly theoretical. The Hansen method is more complicated than the other methods, with less construction experience to validate it, and is not considered further in this text. Belz (1970) and Dismuke (1975) provide a summary of the several design methods in use in the United States. TVA currently uses the Cummings procedure as part of their cofferdam design.

TVA Method of Cellular Cofferdam Design

Terzaghi (1945) presented a paper in cellular cofferdam design in which the methods used by TVA since about 1935 were outlined. TVA (1966) later published a monograph, with the first printing in 1957 outlining in some detail their design methods.

The TVA method considers the following (refer to Fig. 15-4):

1. *Sliding stability.* A cofferdam must provide adequate resistance to sliding on the base caused by the unbalanced hydrostatic pressure. Sliding stability (with $F = 1.0$) is satisfied (Fig. 15-4a), neglecting any active soil pressure, if

$$P_d = P_p + P_f$$

The safety factor for the general case is

$$F = \frac{P_p + P_f}{P_d} \tag{15-1}$$

where P_d = driving force

P_f = developed friction resistance = fW. For soil-to-soil sliding, f can be taken as tan ϕ; for soil on smooth rock TVA uses $f = 0.5$. On rough rock $f = \tan \phi$ is satisfactory. It will be necessary to study the given case to estimate a reasonable value of coefficient of friction.

P_p = developed passive resistance. Depends on depth of embedment of the sheetpiling. With no embedment as on rock, this term is zero

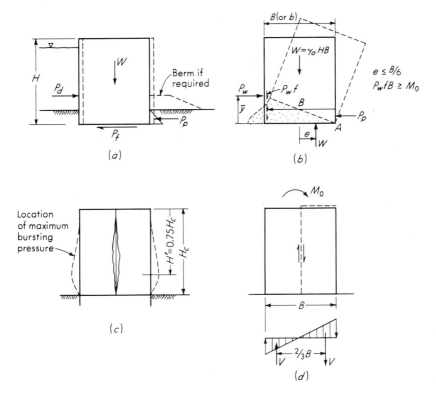

Figure 15-4. Stability of cofferdams. (*a*) Sliding resistance; (*b*) overturning resistance; (*c*) bursting failure by hoop tension $T = q_a r/12$; (*d*) shear along centerline of cell.

The safety factor should be greater than 1.10. A value of at least 1.25 is recommended if this analysis controls the size of the cell.

2. *Overturning stability.* The cofferdam must be stable against overturning. Two possibilities, or types of analysis, can be made when considering this type of stability. To avoid overturning, and reasoning that soil cannot take tension forces, the resultant weight should lie within the middle one-third of the base.

$$e = \frac{P_d \bar{y}}{\gamma H B} \leq \frac{B}{6} \qquad (a)$$

Thus the higher the cell, the wider the cell width B must be. In this equation and those to follow, the width of the cell B is understood to be the average width of the cell, as shown in Fig. 15-3.

Alternatively, one may reason that as the cell tends to tip over, the soil will pour out at the heel. For this to occur the friction resistance of the steel piling on the cell fill is developed. On this side of the cell the water pressure P_w is pushing the piling against the fill so that the friction force per foot of cell width is fP_w. Summing moments about the toe of the cell (point A of Fig. 15-4b):

$$BfP_w = P_w \bar{y}$$

or

$$B = \frac{\bar{y}}{f} \tag{b}$$

where this value of f is the coefficient of friction of the cell fill against the sheetpiling $(f < \tan \phi)$, and the other terms are as identified in the figure. The safety factor is computed as

$$F = \frac{Bf}{\bar{y}} \tag{15-2}$$

A value of $F = 1.1$ to 1.25 is desirable.

If the sheetpiling is embedded to some depth in the soil, the effects of the active and passive soil pressures on the overturning moment and friction resistance should be included in summing moments about point A in equation (b) above. This procedure is not now used by TVA (1966, see Foreword).

3. *Cell shear.* Shear along a plane through the centerline of the cell is another possible mode of failure or excessive cell distortion (Fig. 15-4d). For stability, the shearing resistance along this plane, which is the sum of soil shear resistance and resistance in the interlocks, must be equal to or greater than the shear due to the overturning effects. Referring to Fig. 15-4d and assuming a linear pressure distribution across the base of the cell,

$$M = \tfrac{2}{3}BV$$

Solving for the overturning shear on the plane through the centerline,

$$V = 1.5\frac{M}{B} \tag{15-3}$$

To be stable, the resisting shear must be equal to or greater than this value. The soil shear resistance can be computed as

$$F_s = \tfrac{1}{2}\gamma H^2 K'_a \tan \phi \tag{15-4}$$

where the active-earth-pressure coefficient K'_a is computed from a Mohr's-circle construction (Fig. 15-5c). This computation is necessary since the lateral pressure is not a principal stress with shear on the plane. This pressure is, however, a normal pressure (stress). This normal pressure is represented by the ordinate AB on the Mohr circle. The stress on a plane $90°$ away (the horizontal plane) is the stress value shown as CD, which is also the vertical pressure $\gamma_e H$. From Fig. 15-5c the radius of the circle is

$$FB = FD = \frac{AB \sin \phi}{\cos^2 \phi} \tag{c}$$

By definition, the ratio of the lateral pressure to the vertical pressure is the active-earth-pressure coefficient

$$K'_a = \frac{AB}{CD} \tag{d}$$

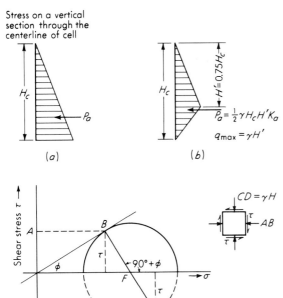

Stress on a vertical
section through the
centerline of cell

H_c

P_o

(a)

H_c

$H' = 0.75H_c$

$P_a = \frac{1}{2}\gamma H_c H' K_a$

$q_{max} = \gamma H'$

(b)

Shear stress $\tau \longrightarrow$

A B

τ

$90° + \phi$

F

ϕ

τ

C

D

(c)

$CD = \gamma H$

τ

AB

τ

Figure 15-5. Internal cell pressures for bursting cell stability. (a) Compute P_a based on K'_a; (b) active-earth-pressure diagram for computing tension in interlocks. Use Rankine or Coulomb value of K_a; (c) Mohr's circle for everlasting K'_a.

From Fig. 15-5c it is evident that

$$AB = CD - 2FD \sin \phi \qquad (e)$$

and by substitution of equations (c) and (e) into equation (d) and solving for K'_a, we obtain

$$K'_a = \frac{\cos^2 \phi}{2 - \cos^2 \phi} \qquad (15\text{-}5)$$

This equation gives progressively smaller values of K'_a as ϕ increases; however, the values tend to be about twice as large as the corresponding Rankine values for the same ϕ. It is convenient, and essentially correct, to use the average unit weight γ_a of the soil in the cell rather than consider part of the soil in a wet state and the remainder submerged. Equation (15-4) is very sensitive to the unit-weight term, but is relatively insensitive to ϕ.

For the cloverleaf cell it will be necessary to use a value of tan δ instead of tan ϕ, since the soil will be acting on the vertical diaphragm wall for a resistance developed of soil on steel rather than soil on soil.

Friction in the interlock joints must be simultaneously overcome for the vertical-shear distortion to take place. This force is computed using the conventional equation for hoop tension,

$$T = q_a r \qquad (f)$$

Since q_a increases with depth, the total hoop-tension force for a cell of total depth H is

$$T = \tfrac{1}{2}\gamma H^2 K_a r \quad \text{using the Rankine or Coulomb value of } K_a \qquad (g)$$

4. *Interlock tension.* Experiences at TVA and elsewhere indicate that during filling the cell, lateral pressures develop during both filling and subsequent consolidation of the fill (order of 10 days or so). This lateral pressure causes the cell to expand in proportion to the lateral pressure and depending on the base restraint as on rock or embedded in the ground. The cell takes on a modified barrel or bulged shape, and field observation finds the bulge most pronounced at from one-fourth to one-third of the free height of the cell above ground or rock. This location has been arbitrarily taken as the one-fourth height as Fig. 15-5b. For a triangular pressure distribution as shown in Fig. 15-5b (we could use a parabolic pressure distribution with a resulting 0.33 increase in computed hoop tension), the hoop-tension force is

$$T = 0.5\gamma H_c(0.75 H_c) K_a r \qquad (h)$$

The resulting resisting friction force is Tf

$$F''_s = Tf = 0.375\gamma H_c^2 K_a r f \qquad (i)$$

The coefficient of friction is usually taken as $f = 0.3$. Tests on high-strength bolted joints have produced friction coefficients from about 0.35 to 0.45; thus $f = 0.3$ seems reasonable.

Since the cell analysis is considering a unit width of cell, the shear resistance is

$$F'_s = \frac{F''_s}{L} = \frac{0.375\gamma H_c^2 K_a \, fr}{L} \qquad (15\text{-}6)$$

where L is the distance between crosswalls for diaphragm cells and $L = r$ for circular cells.

The value of F'_s for a cloverleaf cell can be computed by the same reasoning used to obtain Eq. (15-6). It should be noted that the hoop tension in the arcs is always Tr. The tension in the cross diaphragms will also be Tr if the arcs intersect at 120°. The value of F'_s will then be computed using a value of L equal to the effective length of the cell divided into $2.732\,Tr$. The effective length is measured in the direction perpendicular to overturning.

The total cell shear resistance F_{st} is the sum of the soil friction Eq. (15-4) and interlock friction Eq. (15-6).

$$F_{st} = F_s + F'_s$$

Or combining equations, one obtains

$$F_{st} = \tfrac{1}{2}\gamma H^2 K'_a \tan \phi + \tfrac{3}{8}\gamma H(H_c) K_a \frac{r}{L} f \qquad (15\text{-}7)$$

The safety factor for this analysis is computed as

$$F = \frac{F_{st}}{V}$$

or
$$F = F_{st}\frac{2B}{3M} \qquad (15\text{-}8)$$

A value of F = 1.1 to 1.25 should be used.

5. *Bursting stability.* The cells must be stable against bursting pressures. The critical locations are in the interlock joints and in the tees or wyes used for the connecting arcs. The bursting pressure at a depth in the cell is

$$q = \bar{q}_h + q_w \qquad (15\text{-}9)$$

where $\bar{q}_h$ = effective lateral pressure due to soil

q_w = hydrostatic pressure

it is evident that a free-draining material should be used in the cell so that it can be dewatered.

It is also evident that the cell location during filling will affect the bursting pressure. A cell hydraulically filled near the shore will, during and shortly after filling, be subjected to a full hydrostatic pressure and effective earth pressure. This is compensated to some extent by the cell's being of less total height. Cells in the water will undergo only active effective earth pressure until dewatering, and it would be expected that cell drainage would develop such that the critical condition of bursting develops with full active effective earth pressure and water at the saturation line. Cells filled by other means or in other locations are analyzed similarly.

The bursting force per inch of height is

$$T = q_a\frac{r}{12} \qquad \text{kips/in} \qquad \text{if } q_a \text{ is in ksf and } r \text{ is in feet} \qquad (15\text{-}10)$$

Allowable values for interlock tension T depend on the size, shape, and steel yield point of the sheet-pile sections. The steel sections generally used in this construction are given in Table 15-1 (see Appendix A for profile and section properties).

Table 15-1. Selected sheetpiling data for cofferdam design. Current designation indicates weight/sq ft of wall, i.e., PSA23 weighs 23 lb/sq ft of wall surface

Former section designation (prior to 1971)	Current AISI designation (after 1971)	Guaranteed‡ interlock tension, kips/in ( ) kN/cm	Suggested factor of safety F interlock	Interlock friction f
MP112 (SP-4)*, MP113 (SP-5)	PSA23, PSA28†	12 (21.0)	4§	0.3
MP101 (SP-6A), MP102 (SP-7A)	PS28, PS32† PSX32, (PSX35)	16 (28.0) 28 (49.0)	1.5–2.0 1.5–2.0	0.3 0.3

* ( ) = Bethlehem Steel designation; other designation is US Steel (F_y = 38.5 ksi).
† Rolled only if sufficient quantity ordered.
‡ Sections produced in ASTM A328 and A572 grades (F_y = 42 to 55 ksi). Normal allowable design stresses for tension and bending usually taken as $0.65F_y$. Value given is guaranteed maximum interlock strength and is used in design with a safety factor.
§ This high value due to these sections being "shallow arch," which may straighten with high interlock-tension values.

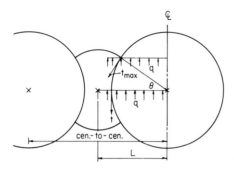

Figure 15-6. Tee stresses according to TVA.

Most cofferdam failures result from failure of the connecting tee from either a fabrication failure or interlock failure [Swatek (1967), Grayman (1970)]. According to TVA (1966) the interlock tension can be computed from the free body of the cell as shown in Fig. 15-6. Summation of forces gives the interlock tension as

$$T' = qL \sec \theta \qquad (15\text{-}11)$$

In this equation $L = 0.5 \times$ center-to-center cell spacing as shown in Fig. 15-6. The maximum interlock tension can be reduced by decreasing θ, which may require use of a wye of 30 to 60° instead of a 90° tee in order to obtain sufficient width of retaining structure in the area of the connecting arc. One may use alternative methods of obtaining the maximum tension force as from a free-body diagram and considering hoop tension in both the main and connecting cells; however, both TVA (1966, p. 112) and Dismuke (1970) show that approximately the same value is obtained using Eq. (15-11).

The Cummings Method (Currently Used by TVA)

Cummings (1960) proposed a method of analysis of cellular cofferdams based on model studies for the tilting of a cofferdam on rock, as shown in Fig. 15-7. The method provides a simple analysis; however, the models were constructed of relatively stiff material for the size of the model, which may not be realistic when related to the flexible sheetpiling sections and dimensions of a field structure. According to TVA, they had made some unpublished model studies similar to and prior to Cummings with essentially the same type of failures observed. It remained for Cummings to develop the analytical method presented here. The method has been successfully used in the design of several cofferdams, and is extremely simple.

The analysis is based on the premise that the cell soil will resist lateral distortion of the cell through the buildup of soil resistance to sliding on horizontal planes (Fig. 15-7b). This resistance will be developed in a triangle as shown, forming an angle of ϕ to the horizontal. The triangle of soil will be in a passive-pressure state and stabilized by the overlying soil which acts as a surcharge. The weight of this soil is termed W_y. The derivation is complete when we can write an expression for the cell resistance in terms of the triangular zone of passive resistance, with shear on the horizontal planes and including the surcharge effect of W_y.

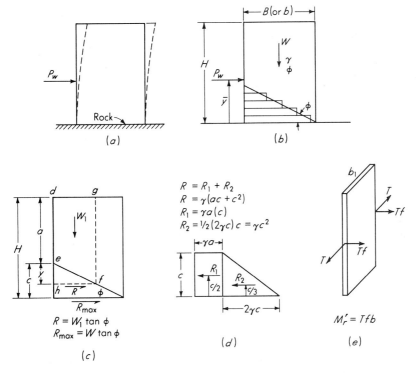

Figure 15-7. Tilt analysis. (*a*) Overturning effect; (*b*) development of internal cell resistance; (*c*) development of sliding (friction) resistance; (*d*) resisting moment due to cell soil; (*e*) resisting moment due to sheet-pile resistance. [*After Cummings (1960)*.]

Referring to Fig. 15-7*c*, the weight of soil overlaying the triangle, zone *defg* yielding W_1, plus the weight of the soil included in the triangle *efh* is

$$W_1 = \gamma(a + y)y \cot \phi \qquad (j)$$

The shear resistance developed by W_1 along the horizontal plane *hf* is

$$R = W_1 \tan \phi = \gamma(ay + y^2) \qquad (k)$$

The maximum value of R occurs when y is a maximum. This occurs when $y = c$, or

$$c = B \tan \phi \qquad (l)$$

The geometry of the problems yields, by inspection,

$$a = H - c \qquad (m)$$

Substituting the values from Eq. (*l*) for y and Eq. (*m*) for a into Eq. (*k*) and using $R_{\max}$ for the maximum force,

$$R_{\max} = \gamma BH \tan \phi \qquad (n)$$

The force R can be interpreted as consisting of two parts, R_1 and R_2 (Fig. 15-7d), referring to Eq. (k). Using $y = c$, these forces are

$$R_1 = \gamma ac \qquad R_2 = \gamma c^2$$

The force R_1 is taken as the area of a rectangle of height c and base γa. Force R_2 is the area of a triangle of height c and base $2\gamma c$. This concept is used so that resisting moments can be computed for these two forces as

$$M_1 = R_1 \bar{y}_1 = R_1 \frac{c}{2} \qquad M_2 = R_2 \bar{y}_2 = R_2 \frac{c}{3}$$

and the total soil resisting moment M_r is

$$M_r = M_1 + M_2$$

Rewriting and substituting, the total soil resisting moment is

$$M_r = \frac{\gamma ac^2}{2} + \frac{\gamma c^3}{3} \tag{15-12}$$

The bending resistance of the piles due to interlock effects (Fig. 15-7e) is computed from the bursting pressure

$$T = \tfrac{1}{2}\gamma H^2 K_a r = Pr$$

This bursting pressure, combined with the interlock friction, provides a couple, as shown in the figure for a pile of width b_1, of

$$M''_r = Tfb_1 = Prfb_1$$

There are n piles in an average width b (or B) of a cell, and since the analysis is for a 1-ft strip of cell, the moment M'_r for a strip is

$$M'_r = \frac{Prfnb_1}{r} \qquad \text{or} \qquad \text{since } nb_1 = b \qquad M'_r = Pfb$$

The total resisting moment M_{tr} developed from the soil and pile resistance is

$$M_{tr} = M_r + M'_r$$

$$= \frac{\gamma ac^2}{2} + \frac{\gamma c^3}{3} + Pfb \tag{15-13}$$

The safety factor against overturning is the ratio of the cell resisting moments M_{tr} to the overturning moments M_0, or

$$F = \frac{M_{tr}}{M_0} \tag{15-14}$$

Stability against sliding in the Cummings method is computed as in Eq. (15-1).
Where cofferdam cells are founded on rock, bearing capacity is not a problem. Where cofferdams are used in construction of power dams, the dam has usually been founded on rock; thus rock was in close proximity to the ground surface and it was

easy to place the cofferdam cells to rock. For many of the Mississippi and Ohio river low-still dams built in conjunction with lock structures to control the river level during floods and allow barge traffic, the dams are founded on piles driven in the sandy riverbed. Cofferdam cell stability in locations such as this or in any situation where the cell piles are not driven to rock will require a bearing-capacity analysis. The analysis may include a deep-seated shear failure if the cell is founded on a firm soil layer overlying a softer clay deposit. In general, the bearing-capacity analysis will follow procedures in Chap. 4 for spread footings and as used for retaining walls in Chap. 12. Consolidation settlements may be a problem if the clay stratum is very susceptible to consolidation, and if the cofferdam is in place a substantial length of time, owing to the differential pressure on the base of the cell from overturning.

15-4. PRACTICAL CONSIDERATIONS IN CELLULAR COFFERDAM DESIGN

The 3/8-in web sheetpiling is widely used for cofferdam design, giving a tension stress for 16 kips/in interlock tension of 42.7 ksi. Since interlock usually controls, and with a nominal $F = 1.5$ giving 10.7 kips/in the corresponding web, tension stress is $10.7/0.375 = 28.4$ ksi. The designer would have to decide if this is satisfactory, since this is slightly more than $0.65F_y$ as given in Table 15-1.

For substantial embedment depths as for cellular cofferdams not on rock, it may be necessary to increase the web thickness to 1/2 in. It is not usually recommended to drive sheetpiling much over 3 to 5 m (10 to 15 ft) with 6 m as an upper practical limit due to driving damage. It is usually desirable to excavate 1 to 2 m of overburden to remove surface stumps, logs, etc., which may damage the sheetpiling if large embedment depths are necessary.

Secondhand sheetpiling is widely used. It may be reused up to about four times, which represents about 25 percent loss from each use. It is for this reason that former as well as current designations for sheet-pile sections are given in Table 15-1.

For dewatering cofferdam cells to reduce the hoop-tension stresses, which usually control the design, it is standard practice to burn holes of about 3.5 to 5 cm diameter on the inside arcs. Practice is to burn the holes at about 1.5- to 2-m centers vertically on every third to sixth sheet pile (weep holes on the same pile result in maximum salvage of piling). Holes are made to the top of the berm or to the inside ground surface if no berm is used. During dewatering operations it is necessary that the drain or weep holes be systematically rodded to maintain drainage. It appears that one cannot rely on drainage through the interlocks to dewater the cells adequately; the interlocks tend to "silt up" during the cell-filling operation. If the dewatering is carefully done and the cell fill is free-draining, current TVA experience (1966, p. 118) indicates that it is satisfactory to use a horizontal saturation line at one-half the free interior cell height. This saturation-line location greatly simplifies the design computations (Fig. 15-8).

Berms are required to provide additional sliding and overturning resistance if the cell is high and sometimes to increase the length of flow path such that the hydraulic gradient does not cause piping or "quick" conditions at the toe of the cells. Berms

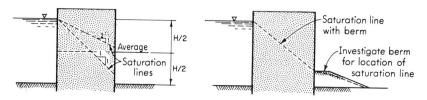

Figure 15-8. Location of the saturation (phreatic line inside cofferdams.

represent internal cofferdam obstruction and additional costs for berm material and are generally undesirable. With the advent of higher-strength interlocks, which were the primary factor in limiting cell diameters (and height) in the past, it is possible that the use of berms will be considerably reduced.

If a berm is required for supplemental sliding and/or overturning resistance, the analysis proceeds by analyzing the berm using the Culmann passive-pressure solution. This value of passive resistance is compared with the total berm sliding ($W_{berm} \times \tan \delta$) and the lesser value is used. Apply the berm resistance at the one-third point. The remaining soil underlying the berm (usually being dissimilar) is analyzed using passive-pressure coefficients—usually Rankine or Coulomb—and treating the berm as a surcharge.

Cofferdams are built with a certain amount of "freeboard" or difference between normal water level and the top of the cell. Freeboard may range from about 1.5 to perhaps 3 m and generally is based on a 5-year flood level. It may be more economical to allow for overtopping the cofferdam during an extreme flood stage than to attempt to design a cofferdam for these conditions—in fact it may not be possible to design for other than minor flood stages owing to the excessive height of cofferdam and the resulting overturning forces due to high water. If it is anticipated that the cofferdam will be overtopped, the fill must be scour-resistant and/or the cell fill capped with a lean concrete mix. The concrete cap will provide resistance to infiltration of rainwater also, especially if it is sloped and a pile is cut on the inside arc to allow the surface accumulation to drain out. Provision should be provided when overtopping is imminent to flood the cofferdam rapidly by means other than overtopping.

Most cofferdam failures appear to be due to failures in the tee of the connecting arc [Grayman (1970)]. The failure may be due to fill saturation after the cell has been dewatered, or inadequate initial design; however, most failures occur in the tee. It appears that many of the past tee failures were associated with welded rather than bolted or riveted fabrication. Current welding state of the art, however, should be sufficient to eliminate welding as a factor in the tee failure.

15-5. DESIGN OF DIAPHRAGM COFFERDAM CELL

This section will consider the design of a diaphragm cell.

Example 15-1. Design a diaphragm cofferdam cell. Assume the cell saturation line to be as shown in the accompanying figure. Other data (see figure):

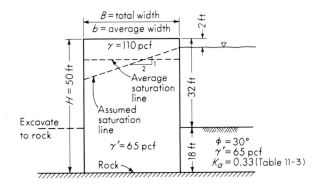

Figure E15-1a.

$$\gamma_{fill} = 110 \text{ pcf clean sand and gravel}$$

$$\gamma'_{fill} = 65 \text{ pcf} \qquad f(\text{fill on rock}) = \tan \phi = 0.57$$

$$\phi(\text{fill}) = 30° \text{ both saturated and damp}$$

Interlock friction = 0.3
Interlock tension = 16 kips/in (Table 15-1)
Use $F = 2$
$\tan \delta$ on sheet pile on fill = 0.4 (Table 11-6)
Steel tensile stress = 25 ksi   A328 steel

SOLUTION. *Step 1. Sliding stability.* The total weight of cell soil to the rock is

$$W = b\left(2 + \frac{b}{4}\right)(0.110) + \left(50 - 2 - \frac{b}{4}\right)(0.065)b$$

$$= b\left(2 + \frac{b}{4}\right)0.110 + b\left(48 - \frac{b}{4}\right)0.065$$

This assumes the saturation line is approximately as shown even at both normal and flood levels. The friction resistance due to the cell weight, neglecting the weight of the steel piling, is

$$F_f = fW = (0.22b + 0.0275b^2 + 3.12b - 0.01625b^2) \tan \phi$$

$$= 0.57(0.0113b^2 + 3.34b)$$

The driving force after inside excavation and overtopping imminent is

$$P_d = 1/2(50)^2(0.0625) + 1/2(0.065)(18)^2(0.33) = 78.1 + 3.5 = 81.6 \text{ kips}$$

For $F = 1.25$,

$$0.57(0.0113b^2 + 3.34b) = 81.6(1.25)$$

Solving for b,

$$b = 46.3 \text{ ft}$$

Step 2. Find width to satisfy overturning ($F = 1.25$).

$$M_0 = P_w(50/3) + P_a(18/3)$$

$$= 78.1(16.7) + 3.5(6) = 1,325.3 \text{ ft-kips}$$

The maximum allowable eccentricity is at the third point, or

$$e = \frac{b}{6} \quad We = M_0 \times F$$

Therefore (and neglecting eccentricity due to actual slope saturation line),

$$(0.0113b^2 + 3.34b)\frac{b}{6} = 1,325.3(1.25)$$

By trial, an approximate value of $b = 50.5$ ft is obtained. Next check overturning from shear of piling on cell fill. Summing moments about the toe, we obtain

$$fb(P_w + P_a) = M_0 \times F$$

$$b = \frac{1,325.3(1.25)}{0.4(81.6)} = 50.8 \text{ ft} \leftarrow \text{controls}$$

Step 3. Check shear along centerline of cell and interlock friction. We shall assume

$$r = L \quad \text{(Fig. 15-3)}$$

The average weight of soil in the cell for a strip 1 ft by b is

$$0.0113(50.8)^2 + 3.34(50.8) = 198.8 \text{ kips}$$

and the average unit weight of the cell soil is

$$\gamma_a = \frac{198.8}{50(50.8)} = 0.078 \text{ kcf}$$

The lateral-pressure coefficient is computed for $\phi = 30°$ and Eq. (15-5) as

$$K'_a = \frac{\cos^2 \phi}{2 - \cos^2 \phi} = \frac{0.75}{2 - 0.75} = 0.60$$

Using the average value of γ_a as a sufficiently precise computation, we obtain for F_s

$$F_s = 1/2(0.078)(50)^2(0.60)(0.577) = 33.8 \text{ kips}$$

For computing the interlock shear F'_s, arbitrarily take

$$H_c = 1/2(50 + 30) = 40 \text{ ft.}$$

$$F'_s = 1/2\gamma H(H_c)K_a f = 1/2(0.078)(50)(40)(0.33)(0.3) = 7.7 \text{ kips}$$

Noting $3/8 = 1/2 \times 3/4$ of Eq. (i), but here we have dimensions and are justified in using the basic computation. The shear on the centerline of the cell due to overturning is

$$V = \frac{1.5M}{b} = \frac{1.5(1,325.3)}{50.8} = 39.1 \text{ kips}$$

The resulting safety factor is

$$F = \frac{F_s + F'_s}{V} = \frac{33.8 + 7.7}{39.1} = 1.06 \; < 1.25 \qquad \text{revise } b$$

An increase of b will decrease V, but other terms do not change:

$$V = \frac{33.8 + 7.7}{1.25} = 33.2$$

and

$$b = \frac{1.5(1,325.3)}{33.2} = 59.9 \leftarrow \text{controls}$$

Step 4. Check interlock tension so that $L = r$ can be determined.

Use $\qquad\qquad H' = H_c, H_w = 40 - 2 - b/2 = 10.3 \qquad$ use 12 ft

$$q_a = 0.078(40)(0.33) + 12(0.0625) = 1.78 \text{ ksf}$$

$$T = \frac{q_a r}{12} = \frac{1.78r}{12} = 0.15r \text{ kips/in}$$

For T less than or equal to 8 kips/in,

$$r = \frac{8}{0.15} = 53.3 \text{ ft maximum radius}$$

It is arbitrarily decided to use $r = L = 40$ ft, resulting in an interlock tension of $0.15(40) = 6.0$ kips/in. The resulting web stress using 3/8-in web is $6.0/0.375 = 16 < 25$ ksi. Using a 120° wye the tension in the diaphragm is the same as in the arc at 6.0 kips/in. The final cell dimensions using PZ-28 piling of $W = 15$ in requires $59.9/1.25 = 47.9$ pieces; use 48 pieces with $b = 60$ ft.

$$r = L = 40 \text{ ft} \qquad b = 60 \text{ ft}$$

The arc layout is as follows (refer to Fig. 15-3a):

$$A + 2r = B \qquad A + 1.82r = b$$

Eliminating A and solving for B, we obtain $B = 67.2$ ft. With $B = 67.2$ ft, the distance A is found to be -12.8 ft. The cell layout is as shown in Fig. E15-1b.

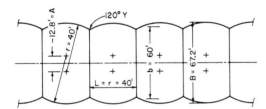

Figure E15-1b.

15-6. CIRCULAR-COFFERDAM DESIGN

This section considers the design of a circular cofferdam on a soil base using both former TVA and Cummings, or current TVA, solution.

Example 15-2. Design a circular-cofferdam cell resting on a riverbed sand stratum approximately 25 m thick.

Other data:
Fill: $\gamma_{wet} = 16.5 \text{ kN/m}^3, \qquad \gamma' = 9.0 \text{ kN/m}^3$
$\qquad \phi = 28°$
Base soil: $\gamma_{sat} = 19.2 \text{ kN/m}^3, \qquad \phi = 34°$
$\tan \delta = 0.38$
Interlock friction 0.3
Interlock tension 28 or 49 kN/m (max. if required)
$H_w = 17$ m; freeboard $= 1.5$ m
Penetration 4 to 5 m
Use $\theta = 45°$ Fig. 15-3a ($b \cong 0.875D$)
All $F \geq 1.25$ neglect dynamic force of river flow.
Saturation line at $H_w/2$ due to free-draining cell fill from river bottom.

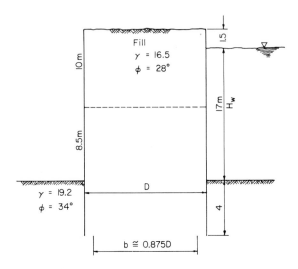

Fill
$\gamma = 16.5$
$\phi = 28°$

10 m

17 m

1.5

H_w

8.5m

D

$\gamma = 19.2$
$\phi = 34°$

4

$b \cong 0.875D$

Figure E15-2a.

Step 1. Compute sliding stability. All computations following are based on a strip 1 m wide unless otherwise indicated.

(a) Weight of cell fill is (Fig. E15-2a) for 4 m of penetration and no change in saturation line for flood or low water level.

$$W = [10(16.5) + 8.5(9) + 4(9.4)]b = 279.1b \text{ kN (effective weight)}$$

(b) Compute net resisting soil pressure $P'(\phi = 34°)$

$$P' = P_p - P_a = 0.5(9.4)(4)^2(3.54 - 0.28) = 245.2 \text{ kN}$$

(c) With water level at flood stage outside and 1 m below soil line inside, the water force is

$$P_w = 0.5(22.5^2 - 3^2)9.807 = 2{,}438.3 \text{ kN}$$

(d) For $F = 1.25$

$$279.1b \tan \phi + 245.2 = 1.25 \, P_w$$

solving

$$b = \frac{1.25(2{,}438.3) - 245.2}{279.1(0.53)} = 18.95 \text{ m}$$

Step 2. Check approximate bearing capacity. Use $B = 19$ m.

$$V = 279.1(19) = 5{,}303 \text{ kN} \qquad H = 2{,}438.3 - 245.2 = 2{,}193 \text{ kN}$$

$$\phi = 34° \qquad N_q = 29.4 \qquad N_\gamma = 34.5$$

$$i_q = \left(1 - 0.5\frac{H}{V}\right)^5 = 0.31 \qquad i_q = \left(1 - 0.7\frac{H}{V}\right)^5 = 0.18$$

$$d_q = 1 + 0.4D/B = 1.08$$

$$q_{ult} = 4(9.4)(29.4)(0.31)1.08 + 0.5(9.4)(19)(34.5)0.18$$

$$= 370.1 + 554.5 = 924.6$$

$$q_a = q_{ult}/2 = 462 \gg 279.1 \qquad O.K.$$

Step 3. Check overturning about inside toe. The overturning moment due to water is

$$M_0 = 0.5\left[(22.5^2)\left(\frac{22.5}{3}\right) - 3^2\left(\frac{3}{3}\right)\right]9.807 = 18{,}573.8 \text{ kN-m}$$

The resisting moment will be made up of $P'\bar{y} + Wb/2 + P_a\,fb$. The friction resistance moment due to P_a on sheetpiling is

$$P_a\,fb = 0.5(9.4)(4)^2(0.28)(0.4)b = 8.4b$$

$$M_r = 245.2\left(\frac{4}{3}\right) + 279.1b\left(\frac{b}{2}\right) + 8.4b$$

For $F = 1.25$

$$139.6b^2 + 8.4b + 326.9 = 1.25(18{,}573.8)$$

$$b^2 + 0.08b = 163.97$$

$$b = 12.8 \text{ m} < 18.95 \qquad O.K.$$

Step 4. Check shear along centerline of cell and interlock friction

$$K_a' = \frac{\cos^2\phi}{2 - \cos^2\phi} = 0.64$$

$$\gamma_{\text{average}} = \frac{279.1}{22.5} = 12.40 \text{ kN/m}^3$$

$$F_s = 0.5(12.4)(22.5)^2(0.64)\tan 28° = 1{,}068.1 \text{ kN}$$

For interlock shear take $H_c = 18.5$ m

$$F_s' = 0.375(12.4)(18.5)^2(0.36)(0.3) = 171.9 \text{ kN}$$

$$M = 18{,}573.8 - 8.4(18.95) - 245.2\left(\frac{4}{3}\right) = 18{,}087.7$$

$$F = \frac{F_s + F_s'}{V} = 1.25$$

$$\frac{(1{,}068.1 + 171.9)b}{1.5(18{,}087.7)} = 1.25$$

$$b = 27.4 \text{ m} > 18.95 \text{ controls}$$

A slight refinement could be made by placing $8.4b$ in numerator to give $b = 27.2$ m.

Step 5. Check interlock tension at soil line with saturation line at $H/2$

$$q_a = 12.4(18.5)(0.36) + 9.807\left(\frac{18.5}{2}\right) = 173.3 \text{ kPa}$$

$$T = \frac{q_a r}{100} \text{ kN/cm} = \frac{173.3r}{100} = 1.73r \qquad \text{kN/cm}$$

For high-strength interlock (28 kips/in or 49 kN/cm) and taking $F = 1.8$

$$r \le \frac{49}{1.73(1.8)} \le 15.74 \text{ m}$$

From step 4, $b = 27.7$

$$D = \frac{27.7}{0.875} = 31.8 \cong 2(15.74) \qquad O.K.$$

$$r = 31.8/2 = 15.9 \text{ m (use this value)}$$

$$T = 1.73(15.9) = 27.5 \text{ kN/cm}$$

Step 6. Check stress in tee connection. Take $L = 1.1r = 17.49$ m

$$T' = \frac{173.3(17.49)(1.4142)}{100} = 42.9 < 49$$

but

$$F = 49/42.9 = 1.14$$

Check web of tee. Use 1/2-in web (1.27 cm) and A572 steel with $F_y = 55$ ksi

$$f_t = \frac{42.9}{1.27} = 33.8 \text{ kN/cm}^2$$

Allowable $F_t \cong 0.6(55)(0.69) = 22.8$ kN/cm² (0.69 converts ksi to metric)
For $f_t = 33.8$ the resulting factor of safety based on F_y (not allowable $= 0.6F_y$) is

$$F = \frac{55(0.69)}{33.8} = 1.12$$

There are three options available:
  (a) Use $\theta = 30°$ (or other value less than 45°).
  (b) Use an inside berm.
  (c) Use this solution recognizing
    (1) Design is for flood condition with overtopping imminent.
    (2) The approximate nature of the tee stress calculation tends to be conservative, and the actual F is probably on the order of 1.3.

Step 7. Design summary (Fig. E15-2b)

$$L = 22.5 \text{ m} \qquad D = 31.8 \text{ m}$$

$$L \text{ of cells} = 2.2r = 34.98 \text{ m}$$

$$H/D = 22.5/31.8 = 0.707$$

r_c connecting cells is

$$BC = 2.2r/2 = 17.49 \text{ m}$$

$$DE = BC - r \cos 45 = 17.49 - 11.24 = 6.25 \text{ m}$$

$$r_c = 6.25/\cos 45 = 8.83 \text{ m}$$

Note final dimensions must be adjusted for an even number of sheet piles and allowing for the special cell connections.

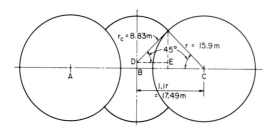

Figure E15-2b.

Example 15-3. Use the data of Example 15-2 and analyze the cell dimensions for stability by the Cummings method.

Fill: $\gamma = 16.5 \text{ kN/m}^3$ \qquad Base soil: $\gamma = 19.2 \text{ kN/m}^3$ (sat)

$\gamma' = 9.00$ \qquad\qquad\qquad $\phi = 34°$

$\phi = 28°$

SOLUTION. Refer to Fig. E15-3 (drawn from final dimensions of Example 15-2). Note the ϕ line of Fig. 15-5c is broken as shown.

$$BI = 4/\tan 34 = 5.93 \text{ m}$$

$$IJ = 8.5/\tan 28° = 15.99$$

$$KL = 5.78 \text{ m}$$

$$FL = 5.78 \tan 28° = 3.07 \text{ m}$$

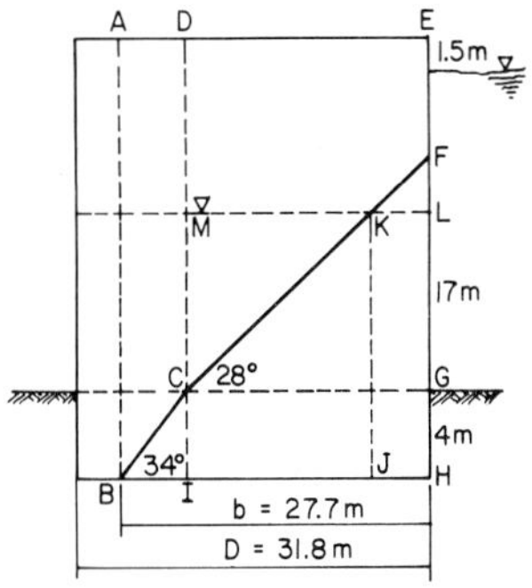

Step 1. Compute resistance of *DCEG*

$$c = B' \tan \phi = 21.77 \tan 28° = 11.58 \text{ m}$$

$$a = 18.5 - 11.58 = 6.92 \text{ m}$$

$$\gamma H = EL \times \gamma + LG \times \gamma'$$

$$18.5\gamma = 10(16.5) + 8.5(9.0)$$

$$\gamma = \frac{241.5}{18.5} = 13.05 \text{ kN/m}^3 \text{ (average)}$$

$$R_1 = \gamma ac = 13.05(6.92)(11.58) = 1,045.7 \text{ kN}$$

$$\bar{y}_1 = c/2 + y_0 = \frac{11.58}{2} + 4 = 9.79$$

$$M_1 = R_1 \bar{y} = 1,045.7(9.79) = 10,237.8 \text{ kN-m}$$

$$R_2 = \gamma c^2 = 13.05(11.58)^2 = 1,750 \text{ kN}$$

$$\bar{y}_2 = c/3 + y_0 = \frac{11.58}{3} + 4 = 7.86$$

$$M_2 = 1,750(7.86) = 13,754.6$$

$$M_T = M_1 + M_2 = 10,237.8 + 13,754.6 = 23,992.4 \text{ kN-m}$$

Step 2. Find M_r of zone *ABDI*

$$\gamma = [10(16.5) + 8.5(9.0) + 4(9.4)]/22.5 = 12.40 \text{ kN/m}^3 \text{ (average)}$$

$$a = 22.5 - 4 = 18.5 \qquad c = 4$$

$$M_r = \gamma \left(\frac{ac^2}{2} + \frac{c^3}{3} \right) = 12.4 \left[\frac{18.5(4)^2}{2} + \frac{4^3}{3} \right] = 2,099.7 \text{ kN-m}$$

Step 3. Find $M'_r = Pfb$; use H at soil line $= 18.5 \text{ m}$

$$P = 0.5\gamma H^2 K_a = 0.5(12.4)(18.5)^2(0.36) = 763.9 \text{ kN}$$

$$Pfb = 763.9(0.3)(27.7) = 6,348.0 \text{ kN-m}$$

Figure E15-3.

Step 4. Compute F against overturning

$$M_w = 0.5(22.5)^2 9.807(22.5/3) = 18,618.0 \text{ kN-m}$$

M_r due to unbalanced soil pressure p' (ex. 15-2, step 1b) and 3 m of water inside cell is

$$M_r = 245.2(4/3) + 44.1(1) = 371.1 \text{ kN-m}$$

$$F = \frac{M_r(\text{total})}{M_0(\text{water})} = \frac{23,992.4 + 2,099.7 + 6,348.0 + 371.1}{18,618.0}$$

$$= \frac{32,811.2}{18,618.0} = 1.76$$

This problem could also have been analyzed by:

1. Considering a series of n vertical sections of say, b/n width each.
2. Computing the a, c dimensions based on ϕ and the width of the section ($b_1 = b/n$, $b_2 = b_1 + b/n$, $b_3 = b_2 + b/n$, etc.).
3. Computing the resisting force of the section using Eq. (n) with $B = b_i$; then $R_i = R_{max} - R_{i-1}$.
4. Placing the force from step 3 at $c/3$ for first value R_1 and at $c/2$ for other values of R_i.
5. With forces and parts of application known, sum moments about toes using statics.

   Normally the Cummings design proceeds by finding b for sliding and bearing capacity as in Ex. 15-2, then making the analysis of this example.

15-7. CLOVERLEAF-COFFERDAM DESIGN

The cloverleaf cell proceeds by making a layout and deciding upon the location of the intersection-cell tees. This step locates the center of the connecting arc and its radius. The connecting arc member may be a wye (30 or 60°) or a 90° tee. Also the dimensions x, y of Fig. 15-3b are determined. The center-to-center spacing is commonly made $L \cong 3.2r$ as shown in the figure instead of the 2.2r of circular cells. The area of the cell (or one-fourth of a cell) is computed, from which the equivalent width of the cloverleaf is

$$B' = \frac{A}{L} = \frac{A}{3.2r}$$

where A is the total area—not the one-fourth computed. Once the equivalent width B' is computed, the analysis proceeds as for a circular cell. The cell is checked for:

1. Sliding.
2. Overturning.
3. Cell shear—when using Eq. (15-4), use tan δ instead of tan ϕ, since the shear force is developed between soil and diaphragm walls.
4. Interlock tension.

PROBLEMS

15-1. For the conditions given in Example 15-2, design a circular cofferdam in which the small connecting circles intersect at 30°.

15-2. Design a diaphragm cofferdam 40 ft high as shown in Fig. P15-1, resting on rock. Coefficient of friction: soil to rock, $f = 0.55$; soil to soil, $f = \tan \phi$; cell wall, $\delta = 22°$.

15-3. Design a diaphragm cofferdam 13.2 m high as shown in Fig. P15-1, resting on rock. Use other data from Prob. 15-2.

15-4. Design a circular cofferdam as shown in Fig. P15-1 if it rests on a sandy-clay soil with $c = 1{,}200$ psf, $\phi = 20°$. Use $\theta = 45°$.

15-5. Design a circular cofferdam as shown in Fig. P15-1 if it rests on a sandy-clay soil with $c = 60$ kPa, $\phi = 20°$. Use $\theta = 45°$.

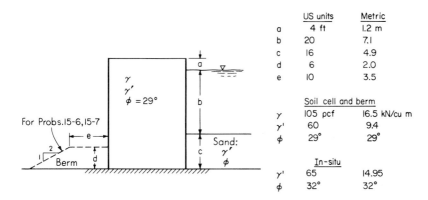

	US units	Metric
a	4 ft	1.2 m
b	20	7.1
c	16	4.9
d	6	2.0
e	10	3.5

Soil cell and berm		
γ	105 pcf	16.5 kN/cu m
γ'	60	9.4
ϕ	29°	29°

In-situ		
γ'	65	14.95
ϕ	32°	32°

Figure P15-1.

15-6. Redo Prob. 15-4 with a berm 6 ft high on the inside (see dotted outline).

15-7. Redo Prob. 15-5 with a berm 2.0 m high (see dotted outline).

15-8. Design a cloverleaf cofferdam if $b = 60$ ft or 19 m. Do not use a berm.

15-9. Make a plot of K'_a vs. ϕ using Eq. (15-5) for a range of ϕ from 25° to 45°. Make appropriate comments.

SIXTEEN

SINGLE PILES—STATIC CAPACITY AND INCLUDING LATERAL LOADS; PILE/POLE BUCKLING

16-1. INTRODUCTION

Piles are structural members of timber, concrete, and/or steel, used to transmit surface loads to lower levels in the soil mass. This load transfer may be by friction, end bearing, or a combination, depending on whether the load is resisted by skin resistance generated along the surface of the pile in the case of *friction piles*, or whether the point rests on a stratum which is firm enough to carry the load in the case of *end-bearing piles*. The pile system may also utilize both end-bearing and skin-resistance components to carry the imposed load. In loose cohesionless soil deposits *compaction piles* may be driven to increase the density (unit weight) of the deposit.

Piles are usually inserted by driving with a steady succession of blows on the top of the pile. The blows may be applied by using a drop hammer, but more commonly with a diesel, steam, or compressed-air-powered hammer. At present the diesel hammer seems to be rapidly supplanting the other types.

Piles have also been driven by the aid of water jets, i.e., displacing the soil at the pile point by using a stream of water under high pressure. Piles are driven also by the use of a vibration device attached to the top of the pile consisting in a pair of eccentric, counterrotating masses.

A pile foundation is generally more expensive than an ordinary shallow foundation, and is used where soil in the shallow-foundation zone is of poor bearing capacity, or settlement problems are anticipated. Therefore, it is absolutely essential for the designer to have an adequate knowledge of the soil conditions at the site when this type of foundation is contemplated.

In pile-foundation analysis it is necessary to make initial estimates of the pile capacity based on a static state of loads and displacements. This is necessary so that the number of piles for a project can be ordered and to the required length. Dynamic analyses can be performed at the same time using the *wave equation* of the next chapter, and other dynamic analyses can be made later during the field operations to establish if the required pile resistance has been obtained so that the next pile can be driven, etc. A load test is usually specified on pile-foundation projects to verify both the static and dynamic analysis.

This chapter will be concerned with the static-analysis methods for pile capacity as well as some introduction to materials and methods used to produce pile members. Methods to compute vertical as well as lateral load capacities will be presented and including some material on pile buckling.

16-2. TIMBER PILES

Timber piles are made of tree trunks with the branches trimmed off and driven with the small end down. Occasionally the large end is driven for special purposes or in very soft soils where the butt end can rest on a firmer stratum. For hard driving the tip may be provided with a metal shoe; otherwise it is either pointed somewhat or cut off square.

Generally there are limitations on the size of the tip and butt end, as well as the magnitude of misalignment; e.g., the Chicago Building Code (1976, Chap. 70) requires that the tip be not less than 6 in in diameter and the butt 10 in if the pile is under 25 ft in length, and with a 12-in butt if the pile is more than 25 ft long. For alignment, the requirement is that a straight line from the center of the butt to the center of the tip shall lie within the pile shaft. The New York Building Code (1969, Art. C26-1109.2) limits timber piles with a uniform shaft taper to an 8-in tip for loads of 50 to 60 kips and a 6-in tip for loads under 50 kips.

Manual 17 published by ASCE (1959) categorizes timber piles as follows:

Class A: To be used for heavy loads and/or large unsupported lengths. The minimum butt diameter is 14 in.

Class B: For medium loads. Minimum butt diameter is 12 to 13 in.

Class C: Use below the permanent water table or for temporary works. Minimum butt diameter is 12 in. Bark may be left on this pile grade.

In addition to minimum dimensions, the ASCE manual also stipulates minimum quality of the timber concerning defects, knots, holes, and type of wood.

If the timber pile is below the permanent water table, it appears that it will last indefinitely. When a timber pile is subjected to alternate wetting and drying, the useful life will be relatively short, perhaps as little as 1 year, unless treated with a wood preservative.

Driving of timber piles usually results in the crushing of the fibers on the driving

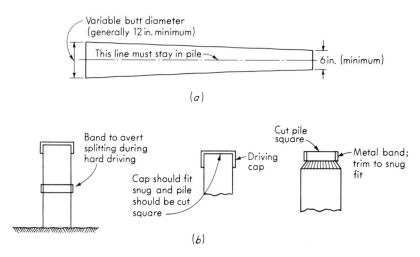

(a)

(b)

Figure 16-1. (*a*) Alignment criteria for timber piles; (*b*) devices to protect pile during driving operations.

end (*brooming*), which can be controlled somewhat by using a driving cap or a metal bond around the butt (Fig. 16-1). Driving may also result in a broken pile in hard soil or soil containing boulders. A sudden increase in pile penetration may be an indication of a broken pile shaft. After driving, the broomed end is cut square, and if previously treated any observed cuts, scars, and holes should be coated with preservative.

Timber-pile splices are undesirable but may be effected as shown in Fig. 16-2. In Fig. 16-2*b* the splice can transmit tension. In both illustrations care should be taken to get good bearing between the two parts of the pile.

Further information on timber piles may be obtained from AWPI (1969, 1967, 1966) and ASTM D25-73 (Part 16).

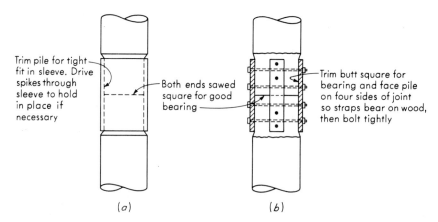

(a)   *(b)*

Figure 16-2. Splices in timber piles. (*a*) Using a metal sleeve with ends carefully trimmed for fit and bearing; (*b*) using splice plates.

The allowable design load for timber piles is computed as

$$P_a = A_p f_a$$

where A_p = average pile cross section at pile cap

f_a = allowable design stresses for the type of wood

Skin-friction coefficients may approach tan ϕ for timber piles in cohesionless soils from a combination of displacement and soil grain penetration of the wood.

16-3. CONCRETE PILES

Table 16-1 indicates that concrete piles may be precast, prestressed, cast in place, or of composite construction.

Precast Concrete Piles

Piles in this category are formed in a central casting yard to the specified length, cured, and then shipped to the construction site. If space is available and a sufficient quantity of piles needed, a casting yard may be provided at the site to reduce transportation costs. Precast piles may be made using ordinary reinforcement as in Fig. 16-3 or prestressed as in Fig. 16-4. Precast piles using ordinary reinforcement are designed for bending stresses during pickup and transport to the site, bending moments from lateral loads, and to provide sufficient resistance to vertical loads and any tension forces developed during driving. The design procedures can be found in any text on reinforced-concrete design; however, handling and driving (tensile) stresses are temporary and allowable concrete stresses are often increased as much as 50 percent to reflect this. The minimum pile reinforcement should be at least 1 percent.

Figure 16-5 illustrates typical bending moments developed during pickup depending on the location of the pickup point. Since the bending moments depend heavily on the location of the pickup point, it should be clearly marked.

Prestressed piles are formed by tensioning high-strength steel, f_{ult} of 250 to 270 ksi, prestressing cables to some value on the order of 0.5 to $0.7f_u$, and casting the concrete pile about the cable. When the concrete hardens, the prestress cables are cut with the tension force in the cables now becoming a compressive stress in the concrete pile as the steel attempts to return to its unstretched length. Some creep and other losses including loss due to axial shortening of the pile under compressive load from the prestress force in the cables occur to reduce the compressive stress in the pile due to the prestress cables. These losses in the absence of refined calculations may be taken as 35,000 psi not including axial-shortening loss due to the applied design loads. Final compressive concrete stresses due to prestressing are usually on the order of 600 to 800 psi.

Table 16-1. Typical pile characteristics and uses*

Pile type	Timber	Steel	Cast-in-place concrete piles (shells driven without mandrel)	Cast-in-place concrete piles (shells withdrawn)
Maximum length	110 ft	Practically unlimited	150 ft	120 ft
Optimum length	30–60 ft	40–160 ft	30–80 ft	25–40 ft
Applicable material specifications	ASTM-D25 for piles. P1-54 for quality of creosote. C1-60 for creosote treatment. (Standards of American Wood Preservers Assoc.)	ASTM A36 for structural sections ASTM A1 for rail sections	ACI Code 318	ACI Code 318
Recommended maximum stresses	Measured at midpoint of length: 600–850 psi for cedar, western hemlock, Norway pine, spruce, and depending on Code 800–1,200 psi for southern pine, Douglas fir, oak, cypress, hickory	$f_s = 9,000$ psi $f_s = 12,000$ $f_s = 20,000$ $f_s = 0.35 f_y$	$0.33f_c'$; $0.4f_c'$ if shell gage ≤ 14; shell stress $= 0.35 f_y$ if thickness of shell ≥ 0.10 in	$0.33f_c'$
Maximum load for usual conditions	60 kips	Maximum allowable stress × cross section	200 kips	300 kips
Optimum-load range	30–50 kips	80–240 kips	100–150 kips	80–200 kips

Disadvantages	Difficult to splice Vulnerable to damage in hard driving Vulnerable to decay unless treated, when piles are intermittently submerged	Vulnerable to corrosion HP section may be damaged or deflected by major obstructions	Hard to splice after concreting Considerable displacement	Concrete should be placed in dry More than average dependence on quality of workmanship
Advantages	Comparatively low initial cost Permanently submerged piles are resistant to decay Easy to handle	Easy to splice High capacity Small displacement Able to penetrate through light obstructions	Can be redriven Shell not easily damaged	Initial economy
Remarks	Best suited for friction pile in granular material	Best suited for end bearing on rock Reduce allowable capacity for corrosive locations	Best suited for friction piles of medium length	Allowable load on pedestal pile is controlled by bearing capacity of stratum immediately below pile
Typical illustrations				

Notes: Stresses given for steel piles and shells are for noncorrosive locations. For corrosive locations estimate possible reduction in steel cross section or provide protection from corrosion.

Table 16-1 (*Continued*)

Pile type	Concrete-filled steel pipe piles	Composite piles	Precast concrete (including prestressed)	Cast-in-place (thin shell driven with mandrel)	Auger-placed pressure-injected concrete (grout) piles
Maximum length	Practically unlimited	180 ft	100 ft for precast 200 ft for prestressed	100 ft for straight sections 40 ft for tapered sections	30–80
Optimum length	40–120 ft	60–120 ft	40–50 ft for precast 60–100 ft for prestressed	40–60 ft for straight 15–35 ft for tapered	40–60
Applicable material specifications	ASTM A36 for core ASTM A252 for pipe ACI Code 318 for concrete	ACI Code 318 for concrete ASTM A36 for structural section ASTM A252 for steel pipe ASTM D25 for timber	ASTM A15 reinforcing steel ASTM A82 cold-drawn wire ACI Code 318 for concrete	ACI Code 318 for concrete	See ACI 318
Recommended maximum stresses	$0.40f'_c$ reinforcement ≤ 30 ksi $0.50f_c$ for core ≤ 25 ksi $0.33f'_c$ for concrete	Same as concrete in other piles Same as steel in other piles Same as timber piles for wood composite	$0.33f'_c$ unless local building code is less $0.4f_y$ if reinforced unless prestressed	$0.33f'_c$; $f_s = 0.4f_y$ if shell gage is ≤ 14 use $f_s = 0.35f_y$ if shell thickness ≥ 0.10 in	0.225 to $0.40f'_c$
Maximum load for usual conditions	400 kips without cores 4,000 kips for large sections with steel cores	400 kips	1,900 kips for prestressed 200 kips for precast	150 kips	160 kips
Optimum-load range	160–250 kips without cores 1,000–3,000 kips with cores	60–160 kips	80–800 kips	60–120 kips	80–120 kips

Disadvantages	High initial cost; Displacement for closed-end pipe	Difficult to attain good joint between two materials	Difficult to splice after concreting; **Redriving not recommended**; Thin shell vulnerable during driving; Considerable displacement	Difficult to handle unless prestressed; High initial cost; Considerable displacement; Prestressed difficult to splice	Dependence on workmanship; Not suitable in compressible soil
Advantages	Best control during installation; No displacement for open-end installation; Open-end pipe best against obstructions; High load capacities; Easy to splice	Considerable length can be provided at comparatively low cost	High load capacities; Corrosion resistance can be attained; Hard driving possible	Initial economy; Tapered sections provide higher bearing resistance in granular stratum	Freedom from noise and vibration; Economy; High skin friction; No splicing
Remarks	Provides high bending resistance where unsupported length is loaded laterally	The weakest of any material used shall govern allowable stresses and capacity	Cylinder piles in particular are suited for bending resistance	Best suited for medium-load friction piles in granular materials	Patented method
Typical illustrations					

Typical illustration labels:

- Grade; 8″ to 36″ dia; Cross section of plain pipe pile; Shell thickness 5/16″ to 1/2″
- Grade; Rock; 12″ to 36″ dia; Cross section of pipe pile with core; Socket required for vertical high loads only; End closure may be omitted
- Typical combinations; Grade; Cased or uncased concrete; Timber; Concrete filled steel shell; Steel pipe concrete filled; BP section; Typical cross sections
- Grade; 12″ to 24″ dia; 12″ to 24″ dia; Note: reinforcing may be pre-stressed; Taper may be omitted; 12″ to 54″ dia; Typical cross sections
- Grade; Grade; Grade; 8″ to 18″ dia; Cross section; Corrugated shell; Thickness 10 ga to 24 ga; Sides straight or tapered; For typical illustration see Fig. for Cast-in-Place Concrete Piles (Shells withdrawn) without pedestal.

* Reprinted and updated from Design Manual, Department of the Navy, Bureau of Yards and Docks, 1971. Concrete and steel stresses from ACI (1973, 1974).

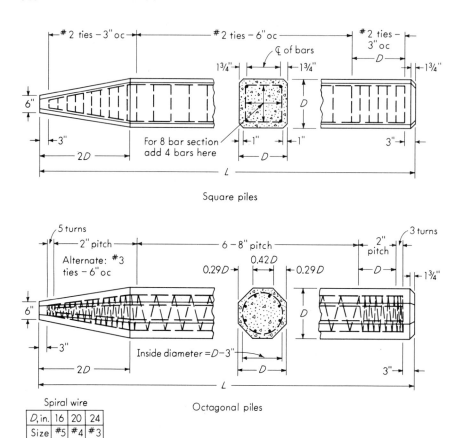

Figure 16-3. Typical details of precast concrete piles. [*After PCA (1951).*]

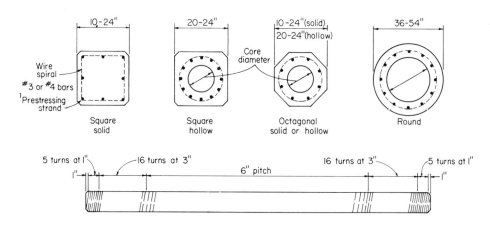

[1]Strand : 1/2 - 7/16 - in diam., f_u = 270 ksi

Figure 16-4. Typical prestressed concrete piles (see also Appendix Table A-5).

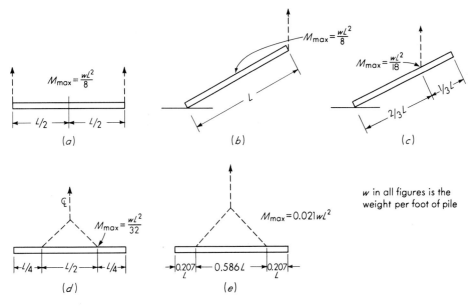

Figure 16-5. Location of pickup points for precast piles, with the indicated resulting bending moments.

The allowable design load P_a for prestressed piles including effect of axial-prestress loss due to load can be computed as

$$P_a = A_g(0.33f'_c - 0.27f_{pe})$$

where A_g = gross concrete area

f_{pe} = effective prestress stress (take 700 psi usually)

Prestressed pile concrete is on the order of $f'_c = 5,000$ to $8,000$ psi. Pickup points should be placed so that the computed bending stress $f_b = Mc/I \leq f_{pe}$, and M is from Fig. 16-5. Prestressing the pile tends to reduce the pile weight, owing to the use of higher-strength concrete, counteracts the tension pickup stresses, and reduces the effect of tension stresses during driving.

Concrete piles are considered permanent; however, certain soils (usually organic) contain materials which may form acids that can damage the concrete, and this possibility should be routinely investigated.

Concrete for structures in fresh or salt water may undergo abrasion from waves, ice, sand, and floating debris. Alternate freezing and thawing may cause concrete damage. In salt water, concrete damage may occur through chemical reaction to the salts. Proposals for nonprestressed concrete to be used in marine structures have been made [Wakeman et al. (1958)] as follows:

1. Use nonreactive aggregates.
2. Use $6\frac{1}{2}$ to $7\frac{1}{2}$ sacks of cement per cubic yard of concrete.
3. Use type V cement (has high sulfate resistance).

4. Use a W/C ratio < 6.0 (i.e., less than 6 gal of water per sack of cement).
5. Use air-entrained concrete in temperate and cold regions.
6. Use 3 in minimum cover on all steel reinforcement.

Cast-in-Place Piles

A cast-in-place pile is formed by making a hole in the ground and filling it with concrete. The hole may be drilled (as in caissons), but more often is formed by driving a shell or casing into the ground. The casing may be filled with a mandrel, after which withdrawal of the mandrel empties the casing. The casing may also be driven with a driving tip on the point, providing a shell that is ready for filling with

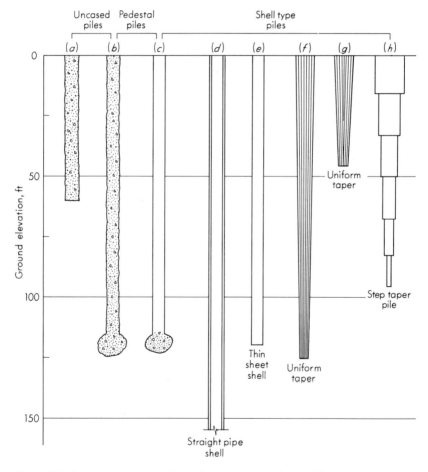

Figure 16-6. Some common types of cast-in-place (patented) piles. (*a*) Western uncased pile; (*b*) Franki uncased-pedestal pile; (*c*) Franki cased-pedestal pile; (*d*) welded or seamless pipe; (*e*) Western cased pile; (*f*) Union or Monotube pile; (*g*) Raymond Standard; (*h*) Raymond step-taper pile. Depths shown indicate usual ranges for the various piles. Current literature from the various foundation equipment companies should be consulted for design data.

concrete immediately, or the casing may be driven open-end, the soil entrapped in the casing being jetted out after the driving is completed.

Various methods with slightly different end results are available and patented. Figure 16-6 indicates some of the commonly available patented cast-in-place piles, and is intended to be representative only. It may be noted that they are basically of three types: (1) shell or cased, (2) shell-less (uncased), or (3) pedestal types.

The allowable design load for all concrete piles (not prestressed) is

$$P_a = A_c f_c + A_s f_s$$

where A_c, A_s = area of, respectively, concrete and steel

f_c, f_s = allowable material stresses

A pile similar in section to that shown in Fig. 16-6a can be formed by using a hollow-stem continuous-flight auger with a diameter of 10 to 16 in. The hole is excavated to the desired elevation, a hose is connected to the auger, and grout (or concrete with small aggregate) is pumped under pressure down the auger stem and out the tip into the cavity formed as the auger is slowly withdrawn. The soil on the auger flights holds the grout from coming up the shaft. Record is kept of the auger depth and quantity of grout pumped so that a check is made that the hole is filled and the auger has not been withdrawn too rapidly and the hole become pinched off from soil collapsing into the cavity. When the hole has been filled, the wet grout, being heavier than soil, will maintain the hole until it sets. If it is desired to provide reinforcement to tie the pile and cap, the reinforcing bars can be carefully inserted into the wet grout to the desired level. The soft layers in layered soils may produce a distorted pile shaft because the grout pressure expands the boring. The principal effect of this is to increase grout quantities.

16-4. STEEL PILES

These members are usually rolled H shapes or pipe piles. Wide-flange beams or I beams may also be used; however, the H shape is especially proportioned to withstand the hard driving stresses to which the pile may be subjected. In the H pile the flanges and web are of equal thickness; the standard WF and I shapes usually have a thinner web than flange. Table A-1 in Appendix A lists the H pile sections produced in the United States. Pipe piles are either welded or seamless-steel pipes which may be driven either open-end or closed-end. Pipe piles are often filled with concrete after driving, although in some cases this is not necessary.

Both open-end pipe and H piles involve small relative volume displacements during driving. In the case of pipe piles, if small boulders are encountered, they may be broken up by a chopping bit or blasting, and removed through the pipe. H piles will either break smaller boulders or displace them to one side. If large boulders are encountered, the possibility of terminating the pile on them can be investigated.

Splices in steel piles are effected in the same manner as in steel columns, i.e., by welding (most common), riveting, or bolting. The design stresses in the splice will depend on the location of the splice and local building codes. An above-ground splice

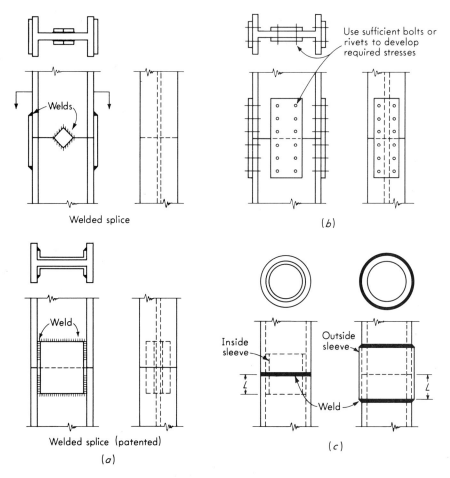

Figure 16-7. Splices for H and pipe piles. (a) Welded pile splices; (b) bolted or riveted splices; (c) splices in pipe piles. Make L adequate for stability.

should be as strong as the pile. Below ground it may only need to be one-third to one-half as strong as the member. The Chicago Building Code (1976) requires that the splice be able to "develop the strength of the pile in compression, tension, bending, and shear" with no qualifications as to location. Figure 16-7 illustrates several pile splices.

Formerly, steel plates were welded on the top of H piles to transfer load from the pile cap to the pile. Research [Ohio (1947)] has indicated that if the top of the pile is adequately embedded (6 in or more) in a concrete pile cap of adequate proportions, which is properly reinforced for the pile reactions, a steel load-transfer cap is not required.

It should be noted that there is little difference in driving a pipe pile with a flat or conical driving point (sometimes called a driving shoe). The reason is that a wedge-shaped zone of soil develops in front of the flat point shaped somewhat like the zone

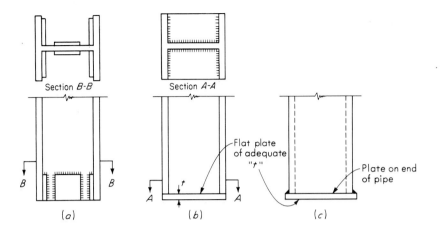

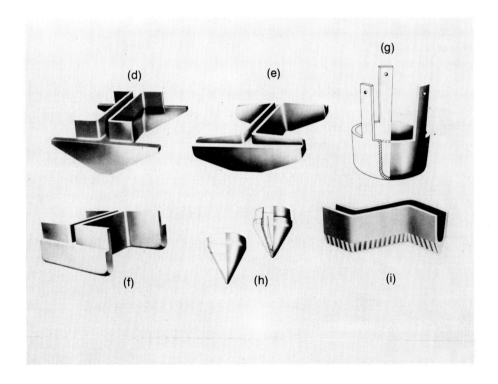

Figure 16-8. Driving points for (a, b) H piles and $(c$ to $e)$ pipe piles. Driving points $(f$ to $i)$ are commercially produced for H pipe, wood and sheet piles (*Courtesy Associated Pile and Fitting Corp.*).

abc of Fig. 4-1 beneath a spread footing. It appears also that open-end pipe piles behave similarly to closed-end piles in driving in that the plug of soil inside the pipe behaves similarly to the driving plate.

H piles tend to form a soil "plug" in the zone between flanges during driving analogous to the soil plug in pipe piles.

H piles and pipe piles may require point reinforcement for driving through dense soil or soil containing boulders. Figure 16-8 indicates some available driving points for these piles. It may not be desirable to use a reinforced pile point for H piles owing to possible reduction in friction capacity near the pile point from the enlarged point opening, not satisfactorily closing back against the pile shaft.

Allowable design loads for steel piles are computed as

$$P_a = A_p f_s$$

16-5. CORROSION OF STEEL PILES

A corrosion study for the National Bureau of Standards [NBS (1962)] on both sheet-pile and bearing-pile substructures indicated that if piles are driven in *undisturbed natural* soil deposits, pile corrosion is not great enough to affect the strength of the piles significantly. This study encompassed soils with pH (a pH less than 7 is an "acid") values from 2.3 to 8.6, and electrical resistivities of 300 to 50,200 ohm-cm, from which it was further concluded that as long as the soil was undisturbed, the soil characteristics and properties are not significant. The substructures studied had been in service from 7 to 40 years.

This study also indicated that piles driven in disturbed or fill soils will tend to undergo relatively more corrosion and may require painting (i.e., paint the pile, then construct the backfill). This was attributed to a higher oxygen concentration in the disturbed soil. Undisturbed soils were found to be oxygen-deficient from a few feet below the ground surface. Piles exposed to sea water or effluents with a pH much above or below 7.0 will require painting or encasement in concrete to resist corrosion. This statement also applies to piles, in general, for the several feet in the zone where the water line fluctuates. As an alternative to painting or concrete encasement, a splice may be made which uses a slightly larger section in the corrosive zone. Some of the newer alloy steels (certain grades of high-strength and copper-alloy steels) may also provide a satisfactory solution, but are generally too recent to have had a service life great enough for drawing valid conclusions.

16-6. SOIL PROPERTIES FOR PILE FOUNDATIONS

Soil properties required for pile-foundation analysis consist in values for the angle of internal friction and cohesion. A value of quake, or elastic recovery from deformation, and damping constants are needed if the wave-equation analysis is used. A value of lateral modulus of subgrade reaction is needed for the laterally loaded pile analysis of Sec. 16-13.

A pile surrounded by soil represents more of a triaxial than plane-strain condition; therefore, the best soil parameters would be obtained from triaxial tests. Normally pile driving will develop excess pore pressures in the vicinity of the pile, and to considerable distances away depending on the soil and the degree of saturation, owing to the rapidity of the driving and the resulting undrained stress conditions. Piles are not normally used to support loads immediately after driving; rather the building loads are applied over a considerable period of time ranging from several months to years. The soil around the pile which has been highly remolded during driving will have regained considerable, all of, or more than the original undrained strength depending on the relative effects of displacement, closing of fissures in stiff clays, consolidation, and densification due to volume change. Under this condition drained tests on remolded samples might be more appropriate.

All types of piles placed in predrilled holes may be subject to drained conditions if one discounts any effects due to soil expansion into the hole and any hole expansion caused by the pile material being placed under pressure. Some question often exists of the effect of wetting the soil at the interface of the soil and pile due to water in the concrete on the soil cohesion. The cement is likely to adhere to many of the soil grains as in pouring a pavement; thus any shear zone will be several millimeters away from the nominal pile dimension and in a zone which will undergo fluctuating water contents normally.

Because of the complexity of predicting the changed soil conditions after driving (or otherwise producing an in situ pile), to prepare laboratory test samples properly, most testing is done on "undisturbed" initial-condition samples. Taking all these factors into consideration as well as costs of laboratory testing, a common practice is to use unconfined-compression (undrained, $\phi = 0$) tests on all cohesive-type soils and penetration tests (SPT) for cohesionless deposits. In some areas the Dutch-cone type of penetrometer may be used instead of the standard penetration test.

The in situ pressuremeter test may be used to advantage for estimating the lateral stress-strain modulus for computing a lateral modulus of subgrade reaction. In general, the in situ soil-parameter tests will not be of particular benefit because the soil properties will change after the pile is driven. These tests may be of considerable value, however, for piles which are not driven.

16-7. PILE CAPACITY IN COHESIVE SOILS

Static pile-capacity computations are necessary to estimate the number of piles for a job and the required pile lengths both for design of the substructure elements and for ordering piles of the correct length from the supplier. For vibrator-driven piles, the dynamic equations do not apply and the static equations must be used to estimate pile-load capacity.

All static pile capacities can be computed by the following equation:

$$P_u = P_{pu} + P_{fu} \qquad (16\text{-}1)$$

where P_u = ultimate static pile capacity

P_{pu} = portion of ultimate pile capacity carried by the pile point in end bearing

P_{fu} = portion of ultimate pile capacity carried in skin resistance (or friction)

Differences in values obtained by the several authorities in the area of pile foundations lie in the methods used to compute the point and skin resistances.

The allowable pile capacity is obtained as

$$P_a = \frac{P_u}{F}$$

where P_a is compatible with the pile material capacity and total loads and the safety factor F is generally taken from 2.5 to 4.0 because of the larger uncertainties in pile design as compared with spread-footing or retaining-wall design.

In general, ultimate point capacity is not developed until after the ultimate skin resistance, and the ultimate skin resistance is not simultaneously developed along the full pile shaft. Those values used in Eq. (16-1) and other equations in this section represent the combined effects of point and skin resistance to obtain an ultimate load. It is academic to consider the timing of the ultimate-resistance development, as the pile is going to behave as a unit to carry some load and the question of interest is the value of the load.

When a pile is driven into a cohesive material, the response will depend on [Tomlinson (1971)] several factors such as:

1. The pile volume displacement. Pipe piles driven "closed-end" and precast concrete piles displace a large volume of soil compared with open-end pipe and H-pile shapes. In clay there may be some plug effect of open-end pipe and H piles, but generally the volume displacement is not likely to be significantly different from the volume of pile material. Volume displacement causes considerable remolding of the soil in the influence zone (some two to five pile diameters) and an increase in pore pressure. Most field observations indicate that the excess pore pressure dissipates and with most of the shear strength recovered in about 30 days. The shear strength (and skin resistance) tends to increase in this zone owing to the consolidation effects obtained when the pore pressure dissipates, and it appears that a soil skin of a thickness of 3 to 25 mm tends to form around the pipe and concrete piles such that one should measure the skin resistance around this outer perimeter rather than using the physical dimensions of the pile. This effect is less certain for H piles because that part of the effective pile perimeter between the flanges is already on a soil zone.
2. The amount and kind of overburden material. Piles penetrating a granular overburden into a cohesive material will tend to drag down granular material into the cohesive material to a depth of about 20 pile diameters based on field observations. The dragdown will increase the skin resistance.
3. Piles penetrating a stiff cohesive soil underlying a soft cohesive layer behave as if a thin skin of soft soil has been dragged down into the driving crack between the pile and clay. This depth of influence also appears to have an upper limit of about

20 pile diameters. This effect is not highly serious, as the soil dragged down tends to be tightly compressed in the crack such that the adhesion is larger than that obtained in the softer upper clay.

4. Displacement piles penetrating a stiff clay tend to form large surface driving cracks alongside the pile and radiating outward from the pile such that adhesion above the zone of about 20 pile diameters is most uncertain. In general, the top 1.2 to 1.8 m should be neglected in computing skin-resistance capacity for piles in stiff clays.

5. Soft clay tends to flow and fill any driving cracks tending to form from driving shock and vibration. For this reason and because of the increased density, consolidation effects, etc., the skin resistance tends to be larger after driving than the initial shear strength. Considering these factors, the α coefficient used later is usually taken as 1.00 to 1.25 in soft clays defined as having a undrained shear strength less than 1,000 psf (50 kPa).

The α and λ coefficients used in this section take into account the factors 1 through 5 cited above so that it is not necessary to estimate the crack depth or any dragdown effects due to overburden.

Using the α factor from Fig. 16-9 or Table 16-2, the equation for *ultimate* pile capacity becomes

$$P_u = cN_c A_p + \int_0^L \alpha(c)(p) \, dL \qquad (16\text{-}2)$$

where N_c = bearing-capacity factor which can be taken as 9 for piles and for undrained ($\phi = 0$) conditions. The N_c value of Table 4-2 can be used when $\phi \neq 0$

A_p = area of pile point

α = coefficient from Fig. 16-9 or Table 16-2

p = pile perimeter

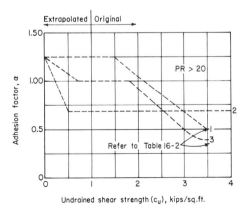

Figure 16-9. Relationship between soil and adhesion factor. Refer to Table 16-2 for curve-number identification. [*After Tomlinson (1971).*]

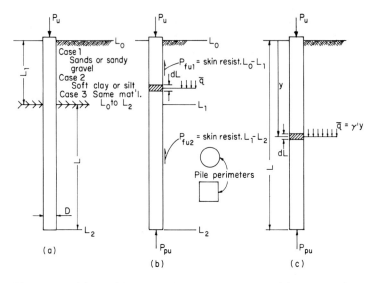

Figure 16-10. (*a*) Conditions for Fig. 16-9 and Table 16-2; (*b*) conditions for Eq. (16-2); (*c*) conditions for Eq. (16-3).

dL = differential of embedment length (Fig. 16-10). This may be replaced by L when the embedment material is the same. It may be replaced with ΔL (finite element) for layered strata

c = *average* soil cohesion (usually undrained values) for the soil layer of interest

Table 16-2. Values of adhesion factors for piles driven into stiff to very stiff cohesive soils for design*

Case	Soil conditions	Penetration ratio†	Adhesion factor, α
1	Sands or sandy gravels overlying stiff to very stiff cohesive soil	< 20	1.25
		> 20	Fig. 16-9
2	Soft clays or silts overlying stiff to very stiff cohesive soil	$8 < PR \le 20$	0.40
		> 20	Fig. 16-9
3	Stiff to very stiff cohesive soils without overlying strata	$8 < PR \le 20$	0.40
		> 20	Fig. 16-9

* After Tomlinson (1971).

† Penetration ratio $PR = \dfrac{\text{depth of penetration into cohesive soil}}{\text{diameter of pile}}$

Where the clay is not saturated, undrained conditions may produce both a ϕ angle and cohesion in triaxial tests. For these conditions Eq. (16-2) must be modified to account for additional skin resistance due to soil friction to obtain

$$P_u = P_{pu} + \int_0^L \alpha(c)(p)\, dL + \int_0^L \bar{q}K \tan \delta\, dL \qquad (16\text{-}3)$$

where the point bearing may also include N_q effects and

$\bar{q}$ = effective vertical stress on element dL

K = coefficient of lateral earth pressure (ranging from about K_0 to 1.75, depending on volume displacement, soil density, ϕ, etc.). Values near K_0 may be most correct because of long-term creep effects, even with large pile-volume displacements

δ = friction angle between soil and pile material (Table 11-6)

The λ Method

Vijayvergiya and Focht (1972) presented an alternative method of obtaining the pile capacity in stiff clay using a λ coefficient obtained from using the Rankine passive earth pressure for undrained ($\phi = 0$) conditions to obtain the skin resistance as

$$f = \lambda(\bar{q} + 2c)$$

where $\bar{q}$ and c have been previously defined in Eqs. (16-2) and (16-3). From a graphical regression analysis of the plot of a large number of pile-load tests, the λ term is as given in Table 16-3. The data shown were taken from the plot cited. According to

Table 16-3. Values of λ obtained from the curves presented by Vijayvergiya and Focht (1972)

L	λ
0	0.50
12.5	0.35
25	0.28
37.5	0.23
50	0.20
62.5	0.18
75	0.16
100	0.15
150	0.135
200	0.12

Vijayvergiya and Focht, this method gives a correlation of ± 10 percent for those piles and soil from which the plot was made. This compares with the Tomlinson data which are more on the order of ± 25 percent. It is not clear why the λ coefficient decreases with increasing pile length unless it is because

1. Longer piles are required in poorer soils and the "undisturbed" sample-recovery procedure deteriorates with increasing depth.
2. Longer piles have considerably more lateral whip during driving, which affects skin contact at the greater depths, resulting in an apparent decrease in skin friction.

Using the λ coefficient, the ultimate pile capacity is computed as

$$P_u = 9cA_p + \int_0^L \lambda(\bar{q} + 2c)(p)\, dL \qquad (16\text{-}4)$$

where terms are defined in Eqs. (16-2) and (16-3) and shown in Fig. 16-10c.

Example 16-1. Estimate the ultimate pile capacity of a 23 m × 30 cm diameter concrete pile for an offshore structure where the average $q_u = 24$ kPa. The submerged unit weight is 8.15 kN/m³. Use both the Tomlinson α coefficient and the λ method.

SOLUTION
(a) Tomlinson Method
Take $\alpha = 1$ (for soft clay see Fig. 16-9)
$A_p = 0.07$ m² $p = \pi D = \pi(0.3) = 0.94$ m
$c = q_u/2 = 12$ kPa
Weight of pile $W_p = 23.5(0.07)(23) = 37.9$ kN

$$P_u = 9(12)(0.07) + \int_0^{23} 1(12)(0.94)\, dL$$
$$= 7.56 + 12(0.94)(23) = 267.0 \text{ kN}$$
$$P_{u(\text{net})} = 267.0 - W_p = 267.0 - 37.9 = 229.1 \text{ kN}$$

(b) λ Method (in soft clay this method may not be strictly applicable)
$\lambda = 0.16$ (Table 16-3)
$P_p =$ same as Tomlinson method
$\bar{q} = 8.15(23/2) = 93.7$ kPa

$$P_u = 7.56 + \int_0^{23} 0.16(93.7 + 24)(0.94)\, dL$$
$$= 7.56 + 0.16(117.7)(0.94)(23) = 414.8 \text{ kN}$$
$$P_{u(\text{net})} = 414.8 - 37.9 = 376.9 \text{ kN}$$

Example 16-2. Estimate the pile length required for an 18-in pipe pile (closed-end) to be filled with concrete in a two-láyer soil system as shown. Use both the Tomlinson and λ methods.

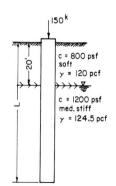

150^k

c = 800 psf
soft
γ = 120 pcf

c = 1200 psf
med. stiff
γ = 124.5 pcf

20'

SOLUTION
(a) Tomlinson Method

Estimate $\alpha_1 = 1.00$ (soft)

$\alpha_2 = 0.70$ stiff (Fig. 16-9)

$A_p = 0.7854(1.5)^2 = 1.77$ ft²

The α coefficient reflects any driving gap in the soil at ground line.

$$P_u = \int_0^{20} (1)(0.80)\pi(1.5)\, dL + \int_0^{L_1} 0.7(1.2)\pi(1.5)\, dL + 9(1.2)(1.77)$$

$$= 0.8(4.71)(20) + 0.7(1.2)(4.71)L_1 + 19.1$$

Figure E16-2.

Use $F = 3$

$$75.4 + 3.96L_1 + 19.1 = 3(150)$$

$$3.96L_1 = 355.5$$

$$L_1 = 355.5/3.96 = 89.7 \quad \text{say 90 ft}$$

$$L = 90 + 20 = 110 \text{ ft}$$

(b) Using λ Method

Estimate a weight factor of 4 on c_2 due to probable penetration depth.

$$c_m = 0.8 + 4(1.2) = (4.8 + 0.8)/5 = 5.6/5 = 1.1$$

Try pile of $L = 100$ ft

$$\lambda = 0.15 \qquad \text{(Table 16-3)}$$

$$P_u = P_p + \int_0^L 0.15(\bar{q} + 2c_m)\pi(1.5)\, dL$$

$$\bar{q} = 20(0.120) + 30(0.062) = 2.4 + 1.86 = 4.26 \text{ ksf}$$

$$P_p = \text{value from } a = 19.1 \text{ kips}$$

Use $F = 3$

$$3(150) = 19.1 + 0.15[4.26 + 2(1.1)]4.71L$$

$$L = 430.9/4.56 = 94.4 \quad \text{say 95 ft}$$

Since this is close to the assumed value of 100 ft and any reduction in depth will result in a reduced $\bar{q}$ and an increase in L, it is not necessary to revise the pile length. Weight of pile is neglected.

Example 16-3. Find the required length of a friction HP10 × 42 pile to carry 450 kN using an $F = 2.0$. Take the average cohesion as 50 kPa.

SOLUTION. *Step 1.* Convert HP10 × 42 to metric but neglect pile weight.

$$A = 12.35 \text{ sq in} = 79.66 \text{ cm}^2$$

$$b = 10.08 \text{ in} = 25.60 \text{ cm}$$

$$d = 9.72 \text{ in} = 24.69 \text{ cm}$$

Minimum shear perimeter is $2b + 2d = 1.006$ m.

Step 2. Find L. Take $\alpha = 1$ due to soft clay and H pile will produce small displacements; also about one-half of perimeter is soil-to-soil.

$$P_{pu} = 9cA_p = 9(50)(79.66 \times 10^{-4})$$

$$= 3.6 \text{ kN} \cong 0 \quad \text{neglect}$$

$$P_{uf} = p\alpha cL = 1.006(1)(50)L = 900$$

$$L = 17.9 \text{ m}$$

16-8. PILES IN COHESIONLESS SOILS

Piles are frequently driven in cohesionless deposits, through loose, cohesionless deposits to rock, or through a cohesive layer into a cohesionless layer for the purpose of controlling settlements when the deposit is loose or may consolidate. Generally cohesionless soils in the vicinity of the pile (three to five diameters) will densify owing to driving vibrations and/or volume displacement.

The ultimate pile capacity in a cohesionless deposit can be computed the same as for a cohesive deposit $(P_u = P_p + P_s)$.

The point bearing capacity can be computed using the Terzaghi bearing-capacity equations of Chap. 4 for round or square piles and using the reduced bearing-capacity factors for N_q as shown in Fig. 16-11 to account for the fact that soil in the zone near the pile tip will fail locally as shown in Fig. 16-12. Experience and observations tend to indicate that the bearing capacity reaches some limiting value rather than increasing infinitely with depth as the $\bar{q}N_q$ term in the Terzaghi bearing-capacity equation tends to indicate.

Again the various authorities are not in general agreement of how to compute the skin-resistance term. In general the skin resistance can be computed as

$$P_s = \int_0^L p(\bar{q})K(z)(\tan \delta) \, dz$$

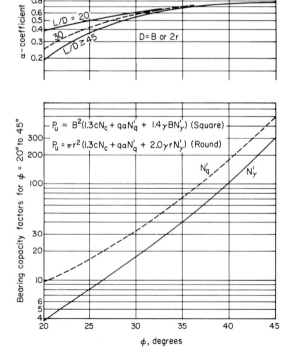

Figure 16-11. Bearing-capacity factors for deep foundations. For values of N_c for all ϕ angles and N'_q and N'_δ for $\phi < 20°$ use Hansen values in Table 4-2. [After Thurman (1964).]

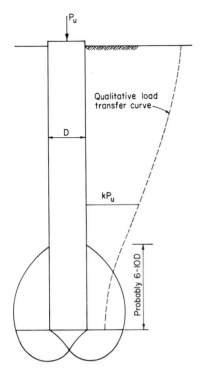

Figure 16-12. Qualitative point-failure zone.

Inspection of the terms in this equation shows that one is simply obtaining the friction resistance due to a lateral force as computed in Chap. 11 on the perimeter of the pile and using the friction coefficient as tan δ. All the authorities use very nearly the same form of this equation. Where the significant difference develops is in the earth-pressure coefficient K. There is not a significant difference in values for tan δ (use Table 11-6). Values of K ranging from the Rankine passive-pressure value to 0.4 to 0.5 have been used. This represents a range of nearly 10 and is much too large. Actually the range of values is now generally accepted to be such that discrepancies on the order of 2 are about maximum.

When a pile is driven into a cohesionless deposit, the displacement and vibration produce densification in the pile area. Increased density produces an increase in the angle of internal friction but probably does not significantly affect the skin-friction angle δ. The increased density also causes lateral pressures to be developed on the order of the passive-earth-pressure value of K_p. It is reasonable to expect, however, that with these high stresses at the pile-soil interface and no appreciable stress changes a short distance (three to five pile diameters) away, stress readjustments and/or creep will gradually reduce the passive pressure to some lateral-pressure value on the order of the original K_0 value to somewhat above this value. From this it follows that a conservative estimate of K is

$$K = K_0 \qquad \text{or} \qquad K = K_0'$$

where K_0' is the at-rest value for the ϕ angle (if it can be determined) resulting from the increased soil density. Larger values of K may be appropriate; however, it is doubtful if K is greater than 1.75, and most likely values of

$$K = 0.60 \text{ to } 1.25$$

will be reasonably correct to use. The lower value(s) are used for silty sands or silty, fine sands and the higher value for other deposits. Tavenas (1971) in a series of pile tests in sands using timber, concrete, and steel piles found K values on the order of 0.5 for steel H piles to 0.7 for the precast concrete. Mansur and Hunter (1970), on a series of pile tests using H piles, steel pipe, and concrete piles, found K values on the order

H piles	$K = 1.4 \text{ to } 1.9$
Steel-pipe piles	$K = 1.0 \text{ to } 1.3$
Precast concrete piles	$K = 1.45 \text{ to } 1.6$
8 tension tests of all types of pile	$K = 0.4 \text{ to } 0.9$

These tension tests produced K values considerably different from Ireland's (1957) tension tests where values of about 1.7 for step-taper piles were obtained. The product of $K \tan \delta$ appears to lie in the range of 0.30 for loose to 1.0 for dense sands.

Nordlund (1963) provided curves based on some field observations of pile performance which tend to account somewhat for driving effects to obtain a K value (K_δ of Fig. 16-13). The Nordlund procedure also attempts to allow for the increased skin resistance which it would appear is possible to obtain using tapered piles.

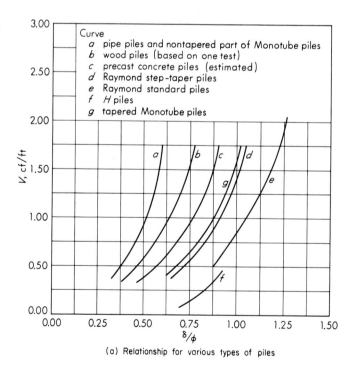

(a) Relationship for various types of piles

Figure 16-13. Method of obtaining the K factor in Nordlund's static pile-capacity equation. [*After Nordlund (1963).*]

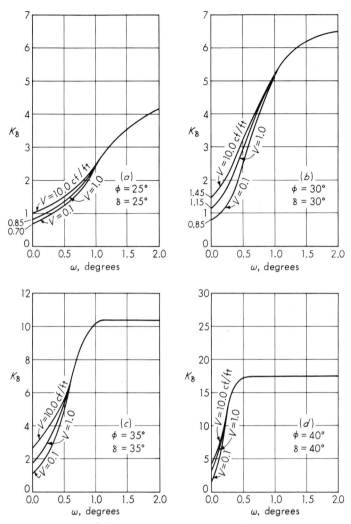

(b) Relationship between K_δ and pile volume for various ϕ-δ-values and pile taper

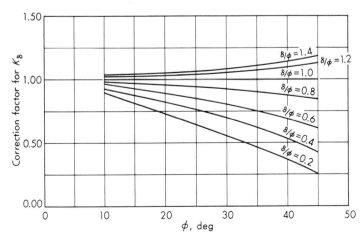

(c) Correction factor for cases where $\delta \neq \phi$

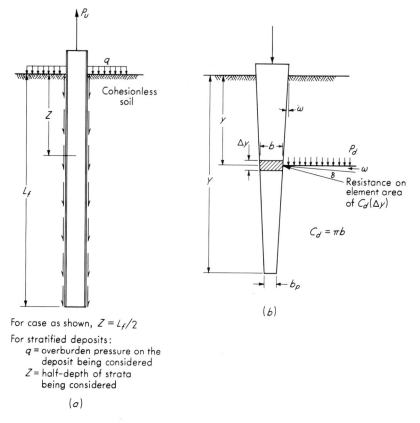

For case as shown, $Z = L_f/2$

For stratified deposits:
   q = overburden pressure on the
      deposit being considered
   Z = half-depth of strata
      being considered

(a)

Figure 16-14. (a) Conditions for Ireland's equation; (b) conditions for Nordlund's equation.

The Nordlund equation is (refer to Fig. 16-14)

$$P_u = A_p \alpha \gamma L N'_q + \int_0^L p K_\delta(\gamma y) \frac{\sin(\omega + \delta)}{\cos \omega} \, dy \qquad (16\text{-}5)$$

where $\alpha N'_q$ = values from Fig. 16-11

     K = lateral-pressure coefficient from Fig. 16-13

     γy = average effective vertical pressure on pile increment dy

Ireland (1957) proposed an equation for friction piles in cohesionless soils using pull-out tests. The form of the equation since has been used for compression piles as well. The equation (refer to Fig. 16-14a) is

$$P_u = P_{pu} + p L_f (\gamma Z + q) K \tan \phi \qquad (16\text{-}6)$$

where Z = depth to center of gravity of embedded portion of pile (or of the corresponding L_f of any stratum in a layered deposit)

     q = surcharge load (for stratified soils, q is the effective pressure of the upper layers on the stratum under consideration)

K = lateral-earth-pressure coefficient [Ireland recommended 1.75 for step-taper piles and lesser values (not specified numerically) for other steel piles in pull-out tests]

p = pile perimeter as used in previous equations; use an average value in the length L_f for tapered piles

The point-bearing term is computed as in the Nordlund equation.

Vesic (1970) reported on a series of pile-load tests in sands and using the idea that the skin resistance reaches a limiting value at a depth of approximately 10 pile diameters in loose sand and 20 pile diameters in dense sand. Tavenas (1971) also reports an indication that in sands the skin resistance reaches a limiting value at some critical depth. Vesic took the relative density as the principal variable and obtained the following equation for the limiting skin resistance as

$$q_s = 0.08(10)^{1.5D_r 4} \tag{16-7}$$

Since the 10 represents the limiting depth for loose sand, it would appear that this factor should be adjusted depending on whether the deposit is loose or dense. Further it has been shown in Chap. 2 that the relative density has little to recommend it, since it is not easily determined. In the Vesic equation above a difference of D_r from 0.4 to 0.5 results in a computed difference factor of $0.0625/0.0256 = 2.4$; driving a displacement pile into a cohesionless deposit could easily change D_r more than this amount.

Using the Vesic skin resistance the ultimate pile resistance is computed as

$$P_u = \text{point bearing} + \text{skin resistance} = P_{pu} + q_s A_s$$

where A_s = pile surface area available for skin resistance

Meyerhof (1956) proposed using the penetration number from the SPT to obtain the pile skin resistance for displacement piles as approximately

$$q_s = 0.04N \quad \text{kips/sq ft} \tag{16-8}$$

For the Dutch-cone penetrometer, Meyerhof proposed

$$q_s = 0.01q_c \quad \text{kips/sq ft} \tag{16-9}$$

For H piles Meyerhof suggested the skin friction be taken as half of either of the above two values. Meyerhof also suggested that q_s be limited to not over 2.0 ksf. For both these equations the total ultimate pile capacity is obtained as

$$P_u = P_{pu} + q_s A_s$$

where the terms are the same as in the Vesic equation. The point bearing capacity can be computed approximately as

$$P_{pu} = 8NA_p \quad \text{(kips) for the SPT test}$$

or $\quad P_{pu} = q_c A_p$ for the Dutch-cone test in units of q_c

For piles in layered deposits consisting in both cohesive and cohesionless soils one must compute the contributions of the separate layers for skin resistance and evaluate the point bearing capacity for that layer in which the point terminates. The total pile capacity is the sum of the resistances from all layers.

Friction piles in silts and loess require additional considerations. For silt:

1. When the cohesion is large, the pile may be designed as in clay, using an appropriate value of cohesion.
2. When there is no cohesion, driving piles may liquefy the soil if it is saturated.
3. It may be necessary to use point-bearing piles which go through the silt layer to a firm underlying stratum.

For loess, Gibbs and Holland (1960) indicate the following criteria:

1. Friction piles may be used if the dry density is 80 to 90 pcf, but driving should be done with the aid of water to consolidate the surrounding material against the pile.
2. Use a spread footing if $\gamma > 90$ pcf, because the bearing capacity may be adequate.
3. Use piles to a lower stratum if $\gamma < 80$ pcf.

The Michigan pile study indicated [ENR (1965, p. 34), Housel (1966)] that if the soil properties are carefully determined, static equations will give results which compare with static load tests within a range of difference of 15 to 20 percent, which is, in general, better correlation than for the dynamic equations. The Michigan study used equations somewhat different from those given in this text,[1] but a study of the report indicates that the equations are similar to either Eq. (16-5) or Eq. (16-6). The basic problem for estimating the static capacity in the study (as was stated earlier) was to arrive at the correct lateral-pressure coefficient and/or a reasonable coefficient of friction or cohesion value.

Example 16-4. Find the length of pile to develop the allowable pile capacity of an HP14 × 89 embedded in the soil stratum shown using Ireland's equation (16-6). Use $f_s = 12$ ksi. SPT numbers averaged 6 in the top 30 ft and approximately 15 for the remainder. Using Table 3-3, ϕ and γ values shown were estimated.

SOLUTION. The pile perimeter (data from Appendix) is $(13.9 + 14.7)2/12 = 4.77$ ft.

$$P_a = f_s A_p = 12(26.19) = 314.3 \text{ kips} \qquad P_u = 3P_a = 943 \text{ kips}$$

Step 1. Find shear resistance of loose layer.
Assume $K = 0.8 \qquad Z = 15$

$$P_u = 4.77(30)(0.105)(15)(0.8) \tan 30° = 104.1 \text{ kips}$$

---

[1] Housel (1966) presents the equation for pile-tip capacity, as well as charts which convert the standard penetration test N values to equation coefficients.

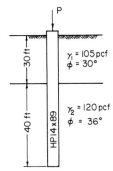

Figure E16-4.

Step 2. Find shear resistance of second layer.

Assume $K = 1.1$, $Z = L/2$, $P_{pu} = 8NA_p = 8(15)(26.19/144) = 22$ kips

$$q = 30(0.105) = 3.15$$

$$P_{u2} = 4.77L\left(0.105\frac{L}{2} + 3.15\right)(1.1) \tan 36°$$

$$= 0.2L^2 + 12.0L$$

$$P_{u2} = 943 - 104 - 22 = 817 \text{ kips}$$

$$0.2L^2 + 12.0L = 817$$

$$L^2 + 60L = 4085$$

$$L = \pm 70.6 - 30 = 40.6 \text{ ft}$$

$$P_{u2} = 820 \text{ kips from back substitution for check}$$

Step 3. Check point capacity. Use Eq. (4-2) and N_q', N_γ' from Fig. 16-11.

$$\gamma D = 30(0.105) + 40.7(0.120) = 8.03$$

For $\phi = 36°$: $N_\gamma' = 48$ $N_q' = 70$ $\alpha = 0.62$

$$A_p = 13.9(14.7)/144 = 1.42 \text{ sq ft}$$

A question arises of whether one should use the pile area or the area of the projected perimeter. Let us use the actual pile area, in which case N_γ is meaningless.

$$P_{pu} = (26.2/144)(8.03)(0.62)(70) = 64 \text{ kips}$$

$$P_u = 104 + 64 + 820 = 988 \text{ kips}$$

$$P_a = 988/3 = 329 \text{ kips} > 314 \qquad O.K.$$

Note that one could cycle to convergence for L and P_{pu}; however, the results are heavily dependent on K, and thus it is not necessary.

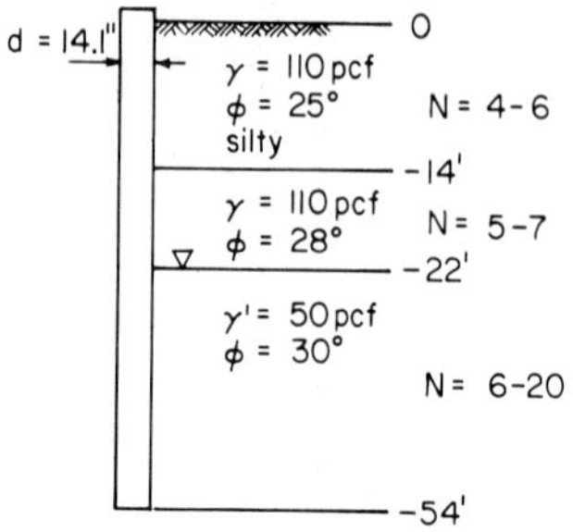

Figure E16-5.

Example 16-5. Compute the allowable pile capacity of the pipe pile (filled with concrete) shown using Nordlund's equation (16-5), $F = 2.0$ and SPT values averaged to give ϕ and γ values used/shown in Fig. E16-5.

SOLUTION. *Step 1.* Ultimate capacity from 0 to 14 ft

$$\phi = 25° \qquad \gamma y \text{ average} = 14(0.110)/2 = 0.77 \text{ ksf}$$

$$V = 0.7854(1.175)^2 = 1.08 \text{ ft}^3/\text{ft}$$

From Fig. 16-12a pile a

$$\delta/\phi = 0.53 \qquad \delta = 0.53(25) = 13.3°$$

From Fig. 16-12b at $\omega = 0$

$$K_\delta = 0.85$$

From Fig. 16-12c at $\phi = 25°$, $\delta/\phi = 0.53$ the correction factor is approximately 0.80 and $K_\delta = 0.85(0.80) = 0.68$

$$P_{u(1)} = \sum_0^{14} 0.77(\pi)(1.175)(14)(0.68) \sin 13.2° = 6.2 \text{ kips}$$

Step 2. Compute capacity from 14 to 22 ft
From Fig. 16-12a for 28° $\delta = 0.53(28) = 14.8°$
From Fig. 16-12b $K_\delta = 1$ and Fig. 16-12c the correction factor is 0.76 and $K_\delta = 0.76(1) = 0.76$
Use the calculus to illustrate an alternative method of solution

$$P_{u(2)} = \int_{14}^{22} (0.76) \sin 14.8°(0.11y)(3.69) \, dy = 0.0788 \left.\frac{y^2}{2}\right]_{14}^{22}$$

$$= 0.0394(22^2 - 14^2) = 11.3 \text{ kips}$$

Step 3. Compute capacity from 22 to 53 ft

$$\gamma y = 0.11(22) + 0.05(31) = 3.20 \text{ ksf} \qquad \delta = 0.53(30) = 15.9°$$

$$K_\delta = 1.1(0.75) = 0.83$$

$$P_{u(3)} = 0.83(3.20)(3.69)(31)(0.274) = 83.2$$

Step 4. Compute point capacity and neglect N'_y term (use Fig. 16-11)

$$P_p = \bar{A}_p(3.97)\alpha N'_q = 1.08(3.97)(0.52)(37) = 82.5 \text{ kips}$$

Step 5. Compute capacity

Weight of pile = 1.08(54)(0.15) = 8.7 kips (neglecting steel shell)

$$P_u = P_{u1} + P_{u2} + P_{u3} + P_p - \text{weight of pile}$$

$$= 6.2 + 11.3 + 83.2 + 82.5 - 8.7 = 174.5 \text{ kips}$$

$$P_a = P_u/2 = 87.2 \text{ kips}$$

16-9. POINT-BEARING PILES

When the material surrounding a pile is soft or has very low friction or adhesion resistance and the point is founded on firm material such as rock or dense sand and/or gravel, the pile is often assumed to derive its total load resistance in end bearing. This is based on the premise that the point would have to displace before skin resistance could develop. Actually the point cannot displace until sufficient load reaches it as the excess over skin resistance.

Based on point capacity alone, the ultimate pile capacity is

$$P_u = A_p q_p$$

where q_p = ultimate bearing resistance of the rock or other firm material on or in which the point terminates.

The bearing pressure of rock can be determined by unconfined or triaxial testing of recovered rock cores. The bearing resistance is often based on coring rock for determining rock quality (cracks, etc.) and for other dense materials the SPT is used and an estimate of bearing capacity is made based on experience.

From instrumented pile tests such as those of D'Appolonia and Hribar (1963) and others which display the loss of pile load with depth due to skin resistance, it can be safely assumed that no pile, in any except the poorest of soils, is in reality a point-bearing pile. In one of the D'Appolonia tests using a step-taper pile to bedrock in point bearing, the actual point load at two times the design load was only about 20 percent of the applied load. It follows from this that some to considerable foundation economy may be obtained by taking into account some skin resistance in nearly all pile-foundation situations where "point-bearing" piles are contemplated. This may be by use of smaller pile cross sections at lower depths or by use of smaller cross sections throughout the pile length. Alternatively, more extensive pile instrumentation should be used on the test pile(s), which are always used to verify pile design on any but the smallest projects, to obtain the skin-resistance contribution and make design revisions based on the actual skin resistance developed.

16-10. BORED OR CAST-IN-PLACE PILES

Piles are often cast in place by drilling a hole which is filled with concrete (Fig. 16-6a), with or without reinforcing. The base may be belled, or enlarged by either placing relatively dry concrete (near zero slump) and ramming it, or otherwise under considerable vertical pressure, causing an enlargement of the base (Fig. 16-6b or c). The hole may not be cased in the case of cohesive soils but usually requires casing with cohesionless deposits, although it may be possible to drill the hole with a clay slurry to hold the sides against caving and use a tremie to place the concrete from the bottom to eliminate weak concrete from intermixing with the slurry. As cited earlier, it is also possible to use hollow-stem augers to build in-place piles.

Piles in the context used here are not over 30 in diameter—larger sizes are classified as caissons, which will be considered in Chap. 19. Bored piles are analyzed using the same static pile equations as for driven piles, the essential difference being that the α coefficient (a λ coefficient is not justified at this time) may be different for bored piles. It appears that for bored piles the coefficient α may not exceed about 0.45; however, if load tests or experience in an area indicate larger values, they should be used. The point capacity is computed similarly as for other static piles. The lateral pressure coefficient K is not likely to exceed K_0 conditions. In the case of the driven pile some densification always occurs. In the case of a bored hole, inward movements can take place even using a slurry, since the density of the slurry is less than the surrounding soil. The concrete, however, has a higher density than the soil and will tend to force the surrounding soil back slightly, resulting in in situ conditions on the order of K_0. Values higher than K_0 do not seem justified unless load tests indicate otherwise. The friction coefficient should approach tan ϕ, since the wet concrete will intermix somewhat with the soil at the interface, forming a skin surface of sufficient roughness that a soil-to-soil slip would be obtained.

16-11. STATIC PILE CAPACITY USING LOAD-TRANSFER LOAD-TEST DATA

Coyle and Reese (1966) presented a method of analysis based on computing the approximate pile movements, and from load-transfer curves of shear strength vs. pile movement (or slip) one could obtain the shear resistance developed by the soil which is compatible with the movement. Basically one can do this as follows:

1. Divide the pile into a convenient number of segments (three are shown in Fig. 16-15a). For very many segments a hand calculation is not practical. Bowles (1974a) used finite elements and matrix notation to develop a computer program to obtain a solution somewhat similar to the Coyle and Reese method, which is available in the cited textbook.
2. Assume a small tip movement Δy_p (zero may be selected, but generally the tip will displace some amount under loads except possibly point-bearing piles).
3. Compute the tip resistance P_p due to this assumed tip movement. Applying a soil "spring" using the concept of subgrade reaction is one method, but the elastic-theory method of Chap. 5 may also be used.
4. Compute the average movement of the bottom segment. For the first trial assume the movement is Δy_p. With this value of "slip" enter the appropriate curve (as Fig. 16-15c) of slip vs. shear strength and obtain the compatible shear strength for

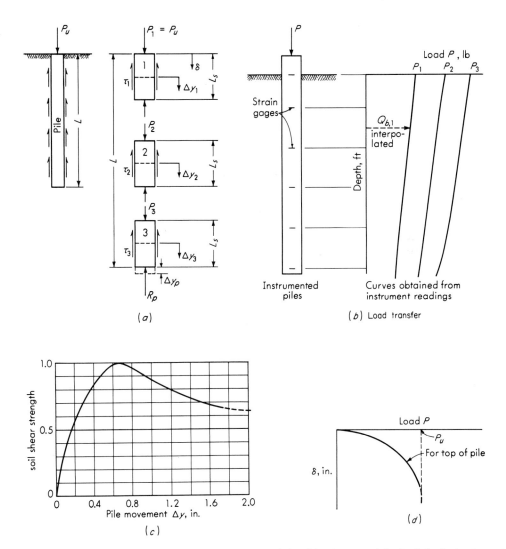

Figure 16-15. Method of computing load-settlement relationships for an axially loaded pile in clay. [*After Coyle and Reese (1966).*]

this slip. The pile load at the top of the segment is the point load + the load carried by skin resistance, or

$$P_3 = P_p + \tau_3 L_3 p$$

Figure 16-15c shows a single curve for the stratum; often there will be a curve for each segment.

5. Recompute the average segment movement until "slip" used and "slip" computed are in satisfactory agreement. Absolute convergence is nearly impossible and certainly of more accuracy than input data will ever justify.

6. With convergence of the bottom segment, proceed to the next segment. As a first estimate of "slip," use the value of $\Delta y_p + \Delta y_n$ from step 5. Obtain the first estimate of shear strength, compute the pile load in the segment $n - 1(P_{n-1})$, and cycle until satisfactory convergence is obtained.

The average load in the bottom segment is

$$P_{av} = \frac{P_t + P_p}{2}$$

and the corresponding average slip is approximately (an exact value requires setting up a differential equation and integrating over the segment length)

$$\Delta y_n = \frac{P_{av}L}{AE} + \Delta y_p$$

Note that $P_t = P_p + \tau_n L_n p$ and τ_n depends on the slip.
   In any other segment the axial deformation is

$$\Delta y_{n-i} = \frac{P_{av}L_i}{AE} + \sum_1^i \Delta y_i + \Delta y_p$$

$$P_{av} = \frac{P_t + P_b}{2}$$

The ultimate load which can be applied at the pile top is

$$P_u = P_p + \sum_1^n \tau_n L_n p$$

16-12. TENSION PILES—PILES FOR RESISTING UPLIFT FORCES

Tension piles may be used beneath buildings to resist uplift forces which may be developed owing to hydrostatic pressures. Tension piles may be used to support structures founded over expansive soils. Tension piles may be used as the foundation element for power-transmission towers. In all these cases a static analysis is necessary. In general the ultimate tension resistance is computed as

$$P_{tu} = P_{fu} + P_{point} + \text{weight of pile}$$

The side resistance is computed similarly to methods presented in the preceding sections, the principal difficulty being the evaluation of the coefficient α, λ, or K. The point-resistance term would be applicable only for those piles which are belled. Belling the base is a common practice for both transmission-tower footings (or piles) and piles in expansive soils. Ireland's (1957) pull-out tests in sands for tapered pile indicated $K \cong 1.75$. He indicated lesser values for other shapes. Sowa (1970) reported on a series of concrete cast-in-place piles in cohesive soils with α ranging from about

0.3 to 0.6 with most values around 0.45. A few piles in cohesionless soils indicated K averages about 0.8 to 1.0.

Transmission-tower footings may be subject to both a compressive force and a tension force depending on the wind direction at a particular time. The use of a concrete shaft of say, 15 to 18 in diameter some 10 to 15 ft long with a belled base to as much as 36 in diameter can provide a footing which can carry a substantial tension force and considerable compression due to both skin resistance and point bearing, since the $\bar{q}N_q$ or the $9cN_c$ term can make a substantial contribution to the bearing capacity of the footing. The pull-out resistance may be evaluated using the method of Sec. 4-10 as an alternative procedure.

Tension piles in expansive soils should preferably have the shaft made as frictionless as possible. This avoids large tension forces in the pile shaft which could separate the bell from the shaft. The shaft will tend to be in tension as the expanding soil compresses around the shaft and expands upward. The tension can be reduced by:

1. Making the shaft as small as practical
2. Leaving an annular space around the shaft which is filled with loose granular material, sawdust, straw, or other material which is compressible but not capable of adhesion to the pile shaft

The pile design proceeds by estimating a depth such that the bell is not affected by the soil expansion. The size of the bell is such that the pile load is carried all in point bearing. The shaft is reinforced (if of concrete, as is the usual case) to carry the estimated compressive and tension forces without failure. The bell is of such dimensions that the tension forces do not separate the shaft and bell or shear the bell from the shaft.

When the pile is at an angle, some modification may be necessary in the computations, and the reader is referred to Hanna (1973), which also contains a bibliography of value for other anchorage resistances.

16-13. LATERALLY LOADED SINGLE PILES

Many piles are subject to lateral as well as axial loads. Lateral loads may be due to lateral earth forces on retaining structures, wind and earthquake loading for buildings, and ship collisions or wave forces for shore and offshore structures.

Laterally loaded piles may be conveniently solved using the finite-element program in the Appendix that has been previously used for beams, rings, and sheet piles, but now $IPILE = 1$. The pile may be fully or partially embedded as shown in Fig. 16-16, which also identifies certain computer-program variables.

The computer-program procedure is identical to the previous problems; i.e., the element EA, ES, ESA^T, and $EASA^T$ matrices are built. The $EASA^T$ is used to construct the global ASA^T, the matrix is inverted to obtain the nodal deflections (translations and rotations), and these are used to obtain the element bending moments for design using the ESA^T, which is recomputed for this step. The user must divide the

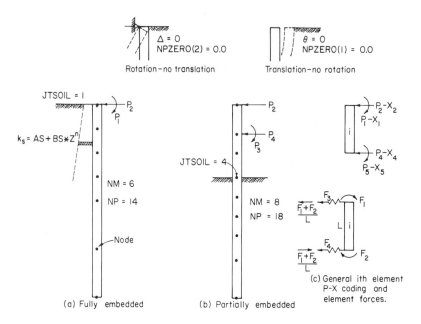

Figure 16-16. Laterally loaded pile using finite elements. Typical loadings shown in (*a*) and (*b*). Note that elements do not have to be same size. Generally use short elements near ground surface and longer elements near pile point where moments are less critical.

pile into a convenient number of finite-element lengths—not necessarily equal—such that loads and boundary conditions fall at nodes.

This program is particularly advantageous in that the following solutions can be readily obtained:

1. Partially embedded piles—input *JTSOIL* as the node number where soil starts.
2. Pile head fixed against rotation—input $NZEROP = 1$ and $NPZERO(1) = 1$ ($1 =$ the *P-X* coding value which is set to zero in the ASA^T).
3. Pile head fixed against translation—input $NZEROP = 1$ and $NPZERO(1) = 2$.
4. Dredge-line translation is zero—as when the pile might be cast in a concrete slab—input $NZEROP = 1$ and $NPZERO(1) = 2$ (*JTSOIL*).
5. More than one displacement specified—input $NZEROP =$ number of displacements specified and $NPZERO(I) =$ the *NP* numbers corresponding to the zero displacements. $XSPEC(I) =$ value of known displacement specified (such as 0.0 or 0.002 rad, 0.0 or 0.1 ft, etc.).

The remaining problem in the lateral-pile solution is to input a reasonable (or correct if known) value of lateral subgrade modulus k_s. Methods previously used for sheet piles may be used as well as the method proposed by the author (Sec. 9-7) of

$$k_s = 36q_a \qquad \text{kcf} \qquad (9\text{-}14)$$

In general as soil surrounds the pile on all sides (the front and sides being of consider-able importance), the above equation should be doubled for piles to give

$$k_s = 72q_a \qquad \text{kcf} \tag{16-10}$$

For undrained-shear-strength data this gives

$$k_s = 72q_u$$

which compares very favorably with a value of $67q_u$ proposed by Terzaghi (1955). Using the Terzaghi or Hansen bearing-capacity equation

$$k_s = 24(cN_c + \gamma DN_q + 0.4\gamma BN_\gamma)$$

which can be rearranged to give

$$k_s = 24(cN_c + 0.4\gamma BN_\gamma) + \gamma N_q D^1$$

The Vesic equation [Eq. (9-13)] may be used and doubled for a condition of pile surrounded by soil to give

$$k_s B = 1.3 \sqrt[12]{\frac{E_s B^4}{E_p I_p}} \frac{E_s}{1-\mu^2} \qquad \text{(units of } FL^{-2}) \tag{16-11}$$

where $E_p I_p$ = flexural rigidity of pile

$\qquad B$ = width of square pile and projected width of round pile

The value of E_s is obtained from

1. Triaxial tests as the secant modulus between 0 and 0.25 and 0.5 the ultimate or peak deviator stress.
2. Borehole pressuremeter tests.
3. SPT test [Yoshida and Yoshinaka (1972)]

$$E_s = 6N \qquad \text{kg/cm}^2$$

   This equation has a maximum error of about 100 percent with the average error close to ±20 percent.
4. Using consolidation test data (cohesive soils) to obtain the coefficient of volume compressibility [Eq. (2-26)] to compute the stress-strain modulus [Francis (1964)] as

$$E_s = \frac{3(1-2\mu)}{m_v} \tag{16-12}$$

   This equation is very sensitive to the value of Poisson's ratio selected.
5. Use the elasticity equation for settlement given in Chap. 5 of

$$S = qB \frac{1-\mu^2}{E_s} I_p$$

Since q/S can be considered to be k_s, using the influence factor for $L/B = 100$ of 3.69, and doubling for soil surrounding the member one obtains

$$k_s \cong \frac{0.6E_s}{B(1 - \mu^2)} \tag{16-13}$$

6. Glick (1948), using the work of others, proposed

$$k_s = \frac{22.24E_s(1 - \mu)}{(1 + \mu)(3 - 4\mu)[2 \ln (2L/d) - 0.443]} \quad \text{kcf} \tag{16-14}$$

where L = pile length (wavelength of a sine curve)

  d = diameter or width of pile

For practical ranges of $2L/d$ of 90 to 120 and Poisson's ratio of 0.2 to 0.4 this equation for practical purposes is

$$k_s \cong 0.8 \text{ to } 1.0E_s \quad \text{kcf} \tag{16-14a}$$

which is not greatly different from Eq. (16-13).

Research by the author indicates that in the same soil, round pile response is different from that for square piles. The response in sands appears to be such that the equation

$$k_s = A + BZ^n$$

should be rewritten as

$$k_s = C_1 A + C_2 BZ^n$$

where $C_1 = C_2 = 1.0$ for square piles (reference lateral soil modulus) and

$$C_1 = 1.3 \text{ to } 1.7$$
$$C_2 = 2.0 \text{ to } 4.4$$

for round piles.

Work by the author indicates the value of subgrade reaction below approximately one-third of the embedded pile length is not critical; however, values much larger than the passive earth pressure (ksf) at this depth divided by, say, 0.01 to 0.02 ft are not realistic. One should use some limiting criteria such as this to establish A, B, and n in the equation

$$k_s = A + BZ^n$$

Table 16-4 gives typical values of lateral modulus of subgrade reaction k_s for several selected soils.

The major advantage of the finite-element method, aside from the greater ease of modeling boundary conditions, is back-computing lateral load test bending moments and lateral deflections measured at selected points along the pile to obtain the likely value of field soil modulus of subgrade reaction.

Table 16-4. Representative range of values of lateral modulus of subgrade reaction (value of A in the equation $k_s = A + Bz^n$)

Soil*	k_s, kcf	k_s, MN/m^3†
Dense sandy gravel	1,400–2,500	220–400
Medium dense coarse sand	1,000–2,000	157–300
Medium sand	700–1,800	110–280
Fine or silty, fine sand	500–1,200	80–200
Stiff clay (wet)	350–1,400	60–220
Stiff clay (saturated)	175– 700	30–110
Medium clay (wet)	250– 900	39–140
Medium clay (saturated)	75– 500	10– 80
Soft clay	10– 250	2– 40

* Either wet or dry unless otherwise indicated.
† MegaNewton = 1,000,000 Newtons.

    Quasi-dynamic analysis of offshore structures subject to wave forces can be obtained by applying the wave force as a time-dependent function on the node(s) and obtaining the resulting pile response at several discrete time intervals using NLC = the number of load conditions or time intervals desired.

    This method of lateral pile analysis may be used to develop the pile-head-response curves to obtain pile constants to use in the pile group analysis of Chap. 18. The procedure is to:

1. Consider the pile head fixed against lateral translation $NPZERO(1) = 2$ and apply a series of moments to the pile at node 1. The computer output gives rotations at node 1, and the unbalanced horizontal force from the soil reactions is

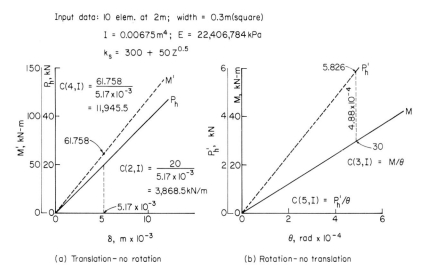

Figure 16-17. Use of the lateral pile analysis to develop "pile constants" for pile group analysis as in Chap. 18, Example 18-7.

the amount of force necessary to apply at node 1 to hold the node to $X(2) = 0.0$. A plot of moment vs. rotation and force required vs. rotation can be made as in Fig. 16-17b.

2. Consider the pile head fixed against rotation $NPZERO(1) = 1$, and apply a series of horizontal forces to the pile at node 1. The computer output gives deflections at node 1 and the fixed-end moment $[F(1)]$ necessary to inhibit rotation. This output may be plotted as force vs. deflection and fixed-end moment vs. deflection as in Fig. 16-17a.

The slope of these four curves provides four of the pile constants needed in Chap. 18.

A more general pile program which can be used for laterally loaded piles, piles partially embedded, piles with axial and lateral forces, and including the effect of lateral deflection producing additional bending ($P - \Delta$ effect) but requiring load-transfer curves as well as lateral subgrade modulus can be found in Bowles (1974a, Chap. 12).

Example 16-6. Given a soft alluvial silt with shear strength 13.8 to 20.7 kPa and a square reinforced-concrete pile 30.5 cm, $I = 0.00774$ m⁴. Take $\mu = 0.4$ and $m_v = 4.679 \times 10^{-4}$ m²/kN. These data are given in Francis (1964) but converted by the author to SI. What is k_s by the Vesic equation and the Bowles method?

SOLUTION. (a) By Vesic equation
Take $E_c = 20,683,200$ kPa

$$E_s = \frac{3[1 - 2(0.4)]}{0.00047} = 1,282.3 \text{ kPa}$$

$$k_s B = 1.3 \sqrt[12]{\frac{1,282.3(0.305)^4}{20.68 \times 10^6(7.74 \times 10^{-3})} \frac{1,282.3}{1 - 0.4^2}}$$

$$= 893.4 \text{ kPa}$$

$$k_s = \frac{893.4}{0.305} = 2,929.3 \text{ kN/m}^3$$

(b) By Bowles method

$$k_s = 2F\frac{q_a}{0.0254} = \frac{6q_a}{0.0254} = 236q_u$$

Use smaller shear strength

$$k_s = 236(13.8) = 3,260 \text{ kN/m}^3$$

or about 11 percent difference.

Example 16-7. Given the same pile as Example 16-6 but sandy clay with shear strength = 2.9 ksf [also from Francis (1964)]. The value of $m_v = 0.286$ ksi and $\mu = 0.4$, 12-in-square concrete pile and $E_c = 3,000$ ksi. Find k_s by Vesic, Eq. (16-14), and Bowles method.

SOLUTION. (a) By Vesic method

$$I = 1,728 \text{ in}^4$$

$$E_s = \frac{3(1 - 2\mu)}{0.286} = 2.097 \text{ ksi} = 302 \text{ ksf}$$

$$k_s = 1.3 \sqrt[12]{\frac{2.097(12)^4}{3,000(1,728)}} \frac{2.097}{1 - 0.16} = 2.178 \text{ ksi}$$

$$k_s = 2.178(144) = 313.8 \text{ kcf}$$

By Eq. (16-14a)

$$k_s = 0.9E_s = 0.9(302) = 272 \text{ kcf}$$

(b) By Bowles equation (assuming shear strength given is half the undrained value q_u)

$$k_s = 72q_u = 72(5.8) = 418 \text{ kcf}$$

Example 16-8. Given the pile shown in the accompanying sketch. Analyze the pile for deflection and bending moment using the finite-element program in the Appendix.

Other data: $I = 50,200 \text{ in}^4 = 2.42091 \text{ ft}^4$

$$E = 30,000 \text{ ksi} = 4,320,000.0 \text{ ksf}$$

$$B = 6.0 \text{ ft (diam)}$$

SOLUTION. The card input data are as follows:

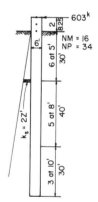

Figure E16-8.

Card No.	Entries							
1	TITLE							
2	FT	KIPS	FT-K	KSF	K/CU FT			
3	NP	NM	NNZP	NLC	IPILE			
	34	16	1	1	1	0	0	0
4	JJS	JTSOIL	NONLIN					
	0	3	0	0	0	0	0	0
5	E	UNITWT	XMAX					
	4,320,000.	0.	3.00					
6	0	2.0	1.0	lateral subgrade modulus				
7	4.125	4.125	5.0	5.0	5.0	5.0	5.0	5.0
8	8.0	8.0	8.0	8.0	8.0	10.0	10.	10.
9	6.0	2.42091	(B, I)					
10	2	1	603.0	(P matrix)				

Comments

1. To define the maximum bending moment reasonably well use smaller divisions in upper part of pile.
2. We will not consider nonlinear effects; make $XMAX = 3.00$, which is large enough not to activate routine ($NONLIN = 0$ causes the $XMAX$ checking routine to be only partially bypassed).

Partial output is as follows:

Member	Bending moments		Node rotation	Node deflection
1	4.00	2,492.00	−0.036 rad	0.976 ft
2	−2,489.00	4,973.00	−0.359	0.827
3	−4,976.00	7,817.00	−0.343	0.681 (D.L.)
4	−7,817.81	9,887.94	−0.031	0.517
5	−9,887.62	10,850.06	−0.027	0.370
6	−10,849.50	10,701.56	−0.022	0.247
7	10,703.00	9,659.38	−0.170	0.149

Note that computer rounding off with IBM 370 series produces slight error in $\sum M = 0$ at the nodes. Node 1 should be $M = 0.00$ but is 4.00. This is a negligible error considering the size of the nodal moments. The bottom moments not shown check quite satisfactorily.

The dredge-line deflection is 0.681 ft, which is certainly a nonlinear deflection, and the problem should be recycled using $XMAX$ on the order of 0.1 ft and $NONLIN = 1$.

16-14. BUCKLING OF FULLY AND PARTIALLY EMBEDDED PILES AND POLES

The author, using a method presented by Wang (1967) for buckling of columns of variable cross section, developed a procedure which can be used to obtain the buckling load for piles either fully or partially embedded. The method is easier to use and considerably more versatile, once a computer program is available, than either the methods of Davisson and Robinson (1965) or those of Reddy and Valsangkar (1970). This method can be used to analyze the buckling load of other pole structures such as steel power-transmission poles [see ASCE (1974) and Dewey and Kempner (1975)].

The method consists in the following steps:

1. Build the ASA^T matrix and obtain the ASA^T inverse of the pile system for whatever the embedment geometry. It is necessary in this inverse, however, to develop the matrix such as shown in Fig. 16-18a. All the rotation P-X are coded first, then the translation P-X's. This results in a matrix which can be partitioned as

$$\frac{P_m}{P_s} = \frac{\begin{array}{c|c} A_1 & A_2 \end{array}}{\begin{array}{c|c} A_2 & A_3 \end{array}} \begin{array}{c} X_R \\ \hline X_s \end{array}$$

2. From the lower right corner of the ASA^T inverse (Fig. 16-18b) take a new matrix called the D matrix (of size $NX_s \times NX_s$), identifying the translation or sidesway X's as X_s

$$X_s = DP_s \qquad (a)$$

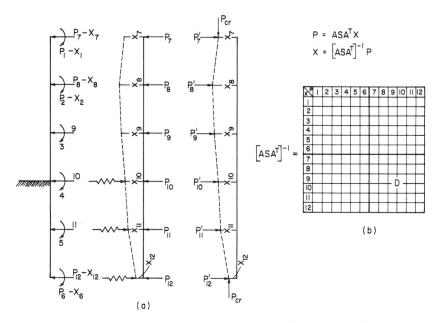

Figure 16-18. (a) General coding and notation used in the pile-buckling problem. The ground line can be specified at any node. Develop the ASA^T, invert it, and obtain the D matrix from the location shown in (b) above.

3. Develop a "second-order string matrix" considering one node deflection at a time as Fig. 16-19a to obtain the G matrix of Fig. 16-19b

$$P'_s = GX_s P_{cr} \qquad (b)$$

4. Since P'_s must be equal to P_s, substitute (b) into (a) and noting that P_{cr} is a column matrix for which the order is not critical to obtain

$$X_s = P_{cr}\{DG\}X_s \qquad (16\text{-}15)$$

This is an eigenvalue problem which can be solved to some predetermined degree of exactness (say, $\Delta X = 0.0000001$) by an iteration process proposed by Wang as follows:

1. Calculate the matrix product of DG (size $NX_s \times NX_s$) and hold.
2. As a first approximation set the column matrix $X_s(i) = 1.00$.
3. Calculate a matrix $X'_s = DG\ X_s$ using the values of 1.00.
4. Normalize the X'_s matrix just computed by dividing all the values by the largest value.
5. Compare the differences of $X_s - X'_s \le \Delta X$ and repeat steps 2 through 5 until the difference criterion is satisfied. On the second and later cycles the current matrix values of X_s are computed from the values of X_s from one cycle back.

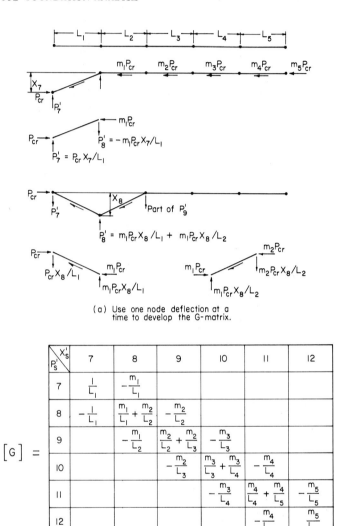

(a) Use one node deflection at a
time to develop the G-matrix.

$[G] =$

X'_s / P'_s	7	8	9	10	11	12
7	$\dfrac{1}{L_1}$	$-\dfrac{m_1}{L_1}$				
8	$-\dfrac{1}{L_1}$	$\dfrac{m_1}{L_1} + \dfrac{m_2}{L_2}$	$-\dfrac{m_2}{L_2}$			
9		$-\dfrac{m_1}{L_2}$	$\dfrac{m_2}{L_2} + \dfrac{m_2}{L_3}$	$-\dfrac{m_3}{L_3}$		
10			$-\dfrac{m_2}{L_3}$	$\dfrac{m_3}{L_3} + \dfrac{m_3}{L_4}$	$-\dfrac{m_4}{L_4}$	
11				$-\dfrac{m_3}{L_4}$	$\dfrac{m_4}{L_4} + \dfrac{m_4}{L_5}$	$-\dfrac{m_5}{L_5}$
12					$-\dfrac{m_4}{L_5}$	$\dfrac{m_5}{L_5}$

(b) The G-matrix for the number of
elements given in (a).

Figure 16-19. The G matrix. For partially embedded piles m will be one until the soil line is encountered.

6. When the convergence criterion has been satisfied, compute the buckling load
using the largest current values in the X'_s and X_s matrix as

$$P_{cr} = \frac{X'_{s(max)}}{X_{s(max)}}$$

This is simply solving Eq. (16-15) for P_{cr} with the left side being the current computation of X_s and the preceding cycle X'_s.

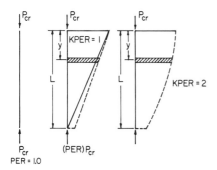

Figure 16-20. Variation of P_{cr} with depth of embedment of the pile or pole. PER = computer variable relating the assumed amount of P_{cr} at the point. $KPER$ = computer variable to specify type of skin-resistance reduction as shown.

If higher buckling modes are desired, and one should always compute at least the first two since this method does not always give the lowest buckling load on the first mode—especially if the values are close together, one may continue steps (1) through (6) using a revised DG matrix for step (1) obtained from the following matrix operation:

$$\{DG\}_{i+1} = \{DG\}_i - \frac{1}{(P_{cr}X_s^TGX_s)_i}(X_s\{GX_s\}^T)_i \qquad (16\text{-}16)$$

where i identifies the current mode and $i + 1$ is the next higher mode. For proof of the validity of Eq. (16-16) see Wang (1967). The values of P_{cr} and X_s are obtained as the values of the ith buckling mode.

Any variation of skin resistance to reduce P_{cr} as illustrated in Fig. 16-20 to develop the string matrix can be used. Note that no skin resistance is used in developing the ASA^T and corresponding D matrix since the assumption of small values of rotation and translation for vertical piles does not produce any skin-resistance effect. Note also that the lateral-soil-resistance effect is included only in the ASA^T matrix and not in the G matrix.

This solution can be readily compared with theoretical solutions by applying one large soil spring at the top and bottom of the pile and no intermediate values (becomes a beam column). It is possible to use a method (similar to that in the included computer program for beams and piles in the Appendix) of zeroing boundary conditions, except that this will not work for the case of a fully embedded pile with top and bottom both specified zero. Satisfactory results can usually be obtained with 8 to 15 finite elements.

PROBLEMS

16-1. A 16-in-diameter pipe pile is driven closed-end 50 ft into a cohesionless soil with an estimated ϕ angle of 32° and a wet unit weight of 105 pcf. The water table is 20 ft below the ground surface, and the submerged unit weight is 55 pcf. Estimate the ultimate pile capacity using (a) Ireland's equation; (b) Nordlund's equation.

16-2. A pile is driven into a cohesionless soil with an estimated ϕ angle of 32° and a wet unit weight of 16.75 kN/m³. The water table is 6.3 m below the ground surface and the submerged unit weight is 9.4 kN/m³. A 40-cm-diameter pipe pile is driven closed-end 17 m into this material. Estimate the ultimate pile capacity using (a) Ireland's equation; (b) Nordlund's equation.

16-3. A pile is driven through a soft cohesive deposit overlying a stiff clay. The average undrained shear strength to the top stratum is 50 kPa and of the lower deposit 165 kPa. The water table is at 6 m below the ground; the stiff clay deposit is at 8 m. The soft clay unit weights are $\gamma = 17.2 \text{ kN/m}^3$ and $\gamma' = 9.7 \text{ kN/m}^3$. The stiff clay unit weight is $\gamma' = 9.9 \text{ kN/m}^3$. What length of 50-cm-diameter pipe pile is required to carry a load of 650 kN?

16-4. Redo Prob. 16-3 for a HP14 × 73. Convert the pile dimensions to metric.

16-5. A J-taper Union Monotube pile with a top diameter of 18 in, a taper of 0.25 in/ft, and a length of 40 ft is driven into a medium stiff clay deposit with an average undrained shear strength of 1.48 ksf. The pile will later be filled with concrete. Estimate the ultimate pile capacity using (a) the α-coefficient method; (b) the λ-coefficient method.

16-6. A Union Monotube F-taper shell is driven into a cohesionless deposit with an average ϕ angle of 33°. The unit weight of the soil is 17.1 kN/m³ above the water table and 9.95 kN/m³ below. The water table is at 6 m below ground surface. The pile top dimension is 45.7 cm and the taper is at 1.17 cm/m. For a 18.3-m pile length, estimate the capacity using (a) Nordlund's equation; and (b) Ireland's equation.

16-7. For the boring logs shown (Fig. P16-7), estimate the pile capacity using Meyerhof's equation and compare with Nordlund's equation. These are actual boring logs. P_u shown are from load tests.

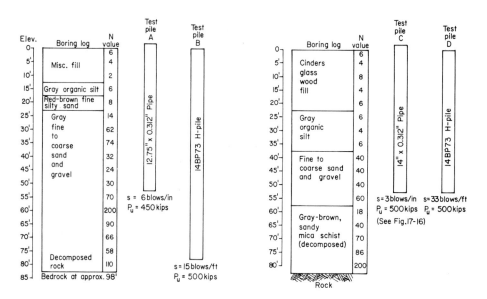

Figure P16-7.

16-8. Verify Example 16-8 using the computer program in the Appendix.

16-9. Redo Example 16-8 using $NONLIN = 1$ and $XMAX = 0.25$ ft, and compare the results with those given in Example 16-8.

16-10. Compute the approximate buckling load for the tapered transmission pole shown in Fig. P16-10.

$$E = 30,000 \text{ ksi for full 160 ft}$$

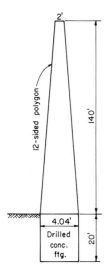

Figure P16-10.

Use 10-ft sections with average section

$$I = 0.07, 0.095, 0.125, 0.155, 0.190, 0.240,$$

$$0.295, 0.350, 0.410, 0.475, 0.550, 0.640,$$

$$0.735, 0.825, 0.870, 0.870 \text{ from top down}$$

$$k_s = 100 + 100z^1$$

Answer: $P_{cr} = 213$ kips.

16-11. Redo Prob. 16-10 if 20-ft embedment is concrete of $E = 3{,}200$ ksi.

SEVENTEEN

SINGLE PILES—DYNAMIC ANALYSIS

17-1. DYNAMIC ANALYSIS

Estimating the ultimate capacity of a pile while it is being driven in the ground at the site has resulted in numerous equations being presented to the engineering profession. Unfortunately, none of the equations is consistently reliable or reliable over an extended range of pile capacity. Because of this, the best means for predicting pile capacity by dynamic means consists in driving a pile, recording the driving history, and load testing the pile. It would be reasonable to assume that other piles with a similar driving history at that site would develop the same load capacity. This chapter will examine some of the driving equations, the load test, and some of the numerous reasons why dynamic pile prediction is so poor. Some of the field problems associated with pile driving such as splicing, redriving, and heave will also be briefly examined.

17-2. PILE DRIVING

Piles are inserted into the ground using a pile hammer resting or clamped to the top of the pile cap, which is in turn connected to the pile. The pile may contain a pile cushion between the cap and pile as shown in Fig. 17-1. The cap usually rests on the pile and may be of, or contain, adequate geometry to effect a reasonably close fit. The pile and hammer is aligned vertically using leads suspended by a crane-type device except for the vibratory hammers, which normally do not use leads. Piles may also be inserted by jetting or partial augering.

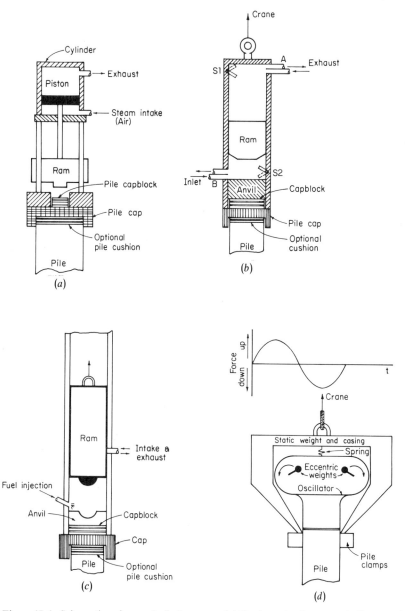

Figure 17-1. Schematics of several pile hammers. (*a*) Single-acting hammer. At bottom of stroke, intake opens with steam pressure raising ram. At top of lift steam is shut off and intake becomes exhaust, allowing ram to fall. (*b*) Double-acting hammer. Ram in down position trips *S2*, which opens inlet and closes exhaust valves at *B* and shuts inlet and opens exhaust at *A*; hammer then raises from steam pressure at *B*. Ram in up position trips *S1*, which shuts inlet *B* and opens exhaust; valve *A* exhaust closes; steam enters and accelerates ram downward. (*c*) Diesel hammer. Crane initially lifts ram. Ram is released and falls; at select point fuel is injected. Ram collides with anvil, igniting fuel. Resulting explosion drives pile and lifts ram for next cycle. (*d*) Vibratory hammer. External power source (electric motor or electric-driven hydraulic pump) rotates eccentric weights in relative directions shown. Horizontal force components cancel—vertical force components add.

Leads provide free travel of the hammer as the pile penetrates the soil and are on the order of 20 ft longer than the pile to provide adequate space for the hammer and other appurtenances.

Mandrels are used to assist in driving pipe piles. These devices fit inside the pipe and rest on the baseplate when the pipe is "closed-end"; they become the pipe point for open-end piles. The mandrel becomes the driving element, which basically drags the pipe down with it during driving so that the thin pipe shell is not damaged.

Spuds are sometimes used in pile-driving operations to penetrate hard strata or seat the pile in rock. The spud may be a separate driving device or simply a heavy driving point attached to pile, especially for H piles seated into rock.

Pile hammers are the devices used to impart sufficient energy to the pile that it penetrates the soil. Several pile hammers are described in the following paragraphs.

Drop Hammers

Drop hammers are still occasionally used for small, relatively inaccessible, jobs. The drop hammer consists in a metal weight fitted with a lifting hook and guides for traveling down the leads with reasonable freedom and alignment. The hook is connected to a cable which fits over a sheave block and is connected to a hoisting drum. The weight is lifted and tripped, freely falling to a collision with the pile. The impact drives the pile into the ground. Principal disadvantages are the slow rate of blows and length of leads required during the early driving to obtain a sufficient height of fall to drive the pile.

Single-Acting Hammers

Single-acting hammers are idealized in Fig. 17-1a. Steam or air pressure is used to lift the ram the necessary height. The ram then drops by gravity onto the anvil, which transmits the impact energy to the capblock, thence to the pile. The hammer is characterized by a relatively slow rate of blows. The hammer length must be such to obtain a reasonable impact velocity (h or height of ram fall), else the driving energy will be small. The blow rate is considerably higher than that of the drop hammer. In general the ratio of ram weight to pile weight including appurtenances should be on the order of between 0.5 to 1.0. Table A-1 in the Appendix gives typical hammer lengths and other useful data.

Double-Acting Hammers

These hammers use steam both to lift the ram and to accelerate it downward. Differential-acting hammers are quite similar except that more control over the steam (or air) is exerted to maintain an essentially constant pressure (nonexpansion) on the accelerating side of the ram piston. This increase in pressure results in a greater energy output per blow than with the conventional double-acting hammer. The blow rate and energy output is usually higher for double-acting or differential hammers (at least for the same ram weight), but steam consumption is also higher than for the single-acting hammer. The hammer length may be several feet shorter for the double-acting hammer than for the single-acting hammer with length ranges on the order of 7 to 15 ft. The ratio of ram weight to pile weight should be on the order of between 0.50 and 1.

Diesel Hammers

Diesel hammers (Fig. 17-1c) consist in a cylinder or casting, ram, anvil block, and a simple fuel-injection system. The ram is raised in the field at the start of operations, fuel is injected near the anvil block, and the ram is released. As the ram falls, the air and fuel compresses and becomes hot because of the compression; when the ram is near the anvil, the heat is sufficient to ignite the air-fuel mixture. The resulting explosion (1) advances the pile, and (2) lifts the ram. If the pile advance is very great as in soft soils, the ram is not lifted by the explosion sufficiently to ignite the air-fuel mixture, requiring that the ram be again manually lifted. It is thus evident that the hammer works most efficiently in hard soils or where the penetration is quite low (point-bearing piles when rock or hardpan is encountered) as maximum ram lift will be obtained.

Diesel hammers are highly mobile, have low fuel consumption (order of 1 to 4 gal/hr), are lighter than steam hammers, and operate efficiently in temperatures as low as 0°C. There is no steam or air-supply generation unit and the resulting hoses. With air there is the additional problem of freezing of the system at temperatures close to freezing. The diesel hammer has a length varying from about 12 to 27 ft (15 to 20 average). The ratio of ram weight to pile weight should be on the order of between 0.25 and 1.0.

Jetting or Preaugering

A water jet is sometimes used to assist in inserting the pile into the ground. This consists in applying a high-pressure stream of water at the pile point to displace the soil. This method may be used to loosen sand or small gravel where for some reason the pile must penetrate to a greater depth in the material than necessary for point bearing. Care must be exercised that the jetting does not lower the point-bearing value. Some additional driving after the jet is halted should ensure seating the point on firm soil.

Preaugering is also sometimes used where a firm upper stratum overlies a compressible stratum which in turn overlies firmer material on which it is desired to seat the pile point to reduce settlement in the compressible stratum. Preaugering will reduce the driving effort through the upper firm material.

For both jetting and preaugering, considerable engineering judgment is required to model the dynamic pile-capacity equations (and static equations) to the field system.

Pile Extraction

Piles may be pulled for inspection for driving damage. Sudden increases of penetration rate may be an indication of broken or badly bent piles. Pile *extractors* are devices specifically fabricated for pulling piles. Double-acting steam hammers may be turned upside down and suitably attached to the pile for the driving impulse and to a hoisting device (crane) to apply a pull at least equal to the weight of the hammer and pile. The hammer impacts loosen and lift the pile, and the crane provides a constant pull to hoist it from the hole.

Vibratory Drivers

A relatively recent (about 1949) method of pile insertion is the use of a vibratory driver. The principle of the vibratory driver is two counterrotating eccentric weights (Fig. 17-1d). The frequency (ranging from 0 to about 30 cycles/sec) is readily computed using equations given in Sec. 20-2. The driver provides two vertical impulses of as much as 160 kips at amplitudes of $\frac{1}{4}$ to 2 in each revolution—one up and one down. The downward pulse acts with the pile weight to increase the apparent gravity force. The pile insertion is accomplished by:

1. The push-pull of the counterrotating weights
2. The conversion of the soil in the immediate vicinity of the pile to a viscous fluid

Best results using vibratory driving are obtained in cohesionless deposits. Results are fairly good in silty and clayey deposits. Driving in heavy clays or soils with appreciable boulders is not possible at present.

Three principal advantages of the vibratory driver (where soils are compatible) are:

1. Reduced driving vibrations—the vibrations are not eliminated but they are less than using impact drivers.
2. Reduced noise.
3. Great speed of penetration—penetration rates of several inches/second are possible.

At present the ultimate pile capacity can only be estimated using static pile methods, although Davisson (1970) developed an equation which purports to estimate the capacity of the patented Bodine Resonant Driver (BRD) used principally by Raymond Concrete Pile company. The other principal vibratory drivers currently used are the patented vibro driver of the L. B. Foster company and a hydraulic-powered device available from McKiernan-Terry Corporation. The BRD equation is

$$R_u = \frac{550 \text{ hp} + 22{,}000 r_p}{r_p + f s_L} \tag{17-1}$$

where r_p = final rate of penetration, ft/sec
f = frequency, Hz
s_L = loss factor, ft/cycle
hp = horsepower delivered to the pile

The loss factors are as follows:

Soil at pile tip	Closed-end pipe	H piles
Loose silt, sand, or gravel	0.0008	−0.0007
Medium dense sand or sand and gravel	0.0025	0.0025
Dense sand or sand and gravel	0.008	0.007

Example 17-1. Use the Davisson equation to solve the pipe pile on the page following in the cited reference (p. 12 of *Foundation Facts*).

$$hp = 414 \quad \text{final penetration} = 240 \text{ sec/ft} = 0.0042 \text{ ft/sec}$$

Soil is dense coarse sand and gravel (based on SPT blow counts); thus $s_L = 0.008$, $f = 126$ Hertz (Hz). Substituting:

$$R_u = \frac{550(414) + 22,000(0.0042)}{0.0042 + 126(0.008)} = 225,046 \text{ lb}$$

The load test indicated $R_u = 550,000$ lb. The pile insertion was terminated nearly on rock for which no s_L was given, and one may debate if that affects the results above. In pile driving, however, piles are often driven until the point reaches approximately refusal—this will always affect the final penetration rate used in Eq. (17-1). It is expected that friction pile computations and load tests might be in closer agreement.

17-3. THE RATIONAL PILE FORMULA

Dynamic formulas have been widely used to predict pile capacity. Some means is needed in the field to determine when a pile has reached a satisfactory bearing value other than by simply driving it to some predetermined depth. Driving the pile to a predetermined depth may or may not obtain the required bearing value because of normal soil variations both laterally and vertically.

It is generally accepted that the dynamic formulas do not provide very reliable predictions. Predictions tend to improve by using a load test in conjunction with the equation to adjust the input variables. Predictions for experienced persons in an area and using certain equipment and with a good knowledge of the input variables of weights, etc., are often considerably better than many of the predictions found in the literature where authors use the reported results of other authors in statistical types of analyses.

The basic dynamic pile-capacity formula termed the *rational pile formula* will be derived in the following material. Nearly all the dynamic pile formulas currently used are based on this equation and by making simplifying assumptions. The rational pile formula depends upon impulse-momentum principles.

For the derivation of the rational pile formulas refer to Fig. 17-2 and the following list of symbols. Applicable symbols from this list are used also with the several

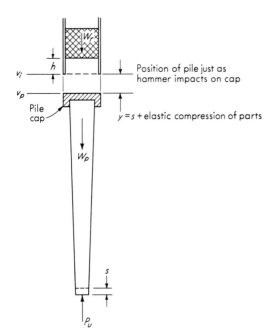

Figure 17-2. Significance of certain terms used in the dynamic pile-driving equations.

pile formulas of the next section and in Table 17-1. The units for the symbol are in parentheses, e.g., (FTL) is the units of force, time, and length.

e_h = hammer efficiency

E_h = manufacturers' hammer-energy rating (LF)

g = acceleration of gravity (LT^{-2})

h = height of fall of ram (L)

I = amount of impulse causing compression or change in momentum (FT)

k_1 = elastic compression of capblock and pile cap and is a form of $P_u L/AE$ (L)

k_2 = elastic compression of pile and is of a form of $P_u L/AE$ (L)

k_3 = elastic compression of soil, also termed quake for wave-equation analysis (L)

L = pile length (L)

m = mass (weight/g) (FT^2L^{-1})

M_r = ram momentum = $m_r v$ (FT)

n = coefficient of restitution

nI = amount of impulse causing restitution (FT)

P_u = ultimate pile capacity (F)

s = amount of point penetration per blow (L)

v_{ce} = velocity of pile and ram at end of compression period (LT^{-1})

v_i = velocity of ram at the moment of impact (LT^{-1})

v_p = velocity of pile at the end of period of restitution (LT^{-1})

v_r = velocity of ram at the end of the period of restitution (LT^{-1})

W_p = weight of pile including weight of pile cap, driving shoe, and capblock (also includes anvil for double-acting steam hammers) (F)

W_r = weight of ram (for double-acting hammers include weight of casing) (F)

At impact, the momentum of the ram is initially

$$M_r = \frac{W_r v_i}{g}$$

End of compression period:

$$M_r = \frac{W_r v_i}{g} - I$$

with a velocity of

$$v_{ce} = \left(\frac{W_r v_i}{g} - I\right)\frac{g}{W_r} \tag{a}$$

If we assume that the pile momentum M_p at this instant is also I, the pile velocity is

$$v_{ce} = \frac{g}{W_p} I \tag{b}$$

Assuming that the pile and ram have not separated at the end of the compression period, the instantaneous velocities of the pile and ram are equal; therefore, combining equations (a) and (b),

$$I = v_i \frac{W_r W_p}{g(W_r + W_p)} \tag{c}$$

At the end of the period of restitution, the momentum of the pile is

$$I + nI = \frac{W_p}{g} v_p \tag{d}$$

and substituting equation (c) for I and solving for the pile velocity,

$$v_p = \frac{W_r + nW_r}{W_r + W_p} v_i \tag{e}$$

At the end of the period of restitution, the momentum of the ram is

$$\frac{W_r v_i}{g} - I - nI = \frac{W_r v_r}{g} \tag{f}$$

Again substituting for I and solving for v_r, we obtain

$$v_r = \frac{W_r - nW_p}{W_r + W_p} v_i \tag{g}$$

The total energy available in the pile and ram at the end of the period of restitution is

$$(\tfrac{1}{2}mv^2)_{\text{pile}} + (\tfrac{1}{2}mv^2)_{\text{ram}}$$

and substituting (e) and (g) and with some simplification one obtains

$$\frac{W_r}{2g} v_r^2 + \frac{W_p}{2g} v_p^2 = e_h W_r h \frac{W_r + n^2 W_p}{W_r + W_p} \tag{h}$$

If the system were 100 percent efficient, the ultimate load P_u multiplied by the point displacement s should be

$$P_u s = e_h W_r h$$

The instant pile top displacement is $s + k_1 + k_2 + k_3$, of which only s is permanent, and the actual input energy to the pile system is

$$e_h W_r h = P_u(s + k_1 + k_2 + k_3) = P_u C$$

Replacing the equivalent energy term with the equivalent from equation (h),

$$P_u = \frac{e_h W_r h}{C} \frac{W_r + n^2 W_p}{W_r + W_p} \qquad (i)$$

Cummings (1940) correctly points out that Eq. (h) already includes the effects of the losses associated with k_i; however, the form of Eq. (i) is generally accepted and used.

The term k_2 can be taken as the elastic compression of the pile $P_u L/AE$ with the corresponding strain energy of $P_u^2 L/2AE$.

Rewriting Eq. (i) and factoring out $\frac{1}{2}$ from all the k terms for strain energy, the Hiley (1930)[1] equation is obtained

$$P_u = \frac{e_h W_r h}{s + \frac{1}{2}(k_1 + k_2 + k_3)} \frac{W_r + n^2 W_p}{W_r + w_p} \qquad (17\text{-}2)$$

For double-acting or differential steam hammers, Chellis (1941, 1961) suggests the following form of the Hiley equation:

$$P_u = \frac{e_h E_h}{s + \frac{1}{2}(k_1 + k_2 + k_3)} \frac{W + n^2 W_p}{W + W_p} \qquad (17\text{-}3)$$

According to Chellis, the manufacturers' energy rating of E_h is based on an equivalent weight term W and height of ram fall h as follows:

$$E_h = Wh = (W_r + \text{weight of casing})h$$

Inspection of the derivation of the Hiley equation indicates the energy-loss fraction should be modified to W as shown in Eq. (17-3) also.

A careful inspection of the Hiley equation or Eq. (i) and a separation of terms results in

Energy in = work + impact loss + cap loss + pile loss + soil loss

$$e_h W_r h = R_u s + e_h W_h \frac{W_p(1 - n^2)}{W_p + W_r} + R_u k_1 + R_u k_2 + R_u k_3$$

Best results from the dynamic formula as a pile-capacity prediction tool are obtained when a careful and separate assessment is made of the several loss factors.

---

[1] Cummings (1940) indicates that Redtenbacher (ca. 1859) may be the originator of this equation.

There may be some question of the correctness of computing the strain energy based on a gradually applied P_u as $P_u^2 L/2AE$ when an impulse-type load is actually applied for which the strain energy is $P_u^2 L/AE$.

It is necessary to use consistent units in Eqs. (17-2) and (17-3) so that the value of P_u is obtained in the force units contained in W_r. For example, if $h =$ ft and $s =$ inches, it is necessary to multiply by 12; if $h =$ meters and $s =$ cm, it is necessary to multiply by 100 to obtain the correct value of P_u.

17-4. OTHER DYNAMIC FORMULAS AND GENERAL CONSIDERATIONS

All the dynamic pile-driving formulas except the Gates formula shown in Table 17-1 are derived from Eq. (17-2) or (17-3) by using various assumptions. The assumptions usually reflect the author's personal experiences and/or attempts to simplify the equation for practical use. Since interpretation of user experience is highly subjective and coupled with wide variability of soils and hammer conditions, the dynamic formulas do not have very good correlation with field experience—especially when used by others in different geographical areas or for statistical comparisons. Statistical comparisons are especially difficult owing to the scarcity of realistic input into the equations of hammer efficiencies, and weights of hammer, and driving equipment such as caps, capblocks, and driving points. For example, Chellis (1961) suggests that pile tips founded on rock or relatively impenetrable material should use a value for pile weight of $W_p/2$. This can make some, to considerable, difference in the loss factor. Also, where is the break point for the factor 2? It would appear that for medium dense materials a factor of 0.75 might be used and gradually increasing to 1.00 for friction piles. Likewise, if the user does not adjust the Hiley equation to include correctly the ram and/or applicable portions of casing and anvil weights, considerable discrepancies can result. Finally, the equations are heavily dependent on hammer efficiency, which must be estimated and which may change during driving operations on the same job.

If we take the impact term in the Hiley equation

$$C_1 = \frac{W_r + n^2 W_p}{W_r + W_p}$$

and rearrange it to

$$C_1 = \frac{1 + n^2 W_r/W_p}{1 + W_r/W_p}$$

and take $n^2 W_r/W_p \cong 0$, we obtain

$$C_1 = \frac{1}{1 + W_r/W_p}$$

which becomes the starting point for the several formula factors.

Table 17-1. Several dynamic pile formulas. Use any consistent set of units

---

Canadian National Building Code (use $F = 3$)

$$P_u = \frac{e_h E_h C_1}{s + C_2 C_3} \qquad C_1 = \frac{W_r + e^2(0.5W_p)}{W_r + W_p}$$

$$C_2 = \frac{P_u}{2A} \qquad C_3 = \frac{L}{E} + 0.0001 .$$

Note that unit of $C_2 C_3$ is same as s

Danish formula [Olsen and Flaate (1967)] (use $F = 3$ to 6)

$$P_u = \frac{e_h E_h}{s + C_1} \qquad C_1 = \sqrt{\frac{e_h E_h L}{2AE}} \text{ (units of } s)$$

Eytelwein formula (use $F = 6$) [Chellis (1941)]

$$P_u = \frac{e_h E_h}{s + 0.1(W_p/W_r)}$$

Gates formula [Gates (1957)] (use $F = 3$)

$$P_u = 0.86\sqrt{e_h E_h} \ (1 - \log s)$$

This gives $P_u =$ kips when $E_h =$ ft-kips and $s =$ inch. In the SI or metric system the equation must be adjusted for the units of E_h and s. This may be done by replacing 0.86 with a coefficient a and 1 by a coefficient b and solving for two known values of P_u

---

Janbu [see Olsen and Flaate (1967), Mansur and Hunter (1970)] (use $F = 3$ to 6)

$$P_u = \frac{e_h E_h}{k_u s} \qquad C_d = 0.75 + 0.15 \frac{W_p}{W_r}$$

$$k_u = C_d\left(1 + \sqrt{1 + \frac{\lambda}{C_d}}\right) \qquad \lambda = \frac{e_h E_h L}{AEs^2}$$

Use consistent units to compute P_u. There is some disagreement of using e_h since it appears to be in C_d; however, a better statistical fit tends to be obtained using e_h, thus it is shown

---

Modified ENR formulas (use $F = 6$)

$$P_u = \frac{1.25 e_h E_h}{s + 0.1} \frac{W_r + e^2 W_p}{W_r + W_p} \qquad \text{[ENR (1965)]}$$

According to AASHO (1973, and no change as of 1976)

$$P_u = \frac{e_h h(W_r + A_r p)}{s + 0.1}$$

For double-acting steam hammers where $A_r =$ ram cross-section area and $p =$ steam (or air) pressure. Use consistent units. Take $e_h \cong 1.0$

Navy-McKay formula (use $F = 6$)

$$P_u = \frac{e_h E_h}{s(1 + 0.3C_1)} \qquad C_1 = \frac{W_p}{W_r}$$

Table 17-1. (*Continued*)

Pacific Coast Uniform Building Code (PCUBC) (from Uniform Building Code, Chap. 28) (use $F = 4$)

$$P_u = \frac{e_h E_h C_1}{s + C_2} \qquad C_1 = \frac{W_r + kW_p}{W_r + W_p} \qquad k = 0.25 \text{ for steel piles}$$

$$= 0.10 \text{ for all other piles}$$

$$C_2 = \frac{P_u L}{AE} \qquad \text{(units of } s\text{)}$$

In general start with $C_2 = 0.0$ and compute value of P_u; reduce value by 25 percent; compute C_2 and a new value of P_u. Use this value of P_u to compute a new C_2, etc., until P_u used $\cong P_u$ computed

---

The *Engineering News-Record* or ENR formula is obtained by lumping all the losses into a single factor and taking $e_h = 1$ to obtain for drop hammers

$$R_u = \frac{W_r h}{s + 1.0} \tag{17-4}$$

and for steam hammers

$$R_u = \frac{W_r h}{s + 0.1} \tag{17-5}$$

A more recent ENR modification (and as used in Table 17-5) is

$$P_u = \frac{e_h W_r h}{s + 0.1} \frac{W_r + n^2 W_p}{W_r + W_p} \tag{17-6}$$

Values of k_1 for use in Eq. (17-2) or (17-3) are presented in Table 17-2. Values of hammer efficiency depend on the condition of the hammer and capblock and possibly the soil (especially for diesel hammers). In the absence of known values the following may be taken as representative of hammers in reasonably good operating condition:

Type	Efficiency e_h
Drop hammers	0.75–1.00
Single-acting hammers	0.75–0.85
Double-acting or differential	0.85
Diesel hammers	0.85–1.00

Chellis (1961) suggests increasing the efficiency 10 percent when using Eqs. (17-2) or (17-3) to compute the driving stresses. Since the reliability of the equations is already with considerable scatter both $(+)$ and $(-)$, it does not appear necessary to make this adjustment.

Table 17-3 presents representative values of coefficient of restitution n. Again the actual value will depend upon the type and condition of the capblock material and whether a pile cushion is used with concrete piles.

Table 17-2. Values for k_1, in inches—temporary elastic compression of pile head and cap*

| | k_1 | | | |
| | Driving stresses P/A on pile head or cap, psi | | | |
Pile material	500	1,000	1,500	2,000
Steel piling or pipe				
Directly on head	0	0	0	0
Directly on head of timber				
pile	0.05	0.10	0.15	0.20
Precast concrete pile with				
3–4 in packing inside cap	0.12	0.25	0.37	0.50
Steel-covered cap containing				
wood packing for steel H				
or pipe piling	0.04	0.08	0.12	0.16
$\frac{3}{16}$-in fiber disk between				
two $\frac{3}{8}$-in steel plates	0.02	0.04	0.06	0.08

* After Chellis (1961).

The term k_2 is computed as $P_u L/AE$, and one may arbitrarily take the k_3 term (quake) as

$$k_3 = 0.0 \text{ for hard soil (rock, very dense sand, and gravels)}$$

$$k_3 = 0.1 \text{ to } 0.2 \text{ for other soils}$$

Equation (17-2) and following must be adjusted when piles are driven on a batter. It will be necessary to compute the axial pile component of $W_r h$ and further reduce this for the friction lost due to the normal component of the pile hammer on the leads or guides. A reasonable estimate of the friction coefficient between hammer and leads may be taken as

$$f = \tan \phi = 0.10.$$

Table 17-3. Representative values of coefficient of restitution for use in the dynamic equations*

Material	n
Broomed wood	0
Wood piles (nondeteriorated end)	0.25
Compact wood cushion on steel pile	0.32
Compact wood cushion over steel pile	0.40
Steel-on-steel anvil on either steel or concrete pile	0.50
Cast-iron hammer on concrete pile without cap	0.40

* After ASCE (1941).

Example 17-2. Estimate the allowable pile capacity of test pile No. 1 reported by Mansur and Hunter (1970, Tables 2, 4, 5, and 6) by the ENR, Janbu, PCUBC, and Hiley equations.

Other data: Hammer = Vulcan 140C; $W_r = 14{,}000$ lb;

$$E_h = 36{,}000 \text{ ft-lb}; \quad e_h = 0.78$$

$$\text{Pile} = 12\text{-in pipe}; \quad A = 17.12 \text{ sq in including instrumentation}$$

$$L = 55 \text{ ft (total)}; \quad E = 29{,}000 \text{ ksi}$$

$$s = 12/16 = 0.75 \text{ in/blow}$$

$$\text{Cap and capblock} = 1{,}710 \text{ lb}$$

$$P_{\text{ult}} \text{ load test} = 280 \text{ kips}$$

SOLUTION. (a) By ENR

$$P_u = \frac{0.78(36)(12)}{0.75 + 0.1} = 396 \text{ kips}$$

$$P_a = 396/6 = 66 \text{ kips}$$

$$\text{Error factor} = \frac{396}{280} = 1.41 \text{ (too large)}$$

(b) By Janbu equation

$$\text{Weight of pile} = \frac{17.12}{144}(490)55 + 1{,}710 = 4{,}914 \text{ lb}$$

$$AE = 17.12(29{,}000) = 496{,}480 \text{ kips}$$

$$C_d = 0.75 + \frac{4{,}914(0.15)}{14{,}000} = 0.80$$

$$\lambda = \frac{0.78(36)(55)(144)}{496{,}480(0.75)^2} = 0.796$$

Note 144 converts sq in to sq ft for units.

$$k_u = C_d\left(1 + \sqrt{1 + \frac{\lambda}{C_d}}\right)$$

$$= 0.80\left(1 + \sqrt{1 + \frac{0.796}{0.80}}\right) = 1.929$$

$$P_u = \frac{e_h E_h}{k_u s} = \frac{0.78(36)(12)}{(1.929)(0.75)} = 233 \text{ kips}$$

$$\text{Error factor} = \frac{233}{280} = 0.83 \quad \text{(too small)}$$

(c) By PCUBC formula

$$k = 0.25$$

$$C_1 = \frac{W_r + kW_p}{W_r + W_p} = \frac{14 + 0.25(4.91)}{14 + 4.91} = 0.80$$

Since we should get the same order of magnitude of P_u, estimate $P_u = 250$ kips

$$C_2 = \frac{250(55 \times 12)}{496,480} = 0.33 \text{ in}$$

$$P_u = \frac{e_h E_h C_1}{s + C_2} = \frac{0.78(36)(12)(0.80)}{0.75 + 0.33} = 251 \text{ kips}$$

Since the value used is adequately close to the assumed value, this is the desired value of P_u

$$P_a = \frac{P_u}{4} = 251/4 = 63 \text{ kips}$$

Error factor $= \dfrac{251}{280} = 0.90$ (too small)

(d) By Hiley equation
For an estimate based on previous steps (a) through (c) of P_u around 200 kips.
Take $n = 0.50$, $k_1 \cong 0.16$ (mandrel driven), $k_3 \cong 0.1$

$$\frac{W_r + n^2 W_p}{W_r + W_p} = 0.80 \text{ as in PCUBC formula}$$

Compute $k_2 \cong \dfrac{200(55)(12)}{496,480} = 0.26$ in

$$P_u = \frac{0.78(36)(12)(0.80)}{0.75 + \frac{1}{2}(0.16 + 0.26 + 0.1)} = 266.9 \text{ kips}$$

Revise k_2; use $P_u = 250$ kips

$$k_2 = \frac{250}{200}(0.26) = 0.32 \text{ in}$$

$$P_u = \frac{269.6}{0.75 + 0.29} = 259.2 \text{ kips}$$

The used and computed values of k_2 are sufficiently close.

Example 17-3. Estimate the ultimate pile capacity of test pile No. 6 (H pile) of Mansur and Hunter (1970) by the ENR, Janbu, and PCUBC equations using metric units. Given data (all converted to metric by the author):

H pile $= 14 \times 73 = 36 \times 1.066$ $L = 40$ ft $= 12.18$ m

Hammer $= 80C$ $E_h = 33.12$ kN-m $W_r = 35.58$ kN $e_h = 0.84$

$W_p = 1.168(40/3.2828) + 1.22 \times 4.447 = 14.22 + 5.43 = 19.65$ kN

$AE = 3,313,000$ kN

$s = 17$ blows/ft $= 12/17 \times 2.54 = 1.79$ cm/blow

$0.1 \times 2.54 = 0.254$

Load-test value $= 280$ kips $= 1,245$ kN
(a) By ENR formula [Eq. (17-5)]

$$P_u = \frac{0.84(33.12)(100)}{1.79 + 0.254} = 1,361.09 \text{ kN}$$

(b) By Janbu formula

$$C_d = 0.75 + \frac{18.41}{35.58}(0.15) = 0.83$$

$$\lambda = \frac{0.84(33.12)(12.18)(10^4)}{3,313,000(1.79)^2} = 0.319$$

$$k_u = C_d\left(1 + \sqrt{1 + \frac{\lambda}{C_d}}\right)$$

$$= 0.83\left(1 + \sqrt{1 + \frac{0.319}{0.83}}\right) = 1.806$$

$$P_u = \frac{0.84(33.12(100)}{1.806(1.79)} = 860.3 \text{ kN}$$

(c) By PCUBC formula

$$C = \frac{P_u L}{AE} \qquad \text{estimate } P_u \text{ approximately 800 kN from } b$$

$$C = \frac{800(12.18)(100)}{3,313,000} = 0.294 \text{ cm}$$

$$P_u = \frac{100 e_h E_h}{s + C} \frac{W_r + kW_p}{W_r + W_p}$$

$$= \frac{100(0.84)(33.12)}{1.79 + 0.294} \frac{35.58 + 0.25(19.65)}{35.58 + 19.65} = 978.7 \text{ kN}$$

17-5. RELIABILITY OF DYNAMIC PILE-DRIVING FORMULAS

Many attempts have been made to improve the reliability of the dynamic formulas. A most comprehensive pile-testing program was undertaken under the direction of the Michigan State Highway Commission (1965). In this program 88 piles were driven and tested as shown in Table 17-4 using the following hammers in the driving operations:

Vulcan No. 1, 50C and 80C
McKiernan-Terry DE30 and DE40
Raymond 15-M
Link-Belt 312 and 520
Delmag D12 and D22

This study found that the true safety factors from using the various dynamic formulas based on pile-load tests are as indicated in Table 17-5. The table indicates reasonable values for the Gates formula in the 0- to 400-kip load range (range in which the formula was derived). The modified *Engineering News-Record* [Eq. (17-6)] formula is reasonably valid over the entire range of load tests. It was proposed from these tests that the modified *Engineering News-Record* formula as given in Eq. (17-6)

Table 17-4. Summary of piles driven in the Michigan State Highway Commission (1965) test program

Pile type	Dimensions	Weight, lb/ft	Manufactured by	Approx. length range, ft	Number driven
H sections CBP124	12-in flange	53	US Steel	44–88	48
12-in-OD pipe piles	0.250-in wall	31.37	Armco	44–178	16
	0.230-in wall	28.98			6
	0.179-in wall	22.60			11
Monotube piles, fluted tapered, F 12-7 (30-in taper section) and an N 12-7 extension	12-in nominal	F 19.63 N 24.50	Union Metal Manufacturing Co.	55–80	5
Step-taper shell with 8-ft sections	$9\frac{1}{2}$-in-OD tip	Varies	Raymond International	58–67	2

be further modified as shown in Table 17-1. This study also brought to light that the amount of energy actually input to the pile for penetration is considerably different from the manufacturers' rating. The actual energy input was heavily dependent on hammer base, capblock, pile cap, and pile cap–pile interfacing. Energy input/E_h was found to range from about 0.26 to 0.65 percent—averaging less than 0.50.

Table 17-5. Summary of safety-factor range for equations used in the Michigan Pile Test Program

Formula	Upper and lower limits of SF = P_u/P_d* Range of P_u, kips		
	0–200	200–400	400–700
Engineering News	1.1–2.4	0.9–2.1	1.2–2.7
Hiley	1.1–4.2	3.0–6.5	4.0–9.6
Pacific Coast Uniform Building Code	2.7–5.3	4.3–9.7	8.8–16.5
Redtenbacher	1.7–3.6	2.8–6.5	6.0–10.9
Eytelwein	1.0–2.4	1.0–3.8	2.2–4.1
Navy-McKay	0.8–3.0	0.2–2.5	0.2–3.0
Rankine	0.9–1.7	1.3–2.7	2.3–5.1
Canadian National Building Code	3.2–6.0	5.1–11.1	10.1–19.9
Modified *Engineering News*	1.7–4.4	1.6–5.2	2.7–5.3
Gates	1.8–3.0	2.5–4.6	3.8–7.3
Rabe	1.0–4.8	2.4–7.0	3.2–8.0

* P_u = ultimate test load.

P_d = *design* capacity, using the safety factor recommended for the equation (values range from 2 to 6, depending on the formula).

Olsen and Flaate (1967) performed a statistical analysis on some 93 other piles which concluded that the Hiley equation [Eq. (17-3)] and the Janbu and Gates formulas (Table 17-1) were with the least deviations or highest statistical correlations. This analysis was based largely on data reported in the literature; thus, some considerable estimating of pile weight, average penetration, pile-cap weight, cap-block weight, and condition (for n) and use of a cushion for concrete piles was required. The hammer condition, which would be particularly critical in obtaining either e_h or E_h, was generally not known.

A further statistical analysis of 30 piles of timber, steel, and concrete was presented by ASCE (1946, p. 28) from an earlier discussion which prompted Peck (1942) to propose a pile formula of $P_u = 91$ tons, which (for the reported data) was statistically as good as any of the several dynamic equations used for computing the pile capacity.

17-6. THE WAVE EQUATION

The wave equation is based on using the stress wave from the hammer impact in a finite-element analysis. This method was first put into practical form by Smith (1962) and later by others. A more detailed discussion of the principles and a reasonably sophisticated computer program is readily available [Bowles (1974a)] and will not be repeated here.

The wave equation can be used to investigate the following problems:

1. Pile capacity—a plot of P_u vs. set is made and the load test plotted on the curve to obtain the correct curve.
2. Equipment compatibility—solutions are not obtained when the hammer is too big or too small for the pile.
3. Driving stresses—plots of stress vs. set can be made to ensure that the pile is not overstressed.

For the discussion to follow, refer to the list of symbols:

A = cross section area of pile
C_m = relative displacement between two adjacent pile elements
D_m'' = element displacement two time intervals back
D_m' = element displacement in preceding time interval DT
D_m = current element displacement
DT = time interval
E = modulus of elasticity of pile material
F_m = element force = $C_m K_m$
F_{am} = unbalanced force in element causing acceleration ($F = ma$)
g = gravitation constant
J = damping constant, use J_s for side value, J_p = point value
K_m = element springs = AE/L for pile segments
K_m' = soil springs = R/quake

L = length of pile element

R_m = side or point resistance including damping effects

R'_m = amount of estimated P_u on each element including the point for 100 percent of P_u on point R_3 through R_{11} of Fig. 17-3b. One zero and $R_{12} = P_u$

v_m = velocity of element m at DT

v'_m = velocity of element m at $DT - 1$

W_m = weight of pile segment m

A pile is formed into a set of discrete elements as shown in Fig. 17-3. The system is then considered in a series of separate time intervals DT chosen sufficiently small that the stress wave should just travel from one element into the next lower element during DT. Practically this is not possible, and DT is taken as a value which usually works as in the following table:

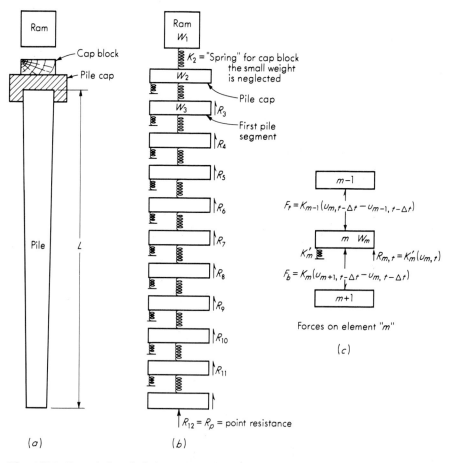

Figure 17-3. Formulation of pile into a dynamic model to solve the wave equation. [*After Smith (1962).*]

Element material	Length	Trial DT
Steel	8–10 ft	0.00025
Wood	8–10 ft	0.00025
Concrete	8–10 ft	0.00033

For shorter lengths, DT should be made correspondingly smaller. The actual time DT can be approximately computed as

$$DT = C\sqrt{\frac{W_m L}{AEg}}$$

where C is 0.5 to 0.75.

The finite-element form of the differential equation used in the wave analysis is

$$D_m = 2D'_m - D''_m + \frac{F_{am}g}{W_m}(DT)^2 \qquad (17\text{-}7)$$

It is not necessary to solve this equation directly, however, since the items of interest for each assumed value of ultimate pile capacity P_u are:

1. Forces in each pile segment
2. Displacement (or set) of the pile point

The instantaneous element displacement is computed as

$$D_m = D'_m + v_m(DT) \qquad (a)$$

With the instantaneous element displacements, the relative compression or tension movement can be computed between any two adjacent elements as

$$C_m = D_m - D_{m+1} \qquad (b)$$

The force in segment m is

$$F_m = C_m\left(\frac{AE}{L}\right)_m = C_m K_m \qquad (c)$$

The soil springs are computed as

$$K'_m = \frac{R'_m}{\text{quake}} \qquad (d)$$

The side or point resistance term is obtained using damping with the side or point value of J and K' as appropriate.

$$R_m = (D_m - D_{sm})K'_m(1 + Jv_m) \qquad (e)$$

The accelerating force in segment m is obtained by summing forces on the element to obtain

$$F_{am} = F_{m-1} - F_m - R_m \qquad (f)$$

The element velocity is computed as

$$v_m = v'_m + \frac{F_{am}g}{W_m}(DT) \tag{g}$$

The wave equation requires the following computation steps:

1. Compute the displacements of each element in turn using Eq. (a) and using consistent units. At $DT = 1$ there is only 1 displacement in the first element; $DT = 2$ there are 2 displacements; $DT = 3$ there are 3 displacements; $DT = m$ computes displacements in all m pile elements.
2. Compute the plastic ground displacements D_{sm}. Values will be obtained only when $D_m >$ quake or elastic ground displacement. This requires 2 SUBROUTINES—one for the point element and one for all other elements.
3. Compute side and the point resistance R_m (use p instead of m for point) using Eq. (e). Use $J =$ side damping for all except the point element; use $J =$ point damping for point element. This requires one equation in a DO loop and a separate point equation.
4. Compute the spring compression in each element C_m using Eq. (b).
5. Compute the forces in each element using C_m and the spring constant AE/L as Eq. (c). Forces in the capblock and pile cap are computed separately using SUBROUTINES because these elements are not usually carrying tension and because of restitution with the dissimilar materials in the capblock and cap cushion (if used).
6. Compute the velocity of each element using Eq. (g).
7. Set the just computed D_m and v_m into storage and reidentify as one time interval back (i.e., become D'_m and v'_m so new values can be computed for D_m and v_m for next DT).
8. Repeat as necessary (generally not less than 40 and not more than 100 iterations unless a poor value of DT is chosen or the pile-hammer compatibility is poor).
9. Program the cycles to stop when
   a. All the velocities become negative
   b. The point-set value becomes smaller than on previous cycle(s)

The wave-equation analysis requires input data as follows:

1. Height of fall of the ram of the pile hammer and weight of ram. This is either given or back-computed as $h = E_h/W_r$. This is needed to obtain the velocity of the pile cap at $DT = 1$ (instant of impact), which is computed as

$$v_1 = \sqrt{e_h(2gh)}$$

2. Weight of pile cap, capblock, pile segments, driving shoe, and modulus of elasticity of pile.
3. Values of capblock and pile-cushion spring constants. Table 17-6 gives values of modulus of elasticity E for several materials used for these elements for computing the spring as $K = AE/L$. Use Table 17-3 for coefficient of restitution.

4. Soil properties of:

Quake (same as k_3 used earlier)
Side damping (usually one-third of point value) J_s
Point damping J_p

Typical values are:

Soil	Quake		Damping constant J_p	
	in	cm	sec/ft	sec/m
Sand	0.05–0.20	0.12–0.50	0.10–0.20	0.33–0.66
Clay	0.05–0.30	0.12–0.75	0.40–1.00	1.30–3.30
Rock	> 0.20	> 0.50		

5. Estimate of percent of the ultimate load carried by the pile point (0 to 100 percent). In general, no pile carries 100 percent of the load on the point, and one should not use more than 80 to 95 percent on the point. Placing 100 percent of load on the point produces a discontinuity in computations, since side load from skin resistance will include damping as shown in Eq. (f), with no side resistance $K'_m = 0.0$.

Plots of P_u vs. blows/inch (or cm) are made by assuming several values of P_u and using the wave-equation computer program to obtain the set. The blows/inch N is obtained as

$$N = \frac{1}{s}$$

For any curve the percent of P_u assumed to be carried by the pile point is held constant as, say, 25, 50, 75, 95 percent.

Plots of $1/s$ (or N) vs. driving stress are obtained for any given P_u by obtaining from the computer output the maximum element force and the corresponding point set for some value of DT. Several other values of maximum element force (not

Table 17-6. Secant modulus of elasticity values for several cap-block and pile-cushion materials. Approximate $A = 12$ in or 30 cm square and $L = A$ unless other data are available to compute spring constant of AE/L

Material	E, ksi	E, kN/cm²
Micarta	450	310.2
Hardwood, oak	45	31.02
Asbestos disks	45	31.02
Plywood, fir	35	24.1
Pine	25	17.2
Softwood, gum	30	20.7

* Data from Smith (1962) and Hirsch et al. (1970).

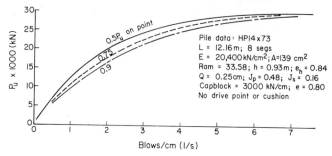

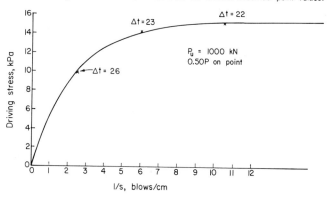

(a) Plot of P_u (assumed values) versus 1/set for several assumed point values.

(b) Plot of driving stress versus 1/set for the assumed value of P_u = 1000kN at Δt values selected from the computer printout for that P_u.

Figure 17-4. The wave equation used to obtain curves of P_u vs. 1/set and driving stresses vs. 1/set for field use.

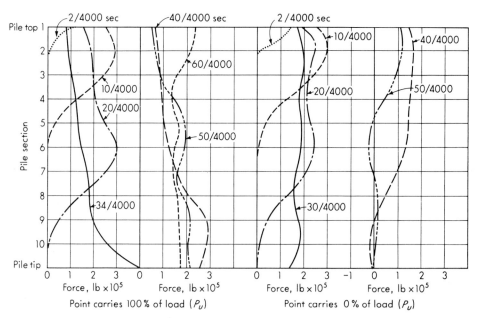

Figure 17-5. Plot of forces computed on pile elements by the wave equation for a HP12 × 53. The plot is shown for selected time intervals and for the case for the point carrying 100 percent of the load and the case for the point carrying no load.

necessarily in the same element) and set at other DT's are also selected so that enough points are obtained to draw a curve. This curve is somewhat erratic, owing to the mathematical model, and must be "faired" through the origin, since it is usually not possible to obtain $1/s$ values as low as 0.5, 1.0 and 1.5 or 2.0. In the region of large $1/s$ it is evident that the curve will approach some asymptotic value of driving stress. Curves of P_u vs. blows/cm and driving stress vs. blows/cm are shown in Fig. 17-4.

Traces of several stress waves down a pile are shown in Fig. 17-5 for a pile with the following data:

$P_u = 200$ kips; HP12 × 53; $L = 100$ ft; 10 pile segments of 10 ft each
$W_r = 5$ kips; $h = 3$ ft; wt. of pile cap = 0.7 kip; $e_h = 0.80$
Wt. of driving point = 0.10 kip; quake = 0.1; side damping = 0.05
Point damping = 0.15; $n = 0.50$ for capblock; no cushion
AE/L for capblock = 2,000 kips/in; $E = 30,000$ ksi

17-7. PILE LOAD TESTS

The most reliable method to determine the load capacity of a pile is to load-test it. This consists in driving a pile and applying a series of loads by some means. The usual procedure is to drive several piles in a group, since no building code will allow a single pile beneath a structure, and use two or more of the adjacent piles for reactions to apply the load. Somewhat similar means are used to test laterally loaded piles. Here the lateral load may be applied by jacking two adjacent piles apart, or suitably connecting several piles for the reaction.

Figure 17-6 illustrates typical data from a pile load test. Figure 17-6a is the usual plot for a load test (the author uses kips where many persons in the field of pile foundations currently use "tons").

The ultimate pile load is taken as the load where the load-settlement curve approaches a vertical condition (as the 225 kips shown in Fig. 17-6a) or the load corresponding to some amount of deflection, say 1 in, and in conjunction with the shape of the load-settlement curve.

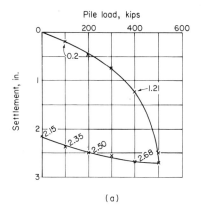
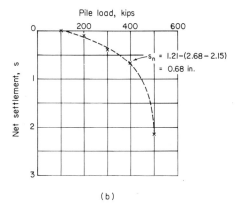

(a)   (b)

Figure 17-6. Pile-load-test data. This is actual pile load test for pile C of home Prob. 16-7 with data shown in Fig. P16-7. Pile is 14-in-diameter pipe with 0.312-in wall and 50 ft long. (a) Usual method of presenting data; (b) plot of load vs. net settlement computed as shown on the figure.

Local building codes usually stipulate how the load test is to be run and interpreted. For example, the Chicago, Atlanta, and New York building codes stipulate the test as follows:

1. Apply load increments of 25 percent of the proposed working load; Atlanta and New York start the first increment at 50 percent. This requires seven or eight load increments.
2. Carry the loading to two times the proposed working load.
3. Apply the loads after specified time lapse or after the settlement rate is some small value.
4. The allowable pile load is taken as one-half that load which causes a net settlement of not more than 0.005 in/kip. For example, referring to Fig. 17-6b, the allowable pile load is about 250 kips ($500 \times 0.005 = 2.50$ in vs. about 2.2 in measured).
5. The building codes limit the minimum value of hammer energy E_h.
6. The codes require a minimum number of test piles per project.

Piles in granular soils may be load-tested as soon as the pile(s) are driven and the load-test arrangements are made. Piles in cohesive soils should be tested after some time lapse to allow the remolding effect to be essentially recovered. This time period should be on the order of 30 days.

17-8. PILE-DRIVING STRESSES

A pile must be adequately proportioned to satisfy both the static and dynamic (driving) stresses. The driving stresses are difficult to evaluate, except as approximations for the same limitations inherent in the dynamic equations as used for computing ultimate resistance.

The wave equation is, in the author's opinion, the best means currently available to obtain driving stresses, both compression for metal and compression and tension for concrete piles. The maximum value of P_u (or the force on a pile element) can be selected from a printout of forces vs. time. This value depends on the amount of side resistance assumed, but in the example used for Fig. 17-5.

$$\text{Maximum pile-element force} = \begin{cases} 302 \text{ kips when } P_p = 0.25P_u \\ 310 \text{ kips when } P_p = 0.75P_u \end{cases}$$

indicating that the ultimate pile load (and pile forces) is not too sensitive to this factor. The solution is more sensitive to the initial assumed value of P_u.

The dynamic equations (ENR and Hiley types) can also be used to evaluate approximately the driving stresses. To use these equations—and using the Hiley equation specifically for illustration—they should be rewritten somewhat. Thus the Hiley (or ENR) equation can be rewritten for a given hammer and pile as

$$P_u = \frac{E_h W_r h}{s + C} \frac{W_r + n^2 W_p}{W_r + W_p} = \frac{K_1 K_2}{s + C}$$

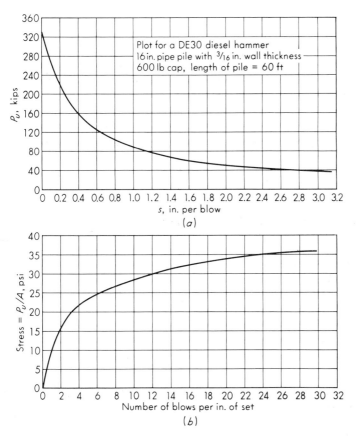

Figure 17-7. Plot of computations of Example 17-4. (a) Plot of s vs. P_u; (b) plot of number of blows for 1 in of set vs. corresponding pile stress.

By assuming values for s and solving for the corresponding P_u for a given hammer-and-pile combination we can establish a curve as in Fig. 17-7a. Since the pile-driving foreman is interested in the stresses vs. number of blows/in, which he can readily measure, it is convenient to assume that a linear relationship exists for the blows vs. set, from which we can plot curve b of Fig. 17-7. The stress ordinate is simply computed as P_u/A. Thus, by having in hand a curve, as shown, which displays the set vs. resistance for the *particular* system being used, the foreman can count the blows for a 1-in penetration, enter the curve with this blow count, and read off the corresponding pile stress. Alternatively, by entering the ordinate of maximum allowable stress, he can read off the maximum number of blows, related to some amount of set beyond which the pile is overstressed.

The set measurement is not a straightforward measurement in that, the instant the hammer strikes the pile, there occurs the term $k_1 + k_2 + k_3$, in addition to the permanent set s.

Attaching a scribing device to the pile and using it to trace on paper or other material the head movement down and up during and after impact, one can obtain the set value reasonably accurately.

There is some diversity of opinion of what the maximum driving stresses can be. Stresses as high as $0.85f'_c$ for concrete piles may result in pile fracture beneath the ground; thus, it would appear that driving stresses should be limited to something on the order of 0.5 to $0.6f'_c$. Driving stresses in timber piles should likewise be limited to something on the order of 0.6 to $0.7f_{ultimate}$ to avoid splits and other damage from driving. Driving stresses for steel piles should be limited to about $0.9f_y$ to avoid yield zones in the pile material below ground which can result in more corrosion from loss of mill scale and formation of Lüder (slip) lines. Some persons are of the opinion that driving stresses for steel piles may be as large as f_y to perhaps 10 to 15 percent above this, since steel has a large residual capacity and because of the "strain-hardening" effect of going through the yield point. In the author's opinion, however, this overlooks the potential corrosion, and thus the driving stresses should be limited.

Example 17-4. Make a set vs. driving resistance curve using the following data:

DE-30 hammer (data as in Appendix)
$W_r = 2,800$ lb
$E = W_r h = 16,800-22,400$ ft-lb   use 20,000 ft-lb
Efficiency e_h taken as 0.85 instead of 1.0
Coefficient of restitution n taken as 0.40 from Table 17-3
$k_1 = 0.10$ from Table 17-2
Pile working load = 210 kips
Driving stresses taken at $F = 1.0$
Curve made for 16.00-in-OD pipe pile, using 600-lb driving cap
Pile filled with concrete after driving

SOLUTION. The approximate area of pile metal is

$$A_s = \frac{210}{F_y} = \frac{210}{36} = 5.34 \text{ sq in} \qquad \text{using A36 steel}$$

A $\frac{3}{16}$-in pipe-wall thickness with a steel area of 9.34 sq in will be selected, providing a nominal driving stress of $f_s = 210/9.34 = 22.7$ ksi. The weight of the pile is 31.7 lb/ft, and the length will be taken as 60 ft (static analysis), resulting in a pile weight $W_p = 31.7(60) = 1,900$ lb. Estimating,

$$k_2 = \frac{P_u L}{AE} = \frac{210(60)(12)}{9.34(29 \times 10^3)} = 0.558 \text{ in}$$

The value of k_3 is estimated at 0.08 (range of 0 to 0.2 given earlier).

$$P_u = \frac{0.85(20)(12)}{s + \frac{1}{2}(0.10 + 0.558 + 0.08)} \cdot \frac{2.8 + (0.4)^2(2.5)}{2.8 + 2.5}$$

$$= \frac{204}{s + 0.369} \cdot \frac{3.2}{5.3} = \frac{123.17}{s + 0.369}$$

from which Table E17-4 can be prepared. Note k_2 is not revised for this computation.

Table E17-4

s	$s + C$	P_u, kips	$1/s = N$	$f_s = P_u/9.34$ ksi
0	0.369	334	∞	35.7
0.05	0.389	294	20	31.5
0.10	0.469	263	10	28.1
0.20	0.569	217	5	23.2
0.80	1.169	105	1.25	11.3
1.00	1.369	90.0	1.00	9.6
2.00	2.369	52.0	0.50	5.6

The data are plotted in Fig. 17-7a and b.

17-9. GENERAL COMMENTS ON PILE DRIVING

Alignment of piles is often difficult to get exactly correct, and often the driven piles will deviate from their computed positions. A tolerance of 2 to 4 in is usually considered allowable. Larger deviations may require additional substructure design to account for eccentricities, or more piles may have to be driven. Alignment of pipe piles may be checked by lowering a light into the tube. If the light source disappears, the alignment is not true. Pile groups should be driven from the interior outward because the lateral displacement of soil may cause excessively hard driving and heaving of already driven piles.

Damage to piles may be avoided or reduced by squaring the driving head with the energy source. Appropriate pile-driving caps and/or cushions should be used. When the required driving resistance is encountered, driving should be stopped. These driving resistances may be arbitrarily taken as

Timber piles	4–5 blows/in
Concrete piles	6–8 blows/in
Steel piles	12–15 blows/in

Driving may require corrective action if the head of a timber pile becomes damaged; e.g., use a cap or metal band or more carefully cut the head of the pile. If during driving any pile changes direction, or the penetration becomes irregular or suddenly increases, the pile may already be broken or bent.

Pile driving may induce heave in saturated fine-grained non-quick-draining soils, where the displaced soil increases the pore-water pressure so that the void ratio cannot rapidly change. As the pore pressure dissipates, the amount of heave may be reduced. Piles already driven in this material may be uplifted, the problem being especially aggravated if the piles are closely spaced [Klohn (1961)]. The problem may or may not be serious, depending on how the heave takes place [Nordlund (1962)], and may be more serious for point-bearing piles if they are driven to refusal and then heave takes place, since excessive settlements may result after the structure is built as the piles reseat themselves. If heave is anticipated, survey benchmarks should be established, and elevations taken on the piles after they are driven and as other piles are driven in the vicinity.

In granular soils a rearrangement of the soil structure from the driving vibrations may result in a subsidence of the adjacent area. Already driven piles may be preloaded to some extent by this phenomenon. A pile driven in a zone within about three pile diameters of an already driven pile will be more difficult to drive because the soil in this zone will be densified.

Continuity of cast-in-place piles is verified by computing the volume of concrete used to fill the pile cavity and comparing this with the theoretical cavity volume.

PROBLEMS

Pile-hammer data in Table A-2 of Appendix.

17-1. A pile load test provides the following data:
Pile = 16-in-diameter pipe; $L = 55$ ft
Hammer = Vulcan 140C; $e_h = 0.78$
Blows for last ft = 38
$A = 23.86$ sq in; $E = 29,000$ ksi; wt = 82.3 lb/ft
Weight includes attachments for instrumentation
Pile cap = 1,710 lb
Find: P_u and P_a by Hiley, ENR, and Janbu equations.
Answer: $P_u = 390$ kips (load test).

17-2. A pile load test provides the following data:
Pile = 40.6-cm-square concrete; $L = 13.71$ m
Hammer = Vulcan 140C; $e_h = 0.81$
Set = 1.38 cm
$A = 1,648.36$ sq cm; $E = 4,343.4$ kN/cm^2
Weight/m = 3.893 kN/m
Pile cap (uses cushion) = 7.604 kN
Find: P_u and P_a by Hiley, ENR, and Janbu equations.
Answer: $P_u = 1,512$ kN (load test).

17-3. A pile load test provides the following data:
Pile = 16-in-square concrete; $L = 55.0$ ft
Hammer = Vulcan 140C; $e_h = 0.81$
Blows for last ft = 48
$E = 6,300$ ksi
Weight of pile cap = 1,710 lb
Required: Compute ultimate and allowable pile capacity using equations from Table 17-1 as assigned.
Answer: $P_u = 480$ kips (load test).

17-4. A pile load test provides the following data:
Pile = timber 1,160 cm^2 butt, 580 cm^2 tip; $L = 12.18$ m
Hammer = Vulcan 65C; $e_h = 0.74$
Set = 1.325 cm/blow
$E = 1,103.1$ kN/cm^2; weight of wood = 20.6 kN/m^3
Weight of pile cap = 4.225 kN
Required: Compute the ultimate and allowable pile capacity using equations from Table 17-1 as assigned.
Answer: $P_u = 712$ kN (load test).

17-5. Plot a curve of P_u vs. $1/s$ and stress vs. $1/s$ for the pile of Prob. 17-1 using the equation assigned by the instructor.

17-6. Plot a curve of P_u vs. $1/s$ and stress vs. $1/s$ for the pile of Prob. 17-2.

17-7. What is the allowable load on the pile of Prob. 17-1 using the PCUBC equation?

17-8. What is the allowable load on the pile of Prob. 17-2 using the PCUBC equation?

17-9. Use the pile data of Prob. 17-1 or 17-3 as assigned and the wave equation. Vary P_u and percent point resistance from 0 to 0.9 and obtain a series of curves of P_u vs. $1/s$. Plot the load test and determine the appropriate curve.

17-10. Plot the assigned load-test data from the following two actual load tests, and select the allowable design load based on pile and load-test data.

	Test No. 1 HP14 × 73, $L = 50$ ft		Test No. 2 32.4 cm × 0.8 cm pipe,* $L = 16.8$ m		
P, kips	Load	Unload	P, kN	Load	Unload
0		0.6 in			2.54 cm
100	0.20 in	0.80	445	0.30 cm	2.92
200	0.35	1.00	890	0.56	3.18
300	0.50	1.15	1,330	1.02	3.43
400	0.80	1.25	1,780	1.65	3.78
500	1.20		2,000	3.18	
	1.30 (24 hr)			3.81 (24 hr)	

* Filled with concrete of $f'_c = 280$ kg/cm².

Use the building code in your area or the Chicago Code method given in Sec. 17-7.

17-11. Compute P_u for the piles shown in Fig. P16-7 using a dynamic equation assigned by the instructor, and compare the solution to the load-test values of P_u shown. The driving hammer in all cases was a Vulcan No. 0 single-acting hammer.

EIGHTEEN

PILE FOUNDATIONS—GROUPS

18-1. SINGLE PILES VS. PILE GROUPS

The preceding two chapters have considered the soil and structural considerations of single piles and including a brief discussion of pile-driving operations. Rarely, however, is the foundation structure likely to consist of a single pile. Generally, there will be a minimum of two or three piles under a foundation element or footing because of alignment problems and inadvertent eccentricities. Building codes[1] may stipulate the minimum number of piles under a building element. The load capacity and settlement associated with pile groups is the topic with which this chapter is concerned. Figure 18-1 presents some typical pile clusters, for illustrative purposes, since the designer must cluster his piles to satisfy the problem.

18-2. PILE-GROUP CONSIDERATIONS

When several piles are clustered, it is reasonable to expect that the soil pressures (either side-friction or point-bearing) developed in the soil as resistances will overlap (idealized in Fig. 18-2). With sufficient overlap, either the soil will fail in shear or the pile group will settle excessively. To avoid the overlap, the spacing of the piles could be increased, but large spacings are impractical, since a pile cap is usually cast over a group of piles for the column base and/or to transmit applied loads to all the piles.

---

[1] The Chicago Building Code (Art. 70.3.3) states: "A column or pier supported by piles shall rest on not less than three piles unless connected to permanent construction which provides lateral support."

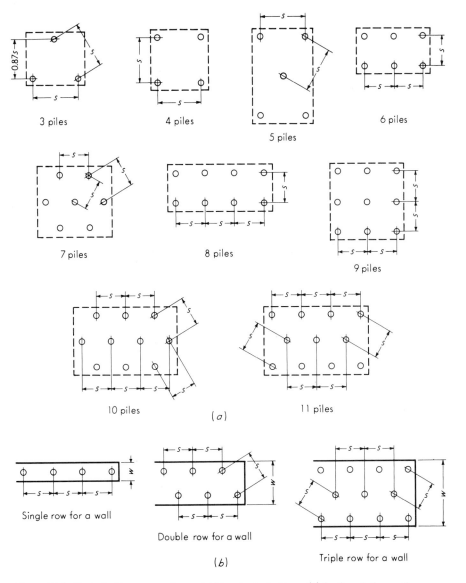

Figure 18-1. Typical pile-group patterns. (*a*) For single footings; (*b*) for foundation walls.

Large spacings will require massive and heavy pile caps, also carried by the piles, unless the cap is in contact with the ground. The minimum allowable spacing of piles is usually stipulated in building codes; e.g., the current National Building Code states that the minimum distance between centers of piles not driven to rock shall be not less than $2D$ for round piles nor twice the diagonal dimension of rectangular-shaped or structural steel piles nor less than $2\frac{1}{2}$ ft. For piles to rock, the spacing shall not be less than $D + 1$ ft for round piles, and the length of the diagonal plus 1 ft for other piles. The current Chicago Building Code has essentially the same requirements.

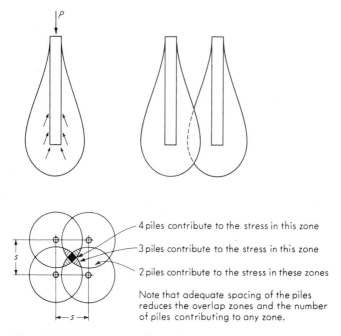

Figure 18-2. Stresses surrounding a friction pile and the summing effects of a pile group.

Generally, the spacing for point-bearing piles (on rock) can be much less than that for friction piles since the high-point-bearing stresses and the superposition effect of overlap of the point stresses will most likely not overstress the underlying material or cause excessive settlements. The spacing s on rock can be computed as

$$s = \sqrt{\frac{\text{load value of pile}}{\text{allowable bearing capacity of rock}}} \qquad (18\text{-}1)$$

with $\qquad\qquad s_{\text{computed}} \geq \text{Code (often } D + 1 \text{ ft)}$

where D = pile diameter or the diagonal dimension of square or H piles

18-3. EFFICIENCY OF PILE GROUPS

From the foregoing discussion it is reasonable to expect that the load-carrying capacity of a group of friction piles will be less than the sum of the individual pile capacities as computed from Chap. 16 or 17. In making this observation, the possible beneficial effect of the pile-driving operations on densifying the soil in and adjacent to the group is neglected. A group of point-bearing piles on rock, with a spacing on the order of 1.75 to 2.5D, should have a group capacity as great as the sum of the individual pile capacities.

The efficiency of a pile group is the ratio of the actual group capacity to the sum of the individual pile capacities. Referring to Fig. 18-3, let the number of rows be n and the number of columns be m and the spacing s. The number of piles is $k = mn$.

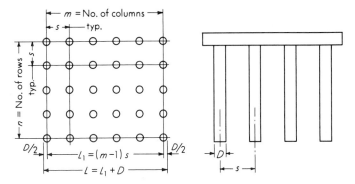

Figure 18-3. Pile-group efficiency.

The perimeter of the group is

$$p = 2[(m - 1)s + s(n - 1)] + 8\frac{D}{2} \tag{a}$$

or
$$p = 2(m + n - 2)s + 4D \tag{b}$$

The group efficiency E_g is the ratio of the skin resistance on the group perimeter pfL_f to the sum of the skin resistances of the individual piles $\pi DkfL_f$, or

$$E_g = \frac{pfL_f}{\pi DkfL_f} = \frac{p}{\pi Dk} \tag{c}$$

and substituting for the group perimeter

$$E_g = \frac{2(m + n - 2)s + 4D}{mn\pi D} \tag{18-2}$$

If we take the efficiency as 100 percent,

$$s = \frac{1.57Dmn - 2D}{m + n - 2} \tag{18-3}$$

or for the individual pile resistances to control the group action, s should be equal to or greater than Eq. (18-3). This equation illustrates a method of computing group efficiencies but is applicable only to the group configuration shown.

Feld (1943) proposed a simple rule-of-thumb method of computing group efficiency by simply reducing pile capacity by 1/16 for each adjacent pile. The Converse-Labarre equation [see Moorhouse and Sheehan (1968)] has been widely used to compute pile-group efficiencies. The Converse-Labarre equation is derived essentially the same as Eq. (18-3) to obtain

$$E_g = 1 - \theta\frac{(n - 1)m + (m - 1)n}{90mn} \tag{18-4}$$

where m, n, D, and s are as defined in Eq. (18-2) and θ is taken as arctan D/s in degrees. Note that both Eqs. (18-3) and (18-4) are based strictly on comparing the skin resistance of a single pile with the block skin resistance.

When a concrete pile cap connecting the piles of a group is poured directly on the ground, one should also consider whether "group efficiency" is the correct parameter to use in describing the group capacity. With the pile cap on the ground (as in the most common case) the group capacity is the block capacity based on perimeter shear + bearing capacity of the block at the pile points regardless of the soil type.

When the pile cap is above ground, as is common for offshore structures, the group capacity will be either:

1. Block capacity based on block perimeter shear + bearing capacity of the block at the pile points for small s/D ratios, or
2. Sum of the capacity of the individual piles $\times$ E_g for larger s/D ratios.

Model pile-group tests in remolded clay by Sowers and Fausold (1961) and Whitaker (1957) for pile caps not in contact with the soil indicated group efficiencies varying from about 0.7 for 16-pile groups at s/D ratios of 1.75 to about 0.9 for larger spacings and smaller pile groups (two to four piles). In all cases the Converse-Labarre efficiency equation predicted group capacities which were on the order of 20 percent too low.

Model pile-group tests by Hanna (1963) and observations by others indicate efficiencies varying from about 0.8 to greater than 1.0 for piles in sands. The lower efficiencies are obtained in dense sand which may be loosened somewhat by pile driving and smaller s/D ratios and the higher values in loose sand which is densified during driving and/or larger s/D ratios.

Example 18-1. Compute the efficiency of the group of friction piles shown by both Eqs. (18-2) and (18-4). Take $D = 16$ in and spacing $s = 3$ ft.

SOLUTION. By Eq. (18-2)

$$m = 5 \qquad n = 3 \qquad k = 3(5) = 15$$

$$E_g = \frac{2(8 - 2)3 + 4(1.33)}{\pi 15(1.33)} = \frac{41.3}{62.7} = 0.66$$

2. By Eq. (18-4)

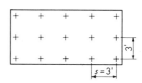

Figure E18-1.

$$\theta = \arctan \frac{D}{s} \, (\text{deg}) = \arctan \frac{1.33}{3.0} = 23.9°$$

$$E_g = 1 - 23.9 \frac{(3 - 1)(5) + (5 - 1)(3)}{90(15)} = 0.61$$

Example 18-2. Assume the pile group of Example 18-1 is founded in a clay soil for which the undrained shear strength is $q_u = 60$ kPa. The piles are 20 m (average) length. Estimate the group capacity. Assume 25 cm of cap projection beyond outer piles all around.

Solution.

Cap dimensions: $L = 4(3)(0.305) + 0.5 = 4.16$ m

$$W = 2(3)(0.305) + 0.5 = 2.33 \text{ m}$$

$$D/B = 20/2.33 = 8.6 > 4 \qquad \therefore N_c = 9.0$$

Perimeter $= 2(4.16 + 2.33) = 12.98$ m

Take $\alpha = 0.9$ since clay is soft

$$Q_u = 9cA + \text{block shear}$$

$$= 9(30)(4.16)(2.33) + 0.9(0.5)(60)(20)(12.98)$$

$$= 2{,}617 + 7{,}009 = 9{,}626 \text{ kN}$$

Load for single pile

$$P_u = 9cA_p + \text{perimeter shear}$$

$$= 9(30)(0.13) + 0.9(0.5)(60)(1.28)(20)$$

$$= 35.1 + 689.9 = 725 \text{ kN}$$

$$\sum P_u = 15(725) = 10{,}875 \text{ kN}$$

$$E_g = \frac{9{,}626}{10{,}875} = 0.89$$

18-4. STRESSES ON UNDERLYING STRATA

The stresses on strata or in the soil underlying a group of piles are not readily evaluated, for several reasons:

1. Influence of pile cap—usually in direct contact with ground except on expansive soils. This results in both the contact soil and the pile carrying the load with the interaction highly indeterminate.
2. The distribution of friction effects along the pile, which are generally not known; hence point load is also not known.
3. The overlap of stresses from adjacent piles, which is difficult to evaluate.
4. The influence of driving the piles on the adjacent soil.
5. Time-dependent effects such as consolidation, thixotropy, varying loads, and change in groundwater level.

Considering all these variables, it is common practice to simplify the stress computations, as illustrated in Fig. 18-4. For friction piles two cases may be considered. In case 1 the load is assumed to spread from a fictitious rigid footing located at the top of the layer, providing friction resistance at a 2 : 1 slope (or 30°). For a homogeneous stratum this is the ground surface. In case 2 the load is placed on a fictitious rigid footing located at $L_f/3$ from the bottom of the piles (average depth), with L_f as in Fig. 18-4b. The spread-out of load is also taken at either 2 : 1 or 30°. Case 1 or 2 should be used, whichever gives the largest computed stresses on underlying strata.

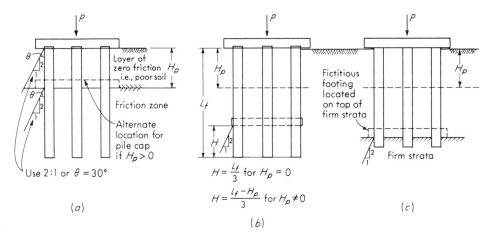

Use 2:1 or $\theta = 30°$

(a)

$H = \dfrac{L_f}{3}$ for $H_p = 0$

$H = \dfrac{L_f - H_p}{3}$ for $H_p \neq 0$

(b)

(c)

Figure 18-4. Simplified computation of soil stresses beneath a pile group. (a) Friction piles; (b) alternative method for stress computations for friction piles; (c) point-bearing piles.

For point-bearing piles in dense sand or sand-gravel deposits, the fictitious footing is placed on the deposit in which the piles penetrate. Again, the load is spread at a 2 : 1 or 30° slope, as shown in Fig. 18-4c.

These analyses are necessary to avoid overstressing the underlying strata. They are also necessary to compute immediate settlements on loose granular deposits or consolidation settlements in clay deposits. As can be seen, a pile group either transmits the load throughout a soil mass of depth L_f for friction piles or to a depth L for an end-bearing pile. The soil at or below these depths must carry the load without excessive deformation, or the load must be transmitted to deeper strata.

An analytical method of evaluating the stresses in the strata underlying a pile group uses an extension of a method proposed by Geddes (1966). This was an adaption by Geddes of the Mindlin (1936a) solution of a point load at the interior of an elastic solid. As with the Boussinesq analysis, the soil is assumed to be semi-infinite, isotropic, homogeneous, and elastic. Soil does not usually fit these assumptions; thus the solutions are in error, but they should be as good as the Boussinesq solutions, which are widely used for footing settlements.

Geddes (1969) later made solutions for the Boussinesq case for subsurface loadings. These are generally less accurate than the Mindlin solution. Poulos and Davis (1968) also used the Mindlin solution to predict settlements. Instead of presenting tables of stress coefficients, they presented charts for settlement-influence factors. Either the Geddes or the Poulos and Davis solutions should provide the same deflection if properly used, since they are both based on the Mindlin solution. The Geddes solution is included since one can easily compute deflections from stresses, but stresses are not as easily back-computed from deflections; stresses may be needed for consolidation settlements.

Geddes developed three cases, as shown in Fig. 18-5. As with the Boussinesq analysis, it is convenient to use stress coefficients which can be evaluated on an

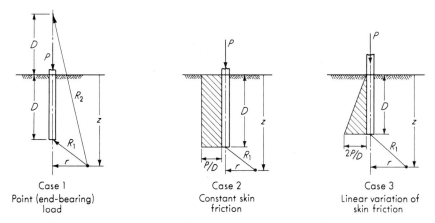

Case 1
Point (end-bearing)
load

Case 2
Constant skin
friction

Case 3
Linear variation of
skin friction

Figure 18-5. Pile-soil system for the evaluation of soil stresses using the Mindlin solution. [*After Geddes (1966).*]

electronic computter. Four stresses can be evaluated for each case (vertical, shear, radial, and circumferential). For case 1 the vertical stress is

$$\sigma_z = \frac{P}{8\pi(1-\mu)} \left\{ -\frac{(1-2\mu)(z-D)}{R_1^3} + \frac{(1-2\mu)(z-D)}{R_2^3} - \frac{3(z-D)^3}{R_1^5} \right.$$
$$\left. - \left[\frac{3(3-4\mu)z(z+D)^2 - 3D(z+D)(5z-D)}{R_2^5} \right] - \frac{30zD(z+D)^3}{R_2^7} \right\} \quad (18\text{-}5)$$

The shearing stress is computed as

$$\tau = \frac{Pr}{8(1-\mu)} \left\{ -\frac{1-2\mu}{R_1^3} + \frac{1-2\mu}{R_2^3} - \frac{3(z-D)^3}{R_1^5} \right.$$
$$\left. - \left[\frac{3(3-4\mu)z(z+D) - 3D(3z+D)}{R_2^5} \right] - \frac{30zD(z+D)^2}{R_2^7} \right\} \quad (18\text{-}6)$$

$$R_1^2 = r^2 + (z-D)^2$$
$$R_2^2 = r^2 + (z+D)^2$$

The other equations are of a similar form, and will not be presented. For computer evaluation they may be expressed in dimensionless form by substituting the following:

$$n = \frac{r}{D} \qquad m = \frac{z}{D} \qquad F^2 = m^2 + n^2$$

$$A^2 = n^2 + (m-1)^2 \qquad B^2 = n^2 + (m+1)^2$$

and introducing a stress coefficient to obtain for the vertical stress

$$\sigma_z = \frac{P}{D^2} K_z \quad (18\text{-}7)$$

The stress coefficient K_z for case 1 is

$$K_z = \frac{1}{8\pi(1-\mu)}\left[-\frac{(1-2\mu)(m-1)}{A^3} + \frac{(1-2\mu)(m-1)}{B^3}\right.$$
$$\left. -\frac{3(m-1)^3}{A^5} - \frac{3(3-4\mu)m(m+1)^2 - 3(m+1)(5m-1)}{B^5} - \frac{30m(m+1)^3}{B^7}\right] \quad (18\text{-}8)$$

For the case of uniform skin friction (case 2) the vertical stress coefficient is

$$K_z = \frac{1}{8\pi(1-\mu)}\left\{ -\frac{2(2-\mu)}{A} + \frac{2(2-\mu) + 2(1-2\mu)\dfrac{m}{n}\left(\dfrac{m}{n} + \dfrac{1}{n}\right)}{B}\right.$$

$$-\frac{(1-2\mu)2\left(\dfrac{m}{n}\right)^2}{F} + \frac{n^2}{A^3} + \frac{4m^2 - 4(1+\mu)\left(\dfrac{m}{n}\right)^2 m^2}{F^3}$$

$$+ \frac{4m(1+\mu)(m+1)\left(\dfrac{m}{n} + \dfrac{1}{n}\right)^2 - (4m^2 + n^2)}{B^3}$$

$$\left. + \frac{6m^2\left(\dfrac{m^4 - n^4}{n^2}\right)}{F^5} + \frac{6m\left[mn^2 - \dfrac{1}{n^2}(m+1)^5\right]}{B^5}\right\} \quad (18\text{-}9)$$

For the case of a linear variation of skin friction (case 3) the vertical stress coefficient is

$$K_z = \frac{1}{4\pi(1-\mu)}\left[-\frac{2(2-\mu)}{A} + \frac{2(2-\mu)(4m+1) - 2(1-2\mu)\left(\dfrac{m}{n}\right)^2(m+1)}{B}\right.$$

$$+ \frac{2(1-2\mu)\dfrac{m^3}{n^2} - 8(2-\mu)m}{F} + \frac{mn^2 + (m-1)^3}{A^3}$$

$$+ \frac{4\mu n^2 m + 4m^3 - 15n^2 m - 2(5+2\mu)\left(\dfrac{m}{n}\right)^2(m+1)^3 + (m+1)^3}{B^3}$$

$$+ \frac{2(7-2\mu)mn^2 - 6m^3 + 2(5+2\mu)\left(\dfrac{m}{n}\right)^2 m^3}{F^3}$$

$$+ \frac{6mn^2(n^2 - m^2) + 12\left(\dfrac{m}{n}\right)^2(m+1)^5}{B^5} - \frac{12\left(\dfrac{m}{n}\right)^2 m^5 + 6mn^2(n^2 - m^2)}{F^5}$$

$$\left. - 2(2-\mu)\ln\left(\frac{A+m-1}{F+m}\frac{B+m+1}{F+m}\right)\right]\cdots \quad (18\text{-}10)$$

Table 18-1a. Stress coefficients for a point load as shown in case 1 of Fig. 18-5. (−) = compression

STRESS COEFFICIENTS FOR POINT LOAD, POISSON RATIO = 0.20
(−) = COMPRESSION

M/N	0.0	0.1	0.2	0.3	0.4	0.5	0.75	1.0	1.5	2.0
1.0		-0.0960	-0.0936	-0.0897	-0.0846	-0.0785	-0.0614	-0.0448	-0.0208	-0.0089
1.1	-17.9689	-3.7753	-0.6188	-0.2238	-0.1332	-0.0999	-0.0659	-0.0467	-0.0222	-0.0099
1.2	-4.5510	-2.7458	-1.0005	-0.3987	-0.2056	-0.1325	-0.0724	-0.0490	-0.0236	-0.0110
1.3	-2.0609	-1.6287	-0.9233	-0.4798	-0.2672	-0.1681	-0.0811	-0.0520	-0.0249	-0.0119
1.4	-1.1858	-1.0382	-0.7330	-0.4652	-0.2926	-0.1930	-0.0905	-0.0555	-0.0263	-0.0129
1.5	-0.7782	-0.7153	-0.5682	-0.4114	-0.2875	-0.2025	-0.0985	-0.0592	-0.0277	-0.0138
1.6	-0.5548	-0.5238	-0.4457	-0.3518	-0.2664	-0.1997	-0.1038	-0.0625	-0.0290	-0.0147
1.7	-0.4188	-0.4018	-0.3569	-0.2984	-0.2399	-0.1893	-0.1061	-0.0651	-0.0303	-0.0156
1.8	-0.3294	-0.3193	-0.2918	-0.2539	-0.2133	-0.1755	-0.1057	-0.0668	-0.0315	-0.0164
1.9	-0.2673	-0.2609	-0.2431	-0.2177	-0.1890	-0.1606	-0.1033	-0.0675	-0.0325	-0.0172
2.0	-0.2222	-0.2180	-0.2060	-0.1883	-0.1676	-0.1462	-0.0995	-0.0673	-0.0334	-0.0179

STRESS COEFFICIENTS FOR POINT LOAD, POISSON RATIO = 0.30
(−) = COMPRESSION

M/N	0.0	0.1	0.2	0.3	0.4	0.5	0.75	1.0	1.5	2.0
1.0		-0.1013	-0.0986	-0.0944	-0.0889	-0.0824	-0.0641	-0.0463	-0.0209	-0.0087
1.1	-19.3926	-3.9054	-0.5978	-0.2123	-0.1287	-0.0986	-0.0668	-0.0475	-0.0222	-0.0097
1.2	-4.9099	-2.9275	-1.0358	-0.4001	-0.2027	-0.1303	-0.0722	-0.0493	-0.0235	-0.0106
1.3	-2.2222	-1.7467	-0.9757	-0.4970	-0.2717	-0.1687	-0.0808	-0.0519	-0.0247	-0.0116
1.4	-1.2777	-1.1152	-0.7805	-0.4891	-0.3032	-0.1974	-0.0908	-0.0555	-0.0260	-0.0125
1.5	-0.8377	-0.7686	-0.6070	-0.4356	-0.3012	-0.2098	-0.0999	-0.0594	-0.0274	-0.0134
1.6	-0.5968	-0.5626	-0.4768	-0.3738	-0.2809	-0.2086	-0.1063	-0.0631	-0.0288	-0.0143
1.7	-0.4500	-0.4312	-0.3819	-0.3177	-0.2538	-0.1988	-0.1094	-0.0661	-0.0302	-0.0152
1.8	-0.3536	-0.3424	-0.3122	-0.2706	-0.2262	-0.1849	-0.1096	-0.0682	-0.0315	-0.0161
1.9	-0.2866	-0.2795	-0.2600	-0.2321	-0.2006	-0.1697	-0.1076	-0.0693	-0.0326	-0.0169
2.0	-0.2380	-0.2333	-0.2201	-0.2007	-0.1780	-0.1547	-0.1039	-0.0694	-0.0336	-0.0177

STRESS COEFFICIENTS FOR POINT LOAD, POISSON RATIO = 0.40
(−) = COMPRESSION

M/N	0.0	0.1	0.2	0.3	0.4	0.5	0.75	1.0	1.5	2.0
1.0		-0.1083	-0.1054	-0.1008	-0.0947	-0.0876	-0.0676	-0.0483	-0.0212	-0.0083
1.1	-21.2910	-4.0788	-0.5699	-0.1970	-0.1228	-0.0970	-0.0680	-0.0486	-0.0223	-0.0093
1.2	-5.3884	-3.1699	-1.0699	-0.4020	-0.1989	-0.1274	-0.0720	-0.0496	-0.0233	-0.0102
1.3	-2.4373	-1.9040	-1.0455	-0.5200	-0.2776	-0.1695	-0.0804	-0.0519	-0.0244	-0.0111
1.4	-1.4002	-1.2179	-0.8438	-0.5208	-0.3173	-0.2032	-0.0913	-0.0554	-0.0256	-0.0120
1.5	-0.9172	-0.8395	-0.6587	-0.4678	-0.3194	-0.2196	-0.1017	-0.0596	-0.0270	-0.0129
1.6	-0.6527	-0.6143	-0.5181	-0.4033	-0.3001	-0.2205	-0.1095	-0.0638	-0.0284	-0.0138
1.7	-0.4915	-0.4705	-0.4152	-0.3435	-0.2724	-0.2116	-0.1138	-0.0675	-0.0300	-0.0147
1.8	-0.3858	-0.3732	-0.3393	-0.2929	-0.2433	-0.1976	-0.1148	-0.0701	-0.0314	-0.0156
1.9	-0.3123	-0.3044	-0.2825	-0.2512	-0.2161	-0.1818	-0.1133	-0.0717	-0.0328	-0.0166
2.0	-0.2590	-0.2537	-0.2390	-0.2173	-0.1919	-0.1659	-0.1098	-0.0722	-0.0340	-0.0174

Table 18-1b. Stress coefficients for constant skin friction as shown in case 2 of Fig. 18-5. (−) = compression

VALUES OF STRESS COEFFICIENTS FOR UNIFORM SKIN FRICTION—POISSON RATIO = 0.20
(−) = COMPRESSION

M/N	0.00	0.02	0.04	0.06	0.08	0.10	0.15	0.20	0.50	1.0	2.0
1.0		-6.4703	-3.2374	-2.1595	-1.6202	-1.2962	-0.8630	-0.6445	-0.2300	-0.0690	-0.0081
1.1	-1.7781	-1.7342	-1.5944	-1.4178	-1.2418	-1.0850	-0.7953	-0.6138	-0.2283	-0.0730	-0.0096
1.2	-0.9015	-0.8789	-0.8576	-0.8269	-0.7882	-0.7446	-0.6317	-0.5307	-0.2231	-0.0759	-0.0111
1.3	-0.5968	-0.5799	-0.5725	-0.5629	-0.5500	-0.5340	-0.4867	-0.4355	-0.2138	-0.0779	-0.0125
1.4	-0.4569	-0.4288	-0.4241	-0.4201	-0.4142	-0.4068	-0.3838	-0.3562	-0.2010	-0.0789	-0.0139
1.5	-0.3482	-0.3359	-0.3334	-0.3313	-0.3282	-0.3242	-0.3113	-0.2952	-0.1862	-0.0790	-0.0152
1.6	-0.2922	-0.2726	-0.2716	-0.2707	-0.2689	-0.2666	-0.2589	-0.2487	-0.1708	-0.0784	-0.0165
1.7	-0.2518	-0.2304	-0.2287	-0.2274	-0.2261	-0.2247	-0.2195	-0.2127	-0.1559	-0.0770	-0.0175
1.8	-0.1772	-0.1953	-0.1949	-0.1942	-0.1936	-0.1925	-0.1891	-0.1844	-0.1420	-0.0750	-0.0185
1.9	-0.1648	-0.1702	-0.1698	-0.1687	-0.1682	-0.1675	-0.1650	-0.1610	-0.1293	-0.0727	-0.0193
2.0	-0.1461	-0.1482	-0.1486	-0.1480	-0.1478	-0.1473	-0.1455	-0.1429	-0.1180	-0.0700	-0.0201

VALUES OF STRESS COEFFICIENTS FOR UNIFORM SKIN FRICTION—POISSON RATIO = 0.30
(−) = COMPRESSION

M/N	0.00	0.02	0.04	0.06	0.08	0.10	0.15	0.20	0.50	1.0	2.0
1.0		-6.8149	-3.4044	-2.2673	-1.6983	-1.3567	-0.8998	-0.6695	-0.2346	-0.0686	-0.0076
1.1	-1.9219	-1.8611	-1.7072	-1.5134	-1.3211	-1.1503	-0.8368	-0.6419	-0.2335	-0.0728	-0.0091
1.2	-0.9699	-0.9403	-0.9166	-0.8825	-0.8400	-0.7922	-0.6688	-0.5588	-0.2292	-0.0760	-0.0105
1.3	-0.6430	-0.6188	-0.6099	-0.5992	-0.5850	-0.5675	-0.5157	-0.4597	-0.2207	-0.0782	-0.0120
1.4	-0.4867	-0.4558	-0.4507	-0.4461	-0.4396	-0.4316	-0.4063	-0.3761	-0.2082	-0.0796	-0.0134
1.5	-0.3766	-0.3561	-0.3533	-0.3510	-0.3476	-0.3432	-0.3291	-0.3115	-0.1934	-0.0800	-0.0148
1.6	-0.3339	-0.2895	-0.2878	-0.2863	-0.2843	-0.2817	-0.2732	-0.2621	-0.1777	-0.0796	-0.0160
1.7	-0.2664	-0.2438	-0.2414	-0.2399	-0.2384	-0.2369	-0.2313	-0.2239	-0.1623	-0.0784	-0.0172
1.8	-0.2025	-0.2065	-0.2054	-0.2044	-0.2038	-0.2026	-0.1989	-0.1938	-0.1479	-0.0765	-0.0182
1.9	-0.1847	-0.1794	-0.1785	-0.1777	-0.1768	-0.1760	-0.1733	-0.1696	-0.1347	-0.0744	-0.0191
2.0	-0.1634	-0.1565	-0.1561	-0.1556	-0.1551	-0.1545	-0.1525	-0.1498	-0.1229	-0.0718	-0.0199

VALUES OF STRESS COEFFICIENTS FOR UNIFORM SKIN FRICTION—POISSON RATIO = 0.40
(−) = COMPRESSION

M/N	0.00	0.02	0.04	0.06	0.08	0.10	0.15	0.20	0.50	1.0	2.0
1.0		-7.2744	-3.6270	-2.4110	-1.8026	-1.4373	-0.9488	-0.7029	-0.2407	-0.0681	-0.0069
1.1	-2.0931	-2.0296	-1.8574	-1.6409	-1.4266	-1.2372	-0.8921	-0.6794	-0.2404	-0.0725	-0.0083
1.2	-1.0486	-1.0209	-0.9947	-0.9567	-0.9091	-0.8556	-0.7181	-0.5964	-0.2373	-0.0760	-0.0098
1.3	-0.6922	-0.6694	-0.6598	-0.6476	-0.6318	-0.6122	-0.5543	-0.4921	-0.2298	-0.0787	-0.0113
1.4	-0.5347	-0.4922	-0.4860	-0.4807	-0.4735	-0.4645	-0.4362	-0.4026	-0.2178	-0.0805	-0.0128
1.5	-0.4020	-0.3823	-0.3798	-0.3771	-0.3734	-0.3684	-0.3527	-0.3332	-0.2029	-0.0813	-0.0142
1.6	-0.3440	-0.3096	-0.3083	-0.3068	-0.3045	-0.3017	-0.2922	-0.2800	-0.1868	-0.0812	-0.0155
1.7	-0.2943	-0.2605	-0.2580	-0.2564	-0.2549	-0.2531	-0.2469	-0.2387	-0.1708	-0.0803	-0.0167
1.8	-0.2114	-0.2195	-0.2189	-0.2181	-0.2174	-0.2161	-0.2119	-0.2063	-0.1558	-0.0787	-0.0178
1.9	-0.1782	-0.1907	-0.1904	-0.1890	-0.1881	-0.1873	-0.1843	-0.1802	-0.1419	-0.0766	-0.0188
2.0	-0.1741	-0.1660	-0.1658	-0.1652	-0.1648	-0.1642	-0.1620	-0.1590	-0.1294	-0.0741	-0.0196

Table 18-1c. Stress coefficients for a linear variation of skin friction as shown in case 3 of Fig. 18-5. $(-)$ = compression

VALUES OF STRESS COEFFICIENTS FOR LINEAR VARIATION OF SKIN FRICTION---POISSON'S RATIO = 0.20
$(-)$ = COMPRESSION

M/N	0.00	0.02	0.04	0.06	0.08	0.10	0.15	0.20	0.50	1.0	2.0
1.0		-11.5315	-5.3127	-3.3023	-2.3263	-1.7582	-1.0372	-0.7033	-0.1963	-0.0018	-0.0082
1.1	-2.8427	-2.7518	-2.4908	-2.1596	-1.8329	-1.5469	-1.0359	-0.7346	-0.2074	-0.0656	-0.0096
1.2	-1.2853	-1.2541	-1.2158	-1.1620	-1.0930	-1.0162	-0.8211	-0.6529	-0.2141	-0.0689	-0.0110
1.3	-0.7673	-0.7753	-0.7585	-0.7420	-0.7195	-0.6928	-0.6142	-0.5312	-0.2139	-0.0717	-0.0123
1.4	-0.5837	-0.5450	-0.5343	-0.5269	-0.5181	-0.5063	-0.4693	-0.4261	-0.2008	-0.0737	-0.0136
1.5	-0.4485	-0.4051	-0.4059	-0.4006	-0.3960	-0.3901	-0.3704	-0.3460	-0.1947	-0.0750	-0.0148
1.6	-0.3635	-0.3201	-0.3226	-0.3183	-0.3154	-0.3123	-0.3008	-0.2861	-0.1803	-0.0754	-0.0160
1.7	-0.3204	-0.2583	-0.2635	-0.2618	-0.2595	-0.2574	-0.2503	-0.2408	-0.1652	-0.0750	-0.0170
1.8	-0.2533	-0.2222	-0.2239	-0.2206	-0.2181	-0.2166	-0.2122	-0.2059	-0.1506	-0.0739	-0.0180
1.9	-0.2382	-0.1761	-0.1855	-0.1880	-0.1878	-0.1853	-0.1827	-0.1782	-0.1371	-0.0722	-0.0188
2.0	-0.1767	-0.1643	-0.1648	-0.1630	-0.1631	-0.1614	-0.1591	-0.1561	-0.1248	-0.0700	-0.0196

VALUES OF STRESS COEFFICIENTS FOR LINEAR VARIATION OF SKIN FRICTION---POISSON'S RATIO = 0.30
$(-)$ = COMPRESSION

M/N	0.00	0.02	0.04	0.06	0.08	0.10	0.15	0.20	0.50	1.0	2.0
1.0		-12.1310	-5.5765	-3.4591	-2.4320	-1.8346	-1.0774	-0.7276	-0.1997	-0.0616	-0.0077
1.1	-3.0612	-2.9620	-2.6751	-2.3119	-1.9547	-1.6433	-1.0908	-0.7680	-0.2115	-0.0654	-0.0090
1.2	-1.3821	-1.3465	-1.3052	-1.2465	-1.1706	-1.0864	-0.8730	-0.6899	-0.2198	-0.0689	-0.0104
1.3	-0.8262	-0.8305	-0.8130	-0.7949	-0.7705	-0.7411	-0.6548	-0.5639	-0.2212	-0.0720	-0.0117
1.4	-0.6194	-0.5827	-0.5722	-0.5630	-0.5540	-0.5410	-0.5005	-0.4530	-0.2150	-0.0744	-0.0130
1.5	-0.5189	-0.4337	-0.4332	-0.4281	-0.4227	-0.4163	-0.3946	-0.3679	-0.2033	-0.0760	-0.0143
1.6	-0.3841	-0.3415	-0.3449	-0.3395	-0.3361	-0.3327	-0.3202	-0.3039	-0.1887	-0.0768	-0.0155
1.7	-0.3337	-0.2784	-0.2810	-0.2782	-0.2760	-0.2739	-0.2660	-0.2556	-0.1732	-0.0762	-0.0166
1.8	-0.2837	-0.2268	-0.2381	-0.2347	-0.2319	-0.2300	-0.2253	-0.2183	-0.1580	-0.0758	-0.0176
1.9	-0.2654	-0.1873	-0.1963	-0.1991	-0.1988	-0.1965	-0.1937	-0.1887	-0.1439	-0.0742	-0.0186
2.0	-0.1872	-0.1730	-0.1744	-0.1732	-0.1725	-0.1714	-0.1684	-0.1651	-0.1310	-0.0721	-0.0194

VALUES OF STRESS COEFFICIENTS FOR LINEAR VARIATION OF SKIN FRICTION---POISSON'S RATIO = 0.40
$(-)$ = COMPRESSION

M/N	0.00	0.02	0.04	0.06	0.08	0.10	0.15	0.20	0.50	1.0	2.0
1.0		-12.9304	-5.9282	-3.6683	-2.5729	-1.9365	-1.1311	-0.7600	-0.2042	-0.0614	-0.0069
1.1	-3.3525	-3.2423	-2.9209	-2.5144	-2.1171	-1.7719	-1.1641	-0.8125	-0.2170	-0.0652	-0.0083
1.2	-1.5030	-1.4712	-1.4255	-1.3588	-1.2742	-1.1800	-0.9422	-0.7394	-0.2274	-0.0689	-0.0096
1.3	-0.8965	-0.9066	-0.8862	-0.8649	-0.8383	-0.8056	-0.7089	-0.6076	-0.2308	-0.0723	-0.0109
1.4	-0.6753	-0.6350	-0.6222	-0.6120	-0.6018	-0.5874	-0.5419	-0.4890	-0.2260	-0.0752	-0.0123
1.5	-0.5629	-0.4718	-0.4712	-0.4641	-0.4584	-0.4511	-0.4270	-0.3971	-0.2147	-0.0773	-0.0136
1.6	-0.4198	-0.3701	-0.3730	-0.3672	-0.3642	-0.3600	-0.3461	-0.3278	-0.1999	-0.0786	-0.0149
1.7	-0.3752	-0.3039	-0.3039	-0.3011	-0.2984	-0.2956	-0.2870	-0.2754	-0.1838	-0.0788	-0.0160
1.8	-0.3158	-0.2496	-0.2575	-0.2530	-0.2497	-0.2479	-0.2427	-0.2349	-0.1680	-0.0782	-0.0172
1.9	-0.2851	-0.2022	-0.2122	-0.2155	-0.2141	-0.2113	-0.2083	-0.2028	-0.1530	-0.0769	-0.0182
2.0	-0.2012	-0.1929	-0.1878	-0.1854	-0.1850	-0.1837	-0.1807	-0.1771	-0.1393	-0.0749	-0.0191

Values in Table 18-1 are not shown for $m < 1.0$ which are stresses above the pile point. The influence values compute as tension when $m < 1.00$ for cases 1 and 3. Tension stresses are not likely to form, as gravity effects would cause the soil to flow downward. It is usually not possible to consider soil weight in these problems, however, as the computations will show deformation due to soil weight. Soil stresses below the pile point cause settlements. Deformation above the pile point will be limited to elastic shortening of the pile in the form of $\alpha P_u L/AE$. In developing the tables, it was necessary to use $n = 0.002^+$ since $n = 0.0$ produces a discontinuity in the computations.

Table 18-1 lists values for K_z for various m and n values and two selected values of Poisson's ratio μ for all three cases. By superposition of effects, these three cases should provide a general solution for the vertical stress at a point for any reasonable type of stress distribution along a pile.

Example 18-3. Compute the vertical stress at point A of the four-pile group shown. Take $\mu = 0.3$. Compare the results with what has been the conventional method of analysis.

SOLUTION:
1. Assume point-loaded piles.

$$r = 2\sqrt{2} = 2.828 \text{ ft}$$

$$m = \frac{z}{D} = \frac{60}{55} = 1.09$$

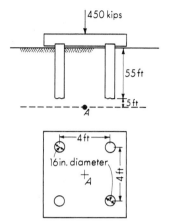

450 kips

55 ft

5 ft

A

4 ft

16 in. diameter

A

4 ft

Figure E18-3.

$$n = \frac{r}{D} = \frac{2.83}{55 \text{ ft}} = 0.0515$$

$$\mu = 0.30$$

From Table 18-1 (case 1),

$$K_z = -12.16 \qquad \text{interpolating}$$

$$\sigma_z = 4\left[\frac{450}{4(55^2)}\right](-12.162) = -1.809 \text{ ksf} \qquad \text{compression}$$

2. Assume one-half of load carried by point and one-half carried by friction, as in result 1 above. For point:

$$\sigma_A = 4\left[\frac{450}{2(4)(55)^2}\right](-12.162) = -0.9 \text{ ksf}$$

For linear variation of skin friction (case 2) and interpolating:

$$\sigma_A = 4\left[\frac{450}{2(4)(55)^2}\right](-2.7) = -0.2 \text{ ksf}$$

$$\sigma_A = 0.0 - 0.2 = -1.1 \text{ ksf} \qquad \text{compression}$$

3. By conventional analysis, what is stress at A?

$$\frac{55}{3} = 18.33 \text{ ft}$$

Therefore, depth to $A = 18.33 + 5 = 23.3$ ft

$$\sigma_A = \frac{450}{(4 + 23.3)^2} = 0.60 \text{ ksf} \qquad \text{compression}$$

This compares with 1.8 ksf for point-load conditions and 1.1 ksf for one-half point, one-half skin friction.

18-5. SETTLEMENTS OF PILE GROUPS

The settlement of a pile group is exactly equal to the displacement of the pile point + the elastic shortening of the pile between cap and point as illustrated in Fig. 18-6. For point-bearing piles the point displacement is relatively small and the principal displacement is the elastic shortening of the pile. For friction piles the point displacement will be the significant quantity causing settlement.

The principal difficulties in computing pile-group settlements are due to:

1. The problem of obtaining the stresses in the strata below the point and the correct elastic properties of the soil so that the point displacement can be computed. Currently the only practical means is to use some type of Boussinesq or Mindlin solution.
2. The determination of the load carried by the piles in the group and the distribution of the load along the pile shaft so that the axial shortening can be computed. When the pile cap is above ground (or water as for offshore structures), the pile loads can be estimated reasonably well. When the cap is of concrete poured directly on the ground as is the usual case, except on expansive soil, the pile load is considerably indeterminate. According to Broms (1972) the modulus of elasticity of concrete piles is not a constant value but deteriorates with time as much as 10 percent. This is not likely to affect the computations to any significant amount, as this change is somewhat speculative for reinforced piles because the transformed section is rarely used and secondly concrete strength gradually improves with age.

This latter modulus reduction, if deemed valid, would also apply equally for wood. The major problem (for all pile materials) is the distribution of load along the pile shaft.

To obtain the pile load (but not its distribution along the shaft—this must be estimated) to make an estimate of pile shortening under load, one can use the mat program in the Appendix and input a composite value of k_s for those mat elements

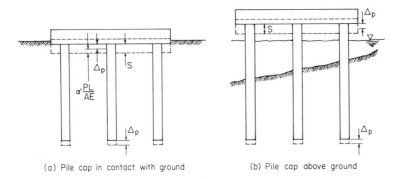

(a) Pile cap in contact with ground   (b) Pile cap above ground

Figure 18-6. Pile cap/group settlement. In (a) the cap-soil interaction introduces considerable difficulty in evaluating the elastic shortening of the pile. In both cases the point-deflection computation is a considerable exercise in engineering judgment.

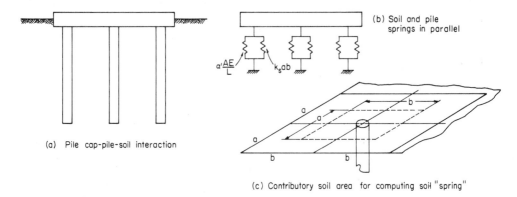

(b) Soil and pile
springs in parallel

$\alpha' \dfrac{AE}{L}$ $k_s ab$

(a) Pile cap-pile-soil interaction

(c) Contributory soil area for computing soil "spring"

Figure 18-7. Method of obtaining equivalent soil modulus for input into "mat" computer program to obtain estimate of the effects of the interaction displayed in (a).

contributing to computation of the soil "spring" at the given pile locations. This method appears to be similar to that of Butterfield and Banerjee (1971).

The composite k_s would be obtained by computing the pile constant for compression (as in Sec. 18-10), which may be of the form $\alpha' AE/L$. The contributory area of the mat for the soil "spring" at the node is obtained as $K_i = k_s(ab)$ as in Fig. 18-7. The equivalent soil modulus is computed by considering the pile and soil springs in parallel (as in Fig. 18-7b) with the same deflection to obtain

$$k_s abX + \alpha' \frac{AE}{L} X = K'X \qquad (a)$$

from which $K' = k_s ab + \alpha' AE/L$.

The equivalent composite soil modulus at that node is $k'_s = K'/ab$. The computer output will give the total nodal force $K'X$, which can be separated into the pile and soil components using Eq. (a) above. It is self-evident that the solution will be only as good as the soil parameter k_s and the pile constant. The α' term used in $\alpha' PL/AE$ is to make allowance for the type of pile and distribution of skin resistance. In any case the computer solution will give values of relative effects which will be useful in estimating pile-group response.

Larger pile groups should settle more than small groups for the same pile loads. This due to the overlapping effect of stresses below the pile point from the additional piles. In addition to the overlap effect the outer piles cause stresses to penetrate to a greater depth L_1 such that integration of strain effects

$$\delta = \int_{L_0}^{L_1} \varepsilon \, dL$$

produces a larger deflection beneath the pile group.

It is usual to assume that the pile cap is rigid such that the cap movements can be described by rigid body translations and rotations.

At present there is not a large body of literature on field measurements of pile groups. There is somewhat more on model groups [Barden and Monckton (1970 with references)], but most of the pile groups have been tested with the cap not in contact with the ground, the principal objective being to obtain efficiency ratios rather than soil-cap interaction.

Pile groups supported by clay soils may produce both elastic, or immediate, and consolidation settlements. The elastic settlements may be the major amount for preconsolidated clays; the consolidation settlements may be the principal value(s) for normally consolidated clays and using friction or "floating" piles. The stress coefficients of Table 18-1 may be used to estimate the stress increase causing consolidation settlements and using the procedures outlined in Sec. 5-8.

Pile groups supported by cohesionless soils will produce only immediate-type settlements, the principal problems being to obtain the correct evaluation of the stress increase in the underlying strata, the depth L_1 through which the stress increase acts, and the elastic properties so that Eq. (5-10) or simply

$$\delta = \frac{\sigma L_1}{E_s}$$

can be used to obtain the point movement δ.

Example 18-4. One of the better-reported series of building and pile settlements available in geotechnical literature was made by Koerner and Partos (1974). From these data the soil profile and typical pile cap on two columns are shown as follows:

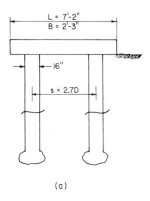

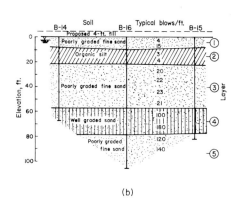

(a)   (b)

Figure E18-4a.   **Figure E18-4b.**

Other data: Pile load = 240 kips (approx.)
Pile length = 25 ft (cased and enlarged base)
Pile diameter = 16 in $f'_c = 5{,}000$ psi
$E_s = 4$ ksi (doubled by Koerner and Partos to
allow for increased density)
$s = 2.7D$
Measured settlements 1.5 to 3.3 in

REQUIRED. Estimate settlement of a typical pile cap. Assume Poisson's ratio = 0.3.

SOLUTION. *Step 1.* Use Table 18-1c, case 3, since load test indicates very little point movement for workingload of 240 kips; therefore, principal load mechanism must be skin resistance.

$$\frac{r}{D} = \frac{1.7D}{25} = 0.09 \qquad \text{use approx. interpolation}$$

$\dfrac{z}{D}$	K_z	$\dfrac{z}{D}$	K_z
1.0	2.30	1.6	0.33
1.1	1.80	1.7	0.27
1.2	1.10	1.8	0.23
1.3	0.76	1.9	0.20
1.4	0.54	2.0	0.17
1.5	0.42		

The average influence value in the zone D to $2D$ using the trapezoidal rule is

$$K_z = \frac{6.665}{10} = 0.67$$

Step 2. Compute average stress in depth D below pile and the corresponding settlement. Assume only two piles.

$$\sigma = \frac{2PK_z}{D^2} = \frac{2(240)(0.67)}{25^2} = 0.515 \text{ ksf}$$

The settlement is

$$\delta = \frac{\sigma L}{E} = \frac{0.515(25)(12)}{4(144)} = 0.27 \text{ in}$$

which compares quite well with the value of 0.21 or 0.22 measured in the load test.

Step 3. Estimate settlement of building.

Method 1: Settlement of cap [Eq. (5-10)] plus point movement.

$$S = qB\frac{1 - \mu^2}{E_s} I_w + \delta$$

$$B = 2.25 \text{ ft} \qquad L = 7.17 \text{ ft} \qquad \text{given}$$

$$L/B = 3 \qquad I_w = 1.5 \qquad \text{plotting and interpolating values from Table 5-4}$$

$$q = 480/(2.25)(7.17) = 29.75 \text{ ksf}$$

$$S = 29.75(2.25)\frac{1 - 0.3^2}{4(144)}(1.5)(12) + 0.27$$

$$= 1.90 + 0.27 = 2.17 \text{ in}$$

Answer is fairly good, but analysis is not realistic in using the cap in this manner.

Method 2: Elastic shortening of pile + δ; assume a linear variation of P from P_0 to $P = P_0 - \Delta P$

$$e = \int_0^L \varepsilon \, dy = \frac{1}{AE} \int_0^L \left(P_0 - \Delta P \frac{y}{L} \right) dy$$

$$e = \frac{1}{AE} \left(P_0 L - \frac{\Delta PL}{2} \right) \qquad \text{take } \Delta P = 0.5 P_0$$

$$e = \frac{L}{AE} \left(P_0 - \frac{0.5}{2} P_0 \right) = \frac{0.75 P_0 L}{AE}$$

Take $E_c = 4{,}060$ ksi $\qquad A = 201$ sq in

$$e = \frac{0.75(240)(25)(12)}{201(4{,}060)} = 0.07 \text{ in}$$

$S = 0.07 + 0.21 = 0.28$ in

It is evident that while this method is correct for the assumed soil deflections, the measured deflections indicate something causing additional settlement.

Method 3: Consider elastic shortening of pile + point movement due to pile load + settlement of 4 ft of fill. Assume the fill produces negative friction such that downward movement is the same as that obtained using Eq. (5-10).
From method 2, movements are $0.07 + 0.21 = 0.28$ in.
q for 4 ft of fill $\cong 0.460$ ksf $\qquad B = 80$ ft
$L/B = 110/80 = 1.4 \qquad$ use 1.5
$I_w = 0.68$ to $1.36 \qquad \mu = 0.3$
Take $E_s = 2{,}000$ instead of 4,000 psi to allow for creep

$$S = qB \frac{1 - \mu^2}{E_s} I_w = \frac{0.46(80)(0.91)(12)}{2(144)} I_w = 1.4 I_w$$

$S = 1.4(0.68) + 0.28 = 1.22$ in (corner)

$S = 1.4(1.36) + 0.28 = 2.18$ in (interior)

This method is most realistic both analytically and by comparison of measured and computed deflections. This computation could be improved by considering the contributions of more adjacent piles to compute a larger point settlement δ and to use a larger surface area for the 4-ft fill zone which would probably be more than 80 ft wide. An additional but also possible factor is the negative skin resistance which may be developed as the fill causes settlement in the compressible layer.

18-6. PILE CAPS

Unless a single pile is used, a cap is necessary to spread the vertical and horizontal loads and any overturning moments to all the piles in the group. The cap is usually of reinforced concrete, poured directly on the ground unless the soil is expansive. Caps for offshore structures are often fabricated from steel shapes. The pile cap has a reaction which is a series of concentrated loads (the pile), and the design considers the column loads and moments, any soil overlying the cap (if it is below the ground surface), and the weight of the cap. It is usual practice to assume that:

1. Each pile carries an equal amount of the load for a concentric axial load on the cap, or for n piles carrying a total load Q, the load P_p per pile is

$$P_p = \frac{Q}{n} \tag{18-11}$$

2. The combined stress equation assuming a planar stress distribution is used for a pile cap noncentrally loaded or loaded with a load Q and a moment, as

$$P_p = \frac{Q}{n} \pm \frac{M_y x}{\sum x^2} + \frac{M_x y}{\sum y^2} \tag{18-12}$$

where M_x, M_y = moments about x and y axes, respectively

x, y = distances from y and x axes to any pile

$\sum x^2$, $\sum y^2$ is the moment of inertia of the group, computed as

$$I = I_0 + Ad^2$$

but I_0 is negligible, and the A term cancels, since it is the pile load desired, and appears in both the numerator and denominator of Eq. (18-12).

The assumption that each pile in a group carries equal load may be nearly correct when the following are all met:

1. The pile cap is in contact with the ground.
2. The piles are all vertical.
3. Load is applied at the center of the pile group.
4. The pile group is symmetrical.

Some model pile-group tests appeared to show that the piles are not all loaded the same; however, most of the model tests did not place the pile cap in contact with the ground. Some persons purport to have developed a theory to show that the piles in a group are not equally loaded; however, most of the theories have enough assumptions to produce any kind of solution the proponent desires. Practically, however, a four-pile symmetrical group loaded with a concentric load will result in each pile and the soil in that quadrant carrying 25 percent of the load. If the cap is rigid, the settlement will be exactly equal to the elastic shortening of the pile and the displacement of the pile point.

The structural design of pile caps of reinforced concrete requires consideration of the following:

1. Bending moment is taken at the same sections as for a reinforced-concrete footing and defined in Art. 15-4 of the ACI Code (also as given in Chap. 8).
2. Pile caps must be reinforced (Art. 15-7).
3. Pile-cap shear is computed at critical sections as shown in Fig. 18-8.
4. Pile caps should end at least 6 in beyond the outside face of exterior piles.

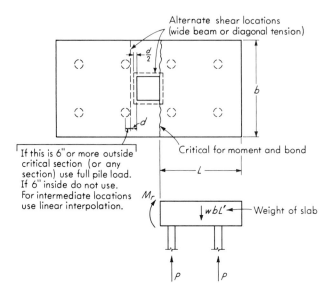

Figure 18-8. Critical pile-cap locations for shear, moment, and bond computations according to ACI 318.

5. Piles should be embedded at least 6 in into the cap. Some building authorities may allow as little as 3 in of pile embedment into the cap. If the embedment is not at least 6 in, the pile should be assumed hinged to the cap.
6. Pile-cap reinforcing bars are placed 3 in above the top of the pile.
7. The minimum thickness of pile cap above the top of the bottom reinforcing bars is 12 in (Art. 15-9).
8. Tension connectors should be attached to the pile to ensure that the pile and cap retain continuity if any of the piles are subjected to tension forces.
9. For fixed-head pile connections, additional reinforcing should be placed around the pile to ensure that the pile does not pull out from or crack the pile cap.

The mat program in the Appendix can be used to investigate the bending stresses in a pile cap and for complicated pile groups will provide bending moments different from those with the simplified ACI method, which is not always conservative. It is necessary when using the mat program to analyze the pile cap to first perform a pile-group analysis using the computer program of Sec. 18-9. The program gives directly the values of forces and indirectly the pile moments at the pile location. Position equilibrium moments are not to be used. The vertical force components and the X and Z pile-head moments are input to the mat program with the vertical load and any X and Z cap moment. It is not necessary to consider rotation about the Y axis nor is it necessary to consider translation forces, as they are accounted for in the solution of the group and do not contribute to bending. With the forces as input in the P matrix it can be conservatively assumed that the pile is supported only by the piles; thus, no soil modulus is used in the problem. This assumption is consistent with the pile-group solution of Sec. 18-9. Program modifications are necessary to investigate the pile-cap-on-ground case with both soil and piles resisting cap deformation.

18-7. BATTER PILES

When large lateral loads are to be resisted by a pile group, it is common practice to use piles driven at a slope with the vertical, i.e., batter piles. It has also been common to assume that the batter piles carry all the lateral loads. All piles have some lateral load-carrying ability dependent on the pile width, the flexural rigidity (EI) of the pile, and the stiffness of the soil in which it is embedded. Early methods of pile-group analysis with both vertical and lateral loads were primarily graphical. These early methods also assumed that the piles were axially loaded, which precluded bending moments being developed. Combining graphical solutions and the assumptions of axial loading, it naturally followed that the lateral loads had to be carried by batter piles.

Modern methods of pile-group analysis use the computer and additionally lateral pile-load tests have verified what the computer solutions illustrate, namely, that all the piles in a group carry lateral load. The graphical solutions are no longer used, since they are obviously incorrect. The computer method of group analysis, being the only practical way of analyzing a group, is the only method presented in this chapter.

Common pile batters range from 1 horizontal to 12 vertical $\triangleleft^{12}_1$ to a batter of $\triangleleft^{12}_5$. When the batter exceeds $\triangleleft^4_1$, the driving may require special equipment, with resulting increased costs.

18-8. NEGATIVE SKIN FRICTION

When a fill is placed on a compressible soil deposit, consolidation of the compressible material will occur. When a pile is driven into the compressible material (either before or after fill placement) before consolidation is complete, the soil will move downward relative to the pile. This relative movement will develop skin friction between the pile and the moving soil termed *negative skin friction*. According to measurements reported by Bjerrum et al. (1969) and Bozozuk (1972), the negative skin friction can exceed the allowable load for pile sections. Fellenius (1972) has also reported large values of measured negative skin resistance.

The principal effect of negative skin resistance is to increase the axial load in the pile. It may result also in increased pile settlements due to the axial shortening of the pile under the increased axial load. Note that in Fig. 18-9 the fill settlement may be such that a gap forms between the bottom of the pile cap and the soil. This will transfer the full cap weight to the piles and may change the bending stresses in the cap.

"Negative" skin friction can produce large tension stresses when the effect is from expansive soils—especially if no, or insufficient, gap is left between soil and pile cap and the soil expands against the cap.

Negative skin friction can be developed from:

1. A cohesive fill placed over a cohesionless soil deposit; the fill develops shear resistance between the soil and pile from lateral pressure/flow effects and drags the

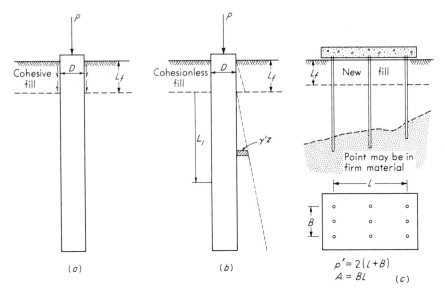

Figure 18-9. Development of negative friction forces on a single pile from a cohesive or cohesionless fill or on a pile group in a cohesive-soil fill.

pile downward as the fill consolidates. Little effect is produced in the underlying soil except that the weight of fill increases the lateral pressure. This provides additional resistance of the pile against further penetration if floating, and raises the center of resistance nearer the cohesive fill for point-bearing piles.

2. A cohesionless fill placed over a compressible, cohesive deposit. In this case there will be some downdrag in the fill zone, but the principal downdrag will occur in the zone of consolidation. For point-bearing piles any settlement of the group will be due to axial shortening of the pile. For floating piles, additional penetration with matching settlement will occur unless the pile is sufficiently long that the bottom portion can develop enough additional positive skin resistance to balance the additional load developed by negative skin resistance. In this case an approximation of the location of the balance, or neutral, point can be made.

The value of negative skin friction (a force) can be computed as follows for a single pile:

For cohesive fill overlying cohesionless soils

$$P_{nf} = \int_0^{L_f} \alpha' p \gamma' z K(dz) = \frac{\alpha' p K \gamma' L_f^2}{2} \tag{18-13}$$

where α' = coefficient relating the effective lateral pressure of $\gamma' L_f K$ to the shearing resistance about the pile perimeter

p = pile perimeter

L_f = depth of fill as shown in Fig. 18-9a

K = lateral-earth-pressure coefficient; use $K = K_0 = 1 - \sin \phi$

For cohesionless fill overlying a compressible-soil deposit, taking the origin of coordinates at the bottom of the fill and neglecting the surcharge effect due to $\gamma_f L_f$, the negative friction force at the depth L_1 is

$$P_{nf} = \int_0^{L_1} \alpha' p \gamma' z K (dz) = \frac{\alpha' p K \gamma' L_1^2}{2} \tag{18-14}$$

Below the neutral point, if there is one, the positive friction is generated as

$$P_{pf} = \int_{L_1}^{L} \alpha_2' p \gamma' z K (dz) = \frac{\alpha_2' p \gamma' K (L^2 - L_1^2)}{2} + P_{np} \tag{18-15}$$

where P_{np} = amount of negative skin resistance carried by the point where point-bearing piles are used

If one assumes the α' coefficients are equal and no allowance is made for increase in soil density from consolidation or depth, for floating piles the depth L_1 to the neutral point is found by equating Eqs. (18-14) and (18-15) to obtain

$$L_1 = \frac{L}{\sqrt{2}}$$

For point-bearing piles, the P_{np} term requires an additional assumption so that the distance L_1 becomes little better than a guess.

When the piles are spaced at small s/D ratios, the negative friction force may act effectively on the block perimeter rather than on the individual piles to obtain two modes of stressing requiring investigation.

1. The negative skin resistance as the sum of the individual piles

$$Q_n = n F_n \tag{18-16}$$

2. The "block" skin resistance

$$Q_n = s L_f p' + \gamma L_f A \tag{18-17}$$

where γ = unit weight of soil enclosed in pile group to depth L_f
  A = area of pile group enclosed in perimeter p', Fig. 18-9
  p' = perimeter of pile group
The maximum value from Eq. (18-16) or (18-17) should be used.

Example 18-5. For the pile group shown, estimate the increase in negative friction (group is square). Characteristics of new fill:

$$\gamma = 110 \text{ pcf} \qquad \phi = 30°$$

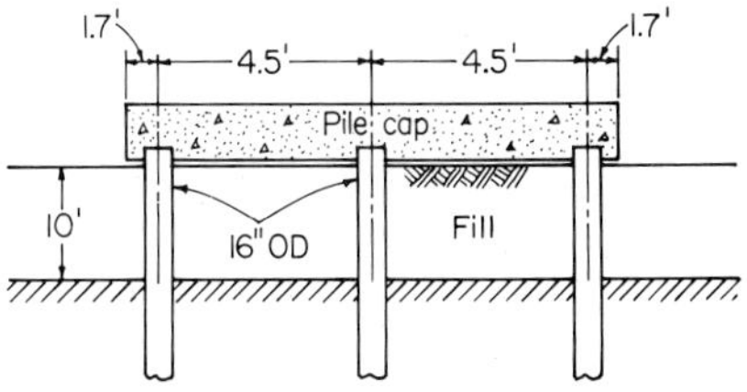

Figure E18-5.

Take

$$f = \tfrac{2}{3} \tan \phi = 0.364 \text{ (soil to pile)}$$

$$F_n = \tfrac{1}{2} L_f^2 pKf\gamma$$

$$K_0 = 1 - \sin 30 = 0.50$$

$$= \tfrac{1}{2}(10)^2 \pi(1.33)(0.110)(0.50)(0.364)$$

$$= 4.2 \text{ kips}$$

$$nF_n = 4.2(9) = 37.7 \text{ kips}$$

Alternative possibility (Eq. 18-17): Take perimeter as c to c of piles

$$F_n = 36(\tfrac{1}{2})(10)^2(0.110)(0.50)(0.364) = 36.0 \text{ kips}$$

$$+ 0.110(10)(81) = 89.1$$

$$\overline{125.1 \text{ kips}} > 37.7 \qquad \text{controls}$$

The increase per pile is $125.1/9 = 13.9$ kips > 4.2.

Example 18-6. Redo Example 18-5 if the fill is 1.5 m deep and the underlying soil is a soft clay. Assume the piles are 25 m long. The clay is $\gamma' = 9.4$ kN/m^3; water table at fill base. It is known that the pile points will not displace. Compute the location of the neutral point and the maximum increase in negative skin resistance. The undrained shear strength $q_u = 80$ kPa.

SOLUTION.

$$L_1 = L/\sqrt{2} = (25 - 1.5)/\sqrt{2} = 16.62 \text{ m}$$

The maximum increase in negative skin resistance is

$$P_n = \alpha' pK\gamma L_1^2/2 = \alpha cL_1 p \qquad \text{(equivalent expression for } \phi = 0)$$

$$p = 1.33(\pi)/3.2828 = 1.28 \text{ m}$$

Take $\alpha \cong 1.00$ as soft clay and consolidating

$$P_n = 1.0(40)(16.62)(1.28) = 848.3 \text{ kN}$$

This is a substantial portion of the allowable capacity of this pile.

18-9. MATRIX ANALYSIS FOR PILE GROUPS

When pile-group loadings consist in vertical loads concentrically placed or with an eccentricity on the order of not more than $0.67s$ and with vertical piles, the pile loads can be predicted with sufficient accuracy using Eq. (18-11) or Eq. (18-12) based on experience and later the model tests like those of Saffery and Tate (1961).

When the pile group is loaded with larger eccentricities, large bending moments and/or horizontal forces, and with both vertical and batter piles the analysis becomes quite complex. Approximate solutions were proposed by Culmann (simple force polygon) and Westergaard (using a center-of-rotation method). Neither of these solutions recognized that vertical piles can carry lateral loads and moments. Later Hrennikoff (1950) proposed a three-dimensional group solution which was simplified by him to place major emphasis on two-dimensional pile groups. This method remained dormant until the early 1960s partly because these analyses are better performed on digital computers. Aschenbrenner (1967) introduced a method of group analysis using pinned pile caps. Saul (1968) introduced a general three-dimensional matrix solution, and Reese et al. (1970) published a similar matrix solution. Bowles (1974a) published a matrix solution similar to the one presented in this section; however, the orientation of the pile forces in the solution presented here makes computation of the direction cosines quite straightforward as compared with the earlier solution.

The matrix solution consists in using the same matrix equations presented earlier in Chap. 9.

$$P = \mathbf{AF} \qquad e = \mathbf{A}^T\mathbf{X} \qquad F = \mathbf{S}e = \mathbf{SA}^T\mathbf{X} \qquad \text{and} \qquad P = \mathbf{ASA}^T\mathbf{X}$$

The essential difference in that solution and the pile-group solution is as follows:

1. The equation $P = AF$ is for a single (ith) pile; thus

    $P = $ that part of the total pile-cap force carried by the ith pile

    $A = $ a complete matrix relating the ith pile forces to the part of the total pile-cap force carried by the ith pile

2. The S matrix introduces the concept of pile constants instead of the familiar $4EI/L$, $2EI/L$, and soil "spring" terms K used in Chap. 9. Here it is necessary to solve a laterally loaded pile to obtain four of the six S-matrix entries and either use a computer program as in Bowles (1974a, Chap. 12) or compute the 5th S-matrix entry as $\lambda AE/L$; the 6th entry is $\psi GJ/L$. A complete 6×6 S matrix is required for each pile.
3. The ASA^T is computed for each pile in the pile group *and summed* into a group ASA^T matrix. For 4 piles, each group ASA^T-matrix value is the sum of 4 individual pile ASA^T values.
4. The pile-group ASA^T matrix (size 6×6) is inverted and the foundation displacements, or X's, are obtained.
5. With the X values the pile-head displacements (e's) are computed using

$$e = A^T X$$

This is necessary because the A matrix (and the A^T) contains entries relating to the pile position, which disallows use of the equation $F = SA^T X$.

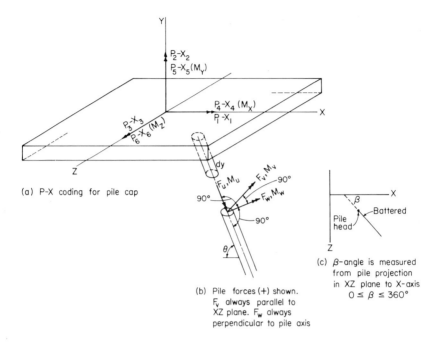

(a) P-X coding for pile cap

(b) Pile forces (+) shown.
F_v always parallel to
XZ plane. F_w always
perpendicular to pile axis

(c) β-angle is measured
from pile projection
in XZ plane to X-axis
$0 \le \beta \le 360°$

Figure 18-10. Coding and pile-force identification for building the A and S matrices.

6. With the pile displacements, the pile forces can be computed as

$$F = Se$$

The matrix solution is completely general, in that six degrees of freedom are used—three translations of X, Y, and Z and three rotations of α_x, α_y, α_z. The principal assumption is that the pile cap is perfectly rigid such that only rigid-body displacements of body translation and rotation with respect to a central set of body axes occur. No bending rotations or cap elongations between pile heads is assumed to take place; e.g., for a given X translation, each pile head has an X-component displacement of the same value, etc.

The A matrix (refer to Fig. 18-10) is built as follows:

1. Note F_v is always parallel to the XZ plane.
2. β = angle of pile projection with X axis.
3. θ = slope of batter pile with horizontal.
4. Pile heads do not have to be at the same elevation.
5. Note that the pile forces act on the cap in the direction opposite to the positive directions shown for the pile.

$$
\begin{bmatrix} F_u \\ F_v \\ F_w \\ M_u \\ M_v \\ M_w \end{bmatrix} =
\begin{bmatrix}
\cos\theta\cos\beta & \sin\beta & \sin\theta\cos\beta & 0 & 0 & 0 \\
-\sin\theta & 0.0 & \cos\theta & 0 & 0 & 0 \\
\cos\theta\sin\beta & -\cos\beta & \sin\theta\sin\beta & 0 & 0 & 0 \\
\begin{array}{l}+Z\sin\theta\\+Y\cos\theta\sin\beta\end{array} & -Y\cos\beta & \begin{array}{l}-Z\cos\theta\\+Y\sin\theta\sin\beta\end{array} & \cos\theta\cos\beta & \sin\beta & \sin\theta\cos\beta \\
\begin{array}{l}+Z\cos\theta\cos\beta\\-X\cos\theta\sin\beta\end{array} & \begin{array}{l}+Z\sin\beta\\+X\cos\beta\end{array} & \begin{array}{l}+Z\sin\theta\cos\beta\\+X\sin\theta\sin\beta\end{array} & -\sin\theta & 0 & \cos\theta \\
\begin{array}{l}-Y\cos\theta\cos\beta\\-X\sin\theta\end{array} & -Y\sin\beta & \begin{array}{l}-Y\sin\theta\cos\beta\\+X\cos\theta\end{array} & \cos\theta\sin\beta & -\cos\beta & \sin\theta\sin\beta
\end{bmatrix}
\begin{bmatrix} P'_x \\ P'_y \\ P'_z \\ M'_x \\ M'_y \\ M'_z \end{bmatrix}
$$

The P_i' and M_i' values are related to the pile-cap forces as follows:

Pile force	Component part of	Pile force	Component part of
P_x'	$P(1)$	M_x'	$P(4)$
P_y'	$P(2)$	M_y'	$P(5)$
P_z'	$P(3)$	M_z'	$P(6)$

$$\sum_1^n P_x' = P(1) \qquad \sum_1^n P_y' = P(2) \qquad \text{etc.}$$

The β angle is zero for vertical piles and varies from 0 to 360° rotated about the Y axis. Refer to Fig. 18-10 for general values of β and θ.

The S matrix is

$$
\begin{bmatrix}
C_1 & 0 & 0 & 0 & 0 & 0 \\
0 & C_2 & 0 & 0 & 0 & C_7 \\
0 & 0 & C_3 & 0 & -C_8 & 0 \\
0 & 0 & 0 & C_4 & 0 & 0 \\
0 & 0 & -C_9 & 0 & C_5 & 0 \\
0 & C_{10} & 0 & 0 & 0 & C_6
\end{bmatrix}
\begin{bmatrix}
e_1 = \delta_u \\
e_2 = \delta_v \\
e_3 = \delta_w \\
e_4 = \alpha_u \\
e_5 = \alpha_v \\
e_6 = \alpha_w
\end{bmatrix}
=
\begin{bmatrix}
e_1 \\
e_2 \\
e_3 \\
e_4 \\
e_5 \\
e_6
\end{bmatrix}
$$

$$C_1 = \frac{\lambda A E}{L}$$ where λ = coefficient so that C_1 is properly computed use $\lambda = 2$ for friction piles

$$C_4 = \frac{\psi G J}{L}$$ use $\psi = 2$ to 3; research by the author using 8-ft model piles fitted with strain gages indicates 2.5 is reasonable

The pile constants C_2 and C_3 are computed by applying a lateral force P_h with no rotation allowed and obtaining a lateral pile solution using the computer program of Chap. 9 (see Fig. 16-17) and contained in the Appendix. The use of I_x or I_y for a pile will produce two different pile constants unless $I_x = I_y$. From the computer output the deflection is used to obtain

$$C_2 \text{ or } C_3 = \frac{P_h}{\delta} = C(2, I)$$

where δ = deflection without rotation. A fixed-end moment M' is computed which is a necessary condition to inhibit pile-head rotation. This moment is the top moment in the first pile segment as shown on the computer output sheet and the pile constants

$$C_7 \text{ or } C_8 = \frac{M'}{\delta} = C(4, I)$$

Pile constants C_5 and C_6 are obtained by applying an end moment to the lateral pile with zero translation to compute a head rotation θ which is used to obtain

$$C_5 \text{ or } C_6 = \frac{M}{\theta} - C(3, I)$$

Likewise the lateral forces on the pile are unbalanced by the amount of force P'_h necessary to produce a zero translation, from which one can compute the constants

$$C_9 \text{ or } C_{10} = \frac{P'_h}{\theta} = C(5, I)$$

In many analyses $C_2 = C_3$; $C_7 = C_8$; $C_5 = C_6$; $C_9 = C_{10}$; thus the S matrix reduces to 6 separate entries instead of 10. The computer program is currently set for reading only 6 pile-constant entries and must be adjusted for reading in the 10 values necessary for a more complete S-matrix formulation. The 6-entry S matrix is as follows:

$$S = \begin{bmatrix} C(1) & 0 & 0 & 0 & 0 & 0 \\ 0 & C(2) & 0 & 0 & 0 & C(4) \\ 0 & 0 & C(2) & 0 & -C(4) & 0 \\ 0 & 0 & 0 & C(6) & 0 & 0 \\ 0 & 0 & -C(5) & 0 & C(3) & 0 \\ 0 & C(5) & 0 & 0 & 0 & C(3) \end{bmatrix}$$

The pile constants reduce to $C1$, $C2$, $C3$, all others are zero, for a pile cap with pin connections to the piles. Partially fixed piles may be simulated by proper adjustment of the S-matrix entries.

No methods currently exist of correctly evaluating the pile constants for a skew pile. No studies have been made of the value of lateral subgrade reaction k_s, for batter piles vs. vertical piles. For nonlinear lateral-subgrade-reaction values the pile constants will be also nonlinear, and it would be necessary to plot curves of forces vs. deformation for the four cases and obtain the slope of the curve in the region of working loads to obtain the best solution. Where under working loads the maximum soil deflection is, say, $\frac{1}{2}$ in or less, it would appear reasonable to assume linear conditions prevail. It should be evident that if linear conditions are used, Maxwell's law of reciprocity applies and the constants $C_7 = C_9$ [in computer program $C(4) = C(5)$].

Example 18-7. Use the computer program in Appendix and find the pile forces for a nine-pile group loaded with loads as shown. The pile group is not very practical; however, it serves to illustrate the input data.

$$k_s = 2,250 + 575z^{0.5}$$

30×30-cm-square concrete piles (see Fig. 16-17)

$L = 20$ m assumed for all piles

$E_c = 22,406,796$ kPa

$I = 0.00675$ m^4

Assume linear soil-structure interaction.
Note that the first soil spring is zero
$(FACP = 0)$ in developing Fig. 16-17a, b.
Compute $C(1,I) = 2AE/L = 2(0.3)^2(22,406,796)/20 = 201,661$ kN/m
Compute $C(6,I) = 3GJ/L$

$$G = \frac{E}{2(1 + \mu)} = \frac{22,406,796}{2(1.15)} = 9,742,085 \text{ kPa}$$

$$J = I_x + I_y = 2I_x = 0.0135 \text{ m}^4$$

$$C(6,I) = 3(9,742,085)(0.0135)/20 = 19,727.7 \text{ kN-m/rad}$$

$$C(2,I) = P_h/\delta = 20/0.00517 = 3,868.5 \text{ kN/m}$$

$$C(4,I) = M'/\delta = 61.758/0.00517 = 11,945.5 \text{ kN-m/m}$$

$$C(3,I) = M/\theta = 30/0.000488 = 61,475.4 \text{ kN-m/rad}$$

$$C(5,I) = P'_h/\theta = 5.826/0.000488 = 11,938.5 \text{ kN/rad}$$

$$\text{Note: } C(4,I) \cong C(5,I)$$

The input data can now be made up as follows:

Card No	Entry
1	TITLE
2	M KN KN-M KPA KN/CU M
3	9 4 1 0
	9 = No. of piles; 4 = No. of nonzero P-matrix entries
4	3.0 2.0 0. 20. 12.
5	201661 3868.5 61475.4 11945.4 11938.5 19727.7
6	0. 2.0 0. 90. 5.
7	Duplicate 5
8	−2. 2.0 0. 0. 3
9	Duplicate 5
10	3. 0. 0. 0. 4.
11	Duplicate 5
12	0. 0. 0. 0. 0.
13	Duplicate 5
14	−2. 0. 0. 0. 0.
15	Duplicate 5
16	3. −2. 0. 340. 12.
17	Duplicate 5
18	0 −2. 0. 270. 5.
19	Duplicate 5
20	−2. −2. 0. 0. 3
21	Duplicate 5

The following 4 cards (NNZP = 4) are the P-matrix entries:

22	1	54.
23	2	−720. (−) = down
24	3	36.
25	6	−63.0 Moment about Z axis use right-hand rule for sign

Partial computer output is as shown in Fig. E18-7b. The foundation displacements are shown on the output with the X displacement = 0.000511 m

$$\alpha_x = -0.000015 \text{ rad about } x \text{ axis using right-hand rule}$$

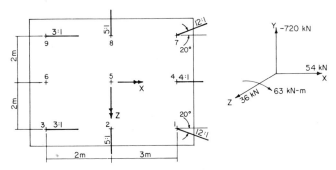

Figure E18-7a

```
J E BOWLES EXAMPLE 18-6    9-PILES

        GENERAL INPUT DATA
PILE NO    X       Y       Z       BETA     BATTER
   1      3.00    2.00    0.00    20.00     12.00
   2      0.00    2.00    0.00    90.00      5.00
   3     -2.00    2.00    0.00     0.00      3.00
   4      3.00    0.00    0.00     0.00      4.00
   5      0.00    0.00    0.00     0.00      0.00
   6     -2.00    0.00    0.00     0.00      0.00
   7      3.00   -2.00    0.00   340.00     12.00
   8      0.00   -2.00    0.00   270.00      5.00
   9     -2.00   -2.00    0.00     0.00      3.00
        THE PILE CONSTANTS ARE
PILE NO    C(1)        C(2)        C(3)        C(4)        C(5)        C(6)

   1    201661.00    3868.50    61475.40   11945.50   11938.50   19727.70
   2    201661.00    3868.50    61475.40   11945.50   11938.50   19727.70
   3    201661.00    3868.50    61475.40   11945.50   11938.50   19727.70
   4    201661.00    3868.50    61475.40   11945.50   11938.50   19727.70
   5    201661.00    3868.50    61475.40   11945.50   11938.50   19727.70
   6    201661.00    3868.50    61475.40   11945.50   11938.50   19727.70
   7    201661.00    3868.50    61475.40   11945.50   11938.50   19727.70
   8    201661.00    3868.50    61475.40   11945.50   11938.50   19727.70
   9    201661.00    3868.50    61475.40   11945.50   11938.50   19727.70

        THE A-MATRIX FOR PILE NO  1***           THE SAT MATRIX--100 FACTORED
 0.0780  0.3420  0.9364  0.0000  0.0000  0.0000    157.4   -2009.6    57.3   114.6  -171.8  -6343.7
-0.9965  0.0000  0.0830  0.0000  0.0000  0.0000     13.2       0.0   -36.4    39.2   119.0     14.3
 0.0284 -0.9397  0.3408  0.0000  0.0000  0.0000     36.2       3.2    13.2   -14.5   -39.6     49.4
 0.0568 -1.8794  0.6817  0.0780  0.3420  0.9364      0.0       0.0     0.0    15.4  -196.6      5.6
-0.0852  2.8191 -1.0225 -0.9965  0.0000  0.0830   -111.8      -9.0   -40.7   128.9   122.1   -383.8
-3.1457 -6.6840 -1.6238  0.0284 -0.9397  0.3408     40.8       0.0  -112.2   351.3   387.6    127.9

        THE A-MATRIX FOR PILE NO  2***           THE SAT MATRIX--100 FACTORED
 0.0000  1.0000  0.0000  0.0000  0.0000  0.0000      0.0   -1977.4   395.5   791.0     0.0     -0.0
-0.9806  0.0000  0.1961  0.0000  0.0000  0.0000     38.7       0.0    -0.0     0.0    23.4     39.8
 0.1961 -0.0000  0.9806  0.0000  0.0000  0.0000      0.0       7.6    37.9   -43.6     0.0     -0.0
 0.3922 -0.0000  1.9612  0.0000  1.0000  0.0000      0.0       0.0    -0.0     0.0  -193.4     38.7
-0.0000  0.0000 -0.0000 -0.9806  0.0000  0.1961     -0.0     -23.4  -117.1   380.6     0.0     -0.0
-0.0000 -2.0000 -0.0000  0.1961 -0.0000  0.9806    119.4       0.0    -0.0     0.0   120.6    364.0

       THE P-MATRIX IS        THE FOUNDATION ASAT MATRIX--100  FACTORED
   1  PX =     54.00       1    884.188  -1959.784     0.000    17.536     0.000  1108.222
   2  PY =   -720.00       2  -1959.784  17458.080     0.000 -1745.420     0.000  6533.256
   3  PZ =     36.00       3      0.000      0.000   503.504 -1053.803  -248.795   -17.536
   4  MX =      0.00       4     17.536  -1745.419 -1053.184  6969.441   783.141  -671.823
   5  MY =      0.00       5      0.000     -0.000  -248.723   783.356  3442.349   -54.609
   6  MZ =    -63.00       6   1107.605   6533.174   -17.536  -671.822   -54.672 83863.230

        THE FOUNDATION DISPLACEMENTS ARE
X =-0.000511  Y=-0.000485  Z= 0.000712  ALPHA X=-0.000015  ALPHA Y= 0.000055  ALPHA Z= 0.000037

        THE FOUNDATION DISPLACEMENTS AND PILE FORCES--**NOTE FU,FV,FW,ETC ARE ACTING ON CAP
PILE    DU      DV      DW     ALPHA U  ALPHA V  ALPHA W     FU       FV      FW       MU      MV       MW
  1  0.0003 -0.0007 -0.0004 -0.0001 -0.0000  0.0000    68.880   -2.610  -1.082   -1.094    2.353   -7.983
  2  0.0006 -0.0006  0.0006 -0.0000 -0.0000  0.0000   122.869   -1.699   2.398   -0.930   -7.779   -4.079
  3  0.0005 -0.0008 -0.0007 -0.0001 -0.0000  0.0000    69.705   -3.028  -2.388   -1.134    6.457   -9.270
  4  0.0002 -0.0005 -0.0006 -0.0001 -0.0000  0.0000    48.164   -2.126  -1.825   -1.136    4.719   -6.596
  5  0.0003 -0.0007 -0.0005 -0.0001 -0.0000  0.0000    97.841   -2.936  -1.533   -1.095    3.818   -9.439
  6  0.0006 -0.0008 -0.0005 -0.0000 -0.0000  0.0000   112.802   -3.365  -1.533   -1.095    3.818  -10.764
  7  0.0003 -0.0004 -0.0006 -0.0001 -0.0000  0.0000    64.970   -1.783  -2.108   -1.135    5.775   -6.054
  8  0.0003  0.0004 -0.0008 -0.0001 -0.0000  0.0000    66.585    1.385  -3.367   -1.217   10.769    3.646
  9  0.0004 -0.0009 -0.0008 -0.0001 -0.0000  0.0000    79.167   -3.265  -1.844   -1.134    4.777  -10.603

        INDIVIDUAL PILE FORCE COMPONENTS TO CHECK SUM OF FORCES ALONG AXES
PILE NO     FX        FY         FZ        MX         MY         MZ
   1       3.4693  -68.7316     4.0398    1.3234   -11.6921  -218.0957
   2      -1.6987 -120.0123    26.4480   45.1175     0.1120    -0.7851
   3      19.7771  -66.8836     3.0276   -3.0979     4.2000    87.7556
   4       9.9111  -47.1685     2.1258   -6.6742    -6.8754  -146.2246
   5      -1.5331  -97.8415     2.9361   -9.4390     1.0948    -3.8184
   6      -1.5331 -112.8020     3.3655  -10.7640     7.8258   221.7856
   7       3.7064  -64.9209     0.5485   -8.8305    -1.0168  -190.6813
   8      -1.3847  -65.9525    -9.7566    8.7437    19.9085    -6.1660
   9      23.2858  -75.6877     3.2652  -16.3792     4.4433   193.1700
TOTAL =   54.0000 -719.9998    35.9999   -0.0002     0.0000   -62.9997
       (  54.000)( -720.000)(  36.000)(  0.000)(   0.000)(  -63.000)
```

Figure E18-7b.

The Y displacement is -0.000485 m downward, as expected from this loading. Pile 1 has an axial displacement of 0.0003 m corresponding to an axial force F_u of

$$F_u = C(\Delta) = 201{,}661(0.0003)$$

$$= 60.5 \text{ kN} \qquad (68.880 \text{ shown})$$

the actual displacement is 0.00034, but output shows only 4 decimals.

Checking F_x, F_y, F_z, and moments of pile 1 as shown in "check"

The batter angle is arcsin $\theta = 85.205°$ $\quad \beta = 20°$ $\quad$ (input)

Thus $\qquad F_x = F_u(\cos \theta)(\cos \beta) + F_v \sin 20 + F_w \sin \theta \cos \beta$

$$= 68.88 \cos \theta \cos \beta - 2.61 \sin \beta - 1.082 \sin \theta \cos \beta$$

$$= 5.41 - 0.89 - 1.01 = 3.51 \ (3.47) \ nK$$

and $\qquad F_y = -68.88 \sin \theta - 1.082 \cos \theta$

$$= -68.64 - 0.09 = -68.73 \ (-68.73) \text{ kN}$$

Note that this check involves using the A-matrix entries and the computed pile forces F_u, F_v, F_w, M_u, M_v, M_w to obtain the components. The moments M_x, M_y, M_z also include position or coordinate-location effects.

The force F_y and the pile-head moments M_v, M_w would be used in a cap analysis (mat computer program) with only the P-matrix entries of $P_y = -720$ kN and $M_z = -63.00$ kN-m.

PROBLEMS

18-1. Compute the efficiency of the 6-pile group of Fig. 18-1a, using the Converse-Labarre equation, and compare with Eq. (18-2), if $s = 80$ cm and the piles are 25 m long and 40 cm in diamter.

18-2. Redo Prob. 18-1 for the 9-pile group of Fig. 18-1a; the 5-pile group.

18-3. A pile group consists of 3 piles as in Fig. 18-1a. The piles are HP14 × 73 used as friction piles. The group is subjected to an axial load of 300 kips. The pile spacing is 2.5 ft. Take $\mu = 0.40$.

(a) Compute the stress, using Table 18-1, in the soil at the center of the group 5 ft below the bottom of the piles, which are 65 ft long.

(b) Compute the stress 5 ft beneath one of the piles.

(c) Compare the stress in (a) and (b) with that obtained using the Boussinesq theory of Chap. 5.

(d) Compare the stresses of (a) and (b) with that obtained from the method shown in Fig. 18-4a and b.

18-4. A clay stratum 13 ft thick with $C_c = 0.56$ is located 13.0 ft beneath the pile group of Prob. 18-3. Compute the expected settlement of the pile group. The soil overlying the clay consists of a sandy material with a unit weight of 115 pcf for the top immediate 12 ft and a buoyant unit weight of 55 pcf to the clay stratum. The unit weight of the clay is 122.5 pcf.

18-5. A pile group consists of 9 piles as in Fig. 18-1a. The piles are 30-cm-square concrete piles, $E_c = 26{,}543{,}400$ kPa. The pile lengths are 32.5 m. The group load $Q = 6{,}475$ kN. The soil has an undrained shear strength $c = 60$ kPa at -3 m to 90 kPa at -40 m. The water table is at elevation -5 m. The cap is poured on the ground. C_c at elevation -15 m is 0.45. The saturated unit weight is 17.53 kN/m³. Estimate the total group settlement.

18-6. Redo Example 18-3 for a linear increase in skin friction.

18-7. Estimate the settlement due to negative skin friction of Example 18-4.

18-8. What width of fill would compute the measured settlement of Example 18-4?
Answer: Approx. 127 ft (interior at 3.3-in value).

18-9. The pile of Prob. 18-5 is in parallel with the soil. The contributory cap area is 2 m × 2 m. The subgrade modulus is 9,500 kN/m^3. Estimate the equivalent soil modulus k_s'.

Answer: Approx. 46,252 kN/m^3 using $\alpha' = 2.0$.

18-10. Assume in Example 18-6 the pile length is 35 m and other conditions the same. What is the negative skin friction? Assume the pipes are filled with concrete of $f_c' = 31,025$ kPa. Can the piles carry this negative friction?

18-11. Redo Example 18-7 if all the piles are vertical.

18-12. Redo Example 18-7 if the front 3 piles are battered at 12 : 1 and skewed as in the example.

18-13. Verify the 4 pile constants from Fig. 16-17a,b.

18-14. Redo Example 18-7 using your estimation of α' and ψ in constants $C(1,I)$ and $C(6,I)$.

NINETEEN

CAISSONS INCLUDING DRILLED PIERS

19-1. TYPES OF CAISSONS

The term *caisson* as used by foundation engineers can have several meanings. For example, a shaft drilled into the soil, which is then filled with concrete, may be termed a caisson. The shaft may be cased with a metal shell to maintain the shaft before the concreting takes place and left as part of the member, or the shell may be gradually withdrawn as the shaft is filled with concrete. The bottom of the shaft may be undercut, underreamed, or belled out to afford a larger end-bearing area. This procedure is almost identical with the construction of cast-in-place piles; therefore, it is evident that we may call the resulting foundation element a pile. Generally, if the diameter of the shaft is large enough for a person to enter for inspection purposes, say, 75 cm or more in diameter [ACI Committee 336 (1972)], the member is termed a caisson. The term *pier* or *drilled pier* also may be used to describe this foundation member. Figure 19-1 illustrates typical drilled caissons. The drilled caisson is used at sites where the soil has low bearing capacity and it is necessary to transmit the loads to a firmer stratum, and pile-driving vibrations are not allowed, pile members are too small for the load, or the larger end-bearing area available with the caisson provides a larger load capacity.

The term *caisson* is also used to describe box-type structures that are used as substructure elements at wet construction sites to transmit large loads through water and poor material to firm strata. The members are filled, or partially filled, with concrete to transmit loads at the base to the underlying soil and to provide structural stability. For convenience, the members will be classified as open-end, closed-end, or floating, as in Fig. 19-2, or pneumatic types (Fig. 19-8), although the member as

Figure 19-1. Typical drilled caissons.

actually constructed may be a combination of these types. These members may be constructed of either concrete or steel, and sometimes a combination. Concrete gives an economical weight advantage which may aid in sinking the element, but steel is easier to modify, should changes be necessary. These substructures may be round, rectangular, or any other convenient geometric shape.

Caissons are used quite extensively for bridge piers and abutments in riers, lakes, and similar marine locations. They may also be used for waterfront structures such as docks, wharves or quays, seawalls, walls, etc. These structures provide a controlled work area for the placing of the foundation on the soil. Generally, caissons (other than drilled caissons) are restricted to major projects because of the cost. A caisson of the open or box type is not generally competitive unless the firm soil stratum is more than 40 ft below the water surface, since a sheet-pile cofferdam will usually be more economical up to these depths.

19-2. OPEN-END CAISSONS

These members may be of any plan shape, as round, oblong, oval, rectangular, etc. (Fig. 19-3), and may be cellular, to provide resistance to external lateral pressures. Cell diaphragm walls may provide considerable economy through bracing the sides

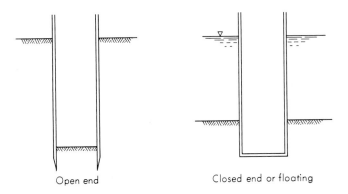

Open end   Closed end or floating

Figure 19-2. Caisson types.

Open wells for dredging or sinking

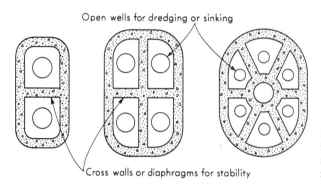

Cross walls or diaphragms for stability

Figure 19-3. Plan views (qualitative) of caissons. Actual dimensions depend on project requirements and designer's ingenuity.

of the caisson, since it is not unusual that dimensions of 30 m or more on a side may be used for bridge-pier foundations or other waterfront structures. The open-end caisson usually has a cutting edge (Fig. 19-4), which, because of its size, may be cast at the site where the caisson is to be located. This may be accomplished by excavating a trench accurately to some grade and fabricating the ring and the first shaft segment in the trench. Excavation then proceeds through the ring to a convenient depth, after which another shaft segment is added and the process continued until the bearing stratum is reached. The open-end allows excavation through the shaft. The method just outlined has primary application to the sinking of a caisson on dry land or by artificially creating an initial dry site, as, for example, by *sand islands*. A sand island is formed by building a cofferdam or sheet-pile perimeter, which is then filled with dredged (sandy) material to above the water level. Sand islands will create an obstruction in river work, and the increased velocity and scour potential should be investigated when contemplating this method of construction. Obstruction problems occurred using this method in the construction of piers for a bridge across the Mississippi River at Baton Rouge, as reported by Blaine (1967). In this case excessive erosion in the stream bed caused a washout of one of the sand islands in a matter of minutes.

It is also possible to construct the cutting edge with a portion (or all) of the shaft at a convenient fabrication location, from which it can be brought to the construction site on a barge, or by plugging the bottom so that it will float, and then towing

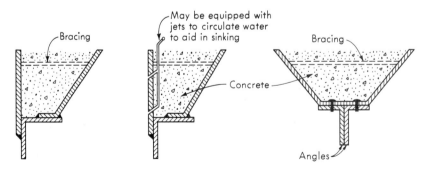

Figure 19-4. Typical fabricated cutting edges for caissons.

the unit. A major problem associated with this method is in obtaining a flat soil base on which to sink or start the caisson sinking operation, so that correct vertical alignment can be maintained.

For both the case of using a cutting edge and where the soil may be dredged to firm strata and a cutting edge is not required, the caisson is worked into correct vertical and horizontal alignment and so that the bottom edge, or rim, is situated on the desired soil stratum. Concrete is then placed under water on the open bottom as a seal to a depth that will contain the hydrostatic uplift pressure so as to avoid a bottom blow-in if or when the caisson is pumped out. When the concrete seal is sufficiently cured, the caisson can be pumped out; however, tension piles or other anchorage may be necessary to prevent flotation of the member until it is sufficiently ballasted by additional concrete or by the use of granular fill.

The concrete seal is tremie-placed under water, which is the least desirable method of concrete placement but is generally unavoidable at wet sites. An additional disadvantage is the lack of inspection of the soil at the bottom of the excavation. The thickness of concrete seal may be computed from theory-of-elasticity principles. For convenience, several caisson-seal conditions are adapted from Den Hartog (1952) and presented as follows:

For round caissons, assuming the edge is fixed (maximum stress at edge)

$$t = 0.87 \sqrt{\frac{q_0 R^2}{f_c}} \qquad (19\text{-}1)$$

For round caissons with simply supported edges (maximum stress at center)

$$t = 1.09 \sqrt{\frac{q_0 R^2}{f_c}} \qquad (19\text{-}2)$$

For rectangular caisson seals with simply supported edges,

$$t = \sqrt{\frac{6\beta q_0 B^2}{f_c}} \qquad (19\text{-}3)$$

For rectangular caisson seals with fixed edges,

$$t = \sqrt{\frac{6\alpha q_0 B^2}{f_c}} \qquad (19\text{-}4)$$

where t = thickness of seal, in consistent units

f_c = allowable concrete stress, use on the order of 0.1 to $0.2f'_c$

q_0 = contact soil pressure or hydrostatic pressure. Inspection of load conditions will indicate appropriate pressure to use.

R = radius of circular caisson

B = width of rectangular caisson

L = length of rectangular caisson

The factors α and β to use in the rectangular equations are given as follows:

L/B	1	1.2	1.4	1.6	1.8	2.0	3	∞*
α	0.051	0.064	0.073	0.078	0.081	0.083	0.083	0.083 at edge in B direction
β	0.048	0.063	0.075	0.086	0.095	0.102	0.119	0.125 at center

* Note limiting value of $\alpha = \frac{1}{12}$ and $\beta = \frac{1}{8}$.

In computing the required thickness of seal, the significance of the q_0 term should be kept in mind. Only a thin seal is necessary if the caisson is filled with sand and then pumped out, or filled with sand and then overlain with a concrete cap, to which the superstructure is attached, rather than having the seal carry the unbalanced hydrostatic pressure occurring if a seal is made and then the caisson is emptied of water.

If the caisson has been placed with the use of a cutting edge, i.e., a combination of dredging, cutting of the soil with the cutting edge, and pressure due to the caisson weight (which may be supplemented with surcharges), the surrounding soil may fit against the shaft tightly enough so that soil adhesion or friction effects may aid in flotation stability. The beneficial effect may be nominal, however, since the soil immediately adjacent to the shaft may be quite remolded and in a nearly liquid state by the time the shaft is sunk. This may be the case especially if the shaft and cutting edge have been constructed with water jets so that water under high pressure can be used as a sinking aid. Laboratory tests may be performed to establish the order of magnitude of skin friction to expect. Skin friction may also be an important factor in sinking the caisson, since this is a resistance which must be overcome in addition to the cutting resistance. In this instance the soil will be less disturbed and the driving resistance will be larger than the adhesion for flotation resistance, with the upper limit approaching the undisturbed shear strength of the soil. Again, soil tests should be performed to obtain reliable information concerning the adhesion and friction values to use.

Example 19-1. A concrete seal is to be poured for the caisson shown. The allowable soil design pressure is 400 kPa. The caisson is to be emptied of water after the seal has cured if possible. Assume that the allowable concrete stress is $0.1f'_c = 2.11$ kg/cm² (based on using $f'_c = 211$ kg/cm² concrete).

SOLUTION. *Step 1.* When the caisson is dewatered, the net upward hydrostatic pressure of the caisson seal is

$$q_0 = 20(9.807) - 23.5t \qquad (\gamma_c = 23.5)$$

$$t = 1.09\sqrt{\frac{(20(9.807) - 23.5t)}{2.11(98.07)}3.4^2}$$

$$t^2 + 1.56t = 13.05$$

$$t = 2.91 \text{ m}$$

This value of t can be justified since the actual required theoretical value of t should be between Eq. (19-2) and Eq. (19-1) and the solution of the latter gives a value of $t = 2.42$ m.

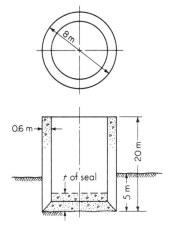

Figure E19-1.

Step 2. Check to see if we can dewater the caisson,

$$\text{Weight of seal is } 2.91(23.5)(0.7854)6.8^2 = 2,483.5 \text{ kN}$$

$$\text{Approx. weight of caisson shell} = (8^2 - 6.8^2)(0.7854)(20)(23.5)$$

$$= 6,555.9 \text{ kN}$$

$$\text{Total weight} = 9,039.4 \text{ kN}$$

$$\text{The buoyant force} = 20(9.807)(0.7854)(8)^2 = 9,859.1 \text{ kN}$$

$$\text{The net force} = 819.7 \text{ kN upward}$$

Therefore, we cannot empty the caisson, or else we must anchor it by perhaps driving two or more tension piles in the bottom or by using 0.7 m additional concrete or about 3.75 m total of concrete seal, for a small factor of safety.

Step 3. Check perimeter shear at the junction of the shell and the seal using $t = 3.75$ m.

$$3.75\pi(6.8)f_c = 9,859.1$$

$$f_c = \frac{9,859.1}{80.11} = 123.07 \text{ kPa} \ll 642 \text{ of wide beam (Table 8-2)} \qquad O.K.$$

Soil pressure when caisson filled with sand at a density 17 kN/m³

$$q_0 = (20 - 3.75)(17) = 276.25$$

$$+ 3.75(23.5) = \underline{88.12}$$
$$\overline{364.37} \text{ kPa} < 400 \text{ kPa} (q_a \text{ of soil}) \qquad O.K.$$

Comments:
1. If caisson were not dewatered, no seal would be required.
2. It would be cheaper to increase seal depth than drive tension piles.

19-3. CLOSED-END, OR BOX, CAISSONS

The *closed-end caisson* differs from the open-end caisson primarily by utilizing a bottom, or base. Provisions may be made by means of *open wells*, or *dredging wells*, through the bottom of the member, to excavate soil from beneath the caisson as

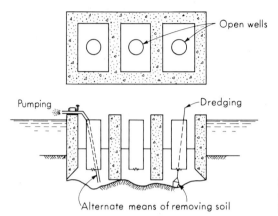

Figure 19-5. Sinking a closed-end caisson, using dredging wells to remove soil from beneath the caisson bottom.

sinking operations begin (Fig. 19-5). This type of caisson is fabricated at a central location and towed to the site and sunk onto a previously leveled soil base (Fig. 19-6), so that the caisson is properly aligned vertically. If this soil is satisfactory in bearing, the caisson is secured into place and ballasted, and the superstructure elements are added as required. If it is necessary to sink the caisson further, it will have been previously equipped with open wells, and often jet openings through which water can be pumped at high pressure, to aid in excavation. Excavation and/or pumping out of the underlying soil allows the caisson to be lowered to the proper elevation.

Many caissons of this type are large enough, for example, so that an entire bridge pier can be placed on them. Thus the member may be more than 30.5 m long by 12 or 16 m wide. These larger members are usually constructed with cells to reduce water-pressure stresses, allow the use of smaller sections or plates, and reduce distortions. Cells also aid in sinking operations and in maintaining vertical alignment,

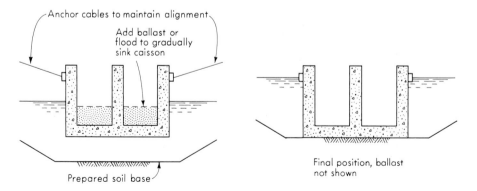

Figure 19-6. Placing a caisson into position.

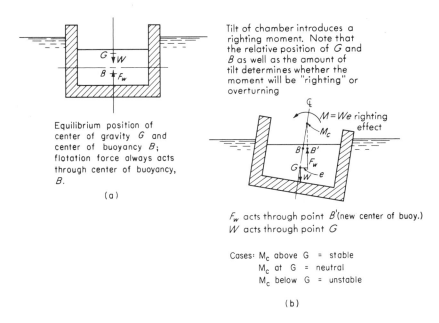

(a)

Equilibrium position of center of gravity G and center of buoyancy B; flotation force always acts through center of buoyancy, B.

Tilt of chamber introduces a righting moment. Note that the relative position of G and B as well as the amount of tilt determines whether the moment will be "righting" or overturning

F_w acts through point B'(new center of buoy.)
W acts through point G

Cases: M_c above G = stable
M_c at G = neutral
M_c below G = unstable

(b)

Figure 19-7. Flotation stability of caissons.

since filling and weight control in the individual cells (using water or other ballast, such as sand) can increase or reduce stresses at certain points. Tilt from an edge hang-up on an obstruction, or refloating the caisson if the horizontal alignment requires adjustment, can be controlled by strategic placement or removal of ballast.

These caissons may be constructed of reinforced concrete or steel plates or a combination of materials, as necessary. The design in any case introduces no new structural analysis techniques the engineer is not already familiar with. The method of excavation and equipment available should be ascertained prior to design, so that the proper size of dredging wells can be provided. Sufficient soil borings should be taken at the site so that the design depth can be reliably established and possible excavation difficulties considered.

Two major precautions with floating caissons are that there be sufficient draft to float the structure to the site and a sufficient freeboard (say, 1.5 to 3 m) so that a small wave does not cause a disaster. It may also be necessary to make a flotation-stability analysis against possible overturning in rough water. Referring to Fig. 19-7, it can be seen that in (b) the moment produces a righting effect, whereas in (a) the couple produced when wave action tips the member tends to turn the caisson over. The position of the center of gravity G (refer to Fig. 19-7) can be obtained by summing moments about the top or base of all elements contributing to the caisson weight. The position of the center of buoyancy B is obtained as the vertical distance from the water surface outside the member to the center of submerged area A_s. The *metacenter* distance $\overline{M_c G}$ is computed as the intersection of the resultant buoyancy force acting through the instantaneous location of B' and the original centerline axis of the caisson. The caisson is stable when M_c is above G and unstable when below.

The necessary values of G, B, and $\overline{M_c G}$ are computed (in most fluid-mechanics texts) as

$$G = \frac{\sum\limits^{n} \gamma A y}{W_t} \qquad (19\text{-}5)$$

$$B = \frac{\sum\limits^{n} A_{s(n)} y_n}{A_{s(\text{total})}} \qquad (19\text{-}6)$$

$$\overline{M_c B} = \frac{I}{V_s} \qquad (19\text{-}7)$$

where Ay = moment of an element n with respect to horizontal axis to which G and B are referred

γ = corresponding unit weight of element n

W_t = total weight of caisson including any water or ballast used

A_s = submerged area

I = moment of inertia of plan area with respect to axis of rotation

V_s = volume of submerged part of caisson

The location of M_c with respect to G is obtained from the relationship that

$$\overline{M_c G} = \overline{M_c B} \mp \overline{GB}$$

using $(-)$ if G is above B.

The box-type caisson just considered could also have been termed a *floating caisson*, since it essentially "floats" in the soil-water system. The logical extension of this analogy would be to use a member of this type beneath buildings located at sites where the soil is of low bearing capacity and where the use of piles may be prohibited because of driving vibrations or where the depth to firm strata is excessive. The early builders recognized, and later measured values [see Casagrande and Fadum (1944)] verified, that a building would "float" in the soil mass without further sinking (settlement) if the weight of the building equaled the weight of excavated soil. Therefore, if a concrete box (steel is more subject to corrosion) of sufficient dimensions is constructed and sunk at the site and the building is constructed on it, the building performance should be satisfactory. This method is not widely used since it is not generally competitive with piles, but in cases where the construction methods are limited, it may provide an economical solution. Golder (1965, 1975) describes uses of this technique in some detail.

Example 19-2. A caisson is constructed as shown in Fig. E19-2. Compute the stability moment if it tilts $10°$ from the vertical while being towed to the site. If necessary for stability add sand ballast at 100 pcf.

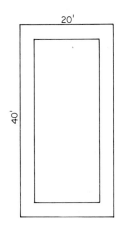

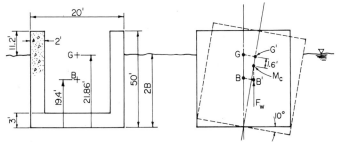

Figure E19-2.

SOLUTION. *Step 1.* Find G.

$$W_t = 50(40)(2)(0.150)(2) \quad = 1,200 \text{ kips} \quad \text{(sides)}$$

$$+ 50(16)(2)(0.150)(2) = \quad 480 \qquad \text{(ends)}$$

$$+ 36(16)(3)(0.150) \quad = \quad \underline{259.2} \qquad \text{(base)}$$

$$= \overline{1,939.2} \text{ kips for a 1-ft-thick}$$
$$\text{slice}$$

From the bottom of the caisson

$$G = \frac{+1,200(25) + 480(25) + 259.2(1.5)}{1,939.2} = 21.86 \text{ ft}$$

The depth of submergence $2B$ is

$$2B = \frac{1,939.2}{0.0625(40)(20)} = 38.8 \text{ ft} \qquad \text{Freeboard} = 50 - 38.8 = 11.2 \text{ ft} \qquad O.K.$$

(Do not use sand ballast at this point.)

$$B = 19.4 \text{ ft in untilted position}$$

Step 2. Compute metacenter distance $\overline{M_c G}$

$$I = \frac{40(20)^3}{12} = 26{,}666.67 \text{ ft}^4$$

$$V_s = 2B(A) = 2(B)(40)(20) = 31{,}040 \text{ ft}^3$$

$$\overline{M_c B} = \frac{26{,}666.67}{31{,}040} = 0.86 \text{ ft}$$

$$\overline{GB} = 19.4 - 21.86 = -2.46 \ (G \text{ above } B)$$

The metacenter is located at $M_c G = 0.86 - 2.46 = -1.6$ ft below G
The overturning couple at a 10° tilt is

$$M = W_t(1.6) \sin 10° = 538 \text{ ft-kips}$$

As a matter of fact, this caisson is inherently unstable for all tilt angles. It can only be made stable by ballast to lower G so that tilt will produce a distance $M_c G$ which falls above G.

19-4. PNEUMATIC CAISSONS

The pneumatic caisson (Fig. 19-8) provides an airtight enclosure and depends on air pressure to maintain a cavity in the excavation area. Its use may be required where an excavation is impossible to maintain because of the soil flowing into the excavation faster than it can be removed, or if it is necessary to maintain the adjacent soil "as is."

The pneumatic caisson is generally unsatisfactory and is a last-resort solution, for the following reasons:

1. Labor costs per unit of excavated material are high.
2. Premium pay becomes necessary because of health hazards.

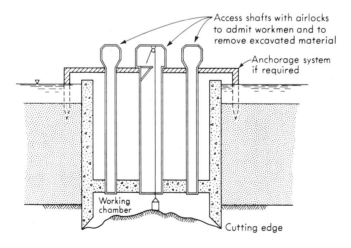

Figure 19-8. Schematic of a pneumatic caisson.

3. Meeting OSHA safety requirements is difficult.
4. Much of the labor period is used up in compression and decompression cycles, which are necessary to avoid the "bends," or the formation of air bubbles under the worker's skin when decompression is too rapid.
5. Insurance rates are apt to be excessive since too rapid decompression can result in death or a permanent disability.

Reportedly, 12 persons died from the bends, or caisson disease, during the construction of the east pier of the Eads Bridge at St. Louis [Steinman and Watson (1941)]. This was the first recorded use of pneumatic caissons in the United States, and took place around 1869. As a consequence of these early experiences on this project, the working conditions are generally outlined by governmental agencies, which usually limit the working pressures in caissons to about 50 psi in excess of atmospheric pressure, amounting to a maximum water depth of about 120 ft. At this maximum pressure some persons may not be able to do any work, and the usable work time for many others may be as low as $\frac{1}{2}$ to 1 hr. Greater working depths may be obtained by lowering the water level outside the caisson.

The pneumatic caisson will require air locks, a working chamber, a decompression chamber, and a means for the workers to get to the working chamber. Means to remove the excavated material must also be provided. At least two power sources must be available so that the air pressure can be reliably maintained. The essential features of the pneumatic caisson are illustrated in Fig. 19-8.

The construction of a pneumatic caisson should start, if possible, with the sinking of an open caisson, which proceeds as deep as practicable. The air shafts which provide entry and egress from the working chamber and through which the excavated material can be removed are then positioned. Either a concrete plug is poured, or a steel diaphragm is lowered and bolted into place above the cutting edge, leaving a head space of 6 to 8 ft, which forms the working chamber. When this is complete and the concrete sufficiently cured (if used), compressed air can be applied, together with any necessary pumping to evacuate the working chamber so that the workers can enter the airshaft and descend to the working chamber to continue the excavation. The excavated material can be hoisted through the material airshaft by bucket devices, or if the material can be liquefied, it can be pumped out. If the soil is loose and sandy, it can be blown out, using air pressure.

The structural design of this type of structure will involve a cutting edge, the roof of the excavation chamber, locating the air locks and material locks, installing a decompression chamber, providing a means to remove the excavated material (as hoists, pumps, etc.), proportioning the members for internal and external pressures, and providing airtight joints at the critical points, such as entrances and exits. Means must also be provided for adding on sections of shaft as the depth of excavation progresses, and to overcome skin friction by installing jets in the shaft walls, so that water under high pressure may be circulated between the shaft and the surrounding soil.

Great care must be continually exercised during the sinking of the shaft to maintain the proper vertical alignment since a tilt may be quite difficult to realign.

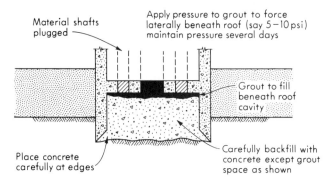

Figure 19-9. Placing concrete seal for pneumatic caisson.

When the excavation reaches the desired stratum, concrete is sent down to the working chamber and carefully placed to fill any base defects. Any voids beneath the cutting edge are also carefully backfilled with concrete. After this careful backfilling the remainder of the chamber, except for a small area beneath the airshaft, is filled with concrete. The remaining area and about 4 to 5 ft of the shaft are then filled with grout, and a pressure of 5 to 10 psi may be applied to force the grout into the underlying concrete and over the concrete to fill any remaining cavities between the concrete and the roof of the working chamber (Fig. 19-9). This pressure should be maintained for several days, while the concrete initially hardens, after which the concrete is allowed additional curing as required, and the superstructure can be removed.

19-5. DRILLED CAISSONS

The drilled caisson is a structural element which has evolved from earlier methods of hand construction, notably the Chicago method (Fig. 19-10a) and the Gow method (Fig. 19-10b). The Chicago method was a hand operation carried on inside a wooden shell consisting of vertically placed boards held in place by circular steel compression rings. The workers excavate a depth (by hand), and the boards are placed vertically around the shaft perimeter and wedged or held in place by the steel rings. Excavation then continues to a depth approximately equal to the length of the next set of boards, which can then be placed. This set of boards may be placed either with a lap on the previously set boards or directly beneath the upper set. In the latter position no reduction in shaft diameter is effected, which is often desirable. The Gow method used telescoping metal shells to maintain the shaft, with each new section of shell of a diameter about 2 in less than the shell just above.

Presently, hand excavation has been supplemented almost entirely by mechanical methods. Hand excavation may be used to clean the bottom of the caisson and in the use of rock drilling, using an electric or air-powered rock drill, into the base of the caisson to verify rock quality during inspection. A common drilling method uses a bucket rig which is a cylindrical device or bucket with cutting vanes on the bottom

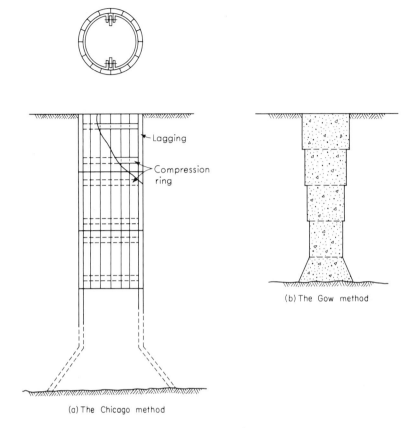

Lagging

Compression
ring

(b) The Gow method

(a) The Chicago method

Figure 19-10. Early methods of caisson construction.

which cuts and forces the soil up into the bucket as it is rotated by the drilling rig at the ground surface. The bucket is periodically withdrawn from the hole and emptied. At the bearing level the bucket is replaced with a belling bucket which is equipped with cutting wings which are hinged to the top and which can be extended to shape the bell by rotating the bucket with the arms progressively further extended.

Another common procedure uses a large auger one or more meters in length. This is turned into the soil until the flights are filled with earth, then withdrawn and spun rapidly to sling the earth off into a spoil pile around the drill hole. The process is repeated until bearing stratum is reached. Either a bell bucket or a rock-coring device is attached to complete the drilling.

The Benoto method is still another method which cuts a steel shell into the soil by turning the shell back and forth and removing the soil and broken rock inside the shell using a tool commonly termed a "grabber." This method is normally used in difficult soil conditions where open augering may cause excessive ground loss or there are large numbers of boulders suspended in the soil mass above the bearing stratum.

A heavy steel shell may be driven to the bearing stratum using pile-driving equipment and the soil and broken rock inside removed. This method is often used where the caisson is to be socketed into rock and/or used with a steel core. The steel core often consists in a very large H-beam section.

Where ground loss, or caving soil, is a problem, it is necessary to case the boring. The casing may be left in place and if the wall is sufficiently thick, say, more than 3 mm (depending on the local building code), it may be considered to carry part of the caisson load. Where the casing is pulled, the cost of installing and pulling the casing may be compared with the cost of using a clay slurry in the boring and placing the concrete by tremie. This method has been used in both England [Palmer and Holland (1966)] and the United States [see O'Neill and Reese (1972)].

If no bell is formed, it may be possible to drill the hole and fill it with concrete before the sides soften (because of exposure) and slough off, thus eliminating the need to use casing, at least in cohesive soils. Depths of 15 or so meters in 15 to 20 min of boring are not uncommon rates of drilling. Belling tools or undercutting devices are available to machine-form the bells, but are much less efficient in operation than shaft drilling. These devices also tend to leave unconsolidated material in the bottom of the hole. Bells can be formed with diameters up to about three times the shaft diameter. The bell can be formed as a dome or on a slope (Fig. 19-1), depending on the equipment used.

Advantages of using caisson or dilled pier foundations include:

1. Economy where caissons can be used.
2. Elimination of pile caps, as a single pier can often be used beneath a column.
3. Absence of vibrations and noise usually associated with pile driving.
4. Heave or ground displacement associated with driven piles. This is especially important if adjacent buildings are close to the foundation location.
5. Ease of providing uplift resistance via belling or underreaming.
6. Ease of providing increased capacity by belling or underreaming.
7. Relative ease of inspection of the sides and bearing surface by sending an inspector down the pier shaft.

Some disadvantages of using caisson foundations include:

1. Operations affected by encountering suspended boulders. Some large rocks may be broken and removed through the shaft.
2. Requires a thorough soil exploration, since it is not usually practical to load-test a caisson. Further, greater reliance is usually placed on bearing and/or skin-resistance soil parameters since caissons carry larger loads than piles.
3. Operations are affected by weather. Drilling and/or concreting during rain is undesirable to impossible.
4. Ground-loss potential where casing is not used.

Baker and Khan (1971) and Woodward et al. (1972) describe in considerable detail various construction and inspection methods to be considered in obtaining satisfactory caissons as well as procedures to correct faultily constructed caissons.

19-6. BEARING CAPACITY AND SETTLEMENTS OF DRILLED CAISSONS

Drilled caissons transfer the load vertically in a soil mass by:

1. Skin resistance (or friction or floating caissons)
2. Point bearing as where the point is founded on dense sand, on sand-gravels, or on (or socketed in) rock
3. A combination of (1) and (2) above

In the United States most caissons appear to be either skin-resistance or point-bearing types. In Europe and much of the rest of the world the combination of skin-resistance and point-bearing-type caisson is widely used. There seems no logical explanation for this except that the end result is likely to be more conservative.

The bearing capacity of any drilled caisson (Fig. 19-11) can be computed as

$$Q = Q_s + Q_p$$

as for a pile. The shaft skin-resistance term for cohesive soils [see Eq. (19-12) for $\phi \neq 0$] is computed as

$$Q_s = A_s \alpha c \qquad (19\text{-}8)$$

and the point-bearing term (assuming round bases) as

$$Q_p = A_p(cN_c + \bar{q}N_q + 0.3\gamma B N_\gamma) \qquad (19\text{-}9)$$

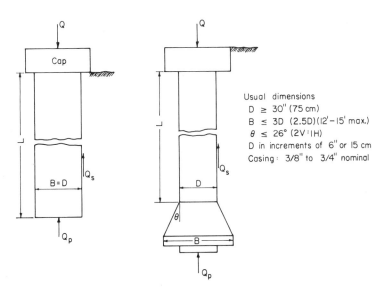

Usual dimensions
D ≥ 30" (75 cm)
B ≤ 3D (2.5D)(12'–15' max.)
θ ≤ 26° (2V:1H)
D in increments of 6" or 15 cm
Casing: 3/8" to 3/4" nominal

Figure 19-11. Capacity of straight-shaft and belled caissons. Also shown are some typical dimensions.

The soil properties (or penetration number) should be taken as the average in a zone from about 1 shaft diameter above to 1.5 shaft or base diameters below the base.

Experiments by Cooke and Whitaker (1961), Whitaker and Cooke (1966), Vesic (1967), and others indicate that the maximum skin resistance is developed at very small shaft movements. These movements are on the order of 3 to about 8 mm. There also appears to be a limiting value of skin resistance that it is possible to obtain, and this value appears to be developed at a depth varying from 10 to not more than 20 shaft diameters.

Experiments by Whitaker and Cooke (1966) and Berezantzev et al. (1961) indicate that the maximum point resistance is mobilized under about the following displacements expressed in terms of a settlement ratio S/B:

S/B	Point resistance
0.05	Full value of skin resistance Q_s is mobilized
0.10–0.15	Bearing-capacity factor $N_c = 9.0$ for belled ends in clay
0.20	Bearing-capacity factor $N_c = 9.0$ for base same size as shaft
	Ultimate bearing capacity developed for bases founded on sand or sand and gravel deposits

Experiments and field tests, principally on piles, tend to show that the shaft resistance increases to a maximum value when the settlement ratio S/B is near 0.05. The shaft resistance then decreases as S/B increases to a relatively constant value corresponding to

$$\alpha = 0.35 \text{ to } 0.40$$

There is not unanimity of agreement on what the value of α should be in computing skin resistance, with values as high as 0.75 to values as low as 0.15 proposed. The best current evidence indicates that except for unusual conditions, one may take

$$\alpha = 0.35 \text{ to } 0.40$$

with a limit on $\alpha\bar{c} \le 100$ kPa as being adequately conservative. There is some opinion that the wet concrete or that the use of a clay slurry would deteriorate the $\alpha\bar{c}$ term. Both O'Neill and Reese (1972) and Palmer and Holland (1966) indicate that this is not a particularly significant factor.

The point-bearing capacity for drilled caissons founded on clay is limited to about

$$Q_p = 9c$$

For clay it is customary to omit the $\bar{q}N_q$ term since the caisson concrete weighs more than the surrounding soil. The overburden term N_q for undrained ($\phi = 0$) conditions is 1, and if one allows for the net weight and similar materials for overburden this becomes

$$\bar{q}(N_q - 1) = 0$$

which introduces an error that is equal to the net weight of concrete and displaced soil as

$$L(\gamma_c - \gamma_s)$$

The value of $N_c = 9$ for $B/D \geq 4$ has been widely used since first proposed by Skempton (1951) for London clay. It appears that $N_c = 9$ is the limiting and widely used value for clay in other geographic locations including the United States, Canada, South America, etc. [De Mello (1969, p. 56)]. The value of $N_c = 9$ is commonly used in Chicago [Baker and Khan (1971)]. Values of N_c measured by O'Neill and Reese (1972) averaged 9.9.

Some persons have reportedly found that the value of N_c may be less than 9 and values ranging from 6.5 to about 8 may be more realistic [Whitaker and Cooke (1966), Burland et al. (1966, p. 55)]. The cause for these differing opinions may be the fact the London clay is badly fissured, causing laboratory values of q_u on solid samples to behave differently from field tests which include the fissures. The value of 9 for N_c is founded on a theoretical analysis.

Berezantzev (1965) has proposed, for drilled foundations resting on sand, that the allowable pressure at the base which limits the relative settlement S/B to about 0.20 be computed as

$$q = \gamma B(B_k) \tag{19-10}$$

where the B_k term is for a settlement ratio $S/B = 0.20$.

$$B_k = 1.667 \left(\frac{D}{B} + \frac{\cos \phi}{2\sqrt{2}} e^{(\pi/4 + \phi)\tan\phi} \right) \frac{\cos\phi(\sin\phi + \cos\phi)}{1 - \sin\phi \cos\phi} e^{2(\pi/4 + \phi)\tan\phi} \tag{19-11}$$

Values of B_k are presented in chart form, for several selected D/B ratios, in Fig. 19-12. For other S/B ratios one may use a linear interpolation. This equation is based on a mathematical analysis of the probable mode of failure, with the value 1.667 included to make the theoretical results correlate with experimental results.

The skin-resistance term for cohesionless soils can be computed in the same manner as for piles, that is

$$Q_s = \int_0^L p(\gamma'z)K(\tan\delta)\, dz \tag{19-12}$$

The terms are defined in Sec. 16-8. If the caisson is cased, the value of K would not be over K_0 and δ would be for soil to casing. If the caisson is uncased, the wet concrete should develop at least K_0 conditions owing to the larger unit weight and pressure head developed during concreting operations. The value of shaft friction angle would be on the order of $\delta = \phi$, since the wet concrete will interface with the soil in the wetted zone.

Long-term settlement analysis on clay soils may be performed using the consolidation theory of Chap. 5. The stresses may be obtained using Table 18-1. In general, it is not expected that long-term consolidation-settlement analyses are necessary since one of the purposes of using caissons is to place the large loads such that bearing capacity and settlements are controlled.

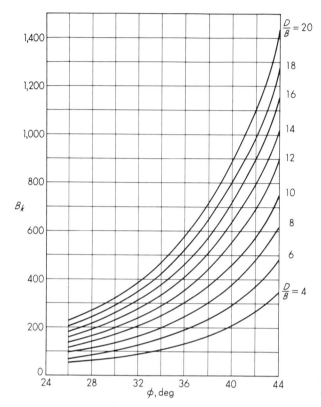

Figure 19-12. Coefficients B_k for Eq. (19-11) based on an S/B ratio of 0.20.

An alternative method of immediate settlement for clay soils was proposed by Burland et al. (1966). This procedure consists in computing the settlement as

$$S = B(K)\frac{q}{q_{ult}} \qquad (19\text{-}13)$$

where the settlement coefficient K can be computed using a modification of Eq. (5-10) to obtain

$$K = \left(N_c + \frac{\gamma' L}{c_u}\right) I_w \frac{1 - \mu^2}{E_s} \qquad (19\text{-}14)$$

For a deep circular plate on an elastic, saturated clay, $N_c = 9$, $I_w = \pi/8$, and $1 - \mu^2 = 0.75$ so that K becomes

$$K = \left(9 + \frac{\gamma' L}{c_u}\right)\frac{0.295 c_u}{E_s}$$

Since the undrained shear strength in a triaxial test is $c_u = \frac{1}{2}(\sigma_1 - \sigma_3)$, and E_s is commonly computed at a deviator stress of one-third to one-half the peak value, one may substitute to obtain

$$K = \varepsilon_1 \left[0.295 \left(9 + \frac{\gamma' L}{c_u} \right) \right] \qquad (19\text{-}15)$$

where ε_1 is the axial strain at a deviator stress of one-third to one-half the peak value.

Settlements on sand will generally not be a problem, since the caisson will normally be founded on dense deposits where the settlements are small. Some of the settlements will be built out of the structure as construction vibrations produce settlements. Settlements on sand may be computed by using Table 18-1 to obtain the stresses in the stratum underlying the base, obtaining or making an estimate of the stress-strain modulus E_s and Poisson's ratio μ, and using Eq. (5-10).

19-7. DESIGN OF DRILLED CAISSONS

The design can proceed once the soil profile, location of water table, and soil characteristics or soil parameters are established. These data are necessary to determine the depth to firm material; whether a bell is needed or whether the soil will stand on an undercut so that a bell can be constructed; and the diameter of the shaft needed, although the shaft diameter will also depend on the applied loading. With these data, the soil bearing capacity and settlements can be estimated. Possible drilling difficulties may also be detected at this stage. The soil and water-table location will determine if casing or drilling mud is required, and how the concrete will have to be placed.

Since the shaft is laterally supported by the surrounding soil, the structural design requirements are satisfied if the bell portion is proportioned so it does not punch out or break off and if the diameter of the shaft is sufficient so that the working stresses of the concrete do not exceed the allowable stresses. A minimum amount of vertical steel should be furnished. Both allowable concrete stresses and minimum vertical-steel requirements may be specified by local building codes. For unreinforced caisson shafts a value of

$$f_c = 0.25 f'_c$$

is often specified in the building codes. This is generally conservative, since we can neglect the steel $(f_s p_g)$ and design the shaft as

$$P = A_g f_c$$

where A_g is the gross area of the caisson shaft. This does not mean that a caisson cannot be designed with reinforcement to take part of the load; but because of possible difficulties in making ties, possible voids, etc., it is more conservative to rely on the concrete to carry the load. In the absence of other requirements, a minimum of 1 percent of steel should be used. Generally, the concrete stresses in members of this type should be kept low, since it is always possible to fill the shaft too rapidly with concrete, thus forming large voids by trapping air or withdrawing the casing too rapidly, allowing soil to intrude into the shaft or effecting a discontinuity in the shaft.

Where steel cores are used in the caisson, building codes may allow up to

$$f_s = 0.50f_y$$

Where the casing is left in place and with a wall thickness greater than about 3 mm, the allowable steel stress is on the order of

$$f_s = 0.35 \text{ to } 0.40f_y$$

Alternatively, the allowable concrete stress may be increased instead of considering the casing steel. Steel reinforcement is usually limited to not over

$$f_s = 0.40f_y$$

The total shaft load with steel reinforcement or casing included is

$$P = A_s \, f_c + A_s \, f_s$$

Example 19-3. Make a tentative design of an uncased caisson for a load of 650 kips with the maximum settlement limited to 1.5 in. It is decided to place the caisson base at elevation -35 ft of the soil profile shown. Use 4,000 psi concrete and a minimum amount of steel reinforcement.

SOLUTION. Find shaft diameter.

$$A_c = \frac{650}{0.25(4)} = 650 \text{ sq in}$$

$$D = \sqrt{\frac{650}{0.7854}} = 28.8 \text{ in} \qquad \text{Try 30 in}$$

Estimate the bearing capacity using Eq. (4-9a). Assume entire load carried on caisson point due to poor soil above elevation -25.

$$q_a = \frac{N}{4}\left(\frac{B+1}{B}\right)^2 K_d S$$

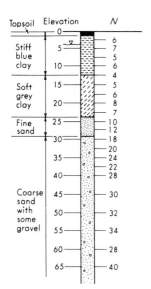

Figure E19-3.

where $N = 28$ as weighted average in zone of $-\dfrac{B}{2} + 2B$ (30 to 50 ft) and not using any type of penetration correction such as Eq. (3-3)

$B = 6$ ft (tentative estimate) $\qquad K_d = 1.3$

$$q_a = \frac{28}{4}\left(\frac{6+1}{6}\right)^2 (1.3)(1.5) \cong 18 \text{ ksf}$$

$$B = \frac{650}{0.7854(18)} = 6.78 \text{ ft} \qquad \text{Try } B = 7.0 \text{ ft}$$

Check using Fig. 19-12

$$D/B = 35/7 = 5.0 \qquad B_k = 120$$

$$q = 0.115(7)(120) = 96.6 \text{ ksf} \qquad \text{for } S/B = 0.2$$

with $S/B = 0.2$; $S = 0.2(7)(12) = 16.8$ in
By proportion for $S = 1.5$ in

$$q_a = 96.6(1.5)/16.8 = 8.6 \text{ ksf}$$

$$B = \sqrt{\frac{650}{0.7854(8.6)}} = 9.8 \text{ ft}$$

Since it is difficult to underream a bell much over 3 × shaft diameter, increase shaft to 36 in and use $B = 9.0$ ft. Rechecking

$$S = 9(0.2)(12) = 21.6 \text{ in} \qquad D/B = 35/9 = 4 \qquad B_k = 100$$

$$q = 0.115(9)(100) = 103.5 \text{ ksf}$$

$$q_a = 103.5(1.5)/21.6 = 7.2 \text{ ksf} \qquad \text{(for } S = 1.5 \text{ in)}$$

q_a based on $B = 9$ ft is

$$q_a = \frac{650}{(0.7854)(9)^2} = 10.2 \text{ ksf}$$

S according to Berezantzev $= 1.5(10.2/7.2) = 2.12$ in. The corresponding factor of safety for bearing-capacity failure is $F = 103.5/10.2 \cong 10$. S according to Eq. (4-11) is computed as follows:

$$q = \frac{28}{4}\left(\frac{9+1}{9}\right)^2 (1.3) = 11.23$$

$$S = \frac{10.2}{11.23}(1) = 0.91 \text{ in}$$

The settlement, which is measured at the top of the caisson, includes both elastic shortening and point displacement.

Axial shortening $= \dfrac{PL}{AE}$

$$E_c = 57,400\sqrt{f'_c} = 3,630 \text{ ksi} \qquad A = 1,018 \text{ sq in}$$

$$e = \frac{650(35)(12)}{3,630(1,018)} = 0.07 \text{ in}$$

This is consistent with assuming point bearing

$$S = 0.91 \text{ to } 2.12 + 0.07 \text{ in}$$

or, say, 1.2 to 1.6 in, which is about as good an estimate as can be made for the information given. Use 1.5 percent steel reinforcing in the caisson

$$A_s = 0.015(1,018) = 15.3 \text{ sq in}$$

Use 15 No 9 bars $\quad A_s = 15(1) = 15$ sq in
Should it prove impractical or impossible to underream at elevation 35, the caisson will have to be placed deeper in the stratum.

Example 19-4. Design underreamed caisson to be founded on firm clay at depth -27 m. The caisson will be 25.9 m long. The top 3.5 m is in water-bearing sand and gravel. The value of q_u at the base will be estimated from the accompanying sketch as 290 kPa. The average q_u for skin resistance will be estimated at $0.6q_{u(\text{point})} = 175$ kPa. Note that $c_u = q_u/2$ is plotted versus depth on Fig. E19-4a.
   The caisson is to carry 10,500 kN. Concrete strength is 350 kg/cm^2. The caisson is to have an overall $F \geq 2.0$. K from Eq. (19-15) is 0.025. Estimate the adhesion coefficient, $\alpha = 0.35$.

SOLUTION. *Step 1.*

$$f_c = 0.25(f_c') = 0.25(350)(98.07) = 8,581 \text{ kPa}$$

$$D = \frac{10,500}{(8,581)(0.7854)} = 1.25 \text{ m} \qquad \text{use } 1.5 \text{ m}$$

Step 2. Find shaft contribution (neglect top 3.5 m)
Assume $D_{\text{bell}} = 3D = 4.50$ m.

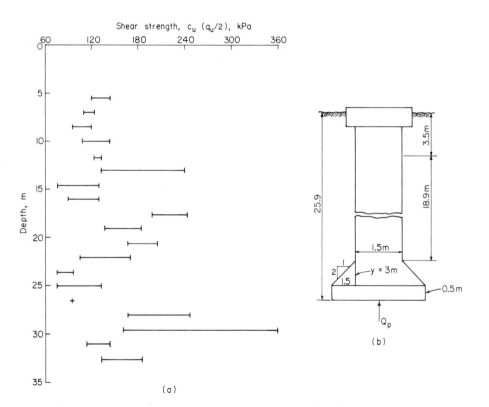

Figure E19-4a.

Figure 19-4b.

Net shaft length at a 26.6° bell angle and 0.5 m edge thickness is

$$\tan 26.6° = \frac{1.50}{y} \qquad y = 3.0$$

$$L = 25.9 - 3.5 - 0.5 - 3.0 = 18.90 \text{ m}$$

$$Q_{ult} = \alpha(18.90)(\pi)(1.5)175/2$$

$$Q_{ult} = 2,728 \text{ kN}$$

Step 3. Find Q_{ult} of base.

$$Q_{ult} = 9cA_b = 9(290/2)A_b$$

$$= 9(145)(15.9) = 20,749 \text{ kN}$$

Step 4. Compute F

$$F = \frac{20,749 + 2,728}{10,500} = 2.23 > 2 \qquad O.K.$$

For the base with full mobilization of shaft resistance

$$F_{base} = \frac{20,749}{10,500 - 2,728} = 2.67 > 2.0$$

Step 5. Estimate settlement using Eq. (19-13)

$$Q = 10,500 - 2,728 = 7,772 \text{ kN}$$

$$S = 4.5(0.025)(7,772/20,749) = 0.04 \text{ m}$$

Total S includes shaft compression

$$P_{av} = (10,500 + 7,772)/2 = 9,136 \text{ kN}$$

$$E_c = 15,100\sqrt{f'_c(98.07)} \cong 28,000,000 \text{ kPa}$$

Take shaft as 24.4 m long

$$\delta = \frac{9,136(24.4)}{28 \times 10^6(15.9)} = 0.0005 \text{ m} \qquad \text{(negligible)}$$

Therefore, $S \cong 0.04$ m.

19-8. LATERALLY LOADED CAISSONS

Laterally loaded caissons may be treated similarly to laterally loaded piles and using the finite-element computer program in the Appendix. There is some opinion, since these members are stiff compared with piles, that the caisson tends to rotate rigidly about some point along the shaft and that the rigid rotation cannot be treated in the same manner as the ordinary laterally loaded pile. The author is in agreement that this may happen; however, lateral loads are not usually of such magnitude as to affect this type of soil-caisson response.

It would seem reasonable to expect that the soil will restrict any rotation or translation at some depth in the mass which depends on the load, soil, and the shaft stiffness of the caisson. Since one does not design these members for large deflections and from observation of large numbers of computations where it is evident that for loads of modest size, the principal load-carrying element is the shaft, the finite-element method as used here is the best method currently available for estimating lateral response of caissons.

ACI Committee 336 (1972) presented essentially the same arguments when the nondimensional solutions (in the form of curves for various boundary conditions) for laterally loaded piles were submitted as a recommended design practice for laterally loaded caissons. Curves are not presented here for the solution of laterally loaded caissons, as the computer program provides a more rapid means of obtaining a solution for various boundary conditions which it is possible to have in any given problem to include:

1. Variation of shaft size and soil modulus along the shaft
2. Fixed or pinned head shaft
3. Both lateral load and moment
4. A substantial part of the shaft above ground

19-9. INSPECTION OF DRILLED CAISSONS

From the foregoing discussion concerning loads and soil properties it is obvious that adequate inspection is necessary. If belled caissons are used, it will be necessary for an inspector to enter the shaft or perhaps pass a TV camera down the shaft to observe the excavation. It is also helpful to obtain penetration readings on the bottom of the shaft (bell) to check the bearing value against the design value. If an inspector does enter the shaft, he should be on guard against possible intrusion of gas or water into the shaft.

When placing the concrete, in addition to the usual concrete testing, the inspector should attempt to compare the poured quantities with the volume of the holes. Peck (1965) has commented on the pitfalls of inadequate inspection, citing an example where the pile had a discontinuity at one point and a bullet nose for a bell. A more recent case has been reported [ENR (1966)] where several caissons for one major structure and one caisson for another major structure had to be repaired. The defect at the site of one of the structures reportedly occurred from pouring the concrete in cold weather, with the resulting vapor (fog) emitting from the shaft, making visibility so poor that the defect was not observed. Settlements began before the structure was completed, and in the resulting investigation it was found that clay had intruded into the shaft. It was also reported that the caissons on this job were cast without using casing to protect the sides. Baker and Khan (1971) discuss the problems of caisson construction and inspection in some detail.

PROBLEMS

19-1. For Example 19-1, what is the theoretical thickness of seal for bursting pressure if the caisson is 8×8 m and 6×12 m in plan?

19-2. How thick will the seal have to be so that the flotation-stability safety factor is 1.2 when the caisson is emptied in Example 19-1, if the soil adhesion on the sides is 40 kPa?

19-3. Referring to Example 19-2, what is the depth of sand in the caisson such that the freeboard is 6.0 ft?

19-4. When the freeboard is 6.0 ft from Prob. 19-3, what is the righting moment?

19-5. Referring to Example 19-4, what is the caisson capacity if the shaft length is increased to 30 m. (*Hint:* Does q_u on point stay the same?)

19-6. Referring to Example 19-4, what size shaft (without a bell) is required to carry the 10,500-kN load?

19-7. For the soil profile shown (Fig. P19-7), design a drilled caisson for a column load of 800 kips. All test values are from undrained tests.

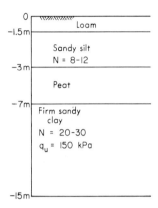

Figure P19-7.   **Figure P19-8.**

19-8. Design a drilled caisson for the conditions in the soil profile given in Fig. P19-8 for a 4,000-kN load. Use a belled caisson is required.

19-9. A circular caisson has an inside diameter of 3 m. How thick must the walls be to overcome a skin friction of 20 kPa if it is 14 m long and of concrete ($\gamma = 23.5$) using its self-weight for sinking?

19-10. A bridge pier for a highway is to be constructed for a roadway width of four 12-ft lanes and a 6-ft center median. The pier load is 4,000 kips. For the soil profile given in Fig. P19-10 make a tentative caisson layout and plan the construction.

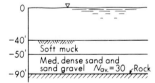

Figure P19-10.

DESIGN OF FOUNDATIONS FOR VIBRATION CONTROL

20-1. INTRODUCTION

Foundations supporting reciprocating engines, compressors, radar towers, punch presses, turbines, large electric motors and generators, etc., are subject to vibrations caused by unbalanced machine forces as well as the static weight of the machine. If these vibrations are excessive, they may damage the machine or cause it not to be able to function properly. Further the vibrations may adversely affect the building or persons working near the machinery unless their frequency and amplitude are controlled.

The design of foundations for control of vibrations was often on the basis of increasing the mass (or weight) of the foundation and/or strengthening the soil beneath the foundation base by using piles. This procedure generally works; however, the designers often recognized that this resulted in considerable overdesign. It was not until the 1950s that a few designers began to use vibration analyses usually based on theory of a surface load on an elastic half-space. The 1960s introduced the lumped-mass approach, refined the elastic half-space theory, and validated both methods.

The principal difficulty which currently exists in vibration analysis consists in determining the necessary soil properties of shear modulus and Poisson's ratio as input into the differential equation describing the vibratory motion. The general design methods to design of foundations subject to vibration and the determination of the soil properties will be taken up in some detail in the following sections.

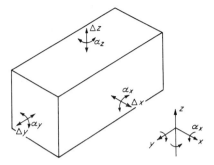

Figure 20-1. Rectangular foundation block with a maximum of six degrees of freedom (three displacements and three rotations).

20-2. ELEMENTARY VIBRATIONS

If one applies three forces and three moments to a rectangular foundation block, as shown in Fig. 20-1, the block will tend to rotate and translate with respect to the three axes; that is, the block has six degrees of freedom. The system may have certain motions restrained, in which case there are fewer than six degrees of freedom; for example, if the block can move only vertically, it has one degree of freedom. The rectangular block in the figure has been chosen for convenience; actually, the member may be of any shape, but regular-shaped figures lend themselves to the formulation of a mathematical model more easily than other shapes.

Now let us consider the motion of an element attached to a spring, as shown in Fig. 20-2, which has been displaced a small distance x from the equilibrium position and released. The differential equation describing the movement of the element after it is released is obtained by summing forces in the vertical direction on the element. If the downward displacement is taken as positive $(+)$, the ΣF_v is

$$-kx - ks + W = \frac{W}{g} a = ma \qquad (a)$$

which yields the following differential equation of motion, since $ks = W$, and using $\ddot{x}$ for the second time derivative,

$$m\ddot{x} + kx = 0 \qquad (b)$$

This is a linear differential equation for which the solution can be found as follows: First divide through by the mass term

$$\ddot{x} + \frac{k}{m} x = 0 \qquad (c)$$

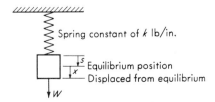

Spring constant of k lb/in.

s Equilibrium position
x Displaced from equilibrium

W

Figure 20-2. Vibrating spring-mass system.

Now, for reasons which become more apparent later, the term k/m will be defined as

$$\omega_n^2 = \frac{k}{m} \quad \text{or} \quad \omega_n = \sqrt{\frac{k}{m}}$$

Rewriting equation (c),

$$\ddot{x} + \omega_n^2 x = 0 \tag{d}$$

which can be written in operator form as

$$(M^2 + \omega_n^2)x = 0$$

and since x is not zero, the part $D + \omega_n^2$ must be the zero term, yielding

$$M = \pm i\omega_n$$

and the solution of equation (c) becomes

$$x = A \sin \omega_n t + B \cos \omega_n t \tag{20-1}$$

The term ω_n is called the *natural circular frequency* (radians/sec) of the vibrating system. If we consider initial conditions (displacement and velocity at $t = 0$ or some other time), we may evaluate the constants A and B. Taking the derivative of Eq. (20-1),

$$\frac{dx}{dt} = \text{velocity} = A\omega_n \cos \omega_n t - B\omega_n \sin \omega_n t \tag{e}$$

and if at $t = 0$, $x = -x_1$ and $V = V_0$, we have $B = -x_1$ and $A = V_0/\omega_n$. Thus Eq. (20-1) becomes

$$x = \frac{V_0}{\omega_n} \sin \omega_n t - x_1 \cos \omega_n t \tag{20-2}$$

Inspection of this equation indicates that the weight W continues to displace or vibrate forever, or the spring-mass system in undergoing *undamped* vibration. If we look at the case of $V_0 = 0$, it can be seen that the displacement x is a negative value x_1 at $\omega_n t = n\pi$, where $n = 0, 2, 4, 6, \ldots$, and is a positive value x_1 at $\omega_n t = n\pi$, where $n = 1, 3, 5, 7, \ldots$, or the cycle repeats at $\omega_n t = 2\pi$ (Fig. 20-3). The value of x_1 is termed the displacement amplitude of the mass. The value of time to repeat the amplitude is termed the period of the cycle T.

$$T = \frac{2\pi}{\omega_n} \tag{20-3}$$

The reciprocal of T is termed the *natural frequency* f_n of the system.

$$f_n = \frac{1}{T} = \frac{\omega_n}{2\pi} \quad \text{cycles/sec or Hertz (Hz)} \tag{20-4}$$

There is damping of some sort involved with all the currently identified vibrations; otherwise, once a system received an initial displacement, it would vibrate forever at the same amplitude and frequency. Experience tells us this does not

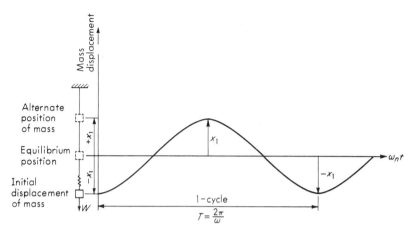

Figure 20-3. Plot of Eq. (20-2) showing the relationship of initial displacement and period T.

happen. Also, we know that some vibrations fade or are damped out more rapidly than others; thus damping is variable for different systems. Considering the simple system we have just analyzed, we shall add a damping device, often termed a "dashpot" (Fig. 20-4). For this case we shall assume that the damping force is proportional to the velocity. The simplest expression we can write for this is

$$F_d = c\dot{x} \qquad\qquad (f)$$

Obviously, other expressions for damping can be written, some of which might describe the damping effects much better than this equation. It might be noted that damping always acts to retard the current direction of motion.

Now referring to Fig. 20-4, we shall give the mass a displacement x from the equilibrium position and sum forces as before.

$$-k(s + x) + W - c\dot{x} = \frac{W}{g}\ddot{x} \qquad\qquad (g)$$

or

$$m\ddot{x} + c\dot{x} + kx = 0 \qquad\qquad (h)$$

The solution of this linear differential equation is obtained as follows:

Let

$$x = Ce^{rt}$$

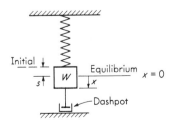

Figure 20-4. Vibrating spring-mass system with damping.

and by successive differentiation obtain

$$\dot{x} = Cre^{rt}$$

$$\ddot{x} = Cr^2 e^{rt}$$

Substitution into equation (h) yields

$$\left(r^2 + \frac{c}{m}r + \omega_n^2\right)Ce^{rt} = 0 \tag{i}$$

where $\omega_n^2 = k/m$ from the preceding undamped case. Since we do not want Ce^{rt} to be zero, the terms in the parentheses must be zero, and

$$r^2 + \frac{c}{m}r + \omega_n^2 = 0$$

Solving for r, we obtain

$$r_{1,2} = \frac{-c/m \pm \sqrt{c^2/m^2 - 4\omega_n^2}}{2}$$

which can be rearranged and placed into the general solution of equation (h):

$$x = C_1 e^{r_1 t} + C_2 e^{r_2 t} \tag{20-5}$$

where

$$r_1 = \frac{-c + \sqrt{c^2 - 4km}}{2m} = \frac{-c}{2m} + \omega_n \sqrt{\left(\frac{c}{2m\omega_n}\right)^2 - 1}$$

and

$$r_2 = \frac{-c - \sqrt{c^2 - 4km}}{2m}$$

For the special case of $c/2\omega_n m < 1.0$, it is preferable to write Eq. (20-5) in the following manner:

$$x = e^{-(c/2m)t}(C_1 \sin \omega_d t + C_2 \cos \omega_d t) \tag{20-6}$$

where ω_d is the circular frequency for free damped vibration,

$$\omega_d = \sqrt{\frac{k}{m} - \left(\frac{c}{2m}\right)^2}$$

The r term contains three possibilities:

1. The radical is real when $c/2\omega_n m > 1.0$ and the solution slowly decays (Fig. 20-5a). This solution is not oscillatory, and need not be considered, because the mass slowly returns to equilibrium.
2. The radical term is zero; then $c/2\omega_n m = 1.0$, indicating that there is no oscillation, but only a rapid return back to the equilibrium position of the mass. The value of c for this condition is termed the critical damping c_c. Equation (20-5) becomes, for this case, $x = (C_1 + C_2 t)e^{-(c_c/2m)t}$.
3. The radical is imaginary when $c/2\omega_n m < 1.0$, and this solution contains the vibrations, which will be further considered (Fig. 20-5c). For this case the roots are complex, and is the reason for introducing Eq. (20-7), following.

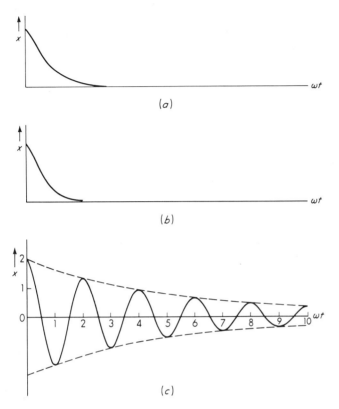

Figure 20-5. Plot of time-displacement curves for three types of damped movement. The plot is relative since the natural frequency is constant and ωt is to the same scale. The variable in the three plots is the damping factor. (*a*) Case 1 (radical positive); (*b*) case 2, critical damping (radical zero); (*c*) case 3, oscillations exist (radical imaginary).

From case 2 the damping factor for critical damping is

$$c_c = 2m\omega_n$$

The period of the damped vibrations is computed similarly as for the undamped case:

$$T = \frac{2\pi}{\omega_d} = \frac{2\pi}{\sqrt{k/m - (c/2m)^2}} = \frac{2\pi}{\sqrt{\omega_n^2 - (c/2m)^2}} \tag{20-7}$$

The coefficients C_1 and C_2 of Eqs. (20-5) and (20-6) can be obtained by differentiating x with respect to time and using the initial conditions $x = x_1$ at $t = 0$ and the velocity $= V_0$ at $t = 0$. For Eq. (20-6) the initial conditions yield

$$x = e^{-(c/2m)t}\left[\frac{V_0}{\omega_d}\sin \omega_d t + x_1\left(\cos \omega_d t + \frac{c}{2m\omega_d}\sin \omega_d t\right)\right] \tag{20-8}$$

For the case of $V_0 = 0$, Eq. (20-8) becomes

$$x = x_1 e^{-(c/2m)t}\left(\cos \omega_d t + \frac{c}{2m\omega_d} \sin \omega_d t\right)$$ (20-8a)

The solution of this equation is shown in Fig. 20-5c.

20-3. FORCED VIBRATIONS FOR A LUMPED MASS

For the case of vibrating machinery on a soil foundation, we have a system which is continually excited with some periodic force (Fig. 20-6); thus, to take into account the exciting force, we must rewrite equation (g) as follows:

$$-k(s + x) + W - c\dot{x} - \frac{W}{g}\ddot{x} = F(t)$$ (j)

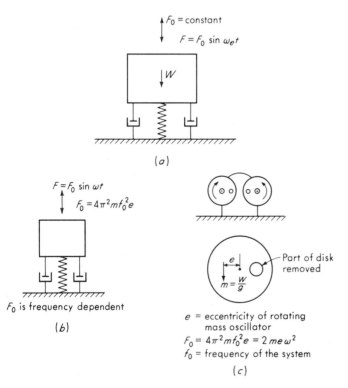

Figure 20-6. Types of foundation exciting forces. (a) Constant-magnitude force; (b) force of varying magnitude, or time-dependent as punch presses or impact machines; (c) rotating-mass oscillator to develop F_0 as for reciprocating machines, gas compressors, etc.

Generally, for oscillating machinery (where the machinery vibrates because an unbalanced rotational force exists), the force can be expressed as a sine or cosine function as (Fig. 20-6)

$$-k(s + x) + W - c\dot{x} - m\ddot{x} = F_0 \sin \omega_e t$$

which simplifies to
$$m\ddot{x} + c\dot{x} + kx = F_0 \sin \omega_e t \tag{20-9}$$

or
$$\ddot{x} + \frac{c}{m}\dot{x} + \frac{k}{m}x = \frac{F_0}{m} \sin \omega_e t \tag{20-9a}$$

It should be pointed out, however, that the force F_0 may be frequency-dependent, as, for example, the oscillator of Fig. 20-6b; it may be due to impulses as from punch presses, etc.

For the solution of the case of $F_0 = $ constant (Fig. 20-6a) we shall assume a solution of the following form, and by successive differentiation obtain

$$x = A \cos \omega_e t + B \sin \omega_e t$$
$$\dot{x} = -A\omega_e \sin \omega_e t + B\omega_e \cos \omega_e t \tag{k}$$
$$\ddot{x} = -A\omega_e^2 \cos \omega_e t - B\omega_e^2 \sin \omega_e t$$

Substituting into Eq. (20-9a), we obtain

$$(-A\omega_e^2 \cos \omega_e t - B\omega_e^2 \sin \omega_e t) + \frac{c}{m}(-A\omega_e \sin \omega_e t + B\omega_e \cos \omega_e t)$$
$$+ \frac{k}{m}(A \cos \omega_e t + B \sin \omega_e t) = \frac{F_0}{m} \sin \omega_e t$$

and equating coefficients, we have for the coefficients of $\sin \omega_e t$

$$-B\omega_e^2 - \frac{c}{m}A\omega_e + \frac{k}{m}B = \frac{F_0}{m}$$

The remaining $\cos \omega_e t$ terms have a zero on the right side of the equation.

$$-A\omega_e^2 + \frac{c}{m}B\omega_e + \frac{k}{m}A = 0$$

Simplifying, we have

$$B\left(\frac{k}{m} - \omega_e^2\right) - A\left(\frac{c\omega_e}{m}\right) = \frac{F_0}{m}$$

$$B\left(\frac{c\omega_e}{m}\right) + A\left(\frac{k}{m} - \omega_e^2\right) = 0$$

The coefficients A and B are obtained from a simultaneous solution of the above, yielding

$$A = -\frac{F_0 c\omega_e}{(k - m\omega_e^2)^2 + (c\omega_e)^2}$$

$$B = \frac{F_0(k - m\omega_e^2)}{(k - m\omega_e^2)^2 - (c\omega_e)^2}$$

This is added to the complementary solution which was obtained before [Eq. (20-8)] to yield the following general solution:

$$x = e^{-(c/2m)t}\left[\frac{V_0}{\omega_d}\sin \omega_d t + x_1\left(\cos \omega_d t + \frac{c}{2m\omega_d}\sin \omega_d t\right)\right]$$
$$+ \frac{-F_0 c\omega_e}{(c\omega_e)^2 + (k - m\omega_e^2)^2}\sin \omega_e t + \frac{F_0(k - m\omega_e^2)\cos \omega_e t}{(c\omega_e)^2 + (k - m\omega_e^2)^2} \qquad (20\text{-}10)$$

In this equation the first term represents displacement due to damped free vibrations, which gradually decays, and will not be further considered. The next two terms represent displacement components due to forced vibrations, with a period of $T = 2\pi/\omega_e$. This is a continuing vibration, which is periodic with the applied force, and is of particular interest. This latter part of Eq. (20-10) can be simplified by interpreting terms as follows:

$$\frac{-F_0 c\omega_e}{(c\omega_e)^2 + (k - m\omega_e^2)^2} = x_2 \cos \Omega$$

$$\frac{F_0(k - m\omega_e^2)}{(c\omega_e)^2 + (k - m\omega_e^2)^2} = x_2 \sin \Omega$$

and placing this into the appropriate part of Eq. (20-10), we obtain the forced vibrating displacement

$$x = x_2(\cos \phi \sin \omega_e t + \sin \phi \cos \omega_e t) = x_2 \sin (\omega_e t + \Omega)$$

The angle Ω is termed the phase angle between the exciting force and the motion of the vibrating mass.

The phase angle Ω can be computed as

$$\tan \Omega = \frac{c\omega_e}{k - m\omega_e^2}$$

The frequency at maximum x, or resonance, can be computed to make the amplification ratio $(x/x_2 = N)$ a maximum

$$f_r = f_n\sqrt{1 - 2(c/c_c)^2}$$

Noting that these x terms represent a pair of vectors which must be added to obtain the displacement, the solution for the displacement due to the forced vibrations of Eq. (20-10) becomes

$$x = \sqrt{A^2 + B^2}\, F_0 \sin (\omega_e t + \Omega) \qquad (20\text{-}11)$$

Using the previously determined values of A and B and making a substitution of $k = \omega_n c_c / 2$, and taking $\omega_n = \sqrt{k/m}$, we obtain

$$x = \frac{F_0}{k} \frac{1}{\sqrt{(2c\omega_e/c_c\omega_n)^2 + [1 - (\omega_e/\omega_n)^2]^2}} \sin{(\omega_e t + \Omega)} \qquad (20\text{-}12)$$

Let us define a damping ratio as

$$D = \frac{\text{actual damping}}{\text{critical damping}} = \frac{c}{c_c}$$

The static spring displacement x_s is

$$x_s = \frac{F_0}{k}$$

and frequency has been defined as

$$f = \frac{\omega_e}{2\pi}$$

with these substitutions

$$x = \frac{F_0}{k} \frac{1}{\sqrt{(2Df/f_n)^2 + [1 - (f/f_n)^2]^2}} \sin{(\omega_e t + \phi)} \qquad (20\text{-}13)$$

But $F_0/k = x_s$, the static displacement of the spring, and the frequency has been defined as $f = \omega_e/2\pi$; therefore

$$\frac{x}{x_s} = \frac{1}{\sqrt{\left(\dfrac{2Df}{f_n}\right)^2 + \left[1 - \left(\dfrac{f}{f_n}\right)^2\right]^2}} \sin{(\omega_e t + \Omega)} \qquad (20\text{-}14)$$

which also may be written as $x/x_s = N \sin{(\omega_e t + \psi)}$. In this form (as earlier) N is seen to be an amplification factor of the displacement amplitude. A plot of the amplification factor N is shown in Fig. 20-7a. The value of interest is the amplification factor N. It is self-evident that at some time the value of $\sin{(\omega_e t + \psi)}$ will be 1. The maximum value of N is obtained at the resonance frequency f_r as

$$N_{\max} = \frac{1}{2D\sqrt{1 - D^2}}$$

When the forced vibration force is frequency-dependent as in Fig. 20-6c, the amplification factor becomes

$$N' = N\left(\frac{f}{f_n}\right)^2 \qquad (20\text{-}15)$$

A plot of this is shown in Fig. 20-7b, and is merely the values obtained from Fig. 20-7a multiplied by the frequency ratio squared.

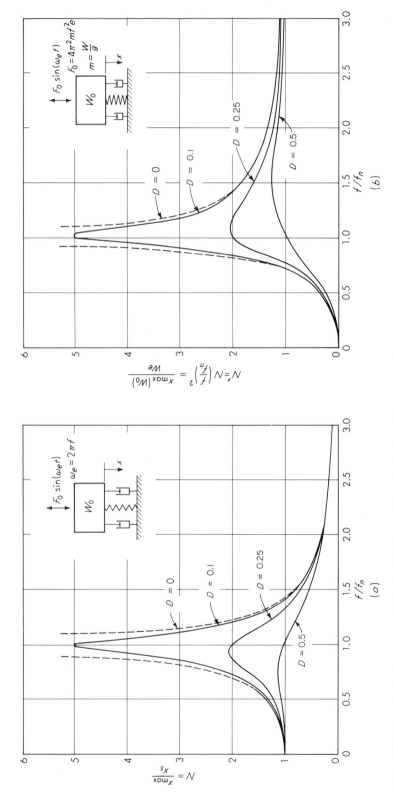

Figure 20-7. Amplitude-frequency relationships for damped forced vibration of a mass-spring system. (a) Constant-amplitude exciting force; (b) exciting force dependent on exciting frequency ω_e. Note in this figure the alternative interpretation of the factor N'.

These two amplification factor vs. frequency ratio plots show the following:

1. Resonance occurs at a frequency ratio of 1 for damping ratios of 0 or no damping.
2. When damping is small, vibration amplitudes are relatively large.
3. For constant force, resonance always occurs at a frequency ratio very nearly 1.0.
4. For frequency-dependent force, resonance always occurs at a frequency ratio greater than 1.0.
5. For frequency-dependent force, damping always shifts resonance away from a frequency ratio of 1.0.

20-4. APPROXIMATE SOLUTION OF VIBRATING FOUNDATION—THEORY OF ELASTIC HALF-SPACE

There are two approaches to the solution of the vibrating foundation. One solution is based on theory of elastic half-space. The first solution appears to have been made by E. Reissner about 1936 for a footing with vertical vibrations. Later solutions have improved on the Reissner solution, and the solution has been extended to include the other five degrees of freedom by Barkan (1962), Sung (1953), Richart (1962), Hsieh (1962), Bycroft (1956), and Arnold et al. (1955).

This section will consider the solution of a footing subject to a vertical vibration to obtain some background for the consideration of the footing subject to several degrees of freedom and using the lumped-mass approach of the next section. The major emphasis in this text is placed on the lumped-mass approach because of its greater simplicity.

If we consider that a rotating-mass oscillator is resting on a semi-infinite body with a circular contact area defined by a radius r_0, the dynamic characteristics of the soil-vibrator system have been found experimentally to depend on:

1. The radius r_0 of the loading area; use $r = r_0$ for circular bases; use $r_0 = \sqrt{A_{base}/\pi}$ for rectangular bases.
2. The mass m_0 of the oscillator and base (W_0/g).
3. The amplitude of the exciting force $F_0 = 2me\omega^2 = 2(We\omega^2/g)$ (Fig. 20-6c).
4. The distribution of the contact pressure.
5. The Poisson's ratio μ, mass density ρ, and shear modulus G of the foundation material.

In making a solution based on theory of elasticity, using Navier's equation, and extending the earlier work of Reissner, Sung (1953) introduced a dimensionless quantity termed the *mass ratio* of the system, which is defined as

$$b = \frac{m_0}{\rho r_0^3} = \frac{W_0}{\gamma r_0^3} = \frac{\pi q_0}{\gamma r_0} \tag{20-16}$$

An interpretation of the mass-ratio factor b is that it is the ratio between the static mass of the oscillator and a cylindrical mass of soil beneath it of radius r_0 and height r_0/π. For submerged soils the saturated unit weight should be used since the water contributes to the mass. Another term used by Sung was the dimensionless frequency factor computed as

$$a_0 = \omega r_0 \sqrt{\frac{\rho}{G}} = 2\pi f r_0 \sqrt{\frac{\rho}{G}} \tag{20-17}$$

A dimensionless amplitude term Z is defined as

$$Z = \sqrt{\frac{f_1^2 + f_2^2}{(1 + ba_0^2 f_1)^2 + (ba_0^2 f_2)^2}} \tag{20-18}$$

The f factors depend on Poisson's ratio and a_0 and are given in Table 20-1 for selected values of μ. An additional factor R is

$$R = \frac{2me\omega^2}{Gr_0} \tag{20-19}$$

These factors are used to compute dimensionless displacement amplitudes as follows:

1. For constant-force excitation ($F = F_0 \sin \omega_e t$)

$$A = \frac{Gr_0}{F_0} X = \frac{Gr_0}{F_0} RZ = Z \tag{20-20}$$

since $Gr_0 R/F_0 = 1.0$

2. For frequency-dependent excitation ($F = 2me\omega^2$)

$$A' = \frac{Gr_0 a_0^2}{2me\omega^2} X = \frac{\rho r_0^3}{2me} X = a_0^2 Z \tag{20-21}$$

One may program Z and $a_0^2 Z$ at sufficiently close intervals of a_0 and selected b values together with the expressions for f_1 and f_2 as given in Table 20-1 for Poisson's ratio μ. This results in plots of the form shown in Fig. 20-8. The values of a_0 corresponding to curve peak and for the given b are plotted in Fig. 20-9. A plot of peak values of A or A' for b is obtained from Fig. 20-8 to plot Fig. 20-10. From Fig. 20-10 one can obtain

Table 20-1. Power series for functions f_1 and f_2 for $a_0 \leq 1.5$

Load condition	$-f_1$	f_2
	Rigid base	
$\mu = 0$	$0.250000 - 0.109375a_0^2 + 0.010905a_0^4$	$0.214474a_0 - 0.039416a_0^3 + 0.002444a_0^5$
$\mu = 0.25$	$0.187500 - 0.070313a_0^2 + 0.006131a_0^4$	$0.148594a_0 - 0.023677a_0^3 + 0.001294a_0^5$
$\mu = 0.50$	$0.125000 - 0.046875a_0^2 + 0.003581a_0^4$	$0.104547a_0 - 0.014717a_0^3 + 0.000717a_0^5$
	Uniform loading	
$\mu = 0$	$0.318310 - 0.092841a_0^2 + 0.007405a_0^4$	$0.214474a_0 - 0.029561a_0^3 + 0.001528a_0^5$
$\mu = 0.25$	$0.238733 - 0.059683a_0^2 + 0.004163a_0^4$	$0.148594a_0 - 0.017757a_0^3 + 0.000808a_0^5$
$\mu = 0.50$	$0.159155 - 0.039789a_0^2 + 0.002432a_0^4$	$0.104547a_0 - 0.011038a_0^3 + 0.000444a_0^5$

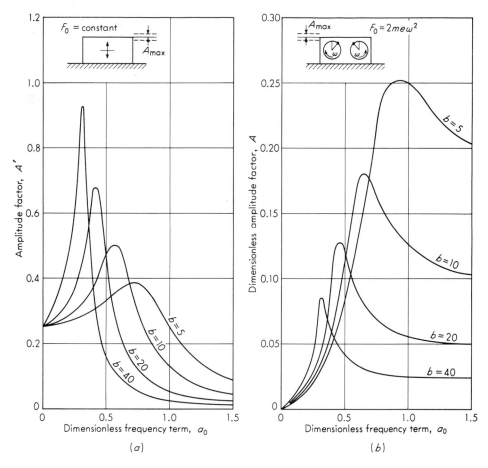

Figure 20-8. Amplitude vs. frequency relationships for a rigid-base oscillator resting on an elastic semi-infinite body. (a) Constant amplitude of exciting force ($\mu = 0.0$); (b) exciting force amplitude dependent upon the exciting frequency ω ($\mu = 0.0$).

the maximum amplitude factor for the b of the system being analyzed to compute the maximum vibration amplitude as

$$X_{max} = \frac{F_0}{Gr_0} A$$

or

$$X_{max} = \frac{2me}{\rho r_0^3} A'$$

The resonance frequency at which X_{max} occurs can be obtained from Eq. (20-17) in terms of either circular frequency or the frequency using the "peak" value of a_0 from the appropriate curve of Fig. 20-9; the frequency at other than peak a_0 is not the

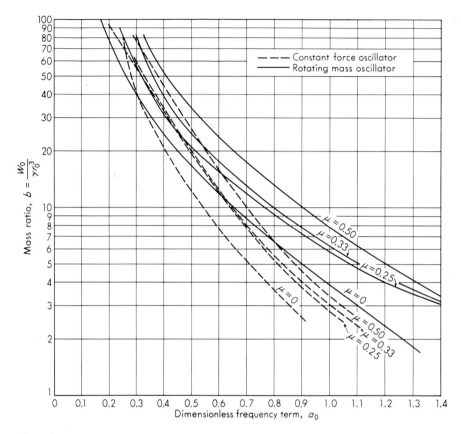

Figure 20-9. Frequency factor vs. mass-ratio relationships for several values of Poisson's ratio.

resonance frequency. The data necessary to obtain the resonance frequency and maximum vibration amplitude are:

1. Soil properties G, μ, γ
2. Unbalanced machine force F_0 or $2me\omega^2$ and operating frequencies
3. Dimensions of foundation block and block weight to compute

$$W_0 = W_{\text{block}} + W_{\text{machine}}$$

The factors f_1 and f_2 in Table 20-1 cannot be used for computing displacement amplitudes for the other five degrees of freedom possible for a vibrating machine foundation. Factors for the other degrees of freedom can be obtained from Hsieh (1962). The range of a_0 in Table 20-1 is from 0 to 1.5. This range of a_0 will usually be adequate for most vibration problems. It is not correct to extend the range of a_0 arbitrarily by using values larger than 1.5 in Table 20-1. The function used to compute f may diverge beyond a_0 of about 1.5 to such an extent that the computed value(s) of f_1 and f_2 could be substantially in error.

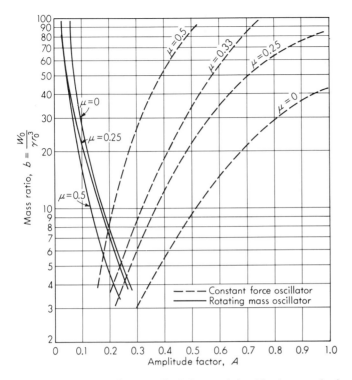

Figure 20-10. Mass ratio vs. amplitude factor relationships for several values of Poisson's ratio, as shown.

Example 20-1. The single-cylinder engine shown in the figure is resting on a foundation block of the given dimensions. The amplitude of vibration should be less than ±0.0015 cm and the resonant frequency should occur at less than 1,000 rpm. Check the system and revise if necessary. The soil is a sandy clay with a unit weight of 17.3 kN/m³; concrete = 22.6 kN/m³. Assume $G = 17,200$ kPa and Poisson's ratio is 0.25. The engine data could be calculated from the given information; however, assume the manufacturer has

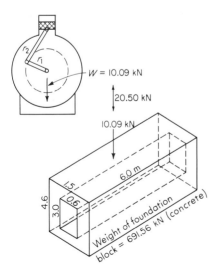

Figure E20-1.

given the following information:

> Bore = 13.00 cm; stroke = 16.51 cm
>
> Rod r_1 = 8.255 cm; r_2 = 27.305 cm
>
> Weight of machine = 10.09 kN
>
> Primary F = 15.74 kN
>
> Secondary F = 4.76 kN
>
> Operating speed = 1,800 rpm

SOLUTION. *Step 1.* Compute weight of block and soil pressure.

$$W = [1.5(4.6) - 0.6(3)](6)(22.6) = 691.56 \text{ kN}$$

$$\text{Static soil pressure} = \frac{691.56 + 10.09}{9} = 77.9 \text{ kPa}$$

Step 2. Compute r_0 and operating frequency.

$$r_0 = \sqrt{\frac{\text{area base}}{\pi}} = \sqrt{\frac{9}{\pi}} = 1.69 \text{ m}$$

$$\omega = \frac{2\pi(1,800)}{60} = 60\pi \text{ rad/sec}$$

Step 3. Compute mass ratio b and the resonant frequency.

$$b = \frac{W_0}{\gamma r_0^3} = \frac{701.65}{17.3(1.69)^3} = 8.4$$

From Fig. 20-9 for $\mu = 0.25$, $b = 8.4$, and rotating-mass oscillator, obtain $a_0 = 0.88$.

$$f_r = \frac{a_0}{2\pi r_0}\sqrt{\frac{Gg}{\gamma}} = \frac{0.88}{2\pi(1.69)}\sqrt{\frac{17,200(9.807)}{17.3}} = 8.20 \text{ cps}$$

$$f_r = 8.2(60) = 491 \text{ rpm} < 1,000 \text{ rpm} \qquad O.K.$$

The angular displacement ω is $8.20(2\pi) = 51.5$ rad/sec and the force is very nearly proportional to the square of rpm to obtain

$$F = (15.74 + 4.76)\left(\frac{491}{1,800}\right)^2 = 1.53 \text{ kN}$$

$$F = 2mew^2$$

$$2me = 1.53/(51.5)^2 = 5.76 \times 10^{-4} \text{ kN-sec}^2$$

From Fig. 20-10 for $b = 8.4$, $A = 0.19$, and

$$X_{max} = \frac{2me}{\rho r_0^3}A = \frac{(5.76 \times 10^{-4})(0.19)(9.807)(100)}{(17.3)(1.69)^3} = 0.0013 \text{ cm} \quad \text{and}$$

$$0.0013 < 0.0015 \qquad O.K.$$

The amplitude of vibration at the operating frequency can be estimated as

$$\frac{f}{f_n} = \frac{1,800}{491} = 3.7$$

From Fig. 20-7b using the alternative definition of N' the factor

$$N' = N\left(\frac{f}{f_n}\right)^2 \cong 1.2 \qquad \text{extrapolating beyond curve range}$$

$$X = N\left(\frac{f}{f_n}\right)^2 \frac{We}{W_0}$$

The term We is also $2me$ previously computed

$$X = 1.2 \frac{5.76 \times 10^{-4}(9.807)(100)}{691.6} = 9.8 \times 10^{-4} \text{ cm}$$

which is less than at resonance, as it should be according to vibration theory.

20-5. LUMPED-MASS SOLUTION OF THE VIBRATING FOUNDATION

The lumped-mass method of obtaining the resonance frequency and amplitude of vibration of a foundation is particularly attractive because of its greater simplicity compared with the elastic half-space solution. The procedure has been and is used along with the elastic half-space method but the validity was not accomplished until relatively recently (about the mid-1960s). Considerable work was done by various research workers to obtain the solution. Extensive bibliographies citing the various work are available in Richart et al. (1970), Novak (1970), and Novak and Beredugo (1972).

   In the lumped-mass approach it is necessary to obtain the static soil displacement, an estimate of the natural frequency of the system, and a term approximating the total vibrating system called the mass ratio. The computations are made as follows:

1. Compute, measure, or estimate the soil properties of shear modulus G, Poisson's ratio μ, and the unit weight. The unit weight is the total weight value including any water present. The shear modulus is a dynamic value which is, in general, considerably different from the static value.
2. Obtain the unbalanced forces causing the foundation to vibrate from the manufacturer if possible; otherwise they must be measured and/or computed. This is normally beyond the scope of the average design office, as dynamic measurements require special equipment.

   It will usually be necessary to obtain (or design) the machine foundation dimensions to determine the center of unbalanced force application with respect to the foundation base to compute any torsional or rocking moments. The dimensions must be such as to satisfy both static and dynamic soil loadings.

3. Compute the combined weight of the machine and foundation so that the mass can be obtained as $m = W/g$.
4. Compute the static soil springs k_i as appropriate for the vibration mode under consideration. These have been obtained from several sources and are given in Table 20-2. The soil springs are used to compute the undamped natural frequency of the foundation system as

$$f_n = \frac{1}{2\pi}\sqrt{\frac{k}{m}}$$

Critical damping can be computed as

$$c_c = 2\sqrt{km}$$

The static foundation displacement is computed as

$$x_s = \frac{F_0}{k_i}$$

Note that x_s may be a rotation depending on F_0 and the soil spring k_i.
5. Compute the equivalent mass radius r_0 using the equations given in Table 20-5. The mass radius is used to compute the mass ratio B_i for the appropriate vibration mode and using equations given in Table 20-4. Compute the damping ratio D_i using the mass ratio and the equations given in Table 20-4.
6. Compute the resonance frequency as
For constant force excitation

$$f_r = f_n\sqrt{1 - 2D^2}$$

For frequency-dependent force excitation

$$f_r = f_n\frac{1}{\sqrt{1 - 2D^2}}$$

With the resonance frequency established, one can compute the frequency ratio f_r/f_n and use Fig. 20-7 to obtain the magnification ratio N or N' to compute the maximum amplitude of vibration as

$$x_{\max} = x_s N \text{ (constant force excitation)}$$

$$x_{\max} = x_s N' \text{ (frequency-dependent force excitation)}$$

The amplitude of vibration at any other frequency ratio can be obtained from Fig. 20-7 as, for example, the vibration amplitude at the operating frequency.

To illustrate the method of obtaining the damping ratio [Lysmer and Richart (1966)], and specifically considering vertical vibrations, a value of damping is obtained as a best fit in a range of $0 \le a_0 \le 1$ to obtain

$$c = \frac{3.4r_0^2}{1 - \mu}\sqrt{\rho G}$$

which is substituted along with the value of the spring constant from Table 20-2 into the general differential equation [Eq. (20-9)] to obtain

$$m\ddot{z} + \frac{3.4r_0^2}{1 - \mu}\sqrt{\rho G}\,\dot{z} + \frac{4Gr_0}{1 - \mu}z = F_0 \sin \omega_e t$$

Since the critical damping was defined as

$$c_c = 2\sqrt{k_z m} = 2\sqrt{\frac{4Gr_0 m}{1 - \mu}} = 8\sqrt{\frac{Gr_0}{1 - \mu}\frac{B_z \rho r_0^3}{1 - \mu}}$$

the critical damping ratio is

$$D = \frac{c}{c_c} = \frac{0.425}{\sqrt{B_z}}$$

The other values of D are obtained in a similar manner, but the reader would have to consult the references to obtain the value of the damping coefficient that is used to obtain the best curve fit.

Example 20-2. Redo Example 20-1 using the lumped-mass method. The following data are from Example 20-1.

$$\omega = 60\pi \text{ rad/sec}$$

$$W_{block} = 691.56 \text{ kN} \qquad W_m = 10.09 \text{ kN} \qquad r_0 = 1.69 \text{ m}$$

$$F_{primary} = 15.74 \text{ kN} \qquad F_{secondary} = 4.76 \text{ kN}$$

$$G = 17{,}200 \text{ kPa} \qquad \mu = 0.25$$

SOLUTION. *Step 1.* Compute f_n

$$k_z = \frac{4Gr_0}{1 - \mu} = \frac{4(17{,}200)(1.69)}{1 - 0.25} = 155{,}029.3 \text{ kN/m}$$

$$m = \frac{W}{g} = \frac{691.56 + 10.09}{9.807} = 71.55$$

$$f_n = \frac{1}{2\pi}\sqrt{\frac{k}{m}} = \frac{1}{2\pi}\sqrt{\frac{155{,}029.3}{71.55}} = 7.41 \text{ Hz}$$

Step 2. Compute x_{max}

$$B_z = \frac{1 - \mu}{4}\frac{W}{\gamma r_0^3} \qquad \text{(Table 20-4)}$$

$$= \frac{1 - 0.25}{4}\frac{701.65}{17.3(1.69)^3} = 1.575$$

$$D = \frac{0.425}{\sqrt{B_z}} = 0.339$$

with $D = 0.339$

$$f_r = f_n \frac{1}{\sqrt{1 - 2D^2}} = 7.41 \frac{1}{\sqrt{1 - 2(0.339)^2}} = \frac{7.41}{0.878}$$

$f_r = 8.44$ Hz vs. 8.2 of Example 20-1

$$\frac{f_r}{f_n} = 1.14 \quad \text{and} \quad N' \cong 1.25 \text{ (interpolating on Fig. 20-7}b\text{)}$$

The unbalanced force F_0 is proportional to the square of the frequency:

$$F_0 = (15.74 + 4.76)\left(\frac{8.44}{30}\right)^2 = 1.62 \text{ kN}$$

$$x_s = 1.62(100)/155,029.5 = 0.00104 \text{ cm}$$

$$x_{max} = x_s(N') = 0.00104(1.25) = 0.0013 \text{ cm}$$

Compares whith 0.0013 of Example 20-1.

Step 3. Check x at operating frequency of $f = 30$ Hz. This is computed the same as in Example 20-1 and will not be repeated.

Example 20-3. Given the machine foundation shown, estimate the vibration amplitudes. The operating frequency is 1,800 rpm. All unbalanced forces are frequency-dependent.

$$G = 25,500 \text{ kPa} \qquad \mu = 0.35$$
$$\text{Silty sand, } \gamma = 18 \text{ kN/m}^3$$
$$B = 2 \text{ m}$$

SOLUTION. The foundation is subject to vertical and horizontal translation and rocking about the Y axis. We will apply superposition and neglect any coupling action. Note that considerable effort could be required to obtain G, μ, W, and unbalanced forces.

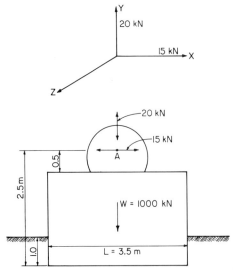

Figure E20-3.

Step 1. Compute r_0 and soil springs for three vibration modes; use Table 20-5.

(a) Vertical: $r_0 = \sqrt{\dfrac{A}{\pi}} = \sqrt{\dfrac{7}{\pi}} = 1.49$ m

(b) Horizontal: $r_0 = 1.49$

(c) Rocking: $r_0 = \sqrt[4]{\dfrac{BL^3}{3\pi}} = \sqrt[4]{\dfrac{85.75}{3\pi}} = 1.74$ m

(d) $I_\psi = m\left(\dfrac{r_0^2}{4} + \dfrac{h^2}{3}\right) = \dfrac{1,000}{9.807}\left(\dfrac{1.74^2}{4} + \dfrac{2^2}{3}\right) = 213.14$

Step 2. Compute soil springs; use Table 20-2 and use *circular base* since L is not much larger than B.

(a) $k_z = \dfrac{4(25,500)(1.49)}{1 - 0.35} = 233,815$ kN/m

(b) $k_x = \dfrac{32(1 - 0.35)(25,500)1.49}{7 - 8(0.35)} = 188,166$ kN/m

use rectangular footing for k_ψ

$$L/B = 1.75 \qquad \text{Use } F_\psi \cong 2.50$$

(c) $k_\psi = \dfrac{2.5(25,500)(2)^3}{1 - 0.35} = 784,615$ kN-m/rad

Step 3. Compute natural frequencies

(a) Vertical: $f_n = \dfrac{1}{2\pi}\sqrt{\dfrac{k}{m}} = \dfrac{1}{2\pi}\sqrt{\dfrac{233,815}{101.97}} = 7.62$ cps (Hertz)

(b) Sliding: $f_n = \dfrac{1}{2\pi}\sqrt{\dfrac{188,166}{101.97}} = 6.84$ Hz

(c) Rocking: $f_n = \dfrac{1}{2\pi}\sqrt{\dfrac{784,615}{101.97}} = 13.96$ Hz

Step 4. Compute mass and damping ratios.

(a) Vertical: $B_z = \dfrac{1 - 0.35}{4}\dfrac{1,000}{18(1.49)^3} = 2.73$

$$D_z = \dfrac{0.425}{\sqrt{2.73}} = 0.257$$

(b) Horizontal: $B_x = \dfrac{7 - 8(0.35)}{32(1 - 0.35)}(16.79) = 3.39$

$$D_x = \dfrac{0.288}{\sqrt{3.39}} = 0.156$$

(c) Rocking: $B_\psi = \dfrac{3(1 - 0.35)}{8}\dfrac{(213.14)(9.807)}{18(1.74)^5} = 1.77$

$$D_\psi = \dfrac{0.15}{(1 + 1.77)\sqrt{1.77}} = 0.041$$

Step 5. Compute resonance frequencies.

(a) Vertical: $f_r = f_n \dfrac{1}{\sqrt{1 - 2D^2}}$

$$= 7.62 \dfrac{1}{\sqrt{1 - 2(0.257)^2}} = 7.62(1.07) = 8.18 \text{ Hz}$$

(b) Horizontal: $f_r = 6.84 \dfrac{1}{\sqrt{1 - 2(0.156)^2}} = 7.01 \text{ Hz}$

(c) Rocking: $f_r = 13.96 \dfrac{1}{\sqrt{1 - 2(0.041)^2}} = 13.98 \text{ Hz}$

From this set of computations it is evident that the maximum vertical and sliding amplitudes occur nearly simultaneously; the maximum rocking amplitude occurs before the maximum sliding and vertical amplitudes; thus it appears that superposition will be satisfactory even if somewhat incorrect.

Step 6. Compute maximum amplitudes of vibration.

(a) Vertical amplitude: $F_{0(z)} \cong 20\left(\dfrac{8.18}{30}\right)^2 = 1.49 \text{ kN}$

$$x_s = \dfrac{1.49(100)}{233,815} = 0.00064 \text{ cm}$$

$$f_r/f_n = \dfrac{8.18}{7.62} = 1.07$$

Interpolating at $D = 0.25$ and $f/f_n = 1.07$, $N' \cong 2.0$ (Fig. 20-7b)

$$x_{max} = 0.00064(2.0) = 0.000128 \text{ cm}$$

(b) Horizontal amplitude:

$$F_{0(x)} \cong 15\left(\dfrac{7.01}{30}\right)^2 = 0.82 \text{ kN}$$

$$x_s = \dfrac{0.82(100)}{188,166} = 0.00044 \text{ cm}$$

$$f_r/f_n = \dfrac{7.01}{6.84} = 1.02$$

For $D = 0.156$ and $f/f_n = 1.02$, $N' \cong 3.3$ (Fig. 20-7b)

$$x_{max} = 0.00044(3.3) = 0.0015 \text{ cm}$$

(c) Rocking amplitude: $F_0 \cong 15\left(\dfrac{13.98}{30}\right)^2 = 3.26 \text{ kN}$

The rocking moment at resonance is

$$M = 3.26(y) = 3.26(2.5) = 8.15 \text{ kN-m}$$

$$x_s = \dfrac{F_0}{k_\psi} = \dfrac{8.15 \text{ kN-m}}{784,615 \text{ kN-m/rad}} = 0.0000104 \text{ rad}$$

$$f_r/f_n = 13.98/13.96 = 1.00$$

For $D = 0.04$ and $f/f_n = 1.00$, $N' \cong 6.0$

　　　Interpolating off scale

$$x_s = 0.0000104(6) = 0.0000624 \text{ rad}$$

(*d*) Total displacement of machine at machine level (point *A* of sketch) is

$$x_{max} = x_x + x_\psi$$

$$= 0.0015 \text{ cm} + 0.0000624(2.5)(100)$$

$$= 0.0015 \text{ cm} + 0.0156 \text{ cm} = 0.0171 \text{ cm}$$

Table 20-2. Spring constants for *rigid foundations*. These constants, in general, are not given directly in the cited sources

Motion	Spring constant	Reference
Rigid circular bases		
Vertical	$k_z = \dfrac{4Gr_0}{1 - \mu}$	Timoshenko and Goodier (1951, p. 368)
Horizontal	$k_x, k_y = \dfrac{32(1 - \mu)Gr_0}{7 - 8\mu}$	Whitman and Richart (1967)
Rocking	$k_\psi = \dfrac{8Gr_0^3}{3(1 - \mu)}$	Borowicka (1943)
Torsion	$k_\theta = \frac{16}{3}Gr_0^3$	Reissner and Sagoci (1944)
Rigid rectangular bases (obtain *F* factors from Table 20-3)		
Vertical	$k_z = F_z \dfrac{G}{1 - \mu}\sqrt{BL}$	Barkan (1962, Chap. 1)
Horizontal	$k_x, k_y = F_x G(1 + \mu)\sqrt{BL}$	Barkan (1962, Chap. 1)
Rocking	$k_\psi = F_\psi \dfrac{G}{1 - \mu} B^3 \quad L/B > 1$	Gorbunov-Possadov and Serebrjanyi (1961)

Table 20-3. *F* coefficients for rectangular footings subject to vibrations indicated

L/B		0.5	1.0	2.0	3.0	5.0	10.0
F_x			2.12	2.18	2.26	2.44	2.82
F_ψ			3.64	2.35	1.82	1.38	1.01
F_x	$\mu = 0.1$	2.08	2.00	2.04	2.10	2.30	2.50
	0.3	1.85	1.74	1.74	1.81	1.90	2.08
	0.4	1.69	1.58	1.57	1.61	1.70	1.88
	0.5	1.54	1.41	1.37	1.40	1.46	1.88

Table 20-4. Mass ratio B and damping factors. Refer to Table 20-5 for equations to compute r_0 and mass moments of inertia I_i

Mode of vibration	Mass ratio	Damping
Vertical	$B_z = \dfrac{1-\mu}{4}\dfrac{W}{\gamma r_0^3}$ Source: Lysmer and Richart (1966)	$D_z = \dfrac{0.425}{\sqrt{B_z}}$
Sliding	$B_x,\,B_y = \dfrac{7-8\mu}{32(1-\mu)}\dfrac{W}{\gamma r_0^3}$ Source: Hall (1967), Richart et al. (1970)	$D_x = \dfrac{0.288}{\sqrt{B_x}}$
Rocking	$B_\psi = \dfrac{3(1-\mu)}{8}\dfrac{I_\psi}{\rho r_0^5}$ Source: Hall (1967)	$D_\psi = \dfrac{0.15}{(1+B_\psi)\sqrt{B_\psi}}$
Torsional	$B_\theta = \dfrac{I_\theta}{\rho r_0^5}$ Source: Whitman and Richart (1967)	$D_\theta = \dfrac{0.50}{1+2B_\theta}$

Table 20-5. Mass radius r_0 and moment of inertia I_i equations

Vibration mode	r_0	I_i
Translation $(x,\,y,\,z)$	$r_0 = \sqrt{\dfrac{BL}{\pi}}$	
Rocking	$r_0 = \sqrt[4]{\dfrac{BL^3}{3\pi}}$	$I_\psi = \left(\dfrac{r_0^2}{4}+\dfrac{h^2}{3}\right)$

r_0 = based on circular area with equivalent moment of inertia
I_ψ = mass moment of inertia of a cylinder of height h about
     an axis through its base

Torsion	$r_0 = \sqrt[4]{\dfrac{BL(B^2+L^2)}{6\pi}}$	$I_\theta = \dfrac{m\pi r_0^4}{2}$

r_0 = same definition as for rocking
I_θ = mass moment of inertia with respect to the vertical
    axis (polar I)

20-6. SOIL PROPERTIES—ELASTIC CONSTANTS

It is evident from Table 20-2 that the spring constants are directly proportional to the value of shear modulus G and also depend on Poisson's ratio μ. The unit weight of the soil also must be obtained for computing $\rho = \gamma/g$. Poisson's ratio is not likely to vary much from 0.25 to 0.35 for cohesionless soils and 0.30 to 0.45 for cohesive soils like those which are likely to be used to support vibrating foundations. A reasonable estimate is

$$\mu = 0.3 \quad \text{(cohesionless soil)}$$

$$\mu = 0.4 \quad \text{(cohesive soil)}$$

The unit weight of the soil is likely to be on the order of 17 to 20 kN/m³ (110 to 125 pcf). A reasonable average estimate in the absence of measured data is 18.5 kN/m³ (about 118 pcf).

The shear modulus can be estimated from resonant-column tests. This involves a special laboratory apparatus consisting in a specially constructed triaxial cell capable of providing a very small amplitude vibration to a soil specimen. The technique is described in some detail in Cunny and Fry (1973) and Hardin and Music (1965).

The value of shear modulus G can be estimated using empirical equations presented by Hardin and Black (1968) and Hardin and Richart (1963) as

$$G = \frac{6{,}900(2.17 - e)^2}{1 + e}\sigma_0^{0.5} \quad \text{kPa} \tag{20-22}$$

for round-grained sands where $e < 80$.

For angular-grained materials with $e > 0.6$ and clays of modest activity the estimate of G is

$$G = \frac{3{,}230(2.97 - e)^2}{1 + e}\sigma_0^{0.5} \quad \text{kPa} \tag{20-23}$$

More recently Hardin and Drnevich (1972) have included the overconsolidation ratio (OCR) to obtain

$$G = \frac{3{,}230(2.97 - e)^2}{1 + e}\text{OCR}^M\sigma_0^{0.5} \quad \text{kPa} \tag{20-24}$$

where e = void ratio (in situ or test sample)
$\sigma_0 = (\sigma_1 + \sigma_2 + \sigma_3)/3$ = mean principal effective stress
(in situ or laboratory triaxial cell), kPa

Values of M in Eq. (20-23) depend on the overconsolidation ratio (OCR) and the plasticity index of the soil as follows:

I_p, %	0	20	40	60	80%
M	0	0.18	0.30	0.41	0.48

Typical values of G as found by several researchers are given in Table 20-6 as a guide or for preliminary estimates of vibration amplitudes. One cannot use static triaxial test values of E_s to compute dynamic values of G. Triaxial values are too small by a factor of 2 or more. This is due primarily to the fact that vibration values of G are at very low strain amplitudes (order of 0.1 to 0.001) of the minimum triaxial value (order of 0.002 to 0.004).

In elastically homogeneous ground stressed at a point near the surface, three elastic waves travel outward at different speeds. Two of the waves are the primary (P wave), or compression, wave and the secondary (S wave), or shear, wave. The third wave is a surface (Rayleigh) wave near and at the ground surface. The velocity of the Rayleigh wave is about 10 percent less than that of the shear wave [Griffiths and King (1965)] and for surface measurements is often used in lieu of the shear wave owing to the complex waveform displayed on the pickup unit (oscilloscope). The

Table 20-6. Representative values of shear modulus G

Material	ksi	kPa × 10^3
Clean dense quartz sand	1.8–3	12–20
Micaceous fine sand	2.3	16
Berlin sand ($e = 0.53$)	2.5–3.5	17–24
Loamy sand	1.5	10
Dense sand-gravel	10^+	70^+
Wet soft silty clay	1.3–2	9–15
Dry soft silty clay	2.5–3	17–21
Dry silty clay	4–5	25–35
Medium clay	2–4	12–30
Sandy clay	2–4	12–30

compression and shear waves are related to the elastic constants of the soil as follows:

$$v_c = \sqrt{\frac{E_s(1 - \mu)}{\rho(1 + \mu)(1 - 2\mu)}} \qquad (20\text{-}25)$$

$$v_s = \sqrt{\frac{G}{\rho}} \qquad (20\text{-}26)$$

The relationship between shear modulus and stress-strain modulus is

$$E_s = 2(1 + \mu)G$$

Also

$$\left(\frac{v_c}{v_s}\right)^2 = \frac{2(1 - \mu)}{1 - 2\mu} \qquad (20\text{-}27)$$

From this last equation it is evident that the shear wave is a little over half as fast as the compression wave, but the ratio is not defined at $\mu = 0.50$ (saturated soil).

The shear modulus can be obtained by obtaining field measurements of the shear-wave velocity and using Eq. (20-26) to obtain

$$G = \rho v_s^2$$

The shear-wave velocity can be obtained in the field using a cross-hole technique described in considerable detail by Stokoe and Woods (1972). Basically the method consists in drilling two boreholes to some depth, say, B to $1.5B$ below the proposed foundation base. A pickup device (velocity transducer) is located in the bottom of one hole and a rod is inserted to the bottom of the other hole. An electric circuit is made between the rod and a hammer (both metal) so that a "storing" oscilloscope is triggered to start recording when the hammer impacts on the rod, causing a shock wave as well as closing the triggering circuit. The oscilloscope trace indicates when the compression and shear waves arrive from the shock. Allowing for the time for the shock wave to travel down the metal rod leaves the time for the shear wave to travel to the pickup unit. Measuring the distance between the two holes as d allows the shear-wave velocity to be computed as

$$v_s = \frac{d}{t}$$

The test can be done several times and at several depth increments, and by using a storage oscilloscope, photographs of the waveform can be made, so that the time of arrival of the shear wave can be reasonably well established. The compression wave always arrives at the pickup first and is indicated on the oscilloscope as the beginning of the wavy trace. The shear-wave arrival is denoted by a much larger trace amplitude. The presence of the water table or saturated soil can be detected by a much larger (and earlier) compression-wave trace amplitude.

The principal source of error in vibration analysis is the soil properties of G and Poisson's ratio μ. Since Poisson's ratio is likely to vary from about 0.3 to 0.5, this establishes the error for this factor. The shear modulus may vary as much as three to five times between field- and laboratory-determined values. The determination of the value of G using field measurements such as the cross-hole method should increase the reliability of G considerably—perhaps to not more than ± 25 percent. A recent study by Cunny and Fry (1973) comparing laboratory and in situ values of shear modulus indicated general reliability on the order of ± 50 percent for laboratory values.

20-7. COUPLED VIBRATIONS

Generally coupled vibration of the rotational and translation mode is of particular interest. Rocking with vertical translation may also be of interest. In general one may write the differential equation of this condition as follows:

$$m\ddot{x} + k_x(x + R\theta) + c_x(\dot{x} + R\dot{\theta}) = F_0(t)$$

where Eq. (20-9) has included two additional terms:

$R\theta$ = additional displacement of point under consideration
    a distance R from cg of foundation for rotational
    angle θ due to rotation
$R\dot{\theta}$ = additional velocity of point under consideration

Note that at the origin of coordinates $R\theta$ and $R\dot{\theta} = 0$. The solution of this equation using the method given in Sec. 20-3 involves four arbitrary constants. This is beyond the scope of this text, and the reader is referred to Hall (1967) for a solution.

Generally, however, as a first estimate one may use superposition, since the displacements are very small. A second approximation is using

$$x_f = \sqrt{x_r^2 + x_t^2} \le x_{\text{allowable}}$$

where x_r, x_t = maximum amplitude of movement due to rotation and translation in consistent units. If either method exceeds the allowable vibration amplitude, one must

1. Redesign based on "superposition," or
2. Evaluate the coupled vibration amplitude (and probably redesign).

20-8. EFFECT OF PILES TO REDUCE FOUNDATION VIBRATIONS

Piles driven into the soil beneath a vibrating foundation may be used to effect changes in vibration amplitudes. The effect is equivalent to increasing the soil stiffness or G. A study of the lumped-mass equations indicates that the use of piles may decrease the amplitude due to larger k and may increase both the natural frequency and damping ratio. In general, the effect may be difficult to quantify. It is possible, however, to estimate the effect of piles driven to rock using the principle of a vibrating rod firmly attached on one end, with a weight on the other. The presence of the surrounding medium is neglected. The basic differential equation is the same as used for wave-equation analysis.

$$\frac{\partial^2 u}{\partial t^2} - \frac{E}{\rho}\frac{\partial^2 u}{\partial y^2} = 0$$

If we define $a = \sqrt{E/\rho}$ and properly manipulate the above equation, we obtain

$$AE\frac{\omega_n}{a}\cos \omega_n \frac{l}{a} = \frac{W}{g}\omega_n^2 \sin \omega_n \frac{l}{a} \qquad (20\text{-}28)$$

where l = length of bar
$\quad A$ = cross-sectional area of bar or pile
$\quad E$ = modulus of elasticity
$\quad W$ = weight of mass on end of bar
$\quad \rho = \gamma/g$

Now, if we further define $\psi = A\gamma l/W$ as the ratio of the weight of the bar to the weight W applied on the end, and further let $\omega_n l/a = \theta$, Eq. (20-28) becomes

$$\psi = \theta \tan \theta \qquad (20\text{-}29)$$

Several values of ψ and θ are given for convenience as follows:

$\psi = 0.01$	0.1	0.3	0.5	0.7	0.9	1.00	1.5	2.0	3.0	4.0	5.0	10.0	30.0	100.0
$\theta = 0.10$	0.32	0.53	0.66	0.75	0.82	0.86	0.98	1.08	1.20	1.27	1.43	1.52	1.565	$\pi/2$

When the weight of the piles is negligible compared with the weight of the attached mass, the quantity ψ is small, and Eq. (20-29) may be rewritten

$$\psi = \theta^2$$

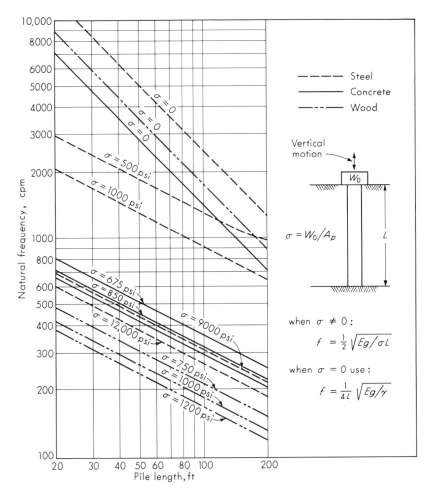

Figure 20-11. Natural resonant frequency for vertical oscillation for an end-bearing pile carrying a static load W_0.

For this case the natural frequency in cycles per time is

$$f = \frac{1}{2\pi} \sqrt{\frac{AEg}{Wl}} = \frac{1}{2\pi} \sqrt{\frac{Eg}{\sigma l}} \qquad (20\text{-}30)$$

These equations have been used to obtain the plot of data, as shown in Fig. 20-11, from which one may arrive at an estimate on the suitability of using piles to change the foundation frequency. This solution has two major errors, in that the effect of the surrounding soil has not been considered, and this plot does not take into account that even if the point is founded on rock, there is some deformation. Fortunately, however, these errors tend to give too high a computed frequency. A value from Fig. 20-11 should set the upper limit of the foundation frequency.

20-9. OTHER CONSIDERATIONS FOR MACHINERY FOUNDATIONS

Foundations for machinery having operating speeds in excess of 1,000 rpm are usually adequate if the natural foundation frequency is one-third to one-half the operating frequency. During starting and stopping operations, however, the foundation will undergo resonance, and this amplitude must be investigated. Heavy low-speed machinery should have a foundation frequency two or more times the operating frequency.

The natural frequency of the foundation can be decreased by increasing the static weight of the foundation, by decreasing the base area of the foundation, reducing the shear modulus G of the soil, or placing the foundation deeper in the soil, although the latter method cannot be reliably analyzed, i.e., increasing m or decreasing k.

Manufacturers of machinery may specify the weight of the foundation block as, say, three to five times the weight of the machine, as a rule of thumb based on experience, to control the vibrations. The weight of the machine block which has been found to perform satisfactorily may simply be specified by the equipment manufacturer.

Embedment of the machine foundation will generally reduce the vibration amplitude and increase the resonant frequency [Novak and Beredugo (1972) and Anandakrishnan and Krishnaswamy (1973)]. Current status of theoretical methods has not advanced to the point of making recommendations. The vibration amplitudes computed for the footing on the surface will in nearly all cases by conservative for embedded footings that are carefully backfilled. One may arbitrarily estimate increases in the resonant frequency of, say, 10 and 20 percent and recompute the vibration amplitudes to see if potential problems exist for these frequencies.

PROBLEMS

20-1. A single-cylinder engine weighs 3.0 kips. Other data are as follows:

Unbalanced vertical forces are: Primary = 3.165 kips

Secondary = 0.94 kips

Soil is a very sandy clay, with a q_u of 4,000 psf. Find the required thickness of a solid concrete foundation block using base-plan dimensions of 5 × 10 ft, so that the amplitude of vibration is less than 0.001 in at either resonance or operating frequency.

20-2. Design the block of Prob. 20-1 so that resonance frequency occurs at 800 rpm (as close as practicable but not to exceed 800 rpm). What is the maximum amplitude of vibration using this foundation block?

20-3. For Example 20-2, if the weight of the foundation block is reduced to 650 kN, compute resonance frequency and the maximum amplitude of vibration.

20-4. Estimate the natural frequency of Example 20-2 if the soil is of such poor bearing quality that concrete piles 60 ft long are used to bedrock.

20-5. Write a computer program to prepare the curves shown in Fig. 20-7a and b for D values of 0.0 to 0.5 in increments of 0.1.

20-6. Prepare a curve similar to those shown in Fig. 20-11 for steel piles, using a steel stress of 9,000 psi, and for concrete piles, using a concrete stress of 1,350 psi.

20-7. Revise Example 20-3 so that the maximum horizontal translation (including rocking) is not more than 0.005 cm. Point A must remain at the same elevation.

20-8. For Example 20-3 assume a horizontal force parallel to both the x and z axis through point A of magnitude shown. Neglecting the rocking amplitude, what is the maximum sliding amplitude? (*Hint:* Vector sum of x and z displacements.)

A

GENERAL PILE-HAMMER AND
PILE DATA TABLES

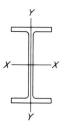

Table A-1. H-pile dimensions and section properties; fps units in dark type, metric units in light type

Desig-nation nominal size wt. in lb/ft cm kN/m	Area of section, in² cm²	Depth of section in cm	Flange		Web thick-ness in cm	Section properties			
			Width in cm	Thick-ness in cm		Axis $X - X$		Axis $Y - Y$	
						I in⁴ cm⁴	S in³ cm³	I in⁴ cm⁴	S in³ cm³
HP 14 × 117	**34.4**	**14.23**	**14.89**	**0.805**	**0.805**	**1,230**	**173**	**443**	**59.5**
36 × 1.71	222	36.14	37.81	2.045	2.045	51,200	2835	18,440	97.5
HP 14 × 102	**30.0**	**14.03**	**14.78**	**0.704**	**0.704**	**1,055**	**150**	**380**	**51.3**
36 × 1.49	194	35.64	37.55	1.788	1.788	43,700	2458	15,800	84.1
HP 14 × 89	**26.2**	**13.86**	**14.70**	**0.616**	**0.616**	**909**	**131**	**326**	**44.4**
36 × 1.30	169	35.20	37.33	1.565	1.565	37,880	2147	13,570	72.8
HP 14 × 73	**21.5**	**13.64**	**14.59**	**0.506**	**0.506**	**733**	**108**	**262**	**35.9**
36 × 1.07	139	34.65	37.05	1.285	1.285	30,550	1770	10,905	58.8
***HP 12 × 117**	**34.3**	**12.77**	**12.87**	**0.930**	**0.930**	**946**	**148**	**331**	**5.15**
30 × 1.71	222	32.46	32.69	2.362	2.362	39,375	2440	13,780	84.2
***HP 12 × 102**	**30.0**	**12.55**	**12.62**	**0.820**	**0.820**	**812**	**129**	**275**	**43.6**
30 × 1.49	194	31.97	32.05	2.082	2.082	33,800	2110	11,500	70.9
***HP 12 × 89**	**26.2**	**12.35**	**12.33**	**0.720**	**0.720**	**693**	**112**	**226**	**36.6**
30 × 1.30	169	31.37	31.31	1.829	1.829	28,850	1840	9,400	60.0
HP 12 × 74	**21.8**	**12.12**	**12.22**	**0.607**	**0.607**	**566**	**93.5**	**185**	**30.2**
30 × 1.08	141	30.78	31.03	1.542	1.542	23,560	1531	7,700	49.5
HP 12 × 53	**15.6**	**11.78**	**12.05**	**0.436**	**0.436**	**395**	**66.9**	**127**	**21.2**
30 × 0.77	100	29.92	30.60	1.107	1.107	16,400	1096	5,290	34.6
HP 10 × 57	**16.8**	**10.01**	**10.22**	**0.564**	**0.564**	**295**	**58.8**	**101**	**19.7**
25 × 0.83	108	25.43	25.97	1.433	1.433	12,280	964	4,200	32.3
HP 10 × 42	**12.4**	**9.72**	**10.08**	**0.418**	**0.418**	**211**	**43.4**	**71.4**	**14.2**
25 × 0.61	80	24.69	25.60	1.062	1.062	8,780	711	2,970	23.3
HP 8 × 36	**10.6**	**8.03**	**8.16**	**0.446**	**0.446**	**120**	**29.9**	**40.4**	**99.1**
20 × 0.53	68.4	20.40	20.72	1.133	1.133	4,995	490	1.680	16.2

* From Algoma Steel Co. (Canadian).

Table A-2. Typical pile-driving hammers from various sources. Consult manufacturer's catalogs for additional hammers, later models, other details

Model No.	Type*	Max rated energy ft-kips	Max rated energy kN-m	Working weight kips	Working weight kN	Ram weight§ kips	Ram weight§ kN	Stroke ft	Stroke m	Blow rate/min	Approx. length, ft
Drop hammers		Variable		0.50–10	2.2–45 kN			Variable‡		Very few	
Vulcan Iron Works West Palm Beach, Fla. 33407											
400C	SA	113.5	153.86	83	369	40	177.9	1.37	0.42	100	17
200C	DA	50.2	68.05	39	174	20	89.0	1.29	0.39	98	13
140C	DA	36.0	48.80	28	125	14	62.3	1.29	0.39	103	12
80C	DA	24.45	33.14	18	80	8	35.58	1.37	0.42	111	12
65C	DA	19.2	26.03	15	67	6.5	28.91	1.29	0.39	117	12
1–106	SA	15.0	20.33	9.7	43	5.0	22.24	3.0	0.91	60	13
7	DA	4.15	5.63	5.1	22.7	0.8	3.56	0.78	0.24	225	6
4N100	D	43.4	58.8	12.8	56.9	5.3	23.5	8.13	2.48	50–60	
1N100	D	24.6	33.4	7.6	33.8	3.0	13.3	8.13	2.48	50–60	
0	SA	24.38	33.04	16.0	71.2	7.5	33.4	3.25	0.99	50	15
McKiernan-Terry, Koehring-MKT Division Dover, N.J. 07801											
MBRS-7000	SA	361.15	489.57	161	712	88.0	391.4	4.10	1.25	40	28
OS-30	SA	90.0	122.0	50.5	225	30.0	133.4	3.0	0.91	60	21
S-20	SA	60.0	81.34	39.0	173	20.0	88.9	3.0	0.91	60	15
S-8	SA	26.0	35.25	18.3	81.4	8.0	35.6	3.25	0.99	53	14
S-5	SA	16.25	22.03	12.5	55.4	5.0	22.2	3.25	0.99	60	13
IHI-J44	D	79.4	107.63	21.5	95.6	9.7	43.2	8.17	2.49	42–70	15
DA55B	D	38.0	51.51	19.6	87.3	5.0	22.2	8.0	2.44	48	17
DE40	D	32.0	43.38	11.2	49.9	4.0	17.8	10.7	3.26	48	15
DE30	D	22.4	30.37	9.1	40.4	2.8	12.4	10.7	3.26	48	15

Table A-2 (*Continued*)

Model No.	Type*	Max rated energy ft-kips	Max rated energy kN-m	Working weight kips	Working weight kN	Ram weight§ kips	Ram weight§ kN	Stroke ft	Stroke m	Blow rate/min	Approx. length, ft
Raymond International, Inc.											
2801 South Post Oak Road, Houston, Tex. 77027											
30X	DA	75.0	107.67	52.0	231.2	30.0	133.4	2.5	0.76	70	19
5/0	SA	56.9	77.10	26.5	117.6	17.5	77.8	3.25	0.99	44	17
150C	DA	48.8	66.09	32.5	144.5	15.0	66.7	1.50	0.46	95–105	16
2/0	SA	32.5	44.06	18.8	83.4	10.0	44.5	3.25	0.99	50	15
80C	DA	24.5	33.14	17.9	79.5	8.0	35.6	1.38	0.42	95–105	12
65C	DA	19.5	26.43	14.7	65.3	6.5	28.9	1.33	0.41	110	12
1	SA	15.0	20.33	11.0	48.9	5.0	22.2	3.0	0.91	60	13
The Foundation Equipment Corp.											
New Commerstown, Ohio 43832 (distributor of Delmag Hammers)											
D55	D	117.175†	158.84	26.3	116.9	11.9	52.8	‡‡		36–47	18
D44	D	87.0	117.94	22.4	99.6	9.5	42.1			37–55	16
D36	D	73.78	100.02	17.8	79.1	7.9	35.3			37–53	16
D30	D	54.2	73.47	12.4	55.1	6.6	29.4			40–60	14
D22	D	39.78	53.93	11.1	49.4	4.8	21.5			40–60	14
D5	D	9.05	12.27	2.7	12.0	1.1	4.9			40–60	13
Link Belt											
Link Belt Speeder Division, FMC Corp., Cedar Rapids, Iowa 52406											
520	D	26.3	35.65	12.6	56.0	5.07	22.55	5.18	1.58	80–84	
440	D	18.2	24.67	10.3	45.8	4.0	17.79	4.35	1.39	86–90	
312	D	15.0	20.33	10.4	46.2	3.86	17.15	3.89	1.18	100–105	
180	D	8.1	10.98	4.6	20.5	1.72	7.67	4.70	1.43	90–95	

L. B. Foster Co. (distributor for Kobe Diesel Hammers)
7 Parkway Center, Pittsburgh, Pa. 15220

K150	D	281.3	381.33	80.5	358.0	33.1	147.2	8.5	2.59	45–60	28
K45	D	91.1	123.51	25.6	113.8	9.9	44.0	9.17	2.8	39–60	19
K42	D	79.0	107.09	24.0	106.7	9.26	41.2	8.5	2.59	45–60	19
K32	D	60.1	81.47	17.8	79.2	7.1	31.4	8.5	2.59	45–60	18
K25	D	50.7	68.73	13.1	58.2	5.5	24.5	9.17	2.80	39–60	18
K13	D	24.4	33.08	8.0	35.6	2.9	12.7	8.5	2.59	45–60	17

Berminghammer Corp., Ltd.
Hamilton, Ontario (Canada)

B500	D	75.0	101.67	16.5	73.4	6.9	30.7	12.0	3.66	40–60	
B225	D	25.0	33.89	6.8	30.2	2.9	12.7	9.7	2.96	40–60	

Mitsubishi International Corp.
875 North Michigan Avenue, Chicago, Ill. 60611

MB70	D	137.0	185.72	46.0	204.6	15.84	70.5	8.5	2.59	38–60	20
M43	D	84.0	113.87	22.6	100.5	9.46	42.1	8.5	2.59	40–60	16
M33	D	64.0	86.76	16.9	75.2	7.26	32.3	8.5	2.59	40–60	16
M23	D	45.0	61.00	11.2	49.8	5.06	22.5	8.5	2.59	42–60	14
M14	D	26.0	35.25	7.3	32.5	2.97	13.2	8.5	2.59	42–60	14

* SA = single-acting; DA = double-acting or differential acting; D = diesel.
† Energy varies from maximum shown to about 60 percent of maximum depending on stroke and soil.
‡ Variable stroke; stroke = energy out/weight of ram.
§ Ram weight or weight of striking part.

Table A-3. Steel sheetpiling sections produced in the United States. Available in ASTM A328 grade or ASTM A572 grades of 45, 50, and 55 ksi yield point steels ($F_y = 38.5$ ksi); fps units in dark type, metric units in light type.

	Section index	Driving distance in cm	Weight lb/ft kN/m	lb/ft^2 kN/m^2	Section modulus§ per pile in^3 cm^3	Moment of inertia§ per pile in^4 cm^4
	PZ-38 MZ-38 ZP-38	**18** 46	**57.0** 0.83	**38.0** 1.82	**70.2** 1,150	**421.2** 17,532 U B
	PZ-32 MZ-32 ZP-32	**21** 53	**56.0** 0.82	**32.0** 1.53	**67.0** 1,098	**385.7** 16,054 U B
	PZ-27 MZ-27 ZP-27	**18** 46	**40.5** 0.59	**27.0** 1.29	**45.3** 742	**276.3** 11,500 U B
	PSX-35*	**15$\frac{1}{4}$** 38.7	**44.5** 0.65	**35.0** 1.67	**3.3** 54	**5.2** B 216
	PSX-32†	**16$\frac{1}{2}$** 41.9	**44.0** 0.64	**32.0** 1.53	**3.3** 54	**5.1** U 212
	PS-32 MP-102 SP-7a	**15** 38.0	**40.0** 0.58	**32.0** 1.53	**2.4** 39.3	**3.6** 150 U B
	PS-28 MP-101 SP-6a	**15** 38	**35.0** 0.51	**28.0** 1.34	**2.4** 39.3	**3.6** 146 U B

Table A-3 (*Continued*)

Section index¶	Driving distance in cm	Weight lb/ft kN/m	lb/ft² kN/m²	Section modulus§ per pile in³ cm³	Moment of inertia§ per pile in⁴ cm⁴
PSA-28‡	**16** 41	**37.3** 0.54	**28.0** 1.34	**3.3** 54	**6.0** 250
MP-113					U
SP-5					B
PSA-23	**16** 41	**30.7** 0.45	**23.0** 1.10	**3.2** 52	**5.5** 229
MP-112					U
SP-4					B
PDA-27	**16** 41	**42.7** 0.62	**32.0** 1.53	**14.3** 234	**53.0** B 2,206
DP-2					
PDA-27	**16** 41	**36.0** 0.53	**27.0** 1.29	**14.3** 234	**53.0** U 2,206
MP-116					
PMA-22	**19⅝** 50	**36.0** 0.52	**22.0** 1.05	**8.8** 144	**22.4** 932
MP-115					U
AP-3					B

U = United States Steel Corporation; B = Bethlehem Steel Corporation.

¶ First group designation is current AISI standard; following are pre-1971 section designations.

* High-strength interlock (28 kips/in), $t_{web} = \frac{1}{2}$ in, same profile as PS-32.

† High-strength interlock (28 kips/in), $t_{web} = 29/64$ in, same profile as PS-32.

‡ Infrequently rolled.

§ Divide driving distance (sheet-pile width) into section modulus or moment of inertia (per pile) to obtain unit-width value.

Following sheet piles interlock with each other:

PSA-28, PSA-23, PDA-27, and PMA-22

PSX-35, PSX-32, PS-32, and PS-28

PZ-38, PZ-32, PSA-28, and PSA-23

PZ-27, PSA-28, and PSA-23

Table A-4. Representative steel pipe sections used for piles and caisson shells. Consult manufacturer's catalogs for more complete size listings

OD in (cm)	Wall thickness		Weight per lin ft, lb	Concrete, cu yd/ft	Area, sq in	
	in	cm			Concrete	Steel
	0.188	0.478	19.65	0.0187	72.7	5.81
10	0.219	0.556	22.85	0.0185	71.8	6.75
(25.40)	0.250	0.635	26.03	0.0182	70.9	7.66
	0.188	0.478	21.15	0.0217	84.5	6.24
10 3/4	0.250	0.635	28.04	0.0212	82.5	8.25
(27.31)	0.307	0.779	34.24	0.0207	80.7	10.07
	0.365	0.927	40.5	0.0203	78.9	11.91
	0.188	0.478	23.65	0.0273	106.1	6.96
12	0.219	0.556	27.52	0.0270	105.0	8.11
(30.48)	0.250	0.635	31.4	0.0267	103.8	9.25
	0.188	0.478	25.16	0.0309	120.3	7.42
	0.250	0.635	33.4	0.0303	117.8	9.82
12 3/4	0.312	0.792	41.5	0.0297	115.5	12.19
(32.38)	0.375	0.953	49.6	0.0291	113.1	14.58
	0.500	1.27	65.4	0.0279	108.4	19.24
	0.219	0.556	32.2	0.0372	144.5	9.48
	0.250	0.635	36.7	0.0368	143.1	10.80
14	0.312	0.792	45.7	0.0362	140.5	13.42
(35.56)	0.375	0.953	54.6	0.0355	137.9	16.05
	0.500	1.27	72.1	0.0341	132.7	21.21
	0.188	0.478	31.7	0.0493	191.7	9.34
	0.250	0.635	42.1	0.0485	188.7	12.37
16	0.312	0.792	52.4	0.0477	185.7	15.38
(40.64)	0.375	0.953	62.6	0.0470	182.6	18.41
	0.500	1.27	82.8	0.0454	176.7	24.35
	0.219	0.556	41.5	0.0623	242.2	12.23
18	0.250	0.635	47.4	0.0619	240.5	13.94
(45.72)	0.312	0.792	59.0	0.0610	237.1	17.34
	0.375	0.953	70.6	0.0601	233.7	20.76
	0.250	0.635	52.7	0.0768	298.6	15.51
20	0.312	0.792	65.7	0.0758	294.9	19.30
(50.80)	0.375	0.953	78.6	0.0749	291.0	23.12
	0.500	1.27	104.1	0.0729	283.5	30.63
	0.250	0.635	63.4	0.1116	433.7	18.7
24	0.312	0.792	79.1	0.1104	429.2	23.2
(60.96)	0.375	0.953	94.6	0.1093	424.5	27.8
	0.500	1.27	125.5	0.1067	415.5	36.9

Table A-4 (*Continued*)

OD in (cm)	Wall thickness		Weight per lin ft, lb	Concrete, cu yd/ft	Area, sq in	
	in	cm			Concrete	Steel
30	0.375	0.953	118.7	0.1728	672.0	34.9
(76.20)	0.500	1.27	157.5	0.1700	660.5	46.3
36	0.375	0.953	142.7	0.2510	975.8	42.0
(91.44)	0.500	1.27	189.6	0.2474	962.1	55.8
42	0.375	0.953	166.7	0.3436	133.0	49.0
(107)	0.500	1.27	221.6	0.3395	132.0	65.2
48	0.375	0.953	190.7	0.4509	175.3	56.1
(122)	0.500	1.27	253.7	0.4462	173.5	74.6
54	0.375	0.953	215	0.573	222.8	63.4
(137)	0.500	1.27	285	0.567	220.6	84.0

Table A-5. Typical prestressed-concrete pile sections*

Pile size diam.* in	Area concrete, sq in	Approx. weight, lb/ft	Minimum prestress force†	Number of Strands/pile‡ 7/16 in	Number of Strands/pile‡ 1/2 in	Inertia, in⁴	Perimeter, in	Pile capacity,§ kips 5,000 psi	Pile capacity,§ kips 6,000 psi
10	100	105	70 kips	4	4	833	40	100 kips	120 kips
12	144	150	101	6	5	1,728	48	144	172
14	196	204	137	8	6	3,201	56	196	235
16	265	276	186	11	8	5,461	64	265	318
18	324	335	227	13	10	8,748	72	324	388
20	400	415	280	16	12	13,333	80	400	480
22	484	505	339	20	15	19,521	88	484	580
24	576	600	403	23	18	27,648	96	576	690
20 HC	305	318	214	13	10	12,615	80	305	366
22 HC	351	365	246	14	11	18,117	88	350	420
24 HC	399	415	280	16	12	25,163	96	400	480
10	83	86	58 kips	4	4	555	33	83 kips	100 kips
12	119	125	83	5	4	1,134	40	119	142
14	162	170	113	7	5	2,105	46	162	194
16	212	220	148	9	7	3,592	53	212	254
18	268	280	188	11	8	5,705	60	268	320
20	331	345	232	14	10	8,770	66	330	396
22	401	420	281	16	12	12,837	73	400	480
24	477	495	334	19	15	18,180	80	477	572
20 HC	236	245	165	10	8	8,050	66	236	283
22 HC	268	280	188	11	8	11,440	73	268	322
24 HC	300	315	210	12	9	15,696	80	300	360

Additional data available from Prestressed Concrete Institute, 20 North Wacker Drive, Chicago, Ill., 60606.

* Voids in 20-, 22-, and 24-in-diameter hollow-core (HC) piles are 11, 13, and 15 in diameter, respectively, providing a minimum 4.5-in wall thickness.
† Minimum prestress force is based on a unit prestress of 700 psi after losses.   ‡ 7/16 and 1/2-in stress relieved strands with $P_{ult} = 31$ and 41.3 kips, respectively.
§ Design capacity based on an allowable point stress of $0.2 f'_c A_c$.

SELECTED COMPUTER PROGRAMS

The following computer programs are included for the convenience of the user. Refer to the comment cards at the beginning or interspersed through the program listings for documentation. Additional aid in using the programs can be obtained by referring to text examples which utilize the program(s) for variable identification and setting up the order of data cards.

The listings which follow were made from operating computer programs; output included in any text examples was obtained from the program as listed here.

B-1. Boussinesq equation for vertical stresses at any depth in a stratum

```
      C          J E BOWLES   PROGRAM TO COMPUTE VERTICAL SOIL PRESSURE
      C               BENEATH ANY SQUARE OR RECTANGULAR SPREAD FOOTING
      C               USING THE BOUSSINESQ EQUATION
      C     VARIABLES ARE AS FOLLOWS:  A,B = LENGTH AND WIDTH RESPECT OF FTG
      C        XO,YO,ZO = COORDINATES OF POINT WHERE PRESSURE IS TO
      C     BE FOUND--YO = VERTICAL (+) = DOWN, XO IS TO RIGHT, ZO IS
      C                  OUT OF PLANE (+)
      C            P = LOAD APPLIED TO FOOTING, DELTAX, DELTAZ = INCREMENT
      C            OF X AND Z RESPECTIVELY IN WHICH DIMENSIONS AX AND BZ ARE
      C              DIVIDED AND MUST BE EVENLY DIVISIBLE BY DELTAX AND DELTAZ
      C     Y-DIRECTION IS VERTICAL, X, Z DEFINE HORIZONTAL PLANE
      C     IDEPTH = NUMBER OF DEPTH INCREMENTS TO OBTAIN VERT PRESS PROFILE
   1         DIMENSION TITLE(20),QV(20)
   2    5000 READ(1,1000,END=150)TITLE
   3    1000 FORMAT(20A4)
   4         READ(1,1001)NOFTG,IDEPTH
   5    1001 FORMAT(16I5)
   6         WRITE(3,1999)
   7    1999 FORMAT('1')
   8         WRITE(3,2000)TITLE
   9    2000 FORMAT(  ////,T5,20A4,//)
  10         READ(1,1004)Y
  11         YO = 0.
  12         DO 130  LL = 1,IDEPTH
  13         YO = YO + Y
  14         DO 110 II=1,NOFTG
  15         IF(LL.EQ.1)READ(1,1004)AX,BZ,XO,ZO,DELTAX,DELTAZ,P
  16    1004 FORMAT(8F10.4)
  17         WRITE(3,2002)
  18    2002 FORMAT(T5,'VERTICAL PRESSURE AT POINT BENEATH SQUARE OR RECT. FOOT
         +ING',/,T10,'USING THE BOUSSINESQ EQUATION--ORIGIN OF COORDINATES I
         2S AT CENTER OF THE FOOTING',//)
  19         WRITE(3,2004)AX,BZ,P,XO,ZO,YO
  20    2004 FORMAT(T5,'A = ',F5.2,' FT', 4X, 'B = ', F5.2,' FT', 5X, 'P = ',
         +F7.2,' KIPS'  ,/, T5, 'COORDS: XO = ',F6.3,' FT', 2X,'ZO = ',F6.3
         1,' FT',/, T13, 'YO = ',F6.3, ' FT BELOW FTG ELEVATION',//)
  21         WRITE(3,2006)DELTAX, DELTAZ
  22    2006 FORMAT(T5, 'DELTAX = ',F5.3, 2X, 'DELTAZ = ',F5.3,///)
  23         R = 0.
  24         PI = 3.141592
  25         AREA = DELTAX*DELTAZ
  26         W = P/(AX*BZ)
  27         Z1= DELTAZ/2.- BZ/2.
  28         JF = BZ/DELTAZ
  29         IF = AX/DELTAX
  30         IK = 0
      C
  31         DO 100  J = 1,JF
  32         X1 = DELTAX/2. - AX/2.
  33         DO 50  I = 1,IF
  34         DR2 = (XO-X1)**2 + (ZO-Z1)**2 + YO**2
  35         DSQRT = SQRT(DR2)
  36         DROOT = 1./(DSQRT**5)
  37         R = R + DROOT
  38         IK = IK + 1
  39      50 X1 = X1 + DELTAX
  40         Z1 = Z1 + DELTAZ
  41     100 CONTINUE
      C
  42         C1 = ((3.*YO**3)/(2.*PI))*W*AREA
  43         QV(II) = C1*R
  44         WRITE(3,2008)QV(II),IK,II,NOFTG
  45    2008 FORMAT(T5,'PRESSURE AT POINT = ',F8.4, ' KSF',/,14X,'NO OF INCREME
         1NT USED = ',I4, 3X, 'FOOTING NO = ',I2, 3X, 'TOTAL NO OF FTGS = ',
         2 I2,/)
  46     110 CONTINUE
  47         QVT = 0.
  48         DO 120  I = 1,NOFTG
  49         QVT = QVT + QV(I)
  50     120 CONTINUE
  51         WRITE(3,2010)QVT,YO,NOFTG
  52    2010 FORMAT(//,10X,' FINAL TOT PRESSURE = ',F8.4,' KSF ',3X,
         +' DEPTH = ',F5.1, ' FT ', 4X,'NO FTGS USED = ',I3,//)
  53     130 CONTINUE
  54         GO TO 5000
  55     150 STOP
  56         END
```

B-2. Boussinesq equation for lateral earth pressure against a wall

```
        C       THIS PROGRAM USES THE BOUSINESQ EQUATION FOR LATERAL STRESSES TO FIND
        C      $THE LATERAL STRESSES ON A VERTICAL LINE IN A SOIL MASS.
        C       EACH FOOTING IS DIVIDED INTO SQUARES AND A DIMENSION GIVEN TO THE
        C      $SQUARE TO EQUAL OUT THE FOOTING DIMENSIONS.
        C
        C       A POINT LOAD IS A ONE BY ONE SQUARE OF SMALL SIZE.
        C       A LINE LOAD IS ANY LENGTH, BUT ONLY ONE SQUARE WIDE.
        C       THE TOTAL LOAD IS DIVIDED BY THE NUMBER OF SQUARES TO GIVE P-SQUARE.
        C       THE BOUSINESQ EQUATION IS THEN SOLVED FOR EACH SQUARE OF THE FOOTING
        C      $FOR ONE DEPTH OF THE VERTICAL LINE.
        C       THE ORIGIN IS TAKEN, LOOKING FROM WALL TO FOOTING, AS THE NEAR, LEFT
        C      $CORNER OF THE FOOTING.  IF ORIGIN TAKEN HERE DOP = = 0 AND SOLVE FOR CORNER
        C           CONTRIBUTION--IF DOP = = 1/2 WIDTH THEN FOOTING IS CENTERED AND BOTH
        C           SIDES CONTRIBUTE
        C
        C       X, Y, R, THETA, AND QH ARE WRITTEN OUT AS CHECKS FOR THE PROCEDURE.
        C       THIS PROGRAM THEN FINDS THE SUM OF QH AT EACH ONE FOOT OF DEPTH, KZ.
        C       THIS CAN BE MODIFIED TO EACH ONE INCH OF DEPTH.
        C       TWO SUCCEEDING  SUM OF QH VALUES ARE THEN AVERAGED AND PLACED ACTING ON
        C      $THE MIDPOINT VALUES OF KZ.
        C       THE TOTAL QH AT EACH DEPTH IS THEN ADDED TO THE OTHER QH'S TO GET THE
        C      $RESULTANT FORCE ON THE VERTICAL LINE.
        C       THE RESULTANT IS IN THE SAME UNITS AS FLOAD.
        C       XBAR IS THEN COMPUTED AS THE POINT OF APPLICATION OF THE RESULT.
        C       THE VERTICAL LINE IS ASSUMED TO BE A ONE FOOT WIDE SECTION OF A WALL.
        C       PROGRAM IS VALID FOR POINT LOAD, LINE LOAD, AND FOOTING LOAD.
        C       READ IN DESCRIPTION OF EACH PROBLEM.
        C
        C       L = NO OF DEPTH INCREMENTS (OR NODES) ON WALL TO BE ANALYZED
        C       DY = DIST ABOVE TOP OF WALL TO LOAD WHERE LOAD IS ON A SLOPE
        C                ABOVE WALL, ETC
        C       HTWALL = WALL HEITHT TO BE ANALYZED
        C      $     NSQW, NUMBER OF ELEMENT SQUARES AWAY FROM VERTICAL WALL.
        C      $     NSQL, NUMBER OF ELEMENT SQUARES PARALLEL TO WALL.
        C      $     WSQ, ELEMENT SQUARE DIMENSION, 1.0 FT., 0.5 FT., ETC.
        C      $     FLOAD, VERTICAL LOAD IN KIPS OR POUNDS.
        C      $     AMU, POISSON'S RATIO.
        C      $     DOP, DISTANCE OF PERPENDICULAR TO THE WALL, FROM THE ORIGIN.
        C      $     PDWF, DISTANCE FROM WALL TO FOOTING EDGE.
        C       LIST = 1 TO INCREASE OUTPUT
        C
        C       FOR 5 FT SQ FTG CENTERED ON WALL DOP = 2.5----IF FRONT EDGE IS 10-FT
        C       FROM WALL PDWF = 10.0
        C
   1            REAL*4 KZ
   2            DIMENSION TITLE(20)
        C
   3     20 READ (1,100,END=25) TITLE
   4    100 FORMAT (20A4)
   5            READ(1,200)UT1,UT3,UT4,UT5
   6    200 FORMAT(4(A4,6X))
        C
   7            READ(1,1004) L,NSQW,NSQL,LIST
   8            READ(1,1006)WSQ,FLOAD,AMU,DOP,PDWF,DY,HTWALL
   9   1004 FORMAT(16I5)
  10   1006 FORMAT(8F10.4)
        C
  11            KZ = L
  12            DDY = HTWALL/(KZ-1)
  13            PI=3.14159265
  14            P=FLOAD/(NSQW*NSQL)
  15            WRITE (3,51)
  16     51 FORMAT (1H1,T5,'BOUSSINESQ EQUATION FOR LATERAL EARTH PRESSURE',/)
  17            WRITE (3,50) TITLE
  18     50 FORMAT (1H0,T5,20A4,/)
  19            WRITE(3,2008)
  20   2008 FORMAT(3X,'DDY',6X,'L', 2X,'NSQW',2X,'NSQL',5X, 'WSQ',5X,'FLOAD',
         16X,'AMU',6X,'DOP',8X,'PDWF',3X,'P-SQUARE',4X,'HTWALL',2X,'LIST',/)
  21            WRITE(3,2010)DDY,L,NSQW,NSQL,WSQ,FLOAD,AMU,DOP,PDWF,P,HTWALL,LIST,
         1DY
  22   2010 FORMAT(1X,F6.3,1X,3I5,  7F10.4,1X,I5,/5X,
         1 'DIST BELOW LOAD WALL STARTS, DY = ',F8.3,//)
        C
  23            QP = 0.
  24            QMP = 0.
  25            CKZ = 0.
  26            TMU = 2.*AMU
        C
  27            DO 8  MM = 1,L
  28            A = MM - 1
  29            KZ = A*DDY + DY
        C
  30            SUMQH = 0.
        C
  31            IF(LIST.EQ.0.OR.MM.GT.2)GO TO 70
  32            WRITE (3,53)
  33     53 FORMAT (1H0,7X,'X',9X,'Y',9X,'R',5X,'THETA',7X,'QH')
  34     70 DO 7 N=1,NSQW,1
  35            DO 7 M=1,NSQL,1
  36            Y=PDWF+(WSQ*(N-0.5))
  37            X=(WSQ*(M-.5))-DOP
  38            IF(ABS(X).LE..01)X = 0.
  39            R=(Y**2+X**2)**.5
  40            IF(KZ.NE.0)THETA = ATAN(R/KZ)
  41            IF(KZ.EQ.0)THETA = PI/2.
  42            F = COS(THETA)
  43            G = SIN(THETA)
  44            IF(KZ.EQ.0)GO TO 44
  45            QH=(P/(2.*PI*(KZ**2)))*((3.*F*F*F*G*G)- ((1.-TMU)*F*F/(1.+F)))
```

B2 (*continued*)

```
46    44  IF(KZ.EQ.0)QH = 0.00
47        IF(QH.LE.0.0) QH=0.0000
48        IF(LIST.EQ.0.OR.MM.GT.2)GO TO 71
49        WRITE (3,45) X,Y,R,THETA,QH
50    45  FORMAT (1H ,5F10.4)
51    71  SUMQH = SUMQH + QH
52     7  CONTINUE
53        IF(MM.EQ.1)GO TO 73
54        CKZ = KZ - .5*DDY - DY
55        QHAVER = (SUMQH + BB)/2.
56        QM = QHAVER*DDY
57        QP = QP + QM
58        QMP = QM*CKZ + QMP
59    73  WRITE(3,54)KZ,UT1,SUMQH,UT5
60    54  FORMAT(//,5X,'DEPTH = ',F7.2,1X,A4,7X, 'SUM-QH AT DEPTH = ',F10.4,
61       1 1X,A4,/)
62        WRITE(3,56)CKZ,UT1,QMP,UT4,QM,UT3
63    56  FORMAT(/,6X,'AVER DEPTH = ',F6.2,1X,A4,3X,'MOMENT FROM TOP = ',F7.
64       12,1X,A4,    /,8X,  'HORIZ FORCE = ',F7.3,1X,A4,//)
65        BB = SUMQH
66     8  CONTINUE
67        XBAR = HTWALL - QMP/QP
68        WRITE(3,55)QP,UT3,XBAR,UT1
69    55  FORMAT( 9X,'TOTAL HORIZONTAL FORCE = ',F7.3,1X,A4   ,5X,'DIST FROM
70       1 BOTTOM OF WALL TO RESULT = ',F7.3,1X,A4,//)
68        GO TO 20
69    25  STOP
70        END
```

B-3. Influence factors F_1 and F_2 for computing the elastic settlement of a footing on a soil stratum of finite thickness

```
      C J E BOWLES INFLUENCE FACTORS FOR SETTLEMENT OF FINITE LAYER
      C
 1          DIMENSION EL(20),AD(20)
 2          WRITE(3,2000)
 3    2000  FORMAT('1',//,5X,'J E BOWLES SETTLEMENT INFLUENCE FACTORS',/)
 4          READ(1,1002)JJ,KK
 5    1002  FORMAT(16I5)
 6          READ(1,1004)(EL(I),I=1,JJ)
 7          READ(1,1004)(AD(I),I=1,KK)
 8    1004  FORMAT(8F10.4)
 9          PI = 3.141593
10          DO 500  I = 1,JJ
11          AL = EL(I)
12          WRITE(3,2002)
13    2002  FORMAT(/,4X,'AL', 6X,'D', 6X, 'F1', 6X,'F2',/)
14          DO 490  J = 1,KK
15          D = AD(J)
16          C1 = (1. + SQRT(AL*AL + 1.))*SQRT(AL*AL + D*D)
17          C2 = AL*(1. + SQRT(AL*AL + D*D + 1.))
18          C3 = (AL + SQRT(AL*AL + 1.))*SQRT( 1.0  + D*D)
19          C4 = AL + SQRT(AL*AL + D*D + 1.)
20          F1 = (1./PI)*(AL*ALOG(C1/C2) + ALOG(C3/C4))
21          IF(D.EQ.0.)F2 = 0.0
22          IF(D.EQ.0.)GO TO 250
23          F2 = (D/(2.*PI))*ATAN(AL/(D*SQRT(AL*AL + D*D + 1.)))
24    250  WRITE(3,2004)AL,D,F1,F2
25    2004  FORMAT(2X,F6.1,2X,F7.1,2(2X,F6.4))
26    490  CONTINUE
27    500  CONTINUE
28          STOP
29          END
```

B-4. Fox equations for influence coefficient to compute settlement of a footing embedded some depth in the soil

```
      C          J E BOWLES PROGRAM FOUNDATION ANALYSIS & DESIGN 2ND EDITION
      C          PROGRAM FOR FOUNDATION SETTLEMENTS OF FOUNDATIONS BENEATH GROUND
      C          SURFACE USING EQUATIONS OF FOX(1948)
      C
      C          XMU = POISSON'S RATIO
      C          C(M) = STRATUM DEPTH:  C/FTG WIDTH = C OF FOX
      C          B(M) = LENGTH OF FOOTING; B/FTG WIDTH = B OF FOX
      C          A = FTG WIDTH TAKEN AS 1.0; THUS RATIO B/A = > 1 AND IS "B" READ IN--
      C          D/B RATIO GREATER OR LESS THAN 1 BUT IS READ IN VALUE OF "C"
      C
   1             DIMENSION XMU(6),C(20),B(20)
   2             WRITE(3,2000)
   3   2000 FORMAT('1',//,4X,'J E BOWLES SETTLEMENT RATIOS',/)
   4             READ(1,1002)II,JJ,KK
   5   1002 FORMAT(16I5)
   6             READ(1,1004)(XMU(M),M=1,II)
   7             READ(1,1004)(C(M),M=1,JJ)
   8             READ(1,1004)(B(M),M=1,KK)
   9   1004 FORMAT(8F10.4)
  10             DO 500  M = 1,II
  11             X = XMU(M)
  12             B1 = 3. - 4.*X
  13             B2 = 5. - 12.*X + 8.*X*X
  14             B3 = -4.*X*(1. - 2.*X)
  15             B4 = -1. + 4.*X - 8.*X*X
  16             B5 = -4.*((1.-2.*X)**2)
  17             DO 490  N = 1,JJ
  18             WRITE(3,2002)
  19   2002 FORMAT(/,2X,'XMU', 5X,'B/A', 5X,'C/B',5X,'WC/WO',/)
  20             W = C(N)
  21             DO 489  L = 1,KK
  22             Z = B(L)
  23             R = 2.*W
  24             P1 = SQRT(1.+R*R)
  25             R2 = SQRT(Z*Z + R*R)
  26             R3 = SQRT(1. + Z*Z + R*R)
  27             R4 = SQRT(1. + Z*Z)
  28             Y1 = ALOG(R4+Z) + Z*ALOG((R4+1.)/Z) - ((R4**3-1.-Z**3)/(3.*Z))
  29             Y2 = ALOG((R3+Z)/R1) + Z*ALOG((R3+1.)/R2) - ((R3**3-R2**3-R1**3+R*
      1*3)/(3.*Z))
  30             Y3 = R*R*ALOG((Z+R2)*R1/((Z+R3)*R)) + (R**2/Z)*ALOG((1.+R1)*R2/((1
      1. + R3)*R))
  31             Y4 = R*R*(R1 + R2 - R3 - R)/Z
  32             Y5 = R*ATAN(Z/(R*R3))
  33             WC = B1*Y1+B2*Y2+B3*Y3+B4*Y4+B5*Y5
  34             WO = (B1+B2)*Y1
  35             RATIO = WC/WO
  36             RR = W/Z
  37             IF(RR.GT.1.0)RR = 1./RR
  38             WRITE(3,2004)X,Z,RR,RATIO,W,Z
  39   2004 FORMAT(1X,F4.2,3X,F6.1,2X,F6.3,2X,F6.4,10X,2F6.1,/)
  40    489 CONTINUE
  41    490 CONTINUE
  42    500 CONTINUE
  43        STOP
  44        END
```

B-5. Finite-element program for solving solving beams, rings, sheet piles, and laterally loaded piles as a beam on an elastic foundation

DOS FORTRAN IV 360N-FO-479 3-8   MAINPGM   DATE 11/28/75   TIME   13.40.17

```
              C   **   J E BOWLES  FINITE ELEMENT PROGRAM FOR BEAMS AND RINGS ON AN
              C        ELASTIC FOUNDATION, LATERALLY LOADED PILES FULLY OR PARTIALLY
              C        EMBEDDED AND SHEET-PILE WALLS - BOTH ANCHORED AND CANTI-
              C        LEVERED -- SOLUTIONS LIMITED TO 2 DEGREES OF FREEDOM
              C
              C        *************** VARIABLE IDENTIFICATION
              C   NP = NO OF P-X CODING       NM = NO OF SEGMENTS        NNZP = NO OF NON-ZERO
              C   P-MATRIX ENTRIES            NLC = NO OF LOADING CONDITIONS
              C   IPILE, IBEAM, ISPILE, IRING, IBRACE = TYPE OF PROBLEM IF > 1--- USE 0 FOR
              C   OTHER VARIABLES NOT USED    NCYC = MAX NO OF ITERATIONS ALLOWED
              C   JJS = NO OF SOIL SPRINGS READ AT #203      JTSOIL = NODE SOIL STARTS
              C   NONLIN = SWITCH TO INCLUDE NON-LINEAR SOIL EFFECTS      IAR = NO OF
              C   NODES WHERE ANCHOR ROD IS LOCATED FOR SHEET PILES      NZEROX = NO
              C   OF SPECIFIED X-DISPLACEMENTS (DEFL OR ROT)AT #97
              C   LIST = SWITCH TO LIST ASAT MATRIX      INERP = SWITCH TO READ VARIABLE
              C   INERTIA FOR PILES      INERB = SWITCH TO READ VARIABLE I FOR BEAMS
              C   IPRESS = SWITCH TO READ NO OF NODAL PRESSURES FOR SHEET PILES AND
              C     BRACED WALLS       E = MODULUS OF ELASTICITY (FT OR M)
              C   UNITWT = UNIT WEIGHT OF BEAM OR RING (K/CU FT)      XMAX = MAX NON-LINEAR
              C     SOIL DEFL (FT OR M)       FACP = DREDGE LINE REDUCTION FACTOR
              C   DEMB = INITIAL DEPTH OF EMBEDMENT OF SHEET PILE WALLS
              C   CONV = CONVERGENCE FACTOR FOR SHEET PILE WALLS       FAC2 = DEPTH INCREM.
              C     FOR SHEET PILES--1 OR 2 FT (.3 OR .6M)      AS,BS,EXPO = VALUES FOR
              C     SOIL MODULUS AS KS = AS + BS**EXPO
              C   H(J) = SEGMENT LENGTHS      B(J) = BEAM WIDTHS,      T(J) = BEAM THICK.
              C   INER(J) = INERTIA VALUES          WIDTH = WIDTH OF CONSTANT WIDTH PILES
              C     OR UNIT OF WIDTH OF SHEETPILE WALLS OR BEAMS      XINER = INERTIA
              C     FOR CONSTANT SECTIONS           NODAR(I) = ANCHOR ROD NODES
              C   ID, OD, DF = DIAMETERS AND THICKNESS OF RING SLAB #187
              C   XSPEC(I) = SPECIFIED VALUES OF KNOWN DISPLACEMENTS #.7
              C   K,J,PR(K,J) = NP,NLC AND P-MATRIX ENTRY #71
              C   PRESS(I) = VALUES OF NODAL SOIL PRESSURES #3050
              C
              C      USE CONSISTENT UNITS OF FT, M, K/SQ FT OF KPA, KCF OR KN/CU M
              C
              C      MISCELLANOUS PROGRAMMING INSTRUCTIONS
              C
              C      READ JTSOIL = 1 FOR RINGS AND BEAMS
              C   USER MUST HAND COMPUTE NODAL SOIL PRESSURES FOR SHEETPILE WALLS
              C     WHERE PHI-ANGLE CHANGES USE H(I), H(I+1) SAME SIZE TO COMPUTE
              C     NODAL FORCES CORRECTLY USING AVERAGE PRESSURE AT THAT NODE
              C          PROGRAM WILL COMPUTE NODAL FORCES FOR P-MATRIX
              C        READ XSPEC(I) IN FT OR RADIANS
              C     XSPEC(I) MAY HAVE SIGN AS WELL AS VALUE USE ZERO (0.0) OR VALUE > 0.
              C   USE JJS FOR SPRINGS ON BRACED WALLS FOR STRUTS
              C   NLC = 1 FOR NONLINEAR OR BRACED EXCAVATIONS
              C   SOIL SPRING FOR END NODES SAME AS INTERIOR NODES
              C     IBRAC = JTSOIL - 1 FOR BRACED WALLS -- IT IS A READ COUNTER AS WELL
              C             AS A PROGRAM SWITCH
              C     IF IBRAC > 0 USE ISPILE > 0 TO READ WIDTH AND XINER
              C
              C
 0001              DIMENSION P(42,3),ASAT(41,41),NPE(20,4),MEMNO(20),H(20),B(20),T(20
                  A),INER(20),EA(4,4),ES(4,4),ESAT(4,4),SK(30),F(4),X(42,3
                  B),SSK(22,2),SUMP(4),JCOUN(10),SOILP(42),NXZERO(42),SK1(20),SK2(20)
                  C,PS(43), XSPEC(42),PRESS(30),PP(42,3),PR(42,3),NODAR(10),TITLE(20)
              C
 0002        REAL*4 INER,ID
 0003        DOUBLE PRECISION UT5, UT6
              C
              C************
 0004        5000 READ(1,1000,END=150)TITLE,UT1,UT3,UT4,UT5,UT6
              C************
 0005        1000 FORMAT(20A4/3(A4,6X),A8,2X,A8)
 0006        WRITE(3,2000)TITLE
 0007        2000 FORMAT('1',//,T5,20A4)
              C************
 0008        READ(1,1002)NP,NM,NNZP,NLC,IPILE,IBEAM,ISPILE,IRING,IBRAC,NCYC
              C************
 0009        READ(1,1002) JJS,JTSOIL,NONLIN,IAR,NZEROX,LIST,INERP,INERB,IPRESS
              C************
 0010        1002 FORMAT(16I5)
              C************
 0011        READ(1,1004)E,UNITWT,XMAX,FACP,DEMB,CONV,FAC2
 0012        XDL1 = 0.0
              C************
 0013        1004 FORMAT(8F10.4)
              C************
 0014        READ(1,1004)AS,BS,EXPO
              C************
```

B5 (*continued*)

```
DOS FORTRAN IV 360N-FO-479 3-8        MAINPGM          DATE  11/28/75     TIME    13.40.17
0015              IF(IBEAM.GT.0)WRITE(3,2222)
0016         2222 FORMAT(//,T5,'SOLUTION FOR BEAM ON ELASTIC FOUNDATION *******',//)
0017              IF(IPILE.GT.0)WRITE(3,2224)
0018         2224 FORMAT(//,T5,'SOLUTION FOR LATERALLY LOADED PILE **********',//)
0019              IF(IRING.GT.0)WRITE(3,2225)
0020         2225 FORMAT(//,T5,'SOLUTION FOR RING FOUNDATION ***********',//)
0021              IF(ISPILE.GT.0)WRITE(3,2226)
0022         2226 FORMAT(//,T5,'SOLUTION FOR SHEET PILE WALL ***************',//)
0023              IF(IBRAC.GT.0)WRITE(3,2228)IBRAC
0024         2228 FORMAT(//,9X,'BRACED WALL COMPUTATIONS FOLLOW ******'3X,'IBRAC =
             1',I3,//)
             C
             C
0025              WRITE(3,2002)E,UT5,UNITWT,UT6,     JJS,JTSOIL,XMAX,UT1,AS,BS,EXPO,
             1UT6,NNZP,NLC,FACP,IPRESS
0026         2002 FORMAT(T5, 'MODULUS OF ELAST = ',F12.1,1X,A4, 3X,'UNIT WT = ',F7.4
             1, 1X,A8,          /, 4X,'NO NODES REQ CORRECT = ',I3, 3X,'NODE SOI
             2L STARTS FOR PILES = ',I3,/,4X, 'MAX NON-LIN SOIL DEFORM = ',F6.3
             3,1X,A4,//, 10X,
             4                       'SOIL MODULUS = ',F10.3,'+ ',F9.3,'*Z**',
             5F5.3,1X,A8,/, T5, 'NO OF NON-ZERO P-MATRIX ENTRIES = ',I3, 3X,'NO
             6OF LOAD CONDITIONS = ',I3,/, 8X, 'GROUND LINE REDUCTION FACTOR = '
             7 , F5.3,4X, 'NO OF NODAL PRESSURE ENTRIES = ',I3,//)
             C
             C            SET ALL ARRAYS TO ZERO WHICH MAY BE TROUBLESOME FOR WORKING
             C                 MORE THAN ONE TYPE OF PROBLEM AT SAME TIME
             C
0027              NMP1 = NM + 1
0028              NPP1 = NP + 1
0029              DO 3001  I = 1,NMP1
0030              H(I) = 0.
0031              T(I) = 0.
0032              B(I) = 0.
0033              INER(I) = 0.
0034              SK1(I) = 0.
0035              SK2(I) = 0.
0036                SSK(I,1) = 0.
0037                SSK(I,2) = 0.
0038         3001 CONTINUE
             C
0039              IF(IRING.GT.0)GO TO 187
             C
             C            NOTE PROVISION TO READ VARIABLE PILE WIDTHS, VARIABLE PILE
             C            MOMENT OF INERTIA AS WELL AS BEAM VARIABLE WIDTH AND MOMENT
             C            OF INERTIA
             C         IF IBRAC > 0 USE ISPILE > 0 TO READ MOMENT OF INERTIA
             C*************
0040              READ(1,1004)(H(J),J=1,NM)
             C
             C            PROVISION TO READ SOIL SPRINGS OF BRACED SHEETING ELEMENTS
             C            IBRAC = JTSOIL - 1 - NOTE SPRING SK2(IBRAC) = 0. IF PILING IS
             C            EMBEDDED DUE TO SPRING IN FRONT OF WALL
             C
0041              IF(IBRAC.GT.0)READ(1,1004)(SK1(I),SK2(I),I=1,IBRAC)
             C
0042              IF(IBEAM.GT.0.OR.INERP.GT.0)READ(1,1004)(B(J),J=1,NM)
             C
0043              IF(IBEAM.GT.0)READ(1,1004)(T(J),J=1,NM)
             C
0044              IF(IBEAM.GT.0.AND.INERB.GT.0.OR.INERP.GT.0)READ(1,1004)(INER(J),J=
             11,NM)
             C
0045              IF((IPILE.GT.0.AND.INERP.LE.0).OR.ISPILE.GT.0)READ(1,1004)WIDTH,XI
             1NER
             C
0046              IF(IAR.GT.0)READ(1,1002)(NODAR(I),I=1,IAR)
             C
             C*************
0047              IF(IBEAM.GT.0)GO TO 15
0048              IF(IPILE.GT.0.OR.ISPILE.GT.0.OR.IBRAC.GT.0)GO TO 17
             C*********************************
0049          15 DO 16  I = 1,NM
0050              IF(INERB.LE.0)INER(I)=B(I)*T(I)**3/12.
0051              SK(I) = AS
0052              IF(IBEAM.GT.0.AND.I.EQ.NM)SK(I+1) = AS
0053              SSK(I,1)=B(I)*H(I)*AS/2.
0054              SSK(I,2)=SSK(I,1)
0055              IF(I.EQ.1)SSK(I,1)=2.*SSK(I,1)
0056              IF(I.EQ.NM)SSK(I,2)=2.*SSK(I,2)
0057          16 CONTINUE
```

B5 (*continued*)

```
DOS FORTRAN IV 360N-FO-479 3-8          MAINPGM          DATE  11/28/75     TIME    13.40.17
0058           IF(IBEAM.GT.0)GO TO 17
         C                    READ RING DATA
         C          NOTE RING ALWAYS USES 20 SEGMENTS
         C               CAN INCLUDE FOOTING WEIGHT IF UNITWT > 0.
         C
         C************
0059     187 READ(1,1004)ID,OD,DF
         C************
0060           RM = SQRT((ID**2 + OD**2)/8.)
0061           AREA = (OD**2 - ID**2)*.785398
0062           AH = RM*.31416
0063           XINER = (OD-ID)*DF**3/24.
0064           RWIDTH = (OD-ID)/2.
0065           SPRING = AS*AREA/40.
0066           PW = UNITWT*AREA*DF/20.
0067           DO 189  I = 1,20
0068           B(I) = RWIDTH
0069           INER(I) = XINER
0070           T(I) = DF
0071           H(I) = AH
0072           SK(I) = AS
0073           SSK(I,1) = SPRING
0074     189 SSK(I,2) = SPRING
0075           WRITE(3,2232)OD,UT1,ID,UT1,RM,UT1,AREA,UT1
0076     2232 FORMAT(//,T5,'OUTSIDE DIAM = ',F8.3,1X,A4,3X,'INSIDE DIAM = ',F8.5
              1,1X,A4,/,T5, 'MEAN RADIUS = ',F8.3,1X,A4,3X,'AREA = ',F9.3,' SQ ',
              2A4,//)
         C
         C                    RING DATA READ AND STORED IN PRECEDING STATEMENTS
0077      17 WRITE(3,2006)UT1
0078     2006 FORMAT(/,T2, 'MEMNO', 3X, 'NP1',3X,'NP2',3X,'NP3',3X, 'NP4',3X,'LE
              1NGTH', 5X, 'WIDTH', 6X,'THICK', 3X, 'INERTIA,',A2, '**4',/)
0079           K = 1
0080           DO 18  I = 1,NM
0081           MEMNO(I) = I
0082           NPE(I,1) = K
0083           NPE(I,2) = K + 1
0084           NPE(I,3) = K + 2
0085           NPE(I,4) = K + 3
0086           IF((IPILE.GT.0.OR.ISPILE.GT.0).AND.INERP.LE.0)B(I) = WIDTH
0087           IF((IPILE.GT.0.OR.ISPILE.GT.0).AND.INERP.LE.0)INER(I) = XINER
0088           IF(IRING.GT.0.AND.I.EQ.NM)NPE(I,4) = NPE(1,2)
0089           IF(IRING.GT.0.AND.I.EQ.NM)NPE(I,3) = NPE(1,1)
0090           IF(IBEAM.LE.0.AND.IRING.LE.0)T(I)=0.
0091      18 K = K + 2
0092     160 DO 19  I = 1,NM
0093      19 WRITE(3,2007)MEMNO(I),(NPE(I,J),J=1,4),H(I),B(I),T(I),INER(I)
0094     2007 FORMAT(5I6,3F10.3,1X,F10.5)
         C                    PROGRAM COUNTERS FOLLOWS
0095           II = 0
0096           NLI = 0
0097           IK = 1
0098           K3 = 1
0099           LISTI = 1
0100                    ISP = 0
0101           JCOUN(K3) = 0
         C
0102           IF(IRING.GT.0)GO TO 238
0103           IF(IBEAM.GT.0)GO TO 170
         C
         C          COMPUTE SOIL MODULUS AT NODES FOR LAT PILES, SHEET PILES
         C               NOTE MODULUS COMPUTED AS BS*(Z**EXPO) NOT AS (BS*Z)**EXPO
         C
         C          RE-ENTER HERE FOR SHEET PILE WALLS WHEN EMBEDMENT DEPTH IS
         C               INCREASED (ISP>0)
         C
0104     5500 CONTINUE
0105           Z = 0.
0106           DO 200  K = JTSOIL,NMP1
0107           SK(K) = AS + BS*Z**EXPO
0108           IF(K.EQ.NMP1)H(K) = 0.
0109     200 Z = Z + H(K)
         C                    COMPUTE SOIL SPRINGS EACH END OF SEGMENT
0110     170 DO 201  K = 1,NM
0111           IF(IBEAM.GT.0)GO TO 22
0112           IF(K.LE.JTSOIL-1)GO TO 172
0113      22 IF(K.EQ.NM)GO TO 23
0114           SSK(K,1) = H(K)*B(K)*(7.*SK(K)+6.*SK(K+1)-SK(K+2))/24.
0115           SSK(K,2) = H(K)*B(K)*(3.*SK(K)+10.*SK(K+1)-SK(K+2))/24.
         C          DOUBLE END SPRING FOR BEAMS
         C          DECREASE 1ST SOIL SPRING BY FACP FOR PILES, SHEET-PILES
```

B5 *(continued)*

```
DOS FORTRAN IV 360N-FO-479 3-8          MAINPGM          DATE  11/28/75     TIME    13.40.17
            C
0116              IF(IBEAM.GT.0.AND.K.EQ.1)SSK(K,1) = 2.*SSK(K,1)
0117              IF(IPILE.GT.0.AND.K.EQ.JTSOIL)SSK(K,1) = FACP*SSK(K,1)
0118              IF(ISPILE.GT.0.AND.K.EQ.JTSOIL)SSK(K,1) = FACP*SSK(K,1)
0119              GO TO 201
0120          23 SSK(K,1) = H(K)*B(K)*(3.*SK(K+1)+10.*SK(K)-SK(K-1))/24.
0121             SSK(K,2) = H(K)*B(K)*(7.*SK(K+1)+6.*SK(K)-SK(K-1))/24.
0122             IF(IBEAM.GT.0.AND.K.EQ.NM)SSK(K,2) = 2.*SSK(K,2)
0123             GO TO 201
0124         172 IF(IBRAC.LE.0)GO TO 201
0125             SSK(K,1) = SK1(K)*H(K)/2.
0126             SSK(K,2) = SK2(K)*H(K)/2.
0127         201 CONTINUE
            C
            C              READ S-MATRIX MODIFICATIONS FOR JJS > 0
            C        TO CORRECT NODES FOR ANCHOR ROD OR SOIL SPRINGS--**EITHER READ A
            C         SPRING FOR BOTH MEMBERS CONNECTING AT A JOINT OR ZERO 1 SPRING AND
            C         READ FULL VALUE FOR OTHER MEMBER
            C
            C
            C
            C        ISP = CYCLE COUNTER FOR SHEET PILE WALLS WHEN EMBEDMENT DEPTH IS
            C             INCREASED AND IS COMPARED TO NO OF CYCLES REQUESTED (NCYC)
            C
0128              IF(ISP.GT.0)GO TO 238
0129              IF(JJS.LE.0)GO TO 238
            C
0130              DO 203  K = 1,JJS
0131         203 READ(1,1007)M,SSK(M,1),SSK(M,2)
0132        1007 FORMAT(I5,2F10.3)
0133             IF(IAR.GT.0)WRITE(3,2328)(NODAR(I),I=1,IAR)
0134        2328 FORMAT(//,9X,'ANCHOR ROD LOCATED AT NODE NOS =',8(2X,I2))
0135         238 WRITE(3,2008)
0136        2008 FORMAT(//,3X,'MEMNO',3X,'SOIL MODULUS',4X,'SPRINGS-SOIL, A.R., OR
                1STRUT',/)
0137             DO 204 K=1,NM
0138             IF(IRING.GT.0.OR.IBEAM.GT.0)SK(K) = AS
0139             IF(JTSOIL.GT.1.AND.K.LT.JTSOIL)SK(K) = 0.
0140             IF(K.GT.NM)SSK(K,1) = 0.
0141             IF(K.GT.NM)SSK(K,2) = 0.
0142         204 WRITE(3,2009)K,SK(K),SSK(K,1),SSK(K,2)
0143        2009 FORMAT(T5,I5,3X,F12.3,3X,2F12.3)
            C
            C        FORM ELEMENT AND GLOBAL MATRICES
            C
0144              DO 20  I = 1,NP
0145                PS(I) = 0.
0146                NXZERO(I) = 0
0147                XSPEC(I) = 0.
0148              DO 20  J = 1,NP
0149          20 ASAT(I,J) = 0.
            C
            C        RE-ENTRY POINT TO COMPUTE EA,ESAT TO FIND F-MATRIX
0150         250 I = 0
0151         240 I = I+1
0152             DO 25  K = 1,4
0153             DO 25  J = 1,4
0154          25 EA(K,J) = 0.
0155             EA(1,1)=1.
0156             EA(2,1) = 1./H(I)
0157             EA(2,2) = 1./H(I)
0158             EA(2,3) = 1.
0159             EA(3,2) = 1.
0160             EA(4,1) = -1./H(I)
0161             EA(4,2) = -1./H(I)
0162             EA(4,4) = 1.
0163             DO 26  K = 1,4
0164             DO 26  J = 1,4
0165          26 ES(K,J) = 0.
0166             ES(1,1) = 4.*E*INER(I)/H(I)
0167             ES(1,2) = .5*ES(1,1)
0168             ES(2,1) = ES(1,2)
0169             ES(2,2) = ES(1,1)
0170             ES(3,3) = SSK(I,1)
0171             ES(4,4) = SSK(I,2)
            C        BUILD SAT MATRIX
0172             DO 28  K = 1,4
0173             DO 28  J = 1,4
0174             ESAT(K,J) = 0.
0175             DO 28  L = 1,4
0176             ESAT(K,J) = ESAT(K,J) + ES(K,L)*EA(J,L)
0177          28 CONTINUE
```

B5 (*continued*)

```
DOS FORTRAN IV 360N-FO-479 3-8        MAINPGM         DATE   11/28/75     TIME    13.40.17
0178              IF(II.GT.0)GO TO 260
          C       BUILD ASAT MATRIX
0179              DO 30  K = 1,4
0180              DO 30  J = 1,4
0181              EASAT(K,J) = 0.
0182              DO 30  L = 1,4
0183           30 EASAT(K,J) = EASAT(K,J) + EA(K,L)*ESAT(L,J)
          C
          C
0184              DO 32  K = 1,4
0185              N1 = NPE(I,K)
0186              DO 32  J = 1,4
0187              N2 = NPE(I,J)
0188           32 ASAT(N1,N2) = ASAT(N1,N2) + EASAT(K,J)
0189          210 IF(I.LT.NM)GO TO 240
          C
          C                    NLI = COUNTER ACTIVATED IN NON-LINEAR SUBROUTINE
          C                       NLI > 0 IF XMAX EXCEEDED
          C
          C                    CHECK FOR ZERO BOUNDARY CONDITIONS
          C
          C        ONLY READ SPECIFIED DISPLACEMENTS ON FIRST CYCLE
          C
0190              IF(NZEROX.EQ.0)GO TO 101
0191              WRITE(3,2235)
0192         2235 FORMAT(//,T5,'PROBLEM BOUNDARY CONDITIONS OF ZERO AT FOLLOWING LOC
                 1ATIONS:',/)
          C                    READ NP-NUMBERS CORRESPONDING TO ZERO ROTAT OR DEFLECTION
0193              DO 99 I = 1,NZEROX
0194              IF(NLI.GT.0.OR.ISP.GT.0)GO TO 97
          C
0195              IF(NLI.EQ.0)READ(1,1002)NXZERO(I)
0196              IF(NLI.EQ.0)READ(1,1004)XSPEC(I)
0197           97 CONTINUE
0198              NPZI = NXZERO(I)
0199              WRITE(3,2236)NPZI
0200         2236 FORMAT(T6,'NP = ',I2)
0201              DO 98  N1 = 1,NP
0202              PS(N1) = PS(N1) - ASAT(N1,NPZI)*XSPEC(I)
0203              ASAT(N1,NPZI) = 0.
0204           98 ASAT(NPZI,N1) = 0.
0205              ASAT(NPZI,NPZI) = 1.
0206              PS(NPZI) = XSPEC(I)
0207           99 CONTINUE
          C
          C                    LIST USED HERE
          C
0208          101 IF(LIST.LE.0)GO TO 51
0209              IF(IK.GT.1)GO TO 51
0210              WRITE(3,2011)
0211         2011 FORMAT(//,T10,'THE ASAT MATRIX CORRECTED FOR ANY BOUNDARY CONDITIO
                 1NS AND 1000 FACTORED',/)
0212         3005 N1 = 1
0213           29 N2 = N1 + 9
0214              IF(N2.GT.NP)N2 = NP
0215              DO 34  I = 1,NP
0216              WRITE(3,2012)I,(ASAT(I,J),J=N1,N2)
0217           34 CONTINUE
0218         2012 FORMAT(T3,I3,1X,-3P10F9.2)
0219              IF(N2.NE.NP)WRITE(3,2011)
0220              N1 = N2 + 1
0221              IF(N2.LT.NP)GO TO 29
          C
          C                    INVERT ASAT MATRIX
0222           51 DO 65  K = 1,NP
0223              DO 61  J = 1,NP
0224           61 IF(J.NE.K)ASAT(K,J) = ASAT(K,J)/ASAT(K,K)
0225              DO 62  I = 1,NP
0226              IF(I.EQ.K)GO TO 62
0227              DO 62  J = 1,NP
0228              IF(J.EQ.K)GO TO 62
0229              ASAT(I,J) = ASAT(I,J) - ASAT(K,J)*ASAT(I,K)
0230           62 CONTINUE
0231              DO 63  I = 1,NP
0232           63 IF(I.NE.K)ASAT(I,K) = -ASAT(I,K)/ASAT(K,K)
0233              ASAT(K,K) = 1./ASAT(K,K)
0234           65 CONTINUE
          C
          C
          C                    END OF ASAT INVERSION
          C
```

B5 (*continued*)

```
DOS FORTRAN IV 360N-FO-479 3-8          MAINPGM          DATE  11/28/75     TIME     13.40.17
              C          TRANSFER CONTROL FOR SHEET PILE OR NONLINEAR ITERATIONS
              C                   USING NLI OR ISP
              C
0235                     IF(NLI.GT.0.OR.ISP.GT.0)GO TO 80
              C
              C          ZERO AND BUILD P-MATRIX
              C          NOTE P-MATRIX ZEROED TO NP + 1 --NOT TO JUST NP
              C
0236                DO 70  I = 1,NPP1
0237                PRESS(I) = 0.0
0238                DO 70  J = 1,NLC
0239                IF(I.EQ.1)SUMP(J) = C.
0240                PR(I,J) = 0.
0241                PP(I,J) = 0.
0242             70 P(I,J) = 0.
              C
              C          NNZP = NO OF NON-ZERO P-MATRIX ENTRIES READ IN
              C          USER MUST COMPUTE NODAL FORCES IF NNZP > 0
              C
0243                     IF(NNZP.LE.0)GO TO 3050
              C
              C*************
0244                DO 71  N = 1,NLC
0245                DO 71  I = 1,NNZP
0246             71 READ(1,1008)K,J,PR(K,J)
0247           1008 FORMAT(2I5,F10.4)
              C*************
              C
              C
              C
              C          USER MUST COMPUTE LATERAL PRESSURE BY HAND THEN READ VALUES IN
              C          IF IPRESS > 0    IPRESS = JTSOIL+1--ALWAYS READ 0.00 FOR LAST VALUE
              C          OF PRESS(I)
              C          LAST P(I,J) IS COMPUTED AT DREDGE LINE--NOT AT JTSOIL + 1 (IPRESS)
              C
              C          READ NODAL PRESSURES IF IPRESS > 0
              C
0248           3050  IF(IPRESS.EQ.0)GO TO 3100
0249                 READ(1,1004)(PRESS(I),I=1,IPRESS)
              C
0250                 L = 0
0251                 LPRESS = IPRESS - 1
0252                 DO 135  I = 1,LPRESS
0253                 L = L + 2
0254                 DO 135  J = 1,NLC
0255                 IF(I.EQ.1)PP(L,J) = H(1)*(2.*PRESS(1) + PRESS(2))/6.
0256                 IF(I.GT.1)PP(L,J) = H(I-1)*(2.*PRESS(I) + PRESS(I-1))/6. + H(I)*(2
                    1.*PRESS(I) + PRESS(I+1))/6.
0257            135 CONTINUE
              C          LOOP TO INCLUDE FOOTING WEIGHT IN P-MATRIX
0258           3100 DO 75  I = 1,NMP1
0259                 IF(IRING.GT.0)GO TO 73
0260                 PW = 0.
0261                 IF(IBEAM.EQ.0)GO TO 73
0262                 IF(I.EQ.1)GO TO 68
0263                 IF(I.EQ.NMP1)GO TO 72
0264                 PW = UNITWT*(H(I)*B(I)*T(I) + H(I-1)*B(I-1)*T(I-1))/2.
0265             68 IF(I.EQ.1)PW = H(I)*B(I)*T(I)*UNITWT/2.
0266             72 IF(I.EQ.NMP1)PW = H(I-1)*B(I-1)*T(I-1)*UNITWT/2.
0267             73 DO 74  J = 1,NLC
0268                 K = NPE(I,2)
0269                 L = NPE(I,1)
0270                 IF(I.EQ.NMP1)K = NPE(I-1,4)
0271                 IF(I.EQ.NMP1)L = NPE(I-1,3)
0272                 IF(IRING.GT.0.AND.I.EQ.NMP1)GO TO 75
              C
              C          SUMM P-MATRIX ENTRIES FROM READ IN VALUES PR(I,J), COMPUTED FROM
              C          PRESSURE VALUES PP(I,J) AND FROM EQUIVALENT ENTRIES FROM XSPEC
              C          BOUNDARY CONDITIONS PS(I)
              C
0273                 P(K,J) = PW + PS(K) + PR(K,J) + PP(K,J)
0274                 P(L,J) = PS(L) + PR(L,J)
0275                 IF(NZEROX.GT.0.AND.NXZERO(I).EQ.K)P(K,J) = PS(K)
0276                 IF(NZEROX.GT.0.AND.NXZERO(I).EQ.L)P(L,J) = PS(L)
0277                 SUMP(J) = SUMP(J) + P(K,J)
0278             74 CONTINUE
0279             75 CONTINUE
              C
              C          CONTROL TRANSFERRED TO HERE WHEN NLI OR ISP > 0
              C
0280             80 DO 76  J = 1,NLC
```

B5 (*continued*)

```
DOS FORTRAN IV 360N-FO-479 3-8          MAINPGM          DATE  11/28/75    TIME   13.40.17
0281          IF('PRESS.LE.0)WRITE(3,2014)UT4,UT3,J
0282     2014 FORMAT(//,T5,'LOAD MATRIX',1X,A4,' OR ',A4,' FOR NLC = ',I2,/)
0283          IF('PRESS.GT.0)WRITE(3,2015)UT4,UT3,UT5 ,J
0284     2015 FORMAT(//,T5,'LOAD MATRIX',1X,A4,' OR ',A4,T45,'WALL PRESSURES,
             1A8,' FOR NLC = ',I2,/)
0285          L = 1
0286          LL = 0
0287          NPO2 = NP/2
0288          DO 76  I = 1,NPO2
0289          LL = L + 1
0290          IF(IPRESS.GT.0.AND.I.LE.IPRESS)WRITE(3,2017)L, P(L,J),LL, P(LL,J),
             1 I,PRESS(I)
0291     2017 FORMAT(4X,I3,2X,F10.4,4X,I3,2X,F10.4,T44,I2,2X,F10.4)
0292          IF(IPRESS.LE.0.OR.I.GT.IPRESS)WRITE(3,2016)L,P(L,J),LL,P(LL,J)
0293     2016 FORMAT(4X,I3,2X,F10.4,4X,I3,2X,F10.4)
0294          L = L +2
0295       76 CONTINUE
         C
         C                       COMPUTE DISPLACEMENT MATRIX
0296          DO 81  I = 1,NP
0297          DO 81  J = 1,NLC
0298          X(I,J) = 0.
0299          DO 81  K = 1,NP
0300          X(I,J) = X(I,J) + ASAT(I,K)*P(K,J)
0301       81 CONTINUE
         C
         C                       COMPUTE ELEMENT FORCE MATRIX   USE ELEMENT SAT*X
         C
0302       82    CONTINUE
0303       77 SUM = 0.
0304          WRITE(3,2018)IK,UT4,UT3
0305     2018 FORMAT(//,T5,'FORCE MATRIX FOR NLC = ',I2,/,T15, 'MOMENTS, ',A4,18
             1X, 'SOIL REACTIONS, ',A4,/)
0306          II = 1
0307          NLI = 0
         C
         C                       REUSES ELEMENT CARD DATA TO RECOMPUTE EA, ESAT FOR F-MATRIX
         C
0308          GO TO 250
0309      260 DO 84  NN = 1,4
0310          F(NN) = 0.
0311          DO 83  KK = 1,4
0312          NS = NPE(I,KK)
0313          IF(NS.GT.NP)X(NS,IK) = 0.
0314       83 F(NN) = F(NN) + ESAT(NN,KK)*X(NS,IK)
0315          IF(NN.EQ.3.OR.NN.EQ.4)SUM = SUM + F(NN)
0316       84 CONTINUE
0317          WRITE(3,2020)MEMNO(I),(F(NN),NN=1,4)
0318     2020 FORMAT(T3,I3,2X,2F12.3,7X,2F12.3)
         C
         C                       REUSE REMAINING ELEMENT CARDS
         C
0319          IF(I.LT.NM)GO TO 240
0320          WRITE(3,2022)SUM,UT3,SUMP(IK),UT3
0321     2022 FORMAT(//,T5,'SUM SOIL REACTIONS = ',F8.3,1X,A4,   3X,'SUM APPLIED
             1 FORCES = ',F8.3,1X,A4,//)
         C
         C           INCREMENT DEPTH OF EMBEDMENT OF SHEET PILES BY 1-FT FOR ANCHORED OR
         C           2-FT FOR CANTILEVER WALLS UNTIL D.L. DEFL CONVERGES
         C
0322          IF(ISPILE.NE.1.OR.NCYC.EQ.1)GO TO 185
0323          ISP = ISP + 1
0324          XDL2 = X(2*JTSOIL,IK)
0325          WRITE(3,2023)XDL1,XDL2,XMAX,ISP,NCYC
0326     2023 FORMAT(//, 5X,'PREVIOUS D.L. DEFL = ',F8.5,2X, 'CURRENT D.L. DEFL
             1 = ',F8.5, 2X,'XMAX = ',F8.5,/,6X,'CURRENT CYCLE = ',I3,4X,'MAX NO
             2 OF CYCLES = ',I2,/)
0327          IF(ABS(XDL2-XDL1).LE.CONV)GO TO 185
0328          XDL1 = XDL2
0329          IF(ISP.GT.NCYC)GO TO 185
0330          FACC = NMP1 - JTSOIL
0331          DEMB1 = 0.
0332          DO 184  I = JTSOIL,NM
0333          H(I) = H(I) + FAC2/FACC
0334      184 DEMB1 = DEMB1 + H(I)
0335          WRITE(3,3043)FAC2,UT1,DEMB,UT1,DEMB1,UT1
0336     3043 FORMAT(//, 6X, 'EMBEDMENT DEPTH INCREASED ',F4.2,1X,A4, 3X, 'ORIGI
             1NAL EMBED. DEPTH = ',F8.4,1X,A4,/,  6X,'NEW EMBEDMENT DEPTH = ',F8
             2.4,1X,A4,/)
         C
0337          II = 0
         C
```

B5 (*continued*)

```
DOS FORTRAN IV 360N-FO-479 3-8        MAINPGM           DATE  11/28/75     TIME    13.40.17

             C
             C
             C       ***************************
0338                 GO TO 5500
             C
0339           185 WRITE(3,2024)UT1,UT5
0340          2024 FORMAT(T5,'NODE NO',3X, 'NODE ROTAT,RAD',3X,'NODE DEFL,',1X,A4,3X,
                  1 'SOIL PRESS,',1X,A4,/)
             C
             C
0341                 K4 = 0
0342                 J = 1
0343                 DO 93  I = 1,NM
0344                 NO = NPE(I,1)
0345                 N1 = NPE(I,2)
0346                 N2 = NPE(I,4)
0347                 SOILP(J) = SK(J)*X(N1,IK)
0348                 IF(X(N1,IK).GT.XMAX)SOILP(J)=SK(J)*XMAX
0349                 IF(IRING.GT.0.AND.SOILP(J).LT.0.)SOILP(J) = 0.
0350                 IF(IBEAM.GT.0.AND.SOILP(J).LT.0.)SOILP(J) = 0.
0351                 GO TO 90
0352           186 IF(I.EQ.NM.AND.N2.EQ.NPE(1,2))GO TO 90
0353                 IF(IBEAM.EQ.0.AND.IRING.EQ.0)GO TO 86
0354                 IF(X(N1,IK).LE.0.)SSK(I,1) = 0.
0355                 IF(X(N2,IK).LE.0.)SSK(I,2) = 0.
0356                 IF(X(N1,IK).LE.0..OR.X(N2,IK).LE.0.)K4 = K4 + 1
0357                 IF(X(N1,IK).LE.0.)NLI = NLI + 1
0358                 IF(X(N2,IK).LE.0.)NLI = NLI + 1
             C
             C       CHECK FOR NON-LINEAR SOIL DEFORM (IF NONLIN > 0)
             C          IF NZEROX > 0   DO NOT CHECK NON-LINEAR SOIL DEFORMATION
             C
0359            86 IF(NONLIN.EQ.0.OR.NZEROX.GT.0)GO TO 90
0360                 IF(I.GT.1.OR.ABS(X(N1,IK)).LE.XMAX)GO TO 87
0361                 IF(P(N1,IK).GE.0.)P(N1,IK) = P(N1,IK) - XMAX*SSK(I,1)
0362                 IF(P(N1,IK).LT.0.)P(N1,IK) = P(N1,IK) + XMAX*SSK(I,1)
0363                 SSK(I,1) = 0.
0364                 NLI = NLI + 1
0365            87 IF(ABS(X(N2,IK)).LE.XMAX)GO TO 89
0366                 IF(I.EQ.NM)GO TO 88
0367                 IF(P(N2,IK).GE.0.)P(N2,IK) = P(N2,IK) - XMAX*(SSK(I,2)-SSK(I+1,1))
0368                 IF(P(N2,IK).LT.0.)P(N2,IK) = P(N2,IK) - XMAX*(SSK(I,2)+SSK(I+1,1))
0369                 SSK(I,2) = 0.
0370                 SSK(I+1,1) = 0.
0371                 NLI = NLI + 1
0372                 GO TO 89
0373            88 IF(P(N2,IK).GT.0.)P(N2,IK) = P(N2,IK) - XMAX*SSK(I,2)
0374                 IF(P(N2,IK).LT.0.)P(N2,IK) = P(N2,IK) + XMAX*SSK(I,2)
0375            89 IF(ABS(X(N1,IK)).GT.XMAX.OR.ABS(X(N2,IK)).GT.XMAX)K4 = K4 + 1
0376            90 WRITE(3,2026)J,X(NO,IK),X(N1,IK),SOILP(J)
0377          2026 FORMAT(T7,I2, T17,F10.7,T33,F10.7,T50,F10.4)
             C
             C
0378                 IF(I.NE.NM.OR.J.GE.NMP1)GO TO 91
0379                 IF(IRING.GT.0.AND.I.EQ.NM)GO TO 93
0380                 NO = NPE(I,3)
0381                 N1 = N2
0382                 J = J + 1
0383                 SOILP(J) = SK(J)*X(N1,IK)
0384                 IF(X(N1,IK).GT.XMAX)SOILP(J)=SK(J)*XMAX
0385                 IF(IRING.GT.0.AND.SOILP(J).LT.0.)SOILP(J) = 0.
0386                 IF(IBEAM.GT.0.AND.SOILP(J).LT.0.)SOILP(J) = 0.
0387                 GO TO 90
0388            91 J = J + 1
0389            93 CONTINUE
0390                 IF(K4.EQ.0)GO TO 95
0391                 IF(K4.GT.0)K3 = K3 + 1
0392                 JCOUN(K3) = NLI-1
0393                 IF(K3.GT.1)WRITE(3,2041)NLI,K3,K4,JCOUN(K3),JCOUN(K3-1)
0394          2041 FORMAT(//,T5,'NLI =',I4,3X,'K3 =',I4,3X,'K4 =',I4,3X,'JCOUN(K3) ='
                  1 ,I4,3X,'JCOUN(K3-1) =',I4,//)
0395                 IF(JCOUN(K3).LE.JCOUN(K3-1))GO TO 149
0396                 IF(JCOUN(K3).GT.JCOUN(K3-1))II = 0
0397                 IF(JCOUN(K3).GT.JCOUN(K3-1))GO TO 238
0398            95   IK = IK + 1
0399                 IF(IK.LE.NLC)GO TO 82
0400           149 GO TO 5000
0401           150 STOP
0402               END
```

B-6. Passive-earth-pressure coefficients using theory of plasticity

```
      C
      C        J E BOWLES   PASSIVE EARTH PRESSURE COEFFICIENTS USING THEORY
      C                OF PLASTICITY--PROGRAM COURTESY OF ROSENFARB AND CHEN WITH
      C                         REVISIONS BY BOWLES
      C
      C            BOWLES PEVISIONS NECESSARY TO MAKE PROGRAM WORK ON IBM-SYSTEMS
      C
      C  COMPUTES NEW PARAMETERS
      C
1          COMMON RALPHA,RPHI,RBETA,X4,S2,S3,X5,X1,S1,X18,X20,X21,X22,X24
          1,CF
2          DOUBLE PRECISION DELT1,DELT4,SUBT1,SUBT2,F,SAVE,FACT
3          DOUBLE PRECISION SAVE1,SAVE4,DD
4          REAL*8 NORM
      C
      C
      C
      C  PASSIVE SANDWICH- TWO PARAMETERS
      C
      C
      C
5          CF=4.*ATAN(1.)/180.
      C          ORIGINAL VALUES K1 = 7,  K2 = 9,  K3 = 4,  K4 = 4      **************
6          K1 = 5
7          K4 = 3
8          CONST1 = .45
9          CONST2 = .15
      C       WALL ANGLE VARIED USING ALPHA----------
10         ALPHA = 70.
11         DO 1005  KL = 1,K4
12         ALPHA=ALPHA+10.
13         RALPHA=ALPHA*CF
14         CONST1 = CONST1 + .05
15         CONST2 = CONST2 + .05
16         WRITE(3,1111)ALPHA
17         WRITE(3,1112)
18         X1=SIN(RALPHA)
19         X2=COS(RALPHA)
      C          ORIGINAL ANGLE PHI = 5. AND VARIED BY INCREMENTS OF 5.
      C  VARIES THE FRICTION ANGLE PHI
20         PHI = 25.
21         DO 1000 I=1,K1
22         PHI = PHI + 5.
23         RPHI=PHI*CF
24         S1=COS(RPHI)
25         X4=TAN(RPHI)
26         S3=1./(1.+9.*X4*X4)
27         X22 = S1
28         X21 = X22
29         X20 = X21
30         X18 = X20
31         X24=SIN(RALPHA+RPHI)
      C               SET PROGRAM OUTPUT TO MATCH LEHIGH PUBLICATION
      C
      C          VARY BACKFILL ANGLE, BETA FROM 0 IN INCREMENTS OF 10 DEG BY K3
      C                    VARY WALL ANGLE DELTA
      C
32         K2 = PHI/5 + 1
33         DO 3000 J=1,K2
34         A = J - 1
35         DEL = A*5.
36         IL=1
37         IF(DEL .GT. PHI) GO TO 3000
      C  CHECKS TO SEE IF SHEARING GOVERNS
38         IF(DEL .EQ. PHI) IL=2
39         RDEL=DEL*CF
40         P3=DEL/PHI
41         X5=TAN(RDEL)
42         Q=COS(RDEL)
43         S2=X5*X2+X1
      C
      C               DETERMINE INITIAL STARTING VALUE
      C          MAKE INITIAL GUESS OF ANGLES--BOWLES' METHOD   NOT LEHIGH
      C
44         K3 = PHI/10
45         IF(K3*10.EQ.PHI)K3 = K3 + 1
46         DO 2000 K=1,K3
47         JJ = 1
48         B = K - 1
49         BETA = B*10.
50         IF(BETA .GT. PHI) GO TO 2000
51         IP=1
52         DD=5.
```

B6 (*continued*)

```
53              IFLAG=1
54              RBETA=BETA*CF
55              P4=BETA/PHI
56              TOTANG = ALPHA + BETA - .01
57                  THETA1 = TOTANG*CONST1
58                  THETA4 = TOTANG*CONST2
59        3999 LK=1
60
61              CALL RAM1(THETA1,THETA4,IL,DELT1,DELT4,F,LK)
        C
        C                     RE-ENTRY POINT TO CHECK FOR MINIMUM VALUE OF F (KP)
62        4000 CONTINUE
63              STOREF = F
64                  IF(IP.LE.41)GO TO 4001
65              THETA2 = 90.-ALPHA+THETA1-PHI
66              THETA3 = THETA2 + THETA4
67                  SUMTH = THETA1 + THETA4
68              APB = ALPHA + BETA
69                      IF(IP.GT.41)WRITE(3,1113)
70              IF(IP.GT.41)WRITE(3,3333)IP,PHI,DEL,BETA,THETA1,THETA4,SUMTH,THETA
               12,THETA3,APB,F
71                  IF(IP.GT.45)GO TO 2002
72        4001 LK = 2
        C COMPUTES THE GRADIENTS
73              CALL RAM2(THETA1,THETA4,IL,DELT1,DELT4,F,LK)
74              SAVE1=DELT1
75              SAVE4=DELT4
76        5555 CONTINUE
77              FACT=DD
78              NORM = DSQRT(DELT1*DELT1 + DELT4*DELT4)
79              SAVE=NORM
80              TEMP1=THETA1
81              TEMP4=THETA4
82              SUBT1 = FACT*DELT1/NORM
83              SUBT2 = FACT*DELT4/NORM
84              THETA1 = THETA1 - SUBT1
85              THETA4 = THETA4 - SUBT2
86          15  THETA2 = 90. - ALPHA + THETA1 - PHI
87              THETA3=THETA2+THETA4
88              IF(THETA4 .LT. 0.) GO TO 6990
89              IF((THETA1+THETA4) .GT. (ALPHA+BETA)) GO TO 6990
90              IF(THETA3 .EQ. BETA) GO TO 6990
        C COMPUTES NEW VALUE
91              LK=1
92              CALL RAM1(THETA1,THETA4,IL,DELT1,DELT4,F,LK)
        C CHECKS TO SEE IF FUNCTION STILL DECREASING
93              IP=IP+1
        C
        C                     MAKE MODIFICATIONS IF KP IS COMPUTING NEGATIVE
94        6990  IF(F.GT.0.)GO TO 7001
95              DD = 6.
96              JJ = JJ + 1
97              IF(JJ.EQ.2)DD = 4.
98              IF(IP.EQ.25)THETA1 = TEMP4
99              IF(IP.EQ.25)THETA4 = TEMP1
100             IF(JJ.LE.3)IP = 1
101             GO TO 3999
        C
        C                 NOTE THAT FOR CERTAIN ALPHA THE VALUE 300. MAY BE TOO SMALL
102       7001 IF(F.LT.300.)GO TO 7002
103             THETA1 = THETA1 - 5.
104             THETA4 = THETA4 - 10.
105             IF(IP.GT.30)THETA4 = THETA4 - 2.
106             IF(IP.GT.30)THETA1 = THETA1 - 1.
107             GO TO 3999
108       7002 IF(F .LT. STOREF) GO TO 4000
109             IF(STOREF.LT.0.)GO TO 4000
110       7003 CONTINUE
111             THETA1=TEMP1
112             THETA4=TEMP4
        C WE ARE NOW VBACK TO PREVIOUS PARAMETERS
113             IF(IFLAG .EQ. 1) GO TO 2001
        C
        C WE HAVE FOUND THE MINIMA
        C
        C                     USE PREVICUS VALUE OF F SINCE IT IS LEAST VALUE
114             F = STOREF/Q
115             THETA2=90.-ALPHA+THETA1-PHI
116             THETA3=THETA2+THETA4
117             SUMTH = THETA1 + THETA4
118             APB = ALPHA + BETA
119             WRITE(3,3333)IP,PHI,DEL,BETA,THETA1,THETA4,SUMTH,THETA2,THETA3,
               1APB,F,ALPHA
120             GO TO 2000
```

B6 (*continued*)

```
121    2002 WRITE(3,7777)
122         GO TO 2000
123    2001 CONTINUE
124         DD=1.
125         IFLAG=2
126         DELT1=SAVE1
127         DELT4=SAVE4
128         NORM=SAVE
129         GO TO 5555
130    2000 CONTINUE
131    3000 CONTINUE
132         WRITE(3,1113)
133    1113 FORMAT(//)
134    1000 CONTINUE
135    1005 CONTINUE
136    1111 FORMAT('1', 30X, 'PASSIVE EARTH PRESSURE COEFFICIENTS, ALPHA = ',
            1F7.2,//)
137    3333 FORMAT(3X,I2,3X,F4.1,4X, F4.1,3X,F4.1, 2X,F5.1,4X,F5.1,2X,F6.1,
            1 3X,F5.1,3X,F5.1,5X,F7.1,7X,G14.5, 9X,F5.1,/)
138    7777 FORMAT(//,9X, 'NO CONVERGENCE AFTER 46 CYCLES',//)
139    1112 FORMAT(/, 3X,'IP', 3X,'PHI',4X, 'DELTA', 3X,'BETA', 2X,'THETA1',2X,
            12X,'THETA4', 2X,'SUMTHET', 2X,'THETA2', 2X,'THETA3', 2X,'ALPH + BE
            2TA',10X,'KP',15X,'ALPHA',/)
140         STOP
141         END
       C
       C
142         SUBROUTINE GRAD(TT1,TT4,IL,G1,G4,F1,LK)
143         COMMON RALPHA,RPHI,RBETA,X4,S2,S3,X5,X1,S1,X18,X20,X21,X22,X24
            1,CF
144         DOUBLE PRECISION F1,P,G1,G4
145         ENTRY RAM1(TT1,TT4,IL,G1,G4,F1,LK)
146         T1=TT1*CF
147         T2=90.*CF-RALPHA+T1-RPHI
148         T4=TT4*CF
149         T3=T2+T4
150         Z6=T2
       C ALL PARAMETERS UNVARIED
151         X6=COS(RALPHA-T1)
152         Q6=X6
153         X7=SIN(RALPHA-T1)
154         Q7=X7
155         X8=SIN(T1)
156         Q8=X8
157         X9=X6
158         Q9=X9
159         X10=SIN(T2+RALPHA+RPHI)
160         Q10 = X10
161         X11=SIN(T2+RALPHA)
162         Q11=X11
163         X12=SIN(T3+RPHI)
164         X13=SIN(T3-RBETA)
165         X14=EXP(T4*X4)
166         X15=X14*X14*X14
167         X16=COS(T4)
168         X17=SIN(T4)
169         X19=SIN(RALPHA-T1+RBETA-T4)
170         X23=SIN(RALPHA+T2+2.*RPHI)
171         Q23=X23
172         GO TO 5001
173         ENTRY RAM2(TT1,TT4,IL,G1,G4,F1,LK)
       C THETA1 VARIED
174         L1=2
175         Z2=T1
176         T1=T1+CF
177         T2=90.*CF-RALPHA+T1-RPHI
178         T3=T2+T4
179         X6=COS(RALPHA-T1)
180         X7=SIN(RALPHA-T1)
181         X8=SIN(T1)
182         X9=X6
183         X10=SIN(T2+RALPHA+RPHI)
184         X11=SIN(T2+RALPHA)
185         X12=SIN(T3+RPHI)
186         X13=SIN(T3-RBETA)
187         X19=SIN(RALPHA-T1+RBETA-T4)
188         X23=SIN(RALPHA+T2+2.*RPHI)
189         GO TO 5001
190    2011 CONTINUE
       C THETA4 VARIED
191         L1=3
192         T1=Z2
193         T4=T4+CF
```

B6 (*continued*)

```
194          T2=Z6
195          T3=T2+T4
196          X6=Q6
197          X7=Q7
198          X8=Q8
199          X9=Q9
200          X10=Q10
201          X11=Q11
202          X12=SIN(T3+RPHI)
203          X13=SIN(T3-RBETA)
204          X14=EXP(T4*X4)
205          X15=X14*X14*X14
206          X16=COS(T4)
207          X17=SIN(T4)
208          X19=SIN(RALPHA-T1+RBETA-T4)
209          X23=Q23
210     5001 CONTINUE
211          B3=X6*(X15*(3.*X4*X16+X17)-3.*X4)
212          B4=X7*(X15*(3.*X4*X17-X16)+1.)
213          WDEXT=S2
214          A=(X11*X11)/(X1*X1*S1*S1)
215          WOAB=(X8*X11)/(X1*X1*X18)
216          WOCD=(X11*X11*X14*X14*X20*X19)/(X1*X1*X18*X18*X13)
217          V1=X1/X10
218          VO1=X6/X10
219          DWALL=X5*VO1
220          IF(IL .EQ. 2) GO TO 7007
221     7008 CONTINUE
222          V3=V1*X14
223          WDOAB=WOAB*V1*X9
224          WOLS=A*S3*V1*(B3+B4)
225          WDOCD=WOCD*V3*X12
226          F1=(1./(WDEXT-DWALL))*(WOLS+WDOAB+WDOCD)
227          GO TO 7009
228     7007 CONTINUE
229          DWALL = 0.
230          V1=(V1*X24*X10)/(X1*X23)
231          VO1=X6/X23
232          GO TO 7008
233     7009 CONTINUE
234          IF(LK .EQ. 1) GO TO 2006
235          IF(L1 .EQ. 2) GO TO 2007
236          GO TO 2008
237     2007 G1=F1-P
238          GO TO 2011
239     2006 P=F1
240          GO TO 4123
241     2008 G4=F1-P
242     4123 CONTINUE
243          RETURN
244          END
```

B-7. Finite-element program for solving mat foundations as a plate on an elastic foundation (may be used for pile caps and counterfort retaining walls)

```
C     J E BOWLES  DIRECT ELEMENT METHOD OF MAT ON ELASTIC FOUNDATION
C     B = MEM WIDTH, T = MEM THICK;  H,V = HOR AND VERT DIST FOR LENGTH
C     PROGRAM COMPUTES I AND J (POLAR I);  E = MOD OF ELAST; G = SHEAR MOD
C     NM = NO OF MEMBERS;  NP = SIZE OF MATRIX;  NNZP = NO OF NON-ZERO P-VALUES
C     NM = NO OF MEMBERS;  NP = SIZE OF MATRIX;  NNZP = NO OF NON-ZERO P-VALUES
C        TO BE READ IN;  NLC = NO OF LOADING CONDITIONS--USE 1 IF IREPT>0
C     NOX,NOY,NNODE = NO OF NODES IN X & Y DIRECTIONS & TOTAL NO OF NODES
C     CAN VARY SOIL MODULUS BUT NOT ACROSS ANY NODE.  READ SK > 0 AND
C        ** USE KSM = 0 AND SK1,SK2 = 0.0 FOR SUPPORTED PLATES
C        USE KSM = 0 AND SK = VALUE FOR CONSTANT SOIL MODULUS AND IT IS
C           NOT NECESSARY TO HAVE SK1,SK2 ON MEMBER DATA CARDS
C           IF USE CARD DEVELOPER PROGRAM READ KSM = 1 AND SK = 0.
C     LAST MEMBER CARD NM+1 IS PUNCHED AS SUCH IN COL 5
C*************
C     GRID MUST BE ORIENTED SO GO ACROSS THEN DOWN--KEEP BAND WIDTH
C        AS SMALL AS POSSIBLE BY ROTATION OF MAT.  NARROW BAND WIDTH IS
C        MAXIMUM MATRIX REDUCTION EFFICIENCY
C        ** FOR ODD SHAPED MATS THE COUNTERS USED FOR SUMMING JOINT MOMENTS
C        MAY NOT COUNT PROPERLY--CHECK OUTPUT UNLESS ELEMENTS ARE HORIZ OR VERT
C        LIST = 1 WILL LIST ASAT MATRIX; LISTB = 1 LISTS BAND MATRIX
C*****************
C        *** NOTE THAT CERTAIN COMBINATIONS OF E,G,T,AND FTG SIZE MAY
C           REQUIRE USING "DOUBLE PRECISION STIFF,P" TO OBTAIN A SATISFACTORY
C           SOLUTION--IF SUM VERT FORCES IS NOT EQUAL TO ZERO WITHIN ABOUT 0.5%
C           ERROR USE DOUBLE PRECISION--AFTER CHECKING MEMBER DATA CARDS
C**************
C     NOZX = NO OF ZERO X-VALUES TO DESCRIBE BOUNDARY CONDITIONS OF ZERO
C           ROTATION OR DEFLECTION
C        ***********
C        ***********      CAUTION    DO NOT USE IREPT AND NLC > 1 AT SAME TIME
C        *************    CAUTION    DO NOT USE NLC > 1 IF PROGRAM RECYCLES
C                           DUE TO NEGATIVE SOIL PRESSURES --PROGRAM MAY OR MAY
C                           NOT FUNCTION PROPERLY IF YOU DO
C*********
C                    IPCAP > 0 SOLVES PILE CAP PROBLEMS FOR BENDING MOMENTS
C                    USE ONLY APPROPRIATE PARTS OF OUTPUT OF 3-D PILE GROUP
C                    AS MAT INPUT TO SOLVE AS MAT PROBLEM
C           ************
C*************
C     IREPT,IREPTM INDEX TO REPEAT PROBLEM IF IREPT>0*** IREPTM IS MAX
C        NUMBER OF REPEAT CYCLES - USE THIS TO COMPARE ROTATION VS NO ROTAT.
C     IF IREPT > 0 MUST READ SEPARATE NOZP DATA, NZEROP DATA AND NEW P-MATRIX EN
C        TRIES TO BUILD P-MATRIX ANEW
C**********
C           ********
C        *** ALPH = COEFF FOR B/A RATIO OF GRID (OR PLATE) = .75 FOR B/A = 1
C     ALPH = 1.00 FOR B/A>1 AND MAY USE 1.05 FOR FIXED EDGE PLATE
C
C
C     ****************** THIS PROGRAM WILL ALSO SOLVE SIMPLY AND FIXED EDGE SUPPORT
C                        PLATES AS FOR EXAMPLE FLOOR SLABS OF CONCRETE
C
C        IF KSM = 0 AND SK1, SK2 ON CARDS WILL ZERO SK1, SK2 (FOR USING SAME MEM
C           CARDS FOR OTHER PROBS)
C        FOR GENERAL SOIL PROBS USING SK1, SK2 ON MEMBER DATA CARDS USE KSM=1
C        IF KSM EQ 0 AND SK > 0 SETS SK1, SK2 = SK WHERE SK = CONSTANT VALUE
C           OF SOIL MODULUS BENEATH FOOTING
C           FOR FLOOR SLABS, PILE CAPS SET KSM = 0 AND SK = 0
C
C
C        ITHICK = SWITCH TO READ VARIABLE THICKNESS FOR VERTICAL WALLS--USE 1
C           FOR WALLS--USE 0 FOR PLATES ON ELASTIC FOUNDATION AS CURRENTLY SET
C
C        I3EDGE = SWITCH FOR PLATES FIXED ON 3-EDGES TO AVOID RECYCLING
C
C
C
C
C
C
C
C
C
1           DIMENSION P(200,2),NPE(6),EA(6,5),ASAT(200),ES(5,5),STIFF(5200),
     1ESAT(5,6),EASAT(6,6),INDEX(10),F(5),                  XX(100),SOILP(100),
     2SUMP(5),NNP(100),ICOUN(10),R(5),S(5),LCOUN(100),MEM(100,4),XMX(110
     3,2),XMY(110,2),FTGWT(200),NZEROX(50),TITLE(20)
2           EQUIVALENCE (XMX(1),STIFF(2200)), (STIFF(2700),XMY(1)),(NNP(1),LCO
     1UN(1)), (STIFF(3201),MEM(1)),(ASAT(1),LCOUN(1))
C
3           DEFINE FILE 10(200,432,U,IR)
4           DOUBLE PRECISION UT5, UT6
C
5     6000 READ(1,1000,END=150) TITLE,UT1,UT3,UT5,UT6
```

B7 (*continued*)

```
  6      1000 FORMAT(20A4/A4,6X,A4,6X,A8,2X,A8)
         C
  7           READ(1,1005)NP,NM,NNZP,KSM,NLC,NBAND,NOX,NOY,NNODE,LIST,LISTB
              1,IREPT
  8           READ(1,1005)NOZX,IREPTM,IPCAP,ITHICK,I3EDGE
         C
  9      1005 FORMAT(16I5)
 10           NMP1 = NM + 1
 11       101 WRITE(3,1002) TITLE
 12      1002 FORMAT('1',//,T5,20A4)
 13           IF(I3EDGE.GT.0)WRITE(3,3102)
 14      3102 FORMAT(//,8X,'SOLUTION FOR PLATE FIXED ON 3-EDGES',/)
 15           IF(IPCAP.GT.0)WRITE(3,3103)
 16      3103 FORMAT(//,8X,'SOLUTION FOR PILE CAP',/)
         C
 17           READ(1,1007)E,G,SK,T,UNITWT,ALPH
 18      1007 FORMAT(8F10.4)
 19           WRITE(3,1008)E,UT5,G,UT5,SK,UT6,UNITWT,UT6,NOX,NOY,NNODE,NP,NOZX,
              1ALPH,NNZP
 20      1008 FORMAT(//T5,'E =',F10.1,1X,A7,5X,'G =',F10.1,1X,A7,5X,'SOIL MODULU
              1S =',F9.2,1X,A7/ T5,'UNIT WT =',F7.3,1X,A7//,T5,'NO OF X-NODES ='
              2,I5,3X,'NO OF Y-NODES =',I5,3X,'TOTAL NO OF NODES =',I6,/, T5,
              3'NO OF NP = ',I4, 3X,'NO OF ZERO NP = ',I3,2X, 'ALPHA COEFF FOR GJ
              4/L = ',F6.2,/, 10X, 'NO OF NON-ZERO P-MATRIX ENTRIES = ',I4,//)
 21           IR = 1
 22           II = 0
 23           III=0
 24           N3=1
 25           IZERO = 1
 26           WRITE(3,104)
 27       104 FORMAT(///,4X,'MEMNO',2X,'NP1',2X,'NP2',2X,'NP3',2X,'NP4',2X,'NP5',
              +2X, 'NP6', 7X,'H',8X,'V',8X,'LEN', 8X,'B', 8X,'T', 6X,'SM1',5X,'SM2',
              A7X,'INERTIA',3X,'POLAR I',/)
         C                RELOOP HERE TO REBUILD SAT
         C                RELOOP HERE TO REBUILD ELEMENT A AND SAT MATRICES TO
         C                     COMPUTE ELEMENT FORCES AFTER ASAT IS INVERTED
 28       106 IF(II.GT.0.OR.III.GT.0)GO TO 11
         C
         C                READ GRID MEMBER DATA AT ONE CARD PER MEMBER
         C
 29           IF(ITHICK.LE.0)READ(1,107)MEMNO,(NPE(I),I=1,6),H,V,B,SK1,SK2
 30           IF(ITHICK.GT.0)READ(1,107)MEMNO,(NPE(I),I=1,6),H V,B,T
         C
 31       107 FORMAT(7I5,3F10.4,2F7.2)
         C
         C        THESE 2 STATEMENTS ALLOWS USE OF SAME MEMBER DATA CARDS WITH NO SOIL
         C        SPRINGS IF KSM = 0 AND SK = 0.0 REGARDLESS OF WHAT IS ON MEM CARD
         C
 32           IF(KSM.EQ.0)SK1=0.                                            $$$
 33           IF(KSM.EQ.0)SK2=0.                                            $$$
 34           BPR = B
 35           AA = AMAX1(BPR,T)
 36           IF(MEMNO.EQ.NMP1)GO TO 21
 37           BB = AMIN1(BPR,T)
 38           C = AA/BB
 39           IF(C.LE.2.)BE = .087*C + .054
 40           IF(C.GT.2..AND.C.LE.4.5)BE = .0288*C + .174
 41           IF(C.GT.4.5)BE = .00218*C + .2902
 42        21 XJ = BE*AA*BB**3
 43           XI = B*T**3/12.
 44           XL = SQRT(H**2+V**2)
 45           IF(MEMNO.NE.NM+1)COSA=H/XL
 46           IF(MEMNO.NE.NM+1)SINA=V/XL
 47           IF(KSM.EQ.0.AND.SK1.EQ.0.)SK1 = SK
 48           IF(KSM.EQ.0.AND.SK2.EQ.0.)SK2 = SK
 49           SM1 = SK1*B*XL/4.
 50           SM2 = SK2*B*XL/4.
 51           FW1 = UNITWT*B*T*XL/4.                                        CHG1
 52           FW2 = UNITWT*B*T*XL/4.                                        CHG1
 53           WRITE(3,110)MEMNO,(NPE(I),I=1,6),H,V,XL,B,T,SM1,SM2,XI,XJ
 54       110 FORMAT(T5,7I5,3X,7F9.3,2X,2F8.5)
         C
         C            WRITE MEMBER DATA ON DISK WORK AREAS 5 AND 6
         C        WORK AREA 6 IS PERMANENT RECORD SO IF NLC > 1 AND MAT-SOIL SEPARATION
         C        HAS OCCURRED THE NODAL SOIL DATA IS NOT LOST FOR REMAINING NLC'S
         C
 55           WRITE (5)MEMNO,(NPE(I),I=1,6),H,V,XL,B,T,SM1,SM2,XI,XJ,COSA,SINA,S
              1K1,SK2
 56           WRITE (6)MEMNO,(NPE(I),I=1,6),H,V,XL,B,T,SM1,SM2,XI,XJ,COSA,SINA,S
              1K1,SK2
 57        11 IF(II.GT.0.OR.III.GT.0)READ(5)MEMNO,(NPE(I),I=1,6),H,V,XL,B,T,
              1SM1,SM2,XI,XJ,COSA,SINA,SK1,SK2
 58           IF(MEMNO.EQ.NM+1)GO TO 301
 59           IF(MEMNO.GT.1.OR.II.GT.0)GO TO 108
 60           DO 75  MM = 1,NP
 61           FTGWT(MM) = 0.                                                CHG1
 62        75 ASAT(MM) = 0.
 63           DO 103  I = 1,NP
 64       103 WRITE(10'IR)(ASAT(MM),MM=1,NP)
 65       108 DO 80  I = 1,6
 66           DO 80  J = 1,5
 67        80 EA(I,J) = 0.
         C        BUILD ELEMENT A MATRIX
 68           EA(1,1) = -SINA
 69           EA(1,3) =-COSA
 70           EA(2,1) = COSA
 71           EA(2,3) = -SINA
 72           EA(3,1) = 1./XL
 73           EA(3,2) = 1./XL
 74           EA(3,4) = -1.
 75           EA(4,2) = -SINA
 76           EA(4,3) = COSA
 77           EA(5,2) = COSA
 78           EA(5,3) = SINA
 79           EA(6,1) = -1./XL
```

B7 (*continued*)

```
250            WRITE(3,3334)LOADC
251     4500   WRITE(3,604)
252      604   FORMAT(//,T5,'MEMNO',9X,'BENDING MOMENTS',T41,'TORSION MOMENT'T57,
               + 'ELEMENT SOIL REACTIONS',//)
253            II = II+1
254            REWIND 5
255            GO TO 11
        C
        C      COMPUTE MEMBER FORCES
256      605   IF(MEMNO.EQ.NM+1)GO TO 957
257            LM = (MEMNO-1)*10
258      606   DO 607 I = 1,5
259            F(I) =0.
260            DO 609 K =1,6
261            N = NPE(K)
262      609   F(I) =F(I) +ESAT(I,K)*P(N,LOADC)
263      607   IF(I.EQ.4.OR.I.EQ.5)SUMR=SUMR+F(I)
264            WRITE(3,608)MEMNO,(F(I),I=1,5)
265      608   FORMAT(T5,I5,5F14.3)
        C                      WRITE SELECTED DATA IN STIFF(I) LOCATION FOR USE
        C                         IN SUMMING MOMENTS AT NODES
        C
266            STIFF(LM+1) = MEMNO
267            STIFF(LM+2) = NPE(1)
268            STIFF(LM+3) = NPE(4)
269            STIFF(LM+6) = H
270            STIFF(LM+7) = V
271            STIFF(LM+8) = F(1)
272            STIFF(LM+9) = F(2)
273            STIFF(LM+10) = F(3)
274            GO TO 11
275      957   WRITE(3,610)SUMP(LOADC),UT3,SUMR,UT3
276      610   FORMAT(//T10,'THE SUM OF VERTICAL CCL LOADS =',F10.4,1X,A4//,T12,
               A'SUM ELEMENT REACTIONS =',F10.4,2X,A4,///)
277            WRITE(3,423)    UT1
278      423   FORMAT(T8,'THE DEFORMATION MATRIX,',1X,A2,' OR RAD--EVERY 3RD = DE
               1FL'//)
279            WRITE(3,426)(I,P(I,LOADC),I=1,NP)
280      426   FORMAT((T4,8(I4,2X,F9.5),//))
        C
        C              NOTE IF IPCAP > 0 TRANSFER CONTROL TO 5426 TO AVOID RECYCLING DUE TO
        C                 NEGATIVE SOIL DEFLECTIONS SINCE NO SOIL IS USED FOR SOIL SPRINGS
        C                         TRANSFER CONTROL TO 5426 FOR PILE CAPS OR PLATES FIXED 3-EDGE
281            IF(IPCAP.GT.0.OR.I3EDGE.GT.0)GO TO 5426
        C
282            JCOUN = 0
283            J = 0
        C                      CHECK FOR MAT - SOIL SEPARATION (- DEFLECTIONS)
        C                         ICOUN(IZERO) = NUMBER OF - DEFLECTIONS FOUND
        C
284            DO 428 I = 1,NP
285            IF(I/3*3.NE.I)GO TO 428
286            J = J+1
287            XX(J) = P(I,LOADC)
288            IF(XX(J).LT.0.)JCOUN = JCOUN+1
289            IF(XX(J).LT.0.)NNP(JCOUN) = I
290      428   CONTINUE
291            IZERO = IZERO + 1
292            ICOUN(IZERO) = JCOUN
293            K = 0
294            M = 0
295            REWIND 5
296            IF(JCOUN.GT.0)REWIND 4
297            DO 431  LK = 1,NMP1
298            READ (5)MEMNO,(NPE(I),I=1,6),H,V,XL,B,T,SM1,SM2,XI,XJ,COSA,SINA,SK
               11,SK2
299            IF(LK.EQ.NMP1)GO TO 408
300            M = M+1
301            IF(H.EQ.0.)M=0
302            IF(H.EQ.0.)GO TO 429
303            IF(M.EQ.1)K=K+1
304            SOILP(K+1) = XX(K+1)*SK2
305            IF(M.EQ.1)SOILP(K) = XX(K)*SK1
306            K = K+1
307      429   IF(JCOUN.EQ.0)GO TO 431
308            DO 430  LM = 1,JCOUN
309            IF(NNP(LM).EQ.NPE(3))SK1 = 0.
310      430   IF(NNP(LM).EQ.NPE(6))SK2 = 0.
311            IF(SK1.EQ.0.)SM1 = 0.
312            IF(SK2.EQ.0.)SM2 = 0.
        C
        C              IF FOOTING SEPARATES, DATA FROM DISK AREA 5 IS WRITTEN ON DIS
        C              IF FOOTING SEPARATES, DATA FROM DISK AREA 5 IS WRITTEN ON
        C                 DISK WORK AREA 4 -- THIS STEP IS NECESSARY SO THAT
        C                 COMPARISON OF CURRENT AND PREVIOUS NUMBER OF SOIL
        C                         SPRINGS CAN BE MADE
313      408   WRITE (4)MEMNO,(NPE(I),I=1,6),H,V,XL,B,T,SM1,SM2,XI,XJ,COSA,SINA,S
               1K1,SK2
314      431   CONTINUE
315            WRITE(3,432)UT5
316      432   FORMAT(//,T8,'THE NODAL SOIL PRESSURE (',A7,')',//)
317            WRITE(3,426)(I,SOILP(I),I=1,J)
318     7513   II = 0
319            IR = 1
320            III = 1
321            WRITE(3,5425)IZERO,ICOUN(IZERO),ICOUN(IZERO-1),JCOUN,N3
322     5425   FORMAT(//,4X, 'IZERO = ',I3,3X, 'ICOUN(IZERO) = ',I3,3X, 'ICOUN(IZE
               1RO-1) = ',I3,3X, 'JCOUN = ',I3,3X, 'CURRENT LOAD CONDIT = ',I2,/)
        C
        C
        C              **************
        C
323            IF(ICOUN(IZERO).LE.ICOUN(IZERO-1))GO TO 5426
```

B7 (*continued*)

```
324              REWIND 4
325              REWIND 5
326              DO 433  M = 1,NMP1
327              READ (4)MEMNO,(NPE(I),I=1,6),H,V,XL,B,T,SM1,SM2,XI,XJ,COSA,SINA,SK
               11,SK2
328         433  WRITE (5)MEMNO,(NPE(I),I=1,6),H,V,XL,B,T,SM1,SM2,XI,XJ,COSA,SINA,S
               1K1,SK2
329              REWIND 5
330              GO TO 11
331        5426  CONTINUE
332              IFF1 = 1
      C*************
      C            **CONVERT ELEMENT FORCES TO BENDING MOMENTS AT NODES IN X & Y DIRECTIONS
      C*************
      C
      C            FIND NUMBER OF MEMBERS FRAMING INTO EACH NODE AND THEIR MEMBER NOS.
      C
333              WRITE(3,7089)
334        7089  FORMAT(///,T5, 'MEMBER NOS AND NO OF MEMBERS AT NODE',//)
335              DO 8011 KO=1,J
336              LCOUN(KO)=1
337              MEM(KO,3)=0
338              MEM(KO,4)=0
339              REWIND 5
340              DO 8010 N=1,NM
341              IF(LCOUN(KO).GT.4)GO TO 8008
342              READ (5)MEMNO,(NPE(I),I=1,6),H,V,XL,B,T,SM1,SM2,XI,XJ,COSA,SINA,SK
               11,SK2
343              IF(NPE(1).EQ.IFF1)MEM(KO,LCOUN(KO)) = N
344              IF(NPE(4).EQ.IFF1)MEM(KO,LCOUN(KO)) = N
345              IF(NPE(1).EQ.IFF1.OR.NPE(4).EQ.IFF1)LCOUN(KO) = LCOUN(KO) + 1
346        8010  CONTINUE
347        8008  LCOUN(KO)=LCOUN(KO) - 1
348              IFF1 = IFF1 + 3
349        8011  CONTINUE
350              WRITE(3,8021)(KO,(MEM(KO,I),I=1,4),LCOUN(KO),KO=1,J)
351        8021  FORMAT(T5,5I6,2X,I6)
      C            FIND NODE MOMENTS EACH SIDE OF NODE—IF ONLY 1 VALUE EXISTS OTHER
      C                VALUE IS MADE 0.0
352              DO 8040 KO=1,J
353              XMX(KO,1) = 0.
354              XMX(KO,2) = 0.
355              XMY(KO,1) = 0.
356              XMY(KO,2) = 0.
357              LM1 = (MEM(KO,1)-1)*10
358              LM2 = (MEM(KO,2)-1)*10
359              LM3 =(MEM(KO,3)-1)*10
360              IF(LCOUN(KO).EQ.2)GO TO 794
361              IF(LCOUN(KO).EQ.3)GO TO 803
362              IF(LCOUN(KO).EQ.4)GO TO 804
363         794  IF(STIFF(LM1+2).EQ.STIFF(LM2+2).AND.STIFF(LM1+6).GE..01)GO TO 793
364              IF(STIFF(LM1+3).EQ.STIFF(LM2+2).AND.STIFF(LM2+6).LE..01)GO TO 795
365              IF(STIFF(LM1+3).EQ.STIFF(LM2+2).AND.STIFF(LM2+6).GE..01)GO TO 796
366              IF(STIFF(LM1+3).EQ.STIFF(LM2+3))GO TO 797
      C          UPPER LEFT CORNER
367         793  XMX(KO,1) = STIFF(LM2+10) + STIFF(LM1+8)
368              XMY(KO,1) = STIFF(LM2+8) - STIFF(LM1+10)
369              GO TO 8040
      C          UPPER RIGHT CORNER
370         795  XMX(KO,1) = STIFF(LM1+9) + STIFF(LM2+10)
371              XMY(KO,1) = STIFF(LM1+10) + STIFF(LM2+8)
372              GO TO 8040
      C          LOWER LEFT CORNER
373         796  XMX(KO,1) = STIFF(LM1+10) - STIFF(LM2+8)
374              XMY(KO,1) = STIFF(LM1+9) - STIFF(LM2+10)
375              GO TO 8040
      C          LOWER RIGHT CORNER
376         797  XMX(KO,1) = STIFF(LM1+10) - STIFF(LM2+9)
377              XMY(KO,1) = STIFF(LM1+9) + STIFF(LM2+10)
378              GO TO 8040
379         803  IF(MEM(KO,2).EQ.MEM(KO,1)+1)GO TO 8001
380              IF(MEM(KO,2).EQ.MEM(KO,1)+1)GO TO 8005
381              IF(STIFF(LM1+3).EQ.STIFF(LM2+2).AND.(STIFF(LM2+2).EQ.STIFF(LM3+2))
               1.AND.(STIFF(LM1+6).LE.0.))GO TO 8004
382              GO TO 8003
      C          BENDING MOMENTS ALONG TOP
383        8001  IF(STIFF(LM1+9).GT.0.)XMX(KO,1) = STIFF(LM1+9) + STIFF(LM2+8)
384              IF(STIFF(LM2+8).GT.0.)XMX(KO,1) = XMX(KO,1) + STIFF(LM2+8)
385              IF(STIFF(LM3+10).GT.0.)XMX(KO,1) = XMX(KO,1) + STIFF(LM3+10)
386              IF(STIFF(LM1+9).LE.0.)XMX(KO,2) = STIFF(LM1+9)
387              IF(STIFF(LM2+8).LE.0.)XMX(KO,2) = XMX(KO,2) + STIFF(LM2+8)
388              IF(STIFF(LM3+10).LE.0.)XMX(KO,2) = XMX(KO,2) + STIFF(LM3+10)
389              XMY(KO,1) = STIFF(LM1+10)+STIFF(LM3+8)-STIFF(LM2+10)
390              GO TO 8040
      C          BENDING MOMENTS ALONG LEFT SIDE
391        8004  IF(STIFF(LM1+9).GT.0.)XMY(KO,1) = STIFF(LM1+9)
392              IF(STIFF(LM2+10).GT.0.)XMY(KO,1) = XMY(KO,1) - STIFF(LM2+10)
393              IF(STIFF(LM3+8).GT.0.)XMY(KO,1) = XMY(KO,1) + STIFF(LM3+8)
394              IF(STIFF(LM1+9).LT.0.)XMY(KO,2) = STIFF(LM1+9)
395              IF(STIFF(LM2+10).LT.0.)XMY(KO,2) = XMY(KO,2) - STIFF(LM2+10)
396              IF(STIFF(LM3+8).LT.0.)XMY(KO,2) = XMY(KO,2) + STIFF(LM3+8)
397              XMX(KO,1) = STIFF(LM1+10)-STIFF(LM2+8)+STIFF(LM3+10)
398              GO TO 8040
      C          BENDING MOMENTS ALONG RIGHT SIDE
399        8003  XMX(KO,1) = STIFF(LM1+10) - STIFF(LM2+9) - STIFF(LM3+10)
400              IF(STIFF(LM1+9).GT.0.)XMY(KO,1) = STIFF(LM1+9)
401              IF(STIFF(LM2+10).GT.0.)XMY(KO,1) = XMY(KO,1) + STIFF(LM2+10)
402              IF(STIFF(LM3+8).GT.0.)XMY(KO,1) = XMY(KO,1) + STIFF(LM3+8)
403              IF(STIFF(LM1+9).LT.0.)XMY(KO,2) = STIFF(LM1+9)
404              IF(STIFF(LM2+10).LT.0.)XMY(KO,2) = XMY(KO,2) + STIFF(LM2+10)
405              IF(STIFF(LM3+8).LT.0.)XMY(KO,2) = XMY(KO,2) + STIFF(LM3+8)
406              GO TO 8040
      C          BENDING MOMENTS ALONG BOTTOM
407        8005  XMY(KO,1) = STIFF(LM1+9) + STIFF(LM2+10) - STIFF(LM3+10)
408              IF(STIFF(LM1+10).GT.0.)XMX(KO,1) = STIFF(LM1+10)
```

B7 (*continued*)

```
409           IF(STIFF(LM2+9).LT.0.)XMX(KO,1) = XMX(KO,1) - STIFF(LM2+9)
410           IF(STIFF(LM3+8).LT.0.)XMX(KO,1) = XMX(KO,1) - STIFF(LM3+8)
411           IF(STIFF(LM1+10).LT.0.)XMX(KO,2) = STIFF(LM1+10)
412           IF(STIFF(LM2+9).GT.0.)XMX(KO,2) = XMX(KO,2) - STIFF(LM2+9)
413           IF(STIFF(LM3+8).GT.0.)XMX(KO,2) = XMX(KO,2) - STIFF(LM3+8)
414           GO TO 8040
415      804  LM4 =(MEM(KO,4)-1)*10
416           S(1) = STIFF(LM1+9)
417           S(2) = STIFF(LM2+10)
418           S(3) = STIFF(LM4+8)
419           S(4) = STIFF(LM3+10)
420           R(1) = STIFF(LM2+9)
421           R(2) = STIFF(LM3+8)
422           R(3) = STIFF(LM4+10)
423           R(4) = STIFF(LM1+10)
424           DO 820  M = 1,3
425           IF(S(M).GT.0.)XMY(KO,1) = XMY(KO,1) + S(M)
426           IF(S(M).LT.0.)XMY(KO,2) = XMY(KO,2) + S(M)
427           IF(R(M).GT.0.)XMX(KO,1) = XMX(KO,1) + R(M)
428      820  IF(R(M).LT.0.)XMX(KO,2) = XMX(KO,2) + R(M)
429           IF(S(4).LT.0.)XMY(KO,1) = XMY(KO,1) - S(4)
430           IF(S(4).GT.0.)XMY(KO,2) = XMY(KO,2) - S(4)
431           IF(R(4).GT.0.)XMX(KO,2) = XMX(KO,2) - R(4)
432           IF(R(4).LT.0.)XMX(KO,1) = XMX(KO,1) - R(4)
433     8040  CONTINUE
434           WRITE(3,8041)
435     8041  FORMAT(///,T5,'NODAL MOMENTS (COLUMN MOMENTS FROM P-MATRIX NOT INC
                1LUDED)',//,T13,'X-1',8X,'X-2',12X,'Y-1', 11X,'Y-2')
436           DO 8042 I=1,KO
437     8042  WRITE(3,8044)I,(XMX(I,J),J=1,2),(XMY(I,J),J=1,2)
438     8044  FORMAT(T3,I5,2X,4F12.4)
        C                      N3 AS LOAD CONDITION COUNTER LS INCREMENTED HERE
439     7551  IF(IREPT.LE.0)N3 = N3 + 1
440           WRITE(3,3457)IZERO,ICOUN(IZERO),ICOUN(IZERO-1),JCOUN,N3
441     3457  FORMAT(//,T5,'IZERO = ',I3,3X,'ICOUN(IZERO) = ',I3,3X, 'ICOUN(IZER
                1O-1) = ',I3,3X,'JCOUN = ',I3,3X,'NEXT LOAD CONDIT = ',I3,//)
        C**********
        C             CHECK ON LOOPING FOR NEGATIVE SOIL PRESSURE, REPEATING PROBLEM (IREPT >
        C               0), CHECK ON NLC
        C**************
442           IF(I3EDGE.GT.0)GO TO 6000
        C
443           IF(N3.LE.NLC.AND.(ICOUN(IZERO-1).LE.0).AND.IREPT.LE.0)GO TO 5500
444           IF(IREPT.GT.0)IREPT = IREPT + 1
445           IF(LOADC.EQ.NLC)GO TO 149
446     9975  FORMAT(7F10.7)
        C
        C             TRANSFER ORIGINAL DATA FROM DISK AREA 6 TO DISK AREA 5
447           REWIND 5
448           REWIND 6
449           DO 434  J = 1,NMP1
450           READ (6)MEMNO,(NPE(I),I=1,6),H,V,XL,B,T,SM1,SM2,XI,XJ,COSA,SINA,SK
                11,SK2
451      434  WRITE (5)MEMNO,(NPE(I),I=1,6),H,V,XL,B,T,SM1,SM2,XI,XJ,COSA,SINA,S
                1K1,SK2
452           REWIND 5
453           REWIND 6
454           II = 0
455           IZERO = 1
456           IR = 1

457           III = 1
458           GO TO 11
459      149  IF(IREPT.GT.1.AND.IREPT.LE.IREPTM)GO TO 9880
460           REWIND 5
461           REWIND 6
462           GO TO 6000
463      150  STOP
464           END
```

B-8. Culmann solution for active or passive earth force against a retaining wall

```
      C     J E BOWLES CULMANN METHOD FOR ACTIVE EARTH PRESSURE WITH OR WITHOUT
      C        COHESION AND CULMANN PASSIVE PRESSURE FOR COHESIONLESS SOIL
      C
      C     *****************   DO NOT USE CONCENTRATED OR OTHER SURCHARGES
      C              UNLESS READ STARTING RHOP > LAST CONC LOAD POINT   *********
      C
      C     CULMANN METHOD FOR ACTIVE EARTH PRESSURE FOR BOTH COHESIONLESS
      C        AND COHESIVE SOILS---ANY SHAPE BACKFILL AND WITH CONCENTRATED
      C        LOADS--READ END COORDS OF BRAKEN BACKFILL TO DEFINE ERREGULAR
      C        SHAPED BACKFILL
      C     XSTART AND YSTART = INITIAL COORDINATES AT BASE OF WALL
      C        XTOP,YTOP = COORDINATES DIRECTLY ABOVE XSTART AND YSTART IF ALPHA = 90
      C     CONCENTRATED LOADS MUST BE LOCATED LESS THAN 85 DEGREES FROM
      C        THE HORIZONTAL
      C     PROGRAM COMPUTES APPROXIMATELY WEDGES FROM 85 DEG TO ABOUT 10 DEG
      C        FROM HORIZONTAL AND SAVES LARGEST VALUE AND ITS LOCATION
      C     ALPHA NOT GREATER THAN 95 DEGREES
      C
      C     FAC = FACTOR FOR WALL ADHESION--USE 0.67 FOR CONCRETE TO SOIL AND
      C        USE FAC = 1.00 FOR SOIL TO SOIL
      C     RHOP = INPUT STARTING RHO-ANGLE FOR PASSIVE PRESS PROBS TO REDUCE ITERAT.
      C
      C        NOTE BACKFILL SLOPE DEFINED BY COORDINATES AND NOT BETA-ANGLE SINCE
      C           BACKFILL MAY BE IRREGULAR SHAPED
      C
      C
      C     ** VIEW WALL SYSTEM WITH SOIL TO RIGHT SIDE OF WALL
      C     NOL = NO OF LINES IN FIGURE NUMBERED LEFT TO RIGHT STARTING
      C     WITH TOP LINE--THEN GO BACK TO STARTING POINT AND NUMBER LINES DOWN
      C     WALL AND TO RIGHT FOR WALLS WITH FOOTINGS
      C
      C     NOCL = NO OF CONCENTRATED LOADS NUMBERED FROM LEFT TO RIGHT IN ORDER
      C
      C
      C     ****** NOTE DO NOT ADJUST INPUT COORDS FOR TENSION CRACK THIS IS DONE
      C           IN PROGRAM
      C
      C
      C     ITENCR = 0 SOLVES COHESIONLESS BACKFILL-- = 1 SOLVES COHESIVE BACKFILL
      C        WITH TENSION CDRACK
      C     WALL COHESION TAKEN AS 0.67 X COHESION OF BACKFILL
      C
      C     COH = COHESION, DELTA = WALL FRICTION ANGLE OR ANGLE OF FRICTION ON
      C        ON SLIP PLANE
      C
      C     WHEN SOLVING PROBLEMS WITH TENSION CRACK WALL MUST BE VERTICAL
      C
      C        SOLVE EITHER STEM FORCE OR TOTAL DRIVING FORCE BY PROPER PLACEMENT
      C     OF XSTART, YSTART, XTOP, YTOP
      C     IHEEL = SWITCH TO INCREMENT LARGER VALUES OF STARTING RHO
      C
      C           IA, IP = 1 FOR ACTIVE OR PASSIVE PRESSURE RESPECTIVELY
      C
      C     EITHER METRIC OR FPS UNITS MAY BE USED DEPENDING ON INPUT DATA
      C
      C        PROGRAM INCREMENTS RHO BY 1 OR 2 DEGS--AT CONC LOADS OR SURFACE
      C        LINE INTERSECTIONS X-COORD IS ADJUSTED IF WITHIN 0.1 TO ACTUAL
      C        COORDINATE--THE Y-COORD IS RECOMPUTED AND THE RHO ANGLE BASED ON
      C        THE S, Y-COORDS  IS COMPUTED--NEXT RHO-ANGLE INCREMENT IS FROM
      C        THE LAST COMPUTED RHO-ANGLE
      C
   1        DIMENSION X(15,2),Y(15,2),PA(120),LOAD(10),XL(10),YL(10),TITLE(20)
   2        DIMENSION YP(15,2),Z(15,2)
   3        DATA AA,PP/'ACT','PASS'/
      C
   4        REAL*4 LOAD
   5        DOUBLE PRECISION UT3,UT4
   6   6000 READ(1,1000,END=150)TITLE,UT1,UT2,UT3,UT4
   7   1000 FORMAT(20A4/2(A4,6X),2(A8,2X))
   8        SMNO = 0.1
   9        BIGNO = 999999999.
  10        WRITE(3,2000)TITLE
  11   2000 FORMAT('1',//,T5,20A4,//)
  12        READ(1,1002)NOL,NOCL,IHEEL,ITENCR,LIST,IA,IP
  13   1002 FORMAT(16I5)
  14        READ(1,1004)G1,PHI,ALPHA,DELTA,COH,FAC,RHOP
  15   1004 FORMAT(8F10.4)
  16        AAG = AA
  17        IF(IP.GT.0)AAG = PP
  18        WRITE(3,2002)AAG
  19   2002 FORMAT(/,4X, 'TYPE OF EARTH PRESSURE PROB = ',A4,//)
  20        WRITE(3,2004)
  21   2004 FORMAT(//,T5, 'LINE NOS AND END COORDS LEFT END FIRST',/,T5, 'LINE
      1 NO',3X,'X',10X,'Y',10X,'X',10X,'Y')
      C
      C        READ END COORDINATES OF ALL TOP LINES AT 1 CARD PER LINE
      C
  22        DO 20 I = 1,NOL
  23        READ(1,1005)LINNO,X(LINNO,1),Y(LINNO,1),X(LINNO,2),Y(LINNO,2)
  24   1005 FORMAT(I5,4F10.4)
  25        Z(LINNO,1) = Y(LINNO,1)
  26        Z(LINNO,2) = Y(LINNO,2)
  27     20 WRITE(3,2006)LINNO,X(LINNO,1),Y(LINNO,1),X(LINNO,2),Y(LINNO,2)
  28   2006 FORMAT(I5,5X,4F10.4)
      C
      C        READ STARTING COORDINATES
      C
  29        READ(1,1004)XSTART,YSTART,XTOP,YTOP
  30        WRITE(3,2008)NOL,NOCL,G1,UT3,PHI,ALPHA,XSTART,YSTART,XTOP,YTOP,DEL
      1TA,COH,UT4,FAC
  31   2008 FORMAT(//,T5, 'NO OF LINES = ',I3,3X,'NO OF CONC LOADS = ',I3,/, T5
      1, 'UNIT WT OF SOIL = ',F8.3,1X,A8, 3X, 'PHI ANGLE = ',F5.1,' DEG',
      2 3X, 'WALL ANGLE, ALPHA = ',F5.1,' DEG'/,T5,'XSTART = ',F8.3,3X,
      3 'YSTART = ',F8.3,3X, 'XTOP = ',F8.3,3X,'YTOP = ',F8.3, 2X,'DELTA =
      5 ',F8.3, ' DEG'/, 6X, 'SOIL COHESION = ',F6.3,1X,A8,3X,'WALL ADH
      6ESION FACTOR = ',F6.3,/)
  32        IF(NOCL.LE.0)GO TO 99
```

B8 (*continued*)

```
33            WRITE(3,2010)
34       2010 FORMAT(//,4X, 'THE CONC LOADS AND COORDS:',/)
         C
         C          READ CONCENTRATED LOADS AND LOAD COORDS--1 CARD PER CONC. LOAD
         C
35            DO 22  J = 1,NOCL
36            READ(1,1008)I,LOAD(I),XL(I),YL(I)
37         22 WRITE(3,2011)I,LOAD(I),XL(I),YL(I)
38       1008 FORMAT(I5,3F10.4)
39       2011 FORMAT(T5,I3,3X,3F10.3)
         C
40         99 PAMAX = 0.0
41            IF(IP.GT.0)PAMAX = 100000.
42            WEIGHT = 0.
43            TAREA = 0.
44            TAM1 = 0.
45            HCRAK = 0.
46            CW = 0.
47            CS = 0.
48            DDA = 0.
49            XI2 = 0.
50            YI2 = 0.
51            XI1 = XSTART
52            YI1 = YSTART
53            MO = 0
54            IF(NOCL.GT.0)MO = 1
55            X3 = XTOP
56            Y3 = YTOP
57            AL = ALPHA/57.2958
58            P1 = PHI/57.2958
59            DE = DELTA/57.2958
60            IF(ITENCR.LE.0)GO TO 24
61            HCRAK = 2.*COH/(G1*TAN(.7854 - P1/2.))
62            WRITE(3,2110)HCRAK
63       2110 FORMAT(//,8X,'HT OF TENSION CRACK = ',F8.4,/, 5X, 'ORIGINAL AND R
              1EVISED Y-COORDS:',/)
64            DO 23  I = 1,NOL
65            DO 18  J = 1,2
66            YP(I,J) = Y(I,J) - HCRAK
67         18 Z(I,J) = YP(I,J)
68            WRITE(3,2111)I,(Y(I,J),J=1,2),(YP(I,J),J=1,2)
69       2111 FORMAT(4X,'I = ',I3,2X, 'Y(I,J) = ',2F8.3,3X,'YP(I,J) = ',2F8.3)
70         23 CONTINUE
         C          ADJUST INPUT VARIABLES FOR COHESION
71            Y3 = Y3 - HCRAK
72            YTOP = YTOP - HCRAK
73            CW = (YTOP - YSTART)*COH*FAC
         C                  NOTE USE OF "FAC" FOR WALL ADHESION
74         24 CONTINUE
         C
         C
         C     START LOOP TO INCREMENT RHO FROM MAXIMUM VALUE TO MINIMUM
         C
75            II = 1
76            J = 1
77            JJ1 = 0
78            IF(ALPHA.GT.90.)RHO = 180.-ALPHA + 2.
79            IF(IHEEL.GT.1.OR.ALPHA.LE.90.)RHO = 90.-4.
80            IF(IP.GT.0)RHO = 90. - 15.
81            IF(RHOP.GT.0.)RHO = RHOP
         C     LIMITED TO 100 ITERATIONS
82            DO 200  I = 1,100
83            RHO = RHO - 1.
84            IF(RHO.LT.(PHI-2.).AND.IA.GT.0)GO TO 204
85            IF(RHO.LT.0..AND.IP.GT.0)GO TO 204
86            RH = RHO/57.2958
87            B = TAN(RH)
88            A = (Z(J,2)-Z(J,1))/(X(J,2)-X(J,1))
89            XI =(YSTART-Z(J,1)-B*XSTART+A*X(J,1))/(A-B)
         C
         C          TEST FOR LINE INTERSECTIONS OR LOAD COORDINATES
90            IF(ABS(XI-X(J,2)).LE.SMNO)XI = X(J,2)
91            IF(ABS(X(J,2)-X3).LE.XI-X3)XI = X(J,2)
92            IF(MO.EQ.0.OR.MO.GT.NOCL)GO TO 104
93            IF(ABS(XI-XL(MO)).LE.SMNO)XI = XL(MO)
94            IF((XL(MO)-X3).LE.(XI-X3))XI = XL(MO)
95        104 CONTINUE
96            IF(ABS(XI-X(J,2)).LE.SMNO.AND.J.LT.NOL)J = J + 1
97        106 IF(J.LE.NOL)YI = A*XI -A*X(J,1) + Z(J,1)
98            IF(XI.EQ.X(J,2))YI = Z(J,2)
99            IF(MO.EQ.0.OR.MO.GT.NOCL)GO TO 108
100           IF(ABS(XI-XL(MO)).LE..0001)YI = YL(MO) - HCRAK
         C
         C          COMPUTE ACTUAL RHO ANGLE AS BASED ON X, Y-COORDINATES
         C
101       108 RHO = (ATAN((YI-YSTART)/(XI-XSTART)))*57.2958
102           RH = RHO/57.2958
103           IF(ITENCR.GT.0)GO TO 116
104           IF(ABS(XI-X(J,1)).LE..001)GO TO 115
105           IF(LIST.GT.0.AND.II.GT.20)GO TO 116
106           IF(I.GT.1.AND.LIST.EQ.0)GO TO 116
107           IF(II.EQ.1.OR.(II/4*4.EQ.II))GO TO 115
108           GO TO 116
109       115 IF(LIST.GT.0)WRITE(3,3999)XI,YI,RHO,II,J,X3,Y3,X(J,2)
110      3999 FORMAT(//,5X,'AT WRITE 115',/,
              A       ,4X,'XI = ',F9.3, 3X, 'YI = ',F9.3,3X, 'RHO = ',F9.4, 3X
              1, 'ITER NO = ',I3,/,6X,'LINE NO = ',I2,3X,'X3 = ',F9.3, 3X,'Y3 =
              2',F9.3,2X,'X(J,2) = ',F8.3)
111       116 CONTINUE
112        41 DAREA = .5*((XSTART+X3)*(YSTART-Y3)+(X3+XI)*(Y3-YI)-(XSTART+XI)*(Y
              1START-YI))
113           TAREA = TAM1 + DAREA
114           TAM1 = TAREA
115           DDA = 0.
116           IF(ITENCR.LE.0)GO TO 36
117           DDA = (XI - X3)*HCRAK*G1
118        36 WEIGHT = DAREA*G1 + WEIGHT+ DDA
         C                  LOOP BACK TO HERE WHEN XI = XL(MO)
```

B8 (*continued*)

```
119      42 IF(JJ1.EQ.2)WEIGHT = WEIGHT + LOAD(MO)
120         IF(ITENCR)47,47,48
121      47 PA(II) = WEIGHT*SIN(RH-P1)/SIN(3.14159-AL+DE-RH+P1)
122         IF(IP.GT.0)PA(II) = WEIGHT*SIN(RH+P1)/SIN(3.14159-AL-DE-RH-P1)
123         GO TO 49
124      48 CONTINUE
125            AA = COS(RH)
126            BB = SIN(RH)
127            CC = COS(RH - P1)
128            DD = SIN(RH - P1)
129         E = SIN(DE)
130         F = COS(DE)
131         AZ = (XI - XSTART)/AA
132         CS = AZ*COH
133         WEIG = WEIGHT - CW
134         PA(II) = ((WEIG/CC) - CS*(BB/CC + AA/DD))/(E/CC + F/DD)
135      49 IF(MO.GT.NOCL.OR.NOCL.EQ.0)GO TO 120
136         IF(NOCL.GT.0.AND.ABS(XI-XL(MO)).LE..0001)GO TO 120
137         IF(NOL.GT.1.AND.XI.EQ.X(J,2))GO TO 120
138         IF(LIST.EQ.0.AND.II.NE.1)GO TO 75
139     120 CONTINUE
140         IF(ITENCR.GT.0)WRITE(3,3399)CW,CS,WEIG,DDA,X3
141    3399 FORMAT(4X,'WALL ADHES = ',F8.2,4X,'SLOPE ADHES = ',F8.2,2X,'WT LES
       1S CW = ',F9.2,/,9X, 'WT TEN CRACK BLOCK = ',F9.3,3X,'PREVIOUS X =
       2',F8.3)
142         LO = 0
143         IF(JJ1.EQ.2)LO = MO
       C        TEST FOR MAXIMUM PA FOR ACTIVE AND MINIMUM PA FOR PASSIVE PRESSURE
       C
144      75 CONTINUE
145         IF(IA.GT.0)GO TO 76
146         IF(PA(II).LT.PAMAX.AND.PA(II).GT.0.)PAMAX = PA(II)
147         IF(PAMAX.EQ.PA(II))RHOMAX = RHO
148         IF(PAMAX.NE.PA(II))GO TO 85

149         GO TO 77
150      76 IF(PA(II).GT.PAMAX)PAMAX = PA(II)
151         IF(PAMAX.EQ.PA(II))RHOMAX = RHO
152         IF(PAMAX.NE.PA(II))GO TO 85
153      77 XMAX = XI
154         YMAX = YI
155         AMAX = TAREA
156         WMAX = WEIGHT
157         CSMAX = CS
158         DMAX = DDA
159         XMAX1 = X3
160         YMAX1 = Y3
161         IMAX = II
       C
       C        CHECK FOR INCREASING OR DECREASING VALUES OF PA
       C
162      85 IF(IA.GT.0)GO TO 87
163         PERC = 1.25*PAMAX
164         IF(NOCL.EQ.0.AND.NOL.EQ.1)PERC = 1.08*PAMAX
165         IF(PA(II).GT.PERC.AND.MO.GT.NOCL)GO TO 204
166         GO TO 86
167      87 PERC = .75*PAMAX
168         IF(NOCL.EQ.0.AND.NOL.EQ.1)PERC = 0.9*PAMAX
169         IF(PERC.LE.0..OR.PA(II).LE.0.)GO TO 86
       C
170         IF(PA(II).LT.PERC.AND.MO.GT.NOCL)GO TO 204
       C
       C        CHECK TO REPEAT PROBLEM IF XI IS AT A CONC LOAD
171      86 IF(NOCL.LE.0.OR.JJ1.EQ.2.OR.MO.GT.NOCL)GO TO 43
172         IF(ABS(XL(MO)-XI).GT..00001)GO TO 43
173         II = II+1
174         JJ1 = 2
175         GO TO 42
176      43 CONTINUE
177         X3 = XI
178         Y3 = YI
179         IF(MO.GT.NOCL.OR.NOCL.EQ.0)GO TO 190
180         IF(NOCL.GT.0.AND.ABS(XL(MO)-XI).LE..00001)MO = MO + 1
181     190 JJ1 = 0
182         II = II+1
183     200 CONTINUE
184     204 CONTINUE
185         WRITE(3,2018)PAMAX,UT2,RHOMAX,IMAX
186    2018 FORMAT(//,T5, 'THE MAXIMUM VALUE OF ACTIVE EARTH PRESSURE = ',F8.3
       1,1X,A4,/,T5, 'THE RHO ANGLE FROM HORIZ = ',F8.3,' DEG',/,T5, 'THE
       2ITER NO = ',I3,///)
187         IF(LIST.LE.0)GO TO 191
188         WRITE(3,2015)II,RHO,WEIGHT,CW,CS,TAM1,DAREA,TAREA
189    2016 FORMAT(//,4X,'LAST COMPUTATION ITER = ',I3,/, 6X, 'LAST RHO = ',F8
       1.3, 3X,'WEIGHT = ',F8.3,3X,'CW = ',F8.3,3X,'CS = ',F8.3,/, 6X,'N-
       2TO-L AREA = ',F8.3, 3X,'DAREA = ',F8.3,3X,'TOTAL AREA = ',F8.3,/)
190         WRITE(3,2012)XMAX,YMAX,AMAX,WMAX,CSMAX,DMAX,HCRAK,CW,XMAX1,YMAX1
191    2012 FORMAT(4X, 'FAILURE WEDGE DATA:',/, 6X,'X-COORD = ',F9.4,3X, 'Y-CO
       1ORD = ',F9.4,/, 6X,'TRIANG AREA = ',F9.4, 3X,'WT OF TOTAL WEDGE =
       2',F9.3,/, 6X,'SLOPE COHES = ',F9.4,3X, 'WT OF TENSION BLOCK AREA
       3= ',F9.4,/,8X,'INIT HT OF TEN CRACK = ',F8.4, 3X, 'WALL ADHES FORC
       4E = ',F8.4,/,8X,'PREVIOUS X-COORD = ',F9.4, 3X,'Y-COORD = ',F9.4)
192     191 GO TO 6000
193     150 STOP
194         END
```

B-9. Retaining-wall solution using ACI 318-71 and USD. Wall may have front- or back-face batter. Program also considers AASHO loss of heel pressure

```
       C
       C         J E BOWLES--RETAINING WALL BY USD 318-71
       C         REVISED PROGRAM 5-18-72     BATTER MAY BE ON EITHER FRONT OR BACK FACE
       C         THIS PROGRAM MAY BE USED TO OBTAIN INITIAL BASE SLAB PROPORTIONS FOR
       C           A COUNTERFORT RETAINING WALL          USE CFORT = 1.0
       C
       C              VARIABLE IDENTIFICATION
       C
       C         HLOSS = 0 FOR NO LOSS OF HEEL PRESSURE;  HLOSS = 1 FOR LOSS OF HEEL
       C            PRESSURE AS FOR AASHO AND AREA METHOD
       C         H = WALL STEM HT;  G1 = UNIT WT BACKFILL;  G2 = UNITWT BASE SOIL
       C         D = DEPTH OF FTG;  COH = COHES OF BASE SOIL;  PHI2 = PHI-ANGLE BASE SOIL
       C         PHI1 = PHI-ANGLE BACKFILL;   SURCHG = SURCHARGE ON BACKFILL;   ** FAC =
       C         LOAD FACTOR FOR USD;      TOP = TOP THICK (IN OR CM)
       C            DC = DEPTH OF CONCRETE;  KPP>0 FOR PASSIVE PRESSURE IN FRONT OF WALL
       C         **    ISLOP = 1 FOR FRONT FACE BATTER,   **   ISLOP = 2 FOR BACK FACE
       C            BATTER AS FOR BRIDGE ABUTMENTS, ETC.
       C
       C         PROGRAM AUTOMATICALLY USES PASSIVE PRESSURE IN FRONT OF WALL IF
       C            FOOTING DEPTH IS MORE THAN 3-FT OR 1-METER
       C
       C         UT1-UT8 DATA CARD ENTRIES    **MOVE FT OR M TO COL 1 ON THE UT CARDS
       C    FT**    IN         KIPS      FT-K      K/SQ FT    K/CU FT     LB/SQ IN   SQ IN
       C     M**    CM         KN        KN-M      KN/SQ M    KN/CU M     KG/SQ CM   SQ CM
       C
       C
       C************* NOTE THAT FU1 - FU15 ARE IN PROGRAM AND NOT ON DATA CARDS
       C*********************************
       C
   1          DIMENSION F(16),FT(16,2),TITLE(20)
       C
       C           THE FOLLOWING PROGRAM DATA ARE USED FOR FU1-FU15
       C
   2          EQUIVALENCE(F(1),FU1),(F(2),FU2),(F(3),FU3),(F(4),FU4),(F(5),FU
      A5),   (F(6),FU6),  (F(7),FU7),(F(8),FU8),  (F(9),FU9),  (F(10),FU10),
      B(F(11),FU11),(F(12),FU12),(F(13),FU13),  (F(14),FU14),(F(15),FU15),
      C(F(16),FU16)
       C
   3          DATA FT/12.,144.,.030,2.00,3.5,1000.,4000.,87000.,100C.,.001,200.,
      1.150,.50,.144,3.0,3.0,        100.,  10000., 4.713, .530, 9.0,70.3,
      2281., 6117.,101.968, .009807, 14.06,23.564,.150,98.07,7.5,1./
       C
   4          DATA UT/'M'/
       C
   5          DOUBLE PRECISION UT5,UT6,UT7,UT8
       C
   6          REAL*4 KA, KP1, NC, NQ, NGAM, IC, IQ, IGAM, KP
       C      ******** BEGIN MAIN PROGRAM
   7   6000   READ(1,1000,END=150)TITLE,UT1,UT2,UT3,UT4,UT5,UT6,UT7,UT8
   8   1000   FORMAT(20A4/(4(A4,6X),4(A8,2X)))
   9          K = 1
  10          IF(UT1.EQ.UT)K = 2
  11          DO 498  I = 1,16
  12    498   F(I) = FT(I,K)
       C
  13    499   READ(1,1004)KPP,ISLOP,H,G1,G2,BETA,PHI1,PHI2
  14   1004   FORMAT(2I5,7F10.4)
       C
  15          READ(1,1005) FIC,FY,TOE,HEEL,DC,D,COH
       C
  16          READ(1,1005)FAC,SURCHG,HLOSS,TOP,CFORT
  17   1005   FORMAT(8F10.4)
       C
  18          WRITE(3,1001)TITLE
  19   1001   FORMAT('1',//,T5,20A4)
  20          IF(HLOSS.LE.0..AND.CFORT.LE.0.)WRITE(3,2008)
  21   2008   FORMAT(T5, 'RETAINING WALL DESIGN BY USD ACI 318-71',// T10, 'GENE
      1RAL INPUT DATA',/)
  22          IF(HLOSS.GT.0.)WRITE(3,208)
  23    208   FORMAT(T5, 'RETAINING WALL DESIGN BY USD ACI 318-71 AND CONSIDERIN
      1G LOSS OF HEEL PRESSURE',//,T10, 'GENERAL INPUT DATA',/)
  24          IF(CFORT.GT.0.)WRITE(3,2009)
  25   2009   FORMAT(4X, 'COUNTERFORT STABILITY BY USD ACI 318-71',//, T10,
      1 'GENERAL INPUT DATA',/)
  26          WRITE(3,2011)G1,UT6,G2,UT6,BETA,PHI1,PHI2,COH,UT5
  27   2011   FORMAT(T5,'SOIL DATA:',T21,'UNIT WT BACKFILL =',F8.3,1X,A7,/,T19,
      1'UNIT WT. BASE SOIL =',F8.3,1X,A7,/,T23,'SLOPE BACKFILL =',F7.3,'
      2DEG',/,T17,'INT FRICT BACKFILL =',F7.3,' DEG',/,T16, 'INT FRICT BA
      3SE SOIL =',F7.3,' DEG',/,T17,'COHES OF BASE SOIL =',F7.2,1X,A7,/)
  28          WRITE(3,2012)H,UT1,FIC,UT7,FY,UT7,FAC,SURCHG,UT5
  29   2012   FORMAT(T7,'WALL HEIGHT ABOVE FTG =',F7.3,1X,A2,/,T9,'28-DAY CONCRE
      1TE STR =',F7.1,1X,A8,/,T6,'YIELD STENGTH OF STEEL =',F8.1,1X,A8,/,
      1 T16,'LOAD FACTOR =',F5.2,/,T18,'SURCHARGE =',F7.3,1X,A7,//)
  30          WRITE(3,201)TOE,UT1,HEEL,UT1,DC,UT1,D,UT1,TOP,UT2
  31    201   FORMAT(T5,'TRIAL FOOTING DIMENSIONS:',T35,'TOE =',F6.3,1X,A2,/,T34
      1,'HEEL =',F6.3,1X,A2,/,T36,'DC =',F6.3,1X,A2,/,T33,'DEPTH =',F6.3,
      2 1X,A2,/,T35,'TOP = ',F6.3,1X,A2,/)
       C
  32          IF(ISLOP.EQ.1)WRITE(3,410)
  33    410   FORMAT(/,5X,'BATTER ON FRONT FACE OF WALL',//, 8X,'WALL COMPUTATIO
      1NS FOLLOW:',/)
  34          IF(ISLOP.EQ.2)WRITE(3,411)
  35    411   FORMAT(/,5X,'BATTER ON BACK FACE OF WALL',//, 8X, 'WALL COMPUTATI
      1ONS FOLLOW:',/)
       C
       C         CALCULATE ACTIVE EARTH PRESSURE BY RANKINE METHOD
       C
  36          I = 1
  37          BETA=BETA/57.296
  38          PHI1 = PHI1/57.296
  39          PHI2 = PHI2/57.296
```

B9 (*continued*)

```
40          A=COS (BETA)
41          B=A**2
42          C = (COS(PHI1))**2
43          KA = A*(A-SQRT(B-C))/(A+SQRT(B-C))
44          WRITE(3,420)KA
45      420 FORMAT(T5,'THE RANKINE EARTH PRESSURE COEFF,KA =', F7.5,//)
        C
        C   COMPUTE STEM THICK. TOP THICK AS READ USING 'TOP'--EFF UNIT WT >.030 KCF
        C
46          QTOP = SURCHG*KA*COS(BETA)
47          IF(CFORT.LE.0.0)GO TO 421
48          Q = (SURCHG + G1*H)*KA*COS(BETA)
49          PWALL = 0.2*Q*H
50          GO TO 422
51      421 QBOTW = (SURCHG*KA + G1*H*KA)*COS(BETA)
52          PWALL = (QTOP+QBOTW)*H/2.
53      422 WRITE(3,2017)PWALL,UT3,G1,UT6
54     2017 FORMAT(T5,'THE HOR SHEAR FORCE BASE OF WALL =',F7.3,1X,A7, /,10X,
        1 'UNIT WT OF BACKFILL USED (IN CASE G1*KA<FU3) = ',F8.3,1X,A7,//)
55          UVC = FU4*.85*SQRT(FIC)
56          UVCK = UVC*FU14
57          TBC = (PWALL*FAC/UVCK + FU5/FU1)*FU1
        C

        C   CALC STEM THICKNESS AT TOP USING BATTER 1:48 ON EITHER FACE
        C   TOP AND BASE STEM THICKNESS ROUNDED TO LARGER INTEGER
        C
58          TT = TBC - H*FU1/48.
59          ITOP = TT
60          ATOP = ITOP
61          IF(ATOP.LT.TT)TT = ATOP+1.
62          IF(TT.LE.TOP)TT = TOP
63          TB = TT + H*FU1/48.
64          ITB = TB
65          ATB = ITB
66          IF(ATB.LT.TB)TB = ATB+1.
        C
        C   CALCULATE ACTUAL SHEAR STRESS AT BASE OF WALL = ULC
        C
67          STRESS = (FAC*PWALL/((TB-FU5)*FU1))*FU9
68          WRITE(3,315)STRESS,UT7,UVC,UT7,TT,UT2,TB,UT2
69      315 FORMAT(T5,'THE ACTUAL CONCRETE SHEAR STRESS =',F7.2,1X,A8,/,T7,'TH
        2E ALLOWABLE SHEAR STRESS =',F7.2,1X,A8,/,T5,'CALC STEM THICK:',T2
        27,'TOP =',F6.2,1X,A2,/, T24,'BOTTOM =',F6.2,1X,A2,/,23X, 'STEM VAL
        3UES ROUNDED TO EVEN UNITS',//)
70          IF(CFORT.GT.0.)GO TO 350
        C
        C   COMPUTE ULT MOMENTS IN STEM, STEEL AREA REQD AND % STEEL
        C
71          WRITE(3,220)UT1,UT4,UT8,UT2
72      220 FORMAT(T5,'ULTIMATE MOMENTS IN STEM AT 0.1 POINTS AND STEEL REQUIR
        1EMENTS PER ',A2,' OF WIDTH',//,T7, 'POINT FROM',T25,'MOMENT',,T41,
        2 'AS',T50, 'PERCENT',T60,'WALL THICK',T74,'WALL THICK FOR AS',T93
        3, 'MAX CODE % STEEL',/,T10, 'TOP', T26,A7,T39,A5,T51,'STEEL', T64,
        4 A2,T78,A2,/)
        C
        C   *** CALCULATE MAX , STEEL PER ACI ART 10.2.7 & ART 10.3.2--.75*PB
        C
73          IF(F1C-FU7)301,301,302
74      302 IDEL = (F1C-FU7)/FU6
75          ADEL = IDEL
76          AKI = 0.85-ADEL*0.05
77      301 IF(F1C.LE.FU7)AKI = 0.85
78      305 PB = .6375*AKI*F1C*(FU8/(FY*(FU8+FY)))
        C
        C   CALC MOM AND REQ'D STEEL @ 0.1 PTS; ASSUME 3.5 IN OR 9 CM TO CGS
        C
79          DY = H/10.
80          G = .5*FY/(.85*F1C*FU2)
81          DO 32  K = 1,11
82          A = K-1
83          M = A
84          UMOM = FAC*KA*(G1*(A*DY)**3 + 3.*SURCHG*(A*DY)**2)/6.*COS(BETA)
85          STEMT = TT + ((TB-TT)/H)*A*DY
86          EDCM = (STEMT-FU5)/FU1
87          F1 = UMOM/(.9*FY*FU10)
88          AS = (EDCM-SQRT(EDCM**2-4.*G*F1))/(2.*G)
89          PER = AS/(EDCM*FU1*FU1)
        C
        C   MIN % STEEL BASED ON ACI 318-71 ART. 10.5.1
        C
90          SMIN = FU11/FY
91          EDCMFT = EDCM*FU1
92          IF(PER-SMIN)304,304,32
93      304 PER = SMIN
94          AS = PER*EDCM*FU1*FU1
95      32  WRITE(3,34)M,UMOM,AS,PER,STEMT,EDCMFT,PB
96      34  FORMAT(T12,I2,T23,F7.3,T39,F7.3,T50,F5.3,T63,F5.2,T77,F5.2,T98,F5.
        13)
        C
        C   NOTE STABILITY COMPUTATIONS BASED ON ACTUAL FORCES AND NOT FORCES X FAC
        C   CALCULATE OVERTURNING STABILITY
        C
97      350 XHT = HEEL*TAN (BETA)
98          HTOT = H+XHT+DC
99          WI=((2.*H)+XHT)/2.*HEEL*G1 + SURCHG*HEEL
100         W2 = TT*H*FU12/FU1
101         W3 = (TB-TT)*H*FU12/(2.*FU1)
        C
        C   ****    NOTE SLIGHT APPROXIMATIONS USED FOR COMPUTING WEIGHT OF
        C   TRIANGULAR WALL ELEMENT AND SOIL IN THIS ZONE WHEN BACK FACE IS
        C   SLOPED--ALSO SLIGHT APPROXIMATION IN COMPUTING MOMENT ARM
        C
102         IF(ISLOP.EQ.2)W3 = 2.*(W3/FU12*(FU12+G1)/2.)+ (TB-TT)/FU1*SURCHG
```

B9 (*continued*)

```
103        W4 = (TOE+HEEL+TB/FU1)*DC*FU12
104        QBASE = SURCHG*KA+GL*HTOT*KA
105        PA = (QTOP/COS(BETA) + QBASE)*HTOT/2.
106        PAV = PA*SIN (BETA)
107        X1 = TOE + TB/FU1 + HEEL/2.
108        X2 = TOE + (2.*TB-TT)/(2.*FU1)
109        X3 = TOE + (TB-TT)*.667/FU1
110        IF(ISLOP.EQ.2)X2 = TOE + TT/(2.*FU1)
111        IF(ISLOP.EQ.2)X3 = TOE + (TT + (TB-TT)/2.)/FU1
112        X4 =(TOE + HEEL + TB/FU1)/2.
113        X5 = 2.*X4
114        RM1 = W1*X1
115        RM2= W2*X2
116        RM3=W3*X3
117        RM4=W4*X4
118        RM5=PAV*X5
119        REMO = RM1+RM2+RM3+RM4+RM5
120        SUMV=W1+W2+W3+W4+PAV
121        IF(I.EQ.1)WRITE(3,2016)PA,UT3,UT3,UT1,UT4
122   2021 FORMAT(    //,9X,'WALL STABILITY FOR PA = ',F8.3,1X,A4,/,
      1                            6X, 'PART', T16, 'WT OF PART,'
      2   , 1X,A4, T35,'ARM', 1X,A2, T46, 'MOMENT',1X,A4,/)
123        IF(I.GT.1)WRITE(3,2021)PA,UT3,UT3,UT1,UT4
124   2016 FORMAT('1',//,9X,'WALL STABILITY FOR PA = ',F8.3,1X,A4,/,
      2                            6X, 'PART', T16, 'WT OF PART,'
      2   , 1X,A4, T35,'ARM', 1X,A2, T46, 'MOMENT',1X,A4,/)
125        WRITE(3,2019)W1,X1,RM1,W2,X1,RM2,W3,X3,RM3,W4,X4,RM4,PAV,X5,RM5
126   2019 FORMAT(T9, '1', T20, F8.3, T36, F6.3, T48, F9.3,/,
      1           T9, '2', T22, F6.3, T36, F6.3, T50, F7.3,/,
      2           T9, '3', T22, F6.3, T36, F6.3, T50, F7.3,/,
      3           T9, '4', T22, F6.3, T36, F6.3, T50, F7.3,/,
      4           T8,'PAV', T22, F6.3, T36, F6.3, T50, F7.3)
127        WRITE(3,2018)SUMV,REMO,UT4
128   2018 FORMAT(T7,'SUM OF WTS =',F9.3, 3X,'SUM RESIST MOM =',F10.3,1X,A4/)
C
C          COMPUTE SF FOR OVERTURNING
C
129        YBAR = (QTOP*HTOT**2/2. + (QBASE-QTOP)*HTOT**2/6.)/PA
130        OTMOM = YBAR*PA*COS(BETA)
131        FSOT = REMO/OTMOM
C
C          COMPUTE ECCENTRICITY OF RESULTANT ON BASE
C
132        XNETM=REMO-OTMOM
133        XBAR=XNETM/SUMV
134        ECCEN = X4-XBAR
135        BASE = TOE + HEEL + TB/FU1
136        EMAX = BASE/6.
137        IF(ECCEN.LE.EMAX)GO TO 46
138   306 CONTINUE
139        I = I+1
140        IF(I/4*4.EQ.I)GO TO 355
141        WRITE(3,47)FU13,UT1
142    47 FORMAT(/,T5,'WALL UNSTABLE--HEEL INCREASED ',F4.2,1X,A2,' ***')
143        HEEL = HEEL + FU13
144        GO TO 357
145   355 WRITE(3,356)FU13,UT1
146   356 FORMAT(/,T5,'WALL UNSTABLE--TOE INCREASED ',F4.2,1X,A2,' *****')
147        TOE = TOE + FU13
148   357 IF(I.LE.8)GO TO 350
149        WRITE(3,41)YBAR,UT1,OTMOM,UT4,FSOT
150    41 FORMAT(T5,'YBAR =',F7.3,1X,A2,/,T5,'OVERTURN MOMENT =',F8.3,1X,A4,
      1  5X,'SF OVERTURN =',F6.3,/)
151        WRITE(3,42)XNETM,UT4,XBAR,UT1,ECCEN,UT1,EMAX,UT1
152    42 FORMAT(T5,'NET RESIST MOM =',F9.3,1X,A4,5X,'COHES RESIST =',F6.3,1X,A2,/,
      1T6,'ECCENTRICITY =',F6.3,1X,A2,5X,'MAX ECCENTRICITY (L/6) ='F6.3,1
      2X,A2,//)
C
C          COMPUTE SLIDING S.F. USING 2/3 FRICT. ANGLE (NOT 2/3TAN (ANGLE))
C          ***NOTE THAT PASSIVE PRESSURE MAY OR MAY NOT BE INCL.
C          NOTE PASSIVE PRESS INCL IF D > 3 FT OR 1 M FOR FU16--PHI OF BASE SOIL USED
C
C          IF SLIDING SF NOT SATISFACTORY ADD HEEL KEY OR RE-PROGRAM WITH
C                          LARGER HEEL DIMENSION BASED ON CURRENT OUTPUT HEEL
C
153        PP = 0.
154        IF(KPP.LE.0.AND.D.LT.FU16)GO TO 35
155        IF(KPP.LE.0)DB = FU16
156        IF(KPP.GT.0)DB = D
157        KP1 = TAN(45./57.2958 + PHI2/2.)
158        KP = KP1**2
159        PP = 0.5*G2*KP*DB*DB + 2.*COH*DB*KP1
160    35 FRIC1 = SUMV*TAN (PHI2)
161        FRIC2 = COH*BASE*0.667
162        TOFR = FRIC1+FRIC2+PP
163        SLIDF = PA*COS (BETA)
164        FSS = TOFR/SLIDF
165        WRITE(3,43)FRIC1,UT3,FRIC2,UT3,PP,UT3,TOFR,UT3,SLIDF,UT3,FSS
166    43 FORMAT(T5,'FRIC RESIS =',F6.2,1X,A4,5X,'COHES RESIST =',F6.2,1X,A
      14,/,T5,'PASSIVE RESIS =',F6.2,1X,A4,5X,'TOTAL SLIDING RESIST =',F7
      2.2,1X,A4,//,T5,'DRIVING FORCE =',F7.2, 1X,A4,5X,'*** SF SLIDING ='
      3,F6.3,//)
C
C          COMPUTATION OF BEARING PRESSURE USING SHAPE, DEPTH AND INCLIN
C          FACTORS OF BRINCH HANSEN (LATEST (1970) REVISIONS)
C          WITH 1-FT STRIP ALL SHAPE FACTORS = 1.   "EXACT" VALUES BEING USED
C
167        IF(PHI2.GT.0.)GO TO 91
168    92 NC = 5.14
169        NQ = 1.
170        GO TO 500
171    91 PI = 3.1416
172        NQ = EXP(PI*TAN(PHI2))*(TAN(PI/4.+PHI2/2.))**2
173        NGAM = 1.5*(NQ-1.)*TAN(PHI2)
174        NC = (NQ-1.)*COTAN(PHI2)
175   500 BPR = BASE -2.*ECCEN
```

B9 (*continued*)

```
176          IQ = (1.-.5*SLIDF/(SUMV+BPR*COTAN(PHI2)*COH))**5
177          IC = IQ - (1.-IQ)/(NQ-1.)
178          IF(PHI2.EQ.0.)IC=0.5-0.5*SQRT(1.-SLIDF/(BPR*COH))


179          IGAM = (1.-.7*SLIDF/(SUMV+BPR*COH*COTAN(PHI2)))**5
180          FDC = 1. + .4*D/BPR
181          IF(PHI2.EQ.0.)FDC = 0.4*D/RPR
182          DQ = 1. + 2.*TAN(PHI2)*(1.-SIN(PHI2))**2*ATAN(D/BPR)
183          IF(PHI2.GT.0.)QULT = COH*NC*FDC*IC + D*G2*NQ*DQ*IQ + .5*G2*BPR*NGA
     1M*IGAM
184          IF(PHI2.EQ.0.)QULT = COH*5.14*(1.+FDC-IC) + D*G2*NQ*DQ*IQ
      C
      C            NOTE SF ON QULT--MAY BE MORE APPROPRIATE TO PUT SF ON COHESION AND PHI
      C
185      61 QALL = QULT/2.
186          IF(COH.GT.0.)QALL = QULT/3.
187      62 WRITE(3,17)NC,NQ,NGAM,IC,IQ,IGAM,FDC,DQ,QULT,QALL,UT5
188      17 FORMAT(T5, 'THE BEARING CAPACITY AND OTHER FACTORS ARE:',/,T6, 'NC
     1 =',F8.3,5X, 'NQ =',F8.3,5X,'NGAM =',F8.3,/,T6,'IC =',F8.3,5X, 'IQ
     3=',F8.3,5X, 'IGAM =',F8.3,/, T6,'FDC =',F8.3, 5X, 'DQ =',F8.3,/, T5
     4,'ULT. SOIL PRESS =',F8.3,3X,'ALLOW SOIL PRESS =',F8.3,1X,A7,//)
189     126 QTOE=SUMV*(1.+6.*ECCEN/BASE)/(BASE*1.)
190          QHEEL=SUMV*(1.-6.*ECCEN/BASE)/(BASE*1.)
191          WRITE(3,48)QTOE,UT5,QHEEL,UT5
192      48 FORMAT(T5, 'ACTUAL SOIL PRESSURES:', T29, 'TOE =', F7.2,1X,A7,/,T2
     19, 'HEEL =',F7.2,1X,A7,//)
193          IF(QTOE.GT.QALL)GO TO 306
      C
      C            COMPUTE WIDE BEAM SHEAR IN TOE & HEEL; COMPUTE TOE & HEEL MOMENTS
      C
      C            COMPUTE SHEAR AT FACE OF STEM; COMPUTE HEEL MOM @ 3.5 IN (9 CM) INSIDE
      C            BACKFACE OF STEM (APPROX CGS); EFFECTIVE D = DC-3.5 IN (9 CM)
      C
194     316 DCE = DC - FU5/FU1
195          V1 = QTOE - FU12*DC
196          V2 = (QTOE-QHEEL)/BASE
197          VTOE = V1*TOE - V2*(TOE**2)/2.
198          TOEMO = (V1*(TOE**2))/2. - (V2*(TOE**3))/6.
199          HEELP = HEEL + FU5/FU1
200          AVEHT = (2.*H +XHT)/2.
201          SOILP = AVEHT*G1 + DC*FU12 + SURCHG
202          V4 = SOILP - QHEEL
203          VHEEL = V4*HEEL - V2*(HEEL**2)/2. + PAV
204          HEELM = V4*(HEELP**2)/2. - V2*(HEELP**3)/6. + PAV*HEELP
205          WRITE(3,155)VTOE,UT3,VHEEL,UT3
206     155 FORMAT(T5,'SHEAR AT STEM FACES:',/,T10,'TOE SHEAR =',F7.2,1X,A4,/,
     1T9, 'HEEL SHEAR =',F7.2,1X,A4,//)
207          VCC = VTOE
208          IF(VHEEL.GE.VCC)VCC = VHEEL
209          VACT = FAC*VCC/(DCE*FU14)
210          IF(CFORT.GT.0.)GO TO 148
211          IF(UVC-VACT)124,125,125
      C
      C            IF SHEAR STRESS EXCEEDS ALLOW FOOT DEPTH INCR 3-IN (7.5 CM)
      C
212     124 WRITE(3,131)FU15,UT2
213     131 FORMAT(T10,'*** SHEAR STRESS EXCEEDED--FTG DEPTH INCR',F5.2,1X,A2
     1,//)
214          DC = DC + FU15/FU1
215          I = 1
216          GO TO 350
217     125 WRITE(3,129)VCC,UT3,VACT,UT7,UVC,UT7
218     129 FORMAT(T5, 'THE ACTUAL SHEAR STRESS IN FTG. BASED ON  A SHEAR OF',
     1F8.3,1X,A4,' IS',F8.3,1X,A8,/,T10,'THE ALLOWABLE SHEAR STRESS IS',
     1F8.3,1X,A8,//)
219          WRITE(3,104)TOEMO,UT4,HEELM,UT4
220     104 FORMAT(T5, 'CONVENTIONAL TOE & HEEL MOMENTS',/,T10,'TOE MOMENT =',

      1F8.2,1X,A4,/,T9,'HEEL MOMENT ='F8.2,1X,A4,//)
      C
      C            ***** ALTERNATIVE CHECK FOR LOSS OF HEEL PRESSURE--ASSUME TOE CARRIES I/3
      C            OF SUMV AS RECTANGULAR PRESSURE BLOCK
      C
221          IF(HLOSS.LE.0.)GO TO 203
222          AVTOE = SUMV*.667
223          AMTOE = AVTOE*TOE/2.
224          AVHEL = (AVEHT*G1 +FU12*DC +SURCHG)*HEEL + PAV
225          AMHEL = (AVHEL-PAV)*HEELP/2. + PAV*HEELP
226          AVCC = AVTOE
227          IF(AVHEL.GT.AVTOE)AVCC = AVHEL
228          AVACT = FAC*AVCC/(DCE*FU14)
      C            ALTERNATE CONDITION CONSIDERS APPROX 10% SF ON ULT SHEAR
229          IF(AVACT-1.55*UVC)52,52,51
230      51 WRITE(3, 107)
231     107 FORMAT(T5,'**** ALTERNATE SHEAR CONDITIONS CONTROL***',//)
232          GO TO 124
      C
233      52 CONTINUE
234          WRITE(3,108)AVTOE,UT3,AMTOE,UT4,AVHEL,UT3,AMHEL,UT4
235     108 FORMAT(T5, 'ALTERNATE TOE, HEEL SHEAR AND MOMENTS',/, T10, 'TOE SH
     1EAR =',F7.2,1X,A4,/,T9,'TOE MOMENT =',F9.2,1X,A7,/,/,'HEEL SHEAR
     1=',F7.2,1X,A4,/,T8,'HEEL MOMENT =',F9.2,1X,A7,//)
236          WRITE(3,112)AVCC,UT3,AVACT,UT7
237     112 FORMAT(T5, 'ALTERNATE SHEAR STRESS IN FTG BASED ON A SHEAR OF', F8
     1.3,1X,A4,' IS',F9.3,1X,A8,//)
      C
      C            IF ALTERNATE MOMENTS ARE ''FAC'' GREATER THAN CONVENTIONAL MOMENT,
      C            THEN ALT. MOMENT CONTROLS FOR STEEL REQUIREMENTS
      C
238          IF(AMHEL.GT.FAC*HEELM)HEELM = AMHEL/FAC
```

B9 (*continued*)

```
239             IF(AMTOE.GT.FAC*TOEMO)TOEMO = AMTOE/FAC
        C
        C       CALCULATE TOE, HEEL STEEL REQUIREMENTS (SQ IN/FT OR SQ CM/M)
240     203 F1 = FAC*TOEMO/(.9*FY*FU10)
241         F2 = FAC*HEELM/(.9*FY*FU10)
242         ASTOE = (DCE - SQRT (DCE**2 -4.*F1*G))/(2.*G)
243         ASHEL = (DCE - SQRT (DCE**2 - 4.*F2*G))/(2.*G)
244         PERTO = ASTOE/(DCE*FU1*FU1)
245         IF(PERTO-SMIN)307,307,308
246     307 PERTO = SMIN
247         ASTOE = DCE*SMIN*FU1*FU1
248     308 PERHE = ASHEL/(DCE*FU1*FU1)
249         IF(PERHE-SMIN)309,309,310
250     309 PERHE = SMIN
251         ASHEL = SMIN*DCE*FU1*FU1
        C
252     310 WRITE(3,145)TOE,UT1,BASE,UT1,HEEL,UT1,DC,UT1,DCE,UT1,TOEMO,UT4,AST
            1OE,UT8,PERTO,HEELM,UT4,ASHEL,UT8,PERHE
253     145 FORMAT(T5, 'THE FINAL SLAB DIMENSIONS ARE:', T38,'TOE =',F6.3,1X,A
            12,T58,'TOTAL BASE WIDTH =',F6.3,1X,A2,/,T37,'HEEL =',F6.3,1X,A2,T5
            28,'THICK BASE SLAB =',F6.3,1X,A2,/,T43,'EFF DEPTH BASE SLAB =',F6.
            33,1X,A2,/,T6,'TOE MOMENT =',F9.2,1X,A7,T37,'STEEL REQD =',F7.3,1X,
            4A7,T68,'% STEEL =',F7.4,//,T5,'HEEL MOMENT =', F9.2,1X,A4,T37,'STE
            5EL REQD =',F7.3,1X,A5,T68,'% STEEL =',F7.4,//)
        C
254         WRITE(3,146)
255     146 FORMAT(10X, 'NOTE ALL FINAL FOOTING DIMENSIONS INCLUDE THE EFFECT
            1OF 3.5-IN OR 9-CM (FU5) OF CONCRETE COVER',/, 12X,'TO C.G.S. WHEN
            2COMPUTING MOMENTS OR SHEAR',/,12X,'AND REQUIRED STEEL AREAS',//)


        C
        C
256     148 GO TO 6000
257     150 STOP
258         END
```

B-10. Finite-element program for pile or pole buckling. Element lengths may be variable; moment of inertia may be variable; pile may be fully or partially embedded

```
        C       J E BOWLES   PILE AND/OR POLE BUCKLING BY FINITE ELEMENT METHOD
        C                    PILE OR POLE MAY BE FULLY OR PARTIALLY EMBEDDED
        C       ********    VARIABLE IDENTIFICATION
        C       XL = LENGTH OF PILE (FT OR M)
        C       ELAS = MODULUS OF ELASTICITY (KSF OR KN/SQ M)
        C       W = PILE WIDTH IF SQUARE OR IF INERTIA READ (FT OR M)
        C       DIAM = DIAMETER OF ROUND PILE (FT OR M)
        C          ** NOTE THAT A ROUND TAPERED PILE CAN ALSO BE READ IN
        C       READ DIAM = 1000. FOR TAPERED PILES & PROGRAM WILL COMPUTE 2 FOR SOLID
        C          SECTIONS, ETC; ALTERNATIVELY MAKE READI < 0. AND READ VALUES IN
        C       READI = VALUE OF MOMENT OF INERTIA (FT OR M**4)--VALUE IS READ IN
        C          WHEN READI < 0.; IS COMPUTED IF READI GE. 0. AND DIAM = 1000.
        C       JTS = TOTAL NUMBER OF JOINTS ON PILE
        C       NPS = NO OF JOINTS IN SIDESWAY--NORMALLY SAME AS JTS
        C       NBMR = NUMBER OF BUCKLING MODES TO BE COMPUTED IN PROBLEM
        C       SOIL MODULUS IS COMPUTED AS  SK(I) = AS + BS*Z**EXPON  (KCF)
        C       PER = PERCENT OF BUCKLING LOAD CARRIED AS POINT LOAD
        C          READ PER AS A DECIMAL
        C       KPER = TYPE OF SKIN RESISTANCE ESTIMATED--USE 1 FOR LINEAR;  USE
        C          2 FOR PARABOLIC DISTRIBUTION
        C       JJS = NUMBER OF NODES REQUIRING SOIL SPRING TO BE READ IN
        C       JSOIL = NODE AT WHICH SOIL STARTS--GREATER THAN 1 FOR PARTIAL
        C          EMBEDMENT OF PILE
        C       ILIST = SWITCH TO WRITE ASAT AND ASAT INVERSE MATRICES
        C       KLIST = SWITCH TO WRITE K-MATRIX
        C       LIST = SWITCH TO WRITEAVERAGED PART OF ASAT MATRIX
        C       NBC = NUMBER OF BEAM STIFFNESS VALUES TO READ IN--USE FOR VARIABLE
        C          MOMENT OF INERTIA AND SEGMENT LENGTHS
        C********   CONTAINS MODIFICATION TO ZERO ASAT MATRIX FOR BOUNDARY
        C          CONDITIONS--JOINT ROTATION OR DEFLECTION = 0.
        C       IZERO = COUNTER OF NUMBER OF NP = 0 IN ASAT MATRIX
        C       UT1 = FT;   UT3 = KIPS; UT4 = KSF; UT5 = K/CU FT OR METRIC EQUIV.
        C       NONCL = SWITCH FOR NON-CONSTANT LENGTH SEGMENTS USE 1; FOR CONSTANT
        C          LENGTH SEGMENTS USE 0
        C       *** TRY TO KEEP RATIO OF ADJACENT MEMBER LENGTHS TO NOT OVER 2.
        C
1           DIMENSION ASAT(40,40), XK(40,40),DK(20,20),CDK(20,20),
            1G(51), X(51), DKX(51), CDKX(51), S(51), PERJT(51), A(51), B(51),
            2C(51),ERNC(51),WDD(51),ZZZ(20),NPZERO(20),XLL(30)
2           DOUBLE PRECISION ASAT,XK,DK,CDK
3           EQUIVALENCE (ASAT(1,1),XK(1,1))
4           DOUBLE PRECISION UT5
        C
5    5000 READ(1,1000,END=6000)ZZZ,UT1,UT3,UT4,UT5
        C
6    1000 FORMAT(20A4/3(A4,6X),A8)
        C
7           READ(1,1005)XL,ELAS,W,DIAM,READI
        C
8    1005 FORMAT(8F10.4)
        C
9           READ(1,1005)FAC,AS,BS,EXPON,PER
10          IF(FAC.EQ.0.)GO TO 410
```

B10 (*continued*)

```
11      9000 CONTINUE
        C
        C                    NOTE READ CONTINUATION CARD
12           READ(1,1006)JTS,NPS,NBMR,KPER,JJS,JSOIL,NBC,ILIST,KLIST,LIST,IZERO
             1, NONCL
        C
        C
        C                    DO NOT USE NPZERO FOR FULLY EMBEDDED PILES TO FIX TOP & BOTTOM
        C
13           IF(IZERO.GT.0)READ(1,1006)(NPZERO(I),I=1,IZERO)
14           IF(IZERO.GT.0)NPZERO(I+1) = 1
15      1006 FORMAT(16I5)
16           IF(JTS.EQ.0)GO TO 410
17           ELAP=ELAS
18           WRITE(3,1001)ZZZ
19      1001 FORMAT('1',//,T5,20A4,//)
20           NB=JTS-1
21           SUME = 0.
        C                    READ SEGMENT LENGTHS IF NONCL > 0;  READ INERTIA VALUES
        C                           IF READI < 0.
22           IF(NONCL.GT.0)READ(1,1005)(XLL(I),I=1,NB)
23           IF(READI.LT.0.)READ(1,1005)(ERNC(I),I=1,NB)
        C
24           IF(NONCL.GT.0)GO TO 99
25           DO 98  I = 1,NB
26           IF(READI.LT.0.)SUME = SUME + ERNC(I)
27        98 XLL(I) = XL/FLOAT(NB)
28           IF(READI.LT.0.)ERN = SUME/FLOAT(NB)
29        99 IF(DIAM.EQ.1000.0)GO TO 18
30           IF(READI.GT.0.)ERN = READI
31           IF(DIAM)102,102,101
32       101 WRITE(3,106) DIAM,UT1
33       106 FORMAT(T10,'DIAMETER OF ROUND PILE = ',F10.5,1X,A2)
34           WIDTH = DIAM
35           IF(READI.EQ.0.)ERN = 3.1416*DIAM**4/64.
36           GO TO 107
37       102 WRITE(3,108) W,UT1
38       108 FORMAT(T10,'WIDTH OF SQUARE OR PIPE PILE = ',F10.4, 1X,A2)
39           IF(READI.EQ.0.)ERN = W**4/12.
40       107 IF(W.GT.0.)WIDTH = W
41           GO TO 15
42        18 READ(1,1005)DIAT,DIAB
43           CTAPER=(DIAT-DIAB)/XL
44           WDD(1) = DIAT
45           XMULT = XLL(1)/2.
46           H = 0.
47           DO 16 I=1,NB
48           H = H + XLL(I)
49           WDD(I+1) = DIAT - H*CTAPER
50           DIAI= DIAT-XMULT*CTAPER
51           IF(READI.GE.0.)ERNC(I) = 3.1416*DIAI**4/64.
52           XMULT = XMULT + XLL(I)/2.
53           SUME=SUME+ERNC(I)
54        16 CONTINUE
55        14 ERN=SUME/NB
56        15 EULER =(3.1416)**2*ELAS*ERN/(XL**2)
57           PER100=PER*100.
58           DO 119 I=1, NB
59           IF(DIAM.LT.1000.AND.READI.GE.0.)ERNC(I) = ERN
60           WRITE(3,2006)I,XLL(I),UT1,ERNC(I),UT1
61      2006 FORMAT(2X,'ELEM NO = ',I2, 3X,'L = ',F10.3,1X,A2, 2X,'INERTIA = ',
             1F10.5,1X,A2,'**4')
62           A(I) = 1./XLL(I)
63           B(I) = 4.*ELAS*ERNC(I)/XLL(I)
64       119 C(I)=B(I)/2.0
        C
        C            READ 4EI/L MODIFICATIONS HERE AS FOR HOLLOW PIPES, ETC
        C            NBC = NUMBER OF B-CHANGES OF 4EI/L
        C
        C
65           IF(NBC.EQ.0) GO TO 116
66           DO 117 I=1, NBC
        C
67           READ(1,1009) J, B(J)
68      1009 FORMAT(I5,F12.2)
69       117 C(J)=B(J)/2.0
70       116 CONTINUE
71           WRITE(3,111)ELAS,UT4, XL,UT1,JTS,NPS,JSOIL,NBMR,PER100,JJS,NBC,EUL
             1ER,UT3,FAC
72       111 FORMAT(//,T10,'MOD OF ELASTICITY = ',F12.2,1X,A4,/,T10, 'LENGTH O
             1F PILE = ',F8.3, 1X,A2,/, T10, 'NUMBER OF JOINTS = ', I4,/,T10,
             2 'NO OF JOINTS IN SIDESWAY = ',I4,/, T10, 'NODE SOIL STARTS = ',I3
             3,/, T10, 'NO OF BUCKLING MODES REQD = ',I3,/, T10, 'PERCENT POINT
             4LOAD = ',F6.2,/, T10, 'NO OF NODES REQUIRING SPRING CORRECTION = '
             5,I3,/,T10, 'NO OF STIFFNESS CHANGES = ',I3,/, T10, '** EULER BUCKL
             6ING LOAD = ',F10.2,1X,A4,/,T5,'1ST NODE REDUCT FACTOR = ',F5.3,
             7//)
73           IF(KPER.EQ.1) WRITE(3,112)
74       112 FORMAT(T5, 'LINEAR SKIN RESISTANCE REDUCTION',/)
75           IF(KPER.EQ.2) WRITE(3,113)
76       113 FORMAT(T5, 'PARABOLIC SKIN RESISTANCE REDUCTION',/)
77           WRITE(3,114) AS,BS,EXPON,UT5
78       114 FORMAT(5X, 'SOIL MOD, KS = ',F10.3,' + ',F10.3,'Z**',F5.3,1X,A7)
79           IF(IZERO.GT.0)WRITE(3,5)(NPZERO(I),I=1,IZERO)
80         5 FORMAT(//,T5,'ZERO NP AT I = ',2X,16I4,//)
        C            ***************************************
        C            CALCULATE SOIL MODULUS AT EACH NODE
        C            ***************************************
81           DO 442 I=1,JTS
82           G(I)=0.0
83       442 S(I)=0.0
84           H = 0.
85           DO 430 I=1,JTS
86           IF(I.LT.JSOIL) GO TO 430
87           S(I)=AS+BS*H**EXPON
88           IF(I.LT.JTS)H = H + XLL(I)
```

B10 (*continued*)

```
 89       430 CONTINUE
 90           WRITE(3,2008)H,UT1
 91      2008 FORMAT(/,T10,'DEPTH OF PILE EMBEDMENT = ',F10.3,1X,A2,/)
 92           IF(DIAM.EQ.1000.0)WIDTH=WDD(JSOIL)
       C********1ST SOIL SPRING REDUCED BY FAC FOR SOIL DISTURBANCE
       C        **** 2ND SOIL SPRING REDUCED BY .25*FAC FOR SOIL DISTURBANCE
       C            IF FAC < 1.
 93           G(JSOIL) = (7.*S(JSOIL)+6.*S(JSOIL+1)-S(JSOIL+2))*XLL(JSOIL)*WIDTH
              1/24.*FAC
 94           IF(DIAM.EQ.1000.0)WIDTH=WDD(NPS)
 95           G(JTS)=(7.0*S(JTS) + 6.0*S(JTS-1) - S(JTS-2))*XLL(NB)*WIDTH/24.
 96           JSP1 = JSOIL+1
 97           FAC2 = 1.
 98           DO 440  I = JSP1,NB
 99           IF(DIAM.EQ.1000.0)WIDTH=WDD(I)
100       440 G(I) = XLL(I-1)*WIDTH*(3.*S(I-1) + 10.*S(I) - S(I+1))/24. + XLL(I)
              1*WIDTH*(3.*S(I+1) + 10.*S(I) - S(I-1))/24.
101           IF(FAC.LT..75)FAC2 = 1. - .25*FAC
102           G(JSOIL+1) = G(JSOIL+1)*FAC2
       C ******** READ SOIL SPRING ADJUSTMENTS HERE FOR NO = JJS
       C
103           IF(JJS.EQ.0) GO TO 517
104           DO 717 I=1,JJS
       C
105       717 READ(1,1009) KC, G(KC)
       C
106       517 CONTINUE
       C    ******************************
       C    BUILDING OF ASAT MATRIX
       C    ******************************
107           M=2*JTS
108           DO 165 I=1,M
109           DO 165 J=1,M
110       165 ASAT(I,J)=0.0
111           DO 110 I=1,JTS
112           K=JTS+I
113           IF(I.EQ.1) ASAT(I,K)=A(I)*(B(I)+C(I))
114           IF(I.EQ.1) ASAT(K,I)=A(I)*(B(I)+C(I))
115           IF(I.EQ.JTS) ASAT(I,K)=-A(I-1)*(B(I-1)+C(I-1))
116           IF(I.EQ.JTS) ASAT(K,I)=-A(I-1)*(B(I-1)+C(I-1))
117           IF(I.EQ.JTS) GO TO 120
118           ASAT(I,K+1)=-A(I)*(B(I)+C(I))
119           ASAT(I,I+1)=C(I)
120           ASAT(K,I+1)=A(I)*(B(I)+C(I))
121           ASAT(K,K+1)=-2.0*A(I)*A(I)*(B(I)+C(I))
122           IF(I.EQ.1)GO TO 131
123       120 CONTINUE
124           ASAT(I,I-1)=C(I-1)
125           ASAT(I,K-1)=A(I-1)*(B(I-1)+C(I-1))
126           ASAT(K,I-1)=-A(I-1)*(B(I-1)+C(I-1))
127           ASAT(K,K-1)=-2.0*A(I-1)*A(I-1)*(B(I-1)+C(I-1))
128           IF(I.EQ.JTS) GO TO 132
129       130 ASAT(I,I)=B(I-1)+B(I)
130           ASAT(K,K)=2.0*A(I-1)*A(I-1)*(B(I-1)+C(I-1))+2.0*A(I)*A(I)*
              1(B(I)+C(I))+G(I)
131           ASAT(I,K)=-A(I-1)*(B(I-1)+C(I-1))+A(I)*(B(I)+C(I))
132           ASAT(K,I)=ASAT(I,K)
133           GO TO 133
134       131 ASAT(I,I)=B(I)
135           ASAT(K,K)=2.0*A(I)*A(I)*(B(I)+C(I))+G(I)
136           GO TO 133
137       132 ASAT(I,I)=B(I-1)
138           ASAT(K,K)=2.0*A(I-1)*A(I-1)*(B(I-1)+C(I-1))+G(I)
139       133 CONTINUE
140       110 CONTINUE
       C*****************
       C            CORRECT ASAT FOR BOUNDARY CONDTIIONS--RCT OR DEFL = 0.
141           INDEX = 0
142           IF(IZERO.EQ.0)GO TO 5499
143           JJ = 1
144           JK = 0
145           DO 1310  IK = 1,M
146           IF(IK.NE.NPZERO(JJ))GO TO 1310
147           IF(IK.EQ.NPZERO(JJ).AND.IK.GT.JTS.AND.JK.EQ.0)JJJ = JJ
148           IF(IK.GT.JTS.AND.JK.EQ.0)JK = 1
149           JJ = JJ + 1
150           DO 1309  K = 1,M
151           ASAT(IK,K) = 0.
152      1309 ASAT(K,IK) = 0.
153           ASAT(IK,IK) = 1.
154      1310 CONTINUE
155      5499 IF(ILIST.EQ.1)CALL WRITEO(ASAT,M,INDEX,JTS)
156      1250 CONTINUE
       C    **********************************
       C    ASAT INVERSION IN PLACE WITHOUT SCALING
       C    **********************************
157           DO 25 K=1,M
158           DO 20 J=1,M
159           IF(J.EQ.K) GO TO 20
160           ASAT(K,J)=ASAT(K,J)/ASAT(K,K)
161        20 CONTINUE
162           DO 26  I = 1,M
163           IF(I.EQ.K)GO TO 26
164           DO 21 J=1,M
165           IF(J.EQ.K) GO TO 21
166           ASAT(I,J)=ASAT(I,J)-ASAT(K,J)*ASAT(I,K)
167        21 CONTINUE
168        26 CONTINUE
169           DO 22 I=1,M
170           IF(I.EQ.K) GO TO 22
171           ASAT(I,K)=-ASAT(I,K)/ASAT(K,K)
172        22 CONTINUE
173           ASAT(K,K)=1.0/ASAT(K,K)
```

B10 (*continued*)

```
174        25 CONTINUE
           C    END OF ASAT INVERSION
           C    *******************************
           C    BUILDING OF SECOND-ORDER STIFFNESS MATRIX
           C    *******************************
175             INDEX = 1
176             IF(ILIST.EQ.1)CALL WRITEO(ASAT,M,INDEX,JTS)
177             DO 452 I=1,JTS
178        452 PERJT(I)=1.0
179             HP1 = 0.
180             IST=JSOIL+1
181             PHI = 1. - PER
182             GO TO (706,707),KPER
183        706 DO 453 I=IST,JTS
184             HP1 = HP1 + XLL(I-1)
185        453 PERJT(I) = 1. - PHI*HP1/H
186             GO TO 456
187        707 CONTINUE
188             DO 454 I=IST, JTS
189             HP1 = HP1 + XLL(I-1)
190        454 PERJT(I) = 1. - PHI*(HP1/H)**2
191        456 CONTINUE
192             WRITE(3,1090)
193       1090 FORMAT(///, 30X,'STIFFNESS',/,10X, 'MEMBER',
                 18X,5H4EI/L,9X,5H2EI/L,4X,6HLENGTH/)
194             DO 128 I=1, NB
195             TEMPL=1.0/A(I)
196             K=I+1
197        128 WRITE(3,1091) I, K, B(I), C(I), TEMPL
198       1091 FORMAT(10X,I2,3H - ,I2,1X,F12.2,2X,F12.2,2X,F8.4)
199             WRITE(3,1018)
200       1018 FORMAT(1H0,9X,35HJOINT      SOIL MOD      SOIL CONST
                 15X, 'FRICTION',/)
201             DO 1501 I=1,JTS
202       1501 WRITE(3,1092) I, S(I), G(I), PERJT(I)
203       1092 FORMAT(10X,I5,3F15.2)
204             DO 135 I=1,JTS
205             DO 135 J=1,JTS
206        135 XK(I,J)=0.0
207             DO 140 I=1,JTS
208             IF(I.EQ.JTS) GO TO 160
209             XK(I,I+1)=-A(I)*PERJT(I+1)
210        160 IF(I.EQ.1) GO TO 170
211             XK(I,I-1)=-A(I-1)*PERJT(I-1)
212        170 IF(I.EQ.1) GO TO 171
213             IF(I.EQ.JTS) GO TO 172
214             XK(I,I)=(A(I-1)+A(I))*PERJT(I)
215             GO TO 173
216        171 XK(I,I)=A(I)*PERJT(I)
217             GO TO 173
218        172 XK(I,I)=A(I-1)*PERJT(I)
219        173 CONTINUE
220        140 CONTINUE
221             INDEX = 2
222             IF(KLIST.EQ.0)GO TO 1350
223             WRITE(3,1344)
224       1344 FORMAT(////,T10,'THE K-MATRIX',//)
225             I1 = 1
226         6 K1 = I1 + 9
227             IF(K1.GT.JTS)K1 = JTS
228             DO 7  K2 = 1,JTS
229             WRITE(3,8)K2,(XK(K2,J2),J2=I1,K1)
230         7 CONTINUE
231         8 FORMAT(T2,I3,12F10.5)
232             I1 = K1 + 1
233             IF(K1.NE.JTS)WRITE(3,9)
234         9 FORMAT(////,T10,'THE MATRIX CONTINUED',//)
235             IF(K1.NE.JTS)GO TO 6
236       1350 CONTINUE
237             JJ=JTS+1

           C***********
           C    MAKE LOWER RIGHT CORNER OF THE ASAT MATRIX SYMMETRICAL
           C    BY AVERAGING VALUES
238             DO 210  I = JJ,M
239             IF(IZERO.EQ.0)GO TO 1460
240             IF(I.EQ.NPZERO(JJJ))ASAT(I,I) = 0.0000001
241             IF(I.EQ.NPZERO(JJJ))JJJ = JJJ + 1
242       1460 DO 210  J = I,M
243             AVER = (ASAT(I,J) + ASAT(J,I))/2.
244             ASAT(I,J) = AVER
245             ASAT(J,I) = AVER
246        210 CONTINUE
247             INDEX = 3
248             IF(LIST.GT.0)CALL WRITEO(ASAT,M,INDEX,JTS)
249       1360 DO 230 I=1,JTS
250             DO 230 J=1,JTS
251             DK(I,J)=0.0
252             DO 230 K=1,JTS
253             II=I+JTS
254             KK=K+JTS
255        230 DK(I,J)=DK(I,J)+ASAT(II,KK)*XK(K,J)
256             DO 240 I=1,JTS
257             DO 240 J=1,JTS
258        240 CDK(I,J)=DK(I,J)
259             NBM=0
           C***********************
           C    LOOP FOR BUCKLING MODES
           C    ***********
260        390 NINTER = 0
261             DO 250 I=1,JTS
262        250 DKX(I)=1.0
263        260 NINTER=NINTER+1
264             IF(NINTER.EQ.1000) GO TO 1111
265             DO 270 I=1,JTS
266        270 X(I)=DKX(I)
```

B10 (*continued*)

```
267          SAVE=0.0
268          DO 290 I=1,JTS
269          DKX(I)=0.0
270          DO 280 J=1,JTS
271      280 DKX(I)=DKX(I)+DK(I,J)*X(J)
     C     OBTAIN COUNTER IS TO USE LATER
272          IF(SAVE.LT. ABS(DKX(I))) IS=I
273      290 SAVE=AMAX1(SAVE, ABS(DKX(I)))
274          DO 300 I=1,JTS
275      300 DKX(I)=DKX(I)/SAVE
276          DO 310 I=1,JTS
277          TEST= ABS(X(I)-DKX(I))
278          IF(TEST.GT.0.000001  ) GO TO 260
279      310 CONTINUE
280     1111 CONTINUE
281          NBM=NBM+1
282          PCR= ABS(X(IS))/SAVE
283          IF(NBM.LE.1) WRITE(3,1500)
284     1500 FORMAT('1',//,T10,'THE MODE INDICATED BY THE ANSWER IS USED AS A C
         1OUNTER -',/,T10,'BY INSPECTION OF THE UNIT DEFLECTIONS THE ACTUAL
         2BUCKLING MODE CAN BE DETERMINED',/,T10,'1000 ITERATIONS IS THE MAX
         2IMUM NUMBER OF ITERATIONS ALLOWED BY THE PROGRAM',//)
285          WRITE(3,628)PCR,UT3,NBM,NINTER
286      628 FORMAT(/,T5,'THE BUCKLING LOAD IS ',F15.3,1X,A4,' FOR MODE ',I2,'
         1AFTER ',I4,' ITERATIONS',//)
287          IF(NBM.GT.NBMR)GO TO 410
     C
288          DO 320 I=1,JTS
289          CDKX(I)=0.0
290          DO 320 J=1,JTS
291      320 CDKX(I)=CDKX(I)+CDK(I,J)*X(J)
292          WRITE(3,1410)
293     1410 FORMAT(//,T15, 'JOINT DEFLECTIONS',/)
294          DO 1400 I=1,JTS
295     1400 WRITE(3,1401) I, CDKX(I), DKX(I)
296     1401 FORMAT(10X,I2,2F10.5)
297          VALUE= ABS(CDKX(IS))
298          DO 330 I=1,JTS
299      330 CDKX(I)=CDKX(I)/VALUE
300      350 XTKX=0.0
301          DO 370 I=1,JTS
302          DO 370 J=1,JTS
303      370 XTKX=XTKX+XK(J,I)*DKX(J)*DKX(I)
304          FATR=1.0/(PCR*XTKX)
305          DO 380 I=1,JTS
306          DO 380 J=1,JTS
307          DO 380 K=1,JTS
308      380 DK(I,J)=DK(I,J)-FATR*DKX(I)*XK(J,K)*DKX(K)
309          IF(NBM.LE.NBMR)GO TO 390
310      360 CONTINUE
311      410 CONTINUE
312          GO TO 5000
313     6000 STOP
314          END

315          SUBROUTINE WRITEO(A,M,INDEX,JTS)
     C*******
     C     ROUTINE TO WRITE VARIOUS MATRICES IN READABLE OUTPUT FORMAT
     C**********
316          DIMENSION A(40,40)
317          DOUBLE PRECISION A
318          MM = M
319          IF(INDEX.EQ.2)MM = JTS
320          LL = 1
321          IF(INDEX.EQ.3)LL = JTS+1
322          IF(INDEX.EQ.0)WRITE(3,22)
323       22 FORMAT('1',//,T5,'THE ASAT MATRIX',//)
324          IF(INDEX.EQ.1)WRITE(3,23)
325       23 FORMAT('1',//,T5, 'THE ASAT INVERSE MATRIX',//)
326          IF(INDEX.EQ.2)WRITE(3,24)
327       24 FORMAT('1',//,T5,'THE K-MATRIX',//)
328          IF(INDEX.EQ.3)WRITE(3,25)
329       25 FORMAT('1',//,T5,'THE LOWER RT CORNER OF ASAT-INVERSE MATRIX AVERA
         1GED',//)
330          I = 1
331          IF(INDEX.EQ.3)I = JTS+1
332        3 K = I + 10
333          IF(K.GT.MM)K = MM
334          DO 20 IK = LL,MM
335          IF(INDEX.NE.0)WRITE(3,40)IK,(A(IK,J),J=I,K)
336          IF(INDEX.EQ.0)WRITE(3,18)IK,(A(IK,J),J=I,K)
337       20 CONTINUE
338       18 FORMAT(T2,I3,11F11.1)
339       40 FORMAT(T2,I3,11F11.6)
340          I = K + 1
341          IF(K.NE.MM)WRITE(3,19)
342       19 FORMAT(////,T10,'THE MATRIX CONTINUED',//)
343          IF(K.NE.MM)GO TO 3
344          RETURN
345          END
```

B-11. Three-dimensional pile-group analysis

```
     C        J E BOWLES  --  3-DIMENSIONAL PILE GROUP ANALYSIS
     C        SIGNS ARE:  +PX = TO RT.  + PY = UP    + PZ = OUT OF PLANE OF PAPER
     C                    +FW IS AT 90 DEG TO THE BATTER
     C                    +FU IS DOWN ALONG THE PILE AXIS
     C                    +FV DEPENDS ON BETA AND IS PARALLEL TO NEG Z AXIS
     C                        WHEN BETA = 0
     C         + MOMENT ACCORDING TO RT HAND RULE
     C         + MOMENT ACCORDING TO RT HAND RULE USING FORCE DIRECTIONS FOR THUMB
     C        BETA = ANGLE OF PILE PROJECTION WITH X-AXIS IN XZ PLANE
     C
     C        BATTER = PILE BATTER WITH VERTICAL AS H:1 WITH 1 = HORIZ.
     C        USE BATTER = 0. FOR VERTICAL PILES -- NOTE THAT BETA DEFINES PILE
     C        BATTER DIRECTION.
     C
     C        BETA = 0. DEG FOR VERTICAL PILES OR PILES BATTERED PARALLEL TO X-AXIS
     C
     C        ** NOTE THAT AS BETA INCREASES FROM ZERO DEG TO 360 DEG THAT
     C           POSITIVE DIRECTIONS OF FW AND FV ALSO ROTATE
     C
     C        ITEST = SWITCH (ITEST>1) TO REDO PILE GROUP WITHOUT CHANGING
     C           OR READING NEW C(I) VALUES SEE LINE 16 JUST FOLLOWING "1006"
     C
  1           DIMENSION XF(6), BETA(30),SUM(6), COMP(30,6),C(30,6),X(30),Y(30),Z
             1(30),          BATTER(30),A(6,6), S(6,6), SAT(6,6),ASAT(6,6),F(6),
             2P(6), ZZ(6), YY(6),XPILE(6),TITLE(20)
  2      5000 READ(1,1000,END=150)TITLE
  3      1000 FORMAT(20A4)
  4           WRITE(3,2000)TITLE
  5      2000 FORMAT(1',//,T5,20A4,//)
  6           READ(1,1004)NPILES,NNZP,LIST,ITEST
  7      1004 FORMAT(16I5)
  8           INDEX = 0
  9           WRITE(3,2002)
 10      2002 FORMAT(T10, 'GENERAL INPUT DATA',/,T5, 'PILE NO',4X, 'X',9X,'Y',9X
             1,'Z', 7X,'BETA', 6X,'BATTER')
 11           DO 20  I = 1,NPILES
 12           READ(1,1006)X(I),Y(I),Z(I),BETA(I),BATTER(I)
 13           WRITE(3,2004)I,X(I),Y(I),Z(I),BETA(I),BATTER(I)
 14      2004 FORMAT(T8,I2,5(3X,F7.2))
 15      1006 FORMAT(8F10.4)
 16           IF(ITEST.GT.0)GO TO 20
 17           READ(1,1006)(C(I,J),J=1,6)
 18        20 CONTINUE
 19           WRITE(3,2006)
 20      2006 FORMAT( T10, 'THE PILE CONSTANTS ARE',/, T5,'PILE NO', 6X,'C(1)',
             A9X, 'C(2)', 9X,'C(3)',9X,'C(4)', 9X,'C(5)', 9X,'C(6)',/)
 21           DO 21  I = 1,NPILES
 22        21 WRITE(3,2008)I,(C(I,J),J=1,6)
 23      2008 FORMAT(T8,I2,3X,F11.2,5(2X,F11.2))
 24           DO 100  I = 1,6
 25           P(I) = 0.
 26           SUM(I) = 0.
 27           DO 100  J = 1,6
 28       100 ASAT(I,J) = 0.
 29           RAD = 180./3.1415927
     C**********
     C        START LOOP DEVELOPING REQ'D MATRICES
     C**********
 30           I = 1
 31        40 BETAR = BETA(I)/RAD
 32           SINB = SIN(BETAR)
 33           COSB = COS(BETAR)
 34           SLOPE = SQRT(1.+BATTER(I)**2)
 35           SINA = BATTER(I)/SLOPE
 36           IF(BATTER(I).EQ.0.)SINA = 1.
 37           COSA = 1./SLOPE
 38           IF(BATTER(I).EQ.0.)COSA = 0.
     C        BUILD A-MATRIX
 39           DO 41  K = 1,6
 40           DO 41  J = 1,6
 41        41 A(K,J) = 0.
 42           A(1,1) = COSA*COSB
 43           A(1,2) = SINB
 44           A(1,3) = SINA*COSB
 45           A(2,1) = -SINA
 46           A(2,3) = COSA
 47           A(3,1) = COSA*SINB
 48           A(3,2) = -COSB
 49           A(3,3) = SINA*SINB
 50           A(4,1) = Z(I)*SINA + Y(I)*COSA*SINB
 51           A(4,2) = -Y(I)*COSB
 52           A(4,3) = -Z(I)*COSA + Y(I)*SINA*SINB
 53           A(4,4) = A(1,1)
 54           A(4,5) = A(1,2)
 55           A(4,6) = A(1,3)
 56           A(5,1) = Z(I)*A(1,1) - X(I)*A(3,1)
 57           A(5,2) = Z(I)*SINB + X(I)*COSB
 58           A(5,3) = Z(I)*A(1,3) - X(I)*A(3,3)
 59           A(5,4) = A(2,1)
 60           A(5,6) = A(2,3)
 61           A(6,1) = -Y(I)*A(1,1) - X(I)*SINA
 62           A(6,2) = -Y(I)*A(1,2)
 63           A(6,3) = -Y(I)*A(1,3) + X(I)*COSA
 64           A(6,4) = A(3,1)
 65           A(6,5) = A(3,2)
 66           A(6,6) = A(3,3)
     C        BUILD S-MATRIX
 67           DO 42  K = 1,6
 68           DO 42  J = 1,6
 69        42 S(K,J) = 0.
 70           S(1,1) = C(I,1)
 71           S(2,2) = C(I,2)
 72           S(2,6) = C(I,4)
 73           S(3,3) = C(I,2)
 74           S(3,5) = -C(I,4)
 75           S(4,4) = C(I,6)
 76           S(5,3) = -C(I,5)
 77           S(5,5) = C(I,3)
 78           S(6,2) = C(I,5)
 79           S(6,6) = C(I,3)
 80           IF(INDEX.GT.0)GO TO 85
 81           DO 60  K = 1,6
 82           DO 60  L = 1,6
 83           SAT(K,L) = 0.
 84           DO 60  M = 1,6
 85        60 SAT(K,L) = SAT(K,L) + S(K,M)*A(L,M)
```

B11 (*continued*)

```
86          IF(LIST.LE.0)GO TO 65
87          WRITE(3,2010)I
88     2010 FORMAT( //,T10, 'THE A-MATRIX FOR PILE NO',I3,'***', T65, 'THE SAT
          1 MATRIX--100 FACTORED')
89          WRITE(3,2011)((A(M,K), K = 1,6),(SAT(M,J), J = 1,6),M=1,6)
90     2011 FORMAT(T5,0P6F8.4,1X,-2P6F9.1)
91     65   DO 70   K = 1,6
92          DO 70   L = 1,6
93          DO 70   M = 1,6
94     70   ASAT(K,L) = ASAT(K,L) + A(K,M)*SAT(M,L)
95          I = I + 1
96          IF(I.LE.NPILES)GO TO 40
       C    END OF ASAT FORMATION
       C    BUILD P-MATRIX
97          READ(1,1010)(J,P(J),K=1,NNZP)
98     1010 FORMAT(I5,F10.4)
99          WRITE(3,2015)
100    2015 FORMAT( ///,T5, 'THE P-MATRIX IS',T26,'THE FOUNDATION ASAT MATRI
          1X--100  FACTORED')
101         WRITE(3,2017)(I,P(I),I,(ASAT(I,J),J=1,6),I=1,6)
102    2017 FORMAT(T5,I3,2X,'PX = ',F9.2, 5X,I3,-2P(6F10.3),/, T5,I3, 2X,'PY =
          1 ',0PF9.2, 5X,I3,-2P(6F10.3),/,  T5, I3,2X,'PZ = ',0PF9.2, 5X,I3,-2
          2P(6F10.3),/, T5,I3,2X,'MX = ',0PF9.2, 5X,I3,-2P(6F10.3),/,T5,I3,2X
          3,'MY = ',0PF9.2,5X,I3,-2P(6F10.3),/, T5,I3,2X, 'MZ = ',0PF9.2, 5X,
          4I3,-2P(6F10.3))
       C    **   INVERT ASAT MATRIX USING IBM STANDARD INVERSION SUBROUTINE  **
103         CALL MINV(ASAT,6,DET,ZZ,YY)
       C    COMPUTE FOUNDATION DISPLACEMENTS XF
104         DO 80   K = 1,6
105         XF(K) = 0.
106         DO 80   J = 1,6
107    80   XF(K) = XF(K) + ASAT(K,J)*P(J)
108         WRITE(3,2020)(XF(J),J=1,6)
109    2020 FORMAT(//,T15,'THE FOUNDATION DISPLACEMENTS ARE',/, T5, 'X =', F9.
          16,2X,'Y=',F9.6,2X,'Z=',F9.6,2X, 'ALPHA X=',F9.6,2X,
          1'ALPHA Y=', F9.6, 2X, 'ALPHA Z=', F9.6)
110         WRITE(3,2022)
111    2022 FORMAT(//,T5, 'THE FOUNDATION DISPLACEMENTS AND PILE FORCES--**NOT
          1E FU,FV,FW,ETC ARE ACTING ON CAP',/,T2, 'PILE',4X,'DU', 6X,'DV',6X
          2, 'DW',3X,'ALPHA U',1X, 'ALPHA V',6X,'ALPHA W',6X,'FU',7X,'FV',7X,
          3'FW',8X,'MU',8X,'MV',8X,'MW')
112         INDEX = INDEX + 1
113         I = 1
114         GO TO 40
115    85   DO 88   J = 1,6
116         XPILE(J) = 0.
117         DO 88   K = 1,6
118    88   XPILE(J) = XPILE(J) + A(K,J)*XF(K)
119    88   CONTINUE
       C    **  COMPUTE PILE FORCES USING F = S(I)*XPILE(I)  **
120         DO 91   J = 1,6



121         F(J) = 0.
122         DO 90   K = 1,6
123    90   F(J) = F(J) + S(J,K)*XPILE(K)
124    91   CONTINUE
125         WRITE(3,2024)I,(XPILE(K),K=1,6),(F(L),L=1,6)
126    2024 FORMAT(T3,I2,2X,F7.4,5(1X,F7.4),2X,F8.3,5(2X,F8.3))
       C ** COMPUTE X,Y,Z FORCE COMPONENTS OF INDIVIDUAL PILES AND SUM AS
       C       STATICS CHECK OF SOLUTION *******
127         DO 95   K = 1,6
128         COMP(I,K) = 0.
129         DO 94   J = 1,6
130    94   COMP(I,K) = COMP(I,K) + A(K,J)*F(J)
131         SUM(K) = SUM(K) + COMP(I,K)
132    95   CONTINUE
133         I = I+1
134         IF(I.LE.NPILES)GO TO 40
135         WRITE(3,2026)
136    2026 FORMAT(//,T10,  'INDIVIDUAL PILE FORCE COMPONENTS TO CHECK SUM OF
          1FORCES ALONG AXES',/,T10,'PILE NO',6X,'FX',10X,'FY',10X,'FZ',10X,
          2 'MX',10X,'MY',10X,'MZ')
137         WRITE(3,2028)(I,(COMP(I,K),K=1,6),I=1,NPILES)
138    2028 FORMAT(T10,I3,3X,6F12.4)
139         WRITE(3,2030)(SUM(I),I=1,6),(P(N),N=1,6)
140    2030 FORMAT(T8, 'TOTAL =',1X,6F12.4,/,T16,'(',F10.3,')(',F10.3,')(',F10
          1.3,')(',F10.3,')(',F10.3,')(',F10.3,')')
141         GO TO 5000
142    150  STOP
143         END
```

REFERENCES

To simplify and condense the reference list, the following abbreviations are used:

AASHO		American Association of State Highway Officials (now Highway and Transportation Officials)
ACI		American Concrete Institute, Detroit, Michigan
	JACI	*Journal of American Concrete Institute* (monthly)
ASCE		American Society of Civil Engineers
	JGED	*Journal of Geotechnical Engineering Division*, ASCE (1974–)
	JSMFD	*Journal of Soil Mechanics and Foundations Division*, ASCE (1955–1973, incl.)
	PSC	*Proceedings of Soil Mechanics and Foundations Division*, ASCE, speciality conferences as follows:

> *1st PSC* = Shear Strength of Cohesive Soils, Boulder, Colo. (1960)
>
> *2d PSC* = Design of Foundations for Control of Settlement, Northwestern University, Evanston, Ill. (1964)
>
> *3d PSC* = Placement and Improvement of Soil to Support Structures, Cambridge, Mass. (1968)
>
> *4th PSC* = Lateral Stresses in the Ground and Design of Earth Retaining Structures, Cornell University, Ithaca, N.Y. (1970)
>
> *5th PSC* = Performance of Earth and Earth Supported Structures, Purdue University (1972)
>
> *6th PSC* = In Situ Measurement of Soil Properties, North Carolina State University, Raleigh, N.C. (1975)

	JSD	*Journal of Structural Division*, ASCE
ASME		American Society for Mechanical Engineers
ASTM		American Society for Testing and Materials, 1916 Race Street, Philadelphia, Pa.
	ASTM STP	*American Society for Testing and Materials Special Technical Publication* (with appropriate number)

AWPI	American Wood Preservers Institute, 2600 Virginia Avenue, Washington, D.C.
CGJ	*Canadian Geotechnical Journal,* Ottawa, Canada
ENR	*Engineering News-Record,* New York (weekly)
ICE	Institution of Civil Engineers (London)
PICE	*Proceedings of Institution of Civil Engineers*
ICSMFE	*Proceedings of International Conference on Soil Mechanics and Foundation Engineering*

1st ICSMFE = Harvard University (1936)
2nd ICSMFE = Rotterdam, Holland (1948)
3d ICSMFE = Zurich, Switzerland (1953)
4th ICSMFE = London (1957)
5th ICSMFE = Paris (1961)
6th ICSMFE = Montreal (1965)
7th ICSMFE = Mexico City (1969)
8th ICSMFE = Moscow (1973)
9th ICSMFE = Tokyo (1977, to be held)

SMFE	*Soil Mechanics and Foundation Engineering*

Geotechnique is published by the Institution of Civil Engineers, London, England.

Highway Research Board, Highway Research Record, etc., published by National Academy of Science, Washington, D.C.

AASHO (1973), "Standard Specifications for Highway Bridges," 11th ed., 469 pp.

Aboshi, H., et al. (1970), Constant Loading Rate Consolidation Test, *Soils and Foundations, Tokyo,* vol. 10, no. 1, March, pp. 43–56.

ACI (1975), Recommended Practice for Design and Construction of Concrete Bins, Silos and Bunkers for Storing Granular Materials, Report of ACI Committee 313, *JACI,* October, pp. 529–565.

ACI (1971), ACI Standard Building Code Requirements for Reinforced Concrete, ACI 318–71, 78 pp.

ACI Committee 336 (1972), Suggested Design and Construction Procedures for Pier Foundations, *JACI,* August, pp. 461–480.

ACI Committee 436 (1966), Suggested Design Procedures for Combined Footings and Mats, *JACI,* October, pp. 1041–1057 (now Committee 336).

AISC (1973), "Steel Construction Manual," 7th ed., American Institute of Steel Construction, 101 Park Ave., New York.

Aldrich, H. P. (1965), Precompression for Support of Shallow Foundations, *JSMFD,* ASCE, vol. 91, SM 2, March, pp. 5–20.

Anandakrishnan, M., and N. R. Krishnaswamy (1973), Response of Embedded Footings to Vertical Vibrations, *JSMFD,* ASCE, vol. 99, SM 10, October, pp. 863–883.

Arman, A., et al. (1975), Study of the Vane Shear, *6th PSC,* ASCE, vol. 1, pp. 93–120.

Arnold, R. N., et al. (1955), Forced Vibrations of a Body on an Infinite Elastic Solid, *Journal of Applied Mechanics,* vol. 77, no. 3, pp. 391–400.

ASCE (1974), Design of Steel Transmission Pole Structures, *JSD,* ASCE, vol. 100, ST 12, December, pp. 2449–2518 (Committee Report).

ASCE (1966), Revised Bibliography on Chemical Grouting, *JSMFD,* ASCE, vol. 92, SM 6, November, pp. 39–66.

ASCE (1959), "Timber Piles and Construction Timbers," Manual of Practice no. 17, 48 pp.

ASCE (1946), "Pile Foundations and Pile Structures," Manual of Practice no. 27, 72 pp. (reprinted 1959).

ASCE (1941), Pile-driving Formulas, *Proceedings ASCE,* vol. 67, no. 5, May, pp. 853–866.

Aschenbrenner, R. (1967), Three Dimensional Analysis of Pile Foundations, *JSD,* ASCE, vol. 93, ST 1, February, pp. 201–219.

AWPI (1969), "Pile Foundations Know-How," 66 pp.

AWPI (1967), "Pressure Treated Timber Foundation Piles for Permanent Structures," 98 pp.

AWPI (1966), "Timber Foundation Pile Study," 46 pp.

Baguelin, F., et al. (1974), Self-boring Placement Method of Soil Characteristics Measurement, *Proceedings Conference on Subsurface Exploration for Underground Excavation and Heavy Construction*, ASCE, pp. 312–332.

Baker, C. N., and F. Khan (1971), Caisson Construction Problems and Corrections in Chicago, *JSMFD*, ASCE, vol. 97, SM 2, February, pp. 417–440.

Balla, A. (1962), Bearing Capacity of Foundations, *JSMFD*, ASCE, vol. 88, SM 5, October, pp. 13–34.

Balla, A. (1961), The Resistance to Breaking Out of Mushroom Foundations for Pylons, *5th ICSMFE*, vol. 1, pp. 569–576.

Ballard, R. J., Jr., and F. G. McLean (1975), Seismic Field Methods for In-Situ Moduli, *6th PSC*, ASCE, vol. 1, pp. 121–150.

Banks, D. C., and B. N. MacIver (1969), "Variation in Angle of Internal Friction with Confining Pressure," U.S. Army Engineer Waterways Experiment Station, Misc. Paper S-69-12 (available from NTIS, Springfield, Va.).

Barber, E. S. (1961), Notes on Secondary Consolidation, *Proceedings Highway Research Board*, vol. 40, pp. 663–675.

Barden, L. (1968), Primary and Secondary Consolidation of Clay and Peat, *Geotechnique*, vol. 18, no. 1, March, pp. 1–24.

Barden, L. (1962), Distribution of Contact Pressure under Foundations, *Geotechnique*, vol. 12, no. 3, September, pp. 181–198.

Barden, L., and A. J. Khayatt (1966), Incremental Strain Rate Ratios and Strength of Sand in the Triaxial Test, *Geotechnique*, vol. 16, no. 4, December, pp. 338–357.

Barden, L., and M. F. Monckton (1970), Tests on Model Pile Groups in Soft and Stiff Clay, *Geotechnique*, vol. 20, no. 1, March, pp. 94–96 (includes most references on model-pile test groups).

Barden, L., et al. (1969), Elastic and Slip Components of the Deformation of Sand, *CGJ*, Ottawa, vol. 6, no. 3, August, pp. 227–240.

Barentsen, P. (1936), Short Description of a Field Method with Cone Shaped Sounding Apparatus, *1st ICSMFE*, vol. 1, pp. 7–10.

Barkan, D. D. (1962), "Dynamics of Bases and Foundations," McGraw-Hill Book Company, New York, 434 pp.

Barron, R. A. (1948), Consolidation of Fine-grained Soils by Drain Wells, *Transactions ASCE*, vol. 113, pp. 718–751.

Bazaraa, A. R. (1967), "Use of the Standard Penetration Test for Estimating Settlements of Shallow Foundations on Sand," Ph.D. Thesis, University of Illinois, Urbana, 379 pp.

Begemann, H. K. (1953), Improved Method of Determining Resistance to Adhesion by Sounding through a Loose Sleeve Placed behind the Cone, *3rd ICSMFE*, vol. 1, pp. 213–217.

Bell, A. L. (1915), The Lateral Pressure and Resistance of Clay, and the Supporting Power of Clay Foundations, in "A Century of Soil Mechanics," ICE, London, pp. 93–134.

Belz, C. A. (1970), Cellular Structure Design Methods, *Proceedings Conference: Design and Installation of Pile Foundations and Cellular Structures*, Lehigh University, pp. 319–338.

Berezantzev, V. G. (1965), Design of Deep Foundations, *6th ICSMFE*, vol. 2, pp. 234–237.

Berezantzev, V. G., et al. (1961), Load Bearing Capacity and Deformation of Piled Foundations, *5th ICSMFE*, vol. 2, pp. 11–15.

Bhatacharya, R. K. (1968), "Stresses and Displacements in Cross-Anisotropic Layered, Elastic Half-Space Due to Axi-Symmetric Loadings on the Top Surface," Ph.D. Thesis, University of Wisconsin, Madison, 162 pp.

Bishop, A. W. (1961), Discussion: Soil Properties and Their Measurement, *5th ICSMFE*, vol. 3, p. 99.

Bishop, A. W., and G. E. Blight (1963), Some Aspects of Effective Stress in Saturated and Partly Saturated Soils, *Geotechnique*, vol. 13, no. 3, pp. 177–197.

Bishop, A. W., and V. K. Garga (1969), Drained Tension Tests on London Clay, *Geotechnique*, vol. 19, no. 2, June, pp. 309–312.

Bishop, A. W., and D. J. Henkel (1962), "The Measurement of Soil Properties in the Triaxial Test," 2d ed., Edward Arnold (Publishers) Ltd., London, 228 pp.

Bjerrum, L. (1972), Embankments on Soft Ground, *5th PSC*, ASCE, vol. 2, pp. 1–54.

Bjerrum, L., and O. Eide (1956), Stability of Strutted Excavations in Clay, *Geotechnique*, vol. 6, no. 1, March, pp. 32–47.

Bjerrum, L., et al. (1969), Reduction of Negative Skin Friction on Steel Piles to Rock, *7th ICSMFE*, vol. 2, pp. 27–34.

Bjerrum, L., and R. Kirkedam (1958), Some Notes on Earth Pressures in Stiff Fissured Clay, *Proceedings Brussels Conference on Earth Pressure Problems*, pp. 15–27.

Blaine, E. S. (1967), Practical Lessons in Caisson Sinking from the Baton Rouge Bridge, *ENR*, Feb. 6, pp. 213–215.

Bond, D. (1961), Influence of Foundation on Size of Settlement, *Geotechnique*, vol. 11, no. 2, June, pp. 121–143.

Borowicka, H. (1943), Über ausmittig belaste starre Platten auf elastischisotorpem Untergrund, *Ingenieur-Archiv*, Berlin, vol. 1, pp. 1–8.

Borowicka, H. (1936), Influence of Rigidity of a Circular Foundation Slab on the Distribution of Pressures over a Contact Surface, *1st ICSMFE*, vol. 2, pp. 144–149.

Bowles, J. E. (1975a), Spread Footings, chap. 15 in "Foundation Engineering Handbook," Van Nostrand Reinhold Co., New York, 751 pp.

Bowles, J. E. (1975b), Combined and Special Footings, chap. 16 in "Foundation Engineering Handbook," Van Nostrand Reinhold Co., New York, 751 pp.

Bowles, J. E. (1974a), "Analytical and Computer Methods in Foundation Engineering," McGraw-Hill Book Company, New York, 519 pp.

Bowles, J. E. (1974b), Foundations for Family Housing, *Technical Report* D-20: Systems Approach to Site Development, Construction Engineering Research Laboratory, Champaign, Ill., 107 pp.

Bowles, J. E. (1970), "Engineering Properties of Soils and Their Measurement," McGraw-Hill Book Company, New York, 189 pp.

Bozozuk, M. (1974), Minor Principal Stress Measurements in Marine Clay with Hydraulic Fracture Tests, *Proceedings Conference on Subsurface Exploration for Underground Excavation and Heavy Construction*, ASCE, pp. 333–349.

Bozozuk, M. (1972), Downdrag Measurements on a 160-ft Floating Pipe Test Pile in Marine Clay, *CGJ*, vol. 9, no. 2, May, pp. 127–136.

BRAB (1968), Criteria for Selection and Design of Residential Slabs-on-Ground, Building Research Advisory Board, *Federal Housing Administration Report* 33, National Academy of Sciences, Washington, D.C., 289 pp.

Brand, E. W., et al. (1972), Load Tests on Small Foundations in Soft Clay, *5th PSC*, ASCE, vol. 1, part 2, pp. 903–928.

Broms, B. B. (1972), Settlements of Pile Groups, *5th PSC*, ASCE, vol. 3, pp. 181–199 (contains extensive bibliography).

Broms, B. B. (1971), Lateral Earth Pressures Due to Compaction of Cohesionless Soils, *4th Budapest Conference SMFE*, Budapest, pp. 373–384.

Brooker, E. W., and H. O. Ireland (1965), Earth Pressures at Rest Related to Stress History, *CGJ*, vol. 2, no. 1, February, pp. 1–15.

Burland, J. B., et al. (1966), The Behaviour and Design of Large Diameter Bored Piles in Stiff Clay, *Proceedings Conference: Large Bored Piles*, ICE, London, pp. 51–71.

Butterfield, R., and P. K. Banerjee (1971), The Problem of Pile Group–Pile Cap Interaction, *Geotechnique*, vol. 21, no. 2, June, pp. 135–142.

Button, S. J. (1953), The Bearing Capacity of Footings on a Two-Layer Cohesive Subsoil, *3rd ICSMFE*, vol. 1, pp. 332–335.

Bycroft, G. N. (1956), Forced Vibrations of a Rigid Circular Plate on a Semi-infinite Elastic Space and on an Elastic Stratum, *Philosophical Transactions of the Royal Society, London*, Series A, vol. 248, pp. 327–368.

Caquot, A., and J. Kerisel (1948), "Tables for the Calculation of Passive Pressure, Active Pressure, and Bearing Capacity of Foundations" (translated by M. A. Bec, London), Gauthier-Villars, Paris.

Carlson, L. (1948), Determination in Situ of the Shear Strength of Undisturbed Clay by Means of a Rotating Auger, *2nd ICSMFE*, vol. 1, pp. 265–270.

Casagrande, A. (1936), The Determination of the Preconsolidation Load and Its Practical Significance, *1st ICSMFE*, vol. 3, pp. 60–64.

Casagrande, A. (1948), Classification and Identification of Soils, *Transactions ASCE*, vol. 113, pp. 901–991.

Casagrande, A., and R. E. Fadum (1944), Application of Soil Mechanics in Designing Building Foundations, *Transactions ASCE*, vol. 109, pp. 383–490, (also see "closure").

Caspe, M. S. (1966), Surface Settlement Adjacent to Braced Open Cuts, *JSMFD*, ASCE, vol. 92, SM 4, July, pp. 51–59.

Chellis, R. D. (1961), "Pile Foundations," 2d ed., McGraw-Hill Book Company, New York, 704 pp.

Chellis, R. D. (1941), Discussion: Pile Driving Formulas, *Proceedings ASCE*, vol. 67, no. 8, October, pp. 1517–1537.

Chen, W. F., and H. L. Davidson (1973), Bearing Capacity Determination by Limit Analysis, *JSMFD*, ASCE, vol. 99, SM 6, June, pp. 433–449.

Chicago Building Code (1976), Obtained from Index Publishing Corp., 308 West Randolph St., Chicago, Ill. 60606 (yearly).

Chowdhury, R. N. (1972), "Deformation Problems in Anisotropic Soil—Application of the Finite Element Method," Conference on Finite Element Method in Civil Engineering, McGill University, Montreal, Canada, pp. 653–675.

Christian, J. T. (1976), Soil-Foundation-structure Interaction, *Proceedings Conference on Foundations for Tall Buildings*, Lehigh University, Envo Publishing Co., pp. 149–180.

Chu, K. H., and O. F. Afandi (1966), Analysis of Circular and Annular Slabs for Chimney Foundations, *JACI*, vol. 63, no. 12, December, pp. 1425–1446.

Chummar, A. V. (1972), Bearing Capacity Theory from Experimental Results, *JSMFD*, ASCE, vol. 98, SM 12, December, pp. 1311–1324.

Clevenger, W. A. (1958), Experiences with Loess as Foundation Material, *Transactions ASCE*, vol. 123, pp. 151–180.

Cooke, R. W., and T. Whitaker (1961), Experiments on Model Piles with Enlarged Bases, *Geotechnique*, vol. 11, no. 1, March, pp. 1–13.

Corps of Engineers (1938), "Report on the Slide of a Portion of the Upstream Face of the Fort Peck Dam," U.S. Department of the Army, Government Printing Office, Washington, D.C. (see also T. Middlebrooks, Fort Peck Slide, *Transactions ASCE*, vol. 107, 1942, pp. 723–764).

Coyle, H. M., et al. (1972), Field Measurements of Lateral Earth Pressures on a Cantilever Retaining Wall, *Research Report* 169-2, Texas Transportation Institute, College Station, Tex., 58 pp.

Coyle, H. M., and L. C. Reese (1966), Load Transfer of Axially Loaded Piles in Clay, *JSMFD*, ASCE, vol. 92, SM 2, March, pp. 1–26.

Crawford, C. B. (1961), Engineering Studies of Leda Clay, in "Soils in Canada," University of Toronto Press, Toronto, Canada, pp. 200–217.

Crawford, C. B., and K. N. Burn (1962), Settlement Studies on the Mt. Sinai Hospital, *Engineering Journal of Canada*, Ottawa, vol. 45, no. 12, December.

Crettaz, P., and H. Zeindler (1974), Penetration Testing in Switzerland, *Proceedings European Conference on Penetration Testing*, Stockholm, vol. 1, pp. 133–136.

Cross, H. (1963), "Arches, Continuous Frames, Columns, and Conduits: Selected Papers of Hardy Cross," The University of Illinois Press, Urbana, Ill.

Cummings, A. E. (1940), Dynamic Pile Driving Formulas, in Contributions to Soil Mechanics (1925-1940), *Journal of the Boston Society of Civil Engineers*, pp. 392–413.

Cummings, E. M. (1960), Cellular Cofferdams and Docks, *Transactions ASCE*, vol. 125, pp. 13–45.

Cunny, R. W., and Z. B. Fry (1973), Vibratory in Situ and Laboratory Soil Moduli Compared, *JSMFD*, ASCE, vol. 99, SM 12, December, pp. 1055–1076.

Cushing, J. J., and R. M. Moline (1975), Curved Diaphragm Cellular Cofferdams, *JGED*, ASCE, vol. 101, GT 10, October, pp. 1055–1059.

Dahlberg, R. (1974), Penetration Testing in Sweden, *Proceedings European Conference on Penetration Testing*, Stockholm, vol. 1, pp. 115–131.

Dakshanamurthy, V., and V. Raman (1973), A Simple Method of Identifying an Expansive Soil, *Soils and Foundations*, Tokyo, vol. 13, no. 1, March, pp. 97–104.

D'Appolonia, D. J., et al. (1968), Settlement of Spread Footings on Sand, *JSMFD*, ASCE, vol. 94, SM 3, May, pp. 735–760.

D'Appolonia, E., and J. A. Hribar (1963), Load Transfer in a Step-Taper Pile, *JSMFD*, ASCE, vol. 89, SM 6, November, pp. 57–77 (see also Discussion, July 1964).

Davisson, M. T. (1970), BRD Vibratory Driving Formula, *Foundation Facts*, published periodically by Raymond Concrete Pile Division of Raymond International, Inc., Houston, Tex., vol. 6, no. 1, pp. 9–11.

Davisson, M. T., and K. E. Robinson (1965), Bending and Buckling of Partially Embedded Piles, *6th ICSMFE*, vol. 2, pp. 243–246.

Dawson, R. F. (1959), Modern Practices Used in the Design of Foundations for Structures on Expansive Soils, *Quarterly of the Colorado School of Mines:* Theoretical and Practical Treatment of Expansive Soils, Golden, Colo., vol. 54, no. 4, pp. 66–87.

De Beer, E. E. (1970), Experimental Determination of the Shape Factors and the Bearing Capacity Factors of Sand, *Geotechnique*, vol. 20, no. 4, December, pp. 387–411.

De Mello, V. F. (1971), The Standard Penetration Test, *4th Panamerican Conference on SMFE*, San Juan, Puerto Rico (published by ASCE), vol. 1, pp. 1–86 (with 353 references).

De Mello, V. F. (1969), Foundations of Buildings in Clay, *7th ICSMFE*, State-of-the-Art Volume, pp. 49–136 (with 344 references).

Den Hartog, J. P. (1952), "Advanced Strength of Materials," pp. 125–133 McGraw-Hill Book Company, New York.

De Ruiter, J. (1971), Electric Penetrometer for Site Investigations, *JSMFD*, ASCE, vol. 97, SM 2, February, pp. 457–472.

Deutsch, G. P., and D. H. Clyde (1967), Flow and Pressure of Granular Materials in Silos, *Journal of Engineering Mechanics Division*, ASCE, vol. 93, EM 6, pp. 103–125.

Dewey, F. B., and L. Kempner, Jr. (1975), Discussion: Design of Steel Transmission Pole Structures, *JSD*, ASCE, vol. 101, ST 11, November, pp. 2439–2441.

Dismuke, T. D. (1975), Cellular Structures and Braced Excavations, chap. 14 in "Foundation Engineering Handbook," Van Nostrand Reinhold Co., New York, 751 pp.

Dismuke, T. D. (1970), Stress Analysis of Sheet Piling in Cellular Structures, *Proceedings Conference: Design and Installation of Pile Foundations and Cellular Structures*, Lehigh University, pp. 339–365.

Drannikov, A. M. (1967), Construction on Loess of Small Thickness, *3rd Asian Regional Conference on SMFE*, Haifa, Israel, vol. 1, pp. 3–4.

Duncan, J. M., and C. Y. Chang (1970), Nonlinear Analysis of Stress and Strain in Soils, *JSMFD*, ASCE, vol. 96, SM 5, pp. 1629–1653.

Dunham, C. W. (1962), "Foundations of Structures," 2d ed., McGraw-Hill Book Company, New York, 722 pp.

Durham, G. N., and F. C. Townsend (1973), Effect of Relative Density on the Liquefaction Susceptibility of a Fine Sand under Controlled-Stress Loading, *ASTM STP* 523, pp. 319–331.

Eden, W. J. (1965), An Evaluation of the Field Vane Test in Sensitive Clay, *ASTM STP* 399, pp. 8–17.

Egorov, K. E. (1965), Calculation of Bed for Foundation with Ring Footing, *6th ICSMFE*, vol. 2, pp. 41–45.

Egorov, K. E., and R. V. Serebrjanyi (1963), Determination of Stresses in a Rigid Circular Foundation, *2nd Asian Regional Conference on SMFE*, Tokyo, Japan, vol. 1, pp. 246–250.

ENR (1966), "Caisson Trouble Hits Chicago," Sept. 29, p. 15 (see also editorial).

ENR (1965), "Michigan Pile Test Program Test Results Are Released," May 20, pp. 26–28, 33–34.

Fahlquist, F. E. (1941), New Methods and Techniques in Subsurface Explorations, in Contributions to Soil Mechanics (1941–1953), *Journal of the Boston Society of Civil Engineers*, pp. 20–44.

Fang, H. Y., and W. F. Chen (1971), New Method for Determination of Tensile Strength of Soils, *Highway Research Record* 345, pp. 62–68.

Feld, J. (1965), Tolerance of Structures to Settlement, *JSMFD*, ASCE, vol. 91, SM 3, May, pp. 63–77.

Feld, J. (1943), Discussion: Timber Pile Foundations, *Transactions ASCE*, vol. 108, pp. 143–144.

Fellenius, B. H. (1972), Down-Drag on Piles in Clay Due to Negative Skin Friction, *CGJ*, Ottawa, vol. 9, no. 4, November, pp. 323–337.

Finn, W. D., et al. (1971), Sand Liquefaction in Triaxial and Simple Shear Tests, *JSMFD*, ASCE, vol. 97, SM 4, April, pp. 639–659.

Fletcher, G. F. (1965), Standard Penetration Test: Its Uses and Abuses, *JSMFD*, ASCE, vol. 91, SM 4, July, pp. 67–75.

Flint, R. F. (1971), "Glacial and Quaternary Geology," John Wiley & Sons, Inc., New York, pp. 251–266.

Fox, E. N. (1948), The Mean Elastic Settlement of a Uniformly Loaded Area at a Depth below the Ground Surface, *2nd ICSMFE*, vol. 1, pp. 129–132.

Francis, A. J. (1964), Analysis of Pile Groups with Flexural Resistance, *JSMFD*, ASCE, vol. 90, SM 3, May, pp. 1–32.

Garbe, C. W., and K. Tsai (1972), Engineering Improvements in Reclaimed Marshland for Housing Project, *Proceedings 2nd International Symposium on Lower-Cost Housing Problems*, University of Missouri–Rolla, pp. 153–157.

Gates, M. (1957), Empirical Formula for Predicting Pile Bearing Capacity, *Civil Engineering*, ASCE, vol. 27, no. 3, March, pp. 65–66.

Geddes, J. D. (1969), Boussinesq-based Approximations to the Vertical Stresses Caused by Pile-Type Subsurface Loadings, *Geotechnique*, vol. 19, no. 4, December, pp. 509–514.

Geddes, J. D. (1966), Stresses in Foundation Soils Due to Vertical Subsurface Loading, *Geotechnique*, vol. 16, no. 3, September, pp. 231–255.

Gibbs, H. J., et al. (1960), Shear Strength of Cohesive Soils, *1st PSC*, ASCE, pp. 33–162.

Gibbs, H. J., and W. Y. Holland (1960), "Petrographic and Engineering Properties of Loess," Engineering Monograph No. 28, U.S. Bureau of Reclamation, Denver, Colo., 37 pp.

Gibbs, H. J., and W. G. Holtz (1957), Research on Determining the Density of Sands by Spoon Penetration Testing, *4th ICSMFE*, vol. 1, pp. 35–39.

Gibson, R. E. (1967), Some Results Concerning Displacements and Stresses in a Non-homogeneous Elastic Half-Space, *Geotechnique*, vol. 17, no. 1, March, pp. 58–67.

Gibson, R. E., and W. F. Anderson (1961), In-Situ Measurement of Soil Properties with the Pressuremeter, *Civil Engineering and Public Works Review*, London, May, pp. 615–618.

Glick, G. W. (1948), Influence of Soft Ground in the Design of Long Piles, *2nd ICSMFE*, vol. 4, pp. 84–88.

Godskesen, O. (1936), Investigation of the Bearing Power of the Subsoil (Especially Moraine) with 25 × 25 mm Pointed Drill Weighted with 100 kg without Samples, *1st ICSMFE*, vol. 1, pp. 311–314.

Golder, H. Q. (1975), Floating Foundations, chap. 18 in "Foundation Engineering Handbook," Van Nostrand Reinhold Co., New York.

Golder, H. Q. (1965), State-of-the-Art of Floating Foundations, *JSMFD*, ASCE, vol. 91, SM 2, March, pp. 81–88.

Golder, H. Q., et al. (1970), Predicted Performance of Braced Excavation, *JSMFD*, ASCE, vol. 96, SM 3, May, pp. 801–815.

Gorbunov-Passadov, M. I., and R. V. Serebrjanyi (1961), Design of Structures on Elastic Foundations, *5th ICSMFE*, vol. 1, pp. 643–648.

Grant, R., et al. (1974), Differential Settlement of Buildings, *JGED*, ASCE, vol. 100, GT 9, September, pp. 973–991.

Grayman, R. (1970), Cellular Structure Failures, *Proceedings Conference: Design and Installation of Pile Foundations and Cellular Structures*, Lehigh University, pp. 383–391.

Griffiths, D. H., and R. F. King (1965), "Applied Geophysics for Engineers and Geologists," Pergamon Press, New York, 223 pp.

Gromko, G. J. (1974), Review of Expansive Soils, *JGED*, ASCE, vol. 100, GT 6, June, pp. 667–687.

Hall, J. R., Jr. (1967), Coupled Rocking and Sliding Oscillations of Rigid Circular Footings, *Proceedings International Symposium on Wave Propagation and Dynamic Properties of Earth Materials*, Albuquerque, N.M., pp. 139–148.

Hamilton, J. J., and C. B. Crawford (1959), Improved Determination of Preconsolidation of a Sensitive Clay, *ASTM STP* 254, pp. 254–271.

Hanna, T. H. (1973), The Influence of Anchor Inclination on Pull-out Resistance of Clays, *CGJ*, Ottawa, vol. 10, no. 4, November, pp. 664–669.

Hanna, T. H. (1963), Model Studies of Foundation Groups in Sand, *Geotechnique*, vol. 13, no. 4, December, pp. 334–351.

Hansen, J. B. (1970), A Revised and Extended Formula for Bearing Capacity, *Danish Geotechnical Institute Bull.* 28, Copenhagen, 21 pp.

Hansen, J. B. (1967), The Philosophy of Foundation Design: Design Criteria, Safety Factors, and Settlement Limits, *Proceedings Symposium on Bearing Capacity and Settlement of Foundations*, Duke University, Durham, N.C., pp. 9–13.

Hardin, B. O., and W. L. Black (1968), Vibration Modulus of Normally Consolidated Clay, *JSMFD*, ASCE, vol. 94, SM 2, March, pp. 27–42.

Hardin, B. O., and V. P. Drnevich (1972), Shear Modulus and Damping in Soils: Design Equations and Curves, *JSMFD*, ASCE, vol. 98, SM 7, July, pp. 667–692.

Hardin, B. O., and J. Music (1965), Apparatus for Vibration of Soil Specimens during the Triaxial Test, *ASTM STP* 392, pp. 55–74.

Hardin, B. O., and F. E. Richart, Jr. (1963), Elastic Wave Velocities in Granular Soils, *JSMFD*, ASCE, vol. 89, SM 1, February, pp. 33–65.

Henkel, D. J. (1970), Geotechnical Considerations of Lateral Stresses, *4th PSC*, ASCE, pp. 1–49.

Henkel, D. J. (1959), The Relationships between the Strength, Pore-Water Pressure, and Volume-Change Characteristics of Saturated Clays, *Geotechnique*, vol. 9, no. 3, pp. 119–135.

Hetenyi, M. (1946), "Beams on Elastic Foundations," The University of Michigan Press, Ann Arbor, Mich., 255 pp.

Hijab, W. A. (1956), A Note on the Centroid of a Logarithmic Spiral Sector, *Geotechnique*, vol. 6, no. 1, March, pp. 66–69.

Hiley, A. (1930), Pile-driving Calculations with Notes on Driving Forces and Ground Resistances, *Structural Engineer*, London, vol. 8, pp. 246–259, 278–288.

Hirsch, T. J., et al. (1970), Pile Driving Analysis by One-Dimensional Wave Theory: State of the Art, *Highway Research Record* 333, pp. 33–54.

Hirst, T. J., et al. (1972), A Static Cone Penetrometer for Ocean Sediments, *ASTM STP* 501, pp. 69–80.

Holtz, W. G. (1973), The Relative Density Approach—Uses, Testing Requirements, Reliability, and Shortcomings, *ASTM STP* 523, pp. 5–17.

Holtz, W. G. (1959), Expansive Clays—Properties and Problems, *Quarterly of the Colorado School of Mines:* Theoretical and Practical Treatment of Expansive Soils, Golden, Colo., vol. 54, no. 4, pp. 89–125.

Hooper, J. A. (1974), Analysis of a Circular Raft in Adhesive Contact with a Thick Elastic Layer, *Geotechnique*, vol. 24, no. 4, December, pp. 561–580.

Hough, B. K. (1969), "Basic Soils Engineering," 2d ed., The Ronald Press Company, New York, 634 pp.

Housel, W. S. (1966), Pile Load Capacity, Estimates and Test Results, *JSMFD*, ASCE, vol. 92, SM 4, July, pp. 1–30.

Housel, W. S. (1929), A Practical Method for the Selection of Foundations Based on Fundamental Research in Soil Mechanics, *University of Michigan Engineering Research Bull.* 13, Ann Arbor, Mich., October.

Hrennikoff, A. (1950), Analysis of Pile Foundations with Batter Piles, *Transactions ASCE*, vol. 115, pp. 351–389.

Hsieh, T. K. (1962), Foundation Vibrations, *PICE*, London, vol. 22, June, pp. 211–226.

Hu, C. Y. (1964), Variable-Factors Theory of Bearing Capacity, *JSMFD*, ASCE, vol. 90, SM 4, July, pp. 85–95.

Hughes, J. M., et al. (1975), A Field Trial of the Reinforcing Effect of a Stone Column in Soil, *Geotechnique*, vol. 25, no. 1, March, pp. 31–44.

Huntington, W. C. (1957), "Earth Pressures and Retaining Walls," John Wiley & Sons, Inc., New York, 534 pp.

Hvorslev, M. J. (1960), Physical Components of the Shear Strength of Saturated Clays, *1st PSC*, ASCE, pp. 169–273 (with extensive references).

Hvorslev, M. J. (1949), "Subsurface Exploration and Sampling of Soils for Civil Engineering Purposes," Waterways Experiment Station (reprinted and available from Engineering Foundation, New York, N.Y.), 521 pp.

Ireland, H. O. (1957), Pulling Tests on Piles in Sand, *4th ICSMFE*, vol. 2, pp. 43–45.

Jaky, J. (1948), Pressure in Silos, *2nd ICSMFE*, vol. 1, pp. 103–107.

James, R. G., and P. L. Bransby (1970), Experimental and Theoretical Investigations of a Passive Earth Pressure Problem, *Geotechnique*, vol. 20, no. 1, March, pp. 17–37.

Janbu, N. (1957), Earth Pressures and Bearing Capacity Calculations by Generalized Procedure of Slices, *4th ICSMFE*, vol. 2, pp. 207–212.

Janes, H. W. (1973), Densification of Sand for Drydock by Terra-Probe, *JSMFD*, ASCE, vol. 99, SM 6, June, pp. 451–470.

Jofriet, J. C. (1975), Design of Rectangular Concrete Tank Walls, *JACI*, July, pp. 329–332.

Johnson, S. J. (1970), "Precompression for Improving Foundation Soils," *JSMFD*, ASCE, vol. 96, SM 1, January, pp. 73–110 (also in *3rd PSC*, ASCE, 1968, pp. 9–39).

Joint ASCE-ACI Committee (1974), The Shear Strength of Reinforced Concrete Members—Slabs, *JSD*, ASCE, vol. 100, ST 8, August, pp. 1543–1591.

Kantey, B. A., and R. K. Morse (1965), A Modern Approach to Highway Materials Sampling, *5th ICSMFE*, vol. 1, pp. 55–58.

Kjellman, W. (1948), A Method of Extracting Long Continuous Cores of Undisturbed Soil, *2nd ICSMFE*, vol. 1, pp. 255–258.

Klohn, E. J. (1961), Pile Heave and Redriving, *JSMFD*, ASCE, vol. 87, SM 4, August, pp. 125–145.

Ko, H. Y., and L. W. Davidson (1973), Bearing Capacity of Footings in Plane Strain, *JSMFD*, ASCE, vol. 99, SM 1, January, pp. 1–23.

Koerner, R. M., and A. Partos (1974), Settlement of Building on Pile Foundation in Sand, *JGED*, ASCE, vol. 100, GT 3, March, pp. 265–278.

Komornik, A., and D. David (1969), Prediction of Swelling Pressure of Clays, *JSMFD*, ASCE, vol. 95, SM 1, January, pp. 209–225.

Kondner, R. L. (1963), Hyperbolic Stress-Strain Response: Cohesive Soils, *JSMFD*, ASCE, vol. 89, SM 1, pp. 115–143.

Kovacs, W. D., et al. (1975), "A Comparative Investigation of the Mobile Drilling Company's Safe-T-Driver with the Standard Cathead with Manila Rope for the Performance of the Standard Penetration Test, School of Civil Engineering, Purdue University, 95 pp.

Kuhn, S. H., and A. B. Williams (1961), Scour Depth and Soil Profile Determinations in River Beds, *5th ICSMFE*, vol. 1, pp. 487–490.

Ladanyi, B. (1963), "Evaluation of Pressuremeter Tests in Granular Soils," 2d Panamerican Conference on SMFE, São Paulo, Brazil, vol. 1, pp. 3–20.

Ladd, C. C. (1966), Shear Strength of Cohesive Soils, *Proceedings of Soil Mechanics Lecture Series: Design and Construction of Earth Structures*, Illinois Institute of Technology, Chicago, Ill., pp. 33–92 (extensive references).

Ladd, C. C., and W. A. Bailey (1964), Discussion: The Behaviour of Saturated Clays during Sampling and Testing, *Geotechnique*, vol. 14, no. 4, December, pp. 353–358.

Ladd, C. C., and T. W. Lambe (1963), The Strength of "Undisturbed" Clay Determined from Undrained Tests, *ASTM STP 361*, pp. 342–371.

Lambe, T. W. (1970), Braced Excavations, *4th PSC*, ASCE, pp. 149–218.

Lambe, T. W. (1967), Stress Path Method, *JSMFD*, ASCE, vol. 93, SM 6, pp. 309–331.

Lambe, T. W. (1964), Methods of Estimating Settlement, *2nd PSC*, ASCE, pp. 47–71.

Lambe, T. W., and R. V. Whitman (1969), "Soil Mechanics," John Wiley & Sons, Inc., New York, 553 pp.

Lambe, T. W., et al. (1970), Measured Performance of Braced Excavation, *JSMFD*, ASCE, vol. 96, SM 3, May, pp. 817–836.

Landau, R. E. (1966), Method of Installation as a Factor in Sand Drain Stabilization Design, *Highway Research Record* 133, Highway Research Board, Washington, D.C., pp. 75–96.

Larew, H. G., and G. A. Leonards (1962), A Repeated Load Strength Criterion, *Proceedings Highway Research Board*, vol. 41, pp. 529–556.

La Rochelle, P., and G. Lefebvre (1970), Sampling Disturbance in Champlain Clays, *ASTM STP 483*, pp. 143–163.

La Russo, R. S. (1963), Wanapum Development—Slurry Trench and Grouted Cut-off, *Proceedings of Symposium: Grouts and Drilling Muds in Engineering Practice*, Butterworths, London, pp. 196–201.

Laursen, E. M. (1962), Scour at Bridge Crossings, *Transactions ASCE*, vol. 127, part 1, pp. 166–209 (including discussions).

Laursen, E. M., and A. Toch (1956), Scour around Bridge Piers and Abutments, *Iowa Highway Research Board Bull.* 4, May, 60 pp.

Lee, I. K. (1974), "Soil Mechanics: New Horizons," American Elsevier Publishing Company, Inc., New York, chap. 5.

Lee, I. K. (1962), Bearing Capacity of Foundations with Particular Reference to the Melbourne Area, *Proceedings Institution of Engineers*, Australia, pp. 283–291.

Lee, I. K., and P. T. Brown (1972), Structure-Foundation Interaction Analysis, *JSD*, ASCE, vol. 98, ST 11, November, pp. 2413–2431.

Lee, K. L. (1970), Comparison of Plane Strain and Triaxial Tests on Sand, *JSMFD*, ASCE, vol. 96, SM 3, May, pp. 901–923.

Lee, K. L., et al. (1973), Reinforced Earth Retaining Walls, *JSMFD*, ASCE, vol. 99, SM 10, October, pp. 745–764.

Leet, L. D. (1950), "Earth Waves," Harvard University Press and John Wiley & Sons, Inc., New York, 122 pp.

Legget, R. F. (1962), "Geology and Engineering," 2d ed., McGraw-Hill Book Company, New York, 884 pp.

Leonards, G. A. (1968), Predicting Settlement of Buildings on Clay Soils, *Proceedings Lecture Series on Foundation Engineering, Northwestern University*, published by Illinois Institute of Technology, Chicago, Ill., pp. 41–51.

Leonards, G. A. (1962), Engineering Properties of Soils, chap. 2 in "Foundation Engineering," McGraw-Hill Book Company, New York.

Leonards, G. A., and P. Girault (1961), A Study of the One-Dimensional Consolidation Test, *5th ICSMFE*, vol. 1, pp. 213–218.

Leonards, G. A., and B. K. Ramiah (1959), Time Effects in the Consolidation of Clays, *ASTM STP* 254, pp. 116–130.

Liepens, A. (1957), "Settlement Analysis of a Seventeen Story Reinforced Concrete Building," S.B. Thesis, Massachusetts Institute of Technology, Cambridge, Mass.

Linnel, K. A., and G. H. Johnston (1973), "Engineering and Design and Construction in Permafrost Regions: A Review," North American Contribution to 2nd International Conference on Permafrost, National Academy of Sciences, Washington, D.C., pp. 553–575.

Littlejohn, G. S. (1970), Soil Anchors, *Proceedings Conference on Ground Engineering*, ICE, London, pp. 33–44.

Lo, K. Y. (1965), Stability of Slopes in Anisotropic Soils, *JSMFD*, ASCE, vol. 91, SM 4, July, pp. 85–106.

Lorenz, H. (1963), Utilization of a Thixotropic Fluid in Trench Cutting and the Sinking of Caissons, *Proceedings of Symposium on Grouts and Drilling Muds in Engineering Practice*, Butterworths, London, pp. 202–205.

Lowe, J. (1974), New Concepts in Consolidation and Settlement Analysis, *JGED*, ASCE, vol. 100, GT 6, pp. 574–612.

Lowe, J., et al. (1969), Controlled Gradient Consolidation Test, *JSMFD*, ASCE, vol. 95, SM 1, pp. 77–87.

Lowe, J., et al. (1964), Consolidation Testing with Back Pressure, *JSMFD*, ASCE, vol. 90, SM 5, pp. 69–86.

Lucas, W. M., Jr. (1970), The Solution of Foundation Mat Problems by Finite Element Methods, *University of Kansas Engineering Bull.* 62, 70 pp.

Lumb, P. (1974), Application of Statistics in Soil Mechanics, chap. 3 in "Soil Mechanics: New Horizons," American Elsevier Publishing Company, Inc., New York.

Lumb, P. (1966), The Variability of Natural Soils, *CGJ*, Ottawa, vol. 3, no. 2, May, pp. 74–97.

Lysmer, J., and F. E. Richart, Jr. (1966), Dynamic Response of Footings to Vertical Loading, *JSMFD*, ASCE, vol. 92, SM 1, January, pp. 65–91.

MacDonald, D. H., and A. W. Skempton (1955), "A Survey of Comparisons between Calculated and Observed Settlements of Structures on Clay," Conference on Correlation of Calculated and Observed Stresses and Displacements, ICE, London, pp. 318–337.

Mackenzie, T. R. (1955), "Strength of Deadman Anchors in Clay," M.S. Thesis, Princeton University, Princeton, N.J.

Mackey, R. D., and D. P. Kirk (1967), At Rest, Active and Passive Earth Pressures, *Proceedings of Southeast Asian Regional Conference on Soil Engineering*, Bangkok, pp. 187–199.

Mackey, R. D., and P. A. Mason (1972), "Pressure Distribution during Filling and Discharging a Silo," 5th European Conference on SMFE, Madrid, vol. 1, pp. 55–62.

McGill Conference (1972), *Proceedings Conference Finite Element Method in Civil Engineering*, McGill University, Montreal, Canada, 1,254 pp.

McLean, F. G., et al. (1975), Influence of Mechanical Variables on the SPT, *6th PSC*, ASCE, vol. 1, pp. 287–318.

Makhlouf, H. M., and J. J. Stewart (1965), Factors Influencing the Modulus of Elasticity of Dry Sand, *6th ICSMFE*, vol. 1, pp. 298–302.

Mansur, C. I., and A. H. Hunter (1970), Pile Tests—Arkansas River Project, *JSMFD*, ASCE, vol. 96, SM 5, September, pp. 1545–1582.

Massarsch, K. R. (1975), New Method for Measurement of Lateral Earth Pressure in Cohesive Soils, *CGJ*, Ottawa, vol. 12, no. 1, February, pp. 142–146.

Massarsch, K. R., et al. (1975), Measurement of Horizontal In-Situ Stresses, *6th PSC*, ASCE, vol. 1, pp. 266–286.

Means, R. E., and J. V. Parcher (1964), "Physical Properties of Soils," Charles E. Merrill Books, Inc., Columbus, Ohio, 464 pp.

Menard, L. (1965), Rules for the Calculation of Bearing Capacity and Foundation Settlement Based on Pressuremeter Tests, *6th ICSMFE*, vol. 2, pp. 295–299 (in French).

Menard, L. (1956), "An Apparatus for Measuring the Strength of Soils in Place," M.Sc. Thesis, University of Illinois, Urbana, Ill.

Meyerhof, G. G. (1974a), Ultimate Bearing Capacity of Footings on Sand Layer Overlying Clay, *CGJ*, Ottawa, vol. 11, no. 2, May, pp. 223–229.

Meyerhof, G. G. (1974b), General Report: Outside Europe, *Proceedings Conference on Penetration Testing*, Stockholm, vol. 2, pp. 40–48.

Meyerhof, G. G. (1972), Stability of Slurry Trench Cuts in Saturated Clay, *5th PSC*, ASCE, vol. 1, part 2, pp. 1451–1466.

Meyerhof, G. G. (1970), Safety Factors in Soil Mechanics, *CGJ*, Ottawa, vol. 7, no. 4, November, pp. 349–355.

Meyerhof, G. G. (1965), Shallow Foundations, *JSMFD*, ASCE, vol. 91, SM 2, March, pp. 21–31.

Meyerhof, G. G. (1963), Some Recent Research on the Bearing Capacity of Foundations, *CGJ*, vol. 1, no. 1, September, pp. 16–26.

Meyerhof, G. G. (1956), Penetration Tests and Bearing Capacity of Cohesionless Soils, *JSMFD*, ASCE, vol. 82, SM 1, pp. 1–19.

Meyerhof, G. G. (1953), The Bearing Capacity of Foundations under Eccentric and Inclined Loads, *3rd ICSMFE*, vol. 1, pp. 440–445.

Meyerhof, G. G. (1951), The Ultimate Bearing Capacity of Foundations, *Geotechnique*, vol. 2, no. 4, pp. 301–331.

Meyerhof, G. G., and J. I. Adams (1968), The Ultimate Uplift Capacity of Foundations, *CGJ*, Ottawa, vol. 5, no. 4, November, pp. 225–244.

Meyerhof, G. G., and J. D. Brown (1967), Discussion: Bearing Capacity of Footings on Layered Clays, *JSMFD*, ASCE, vol. 93, SM 5, part 1, September, pp. 361–363.

Michigan State Highway Commission (1965), "A Performance Investigation of Pile Driving Hammers and Piles," Lansing, Mich., 338 pp.

Mikhejev, V. V., et al. (1961), Foundation Design in the USSR, *5th ICSMFE*, vol. 1, pp. 753–757.

Milovic, D. M. (1965), Comparison between the Calculated and Experimental Values of the Ultimate Bearing Capacity, *6th ICSMFE*, vol. 2, pp. 142–144.

Minard, J. P., and J. P. Owens (1962), Application of Color Aerial Photography to Geologic and Engineering Soil Mapping, *Highway Research Board Bull.* 316, pp. 12–22.

Mindlin, R. D. (1936a), "Discussion: Pressure Distribution on Retaining Walls," *1st ICSMFE*, vol. 3, pp. 155–156.

Mindlin, R. D. (1936b), Force at a Point in the Interior of a Semi-infinite Solid, *Journal of the American Institute of Physics (Physics)*, vol. 7, no. 5, May, pp. 195–202.

Mitchell, J. K., and W. S. Gardner (1975), In-Situ Measurement of Volume Change Characteristics, *6th PSC*, ASCE, vol. 2, pp. 279–345.

Moorhouse, D. C., and G. L. Baker (1968), Sand Densification by Heavy Vibratory Compactor, *3rd PSC*, ASCE, pp. 379–391.

Moorhouse, D. C., and J. V. Sheehan (1968), Predicting Safe Capacity of Pile Groups, *Civil Engineering*, ASCE, vol. 38, no. 10, October, pp. 44–48.

Morgan, J. R., and C. M. Gerrard (1971), Behavior of Sands under Surface Loads, *JSMFD*, ASCE, vol. 97, SM 12, December, pp. 1675–1699.

Morgenstern, N. R., and Z. Eisenstein (1970), Methods of Estimating Lateral Loads and Deformations, *4th PSC*, ASCE, pp. 51–102.

Myslivec, A. (1972), Pressure at Rest of Cohesive Soils, *Proceedings 5th European Conference SMFE*, vol. 1, pp. 63–67.

NAFAC (1971), "Design Manual: Soil Mechanics, Foundations, and Earth Structures," DM-7, Department of the Navy, Washington, D.C., 223 pp.

Nash, J., and G. K. Jones (1963), The Support of Trenches Using Fluid Mud, *Proceedings of Symposium: Grouts and Drilling Muds in Engineering Practice*, Butterworths, London, pp. 177–180.

NBC (1967), "National Building Code," American Insurance Association, 120 South LaSalle Street, Chicago, Ill. 60603.

NBS (1962), "Corrosion of Steel Pilings in Soils," Monograph 58, National Bureau of Standards, U.S. Department of Commerce, Washington, D.C., 22 pp.

NCHRP (1970), "Scour at Bridge Waterways," Synthesis of Highway Practice No. 5, National Academy of Sciences, Washington, D.C., 37 pp.

Neely, W. J., et al. (1973), Failure Loads of Vertical Anchor Plates in Sand, *JSMFD*, ASCE, vol. 99, SM 9, September, pp. 669–685.

Newmark, N. M. (1942), Influence Charts for Computation of Stresses in Elastic Foundations, *University of Illinois Engineering Experiment Station Bull.* 338 (reprinted as vol. 61, no. 92, June 1964).

New York Building Code (1969), Obtained from The City Record, 2213 Municipal Building, New York, N.Y., 10007.

Noorany, I., and I. Poormand (1973), Effect of Sampling on Compressibility of Soft Clay, *JSMFD*, ASCE, vol. 99, SM 12, December, pp. 1184–1188.

Nordlund, R. L. (1963), Bearing Capacity of Piles in Cohesionless Soils, *JSMFD*, ASCE, vol. 89, SM 3, May, pp. 1–36 (see also closure July 1964 with errata).

Nordlund, R. L. (1962), Discussion: Pile Heave and Redriving, *JSMFD*, ASCE, vol. 88, SM 1, February, p. 77.

Novak, M. (1970), Prediction of Footing Vibrations, *JSMFD*, ASCE, vol. 96, SM 3, May, pp. 837–861.

Novak, M., and Y. O. Beredugo (1972), Vertical Vibration of Embedded Footings, *JSMFD*, ASCE, vol. 98, SM 12, December, pp. 1291–1310.

Ohio (1947), Investigation of the Strength of the Connection between a Concrete Cap and the Embedded End of a Steel H-Pile, *Department of Highways Research Report* 1, December.

Olsen, R. E., and K. S. Flaate (1967), Pile Driving Formulas for Friction Piles in Sand, *JSMFE*, ASCE, vol. 93, SM 6, November, pp. 279–296.

O'Neill, M. W., and L. C. Reese (1972), Behavior of Bored Piles in Beaumont Clay, *JSMFD*, ASCE, vol. 98, SM 2, pp. 195–213.

Oosterbaan, M. D., and D. G. Gifford (1972), A Case Study of the Bauer Earth Anchor, *5th PSC*, ASCE, vol. 1, part 2, pp. 1391–1401.

Ovesen, N. K. (1962), Cellular Cofferdams, Calculation Methods and Model Tests, *Danish Geotechnical Institute Bull.* 14, Copenhagen.

Palmer, D. J., and G. R. Holland (1966), The Construction of Large Diameter Bored Piles with Particular Reference to London Clay, *Proceedings Conference: Large Bored Piles*, ICE, London, pp. 105–120.

Palmer, D. J., and J. G. Stuart (1957), Some Observations on the Standard Penetration Test and a Correlation of the Test with a New Penetrometer, *4th ICSMFE*, vol. 1, pp. 231–236.

Parry, R. H. (1971), Stability Analysis for Low Embankments on Soft Clays, *Proceedings of Symposium on Stress-Strain Behavior of Soils*, Cambridge University, England, pp. 643–668.

PCA (1951), "Concrete Piles: Design, Manufacture, Driving," Portland Cement Association, Chicago, Ill., 80 pp.

PCI (1974), "Tentative Recommendations for Prestressed Rock and Soil Anchors," Prestressed Concrete Institute, Chicago, Ill., 32 pp.

Peck, R. B. (1969), Deep Excavations and Tunneling in Soft Ground, *7th ICSMFE*, State-of-the-Art Volume, pp. 225–290.

Peck, R. B. (1965), Pile and Pier Foundations, *JSMFD*, ASCE, vol. 91, SM 2, March, pp. 33–38.

Peck, R. B. (1943), Earth Pressure Measurements in Open Cuts, Chicago (Ill.) Subway, *Transactions ASCE*, vol. 108, pp. 1008–1058.

Peck, R. B. (1942), Discussion: Pile Driving Formulas, *Proceedings ASCE*, vol. 68, no. 2, February, pp. 323–324.

Peck, R. B., and A. S. Bazaraa (1969), Discussion: Settlement of Spread Footings on Sand, *JSMFD*, ASCE, vol. 95, SM 3, May, pp. 905–909.

Peck, R. B., and H. O. Ireland (1961), Full-Scale Lateral Load Test of a Retaining Wall Foundation, *5th ICSMFE*, vol. 2, pp. 453–458.

Peck, R. B., et al. (1948), A Study of Retaining Wall Failures, *2nd ICSMFE*, vol. 3, pp. 296–299.

Peck, R. B., and W. C. Reed (1954), Engineering Properties of Chicago Subsoils, *University of Illinois Engineering Experiment Station Bull.* 423, Urbana.

Polshin, D. E., and R. A. Tokar (1957), Maximum Allowable Non-Uniform Settlement of Structures, *4th ICSMFE*, vol. 1, pp. 402–405.

Poulos, H. G., and E. H. Davis (1974), "Elastic Solutions for Soil and Rock Mechanics," John Wiley & Sons, Inc., New York, 411 pp.

Poulos, H. G., and E. H. Davis (1968), The Settlement Behaviour of Single Axially Loaded Incompressible Piles and Piers, *Geotechnique*, vol. 18, no. 3, September, pp. 351–371.

Prescott, D. M., et al. (1973), Field Measurements of Lateral Earth Pressures on a Pre-Cast Panel Retaining Wall, *Research Report* 169-3, Texas Transport Institute, College Station, Tex., 57 pp.

Purushothamaraj, P., et al. (1974), Bearing Capacity of Strip Footings in Two Layered Cohesive-Friction Soils, *CGJ*, Ottawa, vol. 11, no. 1, February, pp. 32–45.

Ramanathan, D., and V. Raman (1974), Split Tensile Strength of Cohesive Soils, *Soils and Foundations*, Tokyo, vol. 14, no. 1, March, pp. 71–76.

Rankine, W. J. M. (1857), On the Stability of Loose Earth, *Philosophical Transactions of the Royal Society*, London, vol. 147.

Raymond, G. P. (1970), Discussion: Stresses and Displacements in a Cross-Anisotropic Soil, *Geotechnique*, vol. 20, no. 4, December, pp. 456–458.

Reddy, A. S., and R. J. Srinivasan (1967), Bearing Capacity of Footings on Layered Clays, *JSMFD*, ASCE, vol. 93, SM 2, March, pp. 83–99.

Reddy, A. S., and A. J. Valsangkar (1970), Buckling of Fully and Partially Embedded Piles, *JSMFD*, ASCE, vol. 96, SM 6, November, pp. 1951–1965.

Reese, L. C., et al. (1970), Generalized Analysis of Pile Foundations, *JSMFD*, ASCE, vol. 96, SM 1, January, pp. 235–250.

Rehnman, S. E., and B. B. Broms (1972), Lateral Pressures on Basement Wall: Results from Full-Scale Tests, *Proceedings 5th European Conference SMFE*, vol. 1, pp. 189–197.

Reissner, E., and H. F. Sagoci (1944), Forced Torsional Oscillations of an Elastic Half-Space, *Journal of Applied Physics (Physics)*, vol. 15, no. 9, September, pp. 662–654.

Rendulic, L. (1936), Discussion: Relation between Void Ratio and Effective Principal Stresses for a Remoulded Silty Clay, *1st ICSMFE*, vol. 3, pp. 48–51.

Richart, F. E. (1948), Reinforced Wall and Column Footings, *JACI*, vol. 45, October–November, pp. 97–127, 237–260.

Richart, F. E., Jr. (1962), Foundation Vibrations, *Transactions ASCE*, vol. 127, part 1, pp. 863–925.

Richart, F. E., Jr. (1959), Review of the Theories for Sand Drains, *Transactions ASCE*, vol. 124, pp. 709–736.

Richart, E. E., Jr., et al. (1970), "Vibrations of Soils and Foundations," Prentice-Hall, Inc., Englewood Cliffs, N.J., 414 pp.

Robinsky, E. I., and K. E. Bespflug (1973), Design of Insulated Foundations, *JSMFD*, ASCE, vol. 99, SM 9, September, pp. 649–667.

Robinson, K. E., and H. Taylor (1969), Selection and Performance of Anchors for Guyed Transmission Towers, *CGJ*, Ottawa, vol. 6, no. 2, May, pp. 119–137.

Rogers, P. (1952), Design of Large Coal Bunkers, *Transactions ASCE*, vol. 117, pp. 579–595.

Roscoe, K. H., et al. (1958), On the Yielding of Soils, *Geotechnique*, vol. 8, no. 1, March, pp. 22–53.

Roscoe, K. H., and H. B. Poorooshasb (1963), A Theoretical and Experimental Study of Strains in Triaxial Compression Tests on Normally Consolidated Clays, *Geotechnique*, vol. 13, no. 1, March, pp. 12–38.

Rosenfarb, J. L., and W. F. Chen (1972), Limit Analysis Solutions of Earth Pressure Problems, *Fritz Engineering Laboratory Report* 355.14, Lehigh University, 53 pp. (see also *Soils and Foundations*, Tokyo, vol. 13, no. 4, December 1973).

Rowe, P. W. (1957), Sheet Pile Walls in Clay, *PICE*, vol. 7, July, pp. 629–654.

Rowe, P. W. (1954), A Stress-Strain Theory for Cohesionless Soil with Applications to Earth Pressures at Rest and Moving Walls, *Geotechnique*, vol. 4, no. 2, June, pp. 70–88.

Rowe, P. W. (1952), Anchored Sheet Pile Walls, *PICE*, vol. 1, part 1, pp. 27–70.

Rowe, P. W., and K. Peaker (1965), Passive Earth Pressure Measurements, *Geotechnique*, vol. 15, no. 1, March, pp. 57–78.

Rutledge, P. C. (1944), Relation of Undisturbed Sampling to Laboratory Testing, *Transactions ASCE*, vol. 109, pp. 1155–1216.

Saffery, M. R., and A. K. Tate (1961), Model Tests on Pile Groups in a Clay Soil with Particular Reference to the Behaviour of the Group When It Is Loaded Eccentrically, *5th ICSMFE*, vol. 2, pp. 129–134.

Sanglerat, G. (1972), "The Penetrometer and Soil Exploration," Elsevier Publishing Company, Amsterdam, 464 pp.

Sangrey, D. A., et al. (1969), The Effective Stress Response of a Saturated Clay Soil to Repeated Loading, *CGJ*, Ottawa, vol. 6, no. 3, August, pp. 241–252.

Saul, W. E. (1968), Static and Dynamic Analysis of Pile Foundations, *JSD*, ASCE, vol. 94, ST 5, May, pp. 1077–1100.

Schleicher, F. (1926), Zur Theorie des Baugrundes, *Bauingenieur*, vol. 7, pp. 931–935, 949–952.

Schmertmann, J. H. (1975), The Measurement of In-Situ Shear Strength, *6th PSC*, ASCE, vol. 2, pp. 57–138.

Schmertmann, J. H. (1955), The Undisturbed Consolidation Behavior of Clay, *Transactions ASCE*, vol. 120, pp. 1201–1233.

Schultze, E. (1961), Distribution of Stress beneath a Rigid Foundation, *5th ICSMFE*, vol. 1, pp. 807–813.

Schultze, E., and A. Horn (1967), The Base Friction for Horizontally Loaded Footings in Sand and Gravel, *Geotechnique*, vol. 17, no. 4, December, pp. 329–347.

Seed, H. B., and I. M. Idriss (1971), Simplified Procedure for Evaluating Soil Liquefaction Potential, *JSMFD*, ASCE, vol. 97, SM 9, September, pp. 1249–1273.

Seed, H. B., and R. V. Whitman (1970), Design of Earth Retaining Structures for Dynamic Loads, *4th PSC*, ASCE, pp. 103–147.

Seely, F. B., and J. O. Smith (1952), "Advanced Mechanics of Materials," John Wiley & Sons, Inc., New York, p. 271.

Sheeler, J. B. (1968), Summarization and Comparison of Engineering Properties of Loess in the United States, *Highway Research Record* 212, pp. 1–9.

Shepard, E. R., and R. M. Haines (1944), Seismic Subsurface Exploration on the St. Lawrence River Project, *Transactions ASCE*, vol. 109, pp. 194–222.

Shields, D. H., and A. Z. Tolunay (1973), Passive Pressure Coefficients by Method of Slices, *JSMFD*, ASCE, vol. 99, SM 12, December, pp. 1043–1053.

Skempton, A. W. (1960), Effective Stress in Soils, Concrete, and Rocks, *Proceedings Conference on Pore Pressure and Suction in Soils*, Butterworths, London, pp. 4–16.

Skempton, A. W. (1954), The Pore Pressure Coefficients A and B, *Geotechnique*, vol. 4, no. 4, December, pp. 143–147.

Skempton, A. W. (1951), The Bearing Capacity of Clays, *Proceedings Building Research Congress*, vol. 1, pp. 180–189.

Skempton, A. W. (1948), Vane Tests in the Alluvial Plain of the River Forth near Grangemouth, *Geotechnique*, vol. 1, no. 2, pp. 111–124.

Skempton, A. W. (1944), Notes on the Compressibility of Clays, *Quarterly Journal of the Geological Society*, London, vol. C, pp. 119–135.

Skempton, A. W., et al. (1955), Settlement Analyses of Six Structures in Chicago and London, *PICE*, part 1, vol. 4, July, pp. 525–544.

Skempton, A. W., and R. D. Northey (1952), Sensitivity of Clays, *Geotechnique*, vol. 3, pp. 40–51.

Skempton, A. W., and V. A. Sowa (1963), The Behaviour of Saturated Clays during Sampling and Testing, *Geotechnique*, vol. 13, no. 4, December, pp. 269–289.

Smith, E. A. (1962), Pile Driving Analysis by the Wave Equation, *Transactions ASCE*, vol. 127, part 1, pp. 1145–1193.

Smith, J. E. (1957), Tests of Concrete Deadman Anchorages in Sand, *ASTM STP* 206, pp. 115–132.

Smith, J. W., and M. Zar (1964), Chimney Foundations, *JACI*, vol. 61, no. 6, June, pp. 673–700.

Smith, R. E., and H. E. Wahls (1969), Consolidation under Constant Rate of Strain, *JSMFD*, ASCE, vol. 95, SM 2, pp. 519–539.

Soderman, L. G., and R. M. Quigley (1965), Geotechnical Properties of Three Ontario Clays, *CGJ*, Ottawa, vol. 2, no. 2, May, pp. 167–189.

Soderman, L. G., et al. (1968), Field and Laboratory Studies of Modulus of Elasticity of a Clay Till, *Highway Research Record* 243, pp. 1–11.

Sokolovski, V. V. (1960), "Statics of Soil Media," 2d ed., Butterworth Scientific Publications, London, 237 pp.

Sowa, V. A. (1970), Pulling Capacity of Concrete Cast In-Situ Bored Piles, *CGJ*, Ottawa, vol. 7, no. 4, November, pp. 482–493.

Sowers, G. B., and G. F. Sowers (1970), "Introductory Soil Mechanics and Foundations," 3d ed., The Macmillan Company, New York, 556 pp.

Sowers, G. F. (1968), Foundation Problems in Sanitary Land Fills, *Journal of Sanitary Engineering Division*, ASCE, vol. 94, SA 1, pp. 103–116.

Sowers, G. F. (1963), Strength Testing of Soils, *ASTM STP* 361, pp. 3–21.

Sowers, G. F. (1954), Modern Procedures for Underground Investigations, *Proceedings ASCE*, separate no. 435.

Sowers, G. F., et al. (1961), The Bearing Capacity of Friction Pile Groups in Homogeneous Clay from Model Studies, *5th ICSMFE*, vol. 2, pp. 155–159.

Spangler, M. G. (1936), The Distribution of Normal Pressure on a Retaining Wall Due to a Concentrated Surface Load, *1st ICSMFE*, vol. 1, pp. 200–207.

Spangler, M. G., and J. Mickle (1956), Lateral Pressure on Retaining Walls Due to Backfill Surface Loads, *Highway Research Board Bull.* 141, pp. 1–18.

Stagg, K. G., and O. C. Zienkiewicz (1968), "Rock Mechanics in Engineering Practice," John Wiley & Sons, Inc., New York (12 contributing authors), 442 pp.

Steinbrenner, W. (1934), Tafeln zur Setzungsberechnung, *Die Strasse*, vol. 1, October, pp. 121–124.

Steinman, D. B., and S. R. Watson (1941), "Bridges and Their Builders," G. P. Putnam's Sons, New York.

Stokoe, K. H., and R. D. Woods (1972), In Situ Shear Wave Velocity by Cross-Hole Method, *JSMFD*, ASCE, vol. 98, SM 5, May, pp. 443–460.

Sung, T. Y. (1953), Vibrations in Semi-infinite Solids Due to Periodic Surface Loading, *ASTM STP* 156, pp. 35–68.

Swatek, E. P., Jr. (1967), Cellular Cofferdam Design and Practice, *Journal of Waterways and Harbors Division*, ASCE, vol. 93, WW 3, August, pp. 109–132.

Swatek, E. P., Jr., et al. (1972), Performance of Bracing for Deep Chicago Excavation, *5th PSC*, ASCE, vol. 1, part 2, pp. 1303–1322.

Tavenas, F. A. (1971), Load Tests Results on Friction Piles in Sand, *Canadian Geotechnical Journal*, vol. 8, no. 1, February, pp. 7–22.

Taylor, D. W. (1948), "Fundamentals of Soil Mechanics," John Wiley & Sons, Inc., New York, 700 pp.

Teng, W. C. (1962), "Foundation Design," Prentice-Hall, Inc., Englewood Cliffs, N.J., 466 pp.

Terzaghi, K. (1955), Evaluation of Coefficient of Subgrade Reaction, *Geotechnique*, London, vol. 5, no. 4, December, pp. 297–326.

Terzaghi, K. (1954), Anchored Bulkheads, *Transactions ASCE*, vol. 119, pp. 1243–1324.

Terzaghi, K. (1945), Stability and Stiffness of Cellular Cofferdams, *Transactions ASCE*, vol. 110, pp. 1083–1202.

Terzaghi, K. (1943), "Theoretical Soil Mechanics," John Wiley & Sons, Inc., New York, 510 pp.

Terzaghi, K. (1934), Large Retaining Wall Tests, *Engineering News-Record*, Feb. 1, pp. 136–140; Feb. 22, pp. 259–262; Mar. 8, pp. 316–318; Mar. 29, pp. 403–406; Apr. 19, pp. 503–508.

Terzaghi, K., and R. B. Peck (1967), "Soil Mechanics in Engineering Practice," 2d ed., John Wiley & Sons, Inc., New York, 729 pp.

Thornbury, W. D. (1965), "Regional Geomorphology of the United States," John Wiley & Sons, Inc., New York, 609 pp.

Thurman, A. G. (1964), Discussion: Bearing Capacity of Piles in Cohesionless Soils, *JSMFD*, ASCE, vol. 90, SM 1, January, pp. 127–129.

Timoshenko, S., and J. N. Goodier (1951), "Theory of Elasticity," 2d ed., McGraw-Hill Book Company, New York, 506 pp.

Timoshenko, S., and S. Woinowsky-Krieger (1959), "Theory of Plates and Shells," 2d ed., McGraw-Hill Book Company, New York, 580 pp.

Tomlinson, M. J. (1971), Some Effects of Pile Driving on Skin Friction, *Proceedings ICE Conference: Behaviour of Piles*, London, pp. 107–114.

Tschebotarioff, G. P. (1973), "Foundations, Retaining and Earth Structures," 2d ed., McGraw-Hill Book Company, New York, 642 pp.

Tschebotarioff, G. P. (1970), Bridge Abutments on Piles Driven through Plastic Clay, *Proceedings Conference: Design and Installation of Pile Foundations and Cellular Structures*, Lehigh University, pp. 225–238.

Tschebotarioff, G. P. (1962), Retaining Structures, chap. 5 in "Foundation Engineering," McGraw-Hill Book Company, New York, pp. 466–468.

Tschebotarioff, G. P. (1949), Large Scale Earth Pressure Tests with Model Flexible Bulkheads, *Final Report to Bureau of Yards and Docks*, U.S. Navy, Princeton University (see also *Geotechnique*, London, 1948, vol. 1, no. 2, pp. 98–111).

Turitzin, A. M. (1969a), Temporary Bulk Grain Storage Structures, *JSD*, ASCE, vol. 93, ST 3, March, pp. 381–397.

Turitzin, A. M. (1969b), Grain Storage Structures for Food Deficient Areas, *JSD*, ASCE, vol. 93, ST 10, pp. 2161–2181.

Turitzin, A. M. (1963), Dynamic Pressure of Granular Material in Deep Bins, *JSD*, ASCE, vol. 89, ST 2, April, pp. 49–73.

TVA (1966), "Cofferdams on Rock," Technical Monograph 75, Tennessee Valley Authority, Knoxville, Tenn., 281 pp.

Underwood, L. B. (1967), Classification and Identification of Shales, *JSMFD*, ASCE, vol. 93, SM 6, November, pp. 97–116.

US Steel (1974), "Steel Sheet Piling Design Manual," United States Steel Corp., Pittsburgh, Pa., 132 pp.

Van Bruggen, J. P. (1936), Sampling and Testing Undisturbed Sands from Boreholes, *1st ICSMFE*, vol. 1, pp. 3–6.

Vesic, A. S. (1973), Analysis of Ultimate Loads of Shallow Foundations, *JSMFD*, ASCE, vol. 99, SM 1, January, pp. 45–73.

Vesic, A. S. (1970), Tests on Instrumented Piles, Ogeechee River Site, *JSMFD*, ASCE, vol. 96, SM 2, March, pp. 561–584.

Vesic, A. S. (1967), Ultimate Loads and Settlements of Deep Foundations in Sand, *Proceedings Symposium: Bearing Capacity and Settlement of Foundations*, Duke University, Durham, N.C., pp. 53–68.

Vesic, A. S. (1961a), Bending of Beams Resting on Isotropic Elastic Solid, *Journal Engineering Mechanics Division*, ASCE, vol. 87, EM 2, April, pp. 35–53.

Vesic, A. S. (1961b), Beams on Elastic Subgrade and the Winkler's Hypothesis, *5th ICSMFE*, vol. 1, pp. 845–850.

Vesic, A. S., and W. H. Johnson (1963), Model Studies of Beams Resting on a Silt Subgrade, *JSMFD*, ASCE, vol. 89, SM 1, February, pp. 1–31.

Vidal, H. (1969), The Principle of Reinforced Earth, *Highway Research Record* 282, pp. 1–16.

Vijayvergiya, V. N., and John A. Focht, Jr. (1972), "A New Way to Predict Capacity of Piles in Clay," OTC Paper 1718, 4th Offshore Technology Conference, Houston, Tex.

Wahls, H. E. (1962), Analysis of Primary and Secondary Consolidation, *JSMFD*, ASCE, vol. 88, SM 6, December, pp. 207–231.

Wakeman, C. M., et al. (1958), Use of Concrete in Marine Environments, *JACI*, vol. 29, no. 10, April, pp. 841–855.

Wang, C. K. (1970), "Matrix Methods of Structural Analysis," 2d ed., Intext Educational Publishers, Scranton, Pa.

Wang, C. K. (1967), Stability of Rigid Frames with Non-Uniform Members, *JSD, ASCE*, vol. 93, ST 1, February, pp. 275–294.

Watt, A. J., et al. (1967), "Loading Tests on Structures Founded on Soft Cohesive Soils Strengthened by Compacted Granular Columns," 3rd Asian Regional Conference on Soil Mechanics and Foundation Engineering, Haifa, vol. 1, pp. 248–251.

Westergaard, H. M. (1938), A Problem of Elasticity Suggested by a Problem in Soil Mechanics: Soft Material Reinforced by Numerous Strong Horizontal Sheets, in *Contributions to the Mechanics of Solids*, Stephen Timoshenko 60th Anniversary Volume, The Macmillan Company, New York.

Whitaker, T. (1957), Experiments with Model Piles in Groups, *Geotechnique*, vol. 7, no. 4, December, pp. 147–167.

Whitaker, T., and R. W. Cooke (1966), An Investigation of the Shaft and Base Resistances of Large Bored Piles in London Clay, *Proceedings Conference: Large Bored Piles*, ICE, London, pp. 7–49.

White, L. S. (1953), Transcona Elevator Failure: Eye-Witness Account, *Geotechnique*, vol. 3, pp. 209–214 (see also Peck and Bryant, in same vol. pp. 201–208).

Whitman, R. V., and F. E. Richart, Jr. (1967), Design Procedures for Dynamically Loaded Foundations, *JSMFD*, ASCE, vol. 93, SM 6, November, pp. 169–193.

Wineland, J. D. (1975), Borehole Shear Device, *6th PSC*, ASCE, vol. 1, pp. 511–522.

Wissa, A., et al. (1971), Consolidation at Constant Rate of Strain, *JSMFD*, ASCE, vol. 97, SM 10, pp. 1393–1411.

Woodward, R. J., et al. (1972), "Drilled Pier Foundations," McGraw-Hill Book Company, New York, 287 pp.

Wright-Patterson AFB (1965, 1968, 1971), *Proceedings (1st, 2nd, 3rd) Conference on Matrix Methods in Structural Mechanics*, Dayton, Ohio (available from NTIS Springfield, Va.).

Wroth, C. P. (1975), In-Situ Measurement of Initial Stresses and Deformation Characteristics, *6th PSC*, ASCE, vol. 2, pp. 181–230.

Wroth, C. P. (1972), "General Theories of Earth Pressures and Deformations," 5th European Conference on SMFE, Madrid, vol. 2, pp. 33–52.

Wu, T. H. (1976), "Soil Mechanics," 2d ed., Allyn and Bacon, Inc., Boston, Mass., 440 pp.

Wu, T. H., and L. M. Kraft (1967), The Probability of Foundation Safety, *JSMFD*, ASCE, vol. 93, SM 5, pp. 213–231.

Yamagughi, H. (1963), "Practical Formula of Bearing Value for Two-Layered Ground," 2nd Asian Regional Conference on SMFE, Tokyo, vol. 1, pp. 176–180.

Yen, B. C., and B. Scanlon (1975), Sanitary Landfill Settlement Rates, *JGED*, ASCE, vol. 101, GT 5, May, pp. 475–487.

Yoshida, I., and R. Yoshinaka (1972), A Method to Estimate Modulus of Horizontal Subgrade Reaction for a Pile, *Soils and Foundations*, Tokyo, Japan, vol. 12, no. 3, September, pp. 1–17.

Zienkiewicz, O. C. (1971), "The Finite Element Method in Engineering Science," McGraw-Hill Book Company, New York, 521 pp.

Name Index

AASHO, 286, 400, 566
Aboshi, H., 51
ACI, 298, 363, 515, 618, 642
Adams, J. I., 137
Afandi, O. F., 318
AISC, 223
Aldrich, H. P., 183
Anandakrishnan, M., 674
Anderson, W. F., 103
Arman, A., 29, 100
Arnold, R. N., 655
ASCE, 189, 509, 550, 568, 573
Aschenbrenner, R., 609
AWPI, 510

Baguelin, F., 101
Bailey, W. A., 31
Baker, C. N., 632, 635, 642
Baker, G. L., 182
Balla, A., 117, 122, 137
Ballard, R. J., Jr., 110
Banerjee, P. K., 599
Banks, D. C., 25
Barber, E. S., 44
Barden, L., 29, 37, 172, 207, 599
Barentsen, P., 88
Barkan, D. D., 361, 655, 667
Barron, R. A., 185
Bazaraa, A. R., 84, 86, 134, 144
Bell, A. L., 323
Belz, C. A., 488
Beredugo, Y. O., 661, 674
Berenzantzev, V. G., 634, 635
Bespflug, K. E., 204
Bhatacharya, R. K., 39
Bishop, A. W., 25, 28, 54, 120
Bjerrum, L., 100, 457, 472–474, 605
Black, W. L., 669
Blaine, E. S., 620
Blight, G. E., 54
Bond, D., 162
Borowicka, H., 207, 667
Bowles, J. E., 16, 25, 28, 30, 51, 130, 133, 209,
   217, 233, 244, 258, 268, 280, 287, 304, 306,
   307, 393, 412, 414, 548, 573, 609
Bozozuk, M., 104, 105, 605
BRAB, 200
Brand, E. W., 121, 145
Bransby, P. L., 340
Broms, B. B., 340, 377, 378, 598
Brooker E. W., 41, 42
Brown, J. D., 128
Brown, P. T., 304
Burland, J. B., 635, 636
Burn, K. N., 161
Butterfield, R., 599
Button, S. J., 128, 129
Bycroft, G. N., 655

Caquot, A., 334
Carlson, L., 99
Casagrande, A., 19, 27, 47, 626
Caspe, M. S., 463
Chang, C. Y., 37
Chellis, R. D., 564–568
Chen, W. F., 25, 117, 334
Chicago Building Code, 509, 520
Chowdhury, R. N., 40
Christian, J. T., 304
Chu, K. H., 318
Chummar, A. V., 120
Clevenger, W. A., 199
Clyde, D. H., 363
Cooke, R.W., 634, 635
Corps of Engineers, 82
Coyle, H. M., 377, 378, 540, 541
Crawford, C. B., 19, 47, 51, 161
Crettaz, P., 88
Cross, H., 263
Cummings, A. E., 564
Cummings, E. M., 488, 494, 495
Cunny, R. W., 669, 671
Cushing, J. J., 487

Dahlberg, R., 88
Dakshanamurthy, V., 192, 193
D'Appolonia, D. J., 162
D'Appolonia, E., 539
David, D., 201
Davidson, H. L., 117
Davidson, L. W., 120
Davis, E. H., 168, 592
Davisson, M. T., 550, 560
Dawson, R. F., 201
De Beer, E. E., 118
De Mello, V. G., 83, 635
Den Hartog, J. P., 621
De Ruiter, J., 106
Deutsch, G. P., 363
Dewey, F. B., 550
Dismuke, T. D., 488, 494
Drannikov, A. M., 199
Drnevich, V. P., 669
Duncan, J. M., 37
Dunham, C. W., 258
Durham, G. N., 16

Eden, W. J., 101
Egorov, K. E., 318
Eide, O., 472–474
Eisenstein, Z., 342
ENR, 536, 642

Fadum, R. E., 19, 626
Fahlquist, F. E., 82
Fang, H. Y., 25
Fausold, M., Jr., 590
Feld, J., 176, 589

Fellenius, B. H., 605
Finn, W. D., 16
Flaate, K. S., 566, 573
Fletcher, G. F., 83
Flint, R. F., 198
Focht, John A., Jr., 527
Fox, E. N., 159
Francis, A. ., 545, 548
Fry, Z. B., 669, 671

Garbe, C. W., 184
Gardener, W. S., 161
Garga, V. K., 25
Gates, M., 566
Geddes, J. D., 592, 593
Gerrard, C. M., 170
Gibbs, H. J., 84, 85, 99, 100, 199, 536
Gibson, R. E., 103
Gifford, D. G., 448
Girault, P., 43
Glick, G. W., 546
Godskesen, O., 88
Golder, H. Q., 456, 626
Goodier, J. N., 309, 667
Gorbunov-Passadov, M. I., 667
Grant, R., 176, 298
Grayman, R., 494, 498
Griffiths, D. H., 669
Gromko, G. J., 200

Haines, R. M., 110
Hall, J. R., Jr., 668, 671
Hardin, B. O., 669
Hamilton, J. J., 47, 51
Hanna, T. H., 543, 590
Hansen, J. B., 117, 118, 120, 121, 124, 142
Henkel, D. J., 28, 59
Hetenyi, M. 273
Hijab, W. A., 354
Hiley, A., 564
Hirsch, T. J., 577
Hirst, T. J., 106
Holland, G. R., 632, 634
Holland, W. Y., 199, 536
Holtz, W. G., 84, 85, 100, 192, 193
Hooper, J. A., 318
Hough, B. K., 51
Housel, W. S., 97, 536
Hrennikoff, A., 609
Hribar, J. A., 539
Hsieh, T. K.. 655, 658
Hu, C. Y., 117
Hughes, J. M., 188
Hunter, A. H., 532, 566, 569, 570
Huntington, W. C., 385, 386
Hvorslev, M. J., 25, 79, 81, 88, 91

Idriss, I. M., 187
Ireland, H. O., 41, 340, 532, 542

Jaky, J., 41
James, R. G., 340
Janbu, N., 334
Janes, H. W., 187
Jofriet, J. C., 407
Johnson, S. J., 183

Johnson, W. H., 280
Johnston, G. H., 205
Joint ASCE-ACI Committee, 300
Jones, G. K., 481

Kantey, B. A., 70
Kempner, L., Jr., 550
Kerisel, J., 334
Khan, G., 632, 635, 642
Khayatt, A. J., 29
King, R. F., 669
Kirk, D. P., 340
Kirkedam, R., 457
Kjellman, W., 79
Klohn, E. J., 583
Ko, H. Y., 120
Koerner, R. M., 600
Komornik, A., 201
Kondner, R. L., 37, 38
Kovacs, W. K., 84
Kraft, L. M., 142
Krishnaswamy, N. R., 674
Kuhn, S. H., 197

Ladanyi, B., 103
Ladd, C. C., 31
Lambe, T. W., 31, 59, 60, 166, 456, 471
Landau, R. E., 185, 186
Larew, H. G., 37
La Rochelle, P., 100
La Russo, R. S., 480
Laursen, E. M., 197
Lee, I. K., 157, 158, 304
Lee, K. L., 29, 120, 377, 408
Leet, L. D., 110
Lefebvre, G., 100
Legget, R. F., 18
Leonards, G. A., 37, 43, 57
Liepens, A., 161
Linnel, K. A., 205
Littlejohn, G. S., 448
Lo, K. Y., 128
Lorenz, H., 480
Lowe, J., 51
Lucas, W. M., Jr., 307
Lumb, P., 142
Lysmer, J., 668

MacDonald, D. H., 172, 176
McGill Conference, 307
MacIver, B. N., 25
Mackenzie, T. R., 449
Mackey, R. D., 340, 363, 366
McLean, F. G., 84, 110
Makhlouf, H. M., 37
Mansur, C. I., 532, 566, 569, 570
Mason, P. A., 363, 366
Massarsch, K. R., 103
Means, R. E., 46
Menard, L., 101
Meyerhof, G. G., 85, 117, 122, 124, 128, 130, 134, 135, 137, 142, 296, 481, 535
Michigan State Highway Commission, 571, 572
Mickle, J., 355
Mikhejev, V. V., 177
Milovic, D. M., 120, 122

Minard, J. P., 70
Mindlin, R. D., 356, 592
Mitchell, J. K., 161
Moline, R. M., 487
Monckton, M. F., 599
Moorhouse, D. C., 182, 589
Morgan, J. R., 170
Morgenstern, N. R., 342
Morse, R. K., 70
Music, J., 669
Myslivec, A., 41

NAFAC, 343
Nash, J., 481
NBC, 361
NBS, 416, 522
NCHRP, 197
Neely, W. J., 448
Newmark, N. M., 149, 150, 357, 358
New York Building Code, 509
Noorany, I., 48
Nordlund, R. L., 532, 583
Northey, R. D., 52
Novak, M., 661, 674

Ohio, 520
Olsen, R. E., 566, 573
O'Neill, M. W., 480, 632, 634, 635
Oosterbaan, M. D., 448
Ovesen, N. K., 488
Owens, J. P., 70

Palmer, D. J., 87, 88, 632, 634
Parcher, J. V., 46
Parry, R. H., 122
Partos, A., 600
PCA, 416
PCI, 448
Peaker, K., 340
Peck, R. B., 19, 51, 83, 84, 86, 134, 162, 297, 340, 377, 456, 463, 464, 573, 642
Polshin, D. E., 177
Poormand, I., 48
Poorooshasb, H. B., 59
Poulos, H. G., 168, 592
Prescott, D. M., 377

Quigley, R. M., 19

Raman, V., 25, 192, 193
Ramanathan, D., 25
Ramiah, B. K., 43
Rankine, W. J. M., 330
Raymond, G. P., 41
Reddy, A. S., 128, 129, 550
Reed, W. C., 19
Reese, L. C., 480, 540, 541, 609, 632, 634, 635
Rehnman, S. E., 340
Reissner, E., 667
Rendulic, L., 59
Richart, F. E., 217
Richart, F. E., Jr., 185, 655, 661, 667-669
Robinsky, E. I., 204
Robinson, K. E., 139, 550
Rogers, P., 366
Roscoe, K. H., 59
Rosenfarb, J. L.. 334

Rowe, P. W., 340, 378, 414, 433, 434
Rutledge, P. C., 47

Saffery, M. R., 608
Sagoci, H. F., 667
Sanglerat, G., 83
Sangrey, D. A., 37
Saul, W. E., 609
Scanlon, B., 203
Schleicher, F., 158
Schmertamann, J. H., 47, 83, 107, 136
Schultze, E., 207, 209
Seed, H. B., 187, 361
Seely, F. B., 308
Serebrjanyi, R. V., 318, 667
Sheehan, J. V., 589
Sheeler, J. B., 199
Shepard, E. R., 110
Shields, D. H., 334
Skempton, W. W., 31, 51, 52, 54, 58, 59, 99, 172, 176, 202, 472, 635
Smith, E. A., 573, 574, 577
Smith, J. E., 448
Smith, J. O., 308
Smith, J. W., 318
Smith, R. E., 51
Soderman, L. G., 19, 37
Sokolovski, V. V., 334
Sowa, V. A., 31, 542
Sowers, G. B., 51
Sowers, G. F., 29, 51, 87, 204, 590
Spangler, M. G., 355
Srinivasan, R. J., 128, 129
Stagg, K. G., 90
Steinbrenner, W., 157
Steinman, D. B., 629
Stewart, J. J., 37
Stokoe, K. H., 670
Stuart, J. G., 87, 88
Sung, T. Y., 655
Swatek, E. P., Jr., 456, 494

Tate, A. K., 608
Tavenas, F. A., 532, 535
Taylor, D. W., 44, 46, 50
Taylor, H., 139
Teng, W. C., 134
Terzaghi, K., 51, 83, 86, 113, 116, 120, 134, 162, 268, 297, 340, 347, 354, 357, 456, 472, 488, 545
Thornbury, W. D., 18
Thurman, A. G., 530
Timoshenko, S., 304, 309, 667
Toch, A., 197
Tokar, R. A., 177
Tolunay, A. Z., 334
Tomlinson, M. J., 524-526
Townsend, F. C., 16
Tsai, K., 184
Tschebotarioff, G. P., 390, 414, 448, 456, 457
Turitzin, A. M., 363, 364
TVA, 488, 490, 494, 497

Underwood, L. B., 21
US Steel, 449

Valsangkar, A. J., 550
Van Bruggen, J. P., 82

Vesic, A. S., 118, 268, 280, 535, 634
Vidal, H., 408
Vijayvergiya, V. N., 527

Wahls, H. E., 51, 172
Wakeman, C. M., 517
Wang, C. K., 276, 317, 550, 553
Watson, S. R., 629
Watt, A. J., 188
Westergaard, H. M., 153
Whitaker, T., 590, 634, 635
White, L. S., 296
Whitman, R. V., 59, 60, 361, 667, 668
Williams, A. B., 197
Wineland, J. D., 29, 107, 108
Winowsky-Krieger, S., 304, 309

Wissa, A., 51
Woods, R. D., 670
Woodward, R. J., 632
Wright-Patterson AFB, 307
Wroth, C. P., 41, 101
Wu, T. H., 57, 142

Yamagughi, H., 130
Yen, B. C., 203
Yoshida, I., 545
Yoshinaka, R., 545

Zar, M., 318
Zeindler, H., 88
Zienkiewicz, O. C., 90, 130, 168

Subject Index

Abutment wingwalls, 398
Active earth pressure, 323
  in cellular cofferdams, 490
  Coulomb theory for, 325
  Rankine theory for, 330
    tables of, 331
  using plasticity theory, 334
    tables of, 336
  wall movement to develop, 324
Adhesion of clay: on base of footings, 115, 119
  on caissons, 634
  on piles, 525
  on retaining walls, 342, 351, 379
Alignment of piles, 583
Allowable loads on piles: based on dynamic
    equations, 566
  based on static equations, 523, 527, 530
Allowable unit stresses in concrete (see
    Reinforced concrete)
Anchorages for sheet piling, 444
  deadman for, 449
  tiebacks, 448
Anchored sheet piling, 426
  flexibility coefficient used for, 433
  free-earth method, 427
  Rowe's moment reduction, 433
  stability number used with, 428
Angle of internal friction, 25
  depends on test, 25
  determination of, 28
  grain for silos, table of, 363
  typical values, 30
  values based on SPT, 85
Angle of wall friction, 342, 414
Anisotropic soil, 38
  stresses due to footings, 168
Area ratio, 80
Artesian water, 91
ASCE Code for timber piles, 509
At-rest earth pressure, 41, 101, 323, 341
Atterberg limits: defined, 13
  liquid limit for consolidation coefficient, 51
  plastic index for volume change, 192
Auger boring, 75

Backfill: for buildings, 182, 194
  compaction of, 181, 377, 408
  drainage of, 342, 397
  for retaining walls, 341
Band matrix solution, 311
Basement walls, 407
Beam of elastic foundation, 271
  finite element solution of, 276
  ring foundation as, 287
Bearing capacity: based on building codes, 139
  based on building codes, 139
  based on settlement, 134
  on cohesive soil, 120, 202
    Skempton's values for depth, 202

Bearing capacity: depth, shape, and
    inclination factors, 118
  drilled caissons, of, 633
  factors for footings, 132
  general-shear failure, 115
  improving soil for increased, 181
  on layered soil, 127
  reduction for eccentric loads, 124
  reduction for water table, 126
  for retaining walls, 389
  for sands, 186
  on slopes, 131
  based on SPT, 133
Bearing capacity equations: Hansen's, 118
  Terzaghi's factors, for, 118
Bearing pressure for mat foundations, 135
Bell, caisson, 632
Belled piers, 201
Borehole shear test, 29, 107
Borings: data presentation of, 94, 96
  depth and number of, 92
Bottom heave of excavation, 472
Boussinesq equations: influence charts, 148,
    150, 151
  lateral pressure, 151
Box caissons, 623
  base preparation for, 624
  stability during floating, 625
Braced cofferdams: components of, 452
  design of, 458, 467
  excavation depth for stability, 472
  finite element analysis, 465
  ground loss behind, 463
  lateral pressure, 455
  pressure diagrams for, 456
  soil properties for, 459
  soldier beams used in, 453
  stability of bottom of excavation in, 472
Building codes: Atlanta, 140
  Chicago, 140
  national, 140
  New York City, 140
  uniform, 140
Bulk modulus, 36
Buoyant weight of soil, 53

Caisson bell, 632
Caissons, 618
  closed-end, 623
  concrete seal for, 621
  cutting edges for, 620
  drilled: bearing capacity for, 633
    design of, 637
    settlements of, 635
  inspection during construction, 642
  laterally loaded, 641
  pneumatic, 629
  stability of floating, 625
    center of buoyancy, 625

Caissons:
    types of, 619
    uses of, 620
Cantilever footing, 241
Cantilever retaining wall, 381, 399
    allowable bearing pressure for, 389
    design example of, 399
    finite element analysis of, 437
Cantilever sheet piling, 413, 418
    safety factor, 418
Capillary water, 56
Cast-in-place concrete piles, 518, 539
Cellular cofferdams, 484
    average width of, 486
    berms in , 498
    cell fill, 487
    design of, 499, 501
    dewatering of, 497
    interlock friction, 492
    interlock tension, 493
    saturation line location of, 497
    stability: by Cumming's method, 494
        by TVA method, 488
    vertical shear on cell, 491
    weep holes, 497
Circular mats, 318
Classification tests (*see* Soil classification)
Clay: bearing capacity of (*see* Bearing
    capacity)
    classification, 27
    consolidation of (*see* Consolidation)
    extrasensitive, 52
    foundations on, 202
    normally consolidated, 48
    particle size range, 18
    piles in, 524, 527
    preconsolidated, 37, 42, 48
    remolded, 48, 51
    shear strength of, 25, 32
    stability of excavation in, 472, 480
Coal bunkers, pressure in, 362
Coefficient: of consolidation, 44
    of earth pressure: active: Coulomb, 327
        Rankine, 331
        to compute for shear in cofferdams, 490
    of horizontal subgrade reaction, 544
    of permeability, 55
    of restitution, 568
    of subgrade reaction (*see* Modulus, of
        subgrade reaction)
Cofferdam instability due to heave, 472
Cofferdams, 452, 483
    braced, 453
    cellular (*see* Cellular cofferdams)
Cohesion: on base of footings, 115
    on base of retaining walls, 342
    of clay, 25
Cohesive soils: bearing capacity of, 147
    compressibility of, 169
    (*See also* Clay)
Column base plates, 222
    AISC design of, 224
Column fixity to footing, 315
Combined footings, 240
    cantilever (strap), 241, 253
    criteria for rigid or flexible footing, 275

Combined footings:
    rectangular, 240
        flexible, 236
        rigid, 241
            design of, 243
    trapezoid shaped, 248
Compacted fill, 175
Compaction tests, 182
Components of retaining walls, 374, 382
Compression index, 50
    empirical equations for, 51
Compression tests, types of, 29
Computer analysis: of beam on elastic
    foundation, 276
    mat foundations, 307
    pile group analysis, 609
    programs, Appendix B, 685
Concrete piles, data, 684
Cone penetrometer, 105
Consolidation: coefficient of, 44
    coefficient of volume compressibility, 49, 169
    compression index, determination of, 45, 50
    partial theory of, 43
    percent of, 44
    preconsolidation pressure in, 45, 47
    primary, 44
    recompression index, 50
    settlements, 49
    stress path for, 63
    time factor used in, 44, 46
Consolidation test, 43
Contraction joints in retaining walls, 397
Core boring, 87
Corrosion of steel piles, 203
Coulomb's earth-pressure theory, 326
    error in use of passive values, 334
    tables of values, 327
    used for retaining walls, 382
    used for sheet piles, 456
Counterfort retaining wall, 375
    design of, 406
    forces on, 384
    Huntington's method of design, 385
    plate fixed on three edges as, 406
Critical circle, location of, 391
Cross-anisotropic soil, 39
Culmann's method of analysis for: active
        pressure on wall, 345, 353
    passive pressure on wall, 349
Cushions: pile driving, 583
    for toe of retaining wall, 389

Damping factor, 649, 653, 668
Deadmen, 447
Density, relative (*see* Relative density)
Depth: of frost penetration, 192
    of tension crack, 339
Depths: of borings, 92
    of footings, 191
Design loads for footings, 142
Dewatering of excavations, 475
Diesel hammer for pile driving, 559
Differential settlements, 176
    on permafrost, 177
Direct shear test, 29
Double-acting hammer, 558

Drainage of retaining wall backfill, 397
Drill rods, size of, 85
Drilled caisson (*see* Caissons)
Drilling: auger used for, 75
  continuous flight auger for, 75
  mud used for, 75
  wash, 75
Driving points, piles, 521
Durability of steel sheet piling, 416, 497
Dutch cone test, 105
  bearing capacity using, 136
Dynamic pile capacity equations, 566

Earth pressure, lateral: acting on sheet-piling
      walls, 419, 429
    active, 339
    braced cofferdams: in clay, 456
      in sand, 456
    during earthquakes, 361
    due to expansive soil, 362
    at rest, 41, 101, 323
    on retaining walls, 339
Eccentrically loaded footings, 124, 255
  effective width of, 124, 257
Effective pressure, 53, 120, 197
Effective stress path, 60
Effective unit weight, 126
Elastic foundation (*see* Beam on an elastic
      foundation)
End-bearing piles, 539
Environment, 20
Expansive soil, 200
Extraction of piles, 559

Factors of safety, 141, 380, 417
Failure plane in retaining wall backfiill, 341, 391
Fill, structures on, 175
Finite beam on an elastic foundation, 273
Finite element solution of beam on elastic
      foundation, 276
    sheet pile walls, 436
Flexibility coefficient for sheet piles, 433
Floating caissons, 625
Flow net, plan, 56, 477
Foil sampler, 81
Footing with a notch, 265 ,
Footings: adjacent to other footings, 193
  classification of, 113, 207
  common types, 195, 208
  depth of, 113, 191
  eccentrically loaded, 124, 234, 255
  location for frost, 191
  with overturning moment, 234
  for residential construction, 232
  on sand, 116
  on a slope, 131
  on a two-layered cohesive soil, 127
Formulas for concrete design (*see* Reinforced
      concrete)
Free-earth method (*see* Anchored sheet piling)
Friction on base of retaining walls, 379
Friction piles, 527
  in clay, 524
  in cohesionless soil, 530
Frost depth, 204

Geophysical soil exploration, 109
Glacial fill, 19
Grain elevators, pressure in, 364
Grain size, 16
  analysis for soil classification, 16, 22
  sieve sizes used in analysis, 23
Gravity retaining wall, 376, 393
Ground loss behind walls, 464
Ground water, 47, 90, 189
Grouting: of sheet piles, 416
  as stabilization agent, 188

H piles, 519
  table of properties, 676
Hansen's bearing-capacity equations, 118
Heave, stability against, in excavations, 472
Hooke's law, 36
Horizontal force on piles, 543
Horizontal subgrade reaction, 545
Huntington's design procedure for counterfort
      walls, 385
Hydraulic gradient, 54
Hydraulics for soils, 52
Hydraulics fracture test, 104

Ice lenses, 361
Immediate settlements, 146, 157
  influence factors: for computation of, 157
    for depth effects, 159
    for flexible footings, 157, 158
      computer program, 688
    for rigid footings, 157, 158
    for rotation of rigid footings, 158
    size effects, 162
Influence chart for lateral pressure, 358
Influence charts for vertical soil pressure:
  by Boussinesq method, 152
  by Westergaard's method, 156
In situ stress, 41, 101
Isotropic soil, 38, 128

Joints in retaining walls, 396

K_0 conditions, 41
  equations for, 42
  measurement in situ, 101
  in pile design, 531

Lateral earth pressure: due to ice formation, 361
    due to line load, 356
    Newmark chart for, 358
    due to surcharges, 357
    due to water in backfill, 342
    (*See also* Earth pressure)
Laterally loaded caissons, 642
Laterally loaded piles, 543
  finite element analysis of, 544
  modulus of subgrade reaction for, 545
    table of values, 547
Line load surcharge, 356
Liquefaction of soils, 187
Liquid limit, 13, 51
Load settlement curve: pile-load test, 579
  plate-load test, 98
Load test, 95
Location of resultant on retaining wall for
      surcharges, 347, 359

Loess, 19
 foundations on, 198
Logarithm spiral, 354

Mat foundations: arrangements, 294
 bearing pressure, 135, 296
 circular, 318
 conventional design of, 298, 300
 differential settlement, 297
 finite element, 307
 method of finite differences, 304
 modulus of subgrade reaction for, 309
 superstructure interaction with, 307
Metacenter, 625
Mindlin solution for stresses in soil, 591
Modified compaction test, 182
Modulus of elasticity: cyclic tests used to
  obtain, 34, 37
 effect of foundation size on, 161
 empirical equations for, 161
 factors depend on, 37
 hyperbolic equation of, 37
 plate load test for, 160
 triaxial test values, validity of, 34, 161
 typical values of, 35
 used for immediate settlements, 157, 161
Modulus of subgrade reaction, 34, 267
 bearing capacity equation for, 269
 determination from plate load tests, 267
 effect of depth on, 268, 545
 effect of footing size on, 268
 empirical equations for, 268
 typical values, 269
 used in continuous footing design, 270
 used for lateral piles, 545
 used in mat foundation design, 270
 Vesic equation for, 268, 545
Mohr's circle, 28
 used for shear stress in cofferdams, 490
Moisture content (*see* Water content)
Moisture-density relationships, 183
Moment reduction for sheet-pile walls, 433
Montmorillonite clay, 18
Mud, drilling, 75

Negative skin friction, 605
Neutral pressure, 52
Newmark's influence charts, 151, 358
Normally consolidated clays, 48

Open-end caissons, 619
Osterberg piston sampler, 81
Overconsolidation ratio (OCR), 37, 42, 48
Overturning: of cellular cofferdams, 489
 of retaining walls, 380, 399

Passive earth pressure, 324
 Coulomb, 328
 Rankine, 331
Pedestal, design of, 225
Penetration test, standard (SPT), 82
 adjusted for depth, 84, 86
 cone, 105
 refusal, 83
 soil parameters based on, 85, 86
 sounding, 87, 88

Percent of sample recovery, 82
Perched water, 91, 205
Permafrost, 204
Permeability, coefficient of, 55
pH, values in pile corrosion, 197, 522
Piers, drilled, 201, 631
Pile: allowable stress in, 512
 driving stresses, 580
Pile capacity: formulas for dynamic, 566
 by static equations, 527, 534
Pile corrosion, 517, 522
Pile-driver data, table of, 677
Pile-driving formulas: dynamic, 564
 Hiley, 564
 table of, 566
 wave equation, 573
Pile-driving points, 521
Pile foundations (*see* Pile groups)
Pile groups: batter piles used in, 605
 caps for, 603
 efficiency of, 589
 negative skin friction, 605
 pile constants for, 547
 settlements of, 599
 spacing of piles in, 587
 stresses on underlying strata, 591
  stress coefficients for, 595
 three-dimensional analysis of, 609
 typical group patterns of, 587
Pile hammers: efficiency, 567
 tables of hammers, 676
 types of, 559
Pile heave, 583
Pile-load tests, 579
Pile spacing, 588
Piles: alignment of, 583
 in clay, 523
 in cohesionless soil, 530
 concrete, 516
  cast-in-place, 518, 539
  pick-up points for, 517
 in loess, 536
 in salt water, 517
 steel, 519
  points for, 521
  tables of data, 676, 682
 timber, 509, 512
Plane strain, 120
 versus triaxial strain, 120
Plastic limit, 13
Plasticity index, 13
 used to estimate volume change, 192
Plate-load tests, 95, 160
 bearing capacity using, 97
 size effects, 161, 162
Pneumatic caissons, 629
Point bearing piles, 539
Poisson's ratio, 34, 36
 typical values of, 35
Pore pressure parameters A and B, 57, 59
Pore water pressure, 20, 53, 57
Porosity, 10
Precast concrete piles, 511
Preconsolidated clay, 4, 48, 128
Preloading site, 184
Pressure bulbs (*see* Influence charts)

Pressure-void ratio curve, 45
Product of inertia, 265

Quick condition, 54, 57

Raft foundation (*see* Mat foundation)
Rankine earth pressures, 330
   tables of values, 331
   used for retaining walls, 381
Recovery ratio of soil samples, 82
Rectangular spread footing, 228
Reinforced concrete: design (USD), 209
   embedment for bond, 211
   maximum percent steel, 210
   moment capacity, 210
   ϕ-factors, 209
   shear stresses, 212
      table of values, 213
   summary of design checks, 213
   ultimate soil pressure for design, 216, 243
Reinforced earth walls, 408
Reinforcing bars, table of US and metric, 236
Relative density, 15, 85, 198
   to control liquefaction, 187
   SPT test used for, 85
   unified classification, used for, 16
Residual soil, 21
Restitution, coefficient of, 568
Retaining walls: allowable pressure for, 389
   common proportions of, 374
   forces acting on, 382
   reinforced earth of, 408
   rigid considerations, 378
   stability of, 379
      Swedish circle analysis, 391
      tilting of, 390
   USD load factor for, 378
Rigid footing criteria, 158, 208, 275
Rock bits, 89
Rock quality designation (RQD), 90
Rocks: bearing capacity of, 143
   properties, table of values, 143
Rotary drill, 75

Safety factor: for bearing capacity, 141
   for bottom heave of cofferdams, 473
   for overturning of cofferdams, 490
   passive pressure coefficients, applied to, 418
   for retaining walls, 141, 380
   for sheet pile structures, 417, 428
   for slurry walls, 481
Sample spoons, 79
   piston-type, 81
   thin-wall, 79
Sampling of soil, 78
Sand, foundations on, 198
Sand drains, 185
Sand islands, 483, 620
Sanitary landfill, 203
Saturation, degree of, 11
Scour from flowing water, 196
Secondary consolidation, 172, 184
Seismic exploration, 108
   velocities in, 110
Sensitive clay, 51
Sensitivity, 51

Settlement: of drilled caissons, 635
   mat foundations, 298
   retaining walls, 389
   theory of elasticity methods for, 157
      for pile foundations, 599
Settlements: bearing capacity limited by, 124
   on clay, 169
   differential, 173, 176
   on fills, 175
   influence factors for (*see* Immediate
     settlements)
   penetration test based on, 134
   secondary consolidation, 172, 178
Shale, 21
Shallow foundations, 113
Shear strength by Coulomb equation, 25
   Mohr's circle used to compute, 32
   using Dutch cone, 137
Shear strength tests, 29
Sheet pile anchors, 445
Sheet pile materials, 415
Sheet pile walls: anchorages for, 445
   earth pressure against, 413
   safety factor for, 418
Sheet piles: shapes available, 680
   of timber, 415
Shelby tube, 79
Shrinkage limit, 14
SI units, 6
Silos, pressure in, 363
Sizes of drill rods, casing, 85
Skin friction of caisson shaft, 622, 633
Slip circle method, 391
Slopes, footings on, 131
Slurry walls, 479
Soil classification, unified system, 26
Soil exploration, 69
   planning for, 72
   summary of methods, 71
Soil pressure: average increase in stratum
     pressure, 164, 167, 170
     by trapezoidal rule, 170
   Boussinesq method, 148
   cause of settlement, 157, 165, 169
   gross value, 195
   by influence chart, 150, 152, 155
   net value, 135, 195
   2:1 method, 147, 169
   Westergaard theory, 153
Specific gravity, 11, 14
   typical values, table, 15
Splices for piles, 510, 520
Split spoon, 79
Stability number, 428
Standard compaction test, 182
Standard penetration test (SPT), 82
   adjustments to, 84, 86
   for bearing capacity, 134
   for consistency, 86
   for piles, 535, 545
   refusal, defined, 83, 87
   for relative density, 85
   for settlement, 65
Strain, soil for computing settlements, 165
Strap footing, 241
Stress in piles: allowable, 512

Stress in piles: driving, 580
  due to negative skin friction, 607
  wave equation for, 577
Stress paths, 59
Stress-strain modulus (*see* Modulus of
  elasticity)
Stresses: beneath a footing, 147
  in anisotropic soils, 168
  average increase in, 164, 167, 169
  methods to compute, 147, 148, 153
  beneath pile foundation: approximate
    method, 592
    Mindlin solution, 593
Subgrade modulus for mats, 270
  (*See also* Modulus of subgrade reaction)
Surcharge load, location of resultant on wall, 346
  types of, 355
Swedish circle method, 391

Tension cracks, depth of, 337
Tension footings, 137
Tension piles, 542
Terra-probing, 187
Thixotropy, 51
Tilting of retaining walls, 391
Timber piles (*see* Piles)
Trench, slurry, 479
Trial-wedge theory, 353
Triaxial tests: to determine coordinates of
    effective stress path, 61
  to determine modulus of elasticity, 34
  to determine pore pressure parameters, 59

Ultimate Strength Design (USD) (*see*
  Reinforced concrete)
Unconfined compression tests, 29
Undisturbed soil samples, 31, 43, 71, 73, 78
Uniformity coefficient, 27
Unit weight of soil, 11
  buoyant, 53
Units, SI, 6
Unsymmetrical footings, 263

Uplift, footings for, 137

Vane shear test, 29, 33
  correction factors for, 101
  field, 99
  laboratory, 33
Velocity of seismic waves, 110
Vibrations: approximate solution for machine
    foundations, 659, 663, 664
  damped, 649
  elastic half-space, 655
  elementary theory of, 646, 653
  lumped mass solution, 661
  piles to reduce, 673
  soil properties for, 669
  table of design factors, 667
Vibratory pile hammers, 560
  equation for pile capacity, 560
Vibroflotation, 186
Void ratio, 10
Volume change related to I_p, 192

Wales, 444
Wall friction: angle of, 343
  for sheet pile walls, 343, 414
  table of values, 343
Wall movement to develop active pressure, 324
Wash boring, 73
Water in tension crack, 332, 339
Water table location in borings, 90
Wave equation, 573
  coefficient of restitution for, 568
  damping coefficients, 577
Wave velocities, seismic, 110
Weep holes: in cofferdams, 497
  in retaining walls, 397
Westergaard influence charts, 155, 156
Wingwalls, abutment, 398
Winkler foundation, 271
Wood piles (*see* Piles)

Zero-air-voids curve, 183

Bon J